the United States

IGNEOUS ROCKS

HISTORICAL GEOLOGY

The Science of a Dynamic Earth

HISTORICAL GEOLOGY

The Science of a Dynamic Earth

LEIGH W. MINTZ

California State University, Hayward

CHARLES E. MERRILL PUBLISHING COMPANY

A Bell & Howell Company

Columbus, Ohio

Published by
Charles E. Merrill Publishing Co.
A Bell & Howell Company
Columbus, Ohio 43216

International Standard Book Number: 0–675–09143–8

Library of Congress Catalog Number: 70–187716

1 2 3 4 5 6 7 8 9 10—77 76 75 74 73 72

Printed in the United States of America

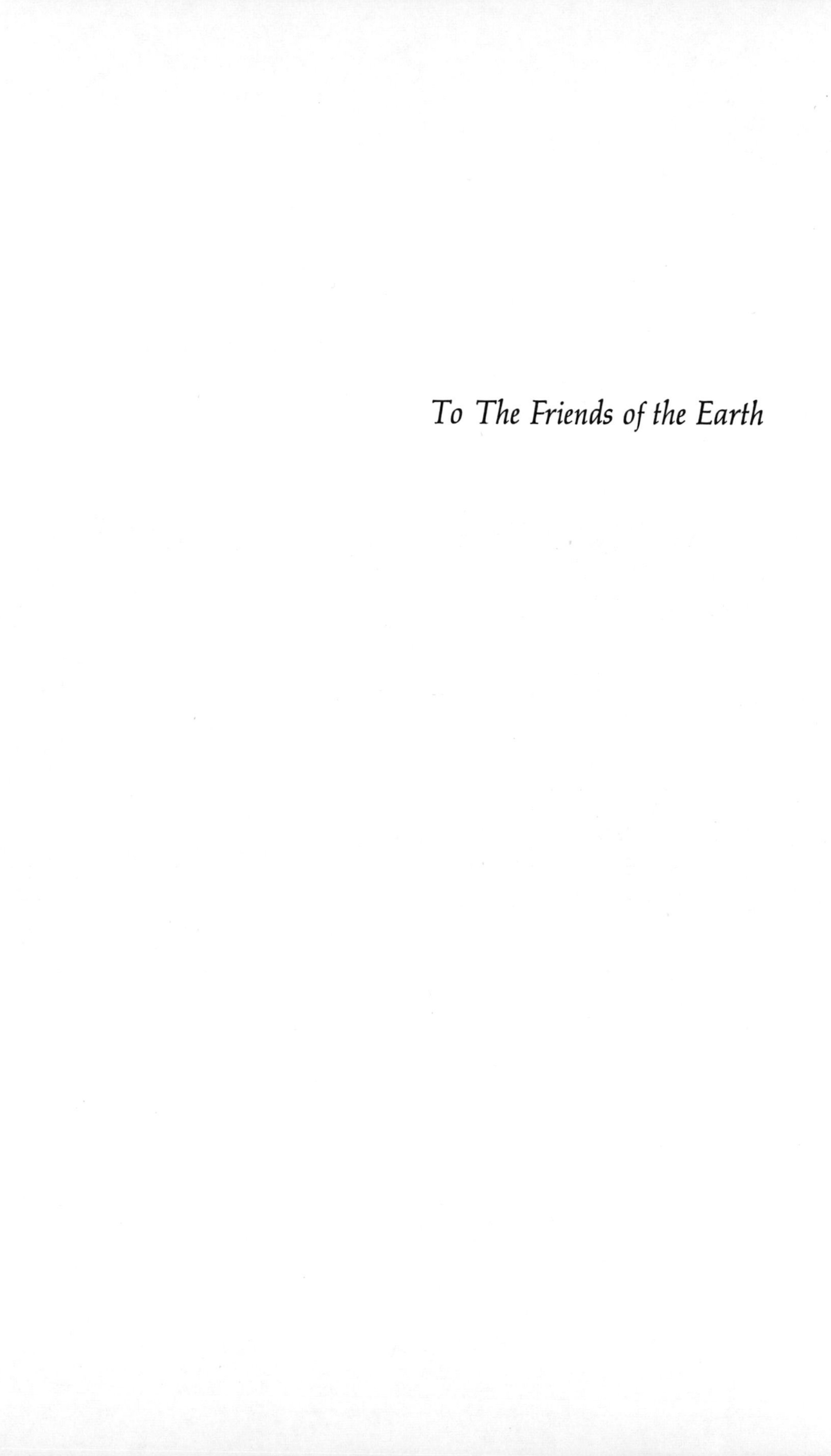

To The Friends of the Earth

Preface

This book has been written with the missionary purpose of corporating the exciting discoveries about global tectonics made in the last ten years into the fabric of an account of earth history. Until now, essentially all historical geology texts discussed global tectonics in the vacuum of a separate chapter, but incorporated little of the subject in their recounting of the earth's evolution. The great impact of plate tectonics on the physical and biological development of the earth has thus been lost.

On the other hand, much of the traditional approach to historical geology, particularly the basic principles of stratigraphy and paleontology, are as important as ever in the process of unravelling the planet's biography. There is a conscious attempt in this book not to submerge the significance these time-proved principles to the nearly overwhelming new discoveries about earth history made in recent years. Both are vital to an understanding of how a geologist interprets earth history and what he has learned.

This book makes no attempt to provide detailed regional stratigraphic histories, but instead concentrates on the broader aspects of plate motions and collisions and the major trends in the fossil record. Rarely are local formation names used and fossil genera are listed only to assist the instructor in providing suitable laboratory material from North American supply houses. The intent is to provide a useful book for students in an introductory one-quarter or one-semester course. Too long a book with too much detail is simply incomprehensible to students in such situations.

Geology is a science that is growing so rapidly that it is hard for any individual to keep abreast of all current hypotheses, theories, and facts. Parts of this book will undoubtedly be superseded even before it is published, but the attempt to be general rather than detailed in the discussion of the earth's evolution should alleviate some of this problem. Several colleagues have read the manuscript and contributed helpful suggestions and criticisms, some of which have been heeded and some have not. All responsibility for factual errors or over enthusiastic interpretations rests with the author, however. The following individuals have read all or most of the manuscript: William B. N. Berry, Rober J. Foster, James Gilluly, Norman J. Silberling, Gary D. Webster, and A. O. Woodford. My colleague at Cal State, Hayward, Alexis Moiseyev, has been an ongoing stimulus for my interest in plate tectonics. As a paleontologist by training, I find it impossible to acknowledge in a limited space all who have stimulated my study of fossils, but Robert V. Kesling and J. Wyatt Durham come most readily to mind. My father provided many of the photographs and my wife, the patience which made this book possible. Finally, I wish to thank Mrs. Anna D. Watson, Mrs. Alice Gould, and Miss Jean McMillan for their many hours of typing and retyping the manuscript.

Contents

HISTORICAL GEOLOGY

The Science of a Dynamic Earth

PART 1

Principles

1

The Present Is the Key to the Past:

Uniformitarianism

The Scientific Method

BASIC TO THE SCIENTIFIC WAY OF VIEWING THE WORLD IS THE SO-CALLED "SCIENTIFIC method." The scientific method accepts as a fundamental postulate, the existence of a real or tangible world that we can perceive with our senses. Debates about the validity of this assumption are left to the philosophers.

To the scientist, the universe consists of two parts, this real world and the conceptual world of the human mind (Fig. 1.1). The boundary between the two is formed by empirical observations. Empirical observations are important to the scientific method at the two ends of the process. First, there is the gathering of data and the formulation of *hypotheses* to explain the data. This process is called *induction.* Using the "laws" of logic and mathematics, the mind discovers that new mental constructs follow from the original hypotheses. These constructs enable the mind to predict certain observations about the real world. This process is called *deduction.* Those hypotheses whose predictions are verified by observing the real world attain the status of *theories.* Notice that to a scientist, the word "theory" is a very strong one implying a high degree of validity, unlike popular usage of the term. After a theory has been tested over a long period of time and found to always make correct predictions within its circumscribed limits of applicability, it becomes a scientific "law," "principle," or "truth."

It is very important to note that no scientific law, principle, or truth is ever considered absolutely true. Although it has always explained and agreed with observations in the past, there is no guarantee that tomorrow someone will not make an observation or perform an experiment that contradicts the law. This then means either that the limits of applicability of the law have to be revised or that the law may have to be discarded. Generally, by the time a hypothesis has become a law, it is the first alternative that actually occurs. This nonabsolutism

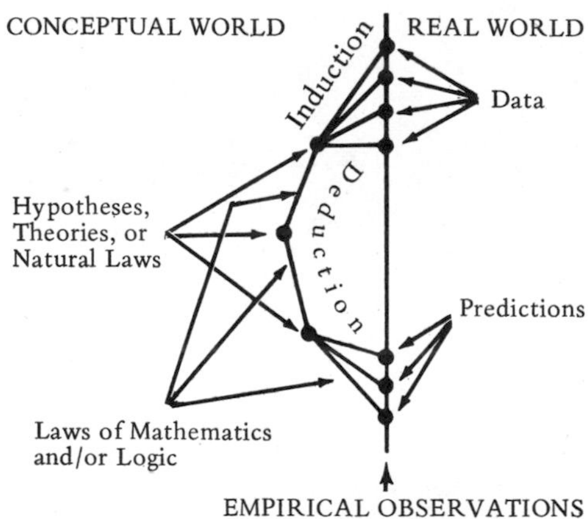

FIGURE 1.1 The Scientific Method. This diagram illustrates the relationships between the real and conceptual worlds as they appear to the scientist.

of science is no doubt surprising to many people who believe that science deals with facts or absolute certainties. It is, however, the construction of theories and laws that truly distinguishes the scientific endeavor and not the collecting of facts. A telephone book is a collection of facts, but it is not scientific. You should also note that the scientific method and, hence, science can be applied to any area of human activity and are not restricted to the pursuits usually considered "scientific." You undoubtedly make use of them in learning and performing most of your daily activities.

In the first half of this text, we will explore the development of those laws and principles which enable geologists to unravel the history of the earth's past. Many will be nothing more than what most of us in this scientific age consider to be "common sense," but the long histories of their implementation testify that this has not always been so. Indeed, to many of the readers of this book who are not familiar with those areas commonly called "sciences," these laws and principles may be anything but intuitively obvious.

Uniformitarianism

Fundamental to all geologic thinking is the principle usually called *uniformitarianism,* but sometimes called *actualism.* The principle states that the physical and biological features of the earth (both past and present) were produced by the same processes that are acting today. This is another way of saying that supernatural causes need not be invoked to explain the real world. To the earliest human intelligence, all of the real world other than his own actions probably seemed inexplicable and, hence, supernatural. Gradually, there must have dawned the notion that some events are "natural" or explicable in terms of the properties of substances and cause-and-effect relationships. This conflict about the supernatural versus the natural has gone on throughout human existence; however, as man's knowledge has expanded, the realm of the supernatural has been constantly decreasing.

Viewed in this context, uniformitarianism ceases to be a principle unique to geology, but becomes the essence of science itself. Uniformitarianism boils down, then, to the statement that the physical and chemical laws of the universe have always been the same. This is implicit in all scientific endeavors, though perhaps the extension of it to the vast immensity of geologic time is truly unique to geology. Without uniformitarianism, history would be impossible for there would be nothing left today that could reliably explain the past. If sandstones or river valleys had not originated in the past as they do now, there would be no way they could tell us anything about the past. To talk of earth history then, is to imply uniformitarianism.

Keep in mind that like all scientific principles, uniformitarianism is an assumption. We can never absolutely prove that our senses are reliable and that we are not being deceived about the reality of the "real world." Such questions are outside the realm of science and in the philosopher's domain.

Historical Recognition

The earliest written records of the application of uniformitarianism are those of the classic *Greeks* of the sixth and fifth centuries B.C. *Xenophanes* (about 600 B.C.), *Xanthus* (about 400 B.C.), and *Herodotus* (about 400 B.C.) all observed fossil seashells in the rocks of mountains high above present sea level. They interpreted the fossil seashells as the remains of once-living marine organisms and their presence in areas well away from and above the level of the present sea indicated that these areas must have been submerged at some time in the past. Similar observations about a changing or dynamic earth were made by the later Greeks Heraclitus, Pythagoras, Empedocles, Democritus, and Aristotle and the Romans Lucretius, Strabo, and Pliny. These ideas were lost in the early Christian era as supernatural explanations of nature dominated the Dark and Middle Ages.

FIGURE 1.2 Leonardo da Vinci (1452–1519).

In the *Renaissance,* the scientific outlook began to reappear again. Around 1500 *Leonardo da Vinci* (Fig. 1.2), that most complete of Renaissance men, revived uniformitarian notions by insisting that fossils were the remains of once-living creatures and that the presence of seashells in the mountains proved that the distribution of lands and seas had changed through time. Unfortunately, da Vinci exerted little influence on his contemporaries. Perhaps because of fear of suffering the fate of others who were imprisoned or killed for similar statements contrary to church decrees, da Vinci wrote all of his notes backwards by using a mirror. As a result, no one could decipher them for several hundred years. From the 1500s through the 1700s only occasional scholars such as the Englishmen *Hooke* and *Burnet;* the Frenchmen *Buffon, Guettard,* and *Desmarest;* and the German *Fuchsel* enunciated uniformitarian thoughts.

The modern demonstration of the validity of uniformitarianism is usually attributed to the Scotsman *James Hutton* (Fig. 1.3) in the late 1700s and the

FIGURE 1.3 James Hutton (1726–1797).

Englishman *Charles Lyell* (Fig. 1.4) in the early to middle 1800s. Both painstakingly gathered together numerous careful observations about the rocks and landforms of Western Europe showing that they could be produced by the same processes that act today if they were allowed plenty of time. Time is the crux of the whole matter. Only tiny amounts of erosion, deposition, uplift, and settling, all apparently without observable effects in the life span of an individual or of a few generations, can be observed today. These can, however, accomplish the wearing away of mountains, the building of deltas, and the shift in position of

FIGURE 1.4 Charles Lyell (1797–1875).

land and seas if only they are given enough time. Once the uniformitarians realized this, it immediately became obvious to them from the detailed rock record that the earth had had a very long and eventful history. As Hutton stated, he saw "no vestige of a beginning—no prospect of an end."

Uniformitarianism and Actualism

Lyell went a bit further in his discussion of uniformitarianism, however. He not only stated that the processes were the same in the past as they are today, but that they have always operated at the same rate and in the same patterns or configurations. Also, there were no additive effects of the same or different processes. In other words, the physical conditions on the earth were never much different than they now are. The positions of lands and seas, mountains and valleys had changed, but the overall average condition had always been similar. Many geologists of his day who accepted the first part of his thesis on uniformitarianism, could not follow Lyell in this second aspect of the principle. One of these men, *Constant Prevost,* coined the term *actualism* for the principle that processes have always been the same, while recognizing that their rates and configurations change through time because of additive effects and chance juxtapositions. Most modern geologists also accept only this part of Lyell's uniformitarianism and because of this, some of them prefer "actualism" to "uniformitarianism." There is evidence that in his later years Lyell also came around to this view of uniformitarianism as evidenced by his acceptance of the Pleistocene Ice Age as a unique event in earth history.

Uniformitarianism versus Catastrophism

In the early 1800s, uniformitarianism was still not accepted by most geologists because it was in competition with a very attractive hypothesis, more acceptable to theologians, known as *catastrophism.* As propounded by its leading exponent,

FIGURE 1.5 Georges Cuvier (1769–1832).

the Frenchman *Baron Georges Cuvier,* (Fig. 1.5) this idea treated the earth's history as a series of catastrophies. Each catastrophe exterminated the existing life, after which a new influx of organisms was either created or migrated in from elsewhere. There were apparently six major catastrophies and these could be inferred to represent the six "days" of Biblical creation if one interpreted "days" to mean "eras" in a loose reading of scripture. Also, the last catastrophe could conveniently be synonymized with Noah's Flood. This made the whole notion very inviting to theologians and to those geologists with theological training or leanings, who were numerous in the 1800s.

Eventually, however, catastrophism passed from the scientific scene as normal deposits in other areas were shown to coincide with the breaks or gaps representing the "catastrophies" in the Paris Basin record studied by Cuvier. Also, some species supposedly exterminated by the catastrophies began to turn up in the rocks overlying these breaks. By mid-century, catastrophism in its original sense was essentially dead in scientific circles, though it persists widely today in fundamentalist religious sects.

Biological Uniformitarianism

Although the uniformity of process in the physical world was largely accepted by 1850, the notion that uniformitarianism also could be applied to biology was not widely recognized at that time. The idea that the slow changes which some scientists saw occurring in organisms today could also produce great change if given enough time seems to follow directly from Lyell's uniformitarianism. It appears remarkable, therefore, that he did not accept this until late in life. If one remembers, however, that Lyell's uniformitarianism was essentially cyclical and non-progressive, whereas evolution is clearly progressive, his reticence is not so surprising.

It remained for another man, *Charles Darwin* (Fig. 4.6), in 1859 to convince the scientific community that the aspect of biologic uniformitarianism called *evolution* also existed. Darwin did for biology precisely what Hutton and Lyell did for geology. He gathered data about the changes in organisms today and indicated what great changes in life they could produce if given great lengths of time. The topic of evolution is very important in earth history and will be discussed at greater length in Chapter 4.

Uniformitarianism Today

"The present is the key to the past" is a widely stated axiom in geology. As we have seen, it has meant different things to different scientists at different times. Today, to most geologists it means only uniformity of processs, physical, chemical, and biologic, and not uniformity of rate. The rates, the times and places of occurrences, the additive effects of occurrences, and the joint occurrences of two or more processes are all variable. They have given rise to the everchanging pageant of earth history we now interpret and not the one of cyclic recurrences envisioned by Hutton and Lyell. Sometimes in the earth's history, continents were higher than at others and, hence, rates of erosion were greater. At other times, when temperatures and precipitation were sufficient, glaciers formed. We have also seen that, in a sense, uniformitarianism is just another way of stating the basic scientific assumption that the physical-chemical-biological laws of the universe are invariable. Geology has shown us how we can use this principle to unravel the past.

Questions

1. What are the differences between hypotheses, theories, and scientific laws or principles?
2. What is the difference between mathematical truth and scientific truth?
3. Cite an example from your everyday life of the application of the scientific method.
4. What is the modern definition of *uniformitarianism?* How does it differ from that of Lyell?
5. How does *catastrophism* differ from *uniformitarianism?*

2

The Time Element:

Superposition, Cross-Cutting Relationships, and Faunal Succession

To the first men who looked at the rocks of the earth's crust, there probably appeared no way of reading a history from their apparent disarray of lithologies and structures. Even today, a nongeologist is often mystified by how a geologist can interpret a time sequence of events by viewing a hillside or roadcut. Although many of the principles involved in making these interpretations could be deduced by simple logic, we have no written record that anyone recognized them until the second half of the seventeenth century.

Superposition

Nicolaus Steno (Fig. 2.1), a physician in Florence, Italy, first formulated the fundamental principle called *superposition*. Simply stated, superposition means that in any sequence of layered rocks undeformed by folding or faulting, each layer was formed after the one below it and before the one above it (Fig. 2.2). It may seem obvious that a layer could not be deposited with nothing beneath it for support, but it was not until 1669 that Steno clearly stated this principle. Notice that superposition does not refer to the age of the particles composing the layer, but to the time of formation of the layer.

FIGURE 2.1 Nicholas Steno (1638–1686).

In Steno's principle there is implied another one, often called the principle of original horizontality. When one observes the formation of layers of sediments at a particular place it usually appears that they form in more or less horizontal bands. Obvious exceptions include foreset beds in deltas, dunes, and alluvial fans, although even here each older layer is beneath each younger layer despite the inclination. The over-zealous application of original horizontality, however, has led to grievous errors in interpreting earth history. Most serious of these is the assumption that layers were truly formed absolutely horizontally and, hence, any widespread layer of the same composition must have formed at the same

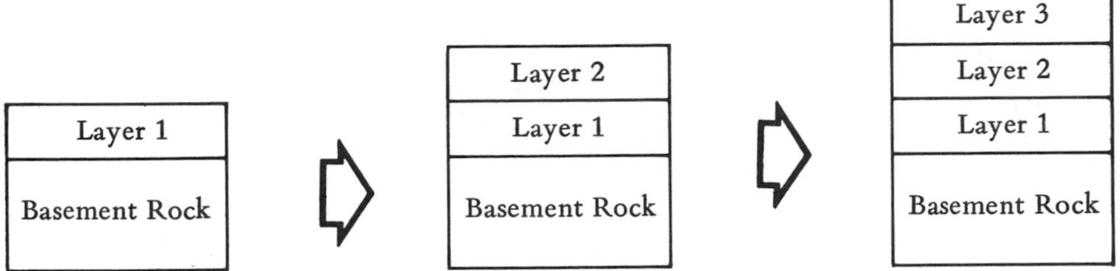

FIGURE 2.2 Superposition. This diagram illustrates the simple, logical reasoning behind the principle of superposition. Each layer was deposited before the one above it.

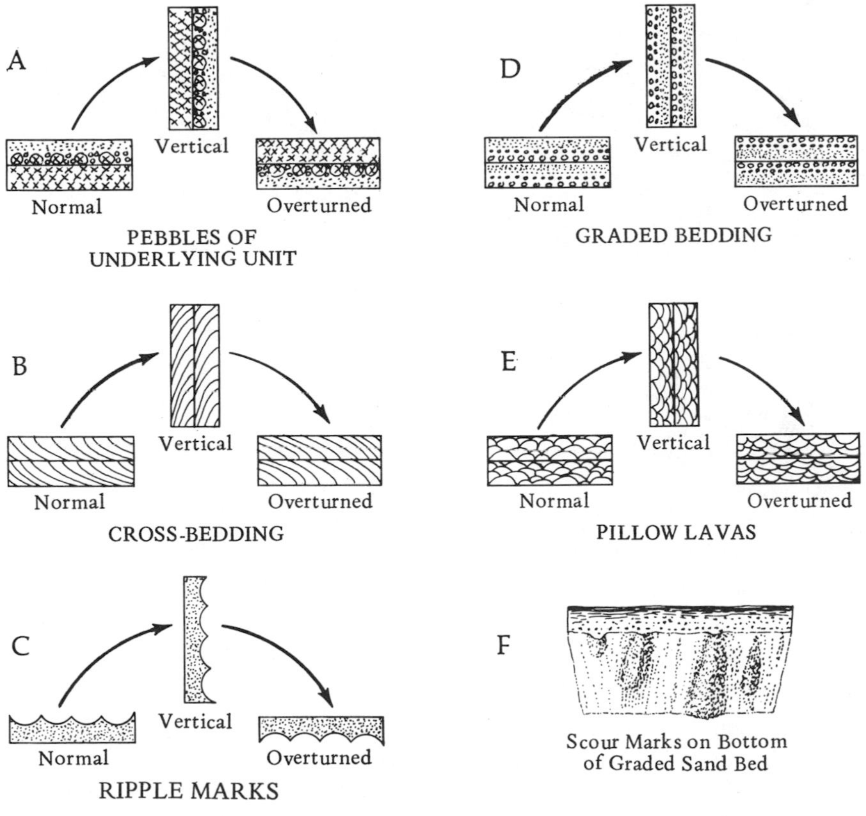

FIGURE 2.3 Tops and Bottoms. Usage of primary sedimentary structures in distinguishing tops from bottoms of rock layers. Each structure has a distinctive shape or orientation relative to the top and/or bottom bedding plane. (A) Pebbles from older bed, (B) Cross bedding, (C) Ripple marks, (D) Graded bedding, (E) Pillows of lava, (F) Linear sole marks.

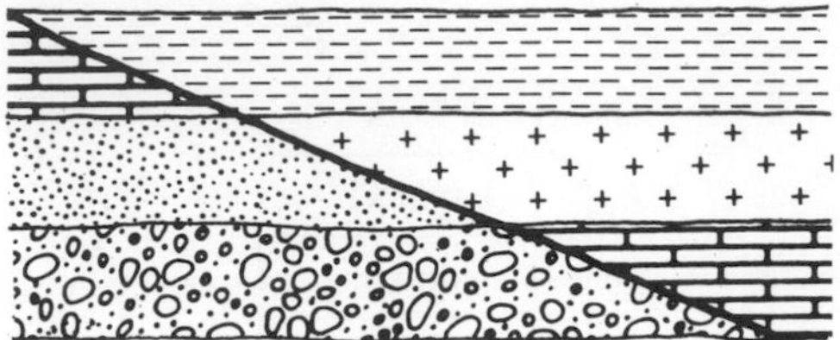

FIGURE 2.4 Recognition of a Fault. Base of units above surface of discontinuity is abruptly truncated. Normally this does not occur along an unconformity as the rock unit overlying the surface of discontinuity parallels the unconformity.

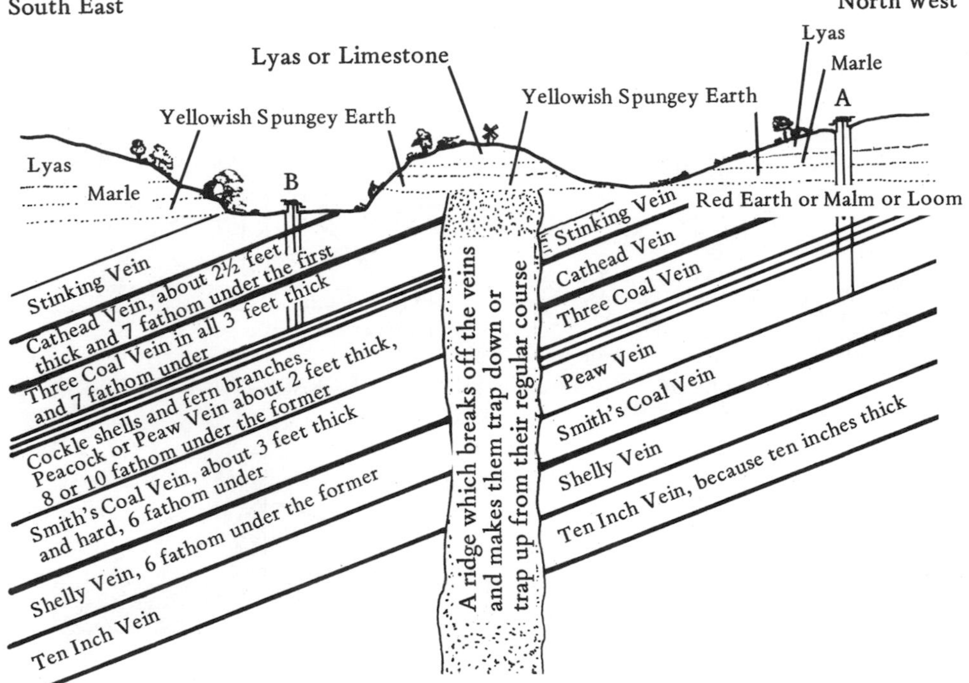

FIGURE 2.5 Strachey's Cross-section. Section drawn across part of Somerset south of Bristol, England in 1719.

time everywhere it is found. The serious consequences of this error will be more fully explored in Chapter 5.

One might logically ask how a geologist deals with rocks deformed by folding and faulting which can overturn the sequence or repeat parts of it. Down through the years, geologists have devised several methods of telling the tops from the bottoms of beds to determine if they are in correct sequence or upside down. These include the distinctive top and bottom shapes of mud cracks, raindrop impressions, oscillation ripple marks, and cross-beds; the gradation from coarse at the bottom to fine at the top of a bed, a feature called graded bedding; and the formation of scour marks and load casts on the bottom of sand layers that made or filled depressions in underlying mud layers. Many of these features and some additional ones are illustrated in Figure 2.3. Sequences inverted or repeated by faulting can often be recognized by the features that occur along fault planes such as gouge and the truncation of strata which lie above a surface of discontinuity (Fig. 2.4). Where these features are not present and where beds have been repeated, superposition alone is usually insufficient to tell the correct age sequence unless the layers can be traced to an area where they are undeformed and in the correct sequence.

Superposition was obviously a very useful tool to tell the *relative* geologic ages of strata (i.e. whether a particular layer was older or younger than another layer in the sequence), but it could have limitations when one dealt with other areas unless one could determine that the same sequence was found worldwide. The 18th century saw the attempt to unravel earth history using only superposition as a guide. We shall now see how well this effort fared.

An Englishman, *John Strachey,* in 1719 and 1725, described and illustrated a cross-section of the strata in Somerset, England (Fig. 2.5). Strachey, using superpositional relationships, worked out a regional succession of strata for this area which contained economically important strata of coal. Note that his diagram clearly portrays an angular unconformity though Strachey apparently did not realize its significance.

A Time Scale Based on Superposition

Geologists then turned their attention to establishing units of wider scope and greater magnitude. *Lazzaro Moro* noted that generally the stratified sedimentary rocks overlie unstratified crystalline rocks and proposed this two-fold subdivision of the rocks of the earth's crust in the early 1700s.

In 1756, *Johann Lehmann,* a German geologist working in the Thuringia area of central Germany (Fig. 2.6), proposed a three-fold subdivision of the rocks of the earth's crust (Fig. 2.7). He called them "Mountains" for their topographic expression. In the cores of the highest mountains and passing superpositionally beneath the younger mountains as one moved outward in either direction were the oldest crystalline (igneous and metamorphic) rocks. He called these the "Ore Mountains" (*Gang-Geburge*) because they were the source of many important metallic ores. They were inferred to have formed at the time of the earth's origin. Outward from the Ore Mountains and superpositionally above them were lower ranges called the "Stratified Mountains" (*Flotz-Geburge*) composed of layered sedimentary rocks with fossils. These limestones, sandstones, and shales were supposed to have formed at the time of the Biblical deluge. Finally, the youngest rocks, forming the lowest foothills of the mountains, were the "Alluvial Mountains" (*Angeschwemmt-Geburge*). These consisted of unconsolidated alluvium, chiefly sands and gravels, and were supposed to have formed since the flood by minor events such as local floods, earthquakes, and volcanic eruptions. To determine the age of a rock (i.e. its time of formation) all one had to do then was determine what type of rock it was.

In 1760, *Giovanni Arduino,* an Italian geologist working in the same general area of northern Italy as Steno, proposed a similar sequence of mountains and their rocks. Arduino's sequence (Fig. 2.8), from oldest to youngest, was: Primitive Mountains composed of crystalline rocks containing metallic ores; Secondary Mountains composed of marbles, limestones and clays which contain fossils;

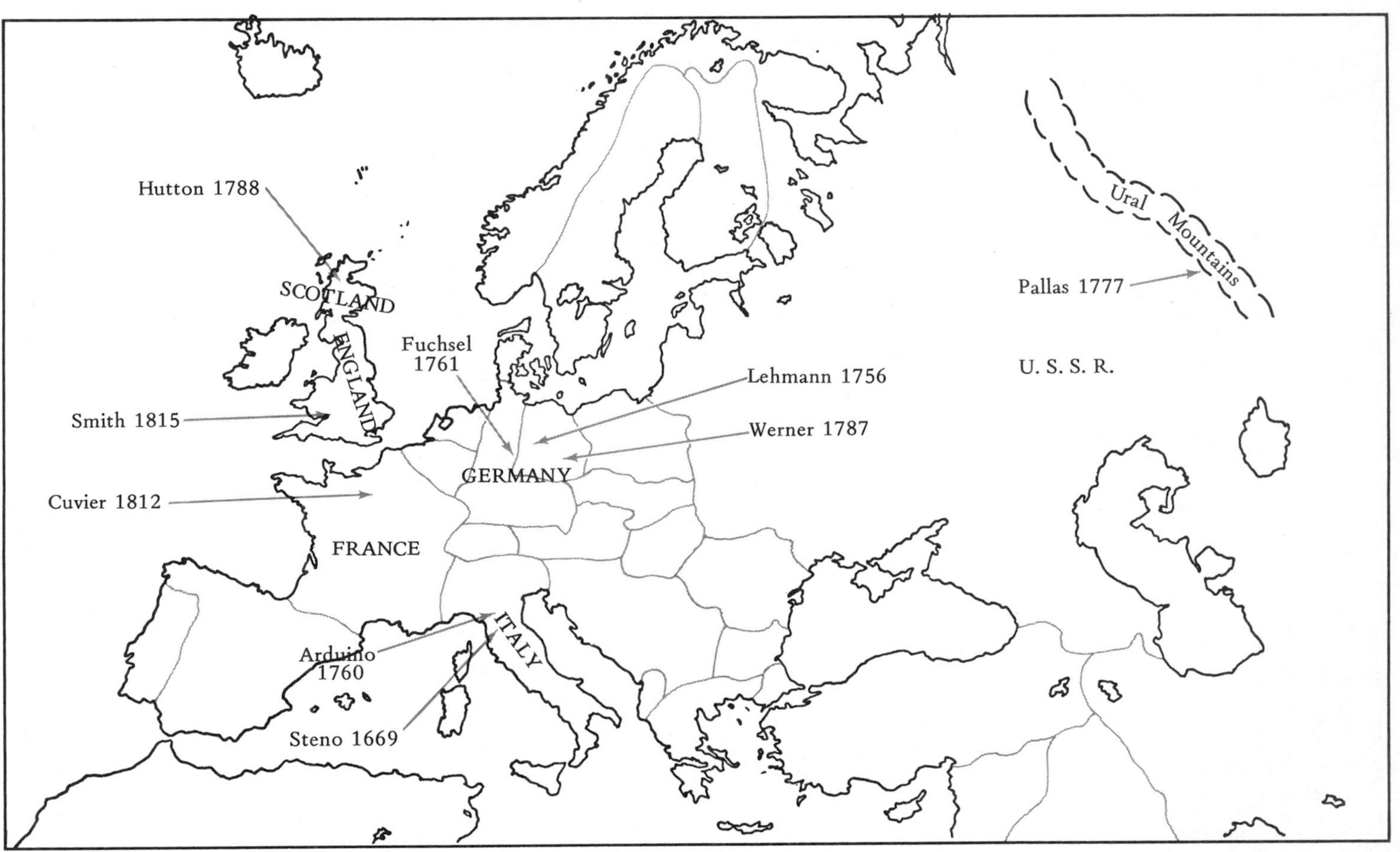

FIGURE 2.6 European Localities of Pioneer Geologic Work. Dates of major publications are indicated.

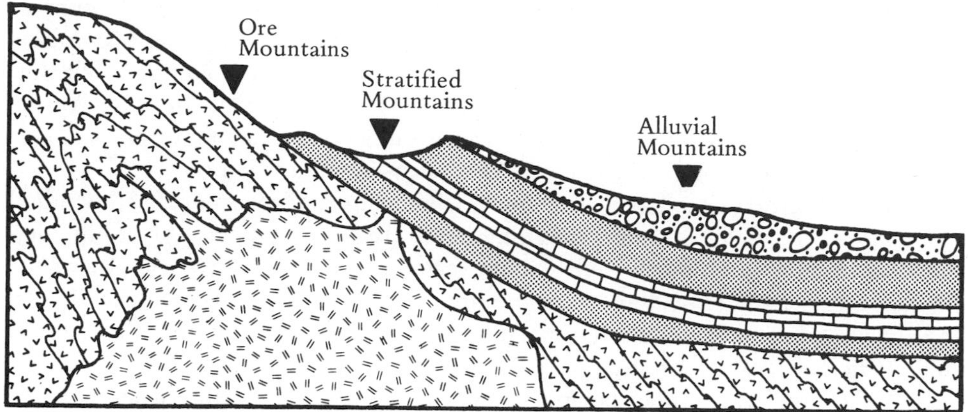

FIGURE 2.7 Lehmann's Subdivisions. A general cross-section showing the superpositional relationships of the various types of rocks in central Germany.

EARLY SUBDIVISIONS				MODERN USAGE			
Arduino 1760	Lehmann 1756 Fuchsel 1760-1773	Werner ca. 1800	English Equivalents	Eras	Periods	Epochs	Alternate Periods
Volcanic	Angeschwemmtgebirge	Angeschwemmtgebirge Neues Flotzgebirge	Alluvium	CENOZOIC Phillips 1841	NEOGENE Hoernes 1853	Holocene Pleistocene	QUATERNARY Desnoyers 1829
						Pliocene Miocene	TERTIARY Arduino 1760
Tertiary			Tertiary		PALEOGENE Naumann 1866	Oligocene Eocene Paleocene	
Secondary	Flötzgebirge	Flötzgebirge	Secondary	MESOZOIC Phillips 1841	CRETACEOUS d'Halloy 1822		
					JURASSIC von Humboldt 1799		
					TRIASSIC von Alberti 1834		
		Ubergangsgebirge	Transition	PALEOZOIC Sedgwick 1838	PERMIAN Murchison 1841		
					PENNSYLVANIAN Williams 1891		CARBONIFEROUS Conybeare & Philips 1822
					MISSISSIPPIAN Winchell 1870		
Primitive	Ganggebirge				DEVONIAN Murchison, Sedgwick 1837		
					SILURIAN Murchison 1835		
					ORDOVICIAN Lapworth 1879		
					CAMBRIAN Sedgwick 1835		
		Urgebirge	Primary	PRECAMBRIAN			

FIGURE 2.8 The Growth of the Relative Geologic Time Scale. The early subdivisions only roughly correspond to the modern ones as they were based on gross lithology rather than fossils.

Tertiary Mountains consisting of gravels, sands, clays, and marls; Alluvium of the plains consisting of unconsolidated debris; and finally, Volcanic Mountains consisting of recent volcanic materials. Arduino, in contrast to Lehmann, felt that this sequence was only a generalization and not universally true. In other words, not all gravels and sands were necessarily younger than all limestones. Geologists had to determine their own local sequences by superposition.

Another German geologist working in Thuringia, *George Fuchsel,* studied Lehmann's Stratified Mountains in some detail. In 1762 and 1773 he subdivided these mountains into nine units he called series, each of which were further subdivided into statumina. He also was the first to erect time units corresponding to the times of formation of these units of rock. In Chapter 5 we will explore the significance of this dichotomy.

In 1777, another German geologist, *Peter Pallas,* extended the ideas of these men to Russia where he recognized Primary, Secondary, and Tertiary subdivisions. His Primary "Mountains" consist of granite, the succeeding Secondary "Mountains" of metamorphic rocks, and the overlying Tertiary "Mountains" of fossiliferous sedimentary rocks. The youngest strata are the unconsolidated sediments of the plains.

Werner

The foremost individual associated with these attempts to determine the earth's history from superposition alone and to tell the age of strata by the type of rock composing them was the German *Abraham Werner* (Fig. 2.9). Werner was the world's leading geologist in the late 1700s and he called his scheme of historical geology the science of geognosy. To Werner, all rocks were the precipitates of a

FIGURE 2.9 Abraham Gottlob Werner (1749–1817).

worldwide ocean which gradually disappeared somewhere. All of the layers precipitated from this ocean were then the same age everywhere and each layer had its own unique composition. At the base were the crystalline Primitive "Mountains" (*Urgebirge*) consisting of granites and gneisses. Above these were the sparsely fossiliferous Transition "Mountains" (*Ubergangsgebirge*) consisting of hard graywackes and limestones. Overlying these were the fossiliferous Stratified "Mountains" (*Flotzgebirge*) consisting of typical sedimentary rocks and, finally above everything else were the Alluvial "Mountains" (*Angeschwemmtgebirge*) consisting of surficial sands and gravels. The earth's crust was like the layers of an onion-skin, except that the outer layers were not continuous because, as the universal sea disappeared, it left islands of older rock which were not overlapped by the younger deposits. Werner only published a brief summary of his work in 1787, but he was a great teacher and produced numerous outstanding students who propagated his work. Werner never left the region, in what is now East Germany (Fig. 2.6), where he taught and, therefore, never checked to see if his sequence was indeed universal or if all rocks formed as precipitates. He and his followers were known as *neptunists* because of this belief about rock formation which brought them into confrontation with those who insisted that some rocks formed by heat.

The *plutonists,* as the latter geologists were called, proclaimed that such rocks as basalt could clearly be seen to be volcanic products and not oceanic precipitates in regions such as the Auvergne of central France. The evidence was so clear that even some of Werner's own students came away from the Auvergne convinced he was wrong. Even more telling a blow to Werner's hypotheses, however, were the works of numerous European geologists who showed that rock units were not worldwide and that each region had its own sequence of strata. Though in a very general, worldwide sense Werner's sequence has some truth to it, it turned out that just about any rock type could occur any place in the superpositional sequence depending on the region.

So as it turned out, superposition was a very useful tool in determining the local sequence of geological events, but it could not be used for matching strata from one region to another where the sequence was different. If rock units had been worldwide, superposition would have been enough to unravel earth history. But they are not and clearly some other principles were needed.

Cross-Cutting Relationships

Some useful lithologic tools were developed by Hutton. Hutton first clearly pointed out another almost intuitively obvious principle which is often called the law of *cross-cutting relationships.* Hutton stated that any body of rock that cut across the boundaries of other units of rock must be younger than those it cuts (Fig. 2.10). How else could it cut them if they were not already there in the first place? This principle led Hutton to recognize what we call today intrusive contacts, where bodies of igneous rock cut across the boundaries of pre-existing rocks. He called them *unconformities.* In defining an unconformity, Hutton noted that in Scotland the upturned edges of the folded Transition rocks were truncated by the essentially flat-lying and overlying Secondary rocks (Fig. 2.11). (This today

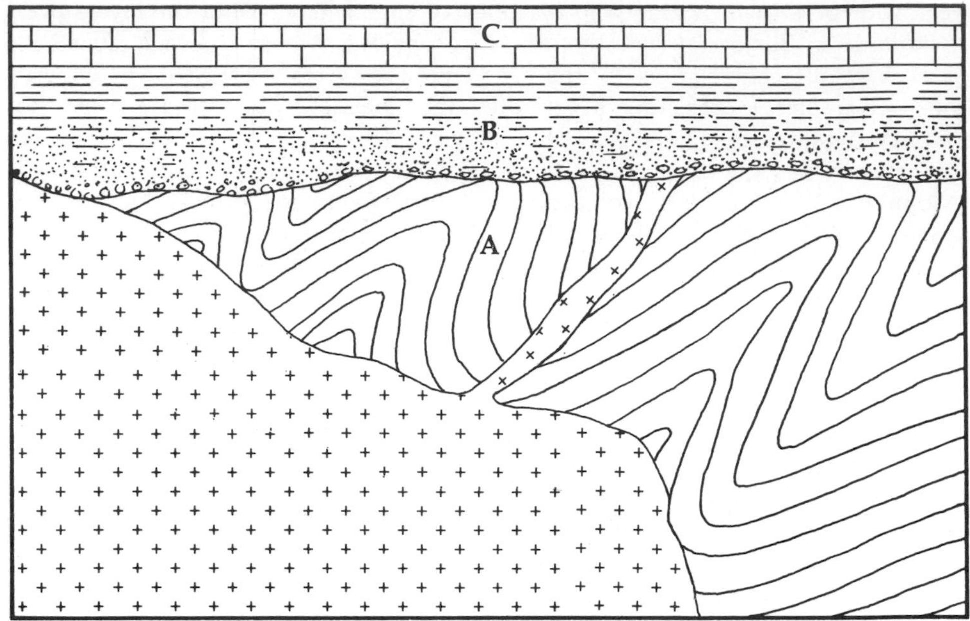

FIGURE 2.10 Cross-Cutting Relationships. The granite cuts Formation A and thus is younger. Formation B cuts both the granite and Formation A and, hence, is younger than both. (Formation C is younger than Formation B by superposition.)

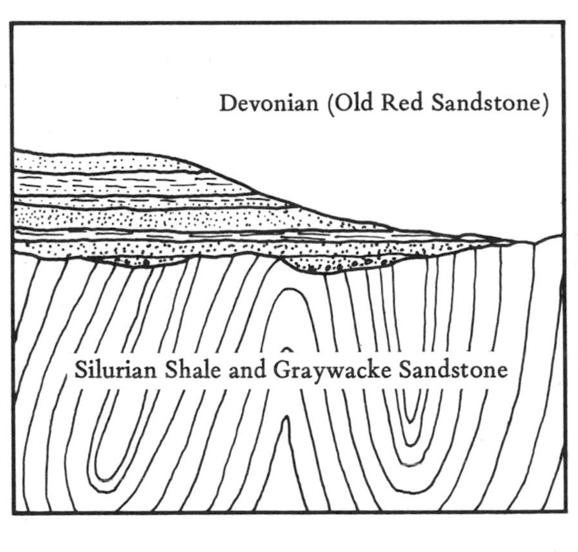

FIGURE 2.11 The Angular Unconformity at Siccar Point, Scotland. Left, diagrammatic cross-section of the unconformity; right, view of the unconformity in the field. (Crown Copyright Geological Survey photograph. Reproduced by permission of the Controller of Her Britannic Majesty's Stationery Office.)

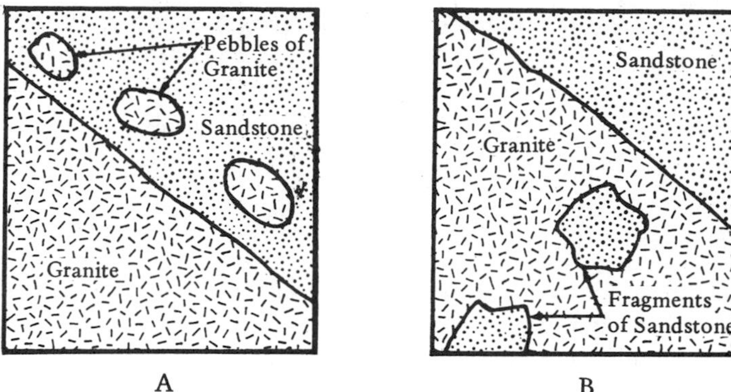

FIGURE 2.12 The Law of Inclusions. The fragments of the older rock occur in the younger rock. (A) An erosional surface, a nonconformity, (B) An intrusive contact between the granite and the intruded rock.

would be called an angular unconformity, only one type of unconformity.) Hutton reasoned that the Transition rocks must have originally been deposited horizontally as are most rocks and then later deformed. Then there must have been a long period of erosion, represented by the unconformable surface, during which time the folded mountains were reduced to a nearly horizontal surface so that when the overlying Secondary rocks were deposited, they were also essentially flat-lying.

Inclusions

A similar principle first clearly formulated by Hutton is the so-called *law of inclusions*. This principle states that a rock body containing fragments of another body of rock must be younger than the unit whose fragments it contains (Fig. 2.12). Again, it seems intuitively obvious that a rock unit could not contain fragments of another rock body unless that other body existed first. But it remained for Hutton's sharp eyes and mind to bring the obvious to the attention of other geologists. Hutton's principles made it possible to determine the proper time of formation of new units or sequences that appeared in an area, but they still did not solve the problem of determining the relative ages of rocks in different areas with different sequences. A much more powerful tool was needed.

The Significance of Fossils

As early as 1777, the churchman *Giraud de Soulavie* had worked out a five-fold subdivision of the rocks in the Vivarais district of southern France and noted that, although the type of rocks were repeated in the succession, the types of fossils were not. There was a definite change in the fossils as one went up the sequence. Fossils more like present forms of life gradually replaced the older more primitive types. Soulavie suggested, therefore, that fossils and not lithology might be the key to determining the relative age of a rock. This was a momentous discovery, but Soulavie was far ahead of his time and it was not until many years later that the principle he suggested became widely accepted. Before we pursue the matter of using fossils to date rocks, however, we must backtrack a bit to explore the history of man's opinions concerning fossils.

The Origin of Fossils

Fossils are usually defined as the remains or traces of prehistoric organisms. (Although the word originally meant anything dug from the earth, through the centuries it has gradually been restricted to its present meaning.) Fossils were noted even by prehistoric men who buried them with their dead, perhaps attaching mystical significance to them. The classical *Greeks* again provide our first written account of opinions concerning fossils. Clearly Xenophanes, Xanthus, and Herodotus recognized that the seashells they found in the mountains of the eastern Mediterranean area were the remains of once-living creatures. Aristotle, though he felt some fossils such as fish bones were the remains of organisms, generally believed that fossils formed in rocks under the influence of celestial bodies.

This sort of geological astrology enjoyed wide popularity in the Dark and Middle Ages when the Christian Church accepted Aristotle's words as dogma. Other explanations arose in the Middle Ages to compete with the idea of celestial influences forming fossils. Some theologians felt fossils grew in the rocks from seeds, solutions, or vapors. Others thought they were unsuccessful creations cast aside by the Creator or tricks of the devil to deceive man about the earth's true history. Another popular notion was that fossils were "sports of nature," creations of Mother Nature in a playful mood. One of the major reasons that fossils could not be accepted as the remains of organisms was because some were clearly unlike living types and the benevolent God who had saved all creatures from the flood in the Ark could not have allowed any species to perish. All of the above ideas persisted in learned circles even until the 1700s and, of course, many persist widely today in the nonscientific community.

In the Renaissance, many original thinkers such as the Italians da Vinci, Fracastoro, and Cesalpino revived the idea of an organic origin of fossils. Their ideas did not enjoy wide circulation, however, because of fear of the persecution that sent others to the dungeon or the stake for heresy.

By the late 1600s and early 1700s, however, it was apparent that fossils were too much like living things to be inorganic. Most scientists, at the urging of *Hooke* in England, *Vallisnieri* in Italy, and *Scheuchzer* in Germany, had come around to accepting fossils as the remains of organisms. Once the scientific tide had swung in this direction, however, the theologians came forward with a new explanation. They then argued that fossils were the remains of creatures killed by the flood. (No one yet would admit the possibility of extinct organisms, however.) Scheuchzer, in particular, championed this view and passed off a fossil giant salamander as the remains of a poor human sinner who had perished in the flood (Fig. 2.13). He also debated with others about whether the Deluge occurred in the spring or the fall because of reproductive structures found on fossil plants.

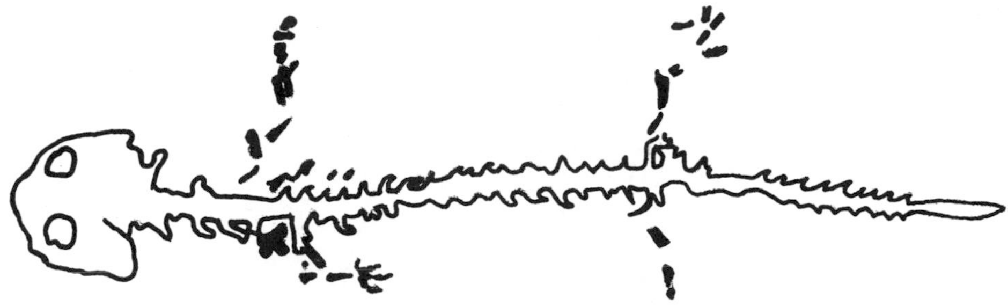

FIGURE 2.13 Scheuchzer's Homo Diluvii Testis.

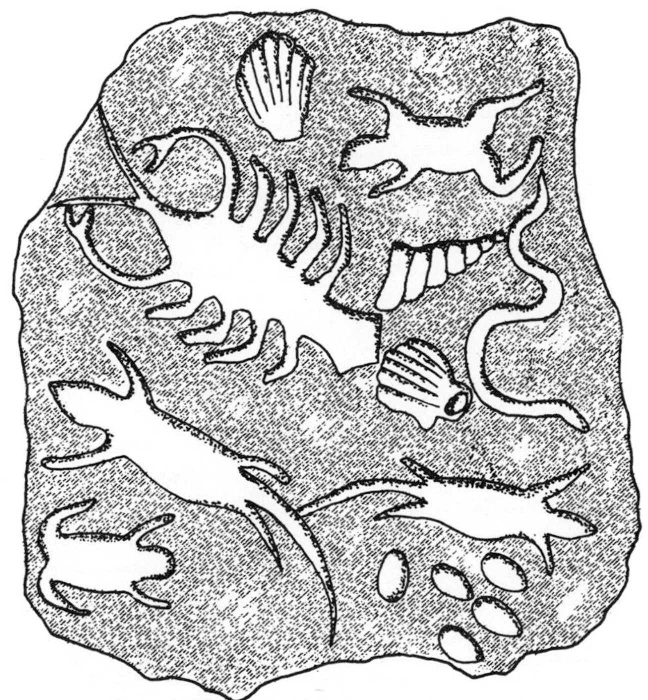

FIGURE 2.14 Beringer's Pseudofossils.

One of the last opponents of the organic origin of fossils was a German professor, Johannes Beringer. Beringer was deceived into believing that a series of carvings made on rocks by another professor were "unique manifestations of nature" (Fig. 2.14). These included not only the images of animals and plants, but Hebrew writing and celestial objects. Beringer described and figured these artifacts in 1726 and only later discovered the hoax, much to his chagrin.

By the time of Soulavie, then, the organic nature of fossils (an aspect of biologic uniformitarianism) was generally accepted in educated circles, although the dispute continued about whether or not they were embedded in the rocks by the Biblical flood. Still, except for isolated individuals such as Soulavie, no one had yet noted anything significant about the pattern of occurrence of fossils in the

FIGURE 2.15 William Smith (1769–1839).

Strata	Thickness	Springs	Fossils Petrifactions, etc., etc.	Descriptive Characters & Situations.
1. Chalk	300	Intermitting on the Downs	Echinites, Pyrites, Mytilites, Dentalia, Funnel-shaped Corals & Madrepores, Nautilites, Strombites, Cochliae, Ostreae, Serpulae	Strata of Silex, imbedded.
2. Sand	70			The fertile vales intersecting Salisbury Plain & the Downs.
3. Clay	30	Between the Black Dog and Berkley		
4. Sand & Stone	30	Hinton, Norton, Woolverton, Bradford Leigh.		Imbedded is a thin Stratum of calcareous Grit. The Stones flat smooth and rounded at the Edges.
5. Clay	15			
6. Forest Marble	10		A Mass of Anomiae and High-waved Cockles with calcareous Cement	The Cover of the upper Bed of Freestone or Oolyte.
7.	60		Scarcely any fossils besides the Coral	Oolyte resting on a thin bed of Coral. Prior Park, Southstoke, Twinny, Winsley, Farley Castle, Westwood, Barfield, Conkwell, Monkton Farley, Coldhorn, Marshfield, Coldashton.
8. Blue Clay	6	Above Bath		
9. Yellow Clay	8			
10. Fullers Earth	6			Visible at a distance by the Slips on the declivities of the Hills round Bath.
11. Bastard Ditto and Sundries	80		Striated Cardia, Mytilites, Anomiae, Pundils and Duck-Muscles.	
12. Freestone	30		Top-covering Anomiae with calcareous Cement, Strombites, Ammonites, Nautilites, Cochliae Hippocephaloides, fibrous Shell resembling Amianth, Cardia, prickly Cockle, Mytilites, lower Stratum of Coral. Large Scollop, Nidus of the Muscle with its Cables	Lincombe, Devonshire Buildings, Englishcombe, Englishbatch, Wilmerton, Dunkerton, Coomhay, Monkton Combe, Wellow, Mitford, Stoke, Freshford, Claverton, Bathford, Batheaston, and Hampton, Charlcombe, Swainswick, Tadwick, Langridge.
13. Sand	30		Ammonites, Belemnites	Sand Burns.
14. Marle Blue	40	Round Bath	Pectenites, Belemnites, Gryphites, High-waved Cockles.	Ore Balls – Mineral Springs of Lincombe, Middle Hill, Cheltenham.
15. Lias Blue	25		Same as the Marl with Nautilites, Ammonites,	The fertile Marl Land of Somersetshire, Twerton, Newton, [illegible], [illegible], Stanton Prior, Timsbury, Paulton, Marksbury, Farmborough, Corston, Hunstreet, Burnet, Keynsham, Whitchurch, Salford, Kelston, Weston, Pucklechurch, Queencharlton, Norton-Malreward, Knowle, Charlton, Kilmersdon, Babington.
16. Ditto White			Dentalia, & fragments of the Enchrinni.	
17. Marlstone, Indigo and black Marls	15		Pyrites and Ochre	A rich Manure
18. Red-ground	180		No fossil known	Pits of Ruddle. Beneath this bed no fossil, shells or animal remains are found; above it no vegetable Impressions. The waters of this Stratum petrify the trunks in which they are conveyed, so as to fill them in about 15 years, with red Matricle, which takes a fine polish. – Highlittleton.
19. Millstone				
20. Pennant Stone			Impressions of unknown plants resembling Equisetum	
21. Grouse				Fragments of Coal & Iron Nodules. – Hanham, Brislington, Mangotsfield, Downend, Winterbourn, Forest of Dean.

FIGURE 2.16 William Smith's Table of Strata near Bath, England dictated to Reverend Townsend (1799).

rocks of earth's crust. The man who finally brought to the attention of the scientific world that fossils did not occur haphazardly but in a definite sequence was *William Smith* (Fig. 2.15).

Smith was an English canal-builder in the late 1700s and early 1800s who traveled widely about Great Britain in his work and carefully collected fossils and noted the rock units from which they came (Fig. 2.16). He discovered that types of fossil organisms occur in the rocks of the earth's crust in a definite and determinable order and, hence, the rocks formed during a particular interval of geologic time can be recognized by their characteristic fossil content. This most significant geologic discovery has become known as the principle or *law of faunal succession.* It is the key that has opened the door to our present understanding of earth history, for here at last, in fossils, are things in the rocks that do not occur in a random fashion but in a regular and irrevocable order. Although a half-century would elapse before Darwin, with his discovery of evolution, would provide the reason for this sequence, the application of faunal succession was immediate. In the 1800s, in relatively rapid succession, the large-scale subdivisions of relative geologic time we now call *periods* and *eras* were established. Many of these were in turn subdivided into still smaller units, all by utilization of this principle.

As so often happens in science, when the general level of knowledge reaches a certain point, several workers independently discover great principles. The discovery of the principle of faunal succession was no exception. At the same time that Smith was making his momentous discovery known in Great Britain, *Cuvier* and *Brongniart,* working in the Paris Basin of France, came to exactly the same conclusion about the time value of fossils (Fig. 2.17). Cuvier, you may recall, turned faunal succession into a line of support for catastrophism. This idea eventually died when it was discovered that, in complete sequences, changes in life were gradual and not abrupt. As we shall see later, abrupt faunal changes do occur in the rock sequence of a particular region, but these are caused either by unconformities or breaks in the geologic record of the region, or by a change in environment in the region carrying with it new organisms adapted to the changed ecology.

The Relative Geologic Time Scale

We shall now chronologically explore the naming and establishment of the geologic periods (Fig. 2.8), and eras that occurred as soon as faunal succession became well established. (Those established before about 1830 received their original names before the discovery of faunal succession, but the present usage follows the implementation of the principle.)

The Periods

The oldest period name that comes down to us today is the *Tertiary* which dates from Arduino's subdivisions of 1760 and thus antedates faunal succession. The Tertiary "Mountains," however, were described sheerly on a lithologic basis (gravels, sands and limestones) and a superpositional basis (between the Secondary and the Alluvial "Mountains"). The type area or region where the unit was first described is in the Apennine Mountains of northern Italy (Tuscany). The usage of the word in its present sense based on fossils dates from the work of Lyell and the French paleontologist (one who studies fossils) *Deshayes* in the 1820s and 30s.

Likewise, the *Jurassic* Period derived its name from a prominent lithologic unit. The German geographer *Alexander von Humboldt,* in 1799, named the great cliff-forming carbonates of the upper Stratified "Mountains" in the Jura Mountains

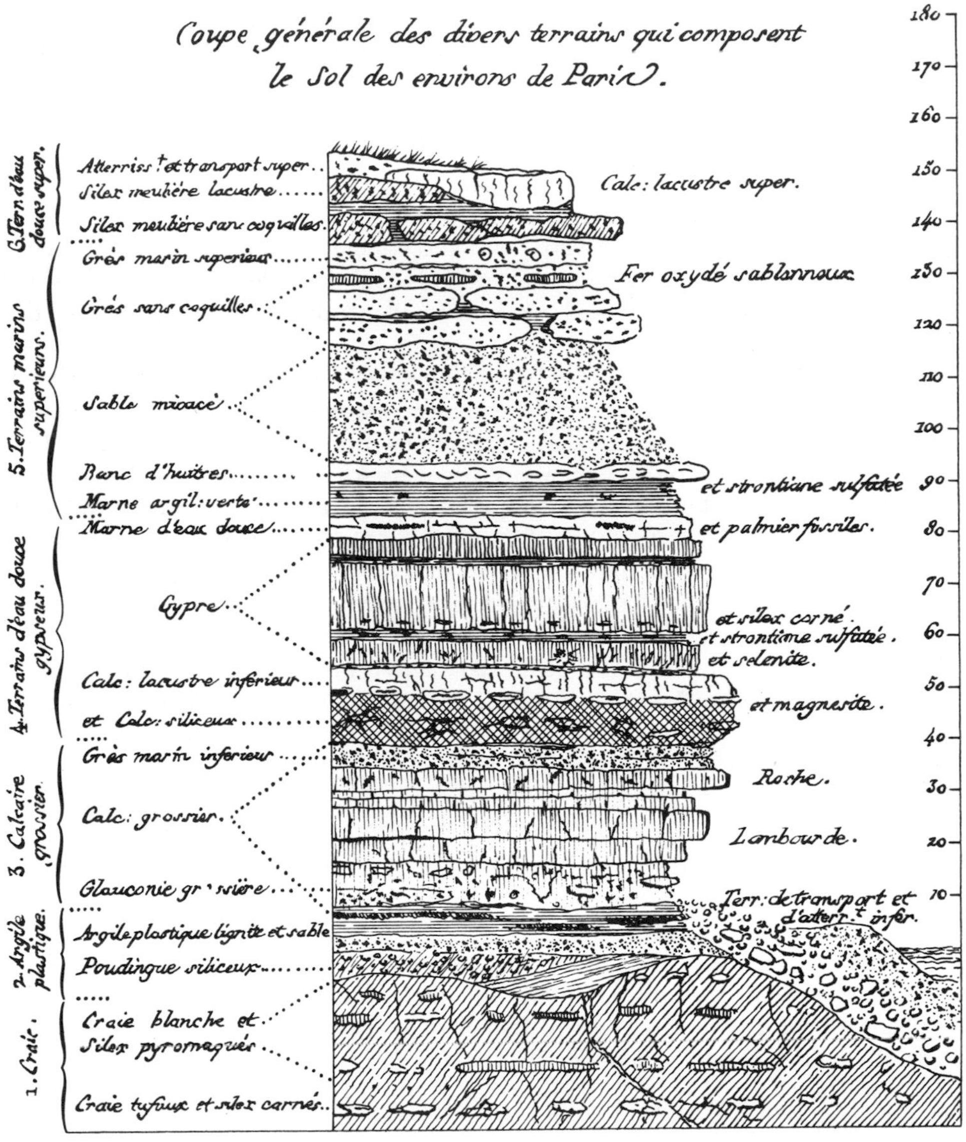

FIGURE 2.17 Cuvier and Brongniart's Table. This cross-section and table of strata in the Paris Basin show the sharp lithologic and faunal changes inferred by Cuvier as supporting his catastrophist theories.

of northwest Switzerland the Jura Limestone. The present usage of Jurassic based on its fossil content dates from the work of the German geologist *Leopold von Buch* in the 1830s.

In 1822, *Conybeare* and *Philips,* two British geologists, published a textbook of British geology in which they named a *Carboniferous* "Order." This subdivision of rocks and the time of its formation was near the boundary between Werner's Transition and Stratified "Mountains." It consisted of the well-known coal-bearing rocks of the British Isles and some underlying sandstones and a limestone. Some mention of the unique fossils found in the unit was made, but it was several years before the detailed studies of fossils both in Britain and western Europe were made that gave the period its present meaning and allowed rocks of that age to be recognized even where there are no coal beds.

Also in 1822, a Belgian geologist, *D'Omalius d'Halloy,* working along the English Channel in northern France, applied the name *Cretaceous* "Terrain" to the prominent chalk beds and underlying greensands, ironsands, and marls exposed there. (*Creta* is the Latin word for "chalk.") These rocks were at the top of the Secondary or Stratified "Mountains" of Werner and other workers. The Cretaceous chalk forms the famous White Cliffs of Dover on the English side of the channel. Again, the original usage was based on rock type and superposition, but application of the principle of faunal succession by many British and French geologists lead to the description of the large and distinctive Cretaceous assemblage of fossils that could be used to recognize rocks of this age whether or not chalk was present. The most famous of these workers was *Alcide d'Orbigny* who published (1840–1855) a monumental work on Cretaceous fossils in which he subdivided the Cretaceous into numerous smaller units which could also be recognized by their fossils.

In 1829, a French geologist, *Paul Desnoyers,* used the term *Quaternary* for the unconsolidated alluvium on the top of the rock record in northern France. The term was proposed as an additional complement to the older Primary, Secondary, and Tertiary units. Quaternary rocks of the type area contain few fossils which could be used to recognize them elsewhere, so the basis for its recognition again was lithologic and superpositional. However, with the discovery that Quaternary at least partly coincided with the Pleistocene described by Lyell (see below) and based on fossils, and with the discovery of the unique ice age that occurred in the period, its recognition became reliable.

Lyell, in addition to his other great achievements in geology, described and named four major subdivisions of the Tertiary which he called periods. Here for the first time, we see fossils used as an essential part of the definition of units of geologic time. In 1833, Lyell proposed the *Eocene* (Dawn of the Recent), *Miocene* (Less Recent), and *Older* and *Newer Pliocene* (More Recent) Periods in order of decreasing antiquity. These subdivisions were based on the percentage of species of living mollusks found in the rocks of the different ages. All identifications were made by the French paleontologist *Deshayes* to insure consistency. The Eocene, which Lyell admitted might be worthy of further subdivision, contained 3½% living species and had its type area in the Paris and London Basins. The Miocene contained 18% of species which survived to the present and had its type area in western France. The Older Pliocene contained 33–50% living species and had its type area in northern Italy, and the Newer Pliocene contained 90% living species and had its type area in southern Italy. Lyell also had a *Recent* Period for the time since the appearance of man. In 1839, Lyell followed his original instincts and separated the Newer Pliocene as a distinct period which he called the *Pleistocene* (Most Recent). The Older Pliocene became simply Pliocene. As time went on, most geologists came to feel that Lyell's periods of the Tertiary were of much smaller magnitude than the other periods of geologic time and they were reduced to smaller scale subdivisions of periods called epochs. In the most-used scheme, the Eocene, Miocene, and Pliocene are epochs of the Tertiary Period, the Pleistocene and Recent are epochs of the Quaternary Period. With the discovery of the ice age that occurred in the Pleistocene, many geologists ignored the original definition of Pleistocene based on fossils and began to consider it synonymous with the glaciation. This switch in definitions, which even Lyell eventually accepted, has caused no end of difficulty in attempting to establish the boundaries of the Pleistocene.

In 1834, a German geologist, *Friedrich von Alberti*, described and named the *Triassic* "Formation," a series of terrestrial redbeds separated by a fossiliferous marine limestone which is prominent in central Germany. Despite the lithologic name, the Triassic was defined on the basis of its fossils which allow strata of this age to be recognized even where there was no tripartite sequence such as the Alps which contain some of the finest Triassic sequences known.

The naming of the next three periods described is inextricably interwoven with the lives of two famous English geologists, *Adam Sedgwick* and *Roderick Murchison* (Fig. 2.18). The two men set out in the 1830s to study the essentially unknown Transition rocks which were known to be well-exposed in Wales. Sedgwick went to the base of the section in the northwest where the rocks are highly deformed and sparsely fossiliferous. Murchison started along the Welsh-English boundary at the top of the Transition (the base of the well-known Old Red Sandstone unit, then placed in the Carboniferous) where the rocks were less deformed and more fossiliferous. In 1835, the two men proposed two new periods

FIGURE 2.18 (A) Roderick Impey Murchison (1792–1871) (B) Adam Sedgwick (1785–1873).

of geologic time, the *Cambrian,* named by Sedgwick for the ancient Roman name for Wales (Cambria), and the *Silurian,* named by Murchison for the Silures, an ancient tribe that lived in the type area. The Cambrian "System" was based solely on lithology and superposition while the Silurian "System" was based on its fossil content which enabled it to be recognized where lithologies were different. Because the two men began at opposite ends of the sequence, it turned out that some of the strata in the middle were claimed by each for his unit. This led to a great feud between the two which affected British geology for decades. Eventually Murchison got the upper hand because he alone had described fossils for his period and so every fossil that was found in the lower Transition rocks gradually became Silurian. With no faunal basis for his period, Sedgwick saw his Cambrian swallowed up by the Silurian, particularly after Murchison became Director of the Geological Survey of Great Britain. It was not until British geologists finally found distinctive fossils in the type Cambrian and a fine fossil assemblage was described from Czechoslovakia by Barrand in the mid-1850s that the Cambrian was salvaged. Final resurrection, however, awaited the resolution of the original problem of overlap, settled in 1879 when this was set aside as the Ordovician Period.

While Sedgwick and Murchison were still friends, they studied the geology of Devonshire and Cornwall in southwestern England. Here were a series of contorted graywackes and limestones similar to the type Cambrian. Both men were ready to call the rocks Cambrian until a local paleontologist, *Lonsdale,* pointed out that the fossils (particularly the corals) were intermediate in structure between those of the Silurian and the Carboniferous. Eventually convinced that fossils were more important than lithology in recognizing time, the two men proposed the *Devonian* "System" in 1839 for the rocks deposited during this interval. They also noted that in the rest of Britain, the rock unit in this interval between the Silurian and the Carboniferous was the Old Red Sandstone, quite different lithologically and paleontologically from the marine rocks of Devonshire. The recognition that different rocks and fossils can be formed at the same time in different environments was an important geologic discovery. Such units of the same age, but from different environments, are now called *facies* and will be discussed in Chapter 3.

In 1840–1841, *Murchison,* by then one of the premier geologists of the world, was invited along with several continental geologists to Russia by the Czar to study the geology of that country. While in Russia, Murchison discovered a flat-lying, fossiliferous sequence of marine rocks between the Carboniferous and Triassic, an interval which was represented in western Europe only by a thick sequence of poorly fossiliferous terrestrial redbeds. These marine rocks were well exposed on the western side of the Ural Mountains in the province of the town of Perm. In 1841, Murchison applied the name *Permian* "System" to rocks of this age as recognized by their abundant fauna.

In 1853, a German geologist named *R. Hoernes* proposed the term *Neogene* System for rocks of Miocene and Pliocene age. Subsequently, the Pleistocene and Recent epochs were also included.

Heinrich von Beyrich, a German geologist, took Lyell at his word that the Eocene Period probably could be split, and erected the *Oligocene* (Little Recent) for the upper third of the Eocene. The type area of the Oligocene is in northern

Germany and contains numerous fossils by which it can be recognized there and elsewhere.

In 1866, another German geologist named *C. F. Naumann* proposed that the Eocene and Oligocene Epochs be grouped together in a *Paleogene* Period. Other Europeans prefer calling the interval the *Nummulitic* after a characteristic type of fossil foram (see Chapter 7) found in it.

Philipp Schimper, a Frenchman, in 1874 completed the process of breaking up Lyell's Eocene by splitting off the lower third as the *Paleocene* (Early Dawn of the Recent). The type area in northern France contains some beds with marine fossils, but consists chiefly of terrestrial beds with plant fossils intermediate in structure between those of the Cretaceous and the Eocene, the basis of its definition. Mammal fossils from these terrestrial beds have been the chief tools for its recognition elsewhere, however.

In 1879 *Charles Lapworth,* a Scotsman, finally resolved the long-standing dispute over the overlap of the Silurian and the Cambrian. Lapworth found that these disputed strata contained an assemblage of fossils different from both the Silurian and Cambrian. He proposed the name *Ordovician* for the rocks, after the Ordovices, an ancient tribe inhabiting the type area in Wales. Although the Ordovician "System" has been widely adopted since Lapworth's time, many Europeans still refer to these rocks as Lower Silurian.

In 1885, *Paul Gervais* proposed the term *Holocene* for Lyell's Recent Epoch at the Third International Geologic Congress. The term gave a certain sense of symmetry to the names of the Tertiary-Quaternary epochs making them all end in the suffix *-cene.* Not all workers accept Holocene and Recent as synonymous, however, reserving Holocene for the time since the retreat of the last ice sheet and Recent for that part of the Holocene since the beginning of written history.

At this point, the standard geologic time scale reached the form it now has in most of the world (Fig. 2.8). At that time, most geologists believed that the periods were "natural" subdivisions of earth history whose boundaries were marked by worldwide mountain-building disturbances that produced the unconformities which bounded most of the periods in their type areas (areas where originally described). Once again, as in Wernerian times, geologists extrapolated from what they saw in a limited area, Europe, to the whole world and once again they were wrong as we shall see later. However, the geologists of the late 1800s and early 1900s operated on this principle and when they left Europe and studied the geology of other continents, they found major unconformities at other positions in the rock sequences of those regions. Reasoning that these also must represent worldwide period boundaries that had been overlooked in Europe, they began the wholesale erection of new periods. North America was particularly fruitful in this regard as the patriotic Americans jumped into the contest with new periods such as Ozarkian, Canadian, and Comanchean. Only two, however, ever enjoyed any widespread acceptance. Those are the Mississippian and Pennsylvanian which roughly correspond to the Early and Late Carboniferous of the remainder of the earth. Only in North America and places where North American geologists have worked extensively are these terms widely used. As geologists realized in this century that the periods are not natural units based on the occurrence of physical events, but instead are arbitrary units whose boundaries are fixed at certain events in the evolution of life, the tendency to propose new periods has ceased.

In 1891, *H. S. Williams* of the U.S. Geological Survey took the lithologic term *Mississippian,* proposed by *Alexander Winchell* in 1870 for the Lower Carboniferous limestones of the Upper Mississippi River Valley, and his own term *Pennsylvanian,* proposed for the overlying coal measures of the state of Pennsylvania, and elevated them to the status of epochs or major subdivisions of the Carboniferous Period. In 1906, *T. C. Chamberlin* and *R. S. Salisbury* published a textbook of geology that was to serve as an American standard for decades. In the book, the authors raised Mississippian and Pennsylvanian to the status of periods and the widespread use of their book has made this terminology nearly universal in North America. Recently, with the recognition of the arbitrary nature of periods, there has been some tendency to revert to the traditional European usage so that everyone has a common terminology.

Before leaving the discussion of periods, one recent development should be noted. For the vast thickness of apparently unfossiliferous and typically highly deformed strata beneath the Cambrian, the name Precambrian has been in long usage. In recent years, many microfossils (fossils so small they must be studied with a microscope) have been found throughout the Precambrian sequence, and in the youngest rocks some large or megafossils have been found. These mega-

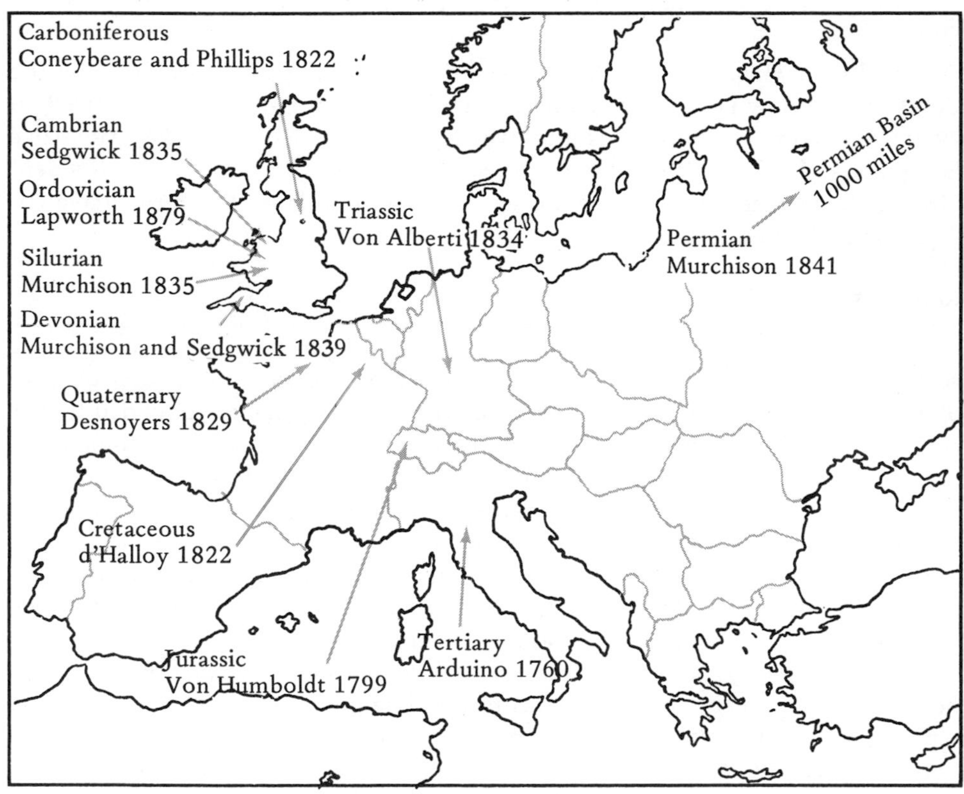

FIGURE 2.19 Type Localities of the Geologic Systems.

fossils are much different from those of the following Cambrian and it has been suggested that a new period, the *Eocambrian,* be established for the period of time in which these organisms lived. This usage is being rapidly adopted by many workers.

The Eras

The periods of geologic time are grouped into larger units called *eras.* The names most widely used for these are *Paleozoic* (Ancient Life) for the Cambrian through the Permian Periods, *Mesozoic* (Middle Life) for the Triassic through the Cretaceous Periods, and *Cenozoic* (Recent Life) for the Tertiary and Quaternary or Paleogene and Neogene Periods. These terms were given their present definition by John Phillips in 1841. Phillips did not coin the word Paleozoic as that term had been proposed earlier by Sedgwick (1838) for the Cambrian-Silurian rocks. Phillips originally spelled the youngest era Kainozoic or Cainozoic as is still often done in Britain. Those geologists who use the Eocambrian Period generally include it in the Paleozoic Era. Several names have been proposed for eras in the Precambrian such as Archeozoic, Archean, Proterozoic, and Algonkian but none have been widely accepted as they were based chiefly on units of rock and not time.

Eons

To avoid the awkward term "Postprecambrian" in talking about the rocks of the Cambrian (or now Eocambrian) onward in which megafossils occur, *Chadwick* in 1930 proposed to group the eras into two larger units called eons. *The Cryptozoic Eon* (Hidden Life) consisted of the Precambrian and the *Phanerozoic Eon* (Evident Life) contained the Paleozoic, Mesozoic, and Cenozoic Eras. The terms have enjoyed some degree of acceptance.

Boundaries

A significant point to note about the original development of the geologic time scale is that the major subdivisions were erected on the basis of major sequences of strata which contain unique fossil assemblages and which are bounded by unconformities representing much missing time. The early geologists assumed that these natural geologic breaks which bounded most of their subdivisions were worldwide. Their illusions were shattered, however, as they explored other regions of the world and found strata with fossils which represent the time missing in the type areas. Clearly it had become necessary to establish arbitrary boundaries for the subdivisions of geologic time. The problem has always been complicated by the fact that the strata which are missing in the type area but found elsewhere are not part of the original definition of any of the units, and, hence, great arguments have arisen as to the time period to which they ought to be assigned. To this day geologists are still involved in the process of fixing the period boundaries. If exact boundaries are to be set for the subdivisions of the relative geologic time scale, some methodology had to be established.

If natural catastrophies had occurred, as workers such as d'Orbigny thought, the task would be easy, but the evidence showed they had not. A German geologist named *Friedrich Quenstedt* suggested in the 1850s that the probable key to the whole problem would be the noting of the precise stratigraphic range of each fossil. In other words, geologists had to record, on a centimeter by centimeter basis, the exact thickness of strata through which a single species of fossil ranged. Quenstedt was convinced that analysis of these precise ranges could provide units with boundaries.

The man who brought this scheme to fruition was another German, *Albert Oppel,* who studied the precise ranges of Jurassic fossils, particularly the ammonoids (relatives of today's pearly nautilus and squid). Using the overlapping ranges of species of fossils, Oppel was able to delineate small-scale units called *zones* which were characterized by the joint occurrence of species not found together above or below their zone (Fig. 2.20). Some species began at the base of the zone, others ended at its top, still others ranged throughout the zone. Here at last were units with boundaries. Ultimately, the boundaries between rock units assigned to one period and those assigned to the overlying period are also the boundaries between successive zones. In other words, the Jurassic-Cretaceous boundary is also the boundary between the highest Jurassic zone and the lowest Cretaceous zone. The principle of zonation based on the overlapping ranges of species of fossils has been successfully applied in this country in our own time by *Robert Kleinpell* and others. Oppelian zones are often called *concurrent range zones* to differentiate them from other types of zones established by geologists.

Recently, some geologists have noted that there is no compelling biologic reason why several species should begin and end at the same point in time. Hence, even zonal boundaries are "fuzzy," that is the boundary of two zones marked by the beginning of two species and the end of two others is in itself a very narrow zone and not a line. Some newer, perhaps more precise, statistical tools have been developed by geologists such as Shaw who is working at such extremely fine levels (Chapter 10).

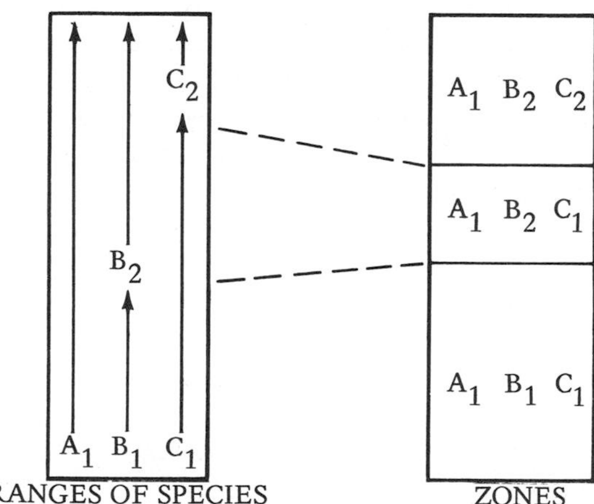

FIGURE 2.20 Oppelian Zones. Each zone is characterized by a unique fossil assemblage that does not occur above or below. Each species in this hypothetical example is designated by a number.

Questions

1. What is the law of superposition and who formulated it?
2. How can oscillation ripple marks be used to tell the tops from the bottoms of beds?
3. What was the basic rock sequence determined by Lehmann, Arduino, and Werner?
4. What was the fundamental conflict between neptunism and plutonism?
5. If a dike cross-cuts a sedimentary layer, which rock unit is older?
6. Why do you suppose medieval men did not recognize the organic nature of fossils?
7. What was the major contribution of William Smith to geology?
8. Name the two geologists who named the Cambrian and Silurian periods and then disputed their boundaries.
9. What is the basis of the relative geologic time scale?
10. Why have geologists had endless debates about system boundaries?

3

The Perversity of Sedimentation and Organisms:

Facies & Biogeography

WE HAVE SEEN HOW GEOLOGISTS ORIGINALLY BELIEVED THE EARTH'S CRUST TO RESEMBLE an onion-skin, composed of several worldwide layers of similar composition, and how this proved to be an illusion. One of the most striking demonstrations of this fact came in the late 1830s when Sedgwick and Murchison discovered that the marine limestones and graywackes of the type Devonian area with their coral and brachiopod fossils occupied the same level in the rock sequence of Great Britain as the terrestrial Old Red Sandstone with its fish fossils found to the north and east.

Facies

Also in the 1830s, a Swiss geologist, *Amanz Gressley*, working in the Jura Mountains, traced several Triassic and Jurassic rock units laterally along the mountains and discovered that both the types of rock and the types of fossils changed. For example, in the hypothetical illustration shown in Fig. 3.1, shale with its characteristic assemblage of fossils gradually changes laterally into a limestone with another characteristic assemblage of fossils. To such lateral variations in both the rocks and fossils of the same age, Gressley applied the term *facies* in 1838. Gressley recognized that these differences were caused by differences in the environment in which the rocks were deposited and the organisms lived, just as there are different environments on the earth's surface today each of which produces its own sediment type and is inhabited by an assemblage of organisms adapted to its conditions (Fig. 3.2).

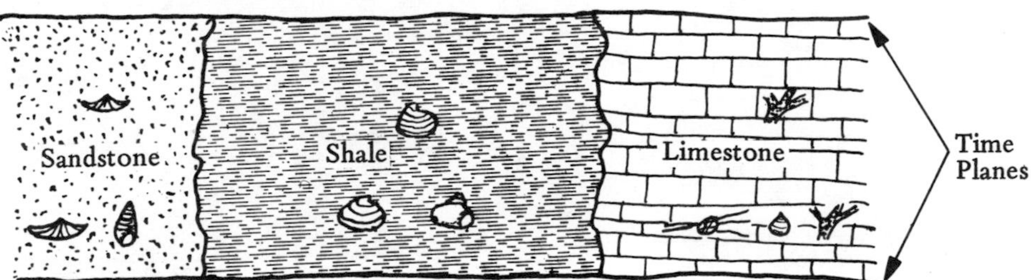

FIGURE 3.1 Facies. This cross-section of a hypothetical region illustrates Gressly's original concept of facies as lateral differences among rocks of the same age. Unfortunately the word has been used in many other ways in stratigraphy since his time.

Gressley and many other geologists experienced a great disillusionment, however, with the realization that species of fossils also were not worldwide and, therefore, fossils also were not perfect tools for dating rocks. Some geologists despaired that unless one could trace strata laterally to determine their time equivalency, there might be no infallible way to date rocks in new areas. The French geologist *Constant Prevost* stated that similar lithologies and fossil assemblages only indicate equivalent environments and not equivalent ages.

Not all geologists dispaired this much, however, for many noted that certain kinds of fossils were found in rocks formed in many different environments. Most notable among these were the ammonites, typically coiled shells with all the loops in one plane and subdivided by internal partitions with crinkly edges (Fig. 7.69). These fossils are very abundant in the widely exposed Mesozoic rocks of western Europe. In 1842, the French geologist *Alcide d'Orbigny* proposed to group together all rocks deposited during the same interval of time, regardless of their lithology and general fossil content, into units called *stages.* Most stages were defined on the basis of widespread ammonite species though some were based on species of other organisms with wide environmental tolerances. D'Orbigny established a sequence of stages for both Jurassic and Cretaceous rocks. In d'Orbigny's work, we see the first formal separation of the terms applied on the basis of their composition and those applied on the basis of the time of their formation. In Chapter 5 we will explore this most important development in some detail.

D'Orbigny's work was refined in the 1850s when *Albert Oppel* established, for the Jurassic, *zones,* still smaller subdivisions of rocks based on their time of formation as defined by species of wide-ranging fossils such as ammonites. Thus by the middle of the last century, Europeans widely recognized the existence of facies, that is, the fact that different types of rocks and fossils could form at the same time and that some separate sort of terminology was needed to differentiate units of rock based on their lithology and units based on their time of formation.

The concept of facies did not achieve widespread acceptance in North America until much later, however. In the late 1800s and early 1900s in North America, an almost Wernerian concept of geologic history prevailed. In this theory of geology, whose leading exponents were Chamberlin, Ulrich, and Schuchert, the periods were natural subdivisions of time based on alternate advances and withdrawals of the seas from the oceans onto the continents. At the beginning of a period, all the continents were high and the seas were marginal; in the middle of the period the continent was flooded; and at the close the sea withdrew. The thin, widespread rock units of the continental interior were easily visualized as the contemporaneous deposits of a sea in full flood. Each rock unit, such as the

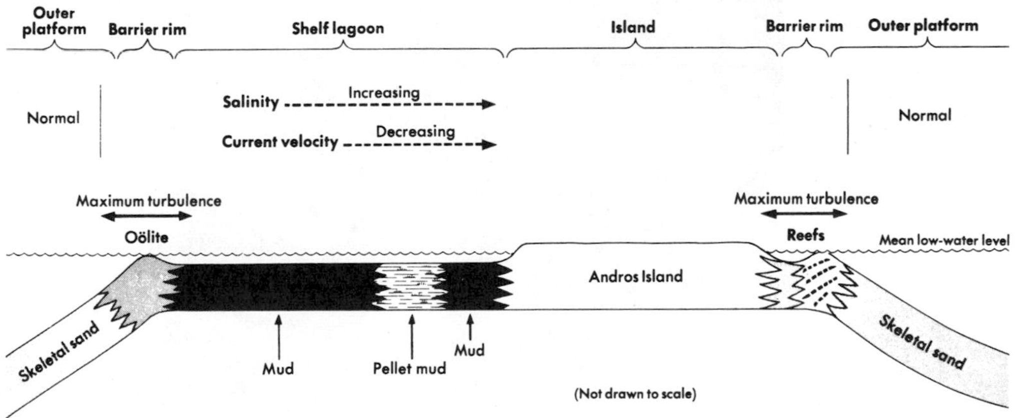

FIGURE 3.2 Cross-Section and Map of Bahama Banks. These illustrations show the distribution of the different sedimentary and biological environments of the Bahama Banks today and some of the factors influencing these patterns. (Con't on pages 42 and 43.) (From Leo F. Laporte, *Ancient Environments,* © 1968 Prentice-Hall, Inc., Englewood Cliffs, N.J.).

Ordovician St. Peter Sandstone (Fig. 3.3A) or the Devonian-Carboniferous Chattanooga Shale (Fig. 3.3B), was considered a deposit of essentially the same age throughout its vast lateral extent. It apparently occurred to few people that these units are of different ages in different places (Fig. 3.4) depending on the geographic location of that environment at various times (see also Chapter 8). Because these deposits covered virtually the whole continental interior and because

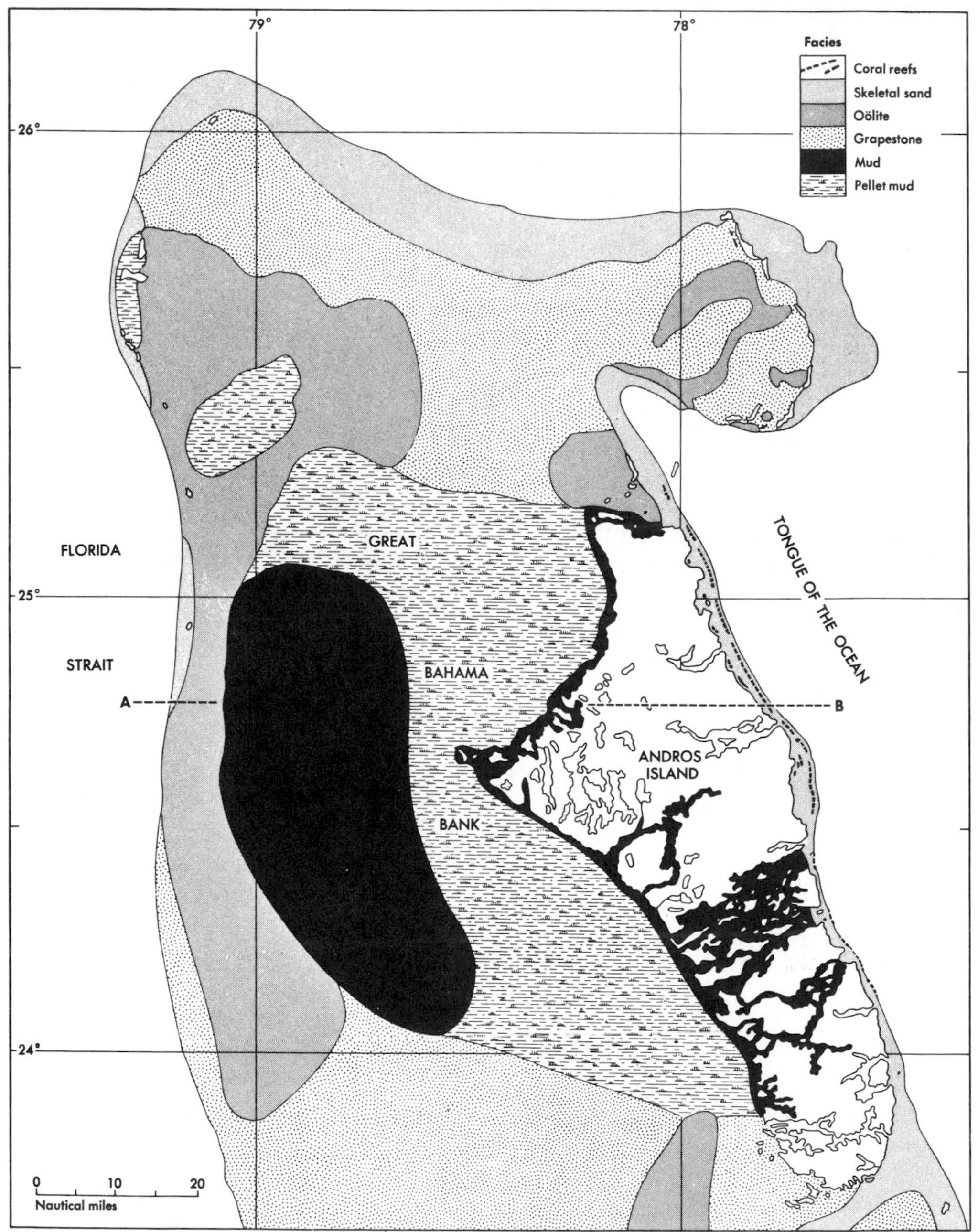

FIGURE 3.2 (con't)

they were assumed to be the same age throughout, it was natural that no one ever found the lateral facies equivalents of these units. Therefore, most American geologists rejected facies out of hand. *E. O. Ulrich*, one of the most important geologists in America in the first decades of this century, was completely convinced of these ideas. His great influence kept in the background the few men who recognized this conceptual error and supported the idea of facies. *Amadeus*

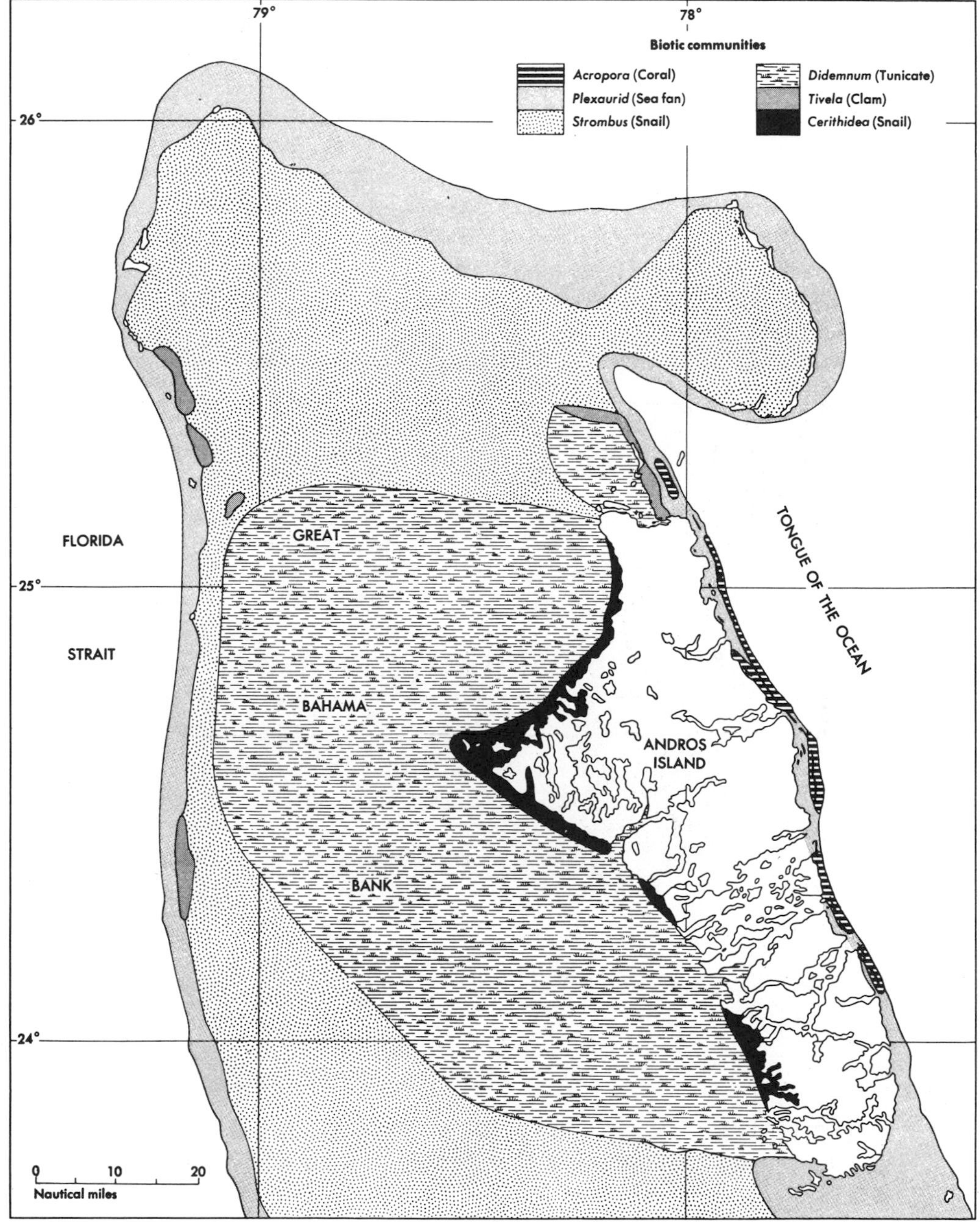

FIGURE 3.2 (con't)

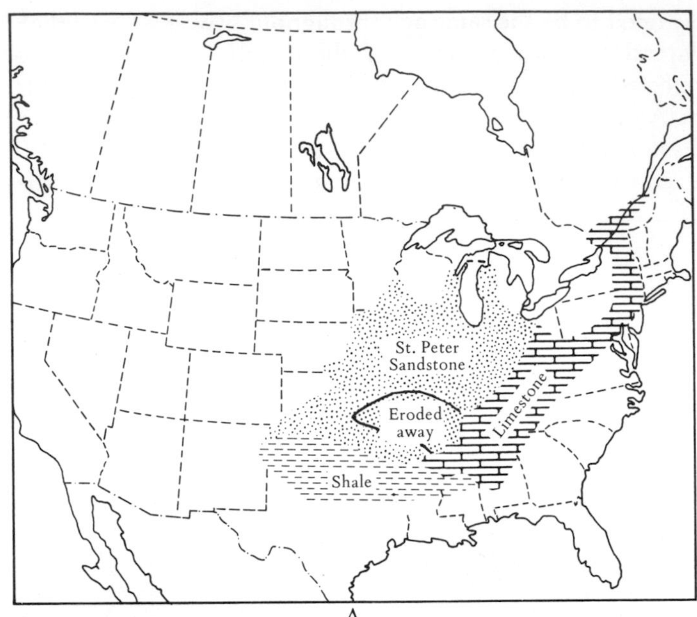

A

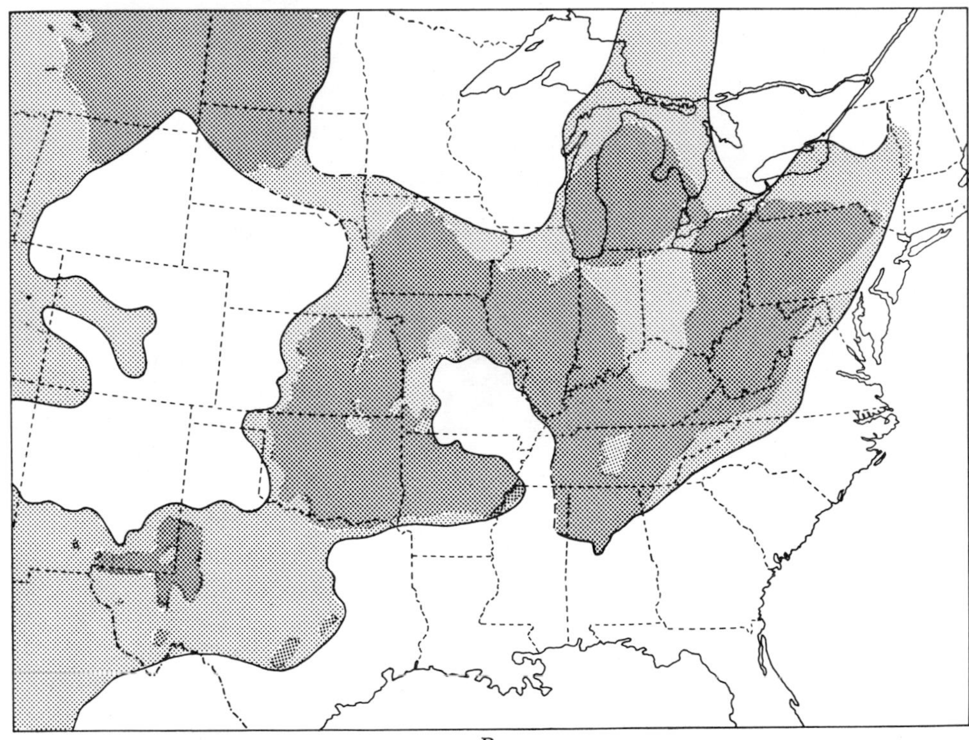

B

FIGURE 3.3 Distribution of (A) St. Peter Sandstone and (B) Chattanooga Shale. These are typical widespread rock units that form on cratons. (B) From Thomas H. Clark and Colin W. Stearn: *Geological Evolution of North America,* Second Ed., Copyright © 1968. The Ronald Press Company, N.Y.)

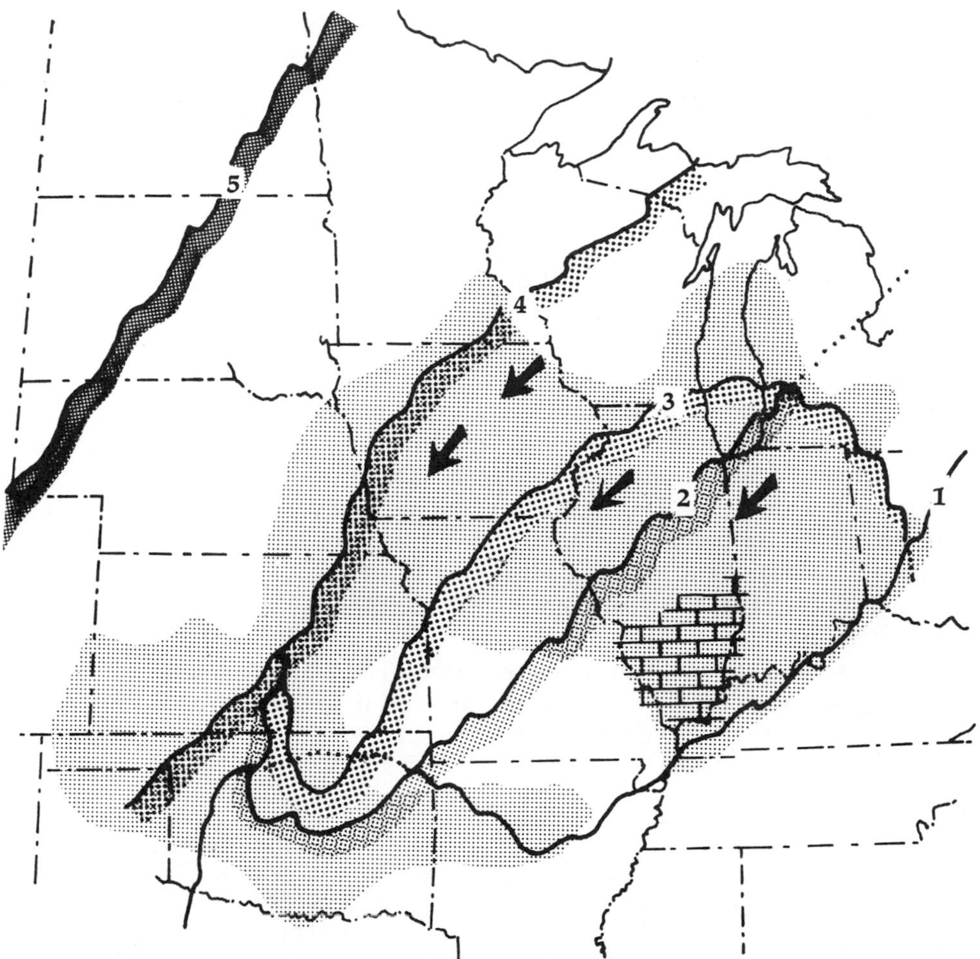

FIGURE 3.4 Different Ages of St. Peter Sandstone. This map illustrates several successive positions of the shoreline during the deposition of the St. Peter Sandstone. Thus, the rock unit is of different ages in different places. (From Edgar Winston Spencer, *Basic Concepts of Historical Geology*, Copyright © 1962. Thomas Y. Crowell Company, Inc. Reprinted with permission of the publisher.)

Grabau, a leading proponent of facies, actually left the country to direct the Geological Survey of China. Beginning in the 1930s, however, the concept of facies enjoyed a renaissance in this country. There are still many geologists, however, who preach facies, but do not practice it.

As we noted above, geologists generally recognize two major subdivisions of facies, those based on the type of rock and those based on the assemblage of fossils. These are known, respectively, as *lithofacies* and *biofacies*. The study of the causes of the lateral variations in rocks formed at the same time is called the science of *sedimentation* and is usually discussed in detail in physical geology texts. Only a brief survey is provided here. The study of the causes of lateral variations in assemblages of organisms living at the same time is called the science of *biogeography.* This topic is seldom covered in geology courses and will be discussed in more detail.

Sedimentation

The major clue to the environment of formation of most sedimentary rocks is their *texture,* the way the grains fit together and their sizes. The two basic textures (Fig. 3.5) of sedimentary rocks are *clastic,* composed of individual fragments, and *crystalline* composed of interlocking crystals that have grown in place.

The major clastic or detrital rocks, in order of decreasing particle size, are conglomerate, sandstone, and shale or mudstone (Fig. 3.6). In addition, there are also many clastic limestones. In clastic rocks, the smaller the particles and the more rounded they are, the longer and generally farther they have been transported. Larger particles cannot be moved much or far without being broken down, and smaller particles, once suspended in a fluid, take longer to settle out and, hence, are carried farther from the source. A particle of any size will become more rounded the more chances it has to collide with other particles. Another aspect of clastic texture is size *sorting* (Fig. 3.5) which refers to the degree to which the sediment consists of particles of uniform size. A well-sorted sediment consists of particles of about the same size, a poorly-sorted one, of

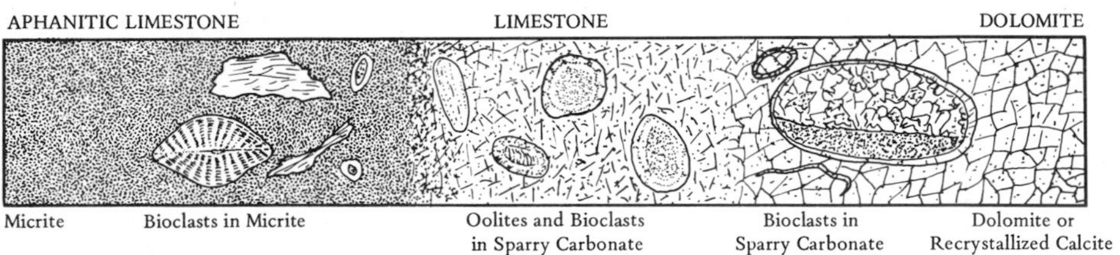

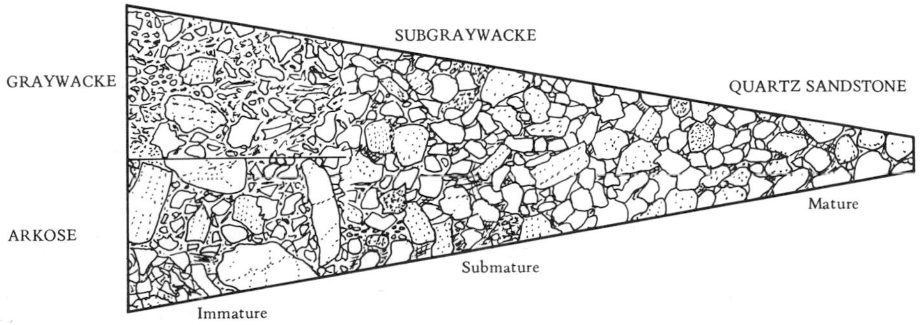

FIGURE 3.5 Textures of Sedimentary Rocks (Carbonates and Sandstones). Note crystalline texture of some carbonates and clastic texture of remaining carbonates and of the sandstones. (After Brice and Levin, 1969.)

particles of numerous sizes. Sorting is also a measure of time of transportation; the poorly-sorted or immature sediments having been formed rapidly with little chance for currents to winnow out the finer grains, while well-sorted sediments have undergone a selective removal of these smaller particles. Clastic limestones (Fig. 3.5) typically consist of whole or partial fossils, the former indicating either quiet or agitated water depending on the fossil and the latter typically indicating agitated water; the tiny spherical, inorganically-formed grains called oolites char-

PRODUCTS OF WEATHERING

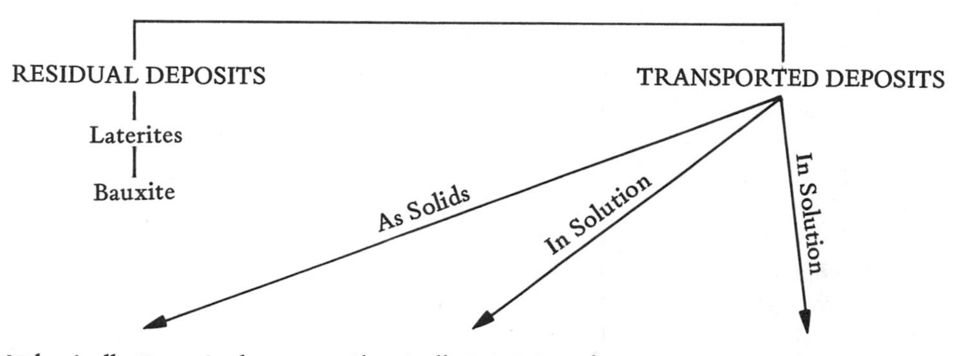

Mechanically Deposited
Detrital Sediments

Rudites (>2 mm.)
a. Conglomerates
b. Breccias

Arenites
(Sandstones)
(Sand Size 1/16 mm–2 mm.)
a. Arkoses
b. Graywackes
c. Subgraywackes
d. Quartz Sandstone

Argillites (<1/16 mm.)
a. Mudstones
b. Shales

Chemically Precipitated
Chemical Sediments

Siliceous
a. Chert
b. Flint
c. Sinter

Carbonate
a. Oolitic Limestone
b. Dolostone
c. Tufa (Springs)
d. Travertine (Caves)
e. Lithographic Limestone
f. Caliche

Ferruginous
a. Ironstones

Other Salts
a. Gypsum
b. Anhydrite
c. Halite
d. Bittern Salts

Organically Extracted
Organic Sediments

Siliceous
a. Radiolarite
b. Diatomite

Calcareous
a. Bioclastic Limestone
b. Chalk
c. Marl

Ferruginous
a. Bog Iron Ore

Phosphatic
a. Phosphorite

Carbonaceous
a. Coal

FIGURE 3.6 A Classification of Sedimentary Rocks. (After Spencer, 1966) acteristic of agitated waters; or limey mud (called micrite when lithified) characteristic of quiet waters.

Crystalline textures (Fig. 3.5) are most commonly found in carbonate rocks and are usually secondary, that is they develop after the original formation of the rock by solution and recrystallization producing crystal growth. Crystalline textures also occur in evaporites such as gypsum and halite which form under conditions of restricted circulation, high salinity, and high evaporation.

Primary sedimentary structures (Fig. 3.7), structures that develop in sediments as they are originally deposited, also furnish evidence of the environment of formation of sedimentary rocks. For example, *cross-bedding* indicates that currents dominated the depositional environment, *ripple marks* commonly indicate shallow water deposition, *mud cracks* indicate alternating wet and dry conditions, and *graded bedding* and *sole marks* indicate deposition by turbidity currents originating on steep submarine slopes.

Color commonly offers environmental clues with reds, yellows, and browns typifying terrestrial rocks; gray and green, marine rocks; and black, stagnant water sediments. There are, however, many exceptions.

Associations (Fig. 3.8) of sedimentary rocks also furnish evidence of regional environments. Shallow water or shelf marine conditions typically produce widespread thin carbonates and well-sorted quartz sandstones, rapidly subsiding marine troughs typically produce narrow but thick sequences of graywacke sandstones with interbedded shales, and rapidly subsiding continental basins typically produce thick accumulations of red arkose sandstone and conglomerate. Sediments also offer climatic evidence. Evaporites today form only in a band varying

FIGURE 3.7 Cross-bedding, Navajo Sandstone, Zion National Park, Utah. (Courtesy of William M. Mintz.)

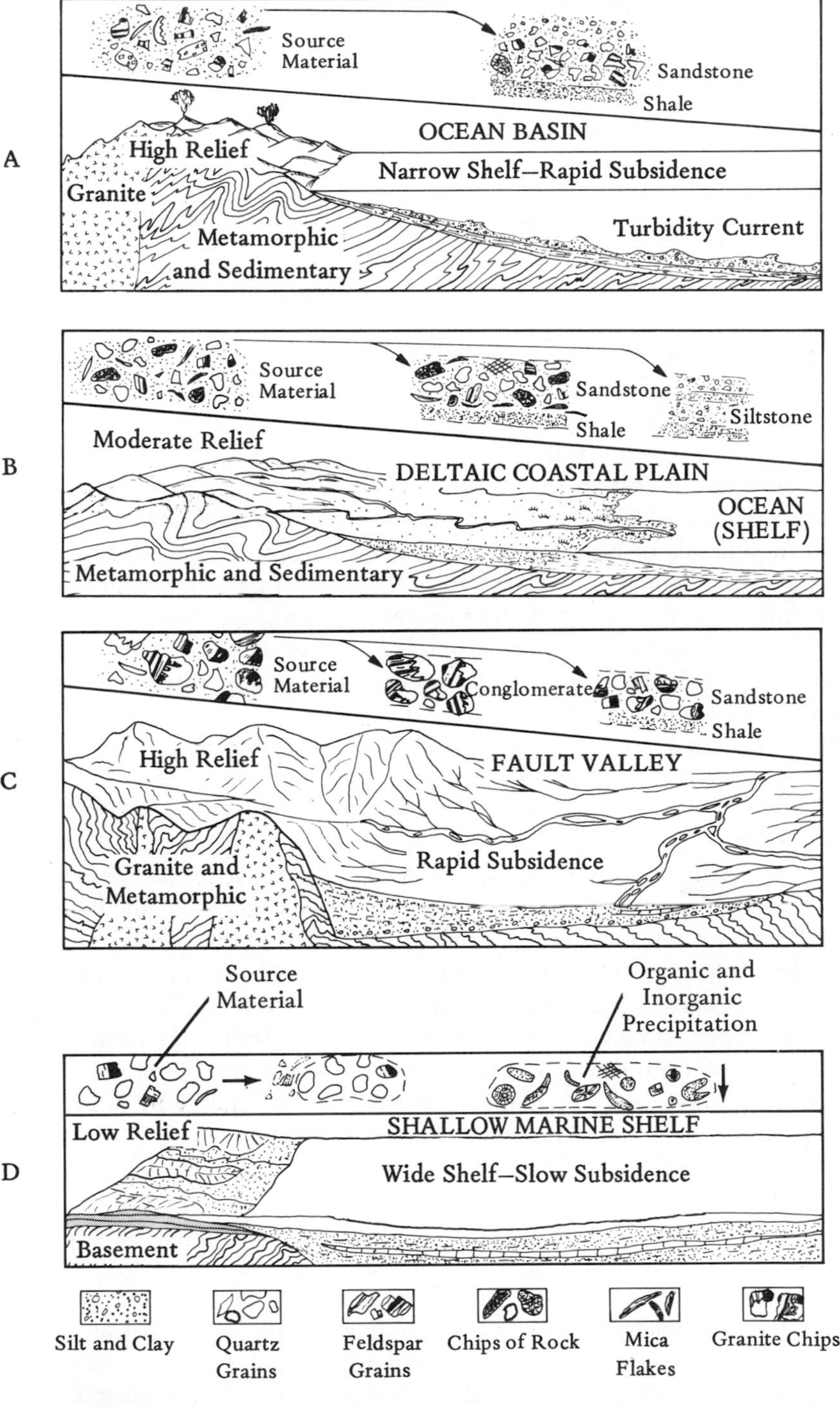

FIGURE 3.8 Environments of Deposition and Associations of Sedimentary Rocks. (A) Graywacke Sandstone, (B) Subgraywacke Sandstone, (C) Arkose Sandstone, (D) Quartz Sandstone and Carbonates. (After Brice, 1960.)

20° north and south of the equator. Red sediments also form today only in the tropics. Coal, the compacted and carbonized remains of plants, presently forms in equatorial regions and in a high latitude zone in both northern and southern hemispheres. Numerous other associations also occur. More will be said about patterns of sedimentation in Chapters 8, 9, and 11.

Biogeography

The science of biogeography can be said to have begun with Ages of Discovery of Western Man, the 1400s through the 1600s, when he began to move into the world beyond Europe. The most startling discovery of these trips for many Europeans was the large number of totally new organisms found on the newly-explored continents. Virtually every continent had a new assemblage of animals and plants and they were nearly all different from those of Europe. These discoveries called into question that most venerable theological tenet, the story of the Biblical Flood and Noah's Ark. It was painfully obvious that no boat could hold two individuals of all the vast number of species of land organisms. As if this were not bad enough, the distribution of land animals and plants did not fit with the story that the center of their present distribution should be Mt. Ararat where the Ark supposedly came aground. Some of the new animals, such as those of Australia, were found only on their own isolated continents and did not occur anywhere even remotely close to Mt. Ararat.

In 1766, the great French naturalist, the Comte de Buffon, prepared a world map showing the distribution and possible migration routes of the four-legged land animals (Fig. 3.9). This was the first real scientific biogeographic work.

In the 1800s, naturalists, particularly from England, fanned out to the distant lands of the earth to study organisms and plot their distribution. Such men as *Sclater, Huxley, Lydekker,* and particularly *Alfred Wallace* defined a series of terrestrial realms, regions, and provinces (in order of decreasing magnitude) based on biogeography. The central unit in this hierarchy, the region, is the most significant. The earth's terrestrial regions (Fig. 3.10) are the *Palearctic* (Europe, Africa north of the Sahara, and Asia north of the Himalayas), the *Nearctic* (North America north of central Mexico), the Oriental (India and southeast Asia), the *Ethiopian* (sub-Sahara Africa), the *Neotropical* (Central and South America), and the *Australian* (Australia and the South Pacific islands).

The determination of the distribution of organisms in the sea began later, in the late 1800s, with the work of *Woodward* and has reached fruition with the research of *Sven Ekman* and others in this century. Much more, however, needs to be done as most of the current data are only on the organisms of the continental shelves (Fig. 3.11). The gathering of data does not a science make, however, and the real scientific effort in biogeography has developed as men have searched for the causes of the patterns of distribution.

From studies of both terrestrial and marine organisms, it has become apparent that there are three basic factors that influence the distribution of organisms. They are the local *environment* (ecology), the presence of *barriers* (geography), and the *geologic history* of the region.

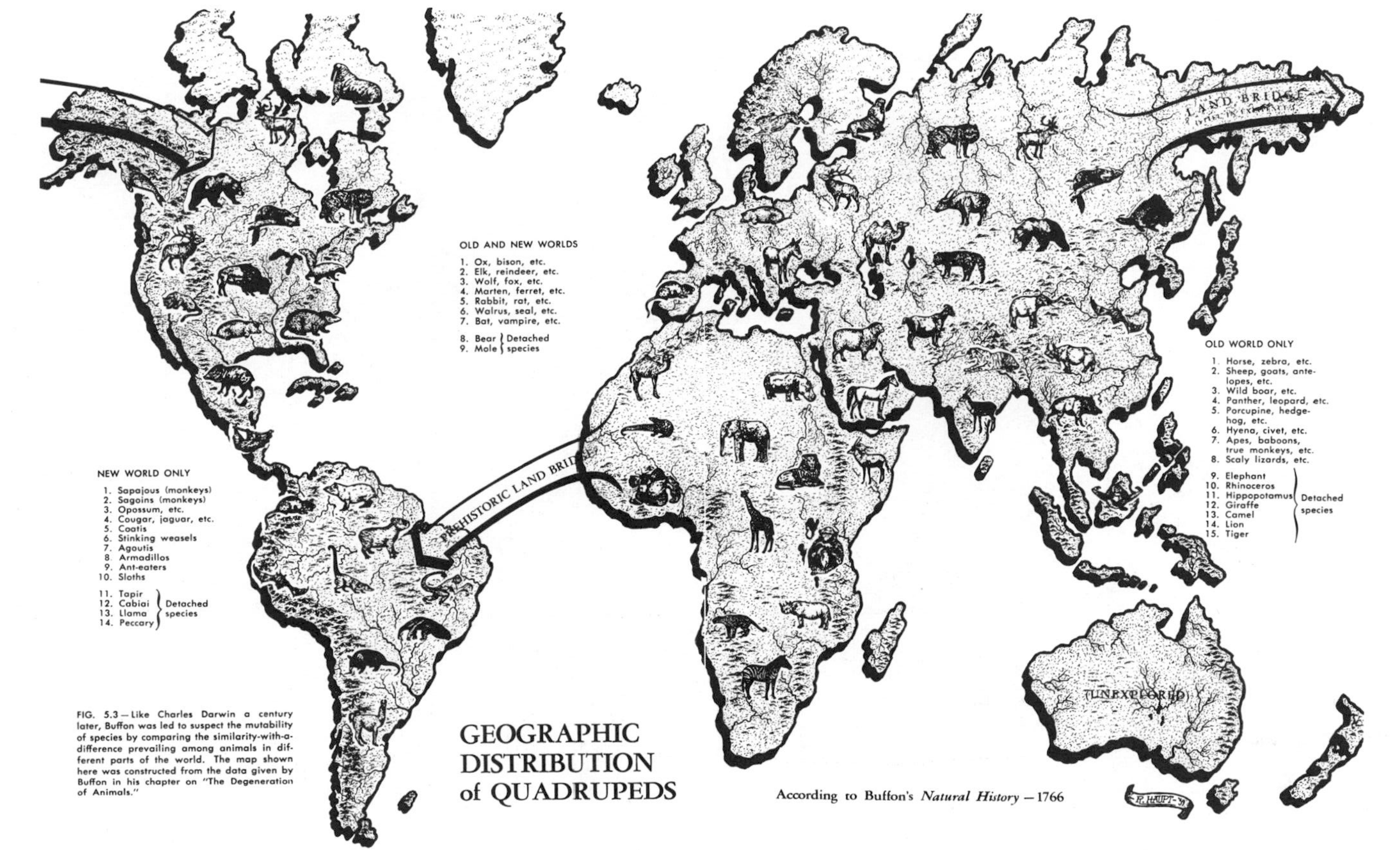

FIGURE 3.9 Geographic Distribution of Quadrupeds According to Buffon's Natural History (1766). (Reproduced by permission from *The Death of Adam: Evolution and Its Impact on Western Thought*, by John C. Greene, © 1959 by the Iowa State University Press, Ames, Iowa.)

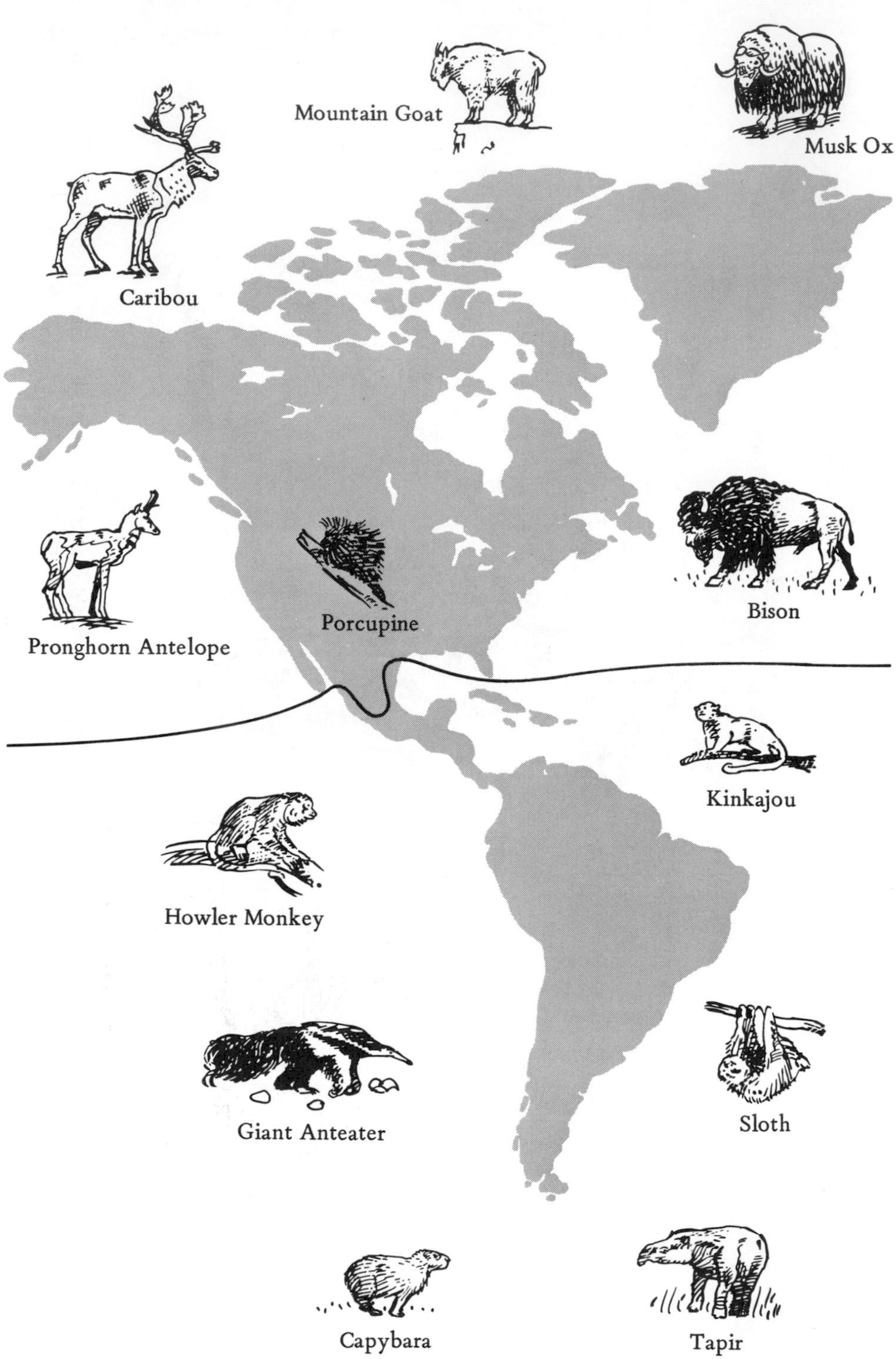

FIGURE 3.10 The Earth's Faunal Regions Today. Characteristic forms are illustrated for each region.

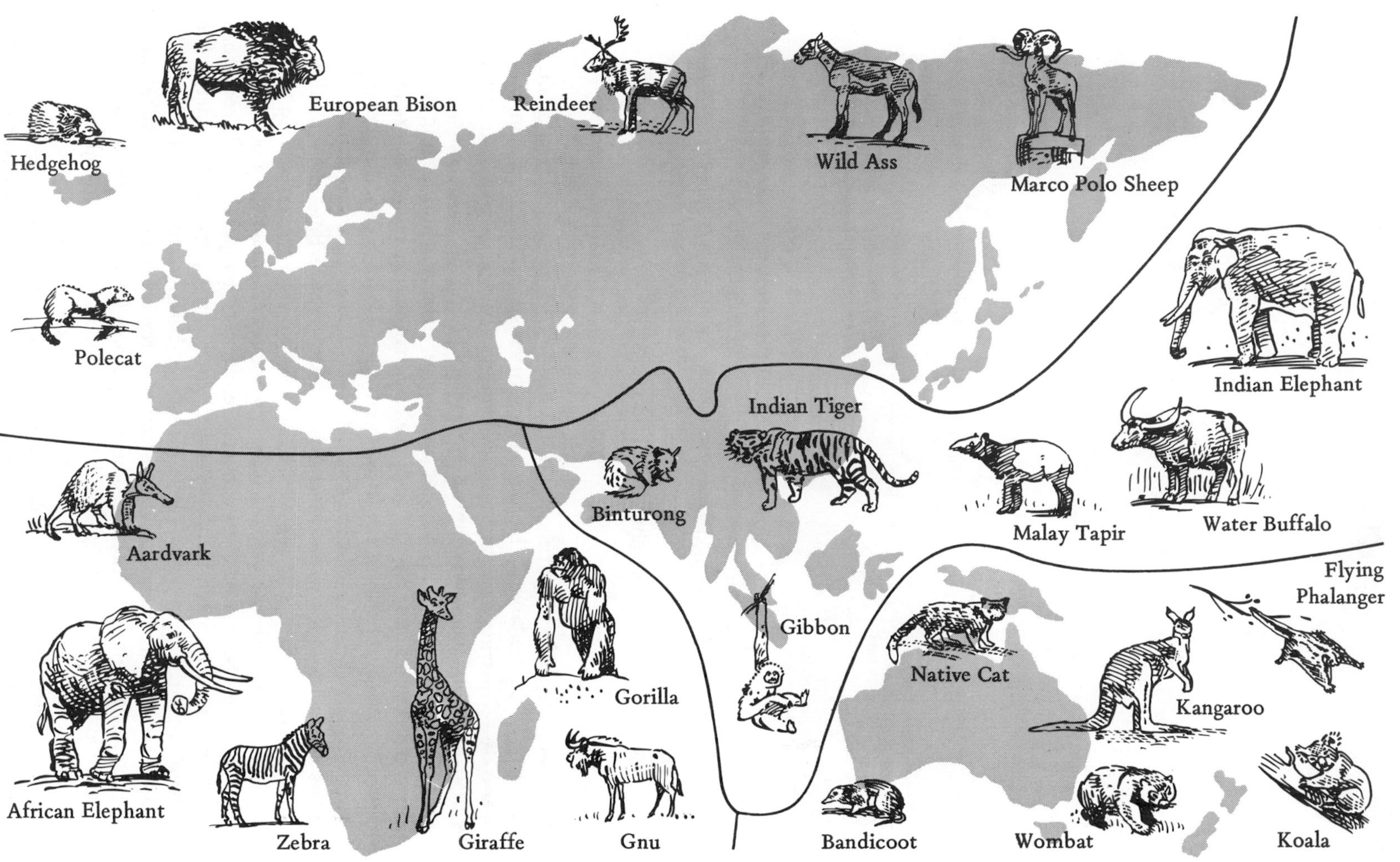

Hedgehog
European Bison
Reindeer
Wild Ass
Marco Polo Sheep
Polecat
Indian Elephant
Indian Tiger
Binturong
Malay Tapir
Water Buffalo
Aardvark
Flying Phalanger
Gibbon
Native Cat
Gorilla
Kangaroo
African Elephant
Zebra
Giraffe
Gnu
Bandicoot
Wombat
Koala

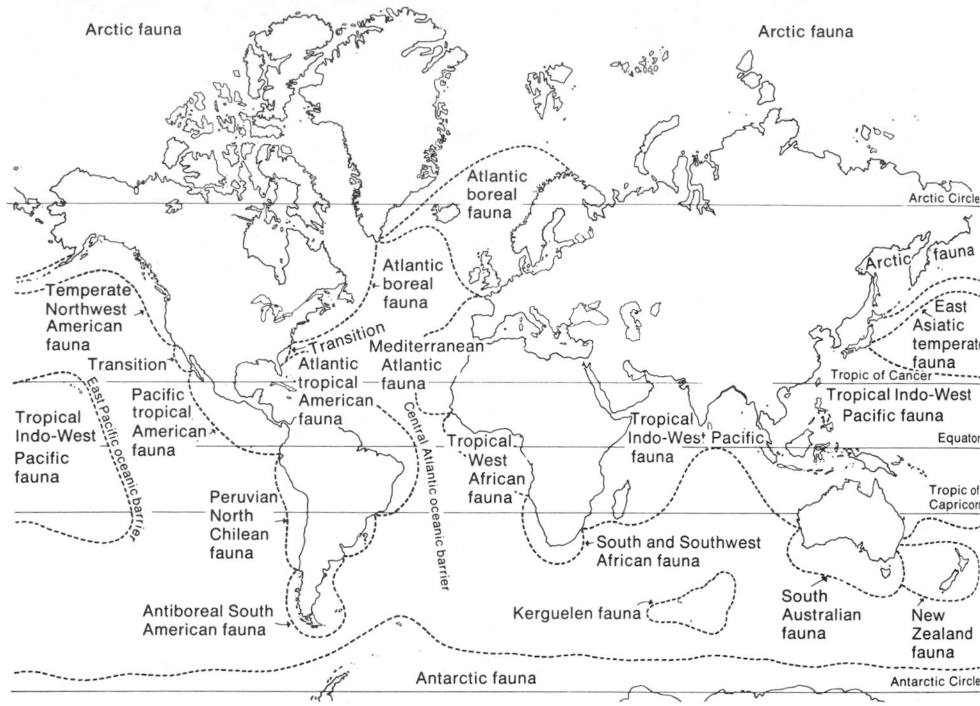

FIGURE 3.11 Zoogeographic Provinces of Shelf Faunas Today. (From *History of the Earth* by Bernhard Kummel, W. H. Freeman and Company. Copyright © 1970, Second Ed.)

Marine Environments

Most fossils are those of marine organisms because the oceans are major sites of deposition, while the land is chiefly a region of erosion. For this reason, we will concentrate on marine environments. The major ecologic factors that influence the distribution of marine organisms are temperature, dissolved gases (chiefly O_2 and CO_2), sunlight, salinity, water turbulence, substrate, and feeding relationships of organisms. Typically, all of these and many other lesser factors interact with one another to produce the local environment (Fig. 3.12). *Temperature* is among the most significant factors as it affects the rate of metabolism of organisms both directly (Fig. 3.13) and indirectly by affecting the solubility of essential gases and salts and establishing convective water movements. In a cross-section of an ocean (Fig. 3.14), the most diverse and abundant life occurs in the shallower waters above 200 meters where warmer temperatures favor more efficient metabolism. Sunlight is also a vital factor as it provides the energy for photosynthesis, the process whereby chlorophyll-bearing organisms ("plants" in the loose sense) convert inorganic substances to the complex organic substances used for metabolism and structure. These organisms form the basis for all life as they are fed upon either directly by herbivores or indirectly through carnivores eating herbivores or other carnivores. Sunlight typically penetrates only the upper 200 meters or so of the ocean and, hence, it is only there that one finds plants or plantlike microorganisms. This is another reason why the shallower seas have the most prolific life.

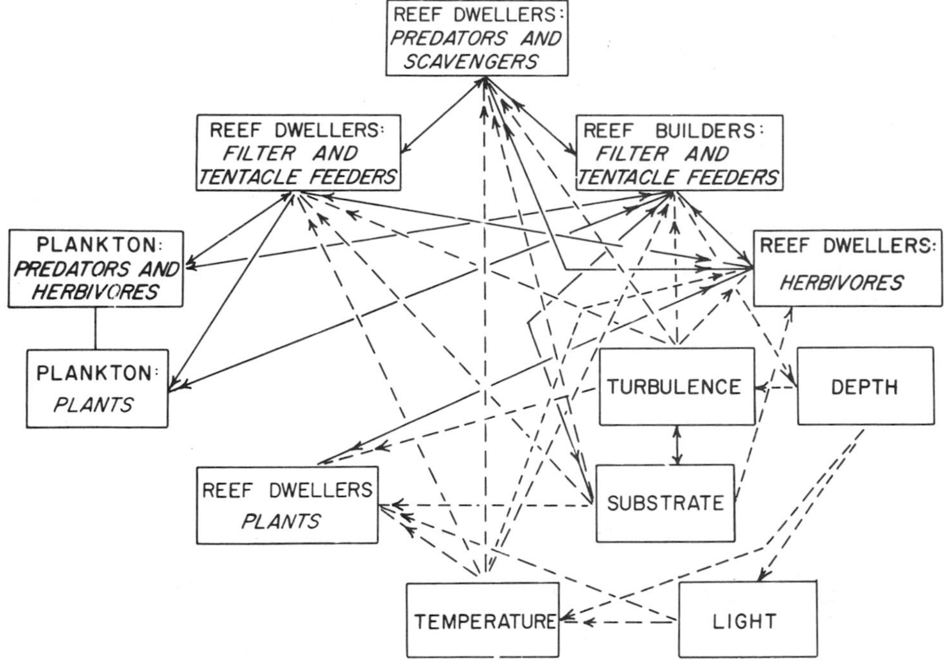

FIGURE 3.12 Complex Interaction of Factors Producing a Silurian Reef Environment. Arrows indicate direction of action. (From James R. Beerbower, *Search for the Past: An Introduction to Paleontology*, Second Ed., © 1968. Reprinted by permission of Prentice-Hall, Inc., Englewood Cliffs, N.J.)

In our cross-section of an ocean (Fig. 3.14), you will note that there are two basic categories of environments, the *benthic* which embraces the surface of the bottom and its substrate, and the *pelagic* which includes all of the ocean waters above the bottom. The benthic environment is subdivided into a series of depth zones which roughly correspond to the physiographic subdivisions listed after their names in parentheses: *littoral* (intertidal), *sublittoral* (continental shelf), *bathyal* (continental slope), *abyssal* (abyssal plains), and *hadal* (trenches). The littoral and sublittoral zones not only have sunlight and more favorable temperatures than the deeper zones, but also have a greater variation in substrates, turbulences, and nutrients which accounts for the plethora of life. A large percentage of the marine rocks on the continents today were deposited in ancient sublittoral environments and fortunately these contain the most abundant benthic fossils. Representative benthic organisms or benthos are illustrated in Figure 3.15.

The pelagic environment is generally subdivided into a thin upper, lighted, *photic* zone in which all the living photosynthesizers and many animals are found and a huge lower *aphotic* zone inhabited only by carnivores, scavengers, and the bacteria which break down dead organisms and return basic mineral salts to the water to be later reused by the photosynthesizers. Pelagic organisms are either floaters, the *plankton*, or swimmers, the *nekton* (Fig. 3.15).

Marine organisms can also be ecologically classified on the basis of their modes of nutrition (Fig. 3.16). Those organisms which can manufacture complex organic molecules from simpler inorganic ones, chiefly the photosynthesizers, are

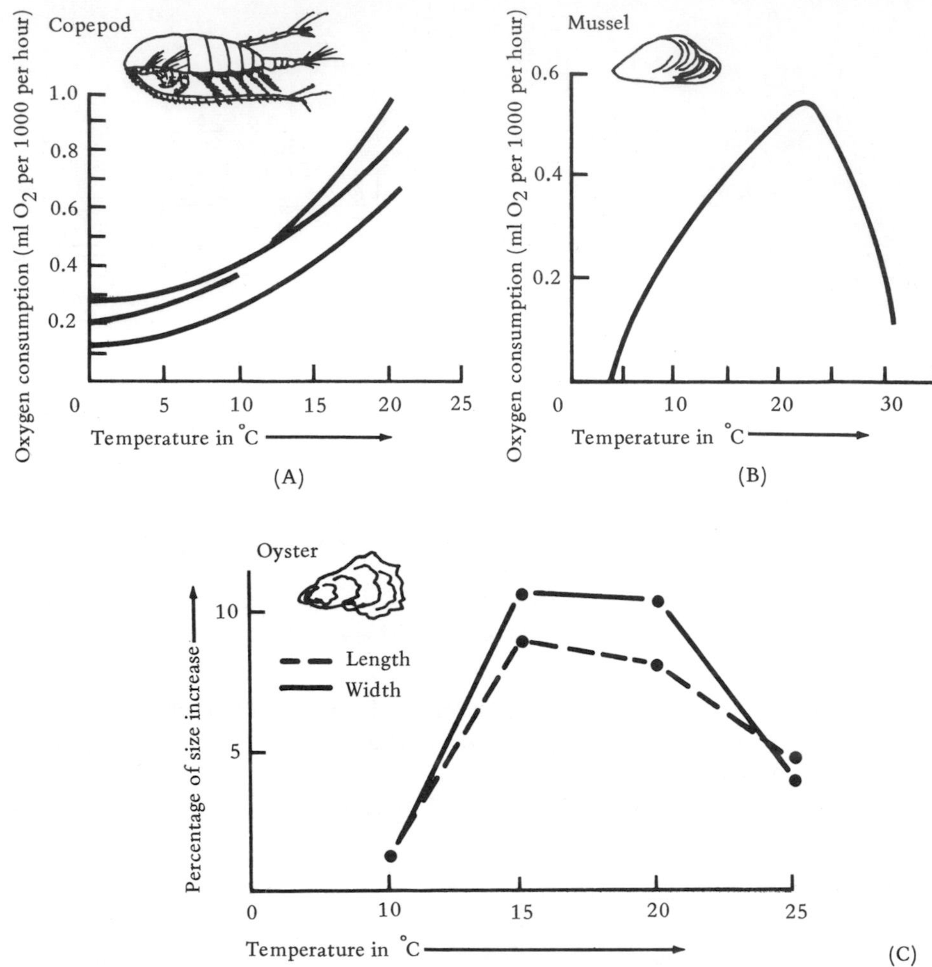

FIGURE 3.13 Effects of Temperature on Metabolism of Organisms. (A–B) Respiration rates increase with increasing temperature until a certain point when they decline sharply. (C) Growth rate curve follows a similar pattern. (From Leo F. Laporte, *Ancient Environments*, © 1968, Prentice-Hall, Inc., Englewood Cliffs, N.J.)

called *producers* or *autotrophs*. These are the true plants and plant-like microorganisms. Those organisms that require an external supply of organic molecules are called *consumers*, *heterotrophs*, or *allotrophs*. These fall into four subtypes, the *herbivores*, which eat producers; the *predators* or *carnivores*, which eat other consumers; *parasites*, which absorb nutrients from other living organisms; and the *saprophytes* and *scavengers*, which obtain their nutrients from dead organic matter. Animals and animal-like microorganisms are found in all the consumer subtypes; the bacteria and fungi are also major contributors to the last two categories.

The interrelationships of organisms through these modes of feeding produce *food chains* and *food webs* (Fig. 3.17) which compose the marine ecosystem

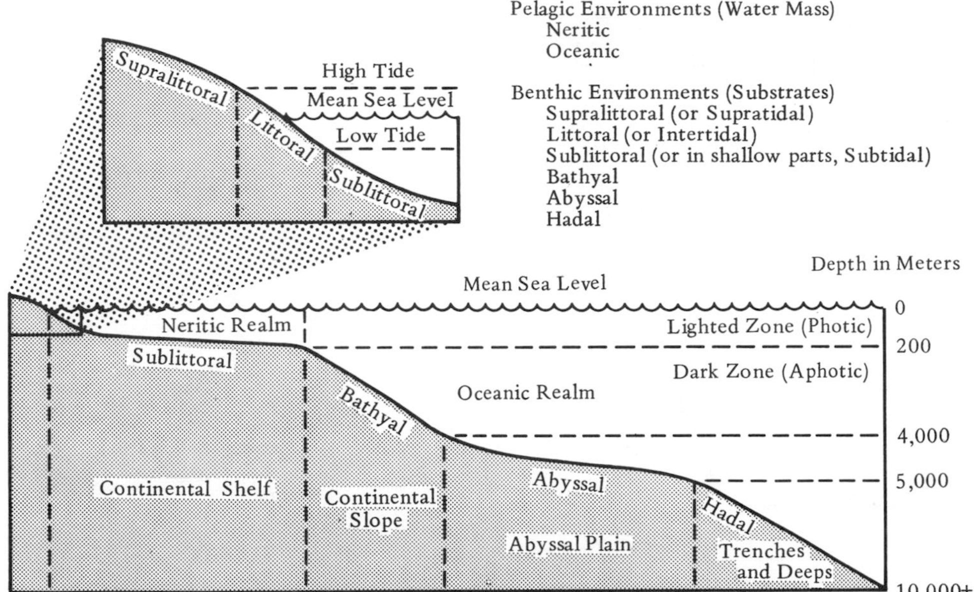

FIGURE 3.14 Partial Cross-section of an Ocean Showing Major Marine Environments. Generally these correspond with major geomorphic zones indicated in diagram.

(Fig. 3.18). In the marine ecosystem, the producers are small floating plant-like microorganisms, the phytoplankton (Figs. 3.15, 3.23). Attached larger algae are only significant in shallow waters where light reaches bottom. Herbivores include the tiny floating animal-like microorganisms and true animals, the zooplankton (Fig. 3.15, 3.23) which actively devour the phytoplankton and the larger groups of benthic animals (Fig. 3.19) which feed on the larger attached algae of shallow waters. The predators come in several levels of successively larger animals (Figs. 3.15, 3.19). Most of these animal groups also contain scavengers. In the sea there are many important fixed or sessile animals that collect masses of plankton by filtering the water (Fig. 3.19). In addition, there are numerous mobile or vagrant benthonic animals (Fig. 3.19) that selectively or nonselectively remove organic detritus from the bottom. The coelenterates are a unique group of attached carnivores (Fig. 3.19). Parasites enter the marine food chains at all levels. The bacteria are the *transformers* which break down organic detritus and return the simple inorganic compounds to the water where they can be reused by the producers.

A cross-section of the oceans only gives part of the ecologic story, however, as there are also the longitudinal differences which affect the basic environmental factors. For example, as one goes north along the North Atlantic coast of North America, the shallow water temperatures decrease and the organic populations change (Fig. 3.20). Some organisms such as reef-building corals (Fig. 3.21) can live only in waters of tropical temperature and shallow depth because the algae that live in their tissues and support their metabolism require sunlight. If reef corals have not changed their environmental requirements throughout geologic history, they should serve as valuable guides in reconstructing ancient environments.

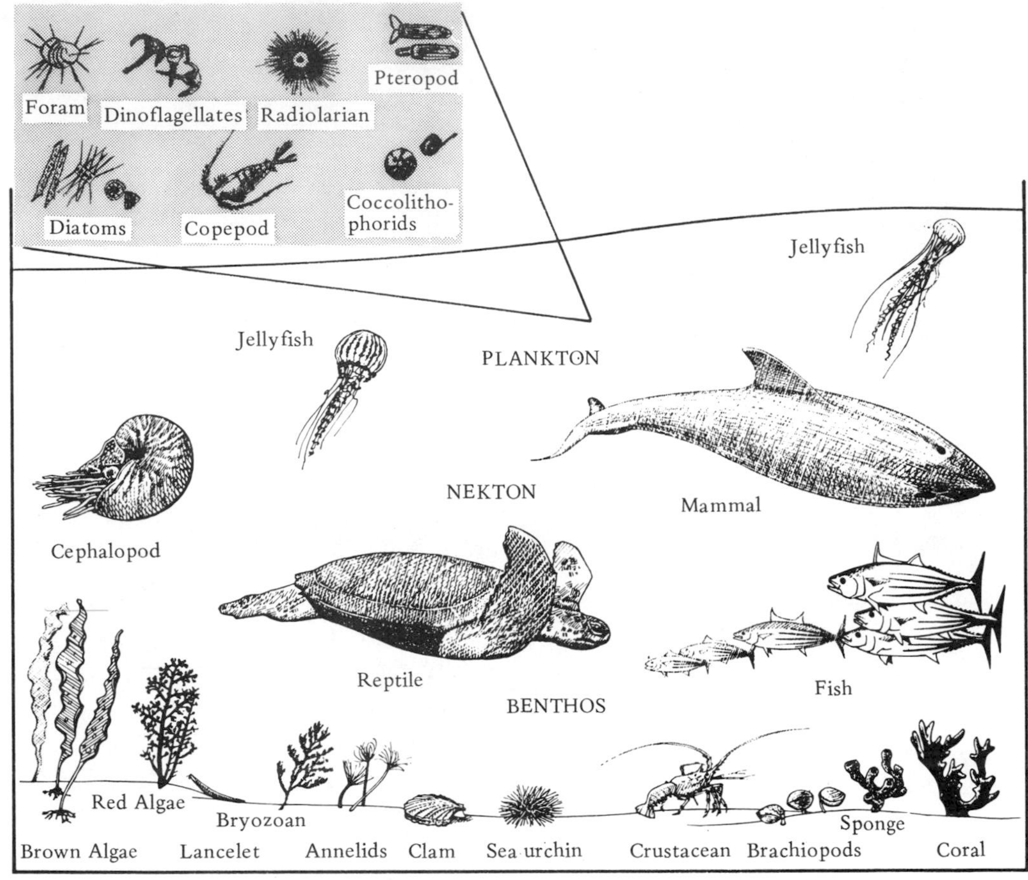

FIGURE 3.15 Life Habits of Marine Organisms. (From A. Lee McAlester, *The History of Life*, © 1968. Reprinted by permission of Prentice-Hall, Inc., Englewood Cliffs, N.J.)

So far, we have seen that all organisms are influenced by their environments and, therefore, if we can compare a fossil organism with a living relative we may have a key to help us understand past environments. This is not the complete picture, however, for many areas have the same ecologic conditions and yet are inhabited by quite different kinds of organisms. For example, the shallow-water, marine molluscan faunas on the two sides of the Atlantic, where ecologies are basically the same, are different (Fig. 3.11). Likewise the mollusks on the east coast and the west coast of North America are different where the environments are the same. Here we see the effects of geographic barriers, deep abyssal waters intervening in the first instance and a land mass, in the second. In still other instances, such as the similarity of the faunas of the Black and Caspian Seas, despite the fact that the two have no present connection, can be explained by the fact that the two were united as part of a larger seaway in the not too distant geologic past.

Terrestrial Environments

In the terrestrial environment, there are some of the same ecologic factors such as temperature, substrate, and feeding relationships that affect marine organisms. Generally, the gases and sunlight are readily available and have little influence. Salinity and water turbulence are not pertinent, though air turbulence may be. Availability of water, of no concern in the marine environment, is of overriding importance in the terrestrial. The terrestrial environment (Fig. 3.22) is considerably more variable than the marine environment, particularly in regard to temperature. Nutritional modes on land (Fig. 3.23) include the same types as in the sea, but the ecosystem is organized differently. Relatively large, immobile plants are the major producers. Herbivorous animals are active, selective seekers of plants and the carnivorous animals, in turn, are selective feeders on the herbivores. Unlike the sea, there are few terrestrial animals that feed on masses of tiny organisms or undifferentiated organic detritus. The bacteria are again the major transformers which recycle the nutrient salts. Geographic barriers significantly affect the distribution of terrestrial organisms as well. These may be features of the land such as belts of unfavorable habitat like mountains or they may be water bodies that separate land masses. Geologic history is important too. For example, South America, which is today in contact with the northern continents through Central America has numerous local organisms not found elsewhere on earth (Fig. 3.10). These are explained by the long isolation of South America throughout most of the Cenozoic.

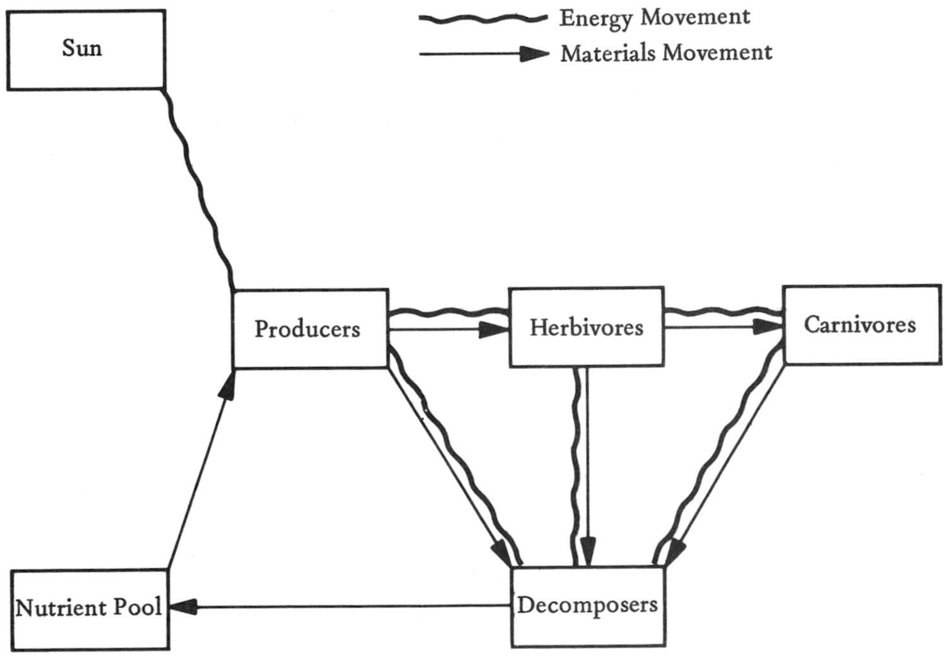

FIGURE 3.16 Interrelationships of Modes of Nutrition in Ecosystems.

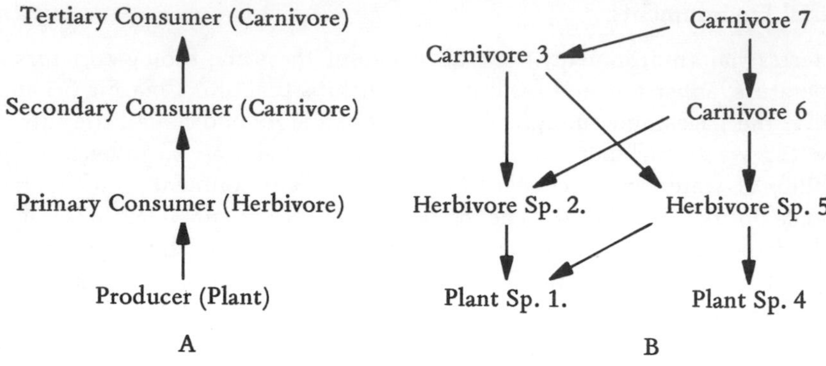

FIGURE 3.17 (A) Food Chain and (B) Food Web.

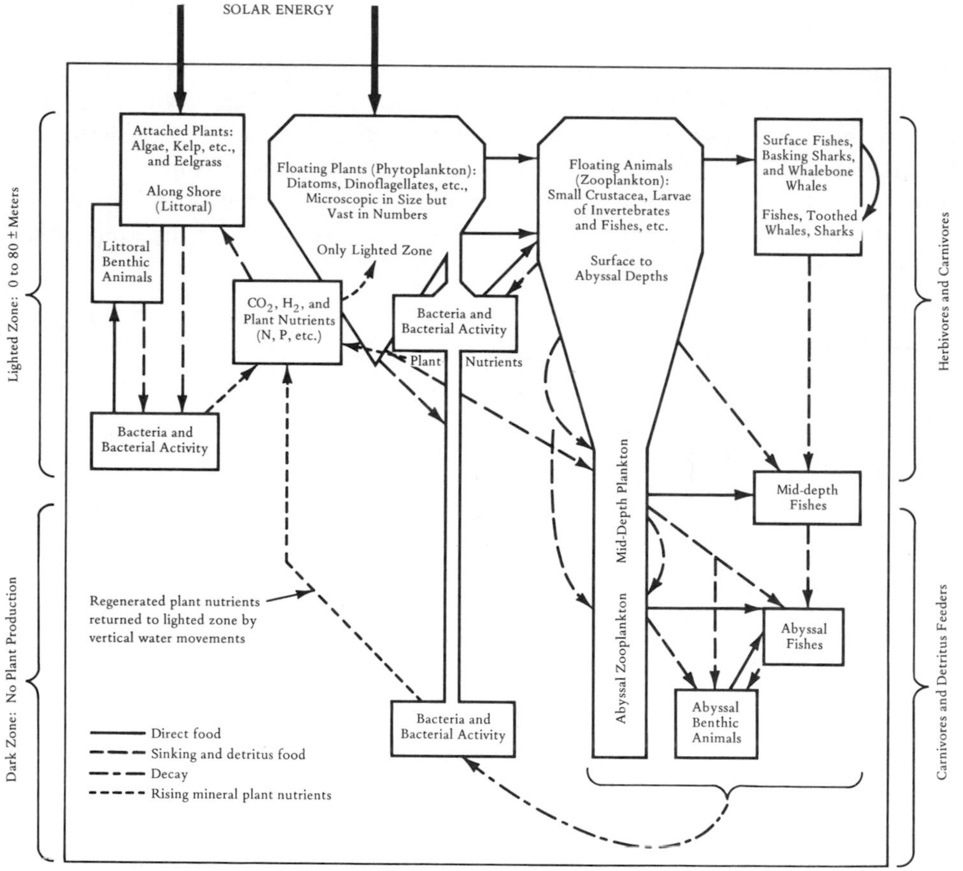

FIGURE 3.18 The Marine Ecosystem. (From *Life: An Introduction to Biology* by George Gaylord Simpson and William S. Beck, © 1957, 1965, by Harcourt Brace Jovanovich, Inc. (Adapted from Allee, *et al.* after H. U. Sverdrup, M. W. Johnson and R. H. Fleming, *The Oceans*—Prentice-Hall, 1942 with permission.) Reproduced by permission of the publishers.)

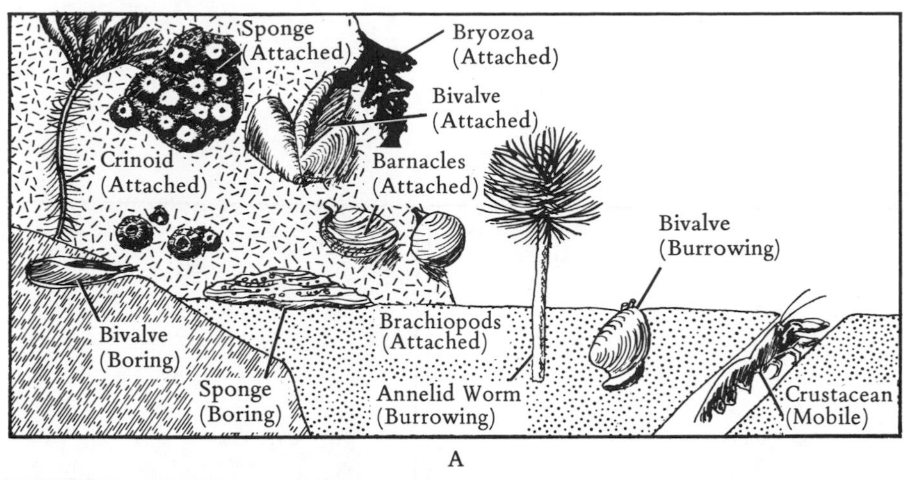

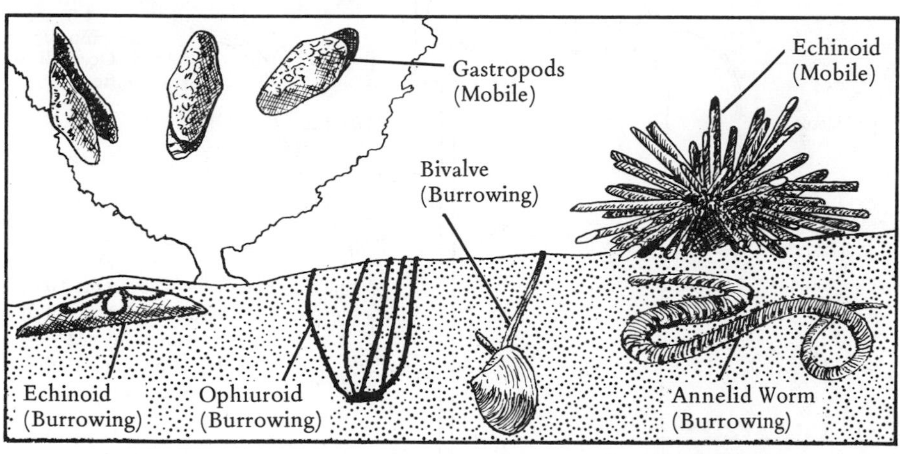

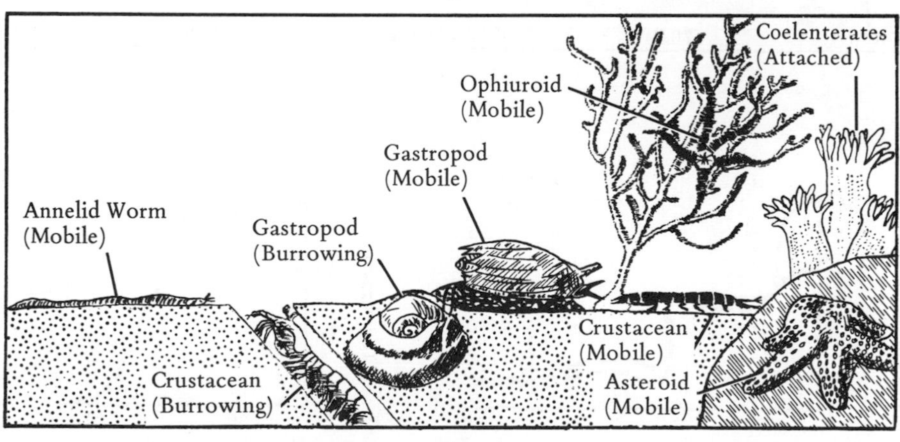

FIGURE 3.19 Major Modes of Nutrition in the Sea. (A) Filter feeders, (B) Sediment feeders, herbivores, (C) Carnivores, scavengers. (From A. Lee McAlester, *The History of Life*, © 1968. Reprinted by permission of Prentice-Hall, Inc., Englewood Cliffs, N.J.)

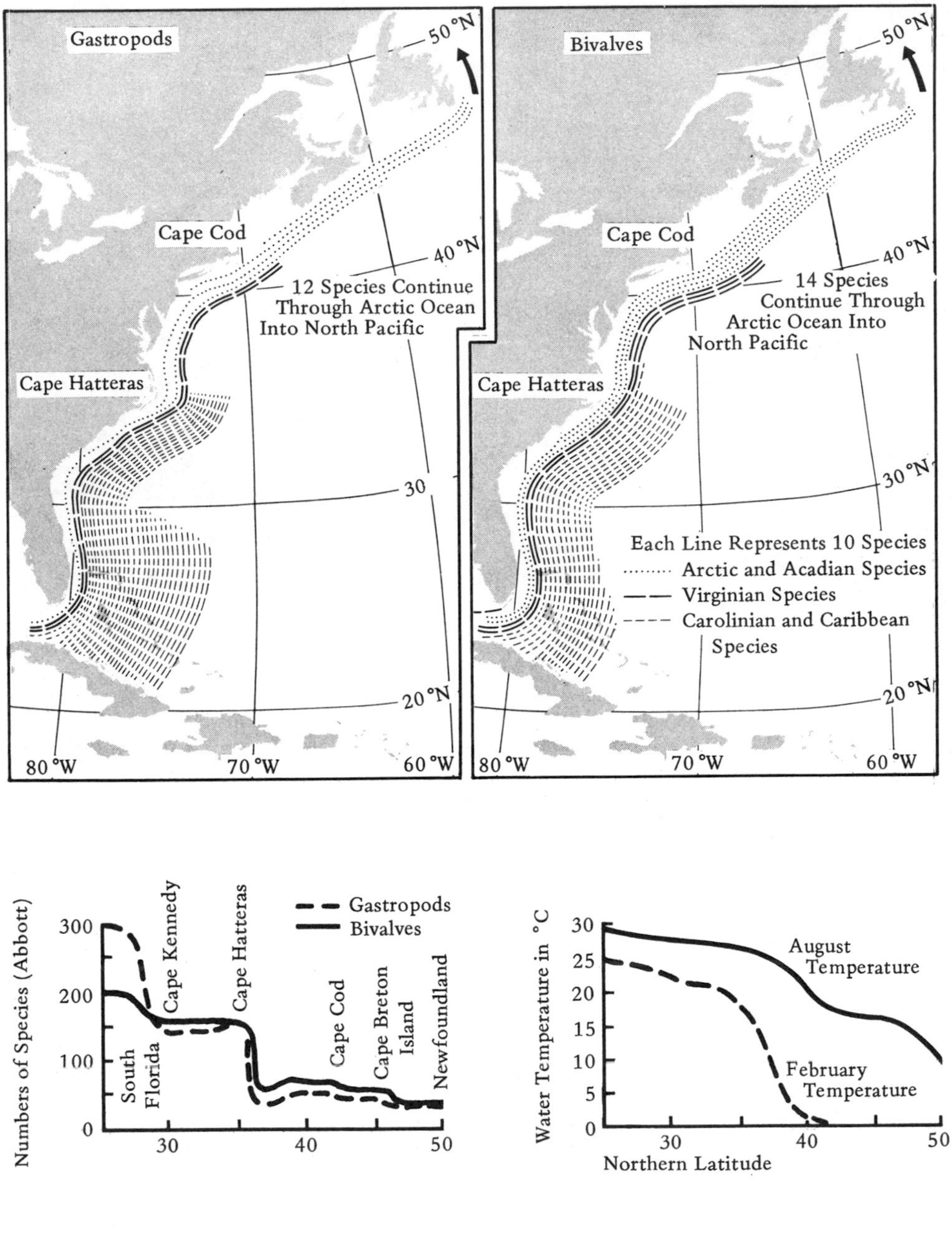

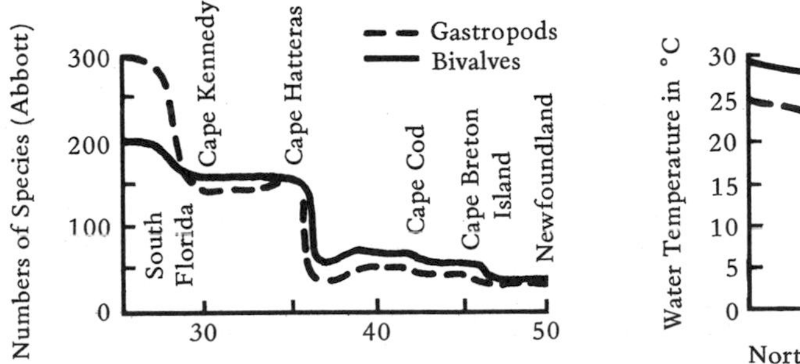

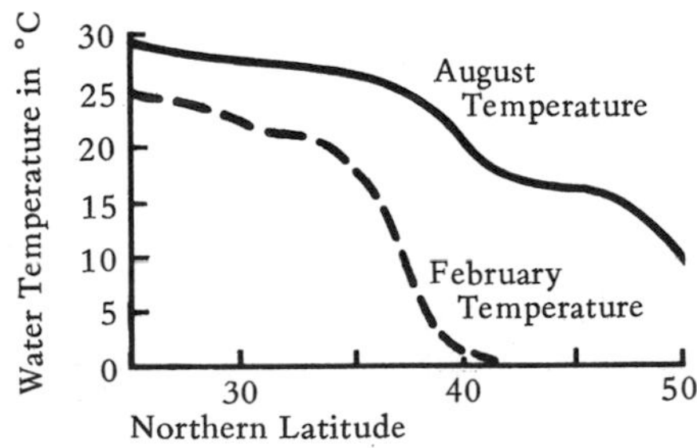

FIGURE 3.20 Longitudinal Differences in Molluscan Diversity. (From Leo F. Laporte, *Ancient Environments*, © 1968, Prentice-Hall, Inc., Englewood Cliffs, N.J.)

FIGURE 3.21
Distribution of Modern
Reef Corals.

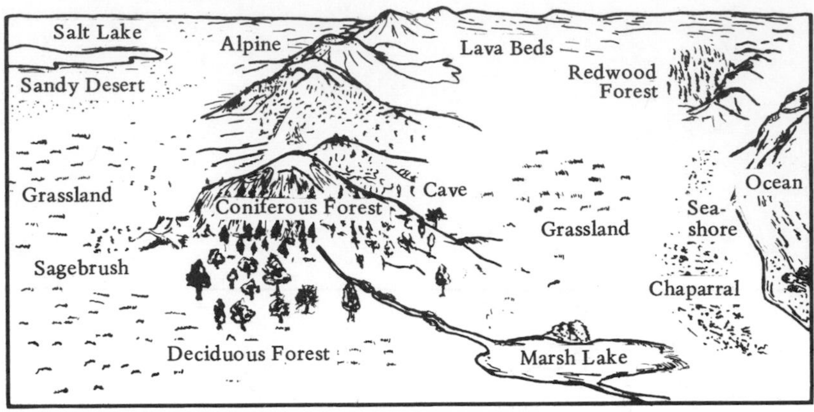

FIGURE 3.22 Representative Terrestrial Environments in the Middle Latitudes of North America.

All of these factors have, of course, also influenced the past distribution of organisms. Thus fossils are guides to the conditions of the past, though until recently there appeared to be some peculiar occurrences because tropical organisms were found in polar regions and polar organisms in the tropics. The discovery of continental drift and sea-floor spreading, discussed in Chapter 11, has explained these anomalous occurrences.

Circumventing the Facies Problem

This brings us back to the original problem of how to circumvent the problem of facies if we wish to determine whether two bodies of rock, deposited in different environments, were deposited at the same time. Our review of life in the sea suggests some answers. You will recall that floating (planktonic) and swimming (nektonic) organisms of the pelagic environment are found in the waters over several different benthic zones (Figs. 3.15, 3.23). When they die, their remains may settle out in several different environments where the benthic organisms and sediment types are different. Indeed, it is the shelled planktonic microorganisms and the nektonic ammonoids that are the most useful in determining the ages of rocks or different facies (Fig. 3.15). This is not to say that pelagic organisms are not also limited by environmental factors but the effects of temperature, sunlight, and substrate which are so prominent in the lives of benthonic organisms are of lesser significance to pelagic organisms latitudinally (not longitudinally, however). In addition, many benthonic organisms have pelagic larva which help overcome the effects of barriers on the distribution of the adults. One might expect the cursorial or running animals and the flyers to help circumvent the facies problem in terrestrial deposits. Cursorial animals often do have wide distributions as long as they have interconnected favorable environments. Flying creatures such as insects and birds are generally too delicate to be preserved commonly enough to be widely used for dating. The pollen and spores of land plants (Fig. 3.24), however, are highly resistant and widely disseminated, making them the most useful of terrestrial fossils for determining ages.

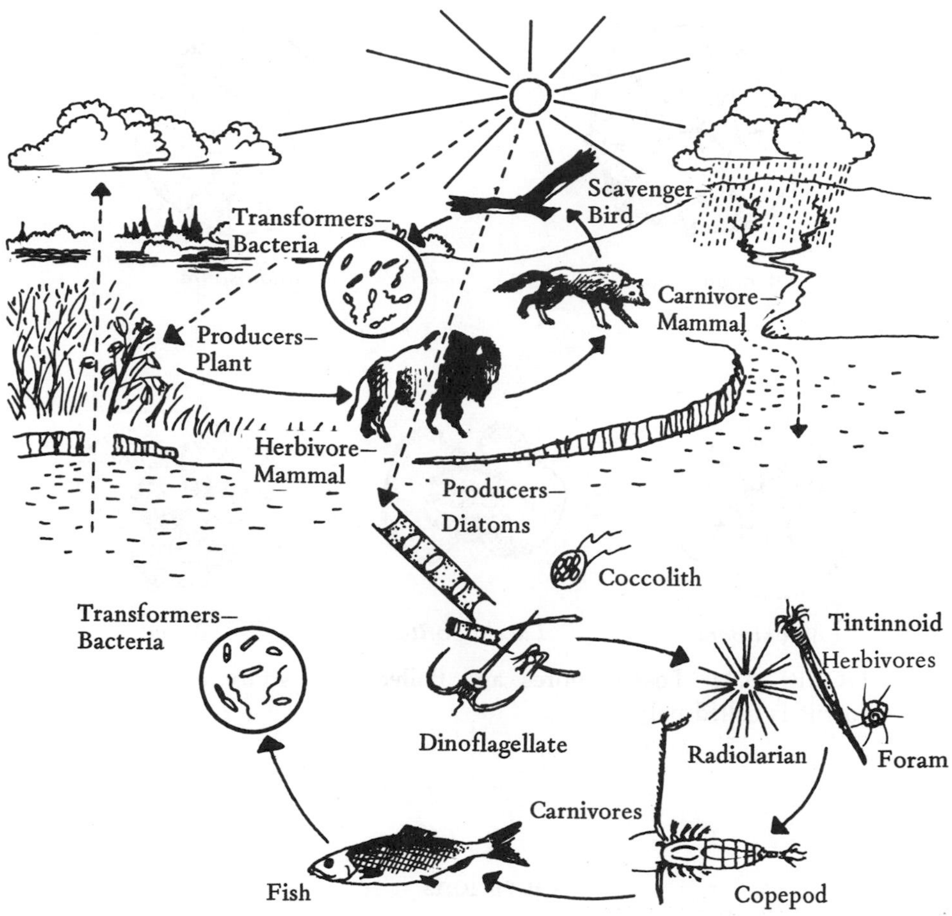

FIGURE 3.23 Terrestrial and Marine Ecosystems.

Before leaving the subject of facies, we should briefly note that the word has been used in several other senses in historical geology since its original definition. Some geologists use facies to refer to the lithogic or biologic record of a particular environment regardless of its age. This is most unfortunate, not only because it distorts a useful concept, but also because there are already words in common geologic usage for those records. The rock record of an environment is called a *formation* and a fossil record of an environment is called an *assemblage zone* or a *zonule.* Other geologists use facies to refer to the lateral variations within formations. In Chapter 5 we will discuss the various types of geologic units and the problems encountered by such misdefinitions. Facies is best left with its original definition of the lateral lithogic and fossil variations within a unit of rocks of the same age.

Grandispora

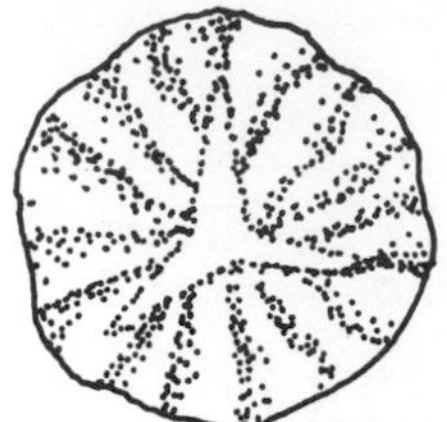

Radiospora

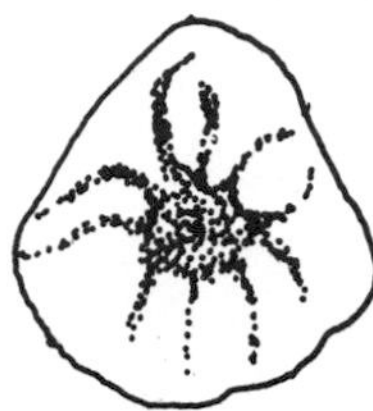

Auroraspora

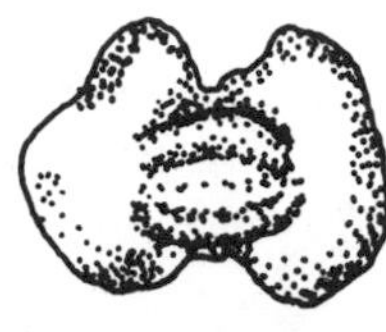

Lueckisporites

Tetrad

FIGURE 3.24 Fossil Spores and Pollen. The study of these fossils is called palynology.

Questions

1. What was the original definition of *facies* and who proposed it?
2. Why did the concept of facies encounter so much opposition in North America?
3. What is the environmental significance of size sorting in sedimentary rocks?
4. Which of the major sedimentary rock associations would you expect to dominate in your state?
5. What are the six major terrestrial faunal regions today?
6. List the three major factors that influence biogeography and give an example of each.
7. What is the difference between benthic and pelagic environments?
8. What are the major producers in the sea? On land?
9. Why are the marine shallow-water faunas different in east Africa and west Africa?
10. What types of organisms are best suited ecologically to circumvent facies problems and date strata?

4

The Key to Faunal Succession and Biofacies:

Evolution

Just as it took man centuries to accept successive changes or evolution of the physical world, the recognition of the evolution of life was a long and difficult process. Both concepts came into widespread use in the mid 1800s as man shifted his notion of a static universe to that of a dynamic one. Biological evolution, in simplest terms, means the change of one kind or species of organism into another, which is another way of saying that all creatures are interrelated. It explains why fossils occur in a definite sequence and why geologic history is important to biogeography.

Historical Development

Once again, it was the classical *Greeks* who left us the first written records of ideas of the transmutation of species. *Anaximander,* who lived in the sixth century B.C., stated that life arose from a mixture of water and mud warmed by the sun and that because of this and the succession of fossils, men had ultimately come from fish. *Empedocles,* who wrote a century later, also spoke of a succession of living things. It is uncertain whether these men had original ideas or derived them from earlier Egyptian scholars.

Aristotle, the greatest Greek scholar, and others developed the idea of a great "*Chain of Being*" which meant that there is a complete gradation in the organic world between the simplest and the most complex organisms. In this scheme, however, it appears that these scholars thought of each species as static on its rung of the ladder and not capable of evolving into anything else. This concept carried on for centuries, even into the time of Linnaeus in the 1700s.

The *Hebrew* creation story, adopted from still earlier legends of other Near Eastern peoples and enshrined in the Book of Genesis, was the dominent view of nature through the Dark and Middle Ages. In this legend, all species were created at once for man's use and each reproduced its own kind without any changes through time. In addition, all species had been saved from the great flood by the benevolent creator and, hence, none had perished.

Organic Diversity and Similarity

With the *Renaissance* and the accompanying *Voyages of Discovery* in the 1400s and 1500s came a renewed interest in observing nature. Disturbing things for the theological viewpoint came to light. We have already spoken of the space problems on the Ark with all the new species and of the fact that the geographical distribution of organisms was not a concentric series of rings around Mt. Ararat. In addition, many creatures were discovered that were intermediate in structure between known species. This meant many new rungs on the "Chain of Being." The biggest surprise, of course, was the discovery of the great apes, creatures that are in many ways intermediate between other animals and men (Fig. 4.1).

Extinct Species

The recognition of *fossils* as organic remains also led to problems for the theological viewpoint. It soon became apparent to forward-thinking men such as *Palissy* (1580), *Hooke* (late 1600s), *Jussieu* (1718), and *Sutton* (1749) that certain species were not alive today, particularly the strange Carboniferous coal age plants. Theologians were able to counter this argument for many years by

FIGURE 4.1 Early Drawing of Chimpanzee. (After Buffon, 1766.)

stating the organisms were living in some as yet unexplored region of the earth. By the close of the eighteenth century, however, it was obvious that certain large species of fossils could no longer remain undiscovered. In 1784, *Suckow* flatly stated that the coal age plant *Calamites* was extinct, and, in 1796, *Cuvier* successfully demonstrated the existence of several species of extinct animals including mastodons and ground sloths (Fig. 4.2).

Classification of Organisms

Another line of scientific advance that paved the way for the recognition of evolution was the development of a formal *classification* of organisms by the great Swedish naturalist Carolus *Linnaeus* in the 1700s. Linnaeus grouped organisms together in his classification (Fig. 4.3) on the basis of the number of characteristics they had in common. Those with numerous characteristics in common were grouped in the lowest and least inclusive categories. As one went up the hierarchy of classification, each level encompassed more creatures with less in common until one reached the highest and most inclusive level (Kingdom) in which the organisms shared few traits. In other words, Linnaean classification brought out the fact that there are degrees of similarity among organisms; some are more alike than others. In his earlier years, Linnaeus was a firm believer in the immutability of species and clearly stated it in his works. As he continued his studies, however, he appears to have recognized that the reason that some organisms are much more alike than others is that they are more closely related. He never publicly spoke for evolution, but he deleted his references to the fixity

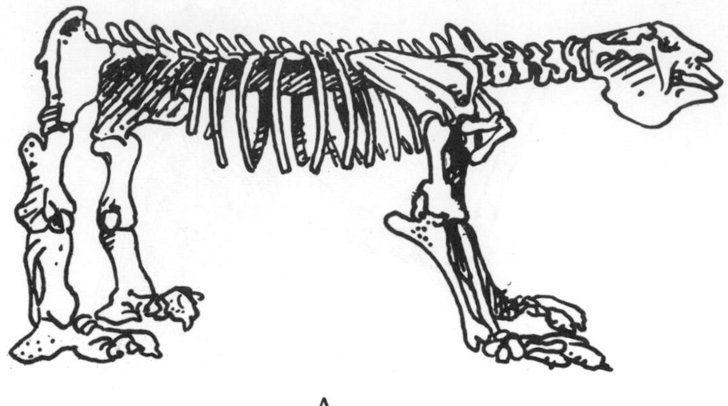

A

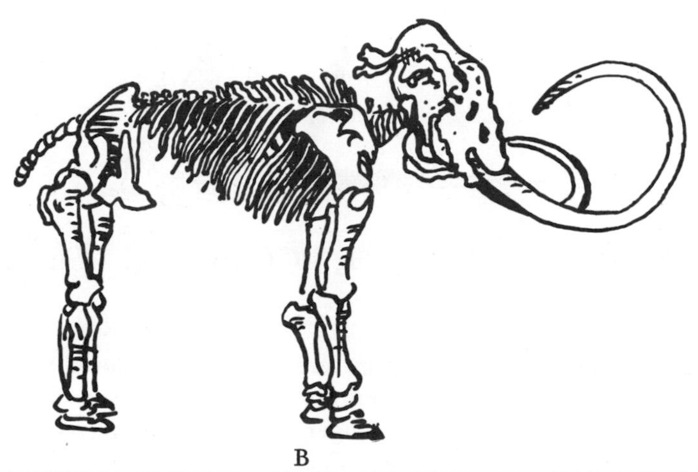

B

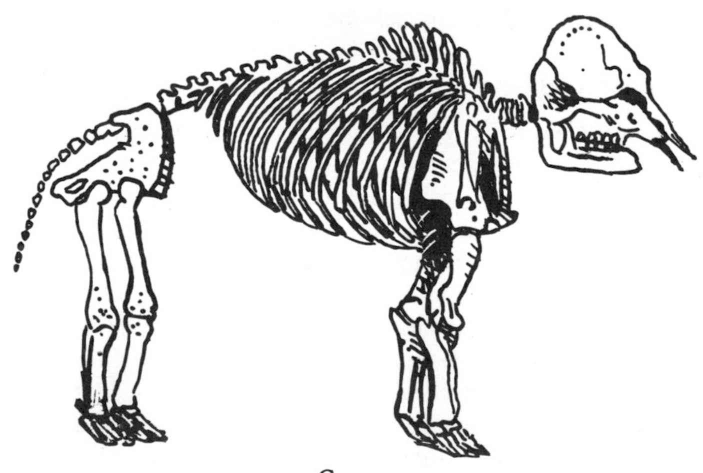

C

FIGURE 4.2 Animals Demonstrated to be Extinct by Cuvier. (A) Giant ground sloth *Megatherium,* (B) Woolly mammoth, (C) American mastodon. (After Cuvier.)

of species from later editions of his works and his correspondence indicates a shift in his ideas. Regardless of Linnaeus' beliefs, the development of a classification clearly pointed up the degrees of resemblances among organisms and was a vital step in the recognition of evolution.

Kingdom
Phylum
Class
Order
Family
Genus
Species

FIGURE 4.3 The Linnaean Hierarchy of Nomenclature.

Geology

The rise of *uniformitarianism* in the late 1700s and early 1800s also prepared the way for evolution, not only because it argued persuasively for evolution in the physical world, but also because it required an immense span of time for the small changes taking place today to produce the great changes seen in the geologic record. Such immense time spans would also be necessary for biological evolution once men began to notice its small, everyday changes. In scientific circles it was becoming obvious that the earth was much older than the Biblical 6000 years. The *stratigraphic succession of fossils* was also discovered in this interval and it indicated that the older rocks contained the simpler organisms and the younger rocks, the more complex ones. This also would fit with a gradual transformation of organisms through time, though few still dared to state the case strongly.

Comparative Anatomy

The rise of *comparative anatomy* with *Cuvier, Lorenz Oken,* and *Geoffrey Saint-Hilaire* was still another significant event at this time. These men noted that certain structures (Fig. 4.4) which have the same position in the structural organization of dissimilar organisms (and hence, are really the same structure) are modified differently by the different modes of life of the creatures in which they perform different functions. Hence, they are superficially unlike. For example, the same bones with somewhat different proportions and shapes are found in the same positions in a man's arm, a bat's wing, and a horse's leg. Again, the solution seems obvious that these similarities were due to a common ancestry, but most scientists did not, at least publicly, recognize it.

Early Evolutionists

Gradually, through the 1700s and early 1800s, it occurred to the more original thinkers in scientific circles that there had been an evolution of the biological, as well as the physical, world. *Benoit de Maillet* (1656–1738) wrote a book discussing the evolution of life from the sea onto the land and anticipated the more advanced ideas of mutations and natural selection which will be discussed

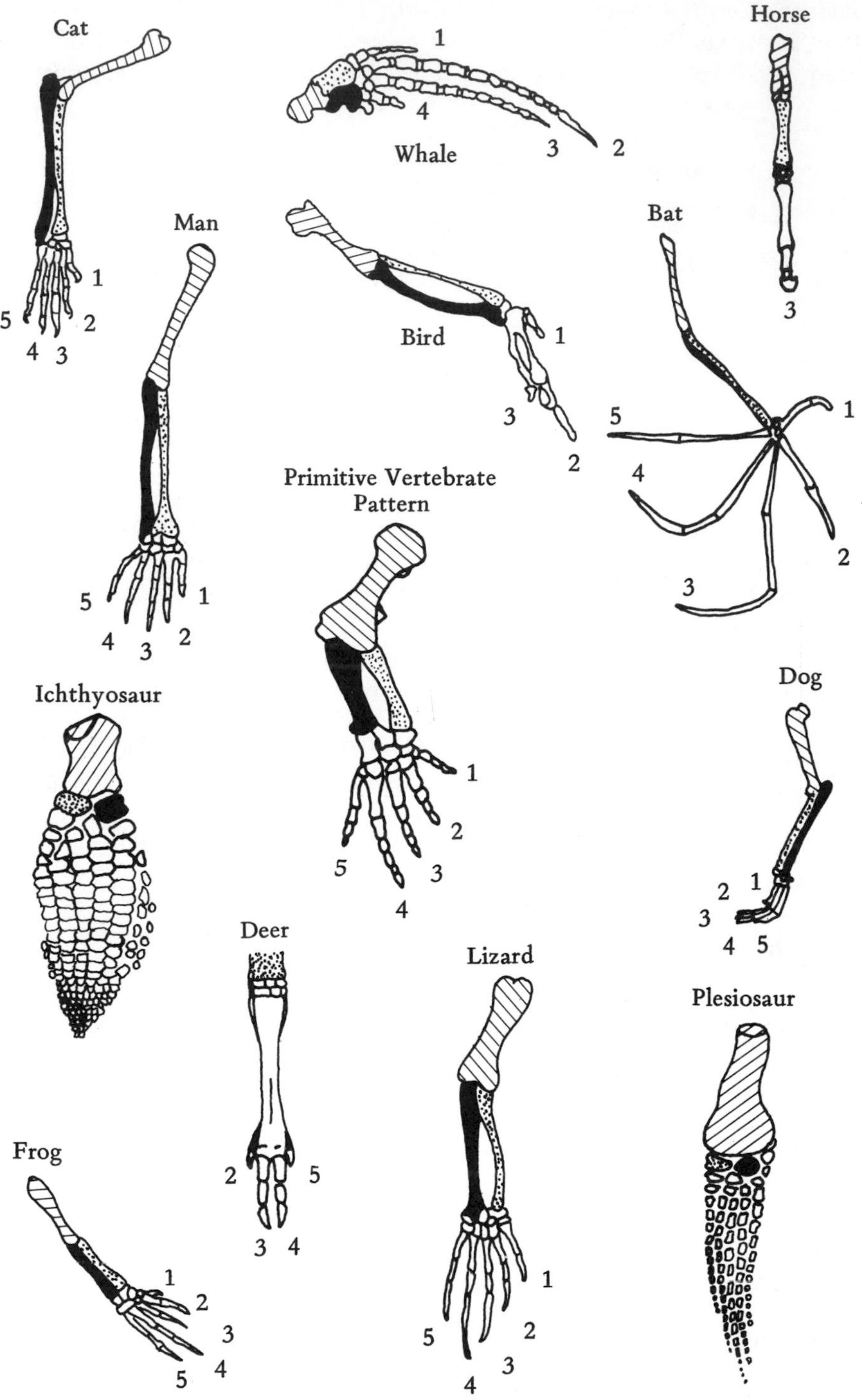

FIGURE 4.4 Homologous Bones in Vertebrate Forelimbs. Homologous bones have same shading or numbering in the various forms.

below. *Georges Buffon* (1707–1788) suggested evolutionary changes in organisms, but was forced to recant by the Church. *Erasmus Darwin* (1731–1802), Charles' grandfather, noted *vestigial structures,* non-functional structures of an organism apparently inherited from an ancestor in which they did function (Fig. 4.5). He called these "the wounds of evolution."

Lamarckism

Finally, around 1800, the French paleontologist *Chevalier de Lamarck* advanced the first coherent theory to explain the evolution of life. To Lamarck, the impulse for evolution came from within an organism. It evolved a structure or improved a structure it needed to survive. And once it had made such an improvement, this advance was passed on to its offspring. If structures were no longer useful, they decayed from disuse. Although Lamarck's theory sounds attractive, particularly from the social viewpoint which has insured its survival in many minds down to the present, much experimental and observational evidence was advanced to refute it. No one has ever seen any organism evolve new structures that were not in its hereditary potential and it is clear that

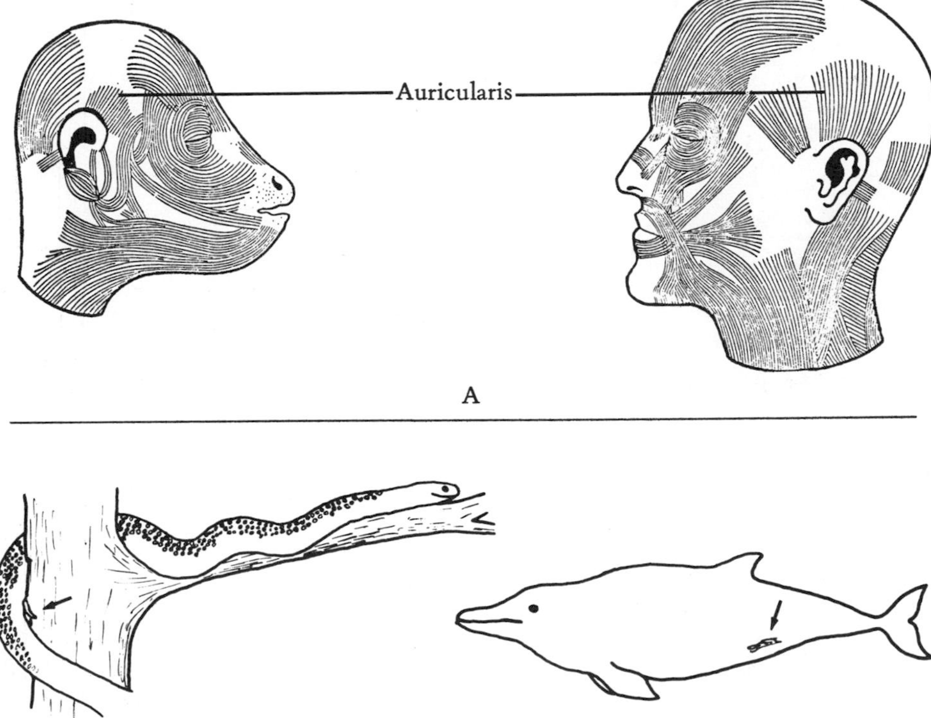

FIGURE 4.5 Vestigial Structures. (A) Auricularis muscles, functional in manipulating ears of monkeys, but not man, (B) Posterior appendages in whales and snakes. (After Elliott, 1957.)

acquired characteristics are not inherited. The son of a steelworker with big biceps will not have his father's big biceps if he becomes an accountant sitting at a desk all day. Despite his erroneous notion of the causes of evolution, Lamarck prepared the way for Darwin by exposing many people to the reality of evolution. Much of the theological opposition to evolution was drained off by Lamarck's work.

Other workers such as *William Wells, Patrick Matthew,* and *Robert Chambers* advanced theories of evolution in the first half of the nineteenth century. Although their works were often a curious admixture of Lamarckian and Darwinian concepts, they did point to lines of evidence supporting evolution such as the modification of species produced by artificial breeding, the reality of natural variation among individuals of the same species, the changes in the fossil record, and the struggle for existence between the members of a species. These men were not taken too seriously as this was the heyday of catastrophism, that brief marriage of theology and geology.

FIGURE 4.6 Charles Darwin (1809–1882).

Darwin

Charles Darwin (Fig. 4.6) was the real giant of science who at last succeeded in winning over the scientific community to the reality of evolution by propounding a consistent theory explaining its causes and supporting it with evidence from many areas of biology and geology. In the 1830s, Darwin traveled around the world on the scientific voyage of the ship *Beagle* (Fig. 4.7). The observations he made on this voyage transformed him from a believer in species fixity to a doubter at the end of the voyage. He spent the remainder of his life following up ideas suggested to him by the voyage which succeeded in convincing him of the reality of evolution. Still, he might never have found time to publish his ideas unless *Alfred Wallace* had not independently reached the same conclusions in the 1850s, an event which stimulated Darwin to action.

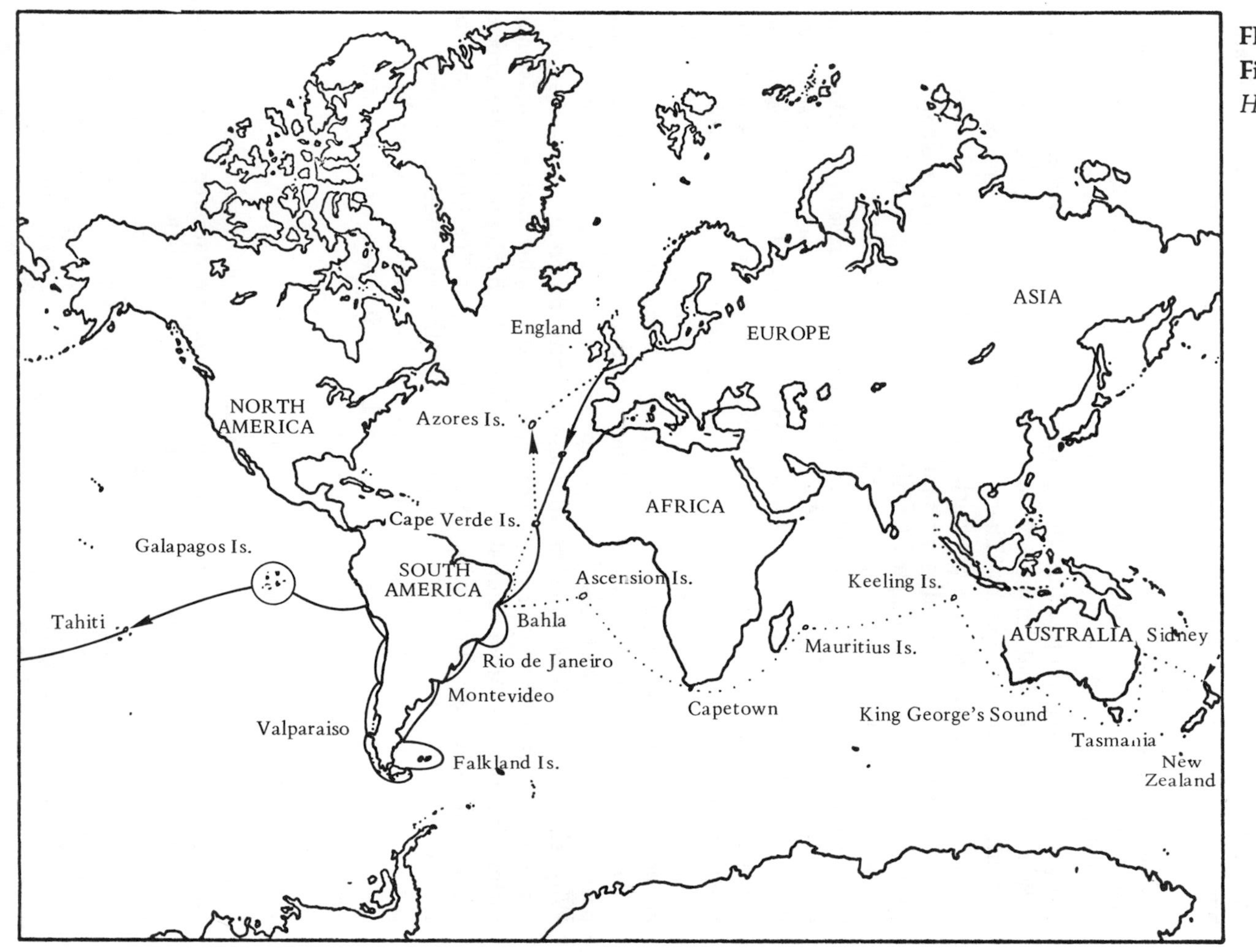

FIGURE 4.7
Five-Year Voyage of
H. M. S. Beagle.

On the *Beagle,* Darwin read Lyell's *Principles of Geology* which spoke consistently of the slow evolutionary changes in the physical world that had produced great cumulative effects. No doubt, this prepared Darwin psychologically to see the evidence of biological change which he would encounter.

During stopovers in South America, Darwin found the fossils of several extinct Cenozoic mammals (Fig. 4.8). To him, they appeared to be forerunners of the mammals inhabiting that continent today. They were certainly unlike any fossil or recent Old World mammals. This suggested to Darwin that perhaps the present South American animals did not migrate in from Mt. Ararat but somehow had descended from ancestors who had lived in South America for a long time.

When he reached the equatorial Galapagos Islands, a volcanic archipelago many hundreds of miles west of South America in the Pacific, Darwin saw many things that really triggered ideas about evolution in his mind. This distant group of islands has no free interchange of species with South America. Instead, only a few species had fortunately survived the great journey because there was little diversity of species. There were, however, several groups of very similar species, most notable of which were the finches (Fig. 4.9). Here were numerous species of birds all very similar in structure, but with slight differences because they were adapted to different modes of life. Surprisingly, most of these species made a living in ways like those of unrelated birds on the mainland and unlike those of the mainland finch species they resembled. One even took up the mode of life of a woodpecker. Could it be, Darwin guessed, that the mainland finch species had gained a foothold here and, in the absence of the competition of other land birds, had evolved to occupy the niches normally filled by other kinds of birds? Certainly it seemed more likely than for the Creator to make all these separate, but similar, Galapagos species when he already had made well-adapted different types of birds on the mainland.

Another observation made by Darwin was that some kinds of animals, particularly the land tortoises, had different species on just about every island. Why should the Creator design all these slightly different separate species reasoned Darwin? Wouldn't it be more logical to imagine that an occasional tortoise or two got from one island to another and that each tortoise had a slightly different hereditary makeup (it was obvious at this time that no two individuals of any species are ever identical) from the remainder of its original population so that its offspring only contained and perpetuated its peculiarities? Eventually its isolated descendants would be so different from the original population they would form a separate species.

Darwin's Theory

These observations and many others set Darwin to exploring further the possibility of evolution upon his return to England. Darwin gathered together data from every conceivable source relative to the problem including information on comparative anatomy, embryology and later developmental stages, vestigial structures, artificial breeding, biogeography, paleontology, and geology. Finally, in 1859 appeared his monumental work, *The Origin of Species.* Darwin not only demonstrated successfully that evolution had occurred, but also offered an explanation. Darwin's basic argument was as follows. All species of organisms have very high reproductive rates, yet populations appear to remain about the same size in the long run. Therefore, there must be a competition for survival among the offspring because most do not survive. Organisms vary; no two have identical characteristics. Therefore, those that survive must have characteristics that give them an edge in the struggle for survival. In Darwin's theory, the organisms do not evolve as Lamarck envisioned, but populations change in composition with time because those individuals with better adaptations stand a better chance of surviving and reproducing. In a way, then, Darwinism is the reverse of Lamarckism. The organisms are, in a sense, passive because all they do is vary, and the impulse for evolution comes from the environment which does the selecting. Darwin called his theory of the cause of evolution the theory of *natural selection*. In the main, his theory is the basis of our own modern one, although it is modified because Darwin did not understand why organisms vary or how characteristics are transmitted from parents to offspring.

Darwin's Evidence

We will only briefly summarize Darwin's evidence here. In *comparative anatomy*, he pointed to homologous structures (Fig. 4.4) which occupy the same position in the organization of different animals, yet are modified differently to perform different functions. As noted above, the best explanation for such structures is that they were inherited from a common ancestry. In embryology and later *ontogenetic* (developmental) stages, Darwin noted that many creatures pass through stages structurally similar to more primitive organisms. For example, the embryo of man (Fig. 4.10) contains gill slits like those of fish, yet they are never used and are soon lost. Darwin stated that the only logical reason for this recapitulation was that man had had fish ancestors at one time and had not yet lost all their traits. *Vestigial structures* (Fig. 4.5) inherited from ancestors where they were functional, could also be best explained by evolution. Their existence was in direct opposition to the theological notion that a Creator made each creature perfectly adapted to its environment for here were useless structures. Darwin pointed to the great changes produced by *selective breeding* of domestic animals such as pigeons, dogs, and sheep (Fig. 4.11). By breeding only individuals with certain desirable traits, man was able to produce structural modifications in an original species as great as those between a chihuahua and a great dane. Just imagine what nature could do over a much longer period even if it were much less selective. *Biogeography* also supported evolution for how else could one explain the fact that in isolated regions such as Australia one group of animals, in this case the pouched marsupials (Fig. 4.12), could fill most of the niches occupied by the typical placental mammals elsewhere. If the Biblical story were true, marsupials should have the center of their distribution at Mt. Ararat like the other mammals and should not be isolated by themselves on one continent. Darwin pointed out that the *fossil record* (Fig. 4.13) showed a progressive change from simpler to more complex organisms as one went up the geologic time scale. That was what one would expect if organisms had evolved, but would not occur if all organisms had been made

FIGURE 4.8 Extinct South American Mammals. Darwin found the remains of some of these forms. (After Marshall Kay, Edwin H. Colbert, *Stratigraphy and Life History,* © 1965, by permission of John Wiley & Sons, Inc.)

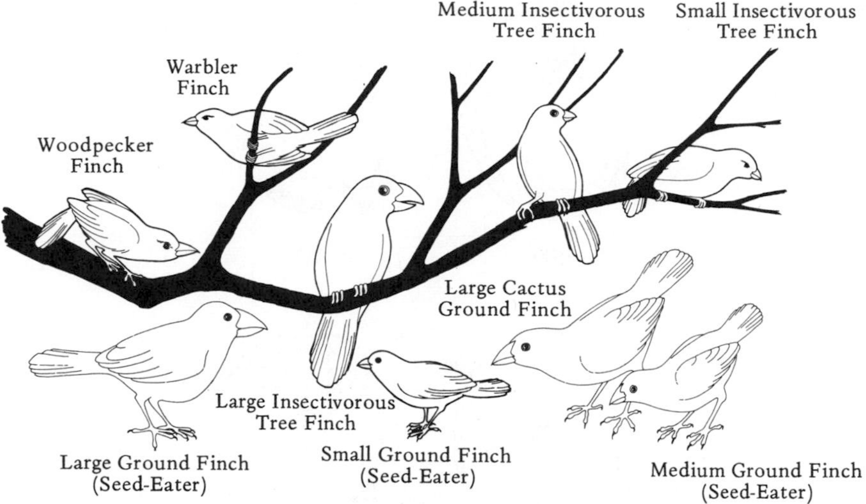

A

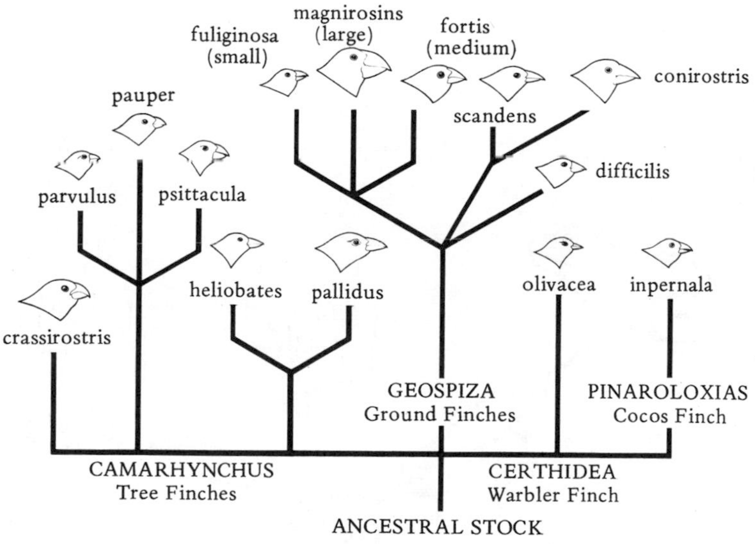

B

FIGURE 4.9 Darwin's Finches. (A) Life Habits, (B) Evolutionary Relationships. (Adapted from E. Peter Volpe, *Understanding Evolution,* © 1970, William C. Brown Co., Publishers, Dubuque, Iowa.)

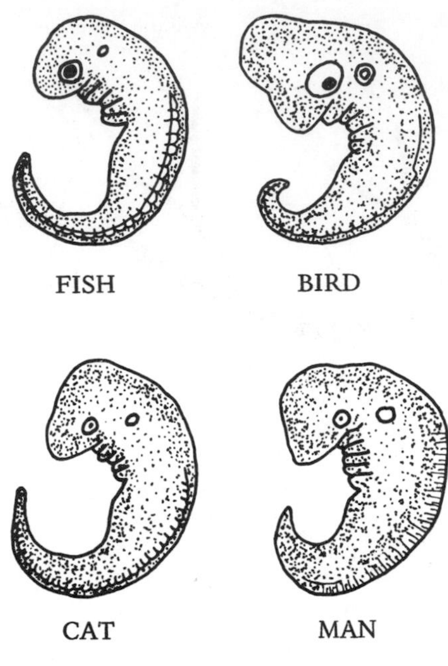

FIGURE 4.10 The Biogenetic Law (A) The early stages of all vertebrates greatly resemble each other. Note the presence of gill slits even in the nonaquatic forms. (B) The development (ontogeny) of a frog duplicates its evolutionary history (phylogeny).

FIGURE 4.11 Selective Breeding. (A) Some of the many varieties of pigeons produced by selective breeding. (B) Some of the many varieties of dogs produced by selective breeding.

FIGURE 4.12 The Marsupial Mammals of Australia. A representative sampling of the unique mammalian fauna of the isolated continent.

at once. Furthermore, the appearance of the different kinds of animals in the geologic record did not match the six-day Biblical sequence. Darwin was pessimistic about the chances of ever finding complete sequences of fossils showing the small changes from one species to the next because of the basically fluctuating nature of deposition and erosion in an area (Fig. 4.14). He drew attention to a second type of unconformity, later called the *disconformity* by Grabau in 1905 (Fig. 4.15), in which the strata above and below are parallel, but a gap in the sequence of fossils indicates much missing time. He felt that these gaps were abundant in the geologic record and precluded the finding of complete evolutionary lineages. As we shall see in a moment, his reasoning about gaps was good, but he was overly pessimistic as several good evolutionary lineages are now known. Lastly, Darwin pointed to the great length of geologic time, indicated by the principle of *uniformitarianism,* as being vital for the slow transformations he envisioned. Although temporarily stymied on this point by the physicists he was later vindicated by the discovery of radiometric dating which proved the earth's great antiquity as we shall see in Chapter 6.

Darwin's Problems

The great stumbling block for Darwin was that he did not understand why organisms varied and how characteristics were transmitted from generation to

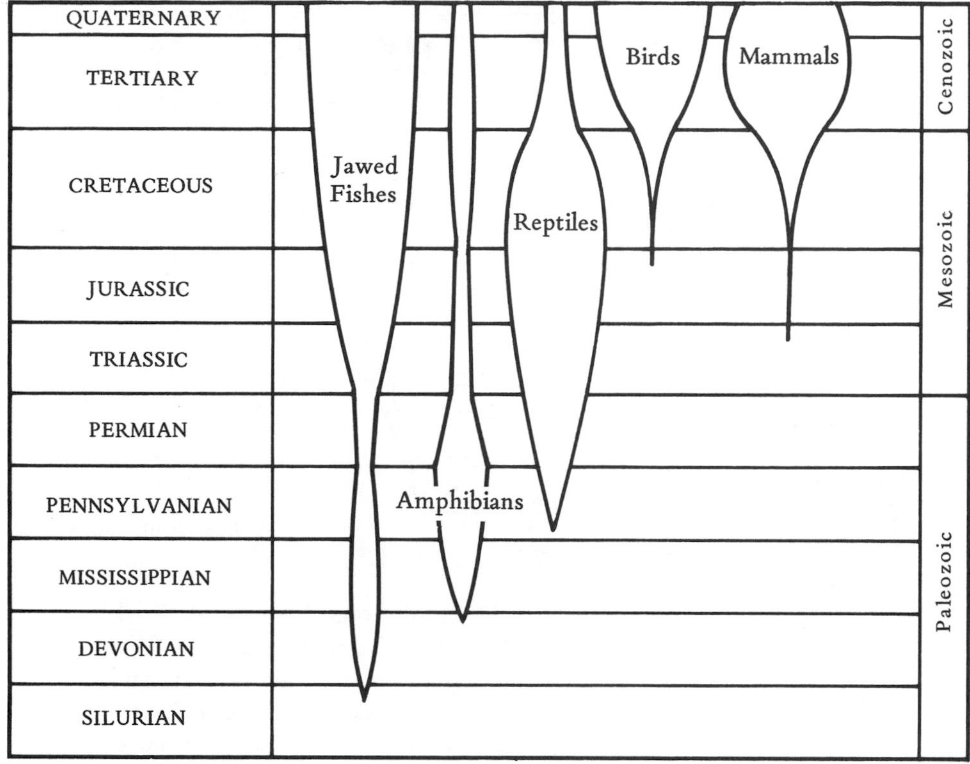

FIGURE 4.13 The Fossil Record of the Vertebrates shows the appearance of more complex types in younger rocks, a fact known even in Darwin's time.

generation. He, like nearly all his contemporaries, envisioned heredity as a blending of parental traits and his critics argued that if this were true, any favorable variation would be swamped out by breeding with those individuals that did not have it. Unless one could envision numerous hereditary changes or *mutations* all in the same direction, no variation could survive. This argument troubled Darwin until his death. At the same time as Darwin's work, an obscure Czech monk, *Gregor Mendel* (Fig. 4.16) had discovered the basic laws of *heredity* by breeding pea plants in a Bohemian monastery. Mendel discovered that the factors which caused individual traits are inherited as separate entities and hence, retain their identities and are not swamped out. But Mendel's work remained unknown until 1900 when its rediscovery vindicated Darwin. In this century the science of heredity, *genetics,* has undergone tremendous advances including the discovery of how spontaneous mutations occur and the chemical basis of inheritance. How far man will go with the control of his own destiny remains to be seen.

Evolution and the Fossil Record

As mentioned above, Darwin was skeptical of the ability of historical geology to vindicate his thesis of slow change from one species into another. In the

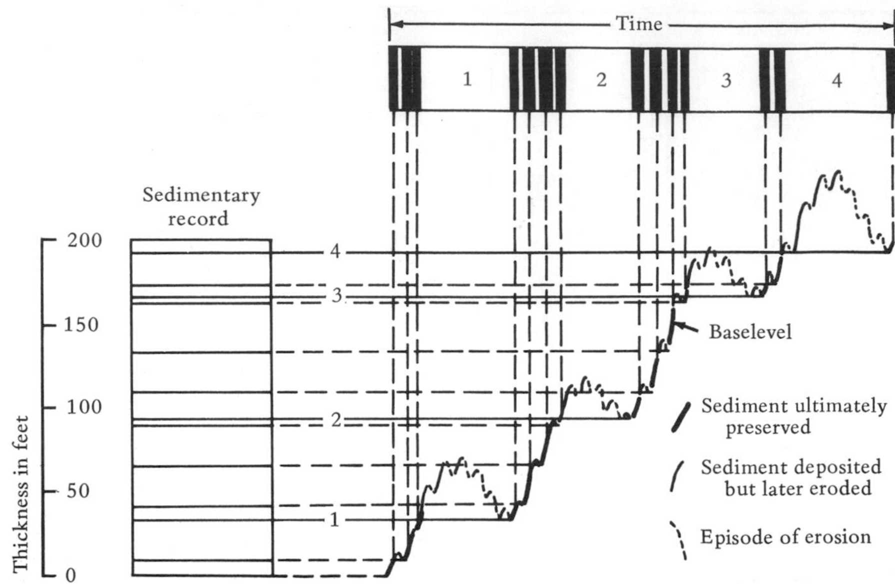

FIGURE 4.14 The Basically Discontinuous Nature of Rock Deposition. Baselevel in a sedimentary basin continually fluctuates. The major trend is one of aggradation, but second- and third-order cycles create a discontinuous record. Where the baselevel rises, sediment accumulates; where it falls, erosion occurs. White portions of the curve are ultimately preserved in the sedimentary record. Total amount of time represented in the stratigraphic record at left is shown in white bars at top. Major diastems are numbered. (From Don L. Eicher, *Geologic Time,* © 1968. Reprinted by permission of Prentice-Hall, Inc., Englewood Cliffs, N.J.)

years after 1859, however, numerous unbroken lineages of fossils began to be unearthed. In 1869, the German paleontologists *Waagen* and *Karpinsky* cited evolutionary sequences in lineages of *ammonites.* (Some examples of ammonite lineages demonstrating evolution are shown in Figure 4.17, although these are not the ones studied by Waagen and Karpinsky.) In the 1870s, *Huxley* in England and *Marsh* in North America discovered the unbroken Eocene to Recent *horse* lineage (Fig. 4.18) which is still perhaps the longest and most complete evolutionary sequence in the fossil record. An Austrian paleontologist, *Neumayr,* in 1875, traced an evolutionary lineage of smooth into ornamented *snails* (Fig. 4.19). In Britain, *Rowe* and others found a series of *heart urchin* lineages (Fig. 4.20) in the Cretaceous chalk. By 1910, *Carruthers* displayed a complete evolutionary sequence of Paleozoic *corals* (Fig. 4.21). The process of finding unbroken evolutionary lineages has continued at an ever-expanding pace down to the present day even in the plant world. Darwin was still basically correct, the geologic record is full of gaps of all sizes in any one region. Even though in the world as a whole there may be strata representing every year of Cambrian to Recent time, clearly every individual cannot have been preserved if for no

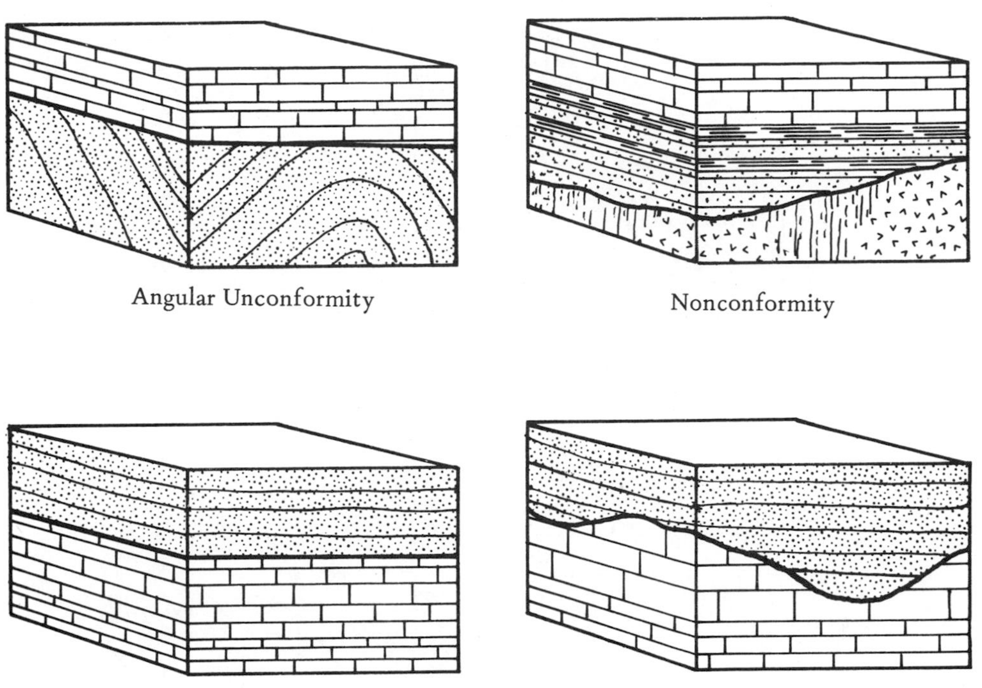

FIGURE 4.15 The Three Major Varieties of Unconformity. Darwin recognized the fact that disconformities exist and are numerous.

FIGURE 4.16 Gregor Mendel (1822–1864).

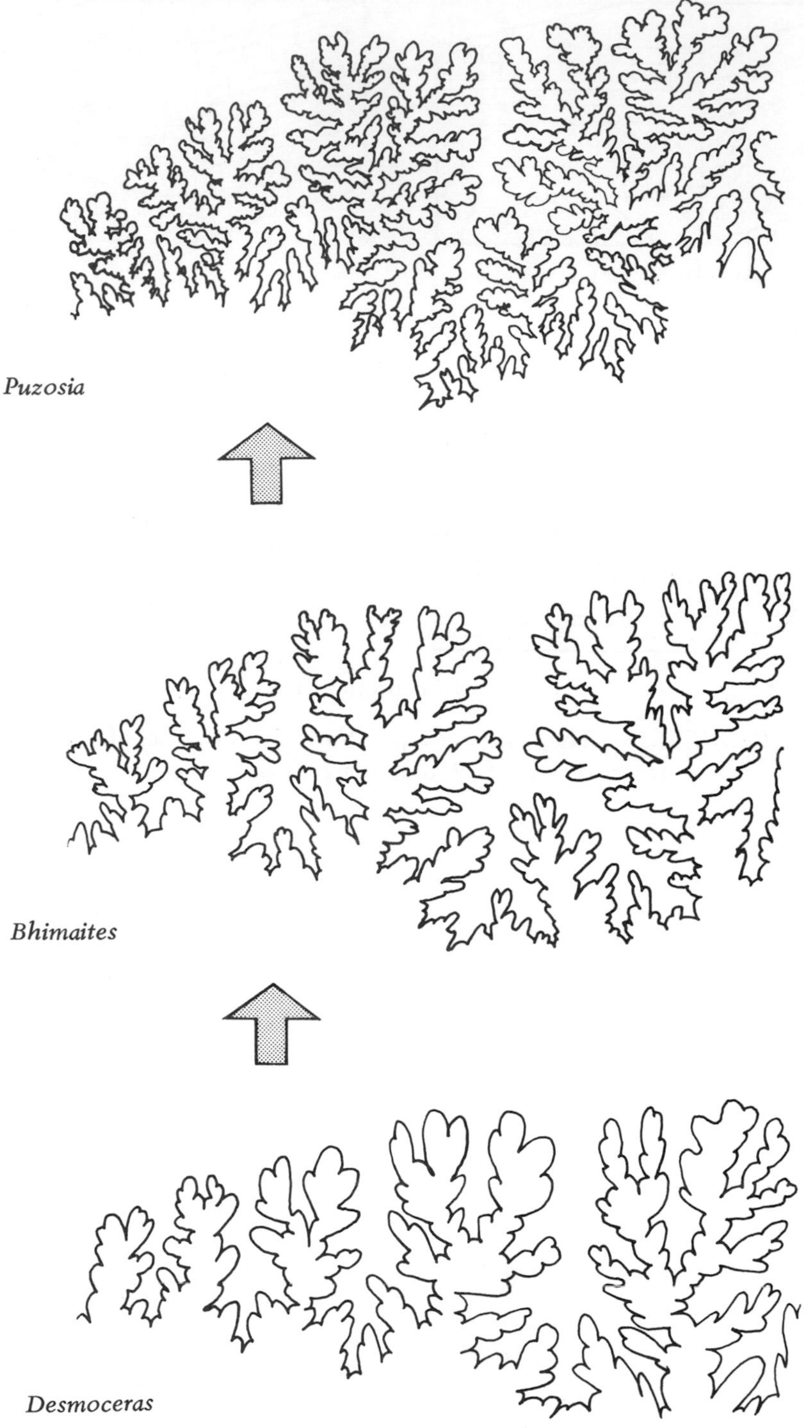

FIGURE 4.17 The Evolution of Ammonoid Sutures. The progressive increase in complexity of these species is very similar to that noted by Waagen and Karpinsky.

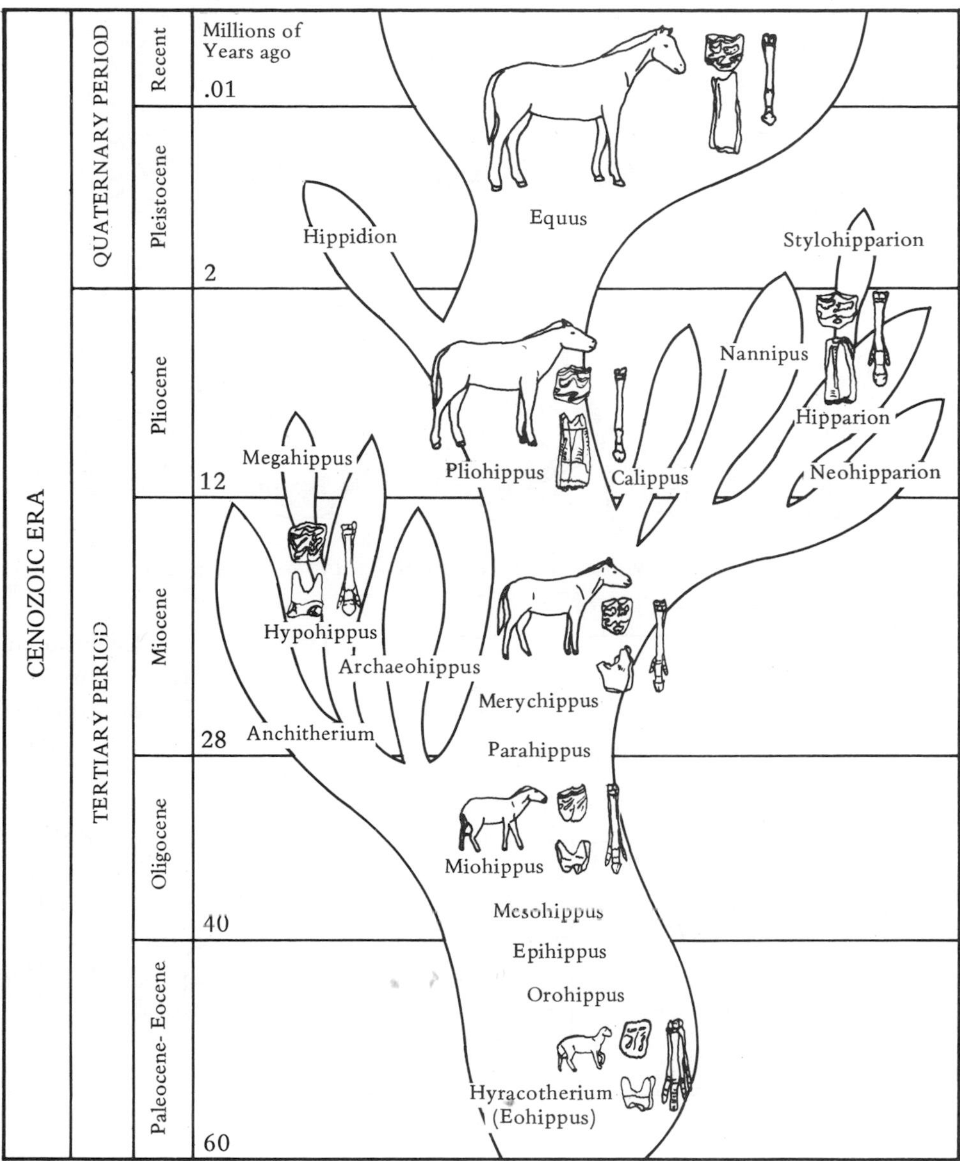

FIGURE 4.18 Evolution of the Horse Family. (After E. Peter Volpe, *Understanding Evolution,* © 1970, William C. Brown Co., Dubuque, Iowa and Bernhard Kummel, *History of the Earth,* W. H. Freeman & Co., San Francisco, Calif.)

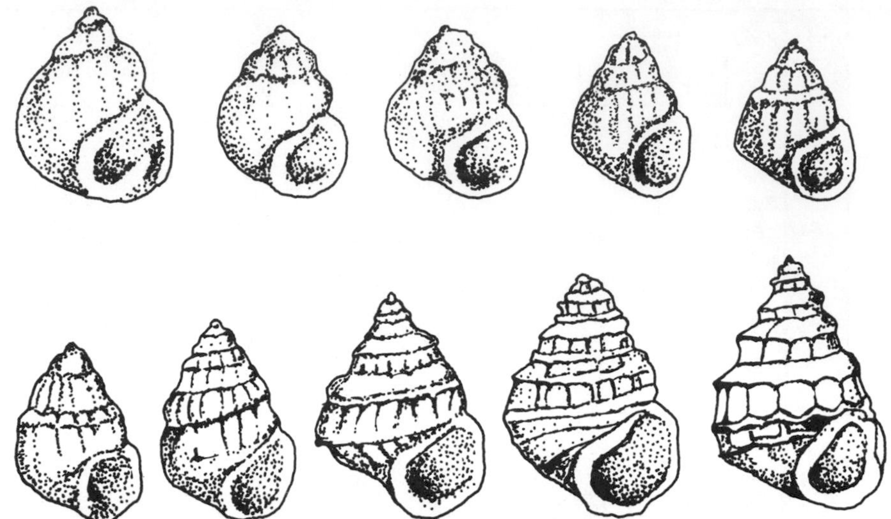

FIGURE 4.19 Evolution of the Snail *Paludina.* This trend was noted by Neumeyr in 1875.

CHANGE IN MICRASTER	TIME ZONES
M. SENONENSIS, M. CORANGUINUM	*Marsupites testudinarius*
	Micraster coranguinum
M. CORTESTUDINARIUM	*Micraster cortestudinarium*
M. CORBOVIS	*Holaster planus*
	Terebratulina lata
M. LESKI	*Cyclothyris cuveri*

FIGURE 4.20 Evolutionary Trends in the Genus *Micraster.* The evolutionary sequence from the Cretaceous Chalk was recognized as early as 1899 by Rowe. (From James R. Beerbower, *Search for the Past: An Introduction to Paleontology,* 2nd Ed., © 1968. Reprinted by permission of Prentice-Hall, Inc., Englewood Cliffs, N.J.)

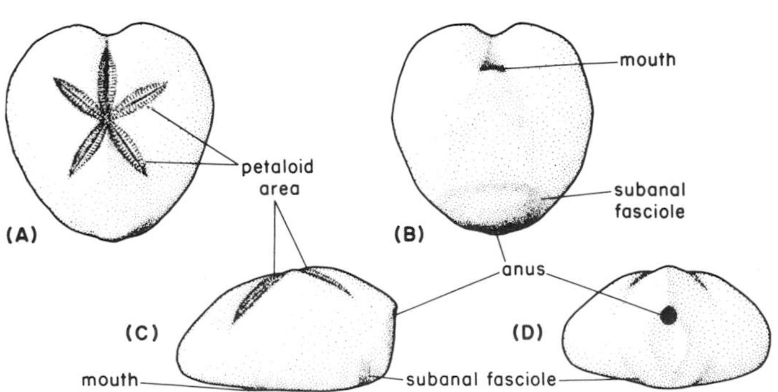

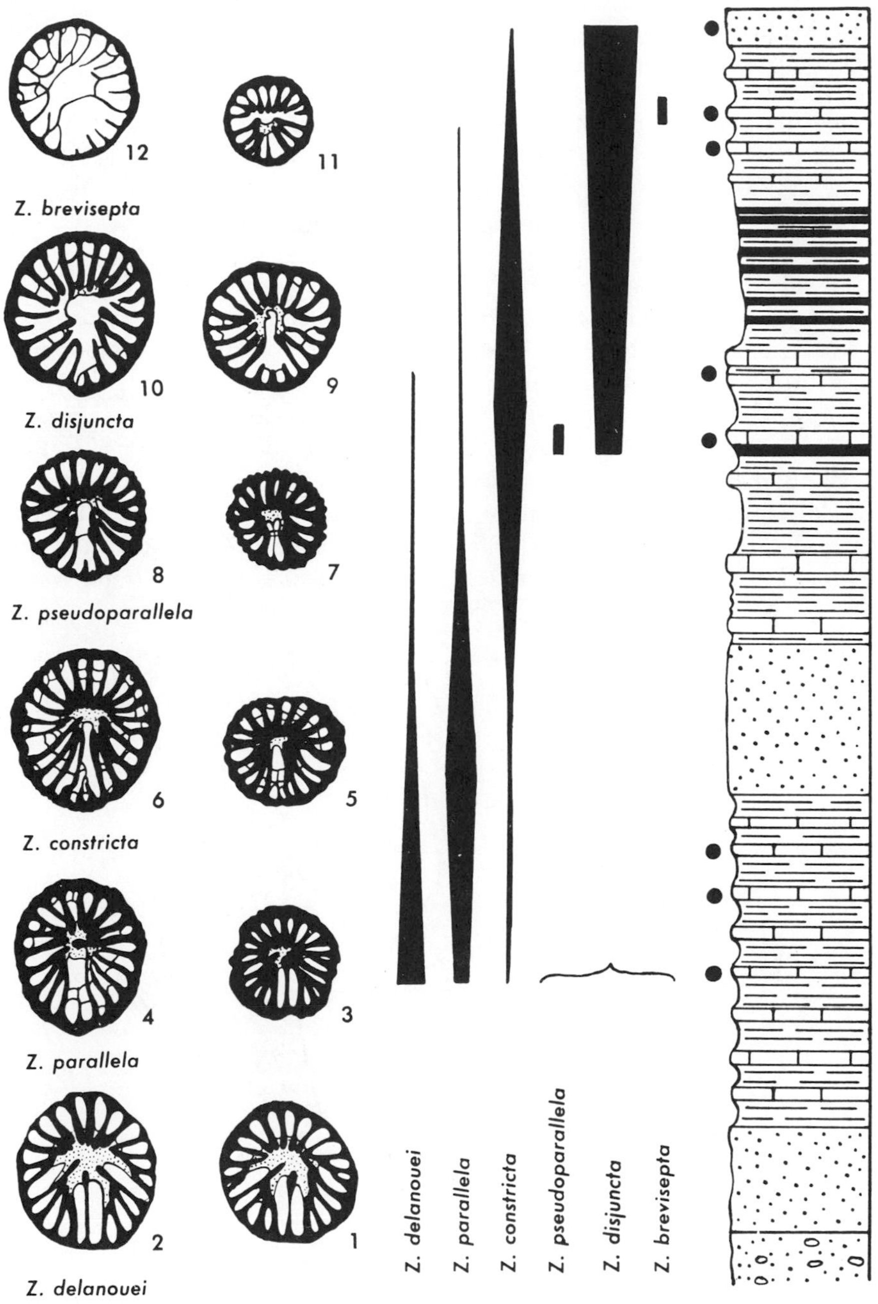

FIGURE 4.21 Evolution and Recapitulation in a Lineage of Fossil Corals. (From *Invertebrate Paleontology*, by William H. Easton, © 1960, Harper & Row Publishers, Inc., Fig. 5.6, "Evolution of *Zaphrentites delanouei*," after Carruthers, 1910, *London Quarterly Journal.*)

other reason than that no species is ubiquitous. As a result, we will probably never know the complete evolutionary lineages of all organisms or even of those likely to be preserved. But enough examples have been found to establish beyond question that the process of evolution occurs.

The Species Problem

One result of the discovery that evolution is indeed recorded in the geologic record by a small series of transformations is that it makes defining a species impossible. Biologists define a *species* as a population or group of populations whose members are capable of interbreeding with one another, but not with members of other species. Such a definition breaks down when the time factor is introduced because then there is no isolation of populations (Fig. 4.22). They are all interconnected through the evolutionary pathway of life. Only when one takes a slice of geologic time such as the biologist is doing with the present, do populations appear isolated. Paleontologists have tried to solve the problem

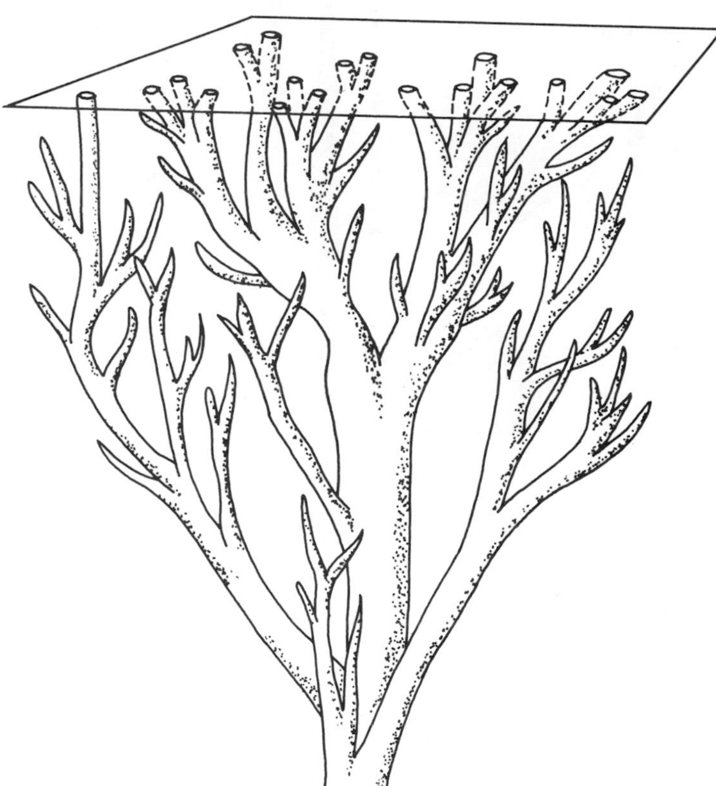

FIGURE 4.22 A Portion of the "Bush of Life" showing the lack of discontinuity in the organic world when viewed from the geologic perspective. The biologist deals only with a slice of time such as the present (top of diagram) and species appear distinct.

by allowing no more variation among the members of a species in a vertical time sense than exists among the members of a "horizontal" contemporaneous population. Even this solution fails when one considers that a population is a mixture of individuals, some of whom are typical of the population for that slice of time, some look like members of ancestral populations, and others look like members of descendant populations. On a strictly structural basis, then, brothers and sisters may be assigned to three separate species.

Clearly there are several ways of looking at fossil populations (Fig. 4.23) and the traditional usage of species is inadequate to cover them all. Someone interested in a slice of geologic time such as a geologist studying the paleocology of a Cretaceous chalk would lump all members of an interbreeding population together in one "species" regardless of their structure. A geologist interested

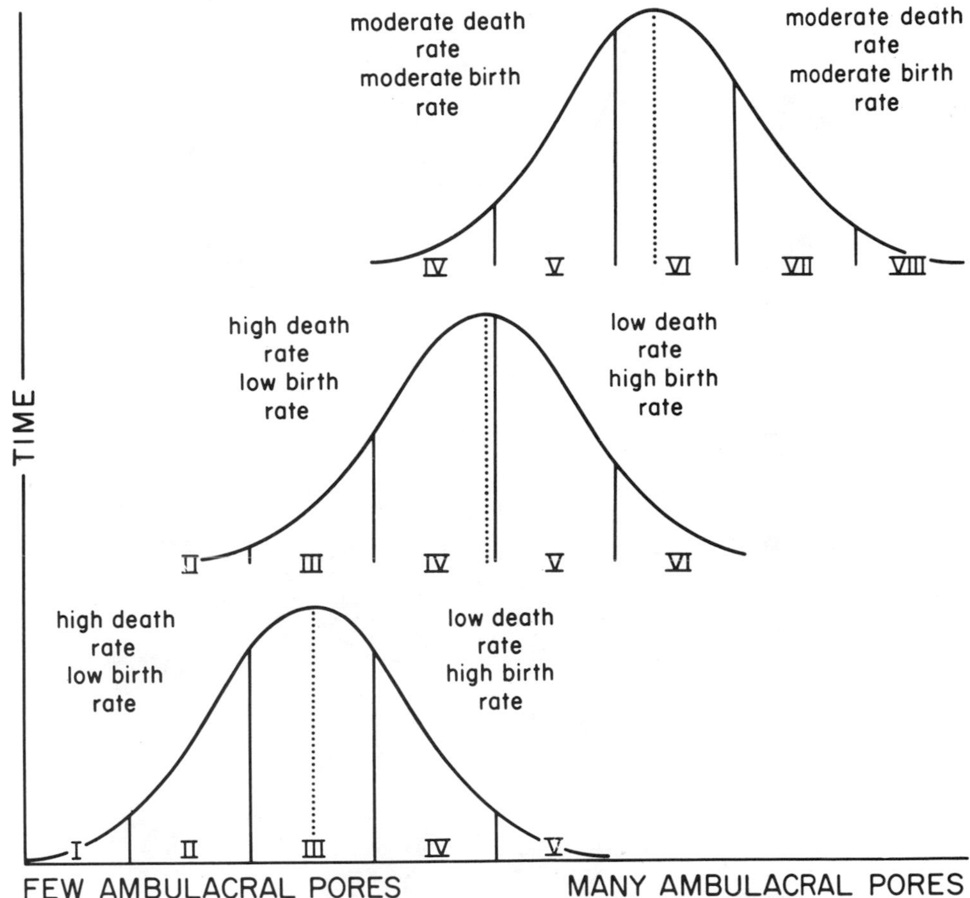

FIGURE 4.23 The Species Problem in Paleontology. Successive populations overlap in morphology. Assignment to a species is purely arbitrary and based on the purposes of a worker as noted in the text. (From James R. Beerbower, *Search for the Past: An Introduction to Paleontology*, 2nd Ed., © 1968. Reprinted by permission of Prentice-Hall, Inc., Englewood Cliffs, N.J.)

in establishing the geological ranges of certain structural types of fossils to zone the Cretaceous would assign the individuals resembling ancestral populations to the ancestral species and those resembling descendant populations to the descendant species. He is concerned with the appearances and disappearances of structural entities for dating purposes and is not concerned about the relationships of individuals at one time. Because of the endless confusion

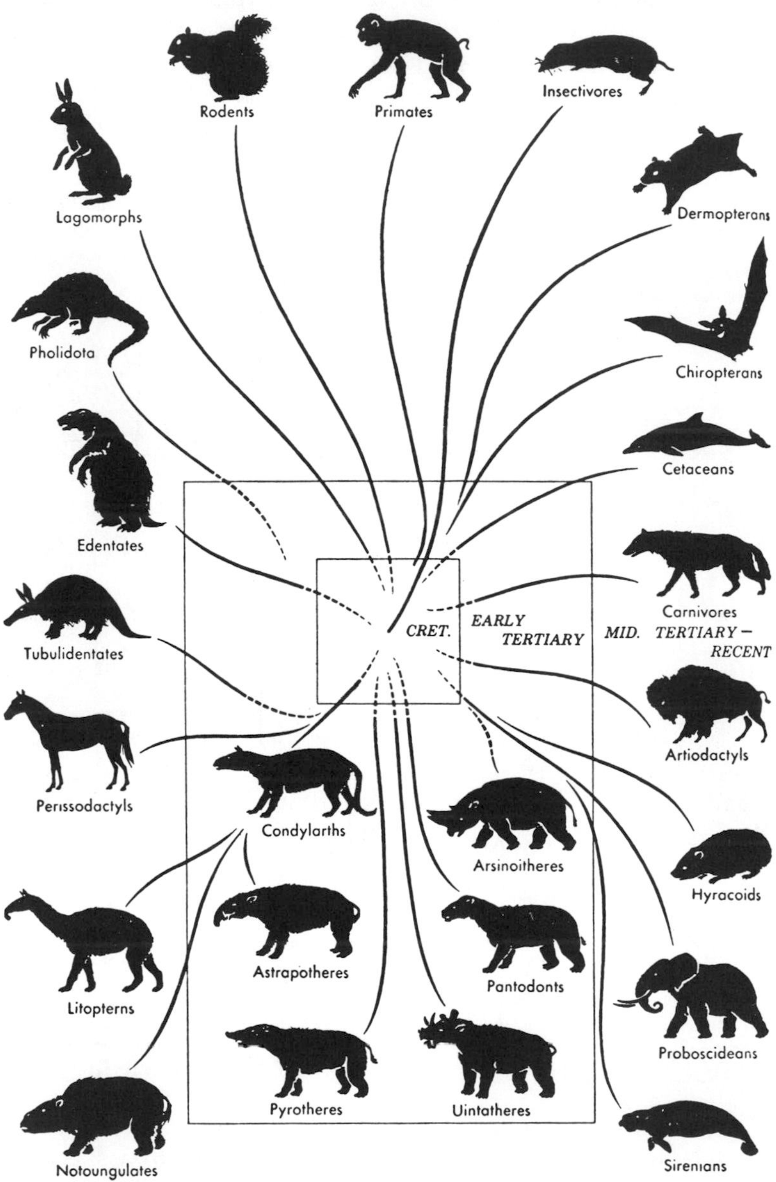

FIGURE 4.24 Adaptive Radiation of the Placental Mammals in the Cenozoic. (From Edwin H. Colbert, *Evolution of the Vertebrates,* © 1969, by permission of John Wiley & Sons, Inc.)

resulting from these different, but useful concepts of species, many biologists and paleontologists suggest we abandon species altogether and simply use computer codes for structures. Time will tell if their efforts will be successful.

Patterns of Evolution

Another result of examining both the fossil record and Recent organisms has been the recognition of patterns of evolution. The more significant ones are adaptive radiation, convergence, parallelism, and extinction.

Adaptive radiation is the process of diversification or multiple-branching of a lineage which develops a significant structural and/or environmental breakthrough. For example, when vertebrates invaded the land they diversified into crawlers, runners, climbers, flyers, and so forth. When the mammalian level of evolution was reached, a variety of types developed ranging from insectivores to herbivores to carnivores, from runners to burrowers to flyers, from terrestrial to aquatic to aerial forms (Fig. 4.24).

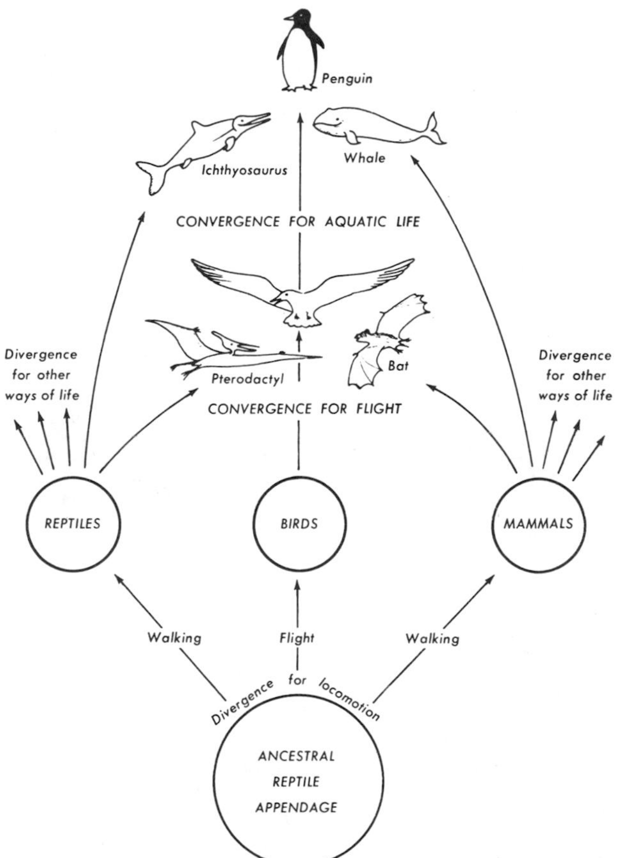

FIGURE 4.25 Convergent Evolution in Certain Vertebrate Groups. (After Alfred M. Elliott, *Zoology*, © 1957, by permission of Appleton-Century-Crofts, Inc.)

In *convergence*, species of distant relationships have evolved similar features and, hence, look similar because they inhabit the same environment which is selecting for generally the same features. For example (Fig. 4.25), fish, whales (mammals) and the extinct icthyosaurs (reptiles) resemble one another as do birds, bats (mammals) and pterosaurs (reptiles), yet all are only distantly related.

Parallelism is similar to convergence in that species of different lineages tend to look alike. But in this case, the lineages are closely related. For another example, the horses and the extinct palaeotheres are members of the same order of mammals and have a close common ancestor (Fig. 4.26). Both lived in the same environment and, hence, went through a series of similar evolutionary changes to better adapt to that environment.

Finally, *extinction*, for the final termination of a lineage, is the lot of most evolutionary lines sooner or later. (We will exclude from extinction the disappearance of a species because it evolved into another species.) Sooner or later, a lineage becomes so highly adapted to an environment or mode of life that a slight change, typically the evolution of a superior competitor by another lineage, occurs too rapidly for the structurally specialized population to evolve

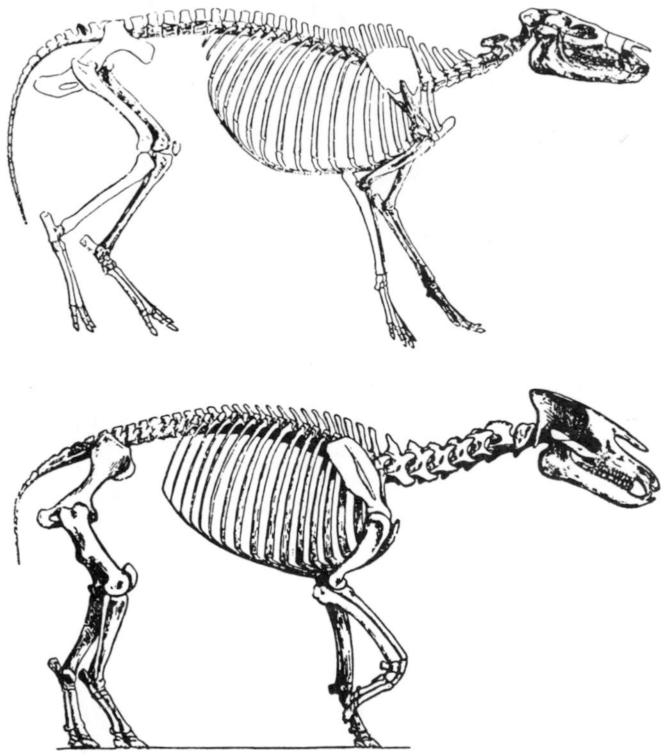

FIGURE 4.26 Parallelism in the Mammals. Horses and Palaeotheres were two lineages of hoofed mammals of the same order that followed similar evolutionary paths in the early Cenozoic. (From A. S. Romer, *Vertebrate Paleontology*, copyright © 1966, by the University of Chicago Press. Used by permission of the publisher.)

into something new and it perishes. This fossil record shows this occurrence over and over again in every major and minor category of organisms (Fig. 4.27). The fact that it has occurred to the most successful of lineages should make man pause to think.

	RADIATION	EXTINCTION
Marine Invertebrates as a Whole	Early Paleozoic	Permian
	Jurassic	Late Cretaceous
Foraminiferans	Silurian	Permian & Triassic
	Mississippian	
	Jurassic-Cretaceous	
Graptolites	Ordovician	Silurian & Devonian
Brachiopods	Ordovician	Devonian & Carboniferous
	Jurassic	Cretaceous
Nautiloids	Ordovician	Silurian
		Mississippian
Ammonoids	Devonian	Early Mississippian
	Early Triassic	Late Triassic
	Early Jurassic	Late Jurassic
	Mid-Cretaceous	Late Cretaceous
Trilobites	Cambrian	Silurian
	Devonian	Carboniferous & Permian
Crinoids	Ordovician	Late Mississippian
	Early Mississippian	Late Permian
Fishes	Devonian	Pennsylvanian
	Cretaceous	
Land Vertebrates as a Whole	Permian	Late Cretaceous
	Jurassic	Pleistocene
	Paleocene-Eocene	
Amphibians	Pennsylvanian	Permian-Triassic
Reptiles	Permian	Late Triassic
	Jurassic	Late Cretaceous
Mammals	Paleocene-Eocene	Pleistocene
Insects	Late Paleozoic	
	Late Cretaceous	
Land Plants	Devonian	Permian
	Pennsylvanian	
	Late Cretaceous	

FIGURE 4.27 Chart of the Main Episodes of Radiation and Extinction in Significant Fossil Groups. (From J. Marvin Weller, *The Course of Evolution*, © 1969, reprinted by permission of McGraw-Hill Book Co.)

Questions

1. What is meant by *organic evolution?*
2. What were the contributions to the recognition of evolution made by the following men: Cuvier, Linnaeus, Erasmus and Darwin?
3. How do Lamarckism and Darwinism differ? Give an example.
4. Discuss any four major lines of evidence used by Darwin in supporting evolution.
5. Briefly describe the theory of natural selection.
6. Cite some examples of evolutionary change documented by the fossil record.
7. Why might two experts on Permian brachiopods assign the same specimen to two different species?
8. What is meant by *adaptive radiation* and *evolutionary convergence?*

5

The Dichotomy Between Rocks and Time:

Formations Versus Stages

In Chapter 3 we noted that it took many centuries for geologists to recognize there were two major ways of subdividing the geologic record. One was on the basis of the types of rocks (lithology) which were present and the other was on the basis of the times of formation of the various rocks. At first, it was not apparent that there was any difference between the two. In Wernerian geognosy, for example, each type of rock formed at one particular time in the past over the entire world and, hence, rock type and time of formation coincided. Geologists soon demonstrated, however, that rock layers were not worldwide. With the recognition of facies by Gressly and others, it became apparent that several varieties of rock could form at the same time. Hence lithology was a guide to environment and not age.

Stages

In the early 1800s, a multiplicity of rock units were named and described on the basis of their lithology. The great French geologist, *Alcide d'Orbigny*, bemoaned this endless terminology and suggested that a universally applicable term was needed to encompass all rocks deposited during the same time interval, regardless of their lithologies. D'Orbigny proposed the term *stage* for this purpose in the 1840s and defined a series of stages for the rocks of Jurassic and Cretaceous ages. These stages were based chiefly on wide-ranging species of ammonoids that were found in numerous facies. D'Orbigny was a catastrophist and believed that the stages were bounded by natural breaks in the record.

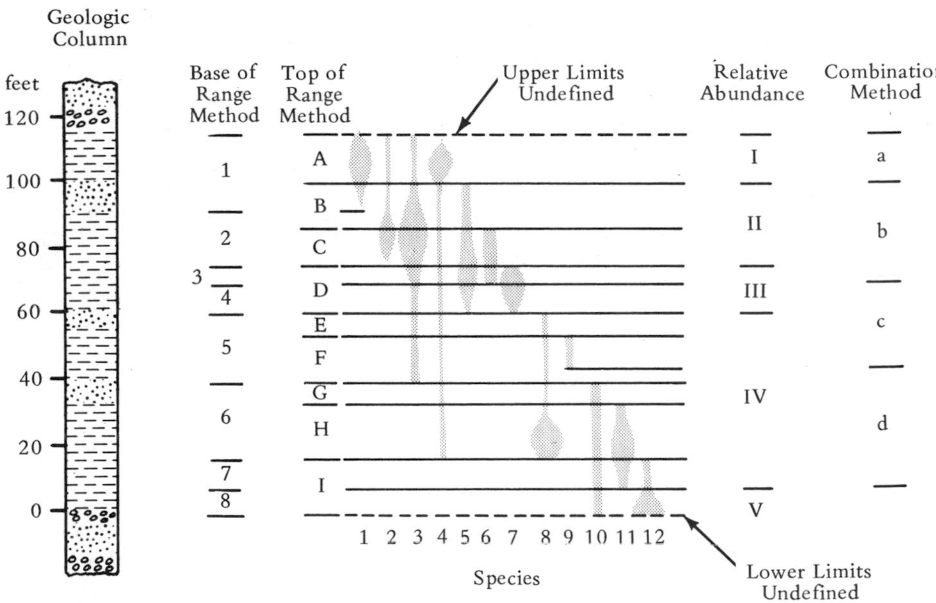

FIGURE 5.1 Concurrent-Range Zones. This diagram illustrates four different ways of subdividing strata into local concurrent-range zones based on overlapping ranges of fossil species. (From Don L. Eicher, *Geologic Time*, © 1968. Reprinted by permission of Prentice-Hall, Inc., Englewood Cliffs, N.J.)

Zones

The German geologist *Albert Oppel* was not a catastrophist, however, and he noted that by careful fossil collecting from numerous facies over a wide area he could piece together a more or less complete time sequence. Species did not all appear and disappear as abruptly as d'Orbigny thought, but instead there was a continuous replacement of old species by new. By utilizing the overlapping ranges of fossil species (Fig. 5.1), Oppel was able to define a continuous sequence of small intervals of strata deposited at the same time and characterized by a unique assemblage of fossils which did not occur earlier or later in the record. He called these units of rock based on their time of formation, *zones*. Several

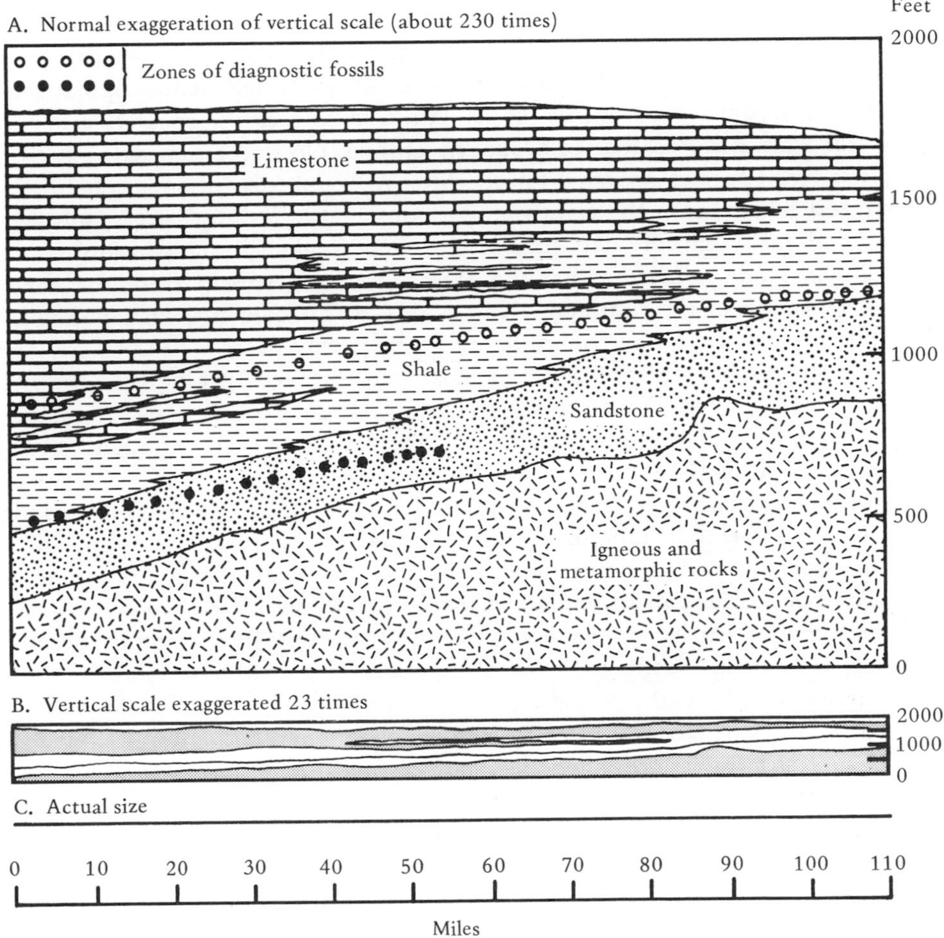

FIGURE 5.2 Effect of Vertical Exaggeration. Cross-sections illustrating the difficulties encountered by geologists in separating lithologic and time aspects of rocks in areas of widespread, nearly flat-lying strata. (From Don L. Eicher, *Geologic Time,* © 1968. Reprinted by permission of Prentice-Hall, Inc., Englewood Cliffs, N.J.)

successive zones that appeared to have much in common paleontologically were assembled into stages.

Rocks Versus Time

In examining a simple cross-section, such as Figure 5.2, showing one type of rock grading laterally into another, one cannot comprehend the great difficulties that still awaited geologists in separating these two aspects of rocks, their lithology and their time of formation as indicated by fossils. In areas such as the great interior plains of North America, which at times in the geologic past were covered by widespread epicontinental or epeiric seas (much like gigantic continental shelves), there are widespread rock units which appear essentially horizontal as one sees them at any one locality. It was very easy for early geologists to envision each of these layers as an essentially contemporaneous deposit. The accurate positions of fossils within these rocks were never reported by careful work such as that of Oppel in Europe. No one checked to see if a diagnostic fossil species occurred lower in the rock unit near the margins of the deposit and higher near the center of the continent; the whole rock unit was one integral entity entirely deposited in the same time span. This hardly seems rational when one considers that the seas must have spread from the edges toward the center of the continent. Thus, any type of sediment, such as a nearshore beach sand, must have only occupied a relatively narrow band at any one time. But such issues did not occur to most mid-continent geologists who proceeded with an almost Wernerian approach to geology. To them, the boundaries of rock units were also the boundaries of time units, for each deposit was contemporaneous throughout.

Proceeding on both observational and logical grounds, *H. S. Williams*, in 1894, pointed out that lithologies were nothing more than the record of a particular geologic environment and migrated wherever these environments went. The boundaries of rock units and time units, therefore, do not correspond. As a hypothetical example, envision a sea transgressing over a gently sloping landmass furnishing clastic particles of all sizes (Fig. 5.3). Nearshore, where wave energy is highest, the coarsest particles, in this case pebbles, occur. Further from shore, the particle size decreases through bands of sand and mud. Finally, beyond the reach of land-derived clasts are chemical percipitates such as limestone. As the sea advances, each band of facies moves progressively inland so that eventually each type of sediment leaves a sheet of sediment behind. Notice, however, that each sheet is of different ages in different places as indicated by the horizontal time lines on the illustration. It is appreciably younger on its seaward margin than at its landward edge. Hence, a rock unit is of slightly different ages as one goes from place to place, unless one is paralleling an ancient shoreline. Instead of the sand being all formed contemporaneously, various parts of it are the same age as adjacent rock types. Note also that with time, each deposit of a transgressing sea is overlapped by the deposits of more offshore sediments. In regressing seas, the reverse is true as the more inshore deposits overlap the more offshore ones. An important generalization, often called *Walther's Law* after the German geologist who proposed it at the turn

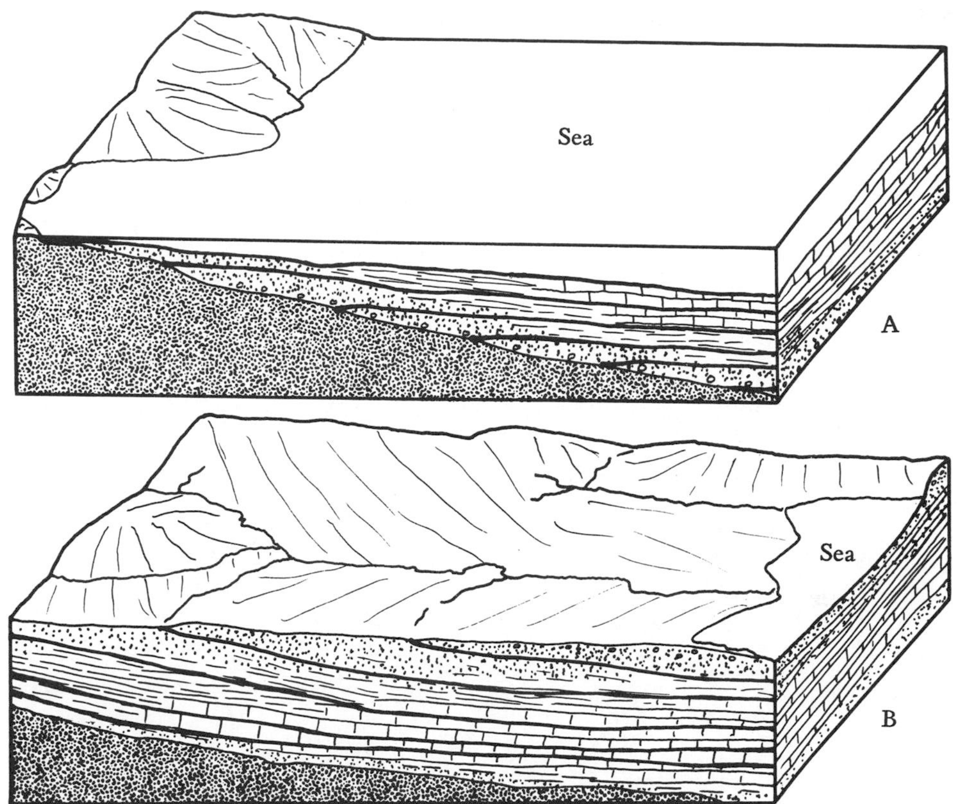

FIGURE 5.3 The Diachronous (Time-Transgressive) Nature of Rock Units. (A) Onlap pattern produced by advancing sea, (B) Offlap pattern produced by retreating sea. Note time lines cross all rock units.

of this century, arises when one looks at these diagrams. In order to see at one locality what the lateral lithologic facies of a rock are, all one has to do is look at the overlying and underlying rock types because they are part of the very same rock units.

A contemporary of Williams, the noted geologist and Colorado River explorer, *John Wesley Powell,* who then directed the United Geological Survey, proposed usage of a separate lithologic term, *formation.* This term was to be utilized to clearly distinguish these units of variable age but constant environment, from stages and other rock subdivisions based on times of formation. As we have seen in an earlier chapter, however, most leading North American geologists of the time would have none of this reasoning. To them, the alternate rise and fall of the seas were worldwide and essentially instantaneous, much like the flushing of a toilet. Thus, every rock unit was the same age everywhere, a throwback to Werner. Once a fossil giving a certain age was found in a rock unit, the age of that entire unit was fixed everywhere. Different lithologic units always meant different ages; they could not have formed contemporaneously in the same region. The key phrase of these men was "diastrophism (earth movement) is the ultimate basis of correlation (the process of determining contemporaneity)."

Codes of Nomenclature

Both international and American geological commissions were established in the late 1800s and early 1900s to bring some order out of this chaos of terminology. The international commission (1900) came up with a dual hierarchy of time terms (era, period, epoch, age, and phase in descending order of magnitude) and rock terms (group, system, series, stage, and zone in order of decreasing magnitude) based on the time formation of the rocks as defined by their fossils. Thus, there was an exact one to one correspondence between the two categories of terms; one set of units was for pure time, the other applied to rocks deposited during that time. For example the Devonian System encompasses all rocks deposited during the Devonian Period. There was no formal terminology for lithologic terms so chaotic usage continued. The American commission (1933) came up with a dual set of terms also, but there was an important difference. On one hand there were units of pure time (era, period, and several levels of epochs) and on the other there were rock units (system, series, group, formation, member, and bed) based on the time of formation of the rocks which was determined by their lithology. In other words, a formation was a lithologic unit that was all deposited during a second-order epoch. Lithology and time were firmly wedded to each other and most American geologists went their own way ignoring Gressly, Williams, Powell, and Grabau.

Such obvious contradictions could not submerge the realities forever, and beginning with the 1930s and continuing down to the present day, the trend has reversed itself. American geologists once again talk of stages and formations as different things even if many of them still do not practice the differentiation. The movement back toward recognizing the fundamental difference between units of rock based on time and those based on lithology had as its fountainhead, Stanford University, and as its leaders, *Hubert Schenck, Siemon Muller*, and their students, the most celebrated of which is *Robert Kleinpell*. In a series of papers in the 1930s which finally culminated in the 1941 work

Geologic-Time Units	Time-Stratigraphic Units	Rock-Stratigraphic Units
Era Period Epoch Age ———	——— System Series Stage Zone	Group Formation Member, etc. Bed, etc.

FIGURE 5.4 Table of Stratigraphic Units Proposed by Schenck and Muller (1941).

of Schenck and Muller, they argued convincingly that at least three separate sets of terms were needed (Fig. 5.4). First there were the units of pure time, such as period, epoch, and age, and corresponding to these exactly were units of rock defined on the basis of their time of formation such as system, series, and stage. (Of course the time-rock units must be recognized first; they are the material basis on which units of time are established.) Written at right angles to these parallel sets of terms were the lithologic units, group, formation, member, and bed, in an effort to clearly denote the fact that these sets of units in no way corresponded to the time-based units, but instead transgress time in virtually all instances. In 1961, the American Commission adopted a new code of units embodying these principles and some additional ones. The International Code of the same year is similar.

Modern Stratigraphic Terminology

The new code recognizes four major categories of terms utilized by those geologists who studied layered or stratified rocks, the stratigraphers. These are the *lithostratigraphic* (rock-stratigraphic) units based solely on lithology and having no necessary time relationship, the *chronostratigraphic* (time-stratigraphic) units based solely on the time of formation of the included rocks, the *geochronologic* (geologic time) units which are subdivisions of pure time recognized on the basis of chronostratigraphic units, and finally, a new set of terms called *biostratigraphic* units.

Biostratigraphic units are units of rock defined solely on their fossil content without any necessary time connotations. This may seem strange at first, when we have been saying all along that fossils are the basis of chronostratigraphic units. There is an important difference, however. The presence of fossils in rocks is an objective thing, just as the presence of certain lithologies defines lithostratigraphic units. Time, however, is an interpretation made from the rocks; it cannot be seen or touched. We have already seen that lithology has no time significance except for certain rock bodies such as ash beds and turbidite beds that represent a single depositional event. In other words, the significance of lithology is in recognizing environments and not time. We have seen that fossils have both environmental and time significance. A certain assemblage of fossil organisms that lived together in an interdependent fossil community has chiefly ecologic significance at first glance. It, like sediment type, will shift about wherever the proper environment is situated for the moment. However, we must not forget the important factor of evolution. No group of populations stays static for very long. Before much time passes, at least one and typically several species in a population evolve into something else or become extinct. Or a new species may migrate in from elsewhere when a distant barrier disappears, thus altering the composition of the community. As a result, there can never be exactly identical communities of organisms living at separate times in the geologic past. Also, it is a statistically impossible event for a species to evolve twice considering the vast number of coincidental hereditary changes that would have to occur. Therefore, the time of existence of a species is a unique biologic event in time. Migrations into an area and extinctions are other time-significant and nonrecurrent biologic events that characterize the appearance and disappearance of fossils, but not rock types.

Thus, the simple occurrence of fossils is an objective thing. The interpretive processes of stripping away the ecologic and other factors that influence the distribution of fossils to get at their time significance produces chronostratigraphic units from this raw data. For example, the occurrence of a species of brachiopod that ranges through two and a half geologic periods has little more than general time significance and just records the presence of the well-adapted species when-

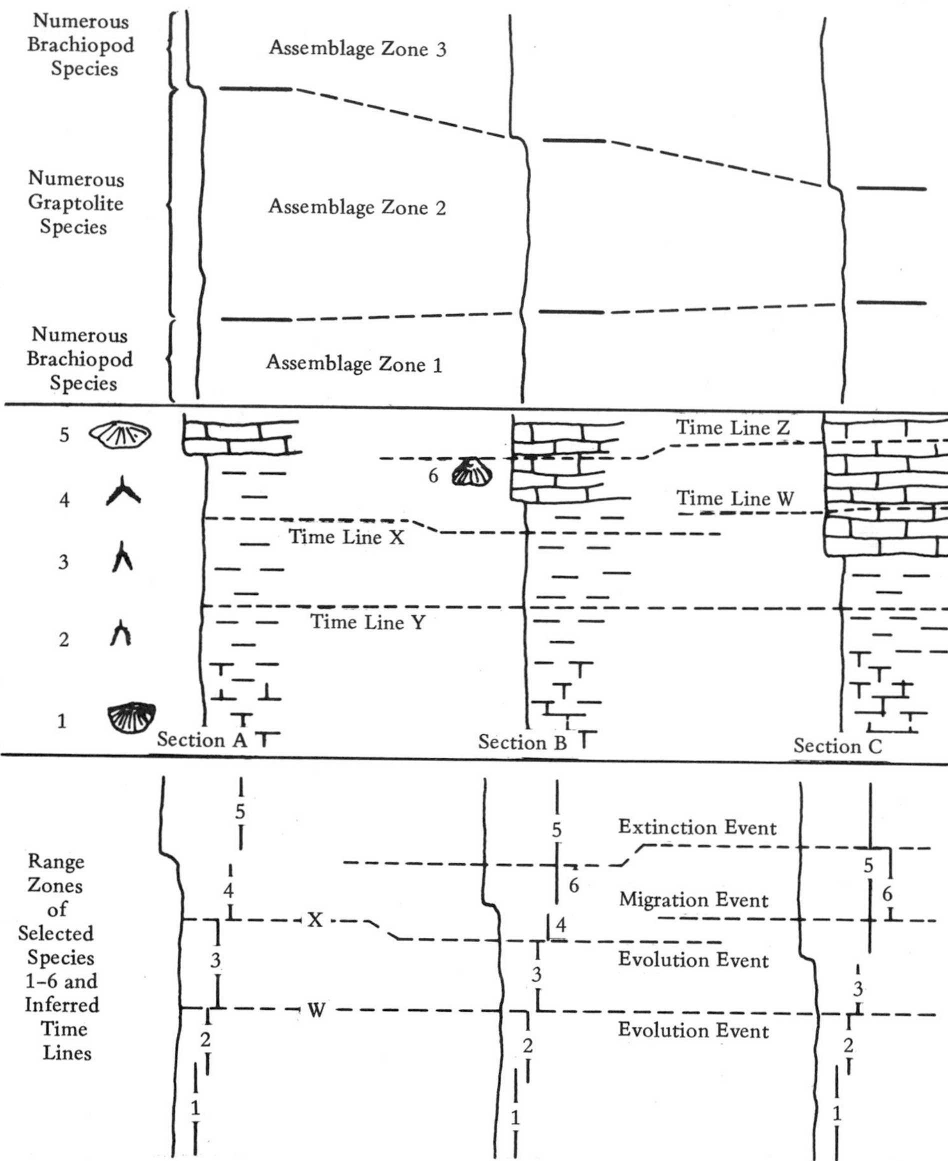

FIGURE 5.5 Diagram Illustrating the Relationships Between Objective Biostratigraphic and Rock-Stratigraphic Terminology and Inferential Time-Significant Events. (From Don L. Eicher, *Geologic Time*, © 1968. Reprinted by permission of Prentice-Hall, Inc., Englewood Cliffs, N.J.)

ever the environment was right. On the other hand, a species of ammonite belonging to a rapidly evolving lineage may dilineate a span of time as short as one sixty-second portion of a period. As another example, we can consider the appearance or disappearance of a species at a formation boundary as being related to a change in the environment of a region and of no time significance. On the other hand, the extinction of a species or its evolution into another species within the same environment is an event of great time significance (Fig. 5.5).

In Figure 5.6 are summarized the geochronologic, chronostratigraphic, lithostratigraphic, and biostratigraphic terms in common usage today. Most of these terms are from the codes, but where the codes do not specify terminology, the most commonly used terms are cited. Note that the word *zone* appears in two places. Zones are biostratigraphic units based on the objective occurrences of fossils without regard to their ranges, whereas chronozones (Oppelian zones) are based on the overlapping ranges of fossil species and are inferred to have time stratigraphic significance.

Geochronologic or Geologic Time	*Chronostratigraphic or Time-Stratigraphic*
Eon	
Era	Erathem
Period	System
Epoch	Series
Age	Stage
Phase	Chronozone

Lithostratigraphic or Rock Stratigraphic	*Biostratigraphic*
Group	
Formation	Zone
Member	Subzone
Bed	Zonule

FIGURE 5.6. Table of Modern Stratigraphic Units. (Modified after American Code of Stratigraphic Nomenclature.)

Questions

1. How did d'Orbigny define a *stage?*
2. Explain the method of establishing a concurrent range zone developed by Oppel.
3. What is Walther's Law?
4. What is the difference between a formation and a stage?
5. What was the major contribution of Schenck and Muller?
6. How do time-stratigraphic and biostratigraphic units differ?
7. List the hierarchy of geochronologic (geologic-time) units now accepted by the code.

6

"No Trace of a Beginning—No Prospect of an End:"

Absolute Time in Geology

With superposition, cross-cutting relationships, inclusion, and faunal succession, geologists had the tools to establish the relative geologic time scale of Precambrian through Recent. They could arrange rocks in the proper sequence and determine the order of events in early history. This task still absorbs the work of numerous geologists worldwide, not only in improving the resolution of detail in known regions, but also in deciphering the history of new regions and incorporating their history into the world pageant. Indeed, for many geologic purposes, this type of dating is all that is necessary. At the same time, however, geologists have also been concerned about absolute time, that is, how many years ago did a particular event occur? Just when did the earth begin? How long was the Devonian Period? How many years does it take one species to evolve into another or for a foot of sandstone to form? And so on.

Early Ideas

The early writings of many religions ascribe various lengths to earth history ranging from relatively very short to rather immense spans of time. The Christian religion is intimately bound with the history of Western Man in whose culture the modern science of geology evolved. To many scholars, the emphasis of Christianity that the earth was made for man to master through knowledge is responsible for the development of that unique Western contribution to civilization we call science. It is natural, then, that Western men first turned to their theologians to answer questions about the antiquity of the earth. In the 1600s, Biblical scholars reasoned that if they worked backward through the genealogies given in the *Bible,* which generally gave the life spans of individuals and sometimes the birth years of their children, they could, with reasonable extrapolations for the gaps, arrive at the time of the creation of Adam and Eve and of the world. In 1644 *Bishop Lightfoot* arrived at the date of September 17, 3928 B.C. at 9:00 A.M. *Archbishop Ussher* in 1658 determined a slightly different date of October 23, 4004 B.C. In general, the dates were in close agreement and indicated that the earth was less than 6,000 years old. Marginal notations of the dates were made in many editions of the Bible making them literally the "gospel truth." Opposition to them was considered heresy.

In the mid 1700s, the great French scientist *Buffon* suggested that the earth must have initially been molten because of the presence of volcanoes and the fact that temperatures increased in deep mines. If so, utilizing the then-known data on the earth's volume, its present temperature and its inferred molten temperature, Buffon reasoned that it must have taken the earth 75,000 years to cool to its present state. Buffon had to recant, and although we would tend to discount his postulates today, he did open the door to consideration that the earth was much older than the Biblical estimate.

The Application of Uniformitarianism

With the rise of uniformitarianism because of Hutton and Lyell's work and the validation of biological evolution by *Darwin,* it became apparent to scientists that vast lengths of time were involved in the geologic past. But how much, no one yet knew. It remained for someone to measure the rates of present processes and then extrapolate these to the history that was recorded in the earth's rock record.

Three basic and classic approaches to the problem were tried in the second half of the nineteenth century and all disagreed for various reasons. One method involved the *salinity* of the oceans. The oceans were assumed to originally have been fresh water. Scientists reasoned that if they could correctly estimate the amount of salt being carried to the oceans each year by rivers and the total amount of salt now in the oceans, they could determine the earth's age by dividing the former figure into the latter. In 1899, *John Joly* made this calculation and came up with an estimate of 90 million years. That was certainly a large figure and lent credence to the uniformitarianism's contention of vast amounts of time. It was, however, less than Darwin had thought and we now know it is far too low because of many unconsidered factors. Joly had no reliable notion of the vast amounts salts lost from the oceans by evaporation of sea water, winds blowing salt inland, the formation of saline sediments, and the uplift of saline rocks to form land again. It also seems possible that the amount of salt now being added to the oceans may be slightly higher than normal for the past because the continents may be higher now than in much of the past and, hence, erosion rates are probably greater. It now appears that the salinity of the oceans has been essentially constant throughout much of geologic history once the factors (Fig. 6.1) reached equilibrium with one another.

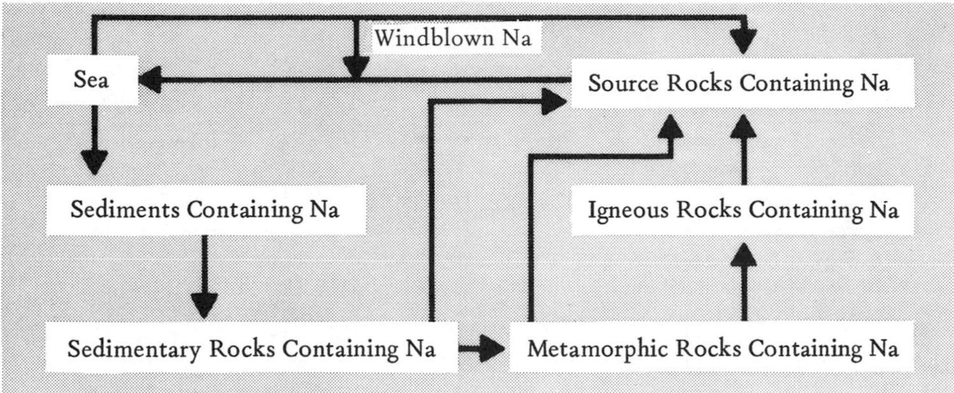

FIGURE 6.1 The Sodium Cycle. This cycle probably attained equilibrium long ago thus invalidating the saltiness-of-the-sea method of dating the earth.

Other geologists reasoned that if they could determine the average rate of accumulation of a foot of *sediment* and then measure the total thicknesses of the geologic record, they could determine the age of the earth by dividing the former figure into the latter. This notion, actually suggested by the ancient Greek Heredotus, was fraught with even greater uncertainties. Geologists recognized that each type of sediment accumulates at different rates, that even one type of sediment accumulates at different rates under different conditions, that the geologic column in any area was full of gaps which may represent as much time as the rocks themselves, and that there was some uncertain amount of compaction which occurred in the conversion of sediment to sedimentary rock. In addition, geologists kept discovering newer and thicker columns in other regions so estimates constantly had to be revised. Between 1860 and 1910, many great geologists made estimates of the length of earth history based on this method (Fig. 6.2). They ranged from a low of three million years to a high of a billion and a half years, but most were between 80 and 100 million years. Again, these figures indicated great lengths for geologic time, but not as much as some uniformitarians had guessed.

It was from the realm of *physics*, however, that the most authoritative estimates of the length of geologic time came in the late 1800s. Physics has always been considered the most basic and precise of the sciences and, therefore, the opinions of its practitioners have always carried heavy weight in scientific circles. *Lord Kelvin*, the foremost physicist of this time, arrived at his estimates of the earth's age using two approaches. One was the old assumption of Buffon that the earth had originally been molten and cooled to its present condition. Kelvin realized the limitations of his data and calculations much better than Buffon, but it was still obvious that the maximum amount of time was less than 100 million years and probably as low as 25 million years. This was much less time than evolutionists and uniformitarians guessed and also indicated that even less time was available for biologic evolution because the crust was too hot early in its history to support life.

DATE	AUTHOR	MAXIMUM THICKNESS (In Feet)	RATE OF DEPOSIT (Years for 1 Foot)	TIME (In Millions of Years)
1860	Phillips	72,000	1332	96
1869	Huxley	100,000	1000	100
1871	Haughton	177,200	8616	1526
1878	Haughton	177,200	?	200
1883	Winchell	——	——	3
1889	Croll	12,000[1]	6000[2]	72
1890	de Lapparent	150,000	600	90
1892	Wallace	177,200	158	28
1892	Geikie	100,000	730–6800	73–680
1893	McGee	264,000	6000	1584
1893	Upham	264,000	316	100
1893	Walcott	——	——	45–70
1893	Reade	31,680[1]	3000[2]	95
1895	Sollas	164,000	100	17
1897	Sederholm	——	——	35–40
1899	Geikie	——	——	100
1900	Sollas	265,000	100	26.5
1908	Joly	265,000	300	80
1909	Sollas	335,800	100	80

After Arthur Holmes, 1913.
[1]Spread evenly over the land areas.
[2]Rate of denudation.

FIGURE 6.2 Estimates of the Earth's Age Based on Rates of Sedimentation. (From Don L. Eicher, *Geologic Time*, © 1968. Reprinted by permission of Prentice-Hall, Inc., Englewood Cliffs, N.J.)

The other method of attacking the problem was based on the assumption that the source of the sun's energy was of a conventional sort; the sun was either a burning incandescent body or released energy by the gravitational contraction of its immense mass. Either way, it could not continue to produce energy at its present fantastically high rate for very long, for its heat would soon be exhausted. The sun, then, could only have illuminated the earth for a few tens of millions of years, and in the past it was much hotter and in the future it would get much cooler. Therefore, Kelvin estimated that at most the earth was inhabitable for organisms between 20 and 40 million years. Evolutionists found it virtually impossible to accept these figures, but all they had were educated guesses in the face of Kelvin's potent mathematics. Darwin and others compromised their original theories in their later years in an effort to reconcile evolution and uniformitarianism with the physicists' estimates. Eventually, however, they were vindicated.

Radioactivity

In all fairness to the physicists, it was one of them, *Henri Becquerel,* who in 1896 made the startling discovery that eventually demolished Kelvin's theories and proved the geologists and biologists right. Becquerel discovered *radioactivity,* the spontaneous decay of certain unstable elements or forms of elements (isotopes), which produces large amounts of energy by the transformation of a tiny amount of mass. Hence, there was no reason to suppose that the earth's crust had been any hotter in the past than now or that it was going to get appreciably cooler in the foreseeable future. The natural decay of the radioactive elements in the earth's crust was chiefly responsible for its heat, not some residual heat left over from an originally molten stage. These calculations, which proved that the heat generated by the amount of radioactive elements in the earth's crust was more than sufficient to provide its heat flow, were also made by a physicist, *R. J. Strutt.* Another significant discovery about radioactivity was that it was apparently the source of the sun's energy. The sun, by the loss of infinitesimal amounts of mass, could produce immense amounts of energy by nuclear reactions and, hence, could keep up its present rate of energy output for very great lengths of time. The sun was not just a big bonfire burning in the sky as Kelvin and others thought, but instead was fueled by this miraculous new energy source, radioactivity.

The most significant of all the wonders that flowed from the discovery of radioactivity, however, was that it at long last provided a reliable means of determining the absolute ages of those rocks which contain radioactive substances. It was found that in the process of radioactive decay, elements were transformed from one to another until a stable form was finally reached. In some cases, an element changed directly from a radioactive to a stable element, in others the transformations were long stepwise processes. In the first decade of this century, an American chemist, *B. B. Boltwood,* found that uranium, a radioactive element, eventually decayed to stable lead and that the ratio of lead to uranium was consistently greater the older the rock. The rate of decay appeared to be constant; half of the original uranium would decay in a certain amount of time, half of the remaining amount would decay after another equal amount of time, and so on. No known physical or chemical process would alter this exponential decay rate. By measuring the amounts of lead and uranium and knowing this so-called *half-life* of the radioactive elements in this transformation, Boltwood reasoned that one could calculate the amount of time that had elapsed since the uranium was first trapped in the crystals of a rock (Fig. 6.3).

To illustrate this method, let us consider a hypothetical example (Fig. 6.4) in which a radioactive element has a half-life of one million years and decays directly to a nonradioactive element. After a million years, then, one half of the original sample consists of the *parent* radioactive element and one half consists of the *daughter* element. After another million years, half of the remaining radioactive element would decay so that now the sample would consist of one-fourth parent and three-fourths daughter element. After another million years, half of the remaining radioactive element would decay so that now the sample

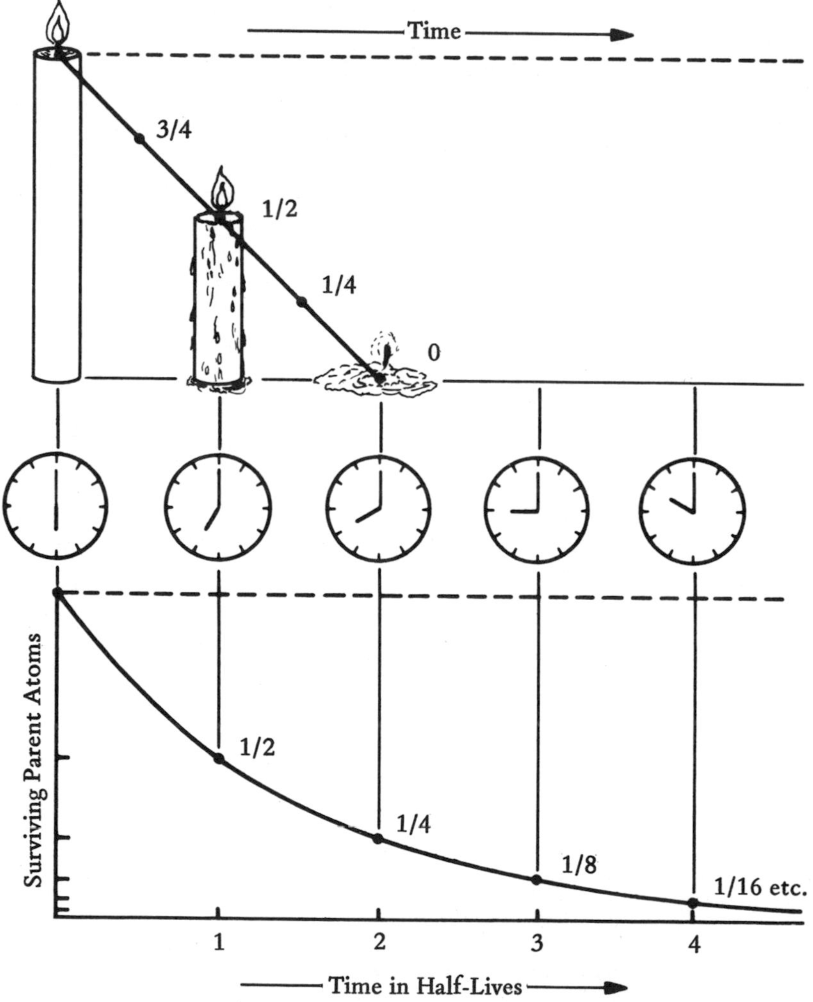

FIGURE 6.3 Diagram Illustrating the Exponential Rate of Radioactive Decay in contrast to normal uniform depletion rates.

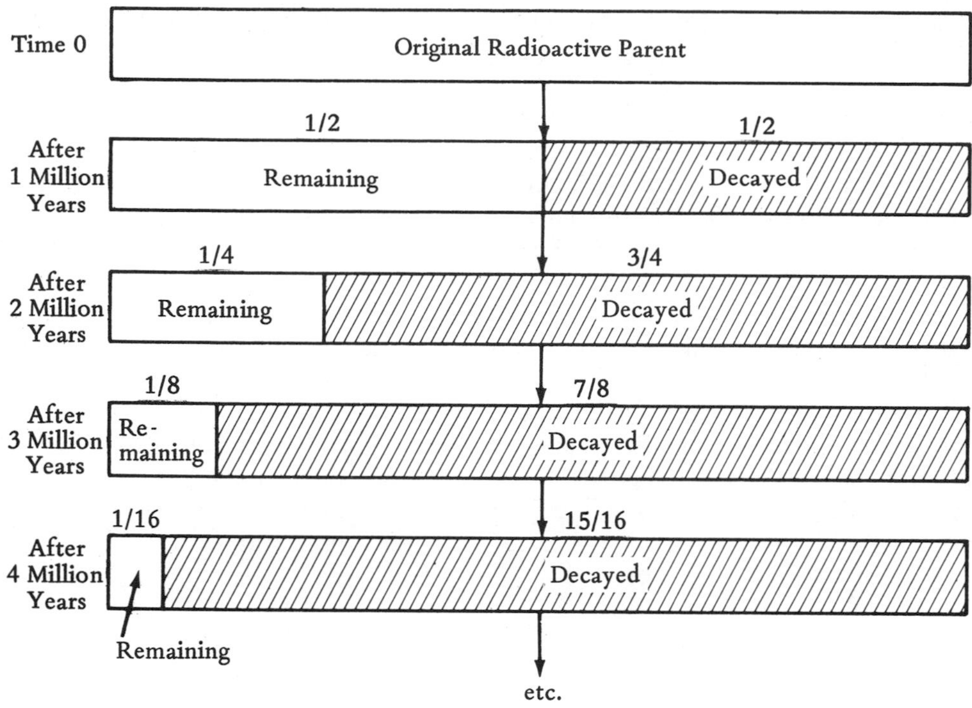

FIGURE 6.4 Diagram of the Radioactive Decay of a Hypothetical Element with a Half-Life of 1 Million Years.

Geological Period	Lead/Uranium Ph/U	Millions of Years
Carboniferous	0.041	340
Devonian	0.045	370
Precarboniferous	0.050	410
Silurian or Ordovician	0.053	430
Precambrian		
a. Sweden	0.125	1025
	0.155	1270
b. United States	0.160	1310
	0.175	1435
c. Ceylon	0.20	1640

FIGURE 6.5 Boltwood's First Radiometric Dates. (After Arthur Holmes, 1911).

would be only ⅛ parent and ⅞ daughter element. Thus if we found a sample that was 15/16 daughter and 1/16 parent element, we would know that it had passed through four half-lives and, hence, four million years had elapsed since this sample of the radioactive element had become isolated in a crystal and started to form its own decay products.

Using this method with the uranium to lead decay series and a crude estimate of the half-lives involved in this multiple-step decay series, Boltwood arrived at the truly astounding figures of hundreds of millions of years ago for the Paleozoic periods to well over a billion and a half years of age for some Precambrian rocks (Fig. 6.5). The uniformitarians and evolutionists had been right after all. The refinement of using radioactive decay to date rocks, the process called *radiometric dating*, has continued with ever increasing success down to the present day. It is now possible to state with some degree of confidence dates of the beginning and ending of the Paleozoic Periods to within a few tens of millions of years, of the Mesozoic Periods within a few million years, and the Cenozoic epochs within a fraction of a million years. With the advent of radiometric dating we have been able to roughly calibrate our relative time scale with an absolute one based on years (Fig. 6.6).

Radiometric Dating Techniques

The major radioactive decay series used in radiometric age determinations are Uranium 238 → Lead 206, Uranium 235 → Lead 207, Potassium 40 → Argon 40, and Rubidium 87 → Strontium 87 (Fig. 6.7). The decay of Carbon 14 → Nitrogen 14 is very useful for dating the late Pleistocene and the Holocene Epochs. Each of these methods deserves some comments. (The numbers refer to the numbers of positively charged protons and uncharged neutrons in the nucleus of an atom, the mass number.)

Uranium-Lead Methods

Naturally-occurring uranium always consists of both varieties or isotopes of the element and, hence, both methods can be used on the same rock and can be cross-checked. U238 has a half-life of 4.51 billion years and U235, a half-life of 713 million years. Each decays through many intermediate steps to the resultant lead 206 and 207. The ratio of the resultant leads provides a further cross-check on the method. Both isotopes occur in the widespread mineral zircon (a zirconium silicate) found in many igneous rocks. In a sample, there is always some original lead 206 and lead 207 which have to be corrected for, but these occur in a constant proportion with ordinary non-radioactively produced lead 204, so once the amount of this isotope is determined, the proper subtraction is readily made. Reheating, such as that produced in metamorphism, and weathering commonly result in the loss of some of the lead in a sample and thus the ages record only the latest episode of lead loss and not necessarily the time of formation of the original igneous body.

FIGURE 6.6 Calibration of Absolute and Relative Geologic Time Scales. G.S.L. is an abbreviation for the Geological Society of London.

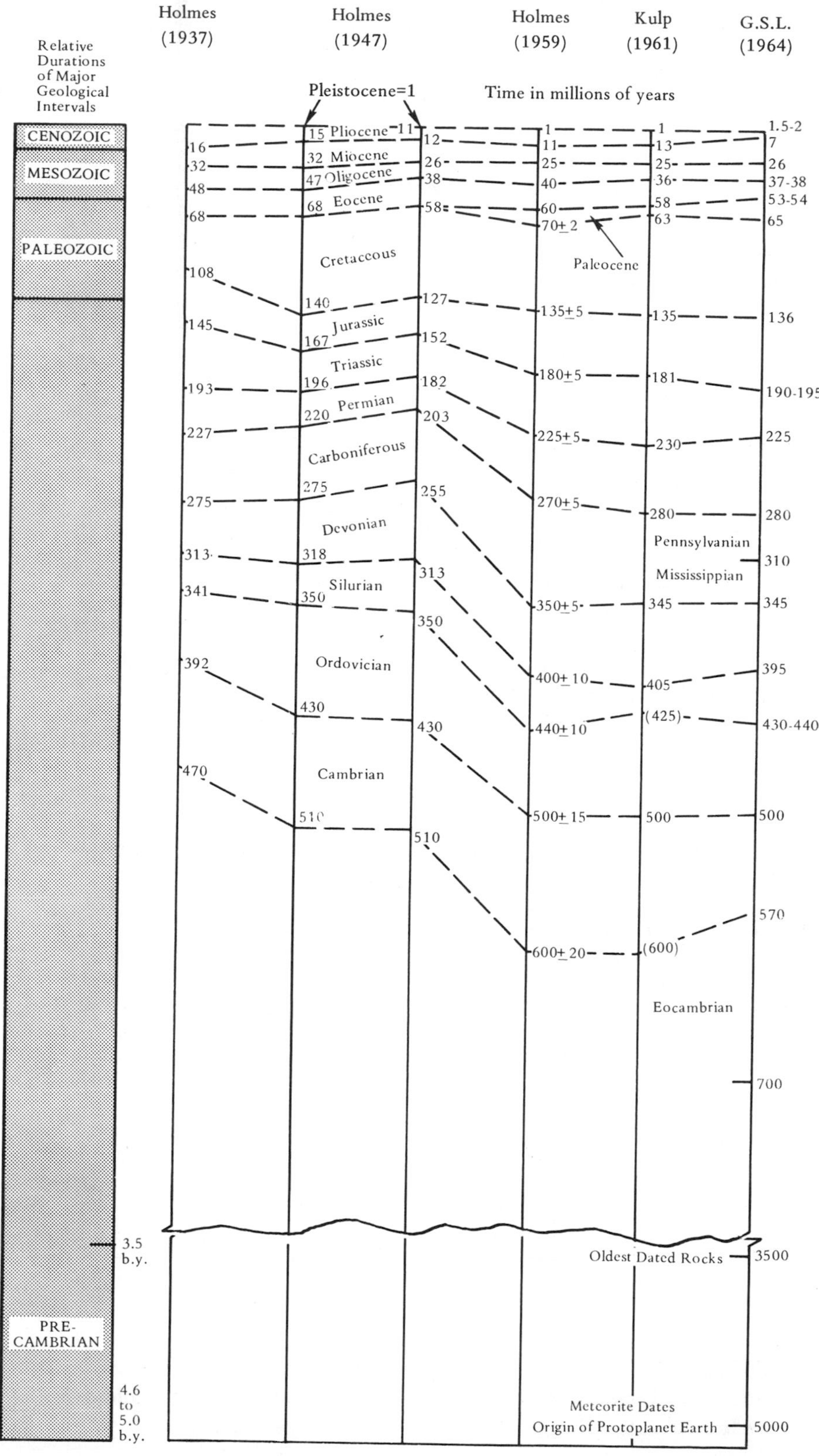
Relative Durations of Major Geological Intervals
Holmes (1937)
Holmes (1947)
Holmes (1959)
Kulp (1961)
G.S.L. (1964)
Pleistocene=1
Time in millions of years
CENOZOIC
MESOZOIC
PALEOZOIC
PRE-CAMBRIAN
3.5 b.y.
4.6 to 5.0 b.y.
Pliocene
Miocene
Oligocene
Eocene
Paleocene
Cretaceous
Jurassic
Triassic
Permian
Carboniferous
Pennsylvanian
Mississippian
Devonian
Silurian
Ordovician
Cambrian
Eocambrian
16
32
48
68
108
145
193
227
275
313
341
392
470
15
32
47
68
140
167
196
220
275
318
350
430
510
11
12
26
38
58
127
152
182
203
255
313
350
430
510
1
11
25
40
60
70± 2
135±5
180±5
225±5
270±5
350±5
400± 10
440±10
500± 15
600± 20
1
13
25
36
58
63
135
181
230
280
345
405
(425)
500
(600)
1.5-2
7
26
37-38
53-54
65
136
190-195
225
280
310
345
395
430-440
500
570
700
Oldest Dated Rocks
3500
Meteorite Dates
Origin of Protoplanet Earth
5000

Parent Nuclide	Half-Life in Years	Daughter Nuclide	Minerals and Rocks Commonly Dated
Uranium-238	4,510 Million	Lead-206	Zircon Uraninite Pitchblende
Uranium-235	713 Million	Lead-207	Zircon Uraninite Pitchblende
Potassium-40	1,300 Million	Argon-40	Muscovite Biotite Hornblende Glauconite Sanidine Whole Volcanic Rock
Rubidium-87	47,000 Million	Strontium-87	Muscovite Biotite Lepidolite Microcline Glauconite Whole Metamorphic Rock

FIGURE 6.7 The Major Methods of Radiometric Age Determination Used by Geologists.

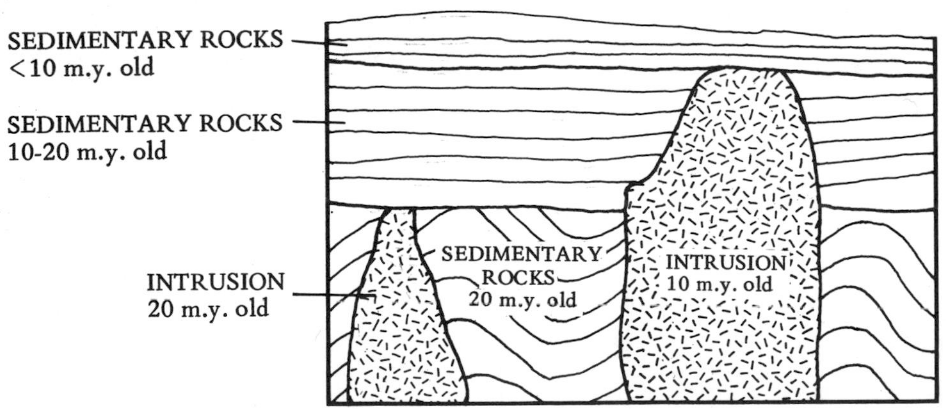

FIGURE 6.8 Use of Cross-cutting Relationships to Date Sedimentary Rocks in Relation to Igneous Rocks Dated by Radiometric Means.

This discussion of the uranium-land dating method points up the two biggest shortcomings of radiometric dating techniques in geology. One is that radioactive elements are typically restricted to igneous rocks, yet we are more concerned with the sedimentary rocks which cover 75% of the earth's land surface and contain the fossils on which our relative time scale is based. Unless we can relate sedimentary rocks to igneous bodies by crosscutting relationships or inclusions it is difficult to date sedimentary rocks. For example, if a sequence of sediments is cut by an igneous intrusion ten million years old, we know that the sedimentary sequence is older than 10 million years old, but how much older we do not know (Fig. 6.8). If, however, these sediments rest unconformably on another sequence of sediments intruded by another igneous body which is truncated at the unconformity and which is twenty million years old, we know that our first sedimentary sequence is between 10 and 20 million years old.

The other difficulty with radiometric dating is that metamorphism, weathering and other processes often result in the loss of parent and/or daughter products and unless this is recognized, erroneous dates can be determined. Enough points on the relative time scale are now calibrated, however, so that such an anomalous date is usually immediately apparent.

Rubidium-Strontium Method

The *rubidium-strontium* method is typically applied to micas and K-feldspars of igneous or metamorphic rocks or to whole metamorphic rocks. Because radioactive potassium also occurs in these minerals, the K-Ar method can commonly be used as a cross-check. The amount of nonradioactively produced strontium 87 in a sample can be determined by its constant ratio to the totally non-radiogenic strontium 86, and a simple subtraction made. Rubidium 87 has a half-life of 47 billion years and, like uranium, is most useful in dating ancient rocks because of its long half-life, but less useful for young rocks because the amounts of strontium generated in a few million years are too small to be measured.

Potassium-Argon Method

The decay of *potassium* 40 to *argon* 40 in one step is the most useful method of radiometrically dating rocks because potassium is such an abundant, widespread crustal element and all natural potassium contains a small amount of the radioactive isotope. Argon, the daughter product, is a gas and easily escapes under conditions of low heat or stress, thus the technique often yields low ages. Unlike the methods described above, there is no way to correct for the amount of nonradioactively produced argon in a crystal. Fortunately, however, there is hardly ever any nonradiogenic argon in the minerals hornblende, biotite, and muscovite, and there are very common minerals in igneous rocks. The K-Ar method can also be used for dating the sedimentary silicate mineral glauconite, providing the enclosing rocks have not been deeply buried or highly weathered. K-Ar dating can also be applied to dating whole volcanic rocks. The half-life of potassium 40 is 1.3 billion years and recent analytic techniques have made it possible to detect the tiny amounts of argon even in rocks as young as 50,000 years, thus making the K-Ar method applicable to rocks of nearly all ages. It is a good method for the most ancient rocks down to the point where the Carbon 14 method becomes effective.

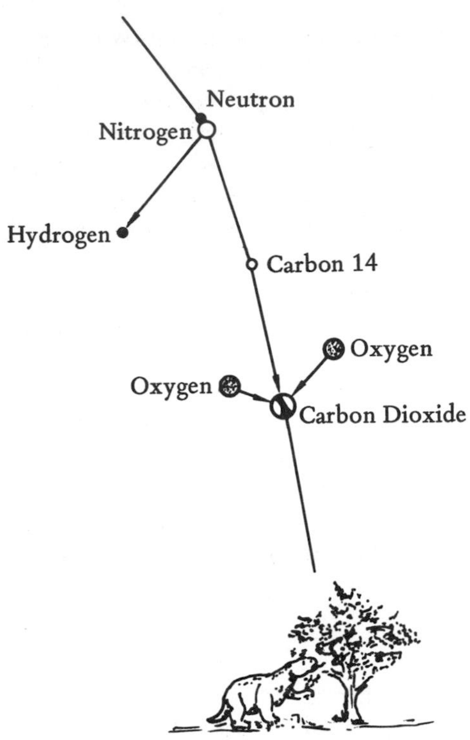

FIGURE 6.9 The Formation of Carbon 14 and Its Uptake by Organisms.

Carbon-14 Method

Carbon 14 is generated continuously in the earth's upper atmosphere by the bombardment of nitrogen 14 by cosmic rays (Fig. 6.9). This isotope of carbon, along with the others, is incorporated into carbon dioxide through which it enters organisms. The half-life of radioactive carbon is very short, only 5,730 years, so that the amount remaining after 50,000 years is negligible. As long as the organism is alive, the decaying Carbon 14 is constantly replaced and the proportions of the radioactive isotope and the normal isotopes remain constant. When the organism dies, however, the proportion of Carbon 14 gradually decreases as it decays to Nitrogen 14. Radiocarbon dates are determined, then, by comparing the proportions of radioactive and normal carbon in a sample. A necessary condition of valid Carbon 14 dating is that the production of C14 in the upper atmosphere has been constant over the last 50,000 years. The dates of known historic events that have been checked by radiocarbon dating clearly indicate that this assumption is valid for historic time though there is presently some reason to believe there were earlier changes of 3–5%. Only Late Pleistocene and Holocene events can be dated by this method because of the short half-life of C14. This is most unfortunate because this method is the most useful one for sedimentary rocks.

Thorium, Protactinium Methods

In recent years, some other methods of radiometric dating have proved useful for dating the Upper Pleistocene and Holocene sediment cores brought up from the

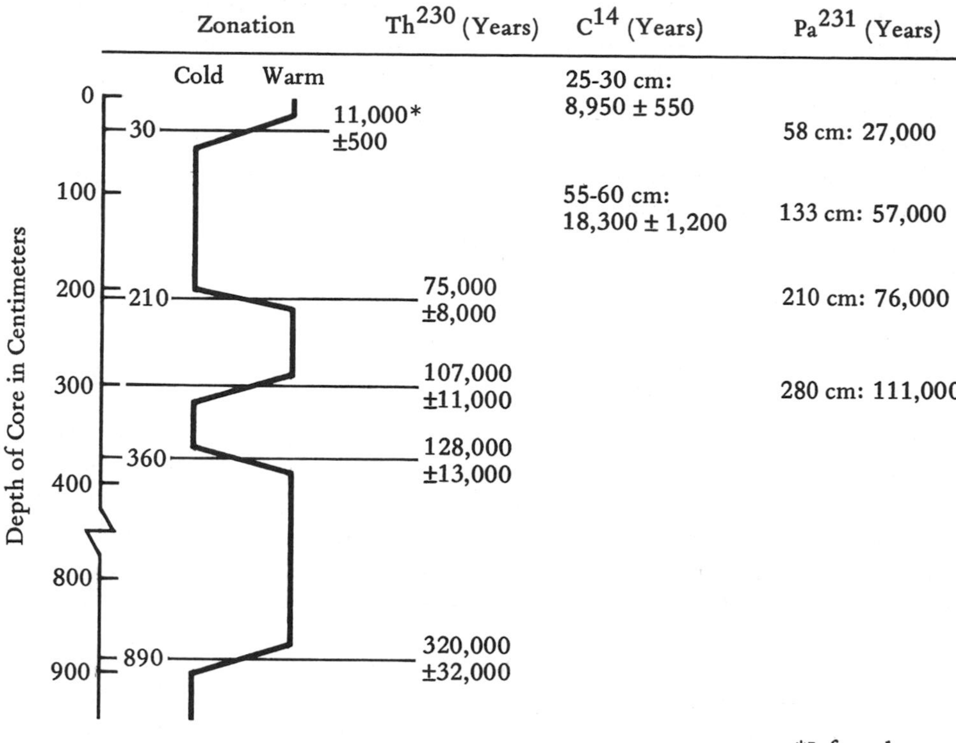

FIGURE 6.10 The Thorium and Protactinium Methods. Dates of an oceanic core utilizing these methods and Carbon 14. (From Karl K. Turekian, *Oceans,* © 1968, by permission of Prentice-Hall, Inc., Englewood Cliffs, N.J.)

ocean floors (Fig. 6.10). Thorium 230, one of the by-products of U238 decay, forms insoluble salts in the ocean and is quickly precipitated. It decays with a half-life of 75,000 years and the amount decreases with the depth of the core, in other words, as one gets down into the older layers. The method is useful for the past several hundred thousand years. Protactinium 231, a by-product of U235 decay, also produces insoluble salts in the sea. Its rate of decay differs from that of Thorium 230 so the ratio of the two can be used to date different layers in cores.

Application of Techniques

For absolute dating of the rocks deposited during most of geologic time, then, geologists must deal with the uranium-lead, rubidium-strontium, and potassium-argon methods. As stated above, only the K-Ar method can be used to directly date sedimentary rocks as it occurs in the complex silicate of sedimentary origin, glauconite. There is almost always some argon loss, however, and such dates are usually minimum ages unless other methods can be used to cross-check them. Interlayed volcanic rocks found in sedimentary sequences can be dated by either the K-Ar or Rb-Sr methods. In volcanic areas, these methods have provided numerous points on the radiometric time scale. Finally, if sedimentary rocks are closely bracketed by igneous intrusives the U-Pb and Rb-Sr methods can be used to provide reference points. Though the radiometric time scale is far from perfect, we now know with the correct order of magnitude the duration of the eras and periods of the relative time scale.

It is another matter, however, when we consider dating an individual rock sample by radiometric means. First of all, it is not often possible to insure that the sample being dated has remained a closed system, in other words, that some of the parent or daughter products have not escaped thus giving erroneous ratios and dates. All radiometric dates are followed by an error figure expressed as plus or minus so many million years or a fraction of a million years. This error figure becomes increasingly important when the amount of the original daughter product present is a large fraction of the radiogenic daughter. In old samples, there is so much radiogenic daughter product that unless the percentage of the original daughter is very large, the percentage error is small. But in younger rocks, where the radiogenic daughter has not been accumulating for very long and, hence, is not present in large amounts, any significant amount of original, nonradiogenic daughter could produce gross errors in dating. The analysis is, therefore, a significant part of radiometric dating with many factors to be considered and weighed.

Radiometric dating has proved, at least until recently, the only practical means of dating rocks from the vast interval of geologic time called the Precambrian. Radiometric dating has shown the Precambrian interval was much longer than anyone anticipated, probably about 4 billion years long or about 90% of the

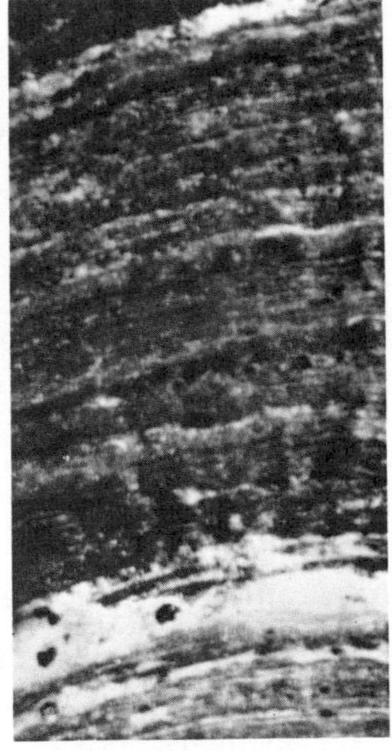

FIGURE 6.11 Daily Growth Lines in Corals. *Holophragma*, a Devonian tetracoral. (Specimens provided by John W. Wells; photography by R. M. Eaton. From *Principles of Paleontology* by David M. Raup and Steven M. Stanley. W. H. Freeman and Company. Copyright © 1971.)

time span since the earth formed. The oldest rocks dated are about three and a half billion years old, but meteorites, which are presumed to be leftover material from the original formation of the solar system, give dates of over four and a half billion years. Similar dates have been obtained from the moon which is inferred to have formed at the same time as the other bodies in the solar system (see Chapter 12). The Paleozoic Era began either .7 or .6 billion years ago depending on whether or not one includes the Eocambrian. In Precambrian rocks excluding the Eocambrian, there are no megafossils (fossils larger than microscopic size) known except for the layered algal mounds called stromatolites. Until recently, no one thought there was much variation in these remains and, thus, there appeared no way to apply the principle of faunal succession to Precambrian rocks. Hence, radiometric dating techniques have provided the only practical method of dating the Precambrian. This is why one sees chiefly radiometric dates in Precambrian time scales in place of eras and periods. In recent years, paleontologists have discovered that the stromatolites are more structurally diverse than thought and successful attempts to zone the later Precambrian have been made. In addition, geologists using sophisticated new techniques have uncovered a wealth of microfossils in Precambrian rocks, even back to the oldest known strata. Powerful new tools, such as the scanning electron microscope, have shown these hitherto uniform-looking fossils also to be diverse and, hence, to have potential for paleontologically subdividing Precambrian time.

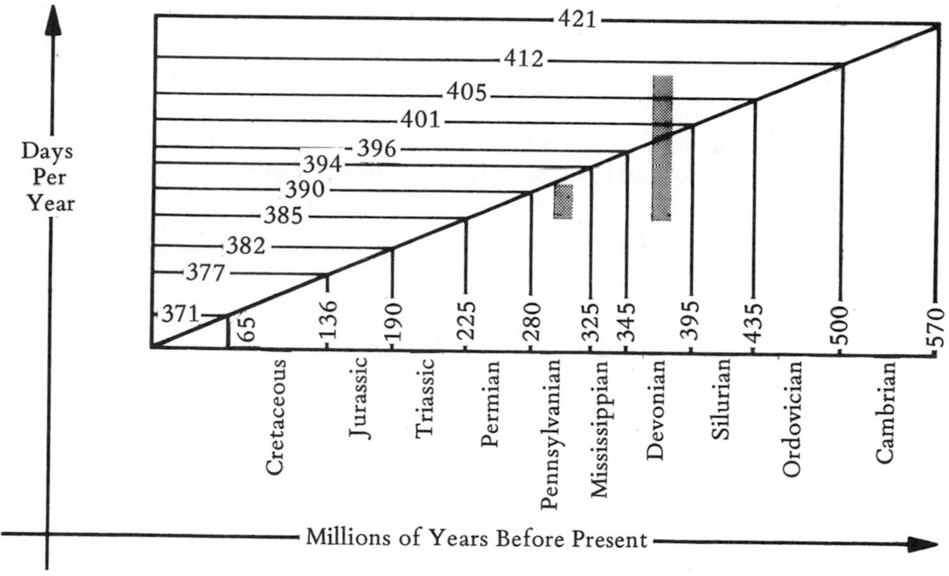

FIGURE 6.12 Predicted Number of Days per Year Plotted Against Quantitative Time Scale. Vertical bars show results from daily and annual growth lines on fossil corals. (From J. W. Wells, 1963.)

Fossils and Absolute Dating

A further development in recent years has been the devising of a method of absolute dating based on the unlikely marriage of paleontology and geophysics, the study of physics applied to the earth. Physicists have long calculated that the earth's rotation rate has been slowing down gradually through geologic time because of tidal friction. The everyday rise and fall of the tides receive their energy from the earth's rotation the rate of which is thus being slowly reduced. This diminution in the rotation is very tiny; the day is increasing in length by only about 2 seconds in 100,000 years. The earth's period of revolution around the sun apparently remains the same and, hence, if the days were formerly shorter than now, there must have been more days in the year in the past than now. Physicists calculated that the day was 21 hours long at the beginning of the Cambrian, and hence, there were 412 days in the year. Until recently, there was no way to confirm these figures.

However, paleontologists have recently begun to note in many organisms that the annual growth rings are in turn subdivided into smaller units which have turned out to be daily growth increments (Fig. 6.11). By counting these daily growth rings in fossil Middle Devonian corals, *Wells* found 385 to 410 of them. In Pennsylvanian corals, he found 385–390 rings. These figures agree closely with the 400 days predicted for the Middle Devonian year and the slightly more than 390 days calculated for the Pennsylvanian year (Fig. 6.12). Similar observations on other types of fossils of other ages have confirmed these findings. Thus, we may be approaching the day when we can tell the age of a well-preserved fossil by counting its growth rings and comparing this number to the scale of days in a year throughout the geologic past. Similar studies on growth rings representing the lunar (synodic) month have led to the development of a graph showing how its length has decreased through time (Fig. 6.13). Refinement of this diagram could also lead to its usefulness in determining rough absolute dates.

Questions

1. What was the rationale behind the oceanic salinity method of dating the earth and why did it fail?
2. Why was there such a large variation in estimates of the earth's age made by the rate of sedimentation method?
3. Summarize Lord Kelvin's arguments about the age of the earth. How was he eventually refuted?
4. What is radioactivity and how can it be used to date rocks?
5. What are the four major radiometric dating techniques useful to geologists?
6. Why is the Carbon 14 method not widely applicable in geology?
7. Briefly summarize the major problems with radiometric dating techniques.
8. Why are radiometric dating techniques not widely used in stratigraphy?
9. What is the apparent age of the earth and how was it determined?
10. What biologic evidence is there for the increase in the length of the day through geologic time?

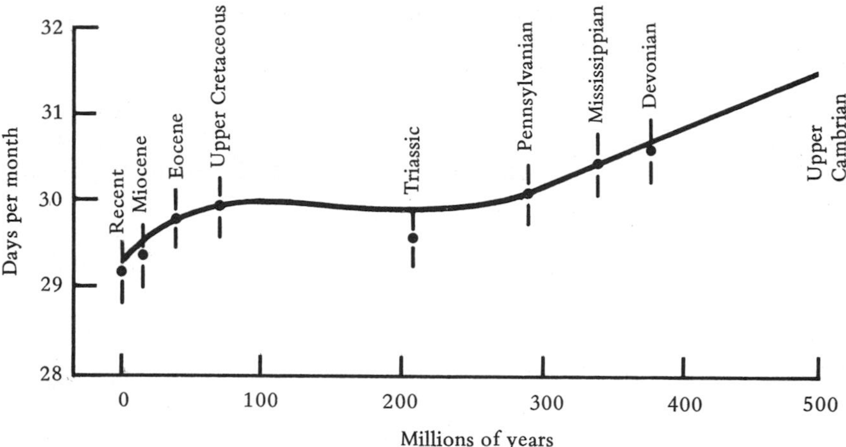

FIGURE 6.13 Variations in the Length of the Synodic Month through Geologic Time as Determined from Count of Growth Lines on Fossil Mollusca (mainly pelecypods). The bars for each point indicate the standard error. (After G. Pannella, C. MacClintock, and M. N. Thompson, *Science*, vol. 162, p. 795, November 15, 1968. Copyright 1968 by the American Association for the Advancement of Science. Used with permission of the authors and AAAS.)

7

The Array of Past Life:

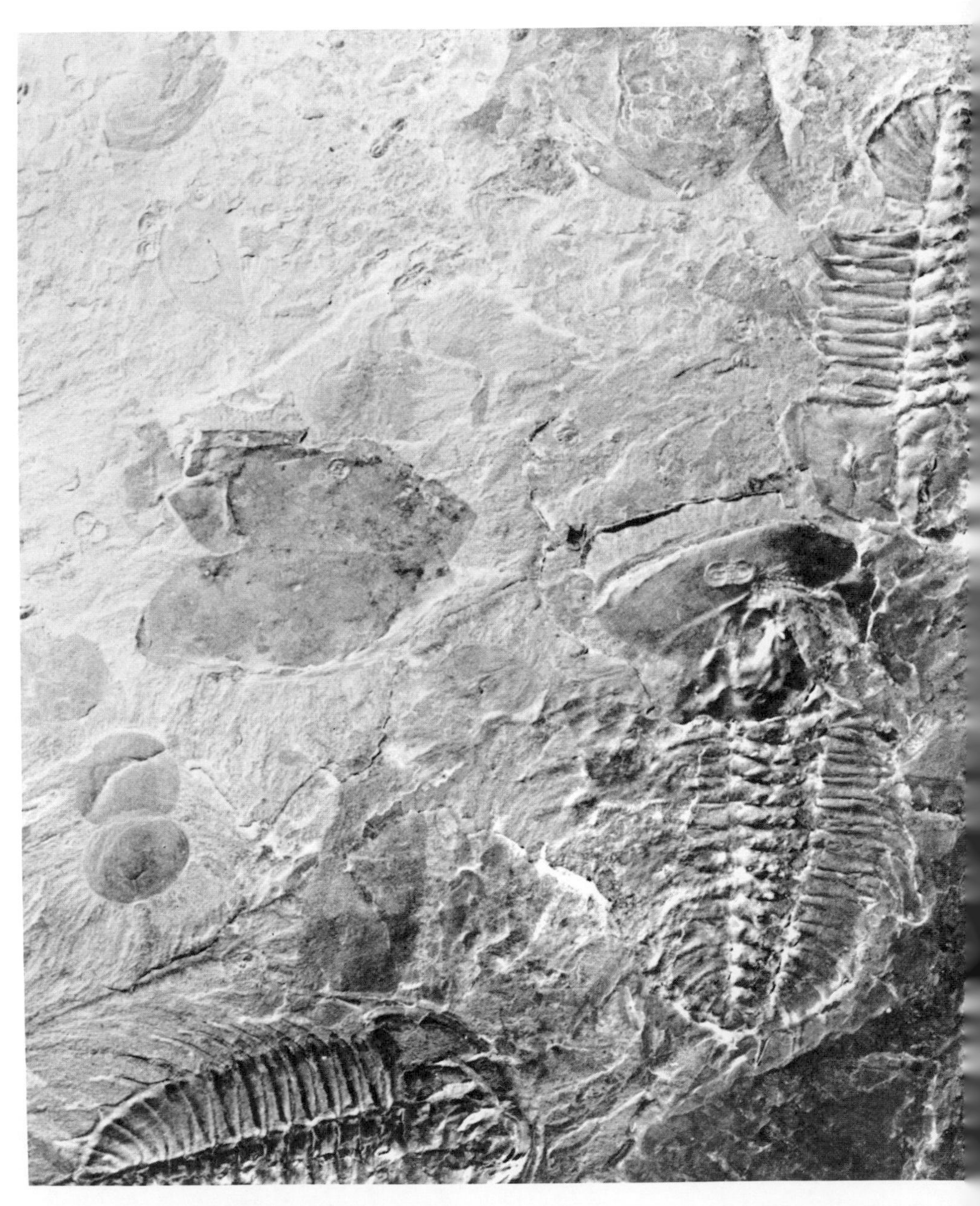

Paleontology

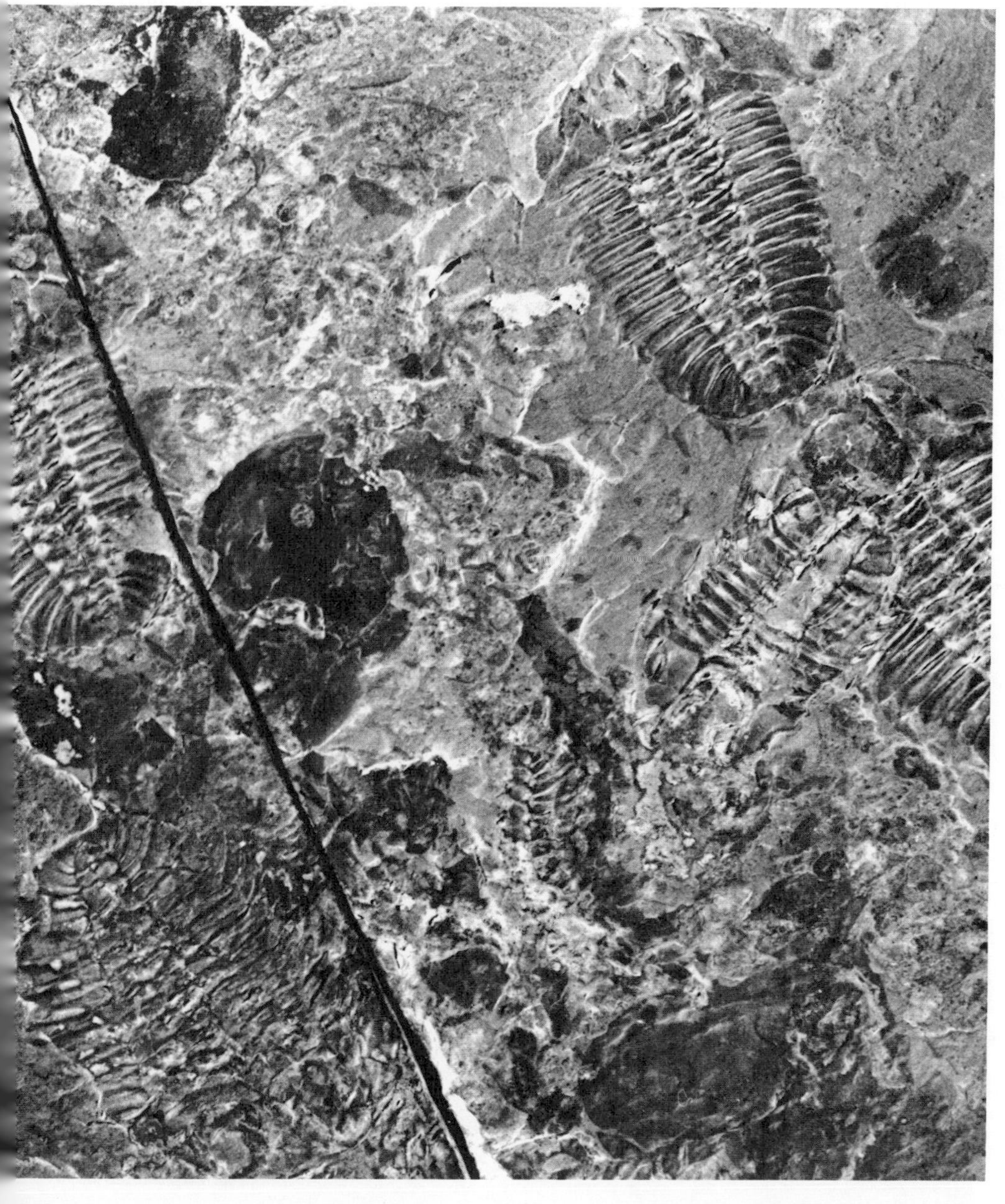

In earlier chapters, we defined fossils as the remains or traces of prehistoric organisms preserved in the earth's crust and pointed out their geologic significance in the principle of faunal successions and their biologic significance in demonstrating the occurrence of evolution. (By prehistoric, we mean older than the dawn of written history 5,000 to 6,000 years ago, though some geologists exclude any Holocene remains.) We shall now turn to a study of the various kinds of fossils and how they are preserved.

Fossilization

Paleontology or paleobiology is the study of fossils and its practitioners are called paleontologists. Paleontologists still do not understand very well the physical-chemical reactions that operate in the process of fossilization, but they do understand some of its major prerequisites. Generally speaking, an organism must have hard, resistant parts such as shell or bone in order to be preserved, for soft, fleshy structures are rapidly destroyed by scavengers and decay bacteria. Furthermore, an organism typically must be quickly buried in some protective medium such as sediment to escape destruction by these agents as well as the geological agents of weathering and erosion which can eliminate even hard parts. Obviously marine organisms are better candidates for burial than terrestrial ones for the oceans are typically sites of sedimentation, while the land is largely a site of erosion.

Organisms with hard parts and which are rapidly buried stand the best chances of being fossilized (Fig. 7.1). For example, most clams have a hard shell and live buried in sand or mud, hence, they have left an excellent fossil record. Organisms of delicate structure which do not live in areas of sedimentation as, for example, butterflies, have left a very poor fossil record. Intermediate cases occur when organisms lack hard parts yet live in favorable environments or have hard parts but do not live in good environments for fossilization. Jellyfish illustrate the first case as they are known only from uncommon impressions of the body, while birds illustrate the latter case for their resistant bones only occasionally collect in areas where they can be rapidly buried. The fossil record is, therefore, very imperfect in the sense that most organisms are not fossilized (and this is fortunate as it keeps the elements recycling), but it does provide us with a good sampling of those organisms with hard parts that lived in areas of sedimentation and an occasional glimpse of the rest.

		Hard Parts	
		Yes	No
Rapid Burial	Yes	GOOD clams	POOR jellyfish
	No	POOR birds	NONE butterflies

FIGURE 7.1 The Likelihood of Fossilization. (From Leo F. Laporte, *Ancient Environments*, © 1968. Prentice-Hall, Inc., Englewood Cliffs, N.J.)

A	PRESERVATION WITHOUT ALTERATION	
	1. Organic Compounds	
	a) Soft parts	Frozen; mummified. Such finds are rare and limited largely to Pleistocene deposits.
	b) Skeletal parts	Organic constituents of bone or shell. Chitin in arthropods, scleroprotein in graptolites, and some other invertebrates. Cartilage in some vertebrates.
	2. Inorganic Compounds	
	a) Calcium carbonate:	
	Calcite	Fairly stable, found in many invertebrate phyla.
	Aragonite	Moderately stable, rare in rocks older than Mesozoic. Corals and molluscs.
	b) Tricalcium phosphate	Brachiopods, arthropods, vertebrates. Quite stable.
	c) Silica (opaline)	Moderately stable. Rare in rocks older than Cenozoic. Sponges and some protozoans.
B	ALTERED IN FOSSILIZATION	
	1. Organic Compounds	
	a) Soft parts	Films of carbon. Rare. Found in fine shales deposited in anerobic environments.
	b) Skeletal parts	Carbonized. Particularly the chitinous skeletons of arthropods and scleroproteins in graptolites.
	2. Inorganic Compounds	
	a) Permineralized	Deposition of minerals in interstices of skeleton. Commonly $CaCO_3$. Less frequently SiO_2,glauconite, iron compounds, etc.
	b) Recrystallized	Less stable inorganic compounds alter in physical form to more stable state without change in chemical composition, e.g., aragonite to calcite. May be very common mode of preservation, but difficult to distinguish from replacement.
	c) Replacement	Removal of original skeleton material by solution and deposition of new compounds, carbonates, silica, iron compounds, etc., in its place. Very common—intergrades with permineralization and recrystallization.
C	PRESERVATION AS MOLDS OR CASTS	
	1. Organic Compounds	
	a) Soft parts	Imprints in fine-grained laminated shales and lithographic limestones.
	b) Skeletal parts	Imprints or casts.
	2. Inorganic Compounds	
	a) Molds	External and internal molds formed by sediment around or within skeletal parts.
	b) Casts	Filling of mold after skeletal parts are dissolved. Intergrades with replacement.
D	EVIDENCES OF ANIMAL ACTIVITIES	
	1. Tracks	Mode of locomotion. Preserved as molds and casts.
	2. Burrows	Animal habitat and behavior. Mode of burrowing. Preserved as molds and casts.
	3. Coprolites	Fossilized excrement. Diet. Structure of gut. May be preserved in any of ways in this table.
	4. Borings and Tooth Marks	Evidence of predation.
	5. Human Artifacts	Evidence of culture.

FIGURE 7.2 Types of Fossil Preservation. (From James R. Beerbower, *Search for the Past; An Introduction to Paleontology,* 2nd Ed., © 1968. Reprinted by permission of Prentice-Hall, Inc., Englewood Cliffs, N.J.)

Types of Preservation

Paleontologists recognize four broad and intergrading categories of types of fossilization, unaltered remains, altered remains, molds and casts, and traces of organic activity (Fig. 7.2). In general, the amount of desirable information decreases in this order with unaltered remains furnishing maximum information and trace fossils, only fragmentary information. But a paleontologist takes what he can find. He will probably ignore molds and casts if he has unaltered or altered remains, but if molds and casts are all that is preserved, they are his only sources of valuable information and he must utilize them. Different types of sedimentary and post-depositional environments tend to favor certain types of preservation. For example, the quiet waters in which muds accumulate are not likely to abrade or break the shells of dead organisms, while the higher energy environments in which sand accumulates often lead to broken and worn fossils. Also, sandstone is much more permeable than shale and, hence, groundwater is more likely to dissolve away the shell leaving only a mold in the sandstone.

Each category of fossilized remains can be further subdivided. *Unaltered* remains can be split into organic, carbon-containing compounds and inorganic, noncarbon-containing compounds. *Organic compounds* are complex mixtures of carbon bonded to hydrogen, oxygen, nitrogen, phosphorus, and sulfur. They occur both in fleshy soft parts and as hard parts of many organisms. *Soft parts* are very rare in the fossil record, as you might expect, for these are very liable to decay and do not survive long under any circumstances. They are restricted to the recent geologic past, late Pleistocene or Holocene, and to very cold and/or dry conditions. The most famous of these fossils are the not uncommon mammoths, extinct hairy elephants, frozen in the permafrost of Siberia and Alaska (Fig. 7.3). Even parasites and the gut contents are preserved in these remarkable

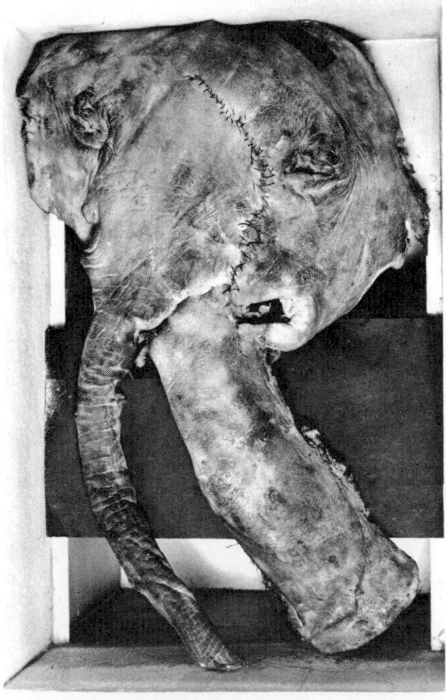

FIGURE 7.3 Young Woolly Mammoth dug from frozen ground in Alaska. (Courtesy of the American Museum of Natural History.)

fossils. A rhinoceros pickled in tar is known from Poland, pieces of ground sloth flesh and fur are known from dry desert caves in both North and South America, mummified clams are found in Greenland, and mummified small crustaceans called ostracods occur in Alaska. This is the entire extent of known fossils of this type of preservation, but degraded organic compounds which cannot be ascribed to any particular creature are known far back into the Precambrian. *Organic skeletal remains* survive only slightly better in the unaltered state. These include *carbohydrates* such as *chitin* (Fig. 7.4) which composes arthropod (crustaceans, insects, etc.) skeletons and worm jaws and spines, and *cellulose* which forms the supporting structures of plants and the shells of the microscopic dinoflagellates; *scleroproteins* which compose the skeletons of certain coral-like groups and the extinct graptolites; and *organic phosphates* such as the cartilage of sharks and other vertebrates. (If you are unfamiliar with any of these groups of organisms, they are discussed and illustrated later in this chapter for reference.) With the exception of the cellulose composing spores and pollen (Fig. 3.24), which are studied by the paleontologists called *palynologists,* these skeletal structures are much more commonly preserved as altered remains.

Inorganic hard parts form the vast majority of unaltered fossils. These are nearly all composed of one of three compounds (Fig. 7.5)—calcium carbonate ($CaCO_3$), calcium phosphate ($Ca_3(PO_4)_2$), and hydrous silicon dioxide ($SiO_2 \cdot nH_2O$). (The carbon in calcium carbonate is part of the carbonate ion and not

FIGURE 7.4 Ant (*Sphecomyrma freyi*) in Cretaceous amber from New Jersey. (Courtesy of Frank M. Carpenter.)

elemental carbon as in organic compounds, so it is classified here.) *Calcium carbonate* is by far the most widely used and abundant skeletal material in the organic world. It occurs in two forms, the minerals calcite and aragonite. Aragonite is less stable and recrystallizes to calcite with time so there are few unaltered aragonite fossils older than Late Mesozoic. *Calcite* composes or contributes to the hard parts of many blue-green algae, charaphytes, many red algae, coccolithophores, most foraminifera, some sponges, stromatoporoids, tetracorals, tabulate corals, many bryozoans, most brachiopods, some snails, many clams, some worm tubes, trilobites, many crustaceans, and echinoderms. *Aragonite* is found in many green algae, hexacorals, many bryozoans, most snails, many clams, most cephalopods, and some worm tubes. *Calcium phosphate* (Fig. 7.5), often in the form of the mineral *apatite,* is much less common, but it does compose the skeletons and teeth of most vertebrates and is therefore very important from our point of view. Other organisms with phosphatic hard parts are the extinct conularoids, some brachiopods, many trilobites, many crustaceans, and the common microfossils called conodonts. *Silicon dioxide* (Fig. 7.5) occurs in organisms in the hydrous, noncrystalline mineral *opal.* With time, opal gradually crystallizes to quartz, so most older sedimentary rocks contain altered siliceous remains. Silica occurs in three abundant groups of microfossils, the silicoflagellates, diatoms, and radiolaria. It also composes the skeletons of most sponges. Unaltered remains occur in rocks of all ages but become more abundant as one approaches the Recent.

Altered remains are also very abundant and come in varying degrees of alteration. Organic remains, both hard and soft, are commonly altered by the process called *distillation* or *carbonization.* In this process, the weight and heat created by overlying rocks force out the volatile carbon compounds in an organism eventually leaving only a black carbon film or residue. Some aquatic vertebrates have had their bodies preserved as sheets of carbon by this process (Fig. 7.6). More common, however, is the distillation of hard parts such as the chitin skeleton of arthropods, the protein skeleton of graptolites, and the cellulose of plants (Fig. 7.7). Indeed, the sedimentary rock coal is a product of the distillation of great thicknesses of plant remains. *Inorganic compounds* are commonly altered by three processes. In *permineralization* porous cavities in hard parts such as bone are filled with the sedimentary matrix of the surrounding rock thus giving the fossil much more strength. *Replacement* is the process whereby the atoms of original material are individually removed and replaced by other ions by percolating solutions. The identical structure is preserved, but the composition is now different (Fig. 7.8). The two commonest replacing substances are calcium carbonate and silica whose stabilities are mutually opposite, the former replacing ions under basic conditions, the latter under acidic conditions. If a calcium carbonate fossil in a limestone is replaced by silica, it can often be easily extracted by dissolving the rock in acid. Many delicate fossils have been

FIGURE 7.5 Distribution of Major Inorganic Skeletal Compounds among Geologically Significant Phyla of Organisms. (From A. Lee McAlester, *The History of Life,* © 1968. Reprinted by permission of Prentice-Hall, Inc., Englewood Cliffs, N.J.)

Phyla		Calcium Carbonate $CaCO_3$		Silica SiO_2	Calcium Phosphate $Ca_5(PO_4)_3OH$
		Calcite	Aragonite		
Schizomycophyta (Bacteria) *Pyrrophyta* (Dinoflagellates)					
Cyanophyta (Blue-Green Algae)		Frequent			
Chlorophyta (Green Algae)			Frequent		
Charophyta (Stone Worts)		Frequent			
Phaeophyta (Brown Algae)					
Rhondophyta (Red Algae)		Common			
Chrysophyta	Diatoms			Common	
	Coccolithophorids	Common			
Mycophyta (Fungi) *Bryophyta* (Mosses) *Tracheophyta* (Vascular Plants)					
Sarcodina	Radiolarians			Common	
	Foraminiferans	Common			
Porifera (Sponges)		Frequent		Common	
Coelenterata (Corals)		Common	Common		
Bryozoa (Bryozoans)		Common	Common		
Brachiopoda (Brachiopods)		Common			Frequent
Mollusca	Snails	Frequent	Common		
	Clams	Common	Common		
	Cephalopods		Common		
Annelida (Segmented Worms)		Frequent	Frequent		
Arthropoda	Trilobites Crustaceans	Common			Common
	Arachnids Insects				
Echinodermata (Echinoderms)		Common			
Chordata	Acorn Worms Tunicates, Lancelets				
	Vertebrates				Common

Common — Frequent — Rare or Absent

FIGURE 7.6 Distilled or Carbonized Fossil Fish, *Diplomystus analis,* from Eocene Green River Shale. (Courtesy American Museum of Natural History.)

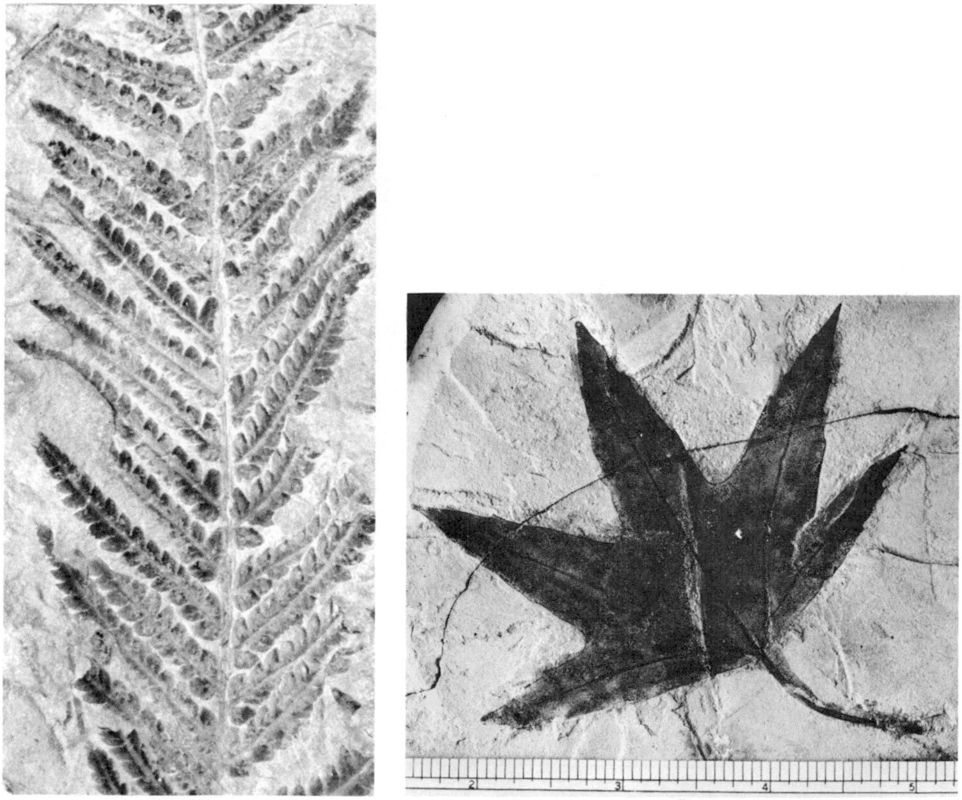

A B

FIGURE 7.7 Distilled or Carbonized Fossil Plants. (A) Carboniferous fern, (B) Cenozoic angiosperm. ((A) Courtesy of T. Delevoryas, Yale Peabody Museum. (B) Courtesy of William L. Stokes, University of Utah.)

FIGURE 7.8 Replaced Fossil Wood. (A) "Old Faithful" petrified log. Petrified Forest National Park, Arizona. (Courtesy of William M. Mintz.) (B) Transverse section of silicified wood. (Courtesy of T. Delevoryas, Yale Peabody Museum.)

obtained this way which could not have been freed from the matrix in any other manner (Fig. 7.9). *Recrystallization or crystallization* is the process whereby an unstable form of a substance changes to a more stable form. This involves an internal shuffling of atoms which distorts or expands the crystals of the original material, thus altering the fossil. The commonest examples are aragonite recrystallizing to calcite and opal crystallizing to quartz.

Molds and *casts* are less desirable fossils than the above because they preserve only the shape of the surface, either inside or outside, of a structure. No internal structure of the shell or bone is preserved. A mold is an impression of a surface. For example (Fig. 7.10), if I were to plunge my fist into a piece of clay and then extract it, the clay would contain a reverse replica of the outside of my hand, an *external mold*. If I were to squeeze a ball of clay in the palm of my hand, the clay would contain a reverse replica of the inside or palm of my hand, an *internal mold*. If a mold is later filled either naturally with sediment or by artificial processes, a cast results. If the mold of my fist were filled with plaster which was allowed to dry and then was removed, the plaster would contain a replica of the surface of my fist, an *external cast*. If the mold of my palm were covered with plaster which was allowed to dry and then, removed from around the mold, a replica of the surface of my palm, an *internal cast*, would result. Molds and natural casts are common fossils and include both soft and hard, organic and inorganic structures. If the molds and casts are so thin that they cannot be distinguished, as in plant leaves, the word "impression" (Fig. 7.11) is often used.

Trace fossils record only the effects of an organism, usually an animal, on its environment. *Trails*, which are records of the passage of a body, and *tracks* (Fig. 7.12), which record the passage of the hands and/or feet of an animal, are commonly preserved as molds and casts. They give some idea of the method of locomotion, the size, and the shape of the animal that made them. Unless actual

FIGURE 7.9 Silicified Fossil Brachiopods. Spiny brachiopods of the Order Productoida from the Permian of the Glass Mountains, Texas. (Courtesy of G. Arthur Cooper and U.S. National Museum.)

remains occur nearby, however, it may be possible to do no more than identify them as "worm trails" or "reptile tracks." *Burrows* which are also preserved as molds and casts, give information about the size and habits of an organism, but are even more difficult to assign to a particular kind of animal. *Coprolites* (Fig. 7.13) are fossilized feces and are preserved by any of the above means. They

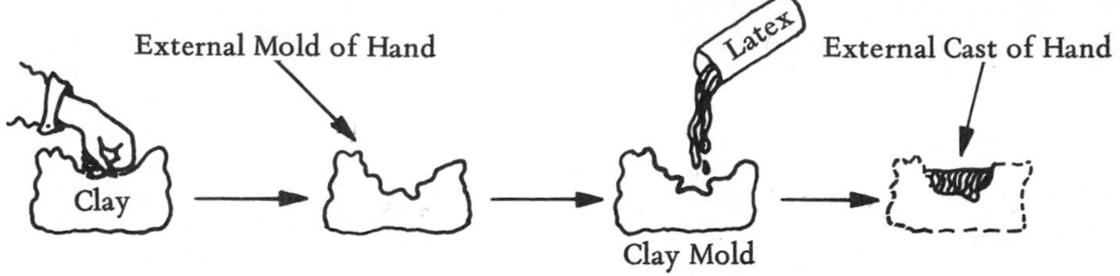

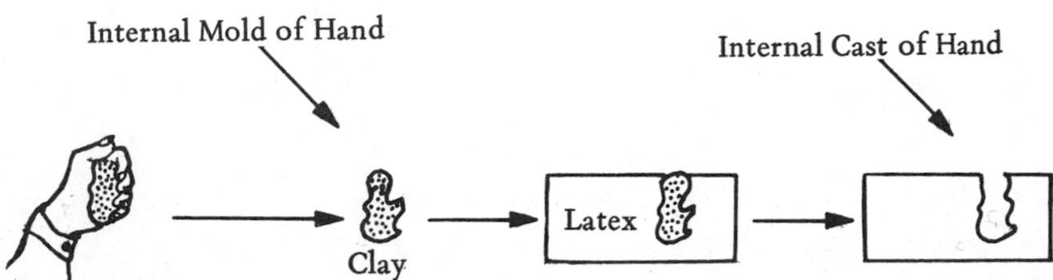

FIGURE 7.10 Diagram Illustrating Formation of External and Internal Molds and Casts.

tell something about the diet and the digestive tract structure of their makers. Small, delicate fossils, not found elsewhere as fossils, often turn up in coprolites. *Borings* in shells and *tooth marks* (Fig. 7.14) on bones and shells offer evidence of who ate whom and how certain organisms fed. Lastly, *human artifacts* made by prehistoric men are trace fossils that offer evidence of the culture of their maker and some evidence of his size and structure.

Geologists and Fossils

In studying a fossil, paleontologists are typically interested in three major tasks: the determination of its position in the evolution of life so it can be properly as-

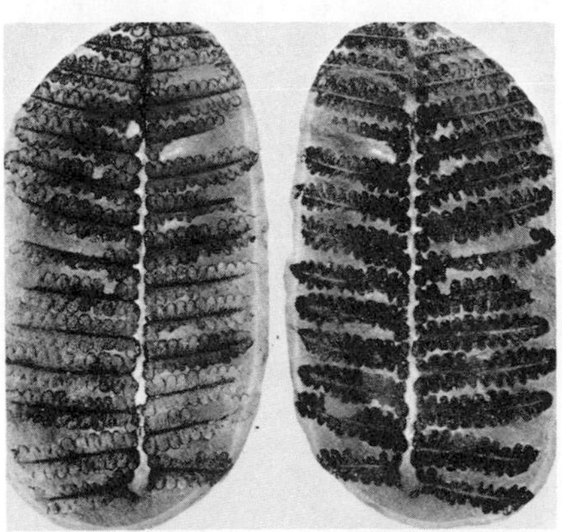

FIGURE 7.11 Fossil Plant Impression. Pennsylvanian, Mazon Creek, Illinois. (Courtesy of T. Delevoryas, Yale Peabody Museum.)

FIGURE 7.12 Fossil Vertebrate Tracks. Reptile tracks in Triassic strata of the Connecticut Valley. (Courtesy of Yale Peabody Museum.)

FIGURE 7.13 Coprolites. Fossilized ground sloth dung, Rampart Cave, Lake Mead National Recreation Area, Nevada. (Courtesy of National Park Service.)

FIGURE 7.14 Fossil Toothmarks of Mosasaur on Ammonoid Shell, Cretaceous Age. (Courtesy Robert V. Kesling and University of Michigan Museum of Paleontology.)

signed in the formal classification, the determination of its ecology so that the environment of the deposit containing it can be ascertained, and the determination of its position in the geologic column so that its age is known. Fossils can be classified in these three ways or any number of other ways depending on one's purposes. The ecology of fossils has already been discussed in Chapter 3, the use of fossils in stratigraphy will be developed in Chapter 10. In this chapter we will pursue the first type of classification.

Classification of Fossils

The formal classification used by paleontologists is the one used by biologists and devised by *Linnaeus* in the 1700s. In this scheme, each type of organism or species is assigned a two-part or binomial species name which is always italicized. The first word, which is capitalized, is the generic name and is usually shared with several other very closely related species. The second word, uncapitalized, is the trivial or species name and is unique in the particular genus, although it may be used for species of other genera. In scientific works, the name of the original author who described the species and the date of his work appear after the name. For example, the horse is *Equus caballus* Linnaeus, 1758. The scientific name of a species is either in Greek or Latin form and although it seems cumbersome to the uninitiated, its universal use eliminates the problems of language barriers and lack of enough everyday words to name every creature. Every species must have a unique binomial name; the name cannot be used twice nor can two names be applied to the same kind of organism.

Another aspect of the Linnaean system is that it establishes a hierarchy of classification terms, that is, each term is less inclusive than the one above it and more inclusive than the one below it. The terms used in the hierarchy today, most of which were coined by Linnaeus, are:

Kingdom
Phylum (Botanists use Division here)
Class
Order
Family
Genus
Species

An order, for example, encompasses several families and it, in turn, belongs with several other orders to a class. There is more diversity recognized in the organic world today than in Linnaeus' time, hence biologists have found it necessary to intercalate several new terms between those of the traditional hierarchy. Most of these have been created by adding the prefixes *super-* and *sub-* to existing categories. For example, a class could be subdivided into several subclasses, each of which could contain several superorders, each of which in turn, could consist of several orders. Sometimes, new words such as *cohort* and *tribe* have been devised where even more subdivisions are needed.

You may recall that Linnaeus initially did not accept evolution, but when he devised his classification it became obvious that the degrees of similarity expressed by his classification could be best explained by evolution. Today, we recognize that the evolutionary relationships of organisms form the basis of our classification. In biological jargon, it is phylogenetic, or at least we try to make it so, though we are not certain of the relationships of many organisms. A necessary implication of such a classification is that each category has only one ancestral species, otherwise we would be classifying together unrelated organisms. For example, birds, bats, and pterosaurs all fly and have numerous similar structural adaptations for flight, but we classify them in separate classes of vertebrates because they are not closely related. This is an obvious example, but there are numerous others, particularly among the invertebrates (animals without backbones) that go undetected.

A Modern Classification of Organisms

In an introductory course in historical geology it is rarely necessary to go below the order level in discussing significant fossil groups. In the succeeding pages, we will examine the major groups of organisms that have left a significant fossil record. In some groups it will be sufficient to discuss the phylum only. In other, more diverse groups we will proceed down to the order level. Zoologists have never standardized the endings for terms above the family level in animals, though this is clearly desirable from the student's viewpoint. In this text, the endings will be standardized as follows: kingdom—*ae*, phylum—*a*, subphylum —*zoa*, class—*ea*, subclass—*ia*, order—*oida*. These are the most commonly used endings, but are not universal so that their application to certain terms such as the Class Gastropodea may seem strange to the biologists reading this text. This

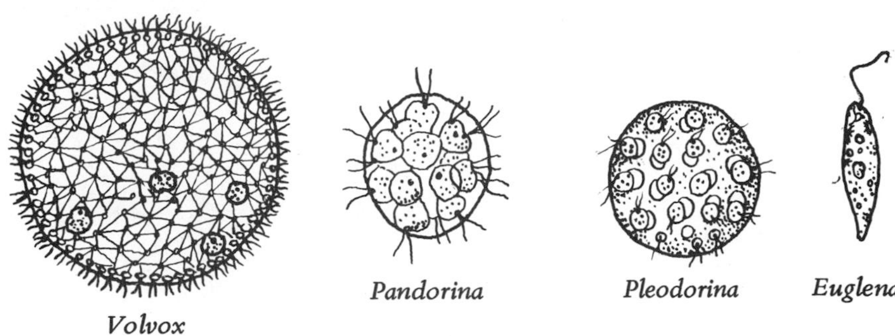

FIGURE 7.15 Animals or Plants? These green flagellates share characteristics of both the traditional kingdoms.

step is taken, however, in the fervent hope that zoologists will finally take this desperately needed action.

Traditional classifications of the organic world have consisted of only two kingdoms, animal and plant. In the last few years, however, there has been a decided shift in biology to recognizing one, two, or even three additional kingdoms. There are two major reasons for this. One is that there are many microorganisms which share features of both traditional animals and plants. The microscopic green flagellates such as *Volvox* and *Euglena* (Fig. 7.15) appear as animals in zoology texts and plants in botany texts. Clearly the traditional kingdoms break down in classifying these creatures. The second reason is that man has traditionally studied the organic world from his anthropocentric viewpoint. It was easy to recognize the major differences among the complex animal groups, but the more removed from man one went, the less obvious significant differences appeared. When one reached the one-celled level this became particularly apparent. Zoologists classed all one-celled "animals" as the Phylum Protozoa and botanists, the simple "plants" in the Division Thallophyta. We have begun to discover the profound differences among these organisms and our classification is now beginning to reflect them. Biologists now generally recognize the following kingdoms: *Monerae, Protistae, Animalae, and Plantae.* Some exclude the fungi from the protistans and add a fifth kingdom, Fungae, but because fungi are relatively rare fossils we need not concern ourselves with the problem here.

Kingdom Monerae

These organisms are distinguished from all others by their lack of a nucleus; the hereditary material is not concentrated and surrounded by a membrane but is diffuse. The evolution of the nucleus was probably one of the most significant evolutionary events in life history and some of its steps are shown in the surviving monerans. Organisms without a nucleus are called procaryotic, those with one (or more), eucaryotic. Monerans are all one-celled or colonies of essentially independent cells. The bacteria (Fig. 7.16), a diverse assemblage of abundant

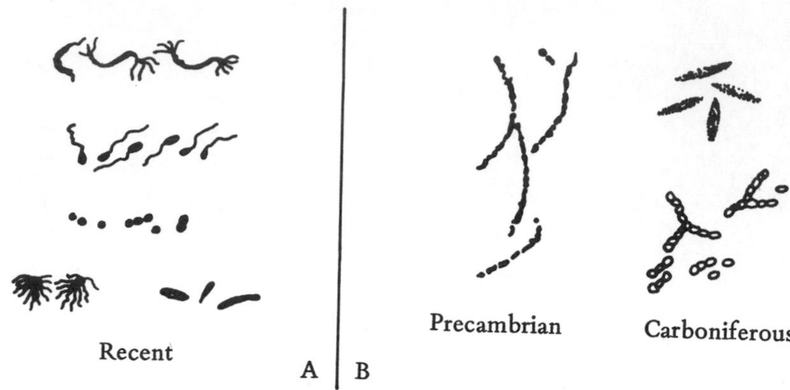

FIGURE 7.16 Recent and Fossil Bacteria. (A) Recent forms, (B) Fossil forms.

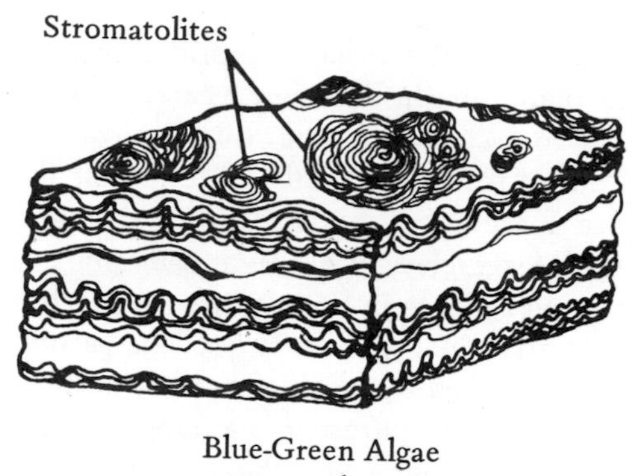

FIGURE 7.17 Stromatolites, the fossil remains of Colonial Blue-Green Algae (Phylum Cyanophyta).

microorganisms, are assigned here, but their geologic importance is presently limited to their appearance among the oldest fossils in rocks over three billion years old and their possible geochemical effects. There is only one geologically significant group of monerans.

Phylum Cyanophyta—The blue-green algae, so named because they contain blue and green (chlorophyll) pigments when alive, are autotrophic, that is they produce their own food by photosynthesis. They have no hard parts, but several of the sheet-like benthonic colonial forms entrap and bind layers of calcium carbonate (calcite) among their mucilaginous filaments. These layered mounds, called stromatolites (Fig. 7.17), range in size from a few inches to many feet across and resemble petrified heads of cabbage. They range throughout the geologic record from Precambrian to Recent, but are most notable in Precambrian rocks in which they are the only megafossils. Most living types of blue-green algae live in fresh water, but marine and terrestrial kinds occur. The fossils lived in all three environments.

Kingdom Protistae

The protistans are the first group of organisms to evolve a nucleus. Most are still one-celled or simple aggregates of cells, but some of the larger algae are multicellular with the cells differentiated. Protistans form their sex cells (gametes) in unicellular structure and do not form embryos. There are nine major fossil groups of protistans: green algae, charophytes, golden algae, diatoms, dinoflagellates, red algae, forams, radiolaria, and tintinnoids.

Phylum Chlorophyta—The green algae are so named because they contain the green pigment chlorophyll which is utilized in photosynthesis. This chlorophyll is the same type as that found in true plants and indicates they probably arose from green algae. Most chlorophytes possess a whip-like flagellum at some time in their life. They range from single cells through colonial cells on up to complex multicellular forms, not all of which may really be closely related. This assemblage, which ranges from Precambrian to Recent, contains only two groups that have left significant numbers of fossils. The *dasycladaceous algae* (Cambrian–Recent) have a central stalk with radiating side branches (Fig. 7.18). The side branches bear facets and supporting structures of calcium carbonate (aragonite and possibly calcite in some fossils). The facets are commonly preserved as fossils and some groups calcify the central axis and/or the lateral branches enabling them to be preserved. In external molds of the body, the passageways originally occupied by the calcified side branches were misinterpreted as pores by many paleontologists and the group was classified with the sponges! Large extinct types called *receptaculitids* (Fig. 7.19) were common in the Ordovician, more typical types abounded in Eurasia in the Triassic and Jurassic. *Codiaceous* algae (Ordovician–Recent) typically have a branching body or thallus which is covered by aragonite in many types (Fig. 7.20). One variety, the jointed *Halimeda* (Fig. 7.21), has been a major contributor to Cenozoic coral reefs. Both types of green algae are marine and have a warm-water preference.

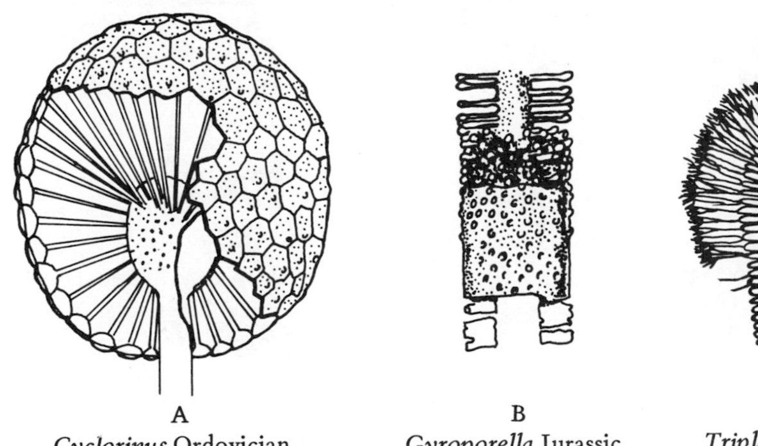

A *Cyclorinus* Ordovician — B *Gyroporella* Jurassic — C *Triploporella* Jurassic

FIGURE 7.18 Dasycladaceous Green Algae. Phylum Chlorophyta. (A) Paleozoic form, (B–C) Mesozoic forms.

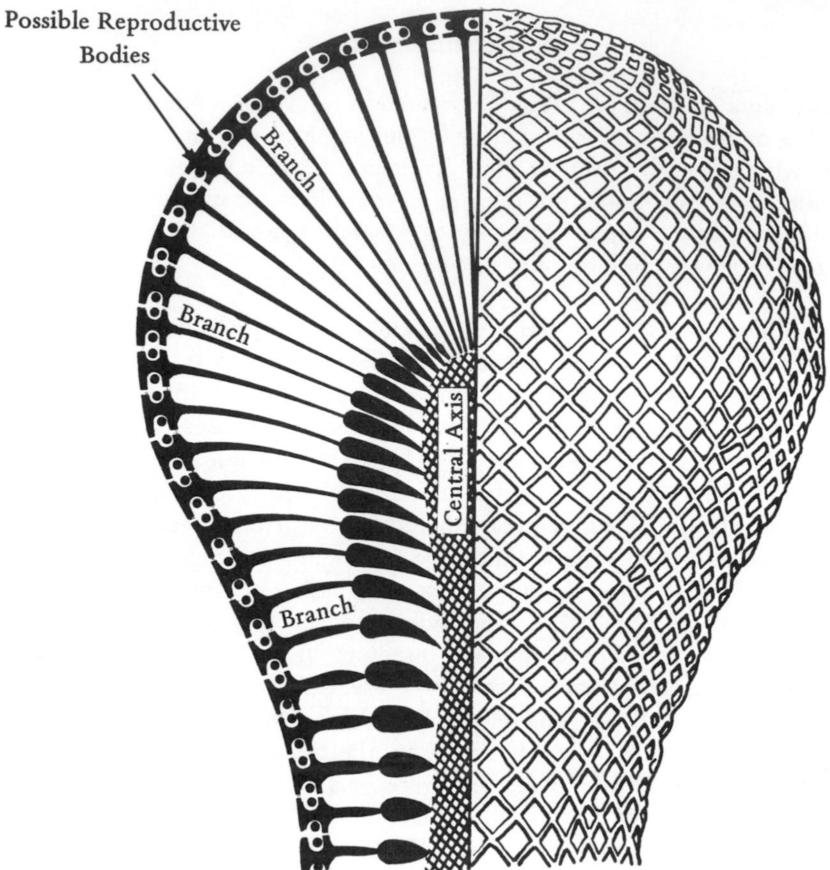

FIGURE 7.19 *Ischadites,* **a Receptaculitid.** These Paleozoic fossils were long mistakenly classified with sponges, but appear to form an extinct group of Dasycladaceous Green Algae.

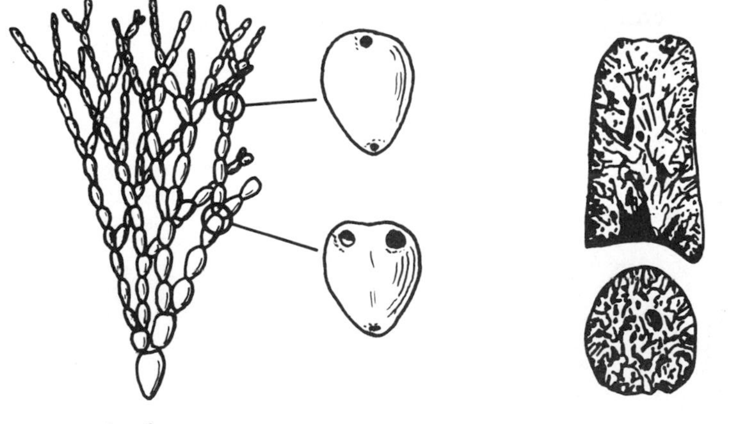

FIGURE 7.20 **Fossil Codiaceous Green Algae** (Phylum Chlorophyta).

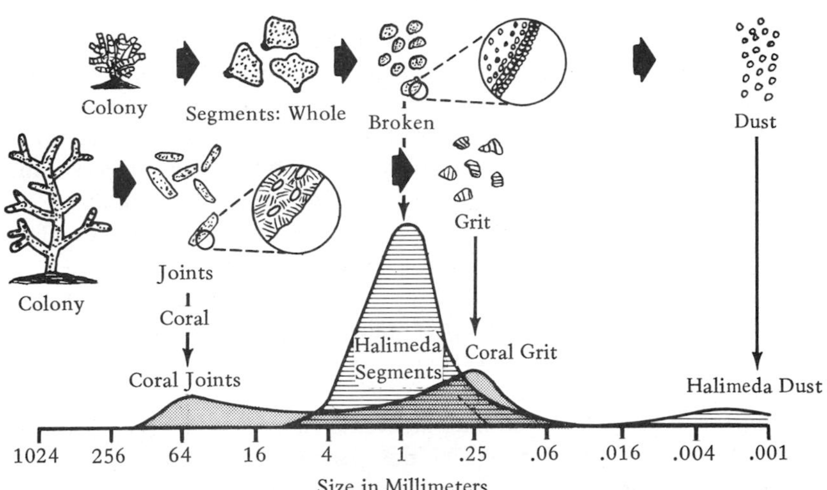

FIGURE 7.21 *Halimeda,* **a Codiaceous Green Algae** important as a sediment-former. (From Leo F. Laporte, *Ancient Environments,* © 1968, Prentice-Hall, Inc., Englewood Cliffs, N.J.)

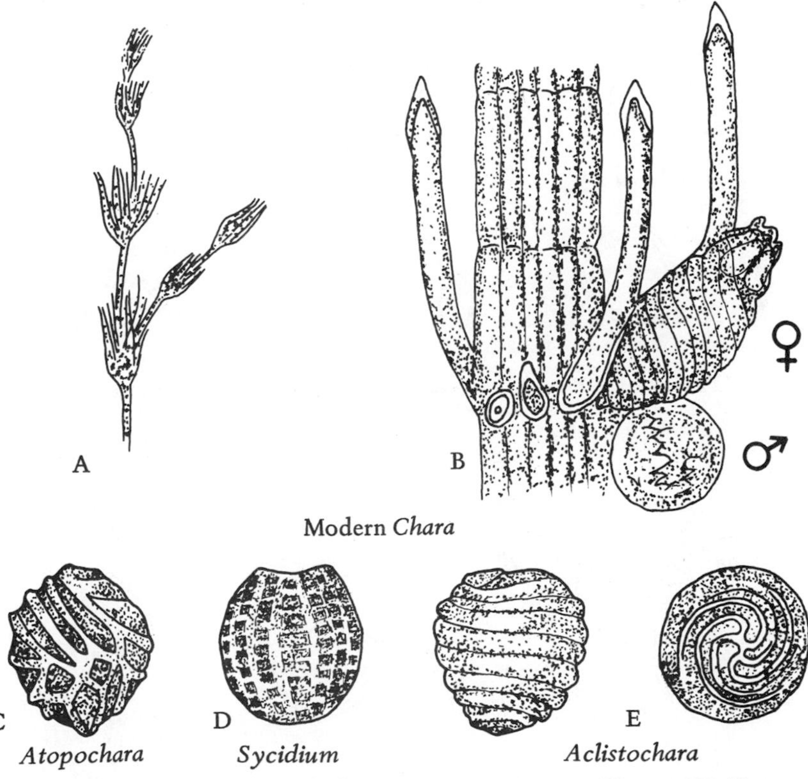

FIGURE 7.22 **Phylum Charophyta.** (A) Recent *Chara,* (B) Detail of reproductive structures on stem, and (C–E) Various fossil oogonia.

Phylum Charophyta—The stoneworts or charophytes are complex, multicellular green algae with structures superficially resembling stems, roots, and leaves (Fig. 7.22). Some believe they are intermediate between chlorophytes and plants. The leaf-like structures occur only at nodes. The reproductive structures are also found at the nodes. In their metabolism, the benthonic charophytes, which inhabit fresh and brackish water, precipitate calcium carbonate over their thalli or bodies. When they rot, a spongy mass of limestone called marl remains. Rarely is there much structure recognized in these fossils, however. The female reproductive structures or oogonia (Fig. 7.22) are tiny, subspherical bodies with a spiral or vertical arrangement of cells which are impregnated with calcite. These microfossils are abundant in fresh water deposits of Mesozoic and Cenozoic age, although they are known as far back as the Devonian.

Phylum Chrysophyta—The golden algae are so named because in life they contain, in addition to chlorophyll, golden pigments. Most are photosynthesizers, but some must eat other creatures to obtain certain nutrients. Most are one celled, though some are colonial, and all possess flagella in life. They are all planktonic. The *chrysomonads* (Cretaceous–Recent) produce siliceous cysts or resting stages (Fig. 7.23). They are chiefly fresh water and are occasionally common fossils. More important are the marine *silicoflagellates*

CHRYSOMONADS

A

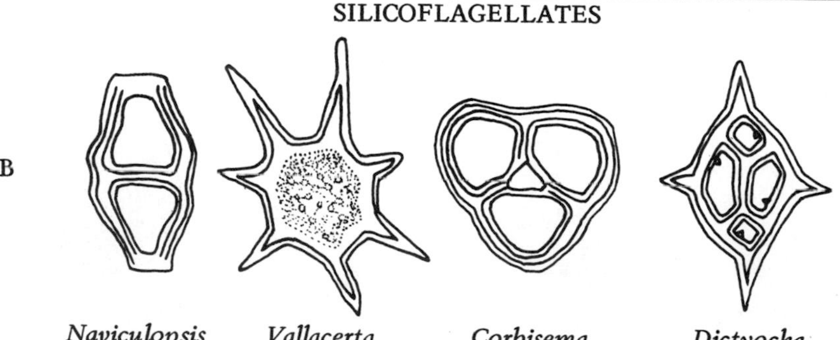

FIGURE 7.23 (A) Chrysomonads and (B) Silicoflagellates (Phylum Chrysophyta).

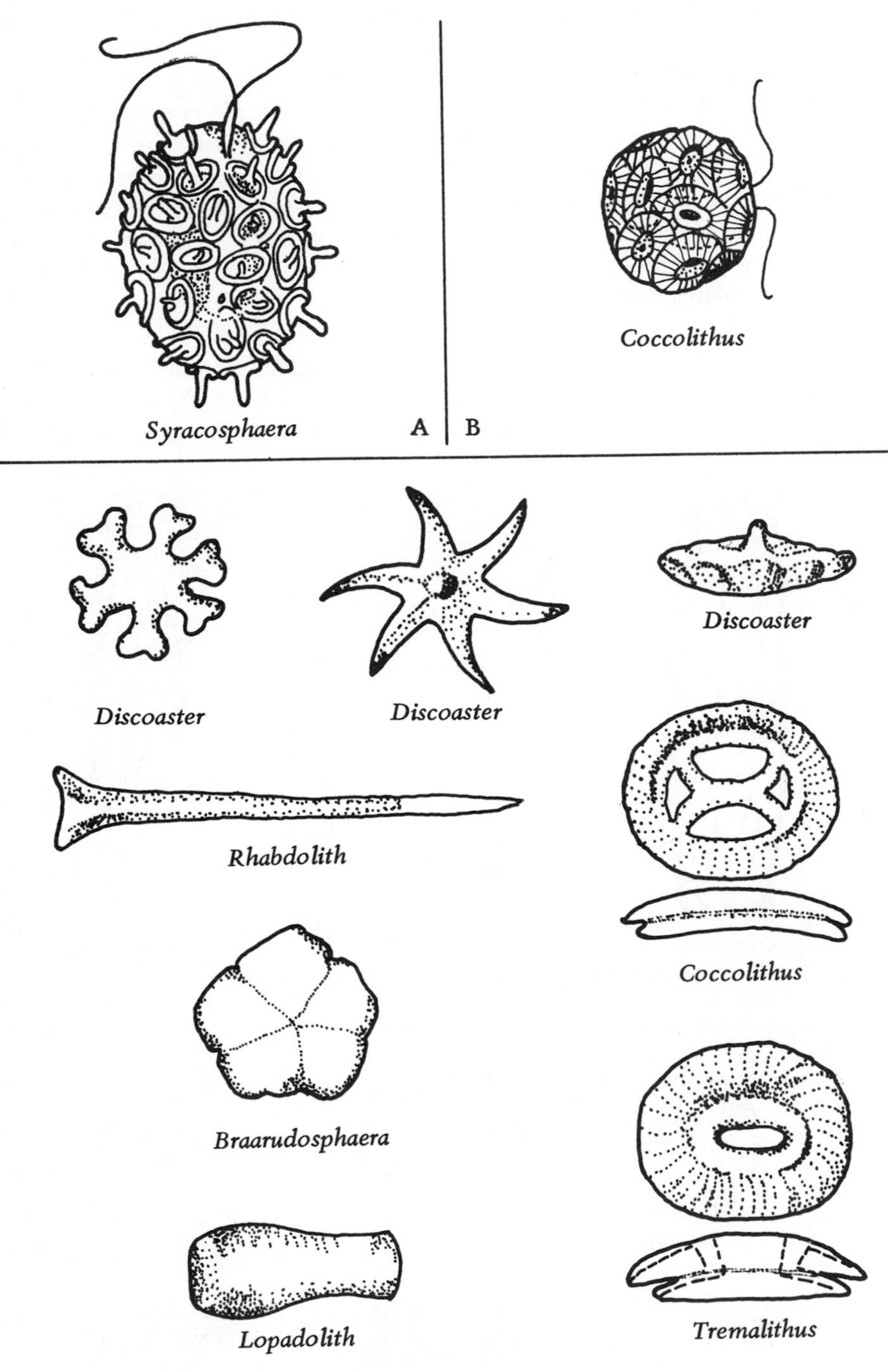

FIGURE 7.24 Coccolithophores. Illustrations of (A) living Coccolithophore, (B) Coccoliths in living position, and (C) Various types of Coccolithophore plates.

(Cretaceous-Recent) which build lattice-like siliceous tests or internal shells (Fig. 7.23) and are abundant in cooler water. The most significant golden algae are the *coccolithophores* (Cretaceous–Recent) which cover their tiny spherical bodies with a cluster of very tiny calcite plates (Fig. 7.24) called coccoliths if they are disc or wheel-shaped, discoasters if star-shaped, rhabdoliths if cylindrical, and lopadoliths if club-shaped. Coccolithophores are the most abundant and important photosynthesizers in the oceans today and prefer warmer waters. Their remains have been so abundant at times in the past that great thicknesses of chalk, a rock composed chiefly of their miniscule plates, have accumulated.

Phylum Bacillariophyta—The diatoms (Fig. 7.25) also have a golden pigment and have chlorophyll for photosynthesis. They differ from the golden algae in several ways including the absence of flagella and the presence of a bi-

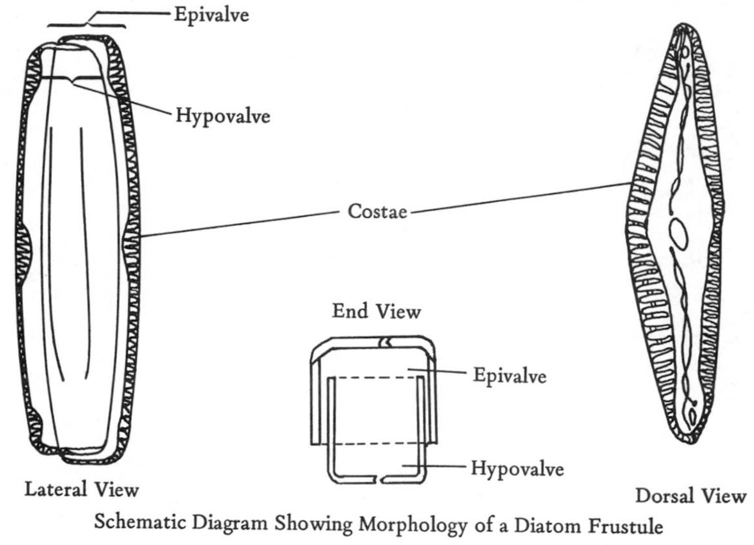

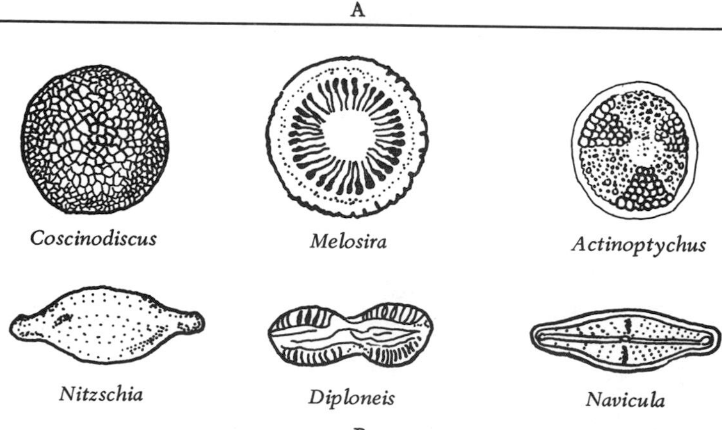

FIGURE 7.25 Diatoms (Phylum Bacillariophyta). (A) General morphology, (B) Fossil forms.

valved siliceous shell or frustule whose two halves fit inside one another like a pill-box. Diatom frustules come in a variety of shapes, but circular disks and spindle-shaped or rod-like forms predominate. Most are single-celled, though many are colonial. Most are marine and pelagic, but several types have inhabited fresh waters since the Miocene and there are several benthonic genera. They range in age from Cretaceous to Recent and have been abundant throughout this interval. Diatoms are among the most important photosynthesizers on earth today.

Phylum Pyrrophyta—The fire algae or dinoflagellates (Fig. 7.26A) also contain golden pigments and most possess chlorophyll for photosynthesis. Those that do not, as well as many of the others, are animal-like in that they eat other organisms for nutrition. They have two flagella in grooves, one meridional and one equatorial. Dinoflagellates are pelagic asymmetrical single cells. The body is encased in a cellulose shell or lorica (impregnated with silica or calcium carbonate in some). This is rarely preserved, but a much thicker cellulose cyst (Fig. 7.26B) which forms around the individual when it reproduces, is a common fossil, particularly the variety known as hystrichospheres which look like spherical pincushions. Dinoflagellates occur as early as the Silurian, but are only common from the late Mesozoic onward.

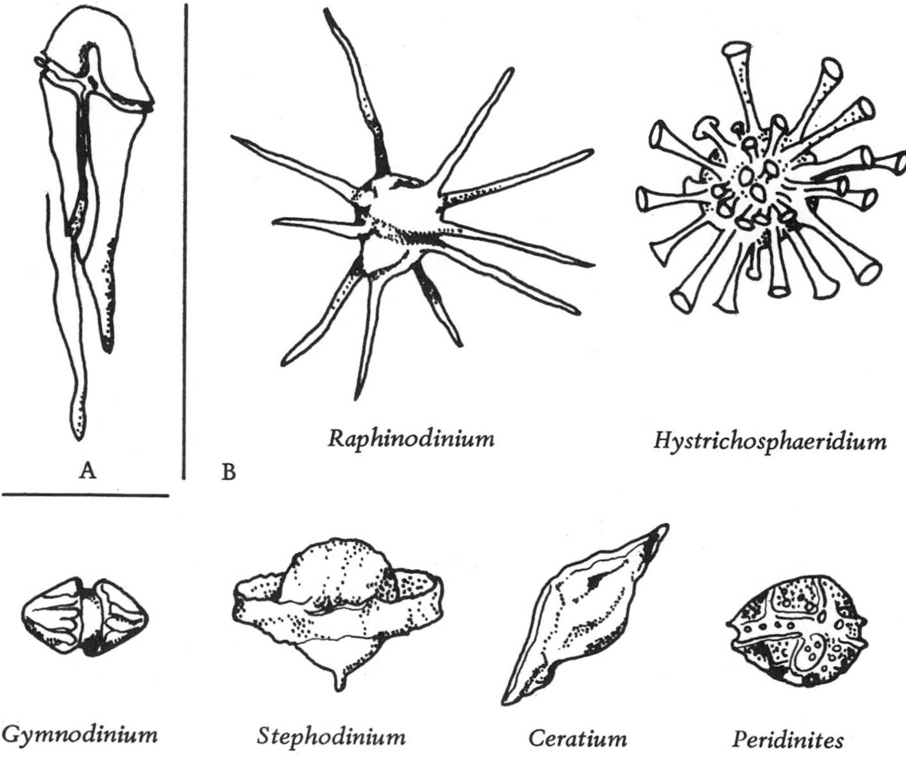

FIGURE 7.26 Dinoflagellates. (A) Living Dinoflagellate *Ceratodinium*, (B) Various fossil dinoflagellate cysts.

Phylum Rhodophyta—The red algae contain a red pigment in addition to chlorophyll for photosynthesis. They lack flagella and are nearly all complex, multicellular benthonic forms. Many are "seaweeds" and have not left fossils, but two types that encrust their thalli with calcite are common fossils. One variety, the corallines (Fig. 7.27A), has delicate branching forms covered with a jointed series of plates. Even more common are *Lithothamnion* and its relatives the nullipores (Fig. 7.27B), which form sheets or lumpy masses that often bind coral reefs together. Red algae are known from the Cambrian to the Recent, but are most abundant in the Cenozoic Era.

Phylum Sarcoda—The protozoans with pseudopods are a diverse lot which may not be a natural group. Pseudopods are fleshy extensions of the one-celled body used in food capture and locomotion. Sarcodine protozoans are all animal-like or heterotrophic, that is they cannot manufacture their own food, but obtain it instead from other organisms. Shell-less forms, such as the amoeba (Fig. 7.28) with its thick pseudopods, are unknown as fossils. The significant fossil groups belong to a separate class characterized by fine pseudopods.

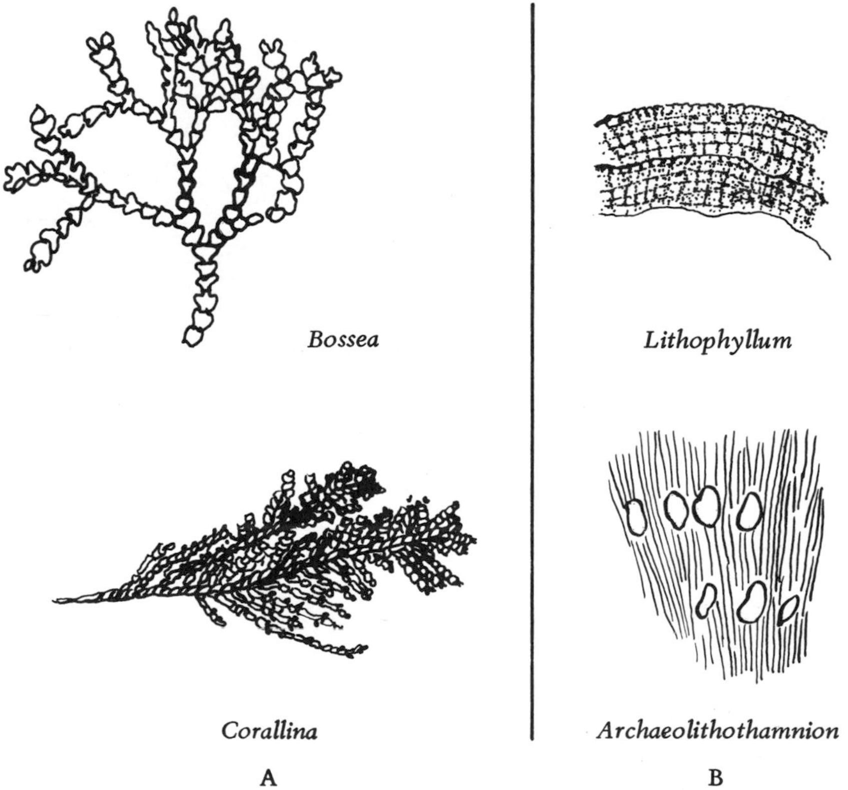

FIGURE 7.27 Red Algae (Phylum Rhodophyta). (A) Living Coralline Red Algae, (B) Cross-sections of fossil Nullipores.

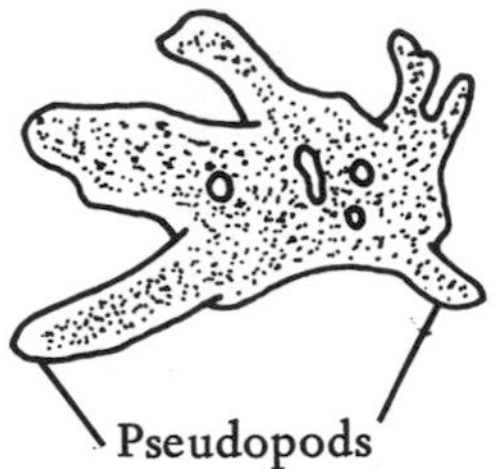

FIGURE 7.28 Amoeba. Living Sarcodine Protozoan of the Class Rhizopodea.

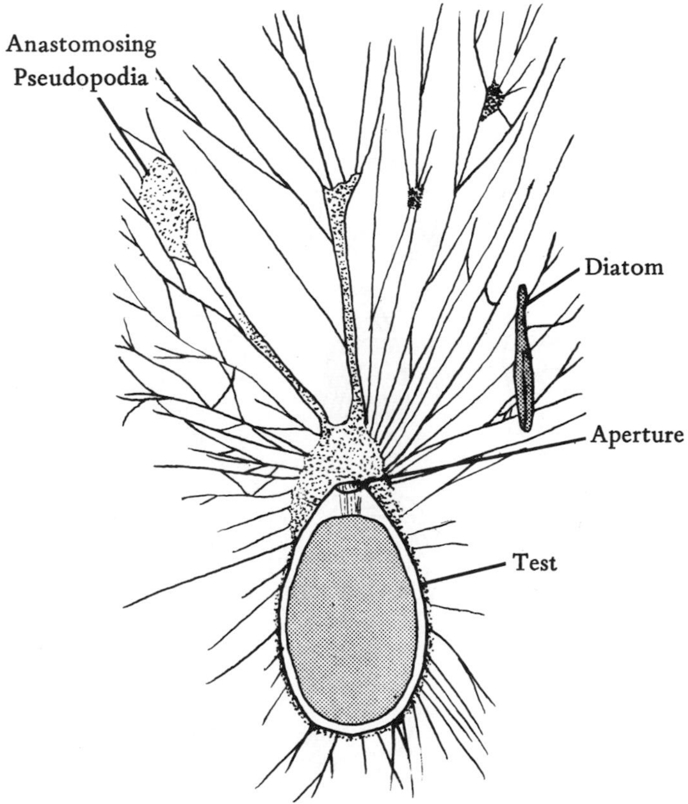

FIGURE 7.29 Order Foraminiferoida. A living foram.

Class Reticularea—These are sarcodines with very fine pseudopods.

Subclass Granuloreticulosia—This group encompasses sarcodines with delicate, anastomosing, unstiffened pseudopods.

Order Foraminiferoida—The forams (Fig. 7.29) are probably still the single most important group of protistans to geology. They possess a test or internal shell (Fig. 7.30) usually composed either of agglutinated mineral particles or, much more commonly, of calcium carbonate (calcite in nearly all). The test of most forams is subdivided internally into a series of chambers, though the simpler varieties are one-chambered. The test shapes are very diverse, ranging from a simple rod to planispiral and conispiral coils, and many others, and the ornamentation is similarly diverse. Most forams are small fossils barely visible to the naked eye, but some very large varieties have evolved. In the Pennsylvanian and Permian, the large spindle or grain-shaped *fusulines* (Fig. 7.31A) are abundant and in the Paleogene, large disk or coin-shaped *nummulites* and *orbitoids* (Fig. 7.31B) abound. The typical smaller forams have been abundant since Jurassic times. Nearly all forams are marine

FIGURE 7.30 Order Foraminiferoida. Illustrations of the diversity in morphology found among the forams.

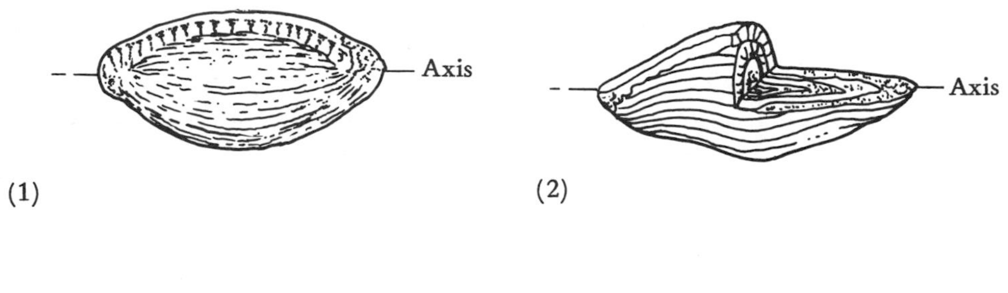

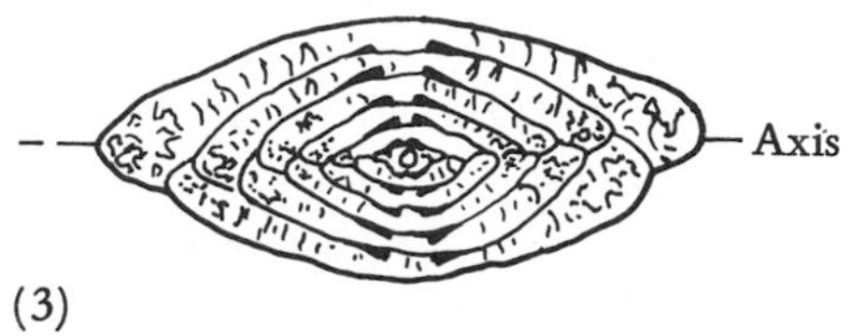

A

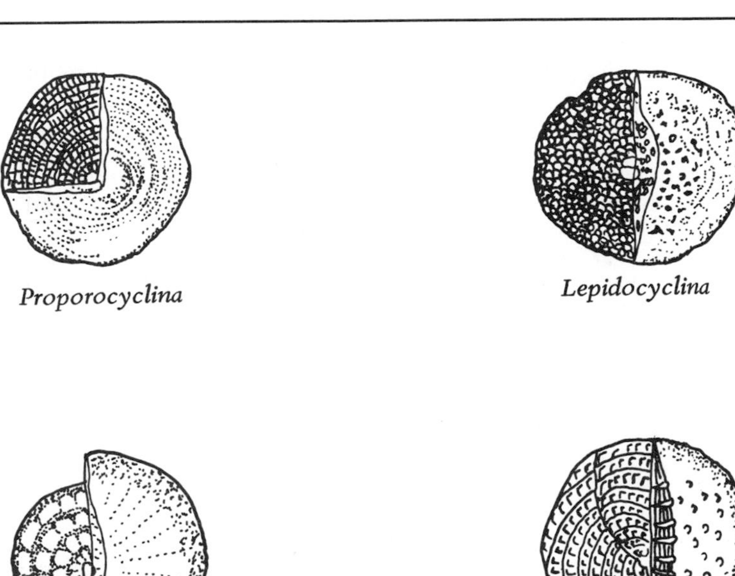

B

FIGURE 7.31 Large Forams. (A) Late Paleozoic Fusulines. External view, partially sectioned shell, cross-section. (B) Cretaceous and Cenozoic Nummulites and Orbitoids.

and most are benthic. A few very common varieties, the *globigerines* (Fig. 7.32), have become planktonic and are major contributors to oceanic oozes. Forams, being small and abundant, are easily recovered from well cores and cuttings and, hence, have been important tools for petroleum geologists.

Subclass Actinopodia—This group encompasses sarcodines with fine, radiating, stiffened pseudopods called axopods (Fig. 7.33A).

Order Radiolaroida—The radiolarians have a spiny, perforate shell or scleracoma (Fig. 7.33B) of siliceous composition. The microscopic scleracomae are spherical to conical in shape and occur in marine rocks from Cambrian to Recent in age, though they are most abundant from the Late Mesozoic onward. They are all planktonic and most frequent the photic zone because of photosynthetic algae in their bodies which are necessary to their metabolism.

Phylum Ciliophora—The protozoans with cilia (Fig. 7.34), tiny hair-like projections used chiefly for locomotion, are typically not fossilizable creatures because most lack hard parts. All are one-celled, but very complex, and all are animal-like, eating other creatures for their nutrition. Many familiar ciliates have uniform cilia over their bodies, but the major fossil group does not belong to that class.

Class Spirotrichea—These are ciliates in which the cilia are restricted to the oral region.

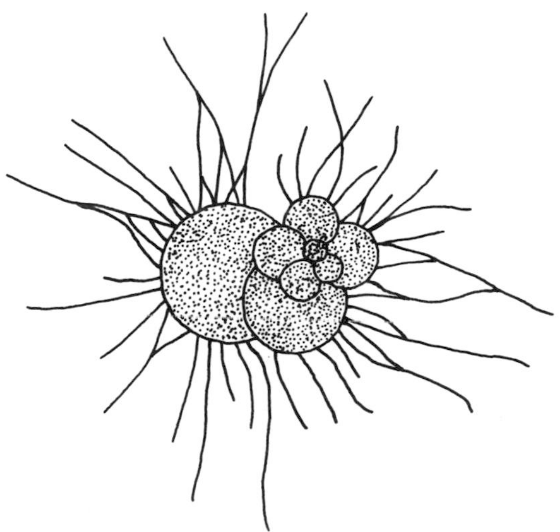

FIGURE 7.32 *Globigerina,* a living planktonic foram.

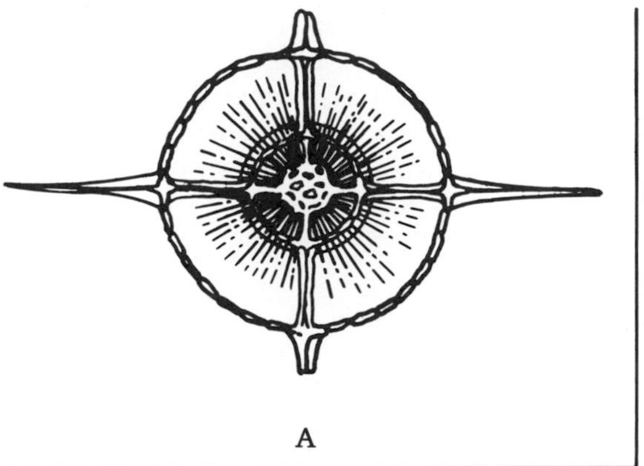

FIGURE 7.33A Subclass Actinopodia. A living radiolarian showing relationship of body and pseudopods to shell or scleracoma.

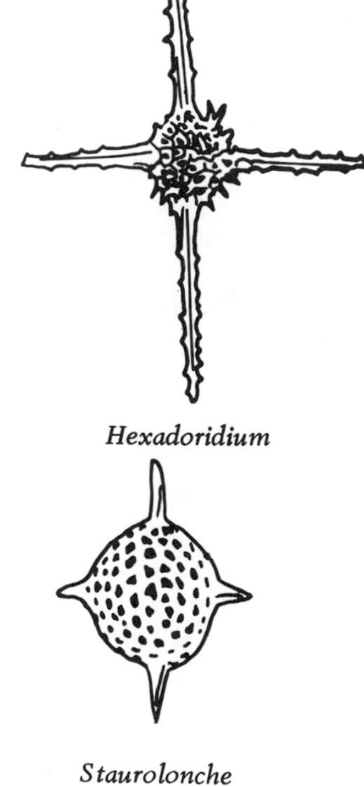

FIGURE 7.33B Order Radiolaroida. Fossil scleracomae.

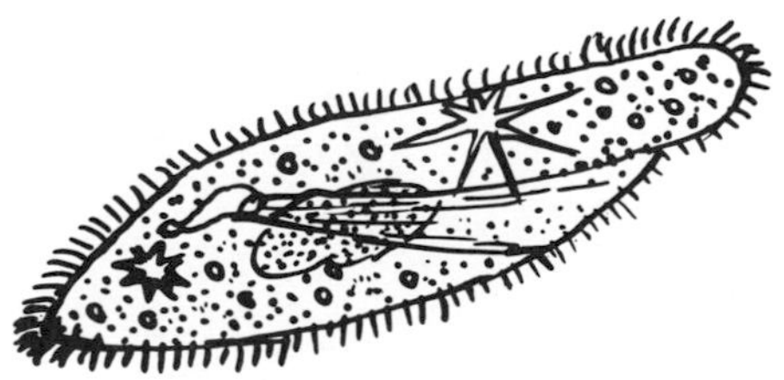

FIGURE 7.34 *Paramecium,* a living ciliate protozoan of the Class Holotrichea.

Order Tintinnoida—Tintinnoids (Fig. 7.35), which compose nearly half of all living ciliates and virtually all known fossil types, possess a conical agglutinated or calcareous shell or lorica. Most are marine and planktonic. They range in age from Ordovician to Recent, but most fossils are from the Jurassic and Cretaceous.

Kingdom Animalae

Animals, like protistans and plants, have a nucleus. All are multicellular and the cells are differentiated for different functions. Animals are unable to synthesize their own food from inorganic substances, and, hence, must obtain their nutrients from other organisms. Their gametes develop in multicellular structures and embryos are formed. There are twelve major animal groups: sponges, archaeocyathans, coelenterates, bryozoans, brachiopods, mollusks, annelid worms, arthropods, echinoderms, graptolites, conodonts, and vertebrates.

Phylum Porifera—The sponges (Figs. 7.36, 7.37), whose relationship to the other animals is uncertain, have a sac-like body perforated by pores for

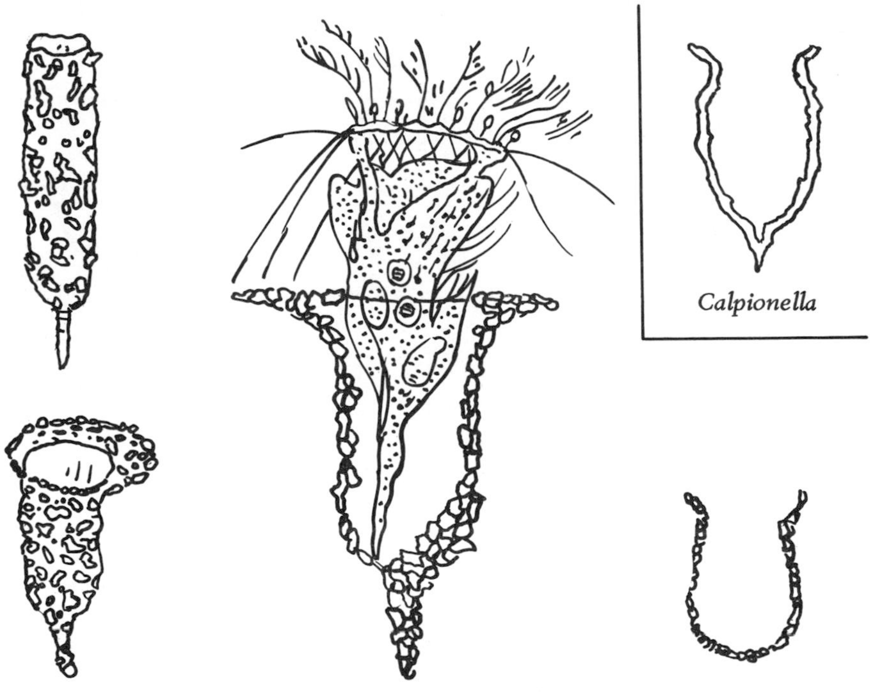

FIGURE 7.35 Order Tintinnoida. Illustration of a living tintinnoid and several fossil loricas.

water intake and a larger opening (osculum) or openings (oscula) for water outflow. The body consists of three weak tissue layers which are hardly more than aggregates of independent cells. The inner layer is composed of collared flagellate cells whose beating sets up the water flow and which also capture the tiny food particles as they flow by. Some flagellate protozoans are very similar and may be their ancestors. Sponge embryology, as well as structure, is unique among animals. The hard parts of most consist of tiny siliceous or calcareous (calcite) elements called spicules (Fig. 7.38) which come in a variety of shapes from simple rods to radiating star-shaped patterns and occur in the middle body layer. Isolated spicules are common fossils, but the major body fossils occur only in certain groups that build a skeleton of rigidly interlocking spicules. Sponges are benthic and nearly all marine. Siliceous sponges were abundant in the Devonian and again in the Jurassic-Cretaceous interval, while the less common calcareous types reached a peak in the latter two periods.

Phylum Archaeocyatha—Archaeocyathans (Fig. 7.39) are one of several common fossil groups that have no living representatives and, hence, their systematic position is uncertain. The basic structure of most is a calcareous (calcite), double-walled cone with the two walls connected by vertical plates called parieties. All body parts are perforated by pores. The general

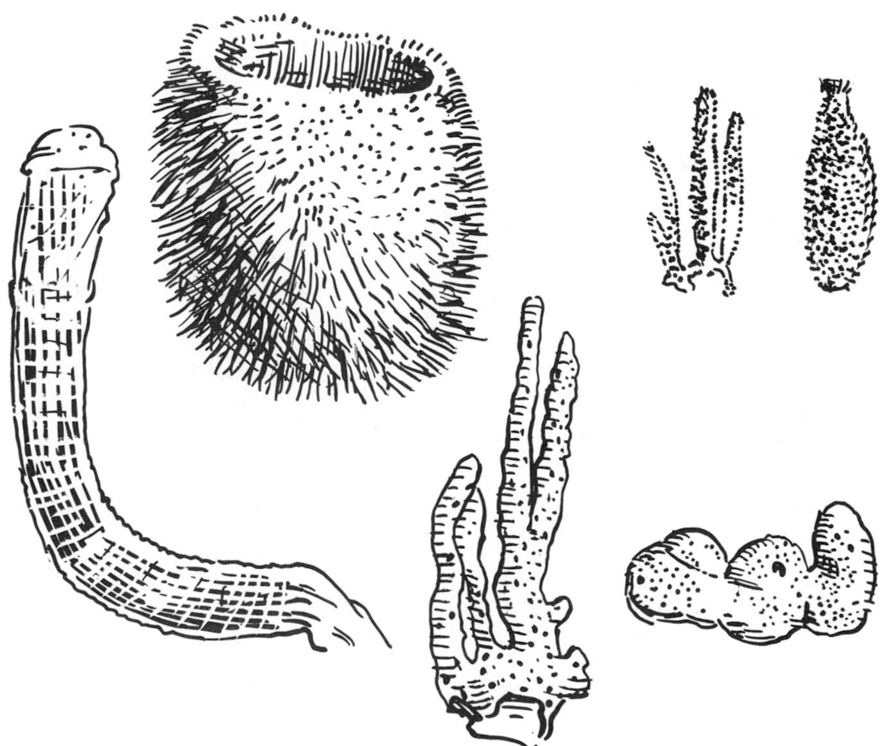

FIGURE 7.36 Sponges (Phylum Porifera). External anatomy of living sponges.

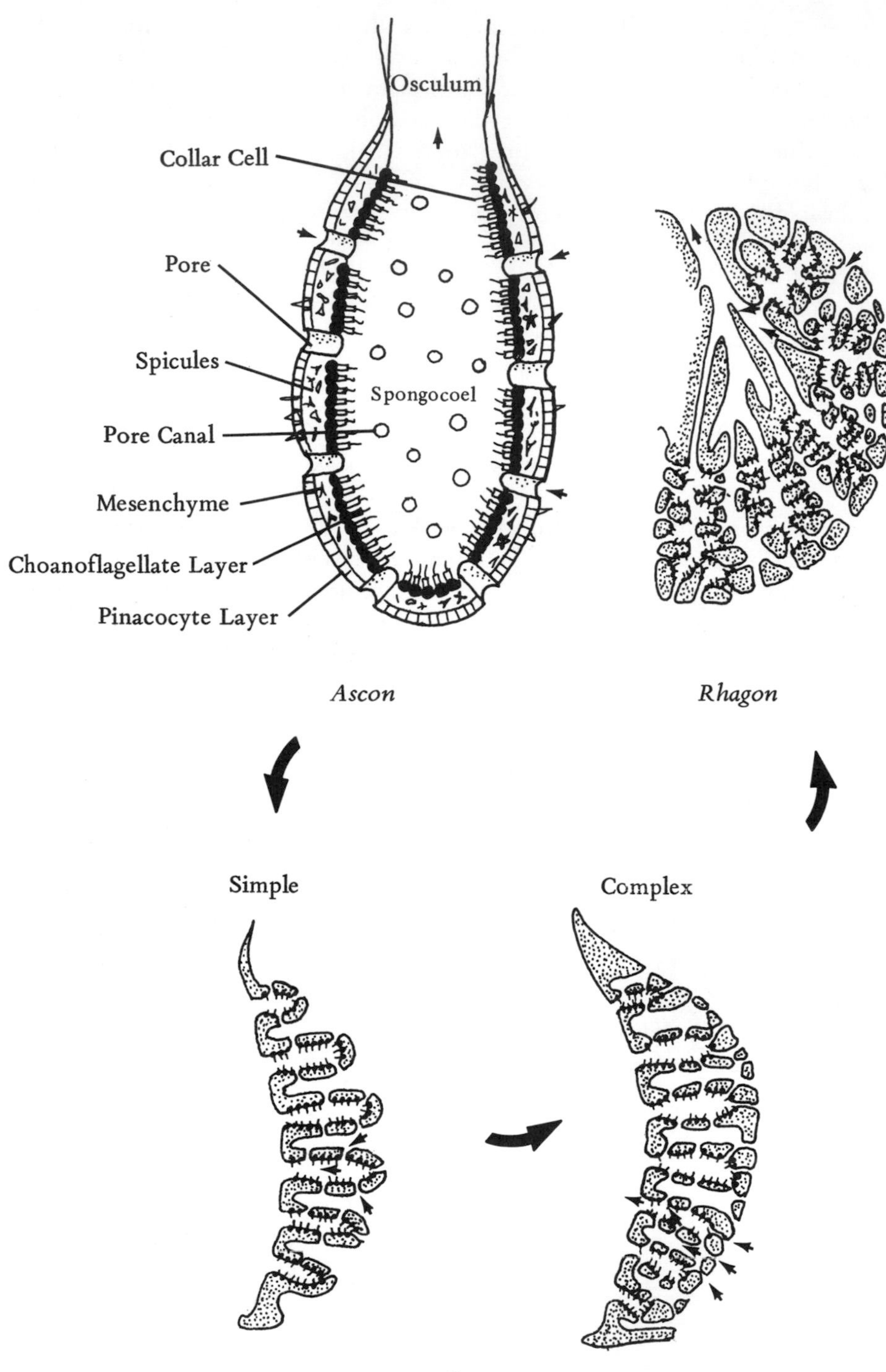

FIGURE 7.37 Sponges (Phylum Porifera). Internal anatomy of living sponges. Small arrows indicate direction of water movement. Large arrows indicate direction of evolution.

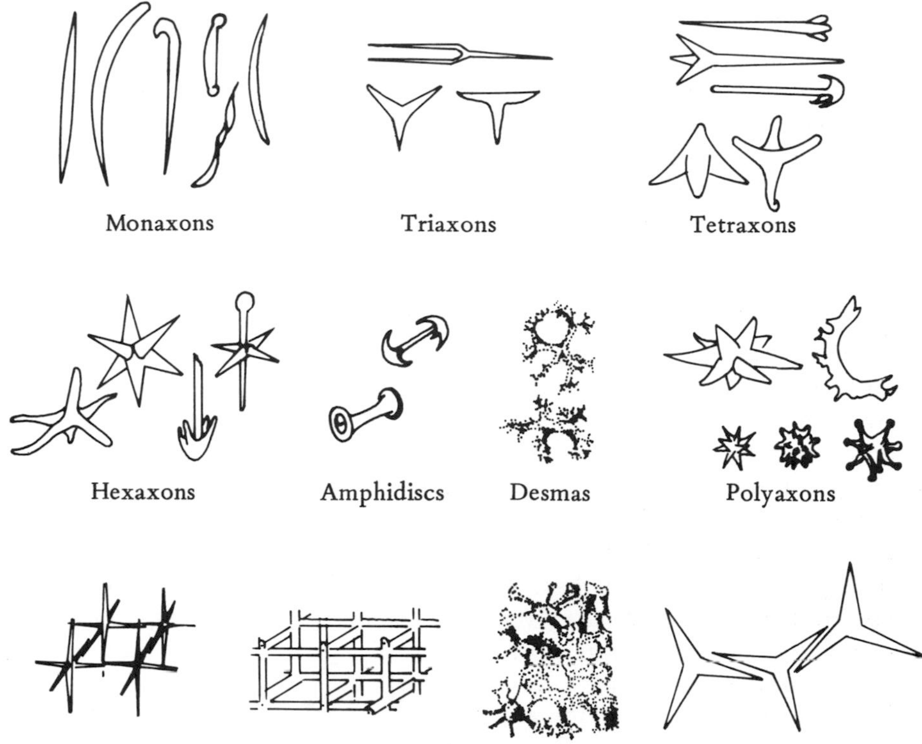

FIGURE 7.38 Sponge Spicules.

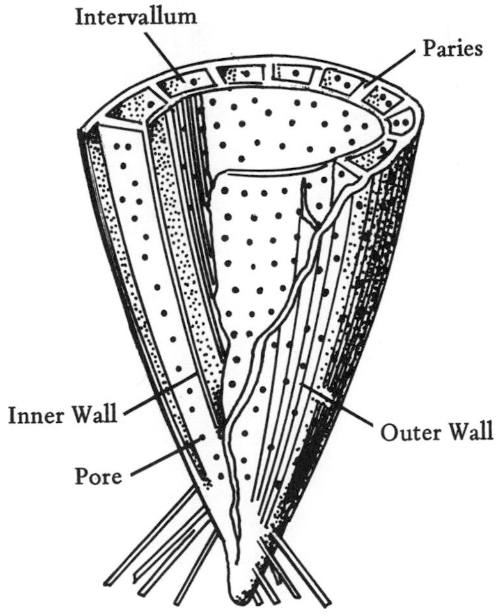

FIGURE 7.39 Phylum Archaeocyatha. Morphology of a typical archaeocyath.

shape and the vertical partitions are like those of corals which, however, lack the inner wall and pores. The perforations suggest sponge affinities, but sponge skeletons consist only of spicules. The general structure of archaeocyathans is also suggestive of dasycladaceous green algae. The consensus of experts places them near the sponges. Archaeocyathans were all marine and commonly built reefs. They are all Cambrian and most are Lower Cambrian.

Phylum Coelenterata—This large and diverse phylum includes hydroids, jellyfish, sea anemones, and corals. The body (Fig. 7.40) is sac-like with only a single opening into the digestive cavity which is surrounded by tentacles for capturing prey. Unlike the majority of animals, most coelenterates are fundamentally radially symmetrical, that is the body parts are disposed in a radial manner about the axis like the rays of the sun, though this symmetry is typically imperfect. Many have two stages in their life cycles (Fig. 7.41), a sessile benthonic polyp attached by its base with tentacles and mouth directed upward, and a pelagic umbrella-shaped medusa or jellyfish with these structures directed downward. Polyps give rise to medusas which then reproduce sexually. Most groups lack hard parts and, hence, are rare fossils, but the stromatoporoids, conularoids, and especially the corals have hard parts and are common fossils. The fossil record of the phylum goes back to the Eocambrian where several jellyfish occur.

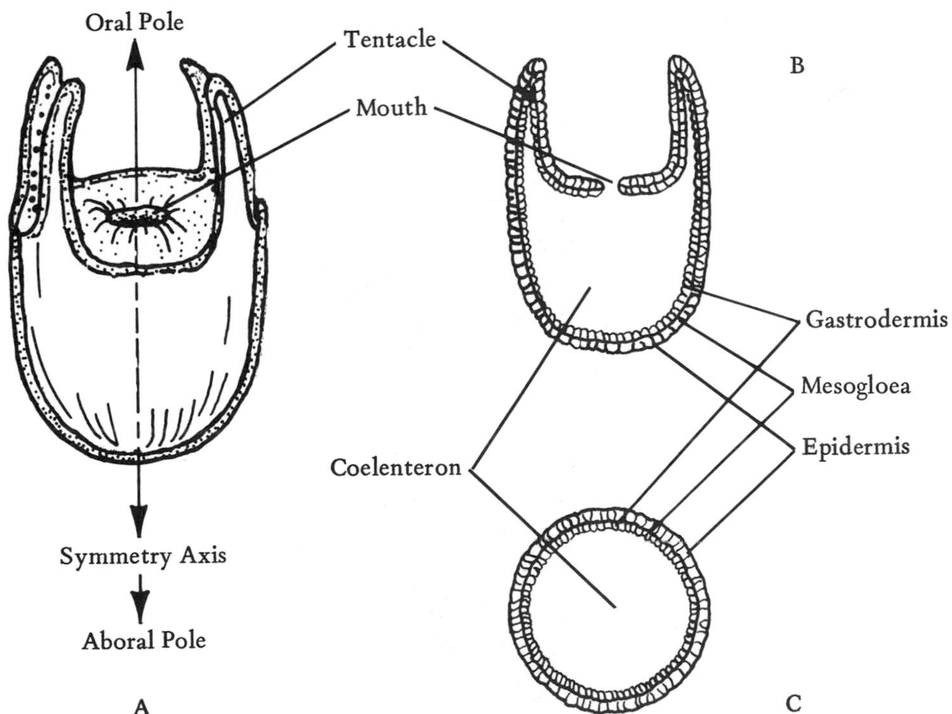

FIGURE 7.40 Basic Coelenterate Body Plan. (A) Generalized perspective view, (B) Longitudinal section, (C) Cross-section.

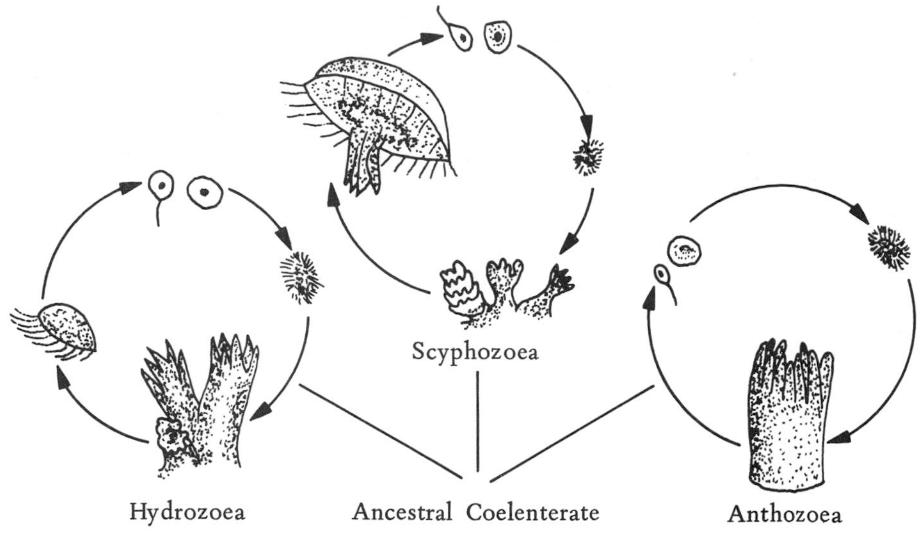

FIGURE 7.41 Life Cycles in the Three Major Coelenterate Groups. Most living and fossil coelenterates are anthozoans.

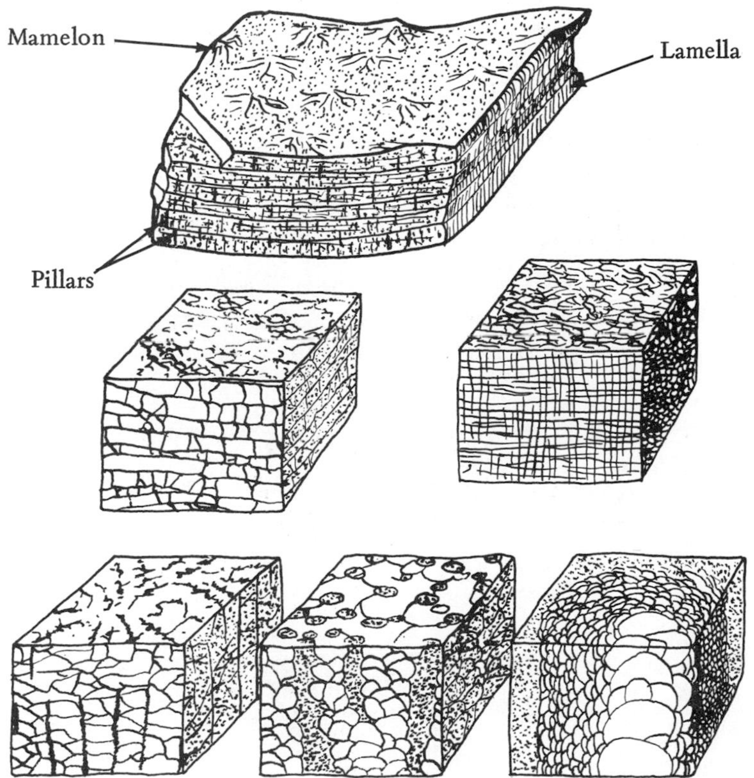

Figure 7.42 Stromatoporoids. An extinct group which may belong to the Porifera or Coelenterata.

Class Hydrozoea—The polyps, which are typically colonial, are the major stage in the life cycle of this group, the medusa being small. The skeleton of most is a tough organic substance rarely preserved, though a few have an aragonite skeleton. The long, sparse fossil record of these forms goes back to the Cambrian and nearly all are marine.

Order Stromatoporoida—The stromes (Fig. 7.42) are another extinct group of uncertain affinities. Their calcite hard parts consist of thin horizontal layers or lamellae penetrated by vertical pillars of varying lengths. Star-shaped structures called astrorhizae occur on the upper surfaces and may have contained polyps. The shape of the marine benthic colonies is variable, but typically stromes form large, lumpy masses. The anatomy is suggestive of some hydrocoral groups, hence, this systematic placement. They also resemble a group of sponges which possess astrorhizae. Stromes range in age from Cambrian to Cretaceous, though some experts restrict the group to Ordovician through Devonian types. They are major reef-building elements in Silurian and Devonian times. A related order, the *spongiomorphoids* (Fig. 7.43), is common in Triassic warm-water deposits.

Class Scyphozoea—The pelagic medusa stage dominates in this class, the polyps being small. The large jellyfish have a four-fold symmetry, are all marine, and are rare fossils because they lack hard parts.

Order Conularoida—These flexible four-sided, conical chitinophosphatic shells (Fig. 7.44) are representatives of a totally extinct group. At first glance they appear most

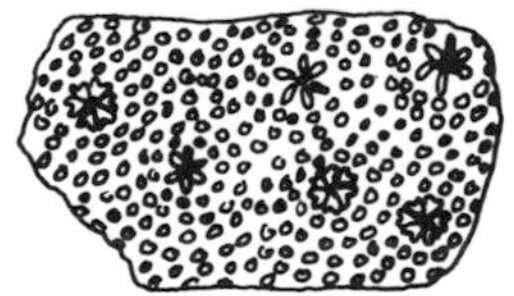

FIGURE 7.43 Spongiomorphoid. Member of an extinct group related to the stromatoporoids.

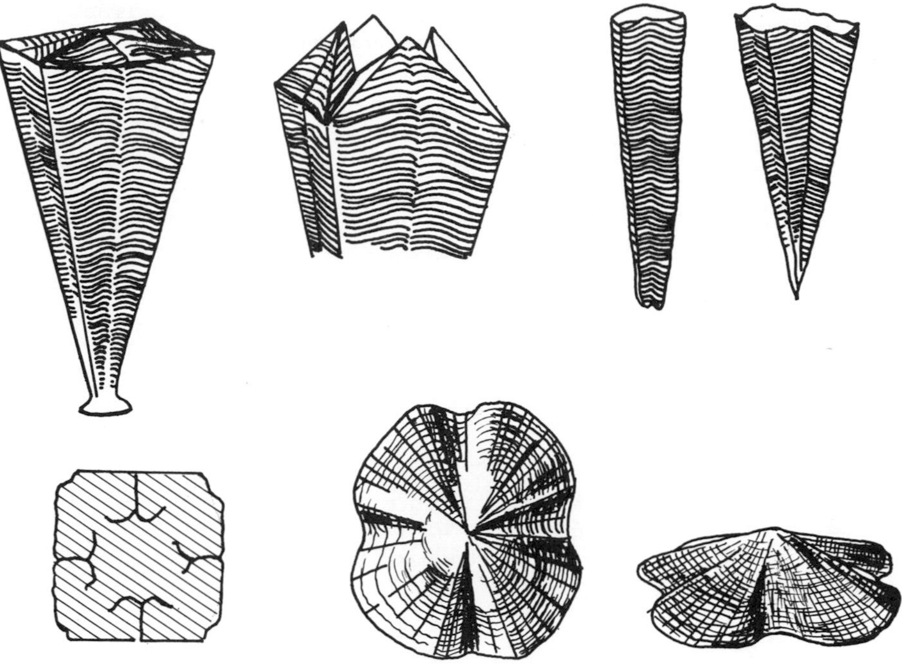

FIGURE 7.44 Conularoids. An extinct group of uncertain affinites, perhaps related to scyphozoan jellyfish.

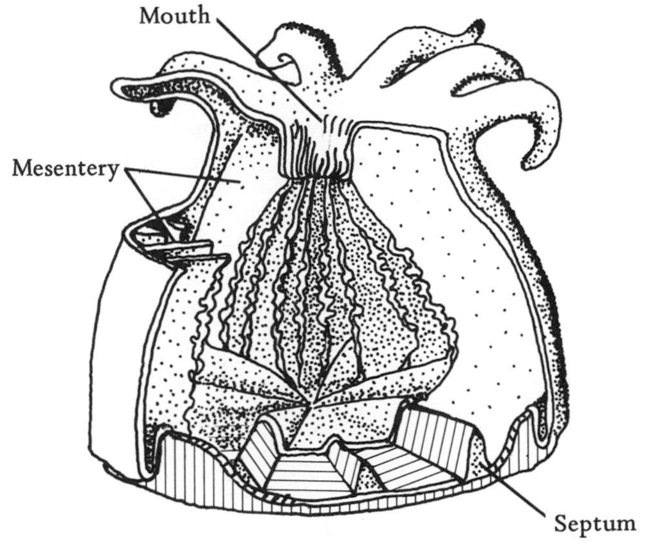

FIGURE 7.45 A Living Coral Polyp, sectioned to show internal anatomy and its relationship to the hard parts.

unlike jellyfish, but they have four internal partitions and tentacles like scyphomedusae and one form is known with a low dome shape similar to that of jellyfish. They are all marine and range in age from Cambrian through Triassic, reaching a maximum in the Upper Paleozoic.

Class Anthozoea—Anthozoans are polyps only; the medusa stage is lacking. The polyps are complex and subdivided internally by vertical partitions or mesenteries (Fig. 7.45). All are marine and benthic. The sea anemones and soft corals lack hard parts and are rare as fossils. The stony corals have a conical or cylindrical skeleton, the theca, subdivided by radiating internal partitions or septa in most cases.

Subclass Zoantharia—This group includes the stony corals and most sea anemones. The theca, when present, begins with six major septa and the orders are based on the positions where new septa are added as the animal grows.

Order Tabulatoida—In this extinct Paleozoic order, the septa have not yet evolved or are very poorly developed. All are colonial and consist of polygonal or round tubes generally closely packed (Fig. 7.46). Many have pores on their walls, a feature not found on other corals. Prominent horizontal partitions or tabulae supported the base of the animal as it grew upward as in other corals. Tabulates range in age from Ordovician through the Permian (a few may have survived later) but were important chiefly in the Lower Paleozoic, especially the Silurian, when they built large reefs. Some workers feel they are really sponges.

Order Rugosoida—Another extinct group, these corals are often called *tetracorals* (Fig. 7.47) because they insert their new septa in only four of the six spaces between the initial septa (Fig. 7.48). They are either solitary or colonial, the former type dominating and having a typically curved theca, hence, the name horn corals. The top usually contains a depression or calyx for the body. Rugose corals were the major reef builders of the Late Paleozoic, particularly the Devonian. They are known from the Ordovician through the very earliest Triassic.

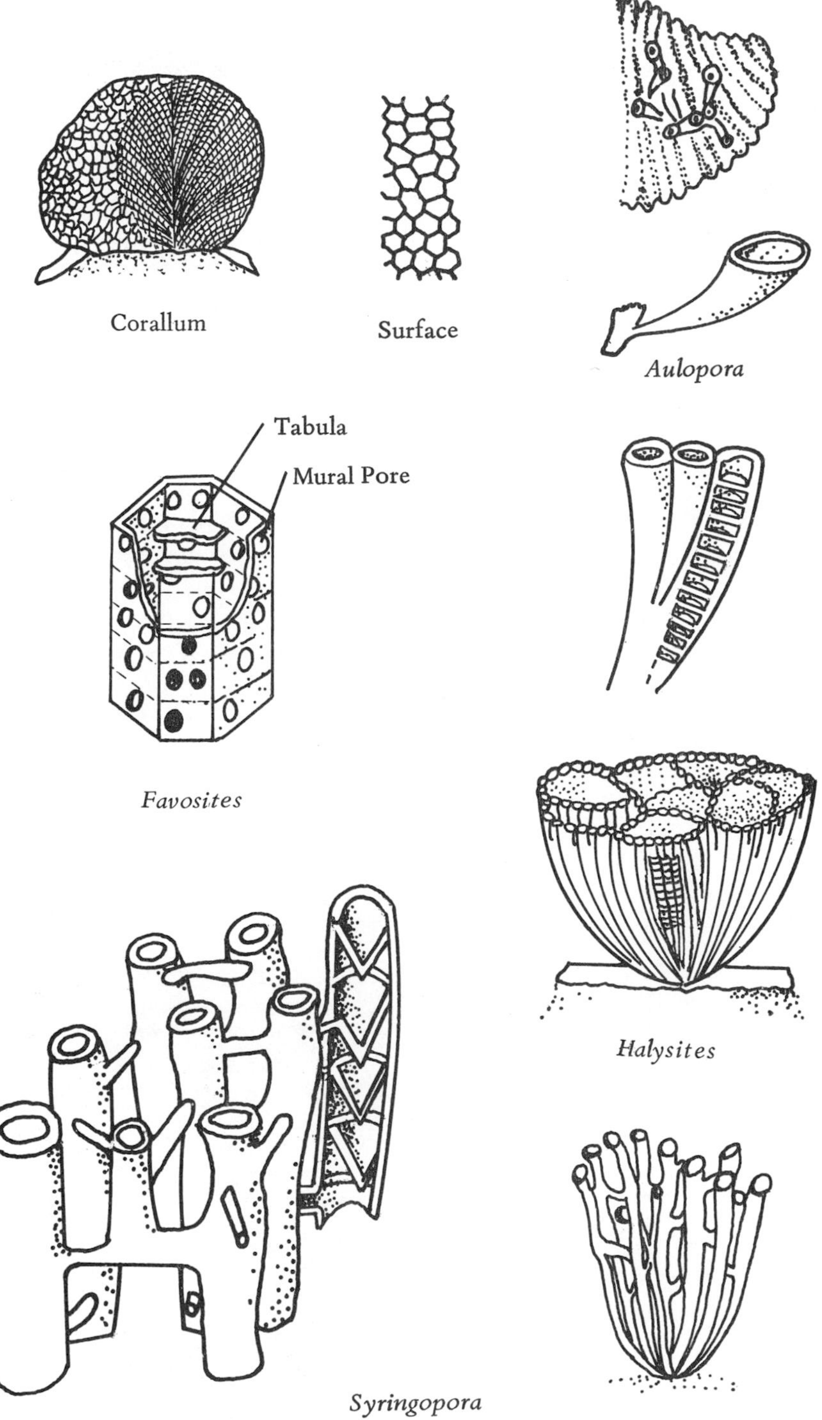

FIGURE 7.46 Morphology of Tabulate Corals.

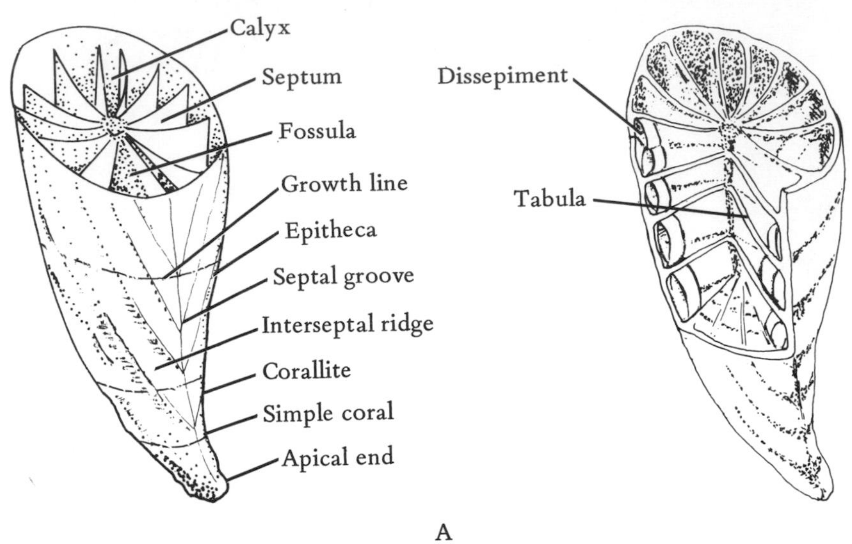
Calyx
Septum
Fossula
Growth line
Epitheca
Septal groove
Interseptal ridge
Corallite
Simple coral
Apical end
Dissepiment
Tabula

A

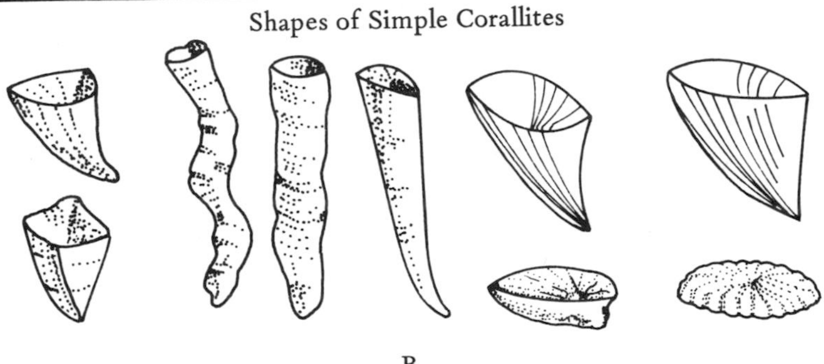
Shapes of Simple Corallites

B

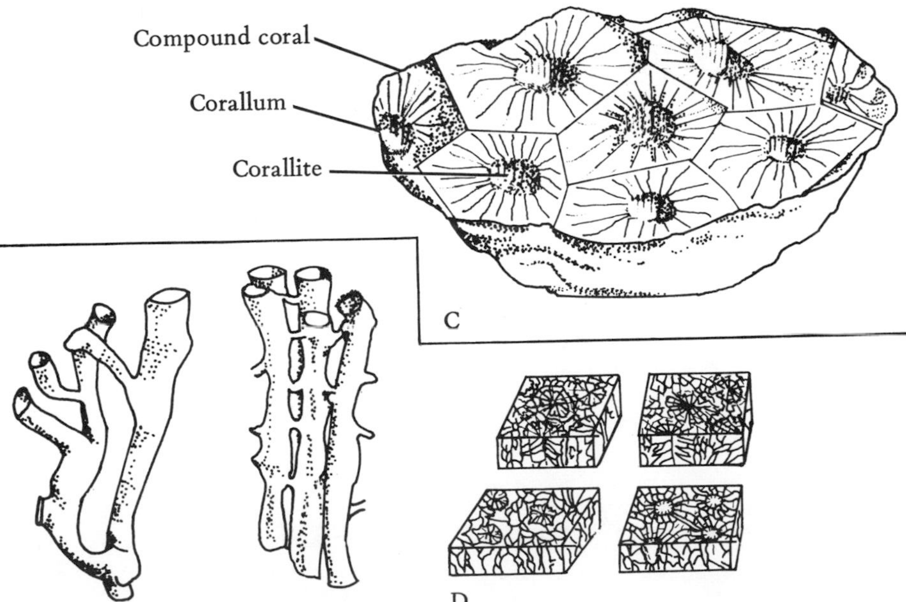
Compound coral
Corallum
Corallite

C

D

FIGURE 7.47 Morphology and Diversity of Rugose Corals (Tetracorals). (A) Simple corallum, (B) Variation of simple coralla, (C) Compound corallum, (D) Variation of compound coralla.

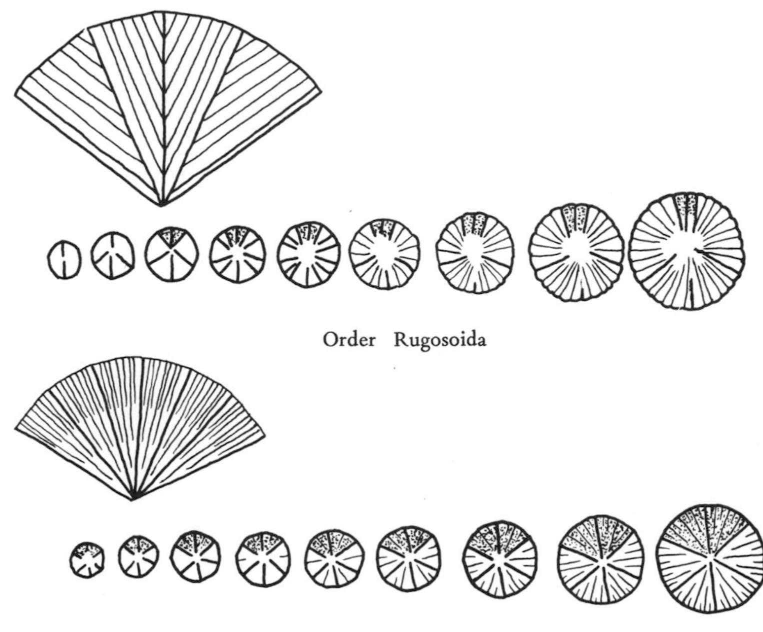

FIGURE 7.48 Patterns of Septal Insertion in Tetracorals and Hexacorals.

Order Scleractinoida—These are usually known as *hexacorals* (Fig. 7.49) because they add new septa in all six spaces between the original six (Fig. 7.48). They are both solitary and colonial with the latter type dominating. The solitary forms are typically uncurved and lack a distinct outer wall, hence, there is usually no calyx. Hexacorals, which range from Triassic to Recent, are the major reef builders of Mesozoic and Cenozoic times. The reef builders are tropical and are restricted to shallow waters because of photosynthetic algae in their tissues which are necessary for their metabolism.

Phylum Bryozoa—The bryozoans or moss animals have been very abundant from the Ordovician to the Recent. Individually, they are microscopic, but all are colonial and the colonies often are many inches across. The tiny

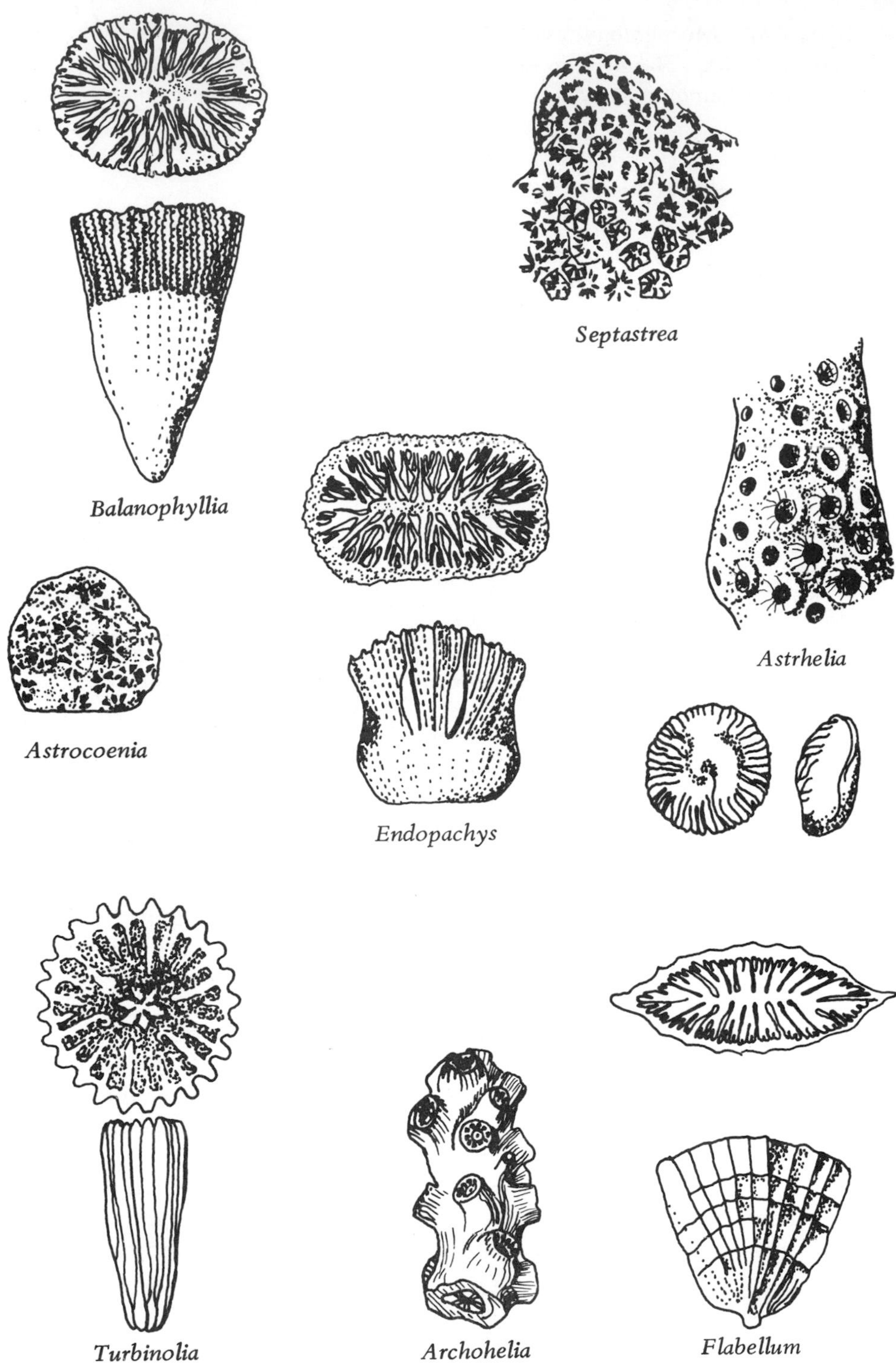

FIGURE 7.49 Diversity of Fossil Scleractinian Coral (Hexacorals).

bodies or zooids (Fig. 7.50) of individual bryozoans contain a retractible portion, the polypide, which has a U-shaped gut and a ciliated tentacle-bearing ridge or lophophore used in capturing microscopic food. The soft parts of most are encased in a calcareous (mixture of calcite and aragonite) skeleton or zooecium, collectively known as the zooarium of the colony. Bryozoans are all benthic and nearly all marine.

Class Stenolaematea—These have tubular zooecia with a terminal aperture.

Order Cyclostomoida—Cyclostome bryozoa (Fig. 7.51) generally lack the internal horizontal partitions of other tubular bryozoans. Most zooaria are smooth and twig-like. They range from Cambrian to Recent, but reached a peak in the Jurassic and Cretaceous and are common in the Cenozoic.

Orders Cystoporoida and Trepostomoida—These extinct stony bryozoa (Fig. 7.52) have typically twig-shaped zooaria with their surfaces covered by mounds and depressions, though the latter order includes many massive forms also. They are complex internally with cross-partitions (diaphragms) and other structures (Fig. 7.52). The orders are distinguished on features of the zooecial walls and need not concern us here. Both are Ordovician through Permian in age, but reached a great peak of abundance in the Ordovician.

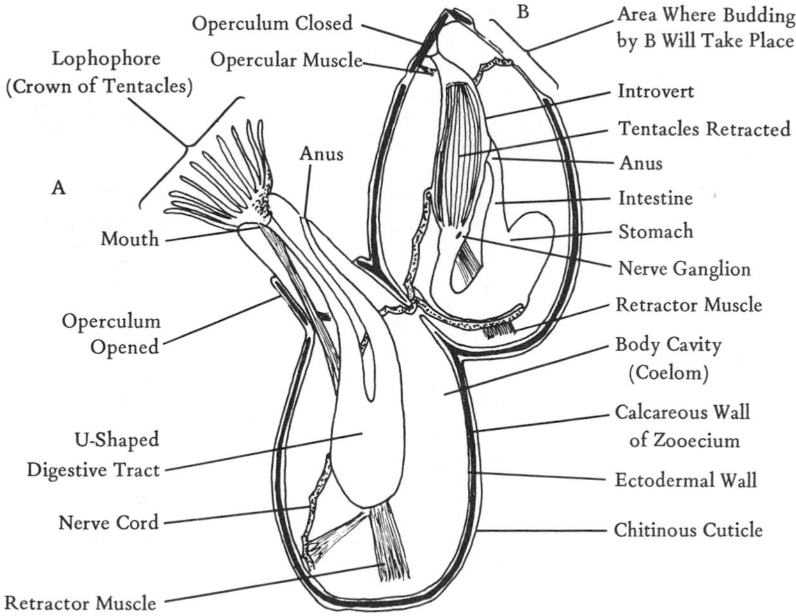

FIGURE 7.50 Structure of a Living Bryozoan (A) with polypide extruded, (B) with polypide retracted.

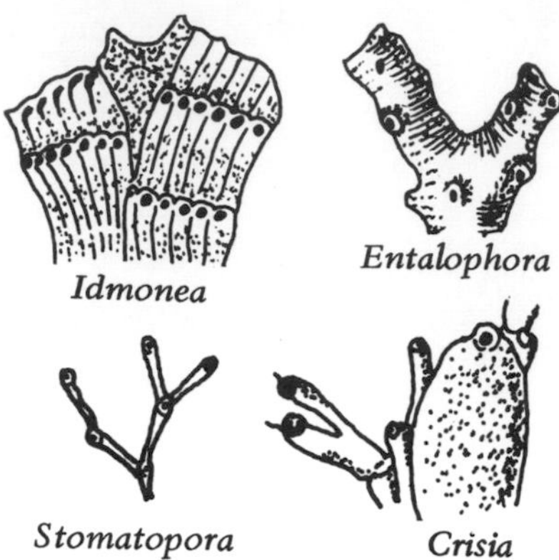

FIGURE 7.51 Cyclostome Bryozoan Fossils.

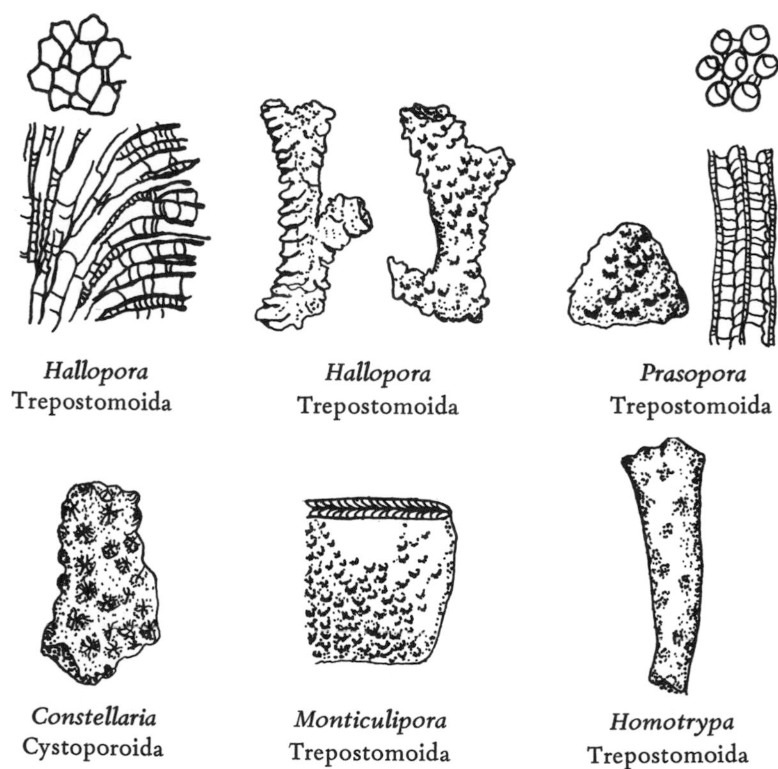

FIGURE 7.52 External and Internal Morphology of Trepostome and Cystoporoid Bryozoans.

Order Cryptostomoida—This extinct group of bryozoans consists chiefly of forms with delicate, lacy zooaria (Fig. 7.53). The openings of individual zooecia occur on the vertical and cross-bars of the fronds. One common genus had a screw-shaped axis to support its lacy fronds. The group ranges in age from Ordovician to earliest Triassic but is most common in the late Paleozoic, particularly the Devonian and Carboniferous.

Class Gymnolaematea—These typically have box-like zooecia with lateral openings (Fig. 7.54).

Order Cheilostomoida—This, the largest group of living bryozoans, appeared in the Cretaceous and has dominated ever since. The zooaria are very diverse, but are always very thin, commonly forming sheet-like incrustations, delicate twigs, or blade-like fronds. Most lack internal partitions. Almost any Cenozoic shell contains the delicate tracery of encrusting cheilostome bryozoans.

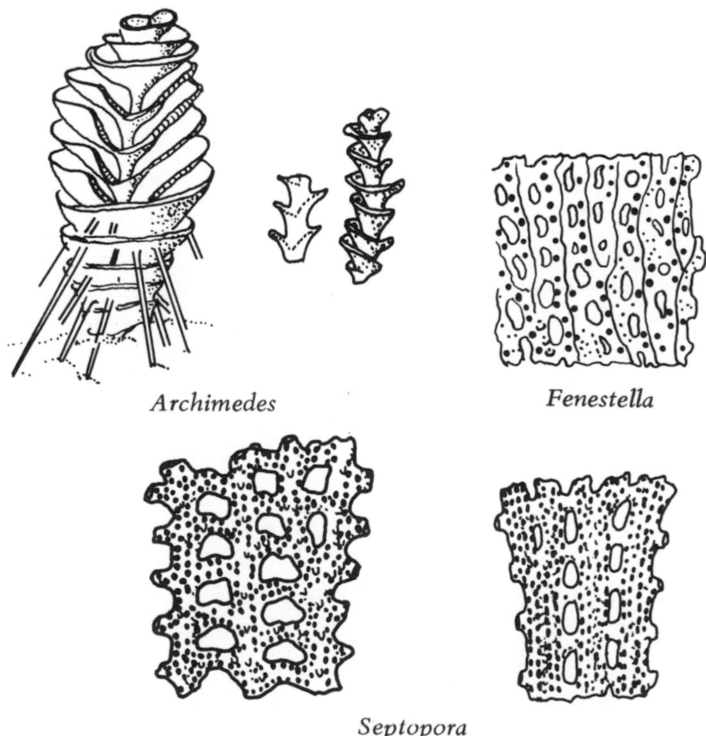

FIGURE 7.53 External Morphology of Cryptostome Bryozoans.

Phylum Brachiopoda—The brachiopods or lamp shells are not abundant today, but they are among the most common megafossils, particularly in Paleozoic rocks. The simple body (Fig. 7.55) contains a large, feathery, two-armed anterior lophophore used for feeding; a fleshy posterior stalk or pedicle for attachment; and flaps called the mantle which secrete a bivalved shell. The shell is either chitinophosphatic or, much more commonly, calcareous (calcite). The two valves are dorsal and ventral in relation to the body. Both are bilaterally symmetrical with one, the valve housing the pedicle and called the pedicle valve, typically larger than the other, called the brachial valve because it supports the lophophore arms. The posterior prolongations of the valves, the beaks, are typically much larger on the pedicle valve because they house the pedicle. Beginners often confuse brachiopods with clams which belong to the much more complex phylum Mollusca and are also bivalved. In clams, however, the valves are typically right and left in relation to the body and the plane of symmetry passes between the valves, each valve not being symmetrical about its midline (Fig. 7.56). Brachiopods are all marine and benthic. They range in age from Cambrian to Recent, but are most abundant in Paleozoic and some Mesozoic rocks. The major types of this most significant fossil group are listed below.

Class Inarticulatea—In the inarticulate brachiopods (Fig. 7.57), the two valves are not hinged and are typically chitino-phosphatic in composition. The inarticulates have a long geologic record from Cambrian to Recent, but were most abundant in the Cambrian, when they were the major brachiopod group, and the Ordovician.

Class Articulatea—Members of this much larger group of brachiopods have hinged, calcareous shells. On the inside of the brachial valve are supports for the lophophore.

Order Orthoida—Orthoids (Fig. 7.58A) have squarish to round biconvex shells. The posterior hinge line is typically wide. The valves are covered with fine radial or-

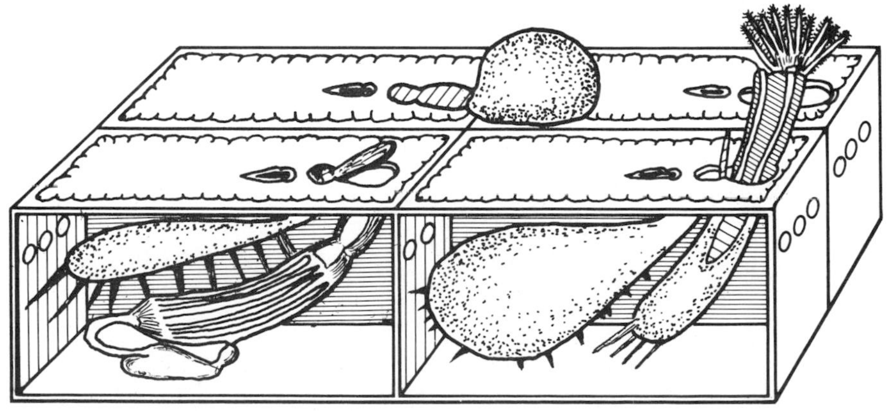

FIGURE 7.54 Morphology of Cheilostome Bryozoans.

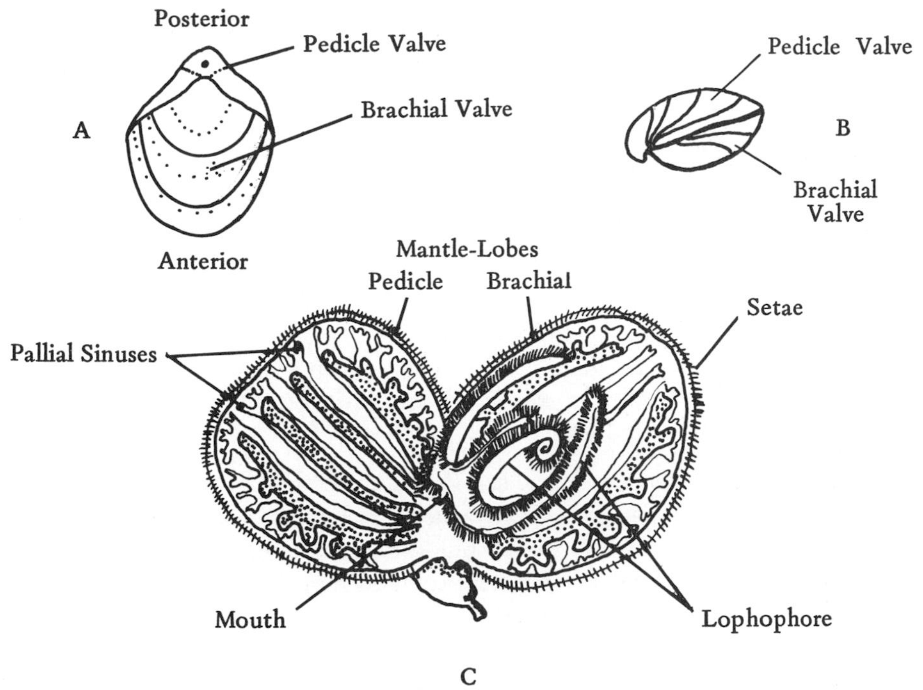

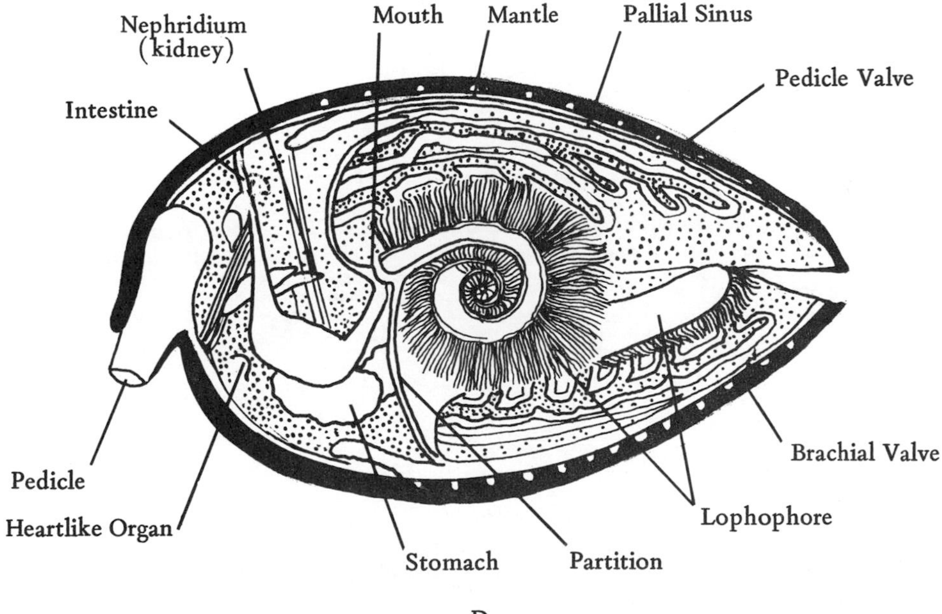

FIGURE 7.55 Phylum Brachiopoda. The shells and soft parts of a Recent Brachiopod. (A) Ventral view, (B) Lateral view, (C) Section along symmetry plane, (D) Body with mantle lobes separated.

namentation or costae in most. Orthoids range from Cambrian through Permian, reaching an Ordovician peak.

Order Strophomenoida—Strophomenoid shells (Fig. 7.58B) are similar to orthoids, but one valve is concave. The pedicle opening is sealed in adults indicating loss of the pedicle. The order ranges from Ordovician through Jurassic, but has an Ordovician peak.

Order Chonetoida—Chonetoids (Fig. 7.58C) are like strophomenoids, but have a row of spines along the posterior margin of their concave-convex shell. They range from Silurian through Permian, reaching a Devonian-Carboniferous peak.

Order Productoida—Productoids (Fig. 7.58D) resemble the preceding two orders in having wide-hinged, concave-convex shells, but differ in having large pedicle valve beaks and a shell covered with spines. Productoids range

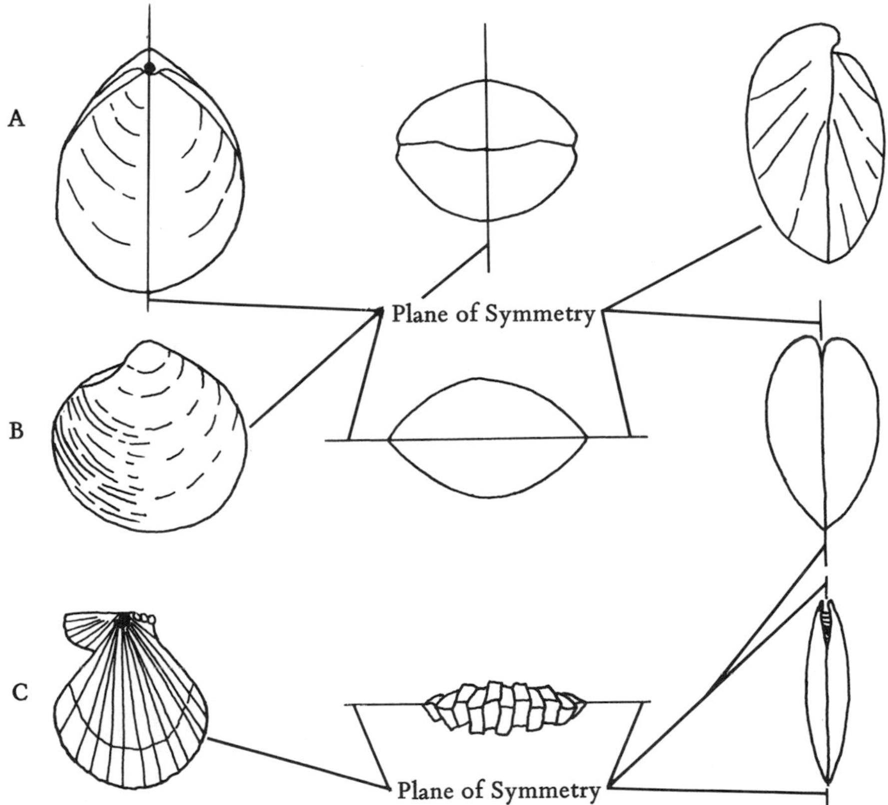

FIGURE 7.56 The Morphology of Brachiopod and Bivalve Shells Contrasted. (A) Typical brachiopods, (B) Typical bivalve, (C) Scallop, a bivalve with symmetry resembling a brachiopod.

from Devonian through earliest Triassic, reaching a Carboniferous-Permian peak.

Order Davidsonioida—This group (Fig. 7.58E) is characterized by a biconvex shell with one valve having a somewhat conical shape and a prominent irregular beak. Davidsonioids range from Ordovician through Triassic and are common in Carboniferous and Permian rocks.

Order Pentameroida—Pentameroids (Fig. 7.58F) have biconvex shells with short hinge lines. Inside the valves they have prominent plates which form areas of muscle attachment. These appear on internal molds as slits (Fig. 7.58F). Pentameroids range in age from Cambrian through Devonian, reaching a Silurian maximum.

Order Rhynchonelloida—Members of this order (Fig. 7.58G), one of the two surviving articulate groups, possess highly corrugated (plicate) biconvex shells with very narrow hinges and sharp beaks. They range from Ordovician to Recent, with Silurian-Devonian and Jurassic peaks.

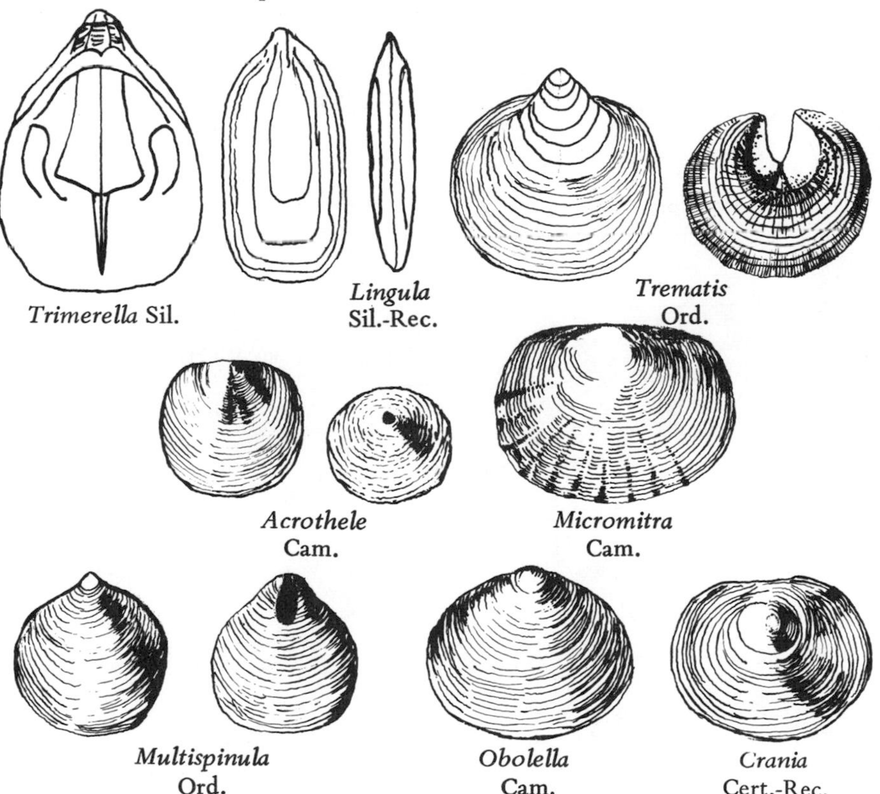

FIGURE 7.57 Class Inarticulatea. Representative fossil inarticulate brachiopods.

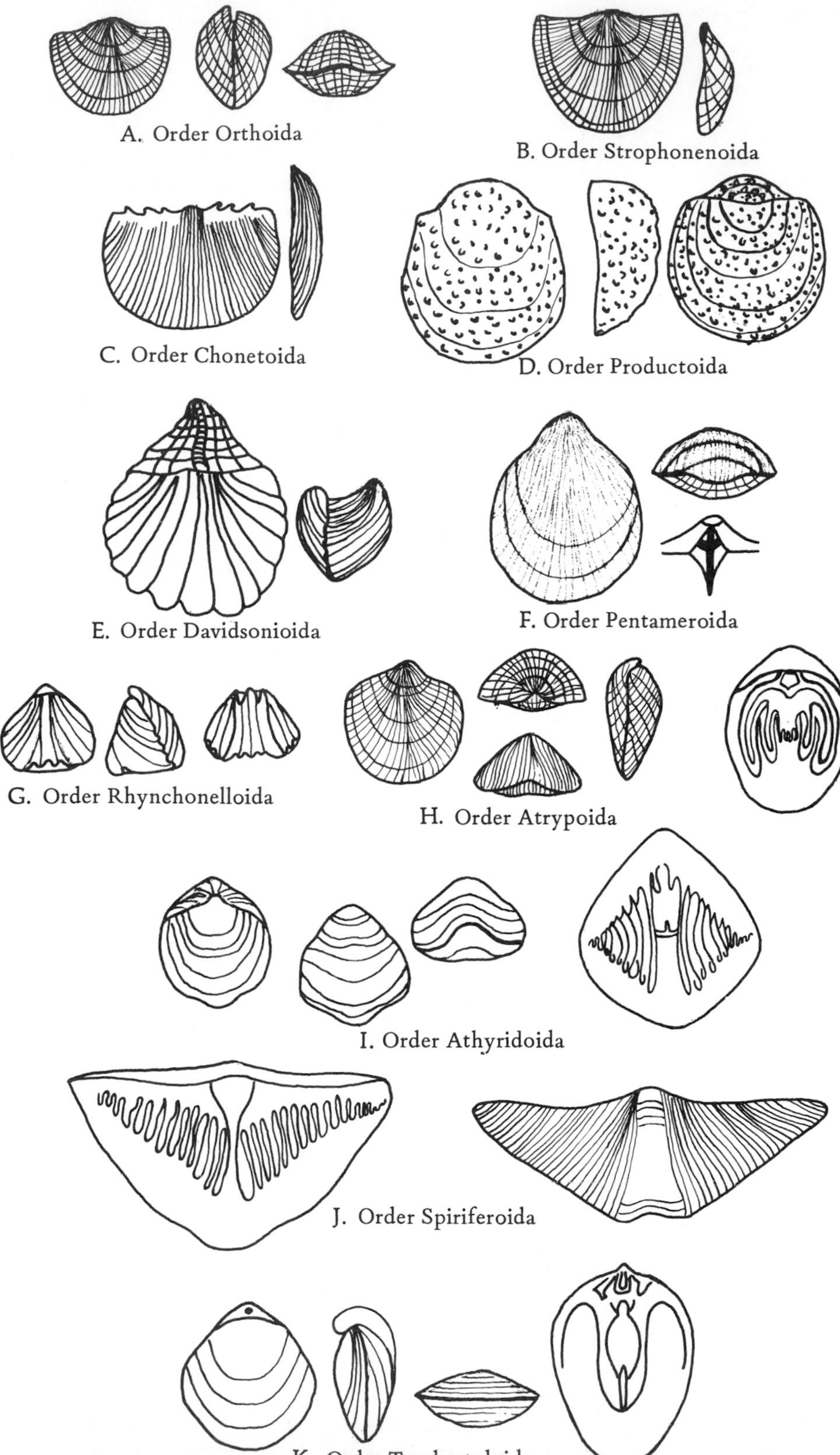
A. Order Orthoida
B. Order Strophonenoida
C. Order Chonetoida
D. Order Productoida
E. Order Davidsonioida
F. Order Pentameroida
G. Order Rhynchonelloida
H. Order Atrypoida
I. Order Athyridoida
J. Order Spiriferoida
K. Order Terebratuloida

FIGURE 7.58 Class Articulatea. Representatives of the major articulate brachiopod orders.

Order Atrypoida—Atrypoids (Fig. 7.58H) have finely costate shells with narrow hinges. One valve, the one with the larger beak, is typically flat, the other convex. Inside the valves are calcified coiled lophophore supports. Atrypoids range from Ordovician through Devonian with a Devonian peak.

Order Athyridoida—Arthyridoids (Fig. 7.58I), are one of two types of brachiopods that lack radial ornamentation. They have a narrow hinge line and an anterior depression (sulcus) and elevation (fold). Calcified coiled lophophore supports are present. They are Devonian through Triassic in age, but reached a Carboniferous peak.

Order Spiriferoida—Spiriferoids (Fig. 7.58J) are distinctive brachiopods recognized by their very wide hinge lines caused by the lateral prolongation of the valves called alae. The valves are costate and there are broad, internal, calcified, coiled lophophore supports. Spiriferoids range from Silurian through Jurassic, but are most abundant in the Upper Paleozoic, particularly the Devonian.

Order Terebratuloida—The other order of surviving articulates, the terebratuloids (Fig. 7.58K), typically have smooth shells lacking the prominent fold and sulcus of athyridoids. The hinge line is narrow and there are calcified, loop-shaped lophophore supports. The group ranges from Devonian to Recent, but is most common from the Jurassic onward.

Above the brachiopods, the tree of life forks into two branches, one containing mollusks, annelids, and arthropods, the other echinoderms, hemichordates, and vertebrates (Fig. 7.59).

Phylum Mollusca—The mollusks (Fig. 7.60) are a large, diverse phylum which has probably left the most impressive fossil record of all. Three major groups, the snails, clams, and cephalopods, and several lesser ones contribute to this array. The phylum is so diverse that it is difficult to find common features and there are exceptions to every rule. Mollusks have a large muscular foot which is flattened for creeping, compressed for burrowing, or modified into tentacles for seizing. Most respire by filamentous gills or ctenidia which lie in a cavity formed by flaps of the body called the mantle. The soft parts of most are unsegmented, but in the small class of monoplacophorans (Fig. 7.61) segmentation occurs and may indicate a relationship with the annelid worms. Most mollusks have a shell covering the outside of the body, though it is lost or internal in some. In most, the shell is of one piece, but in the clams it has two parts and in the minor groups of the

FIGURE 7.59 An Inferred Phylogeny of the Kingdom Animalae. (Reproduced with permission of A. Heintz and L. Stormer.)

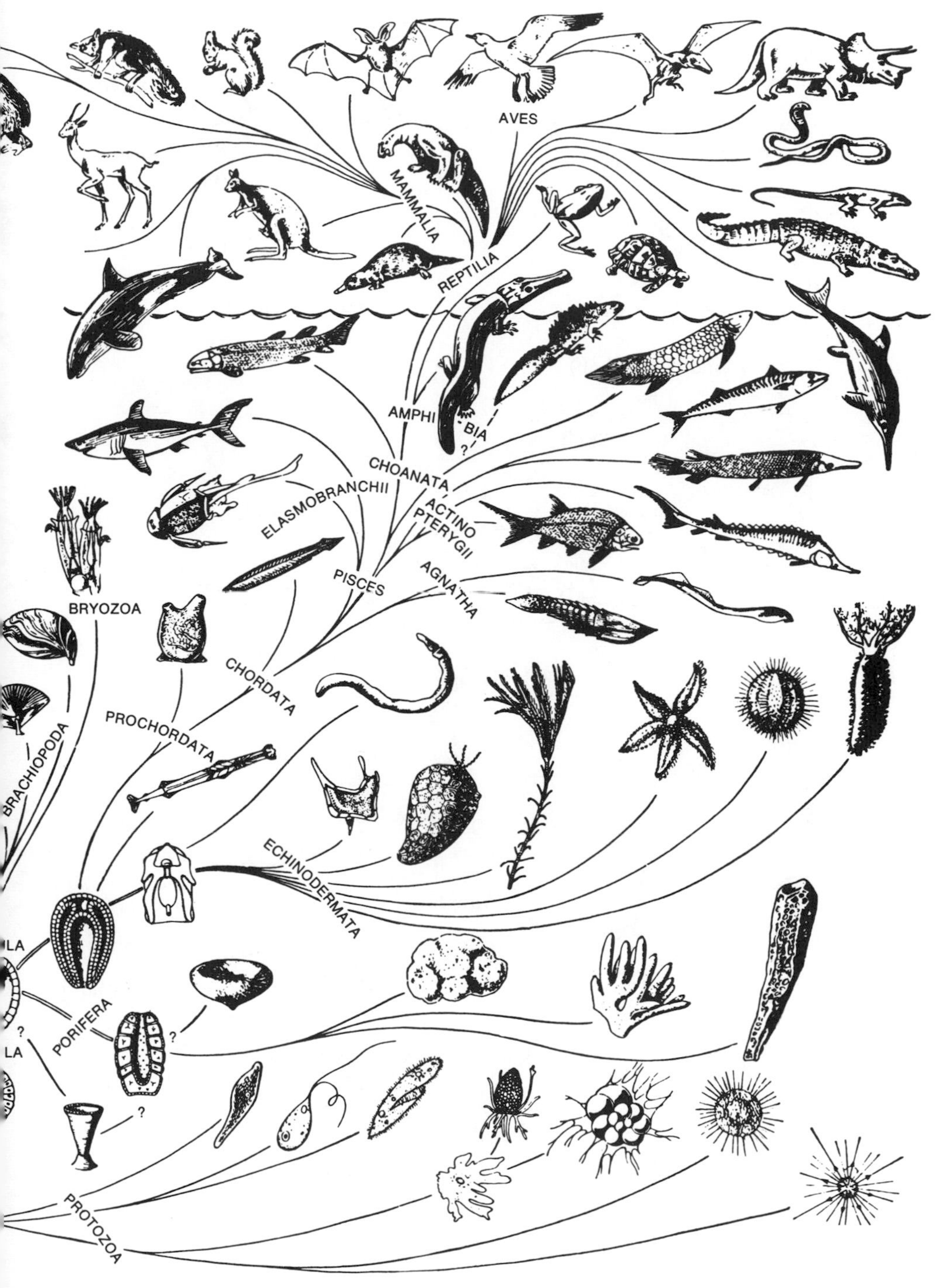
AVES
MAMMALIA
REPTILIA
AMPHI
BIA
CHOANATA
ELASMOBRANCHII
ACTINO
PTERYGII
PISCES
AGNATHA
BRYOZOA
CHORDATA
PROCHORDATA
BRACHIOPODA
ECHINODERMATA
PORIFERA
PROTOZOA

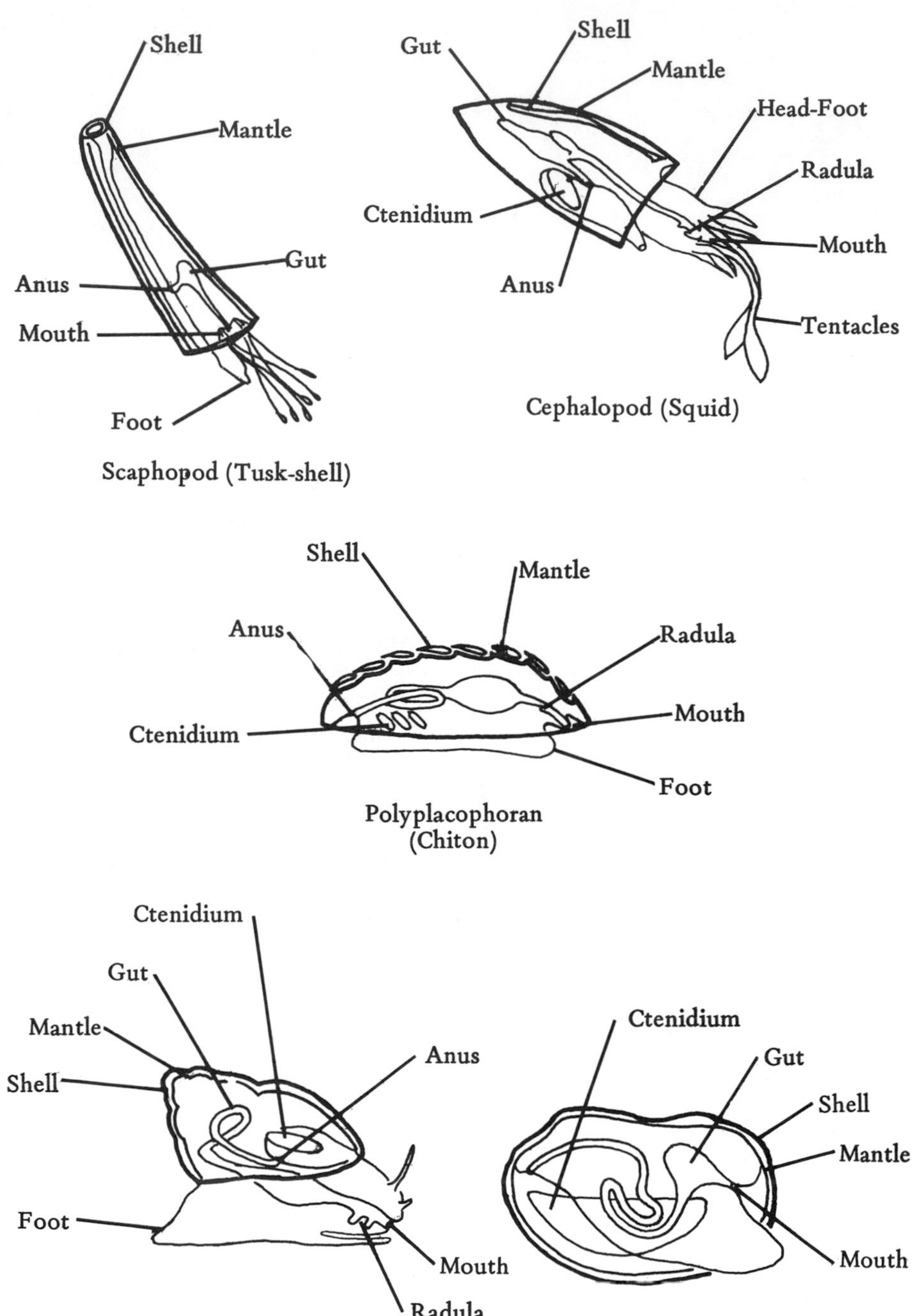

FIGURE 7.60 Phylum Mollusca. The major classes showing the shell and its relationships to the soft parts.

chitons it consists of eight overlapping valves. The shell is composed of aragonite in most cases, but calcite dominates in many clams. Mollusks are chiefly benthic, but several snails and many cephalopods are pelagic. Most are also marine, but snails and clams have invaded fresh water and the snails, the land.

Class Gastropodea—The snails (Fig. 7.60) and their shell-less relatives, the slugs, compose this largest of molluscan classes. Most have a head with tentacles, a flat foot for creeping, and the posterior of the body arched up over the back (and twisted) so the mantle cavity faces forward. Snails have tiny teeth they use in eating plants or animals. The shell (Fig. 7.62) is of one piece and lacks internal partitions. Most are conispirally coiled, but some are planispiral and some, not coiled at all. Snails range from Cambrian to Recent and have been at their peak from the Cretaceous onward.

Class Bivalvea or Pelecypodea—The clams (Fig. 7.60) have a laterally compressed, headless body and foot. The mantle cavity is large and extends forward from the posterior along the sides of the body. It contains the large gills which are the chief feeding organs, straining out microscopic food particles from water which enters and leaves by a pair of posterior siphons in most clams. Clams are typically burrowers though many are attached or free on the surface. The shell (Fig. 7.63) consists of two valves that hang down on either side of the body and which are mirror images of one another except in those that lie on one side like the scallops, oysters, and the extinct coral-shaped rudists. Clams range from Cambrian to Recent, but, like snails, they dominate Cretaceous and Cenozoic aquatic environments.

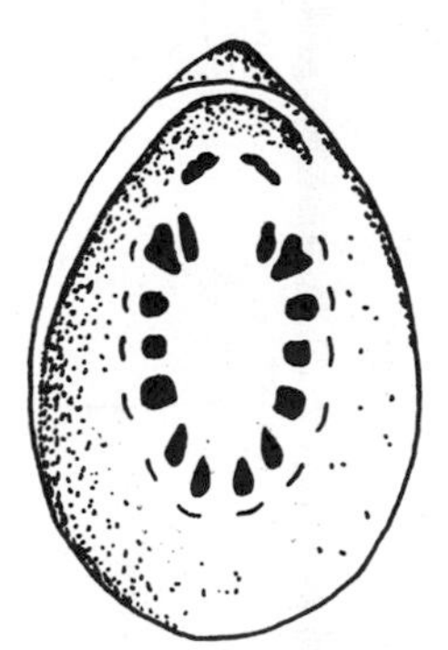

FIGURE 7.61 *Pilina,* **a Monoplacophoran Mollusk.** Exterior and interior views of fossil shell.

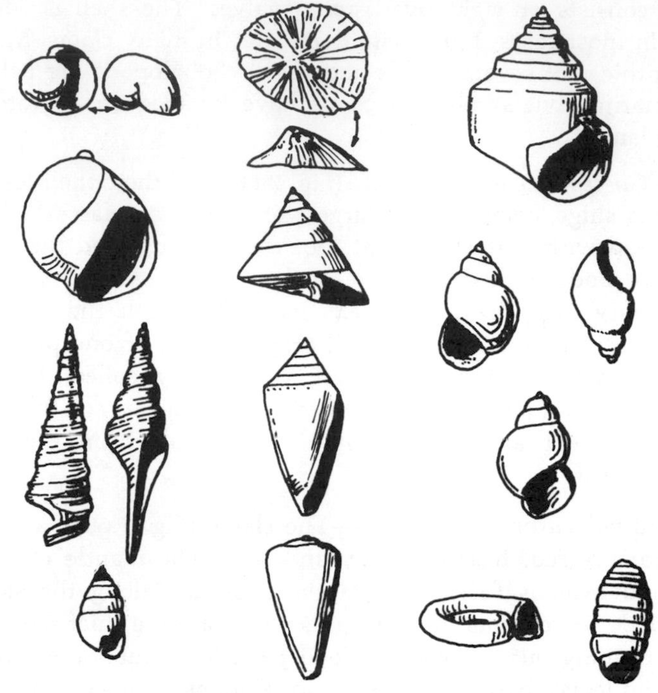

FIGURE 7.62 Variation Among Gastropod Shells.

FIGURE 7.63 Variation Among Bivalve Shells.

Class Scaphopodea—The curved, conical shells (Fig. 7.64) of this small group, the tusk or tooth shells, are open at both ends. The animal lives buried in the bottom with only the narrower top protruding. Water currents enter and leave this opening, tiny tentacles gather food at the other end. Scaphapods range from Ordovician to Recent and are common elements in Cenozoic faunas.

Class Cephalopodea—This group includes the living squids, cuttlefish, octopuses, and the pearly nautilus (Fig. 7.60) which is more like the host of fossil types. The cephalopod foot is modified to tentacles for seizing prey and a funnel used for jet-propelled locomotion. Cephalopods are chiefly active, pelagic mollusks and have highly developed nervous and circulatory systems. Most have a one-piece shell divided into a series of internal chambers by cross-partitions called septa (Fig. 7.65). The lines of junction between septa and shell are called sutures and their structure is the basis for classifying most cephalopods. The shell was external in most fossils, as it is in the modern pearly nautilus, but it has become internal and is reduced or lacking in the other modern types. The shells are typically either straight or planispirally coiled, though some are bizarrely coiled. The fossil record extends from Cambrian to Recent and the group has been abundant and useful in many of the intervening periods because of its nektonic mode of life which often transcends facies control.

Subclass Nautiloidia—In nautiloids, the suture (Fig. 7.66), when unrolled, forms a straight or slightly wavy line in all but a few modern types such as the living nautilus. The shell (Fig. 7.67) is straight, curved, or planispirally

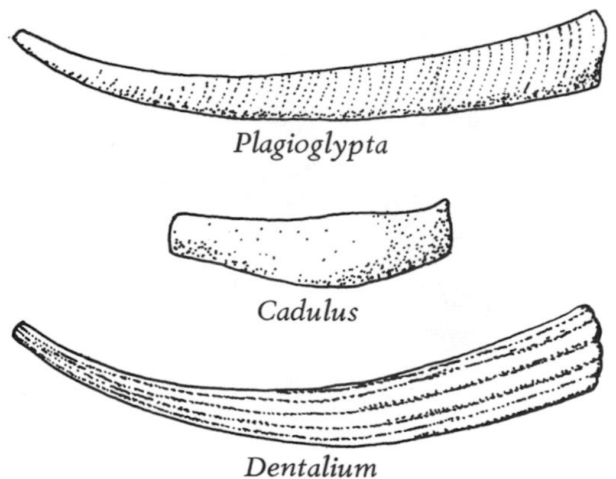

FIGURE 7.64 Class Scaphopodea. Representative fossil tusk or tooth shells.

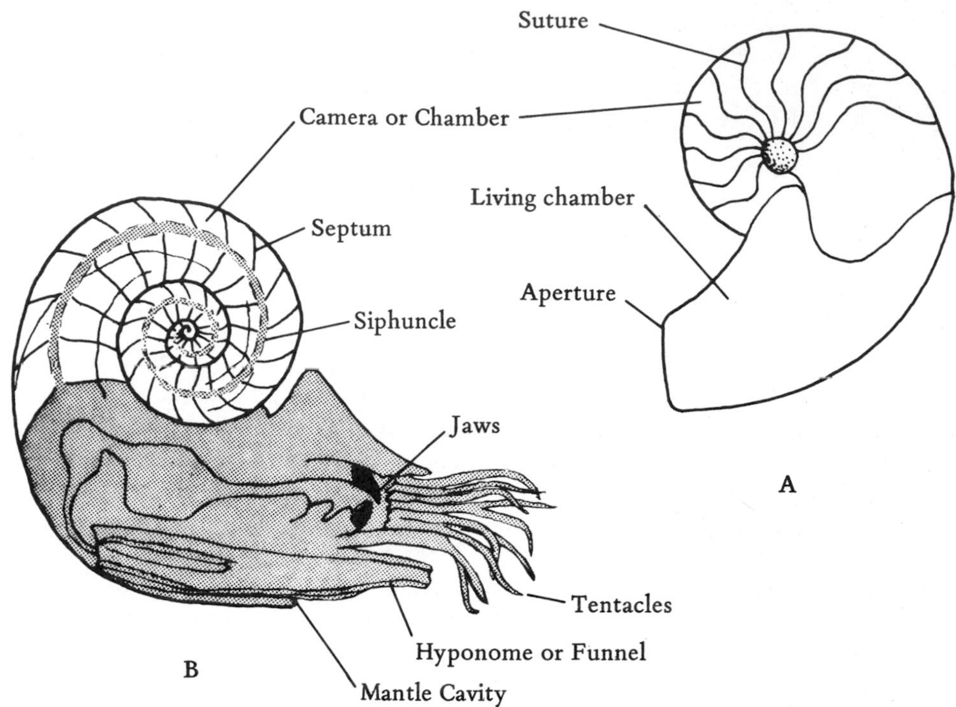

FIGURE 7.65 The Living Pearly *Nautilus.* (A) Internal mold of shell, (B) Section through living organism.

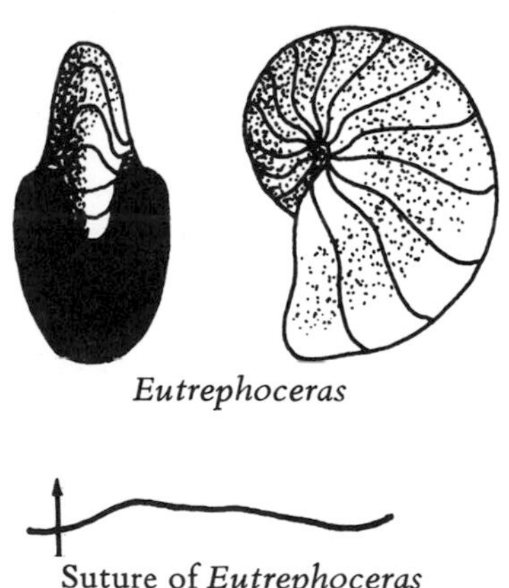

FIGURE 7.66 Coiled Nautiloid Shell and Nautiloid Suture.

coiled in nearly all members. The group ranges from Cambrian to Recent, but reached a peak in the Ordovician.

Subclass Coleoidia—This group includes nearly all living cephalopods and is characterized by an internal, reduced, or lost shell. In the early fossil types, however, a heavy shell was formed. Coleoids were derived from straight nautiloids.

Order Belemnitoida—In belemnoids (Fig. 7.68), the chambered part of the shell is reduced, but the outer walls are tremendously thickened to form a heavy cigar-shaped guard or rostrum. Belemnites range from Carboniferous through Paleogene, but are abundant only in the Jurassic and Cretaceous. A similar and closely related group, the *aulacoceroids,* is common in the Triassic.

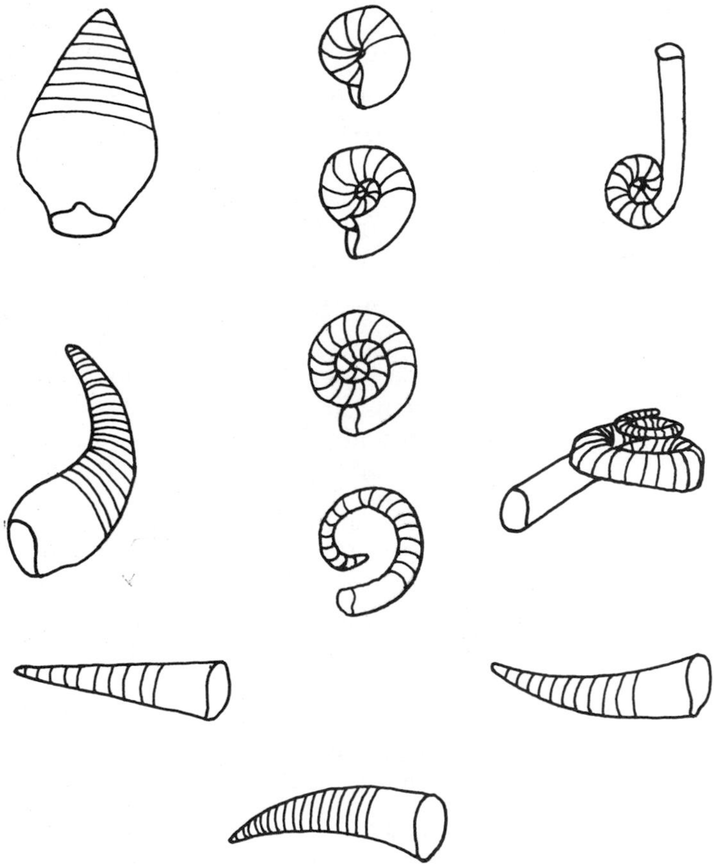

FIGURE 7.67 Variation Among Fossil Nautiloid Shells.

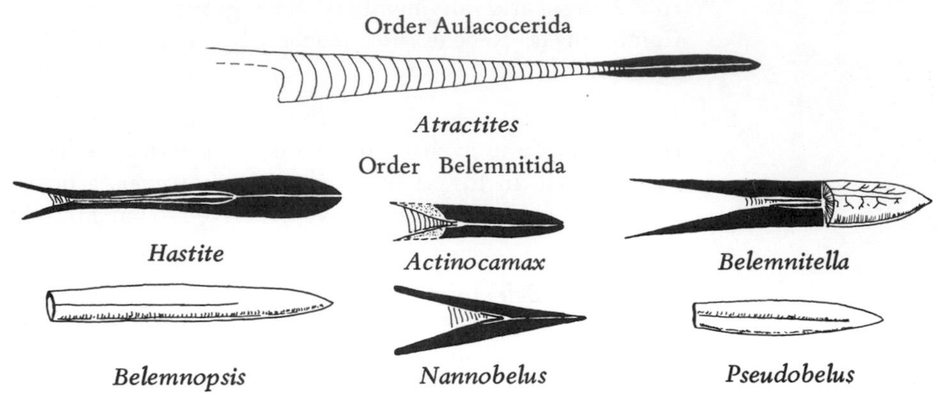

FIGURE 7.68 Shells of Fossil Belemnoids and a Related Aulacoceroid (*Atractites*).

Subclass Ammonoidia—The extinct ammonoids are the most geologically significant cephalopods. They have complex sutures and chiefly planispiral shells (Fig. 7.69). They range from Devonian through Cretaceous.

Order Goniatitoida—These chiefly have goniatite sutures (Fig. 7.70), that is the flexures toward the aperture (saddles) and those away from the aperture (lobes) are not crenulate. They range from Devonian through earliest Triassic with a Carboniferous and Permian peak.

Order Ceratitoida—These chiefly have ceratite sutures (Fig. 7.70), that is the lobes, but not the saddles, are crenulate. They range from Devonian through Triassic, though ceratite sutures did not appear until the Mississippian. Most are Triassic in age.

Order Ammonitoida—These chiefly have ammonite sutures (Fig. 7.70), that is both lobes and saddles are crenulate. Most are highly ornamented. Some, called *lytocerines,* have very bizarre patterns of coiling or uncoil (Fig. 7.71). They range from Triassic through Cretaceous and have a Jurassic-Cretaceous peak. (Ammonitic sutures appear in the Permian in advanced members of the other orders.)

?Class Hyolithea—This is another extinct group of uncertain relationships. Hyolithans have a three-sided, conical, calcite shell (Fig. 7.72). They range from Cambrian through Permian in age but are most abundant in the Cambrian.

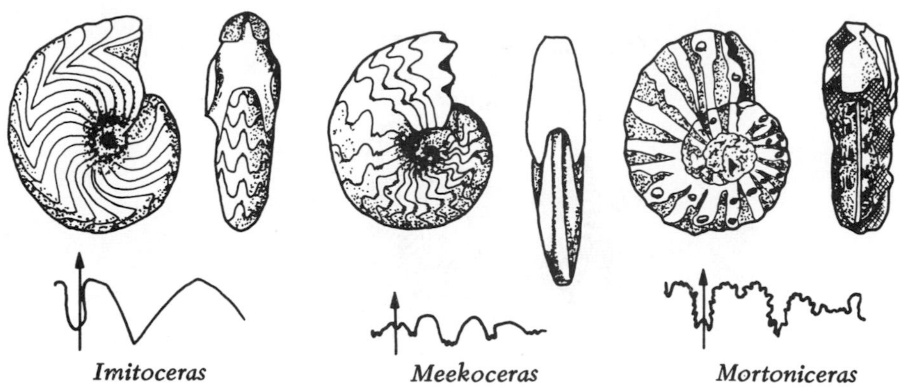

FIGURE 7.69 Ammonoids of Each of the Three Orders showing shell shape and sutures.

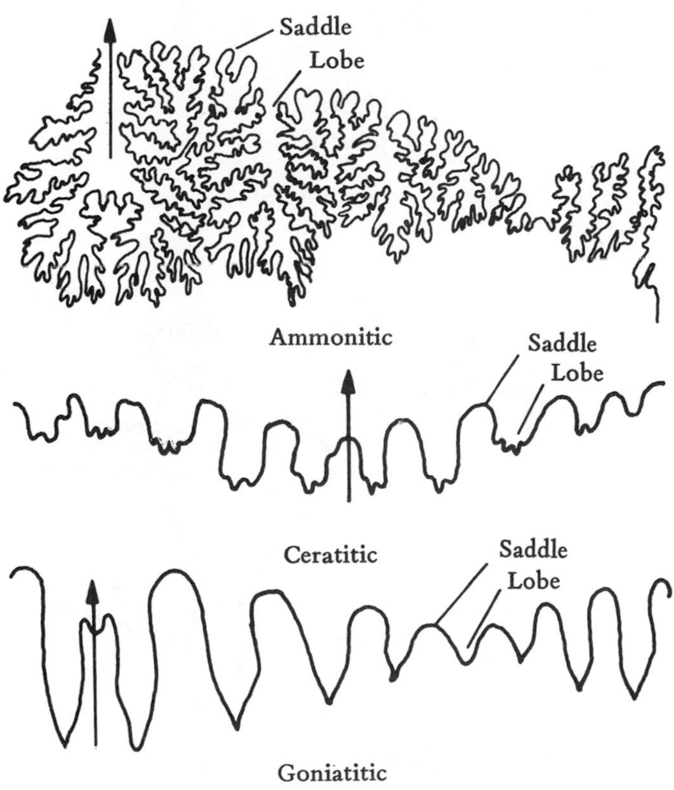

FIGURE 7.70 Types of Ammonoid Sutures.

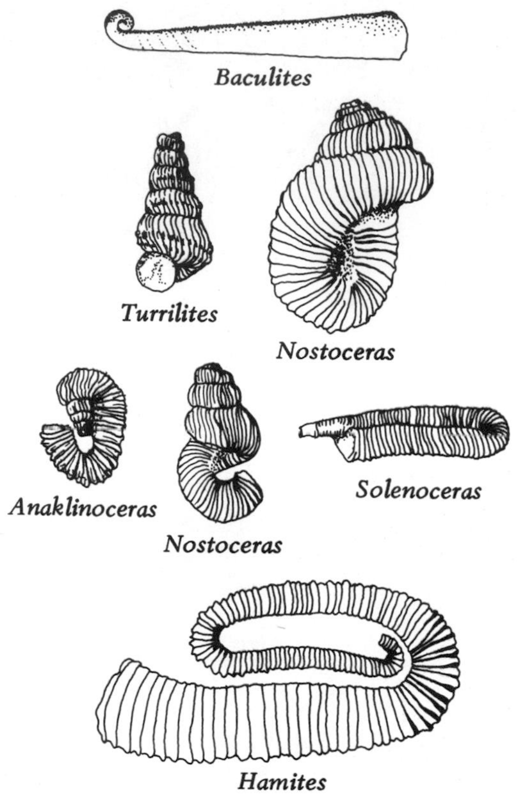

FIGURE 7.71 Lytocerine Ammonoid Shells.

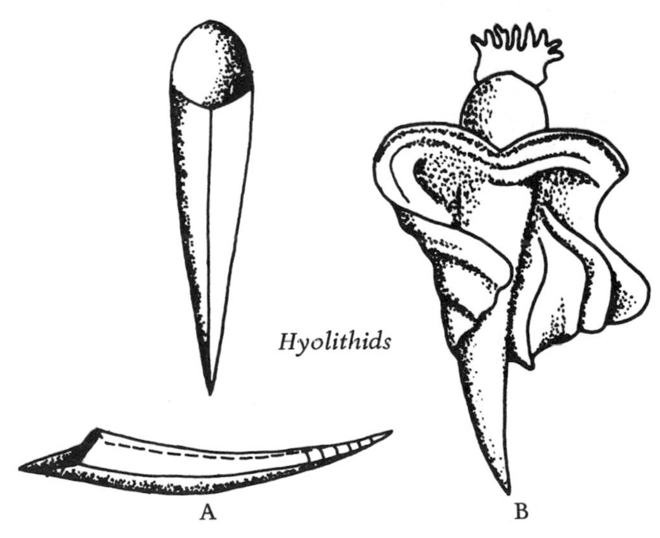

FIGURE 7.72 Class Hyolithea (A) Fossil shells, (B) Restoration.

?Class Tentaculitea—Still another extinct group, the tentaculites have conical ringed shells (Fig. 7.73) of circular cross-section. They range in age from Ordovician through Devonian and are common throughout this interval.

Phylum Annulata—The segmented worms or annelids (Fig. 7.74) have a body characterized by the repetition of similar parts in a series. Only the marine group, the polychaetes, have left much of a record. The mobile forms have few hard parts except for their chitinous jaws or scolecodonts (Fig. 7.75). Consequently their fossil record consists of some body impressions and occasionally common scolecodonts. Many of the attached or burrowing forms build tubes (Fig. 7.76) which are typically agglutinated, but some build calcite or aragonite tubes and these are common fossils. Annelids range from Eocambrian to Recent with identifiable tubes appearing in the Cambrian and scolecodonts, in the Ordovician.

Phylum Arthropoda—This is the largest phylum of living organisms. The body is segmented, suggesting an annelid relationship, but there is a fusion and specialization of groups of segments in arthropods giving rise to a regionally differentiated body (Fig. 7.77). Paired, segmented appendages are typically present on most segments and those are modified for numerous functions such as sensing, feeding, walking, and reproducing. They are also specialized on various regions of the body. Being active creatures, arthropods have a well-developed nervous system. The body is covered by a chitin exoskeleton which covers the outside of the body and the appendages. In some groups, it is mineralized with calcite and/or calcium phosphate which greatly im-

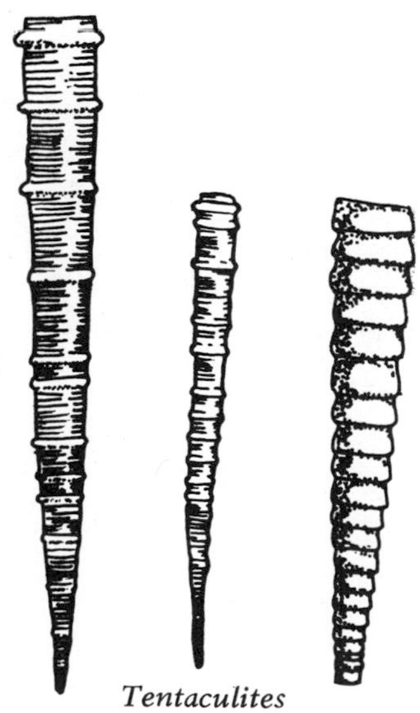

FIGURE 7.73 Class Tentaculitea.

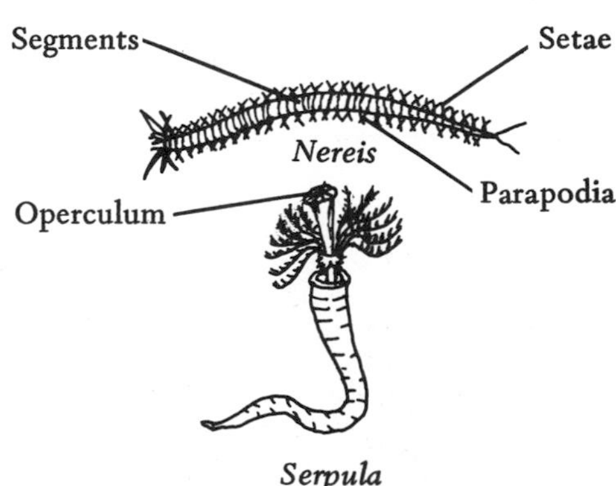

FIGURE 7.74 Living Polychaete Annelids.

proves its chances of fossilization. Because an arthropod has an outside skeleton, it must shed it periodically in order to grow and this means an individual may leave several fossils. The fossil record ranges from Cambrian (possibly Eocambrian) to Recent with trilobites dominating the Lower Paleozoic and crustaceans taking over thereafter, with insects spasmodically common. Arthropods live in marine, fresh water, and terrestrial environments with the greatest numbers living in the last, but the fossil record biased toward the first.

Subphylum Trilobitazoa—Trilobites and some related forms are placed here.

Class Trilobitea—Trilobites (Fig. 7.78) are an extinct group of arthropods whose members are longitudinally trilobed, with an axis and two pleurae on either side. Their body is also transversely specialized into a cephalon or head, a thorax or body, and a pygidium or abdomen. The first and last consist of fused segments and are the commonest fossils. The axial part of the cephalon is the glabella, the pleural parts are the genae or cheeks. The cheeks are usually crossed by a facial suture along which the trilobite splits its exoskeleton when molting. The

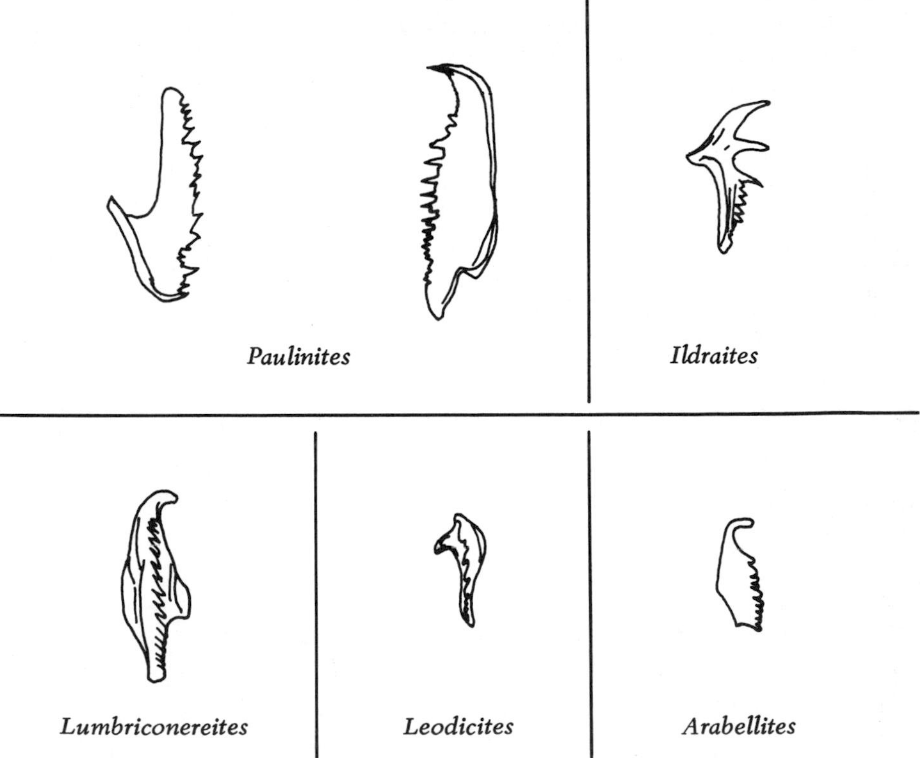

FIGURE 7.75 **Scolecodonts** or fossil worm jaws.

first pair of appendanges are antennae, the remainder are all alike (except in size) with one branch for walking and one which bears the gill. Trilobites were all marine and chiefly benthic. Lacking jaws, they were probably scavengers and detritus feeders. Trilobites range from Cambrian through Permian but were most abundant in the Cambrian and Ordovician. Several different kinds were important.

Subclass Opisthoparia—This, the largest group of trilobites, is characterized by sutures (Fig. 7.79) that intersect the back of the head on the inside of the back corners (genal angles). They range from Cambrian through Permian.

Order Redlichioida—These strictly Cambrian forms (Fig. 7.80A) have large semicircular cephalons, large crescentic eyes, genal spines, numerous thoracic segments, and a small pygidium. Most have opisthoparian sutures, but they atrophied in one group.

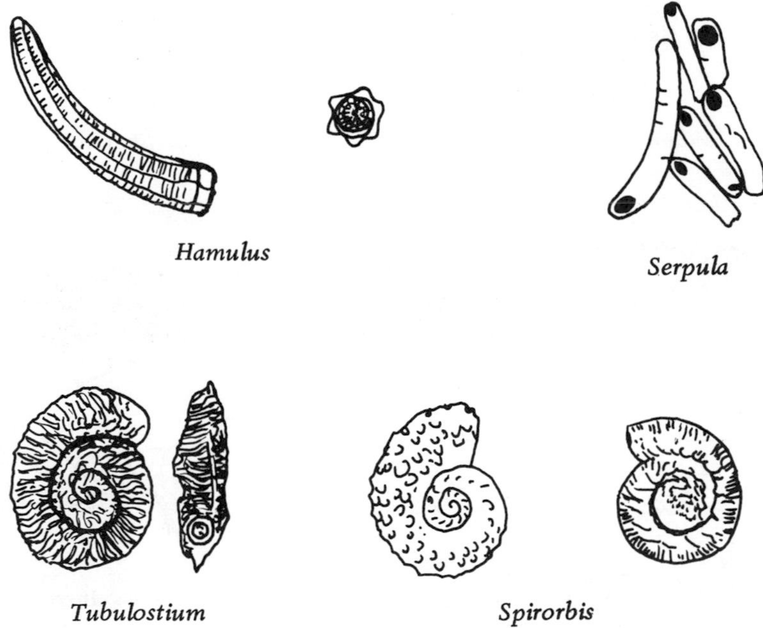

FIGURE 7.76 Fossil Annelid Worm Tubes.

Order Corynexochoida—Another group restricted to the Cambrian, corynexochoids (Fig. 7.80B) have an elongate, subelliptical body, genal spines (in most), and a large pygidium.

Order Ptychoparioida—Ptychoparioids (Fig. 7.80C) are found in Cambrian

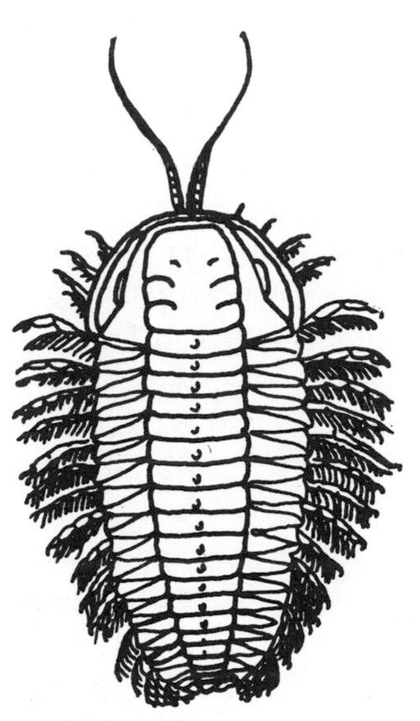

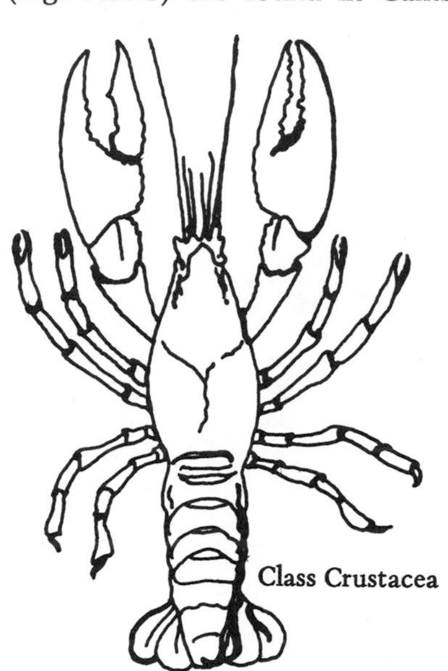

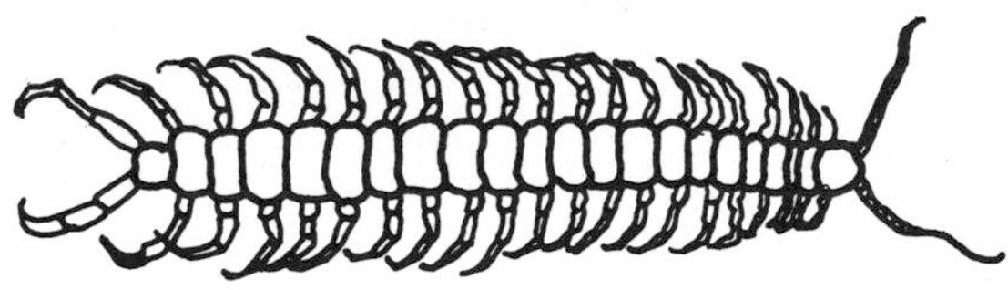

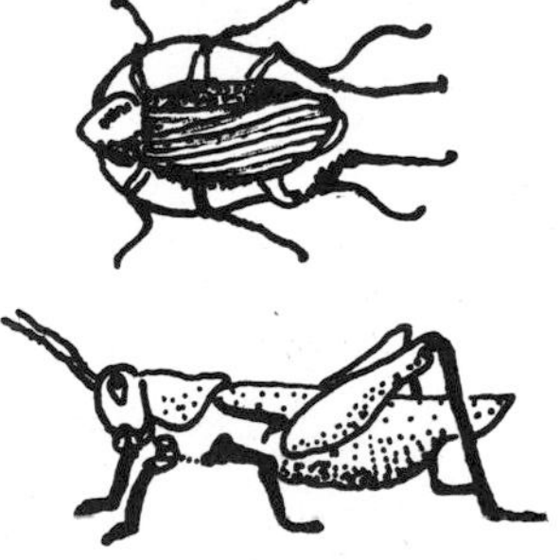

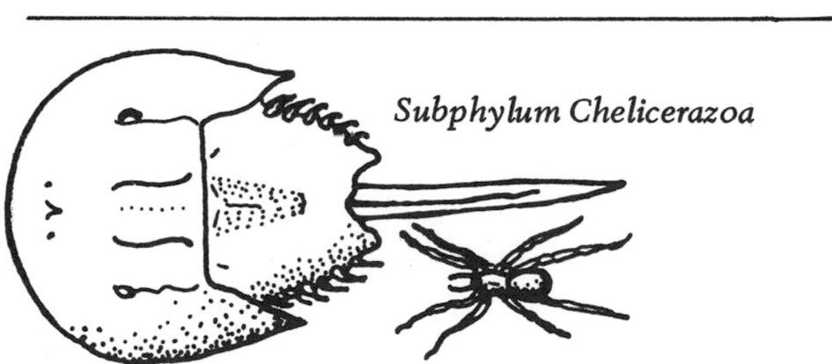

FIGURE 7.77 Phylum Arthropoda. Representatives of the three major subphyla.

and Ordovician rocks and are common in both. They are very generalized trilobites with a large preglabellar area and a moderate pygidium.

Order Asaphoida—Asaphoids (Fig. 7.80 D) have large, smooth, subequal, subtriangular cephalons and pygidia. The eyes are tuberculate. Asaphoids range from Cambrian through Ordovician, but most are Ordovician in age.

Order Illaenoida—Illaenoids (Fig. 7.80E) have a large, nearly smooth glabella reaching almost to the front of the cephalon and a large pygidium. They are Cambrian through Permian in age, but are most common in Ordovician and Silurian rocks.

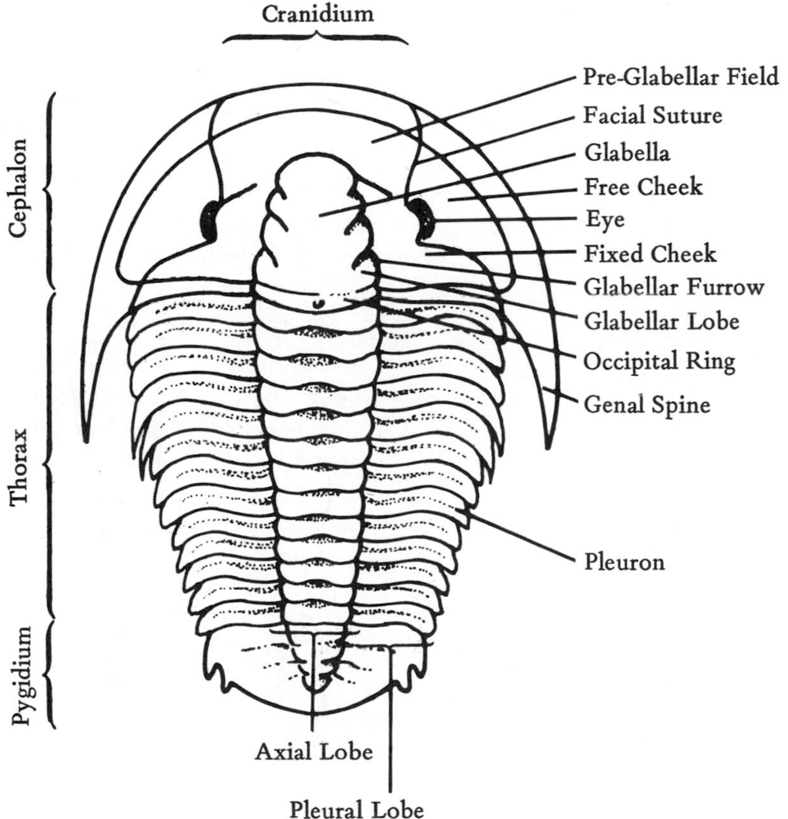

FIGURE 7.78 Trilobite Morphology. Dorsal view.

Order Trinucleoida—The suture has become marginal in these small trilobites (Fig. 7.80F) with a large cephalon. The most distinctive features are the pitted anterior of the cephalon and the high glabella. The group occurs in both Ordovician and Silurian rocks, but peaks in the former.

Subclass Proparia—Proparian trilobites have sutures (Fig. 7.79) that end posteriorly either at the genal angle or outside it. They are Ordovician through Devonian in age.

Order Cheiruroida—These (Fig. 7.80G) have a glabella that expands anteriorly, prominent genal spines, sutures that end outside the genal angle, and a small pygidium. They range from Ordovician through Devonian with an Ordovician peak.

Order Phacopoida—Phacopoids (Fig. 7.80 H) have large multifaceted eyes, a very large gabella expanding anteriorly, sutures that end outside the genal angle, and a moderate to large pygidium. They are Ordovician through Devonian in age, but are most common in the Devonian.

Order Calymenoida—Calymenoids (Fig. 7.80I) have a deeply furrowed glabella that tapers anteriorly, sutures that end at the genal angle, no genal spines, and a moderate to large pygidium. Of Ordovician through Devonian age, they peak in the Silurian.

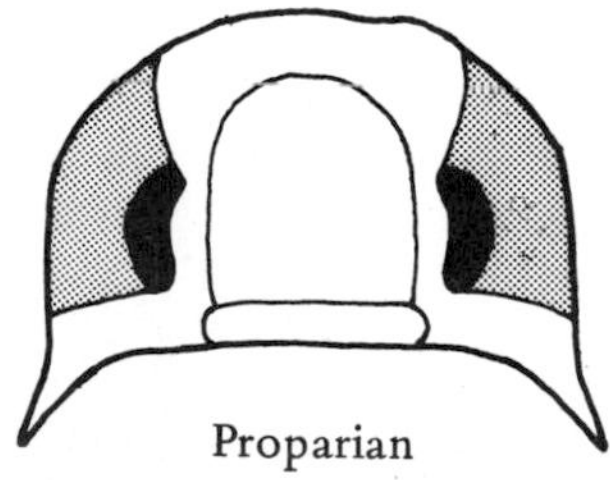

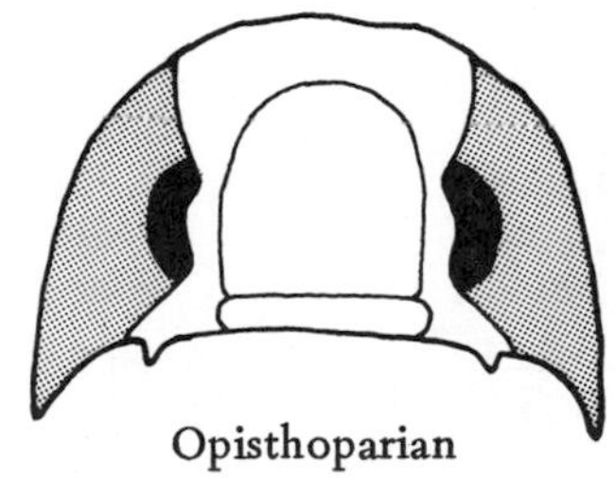

FIGURE 7.79 Trilobite Facial Sutures. The two major varieties.

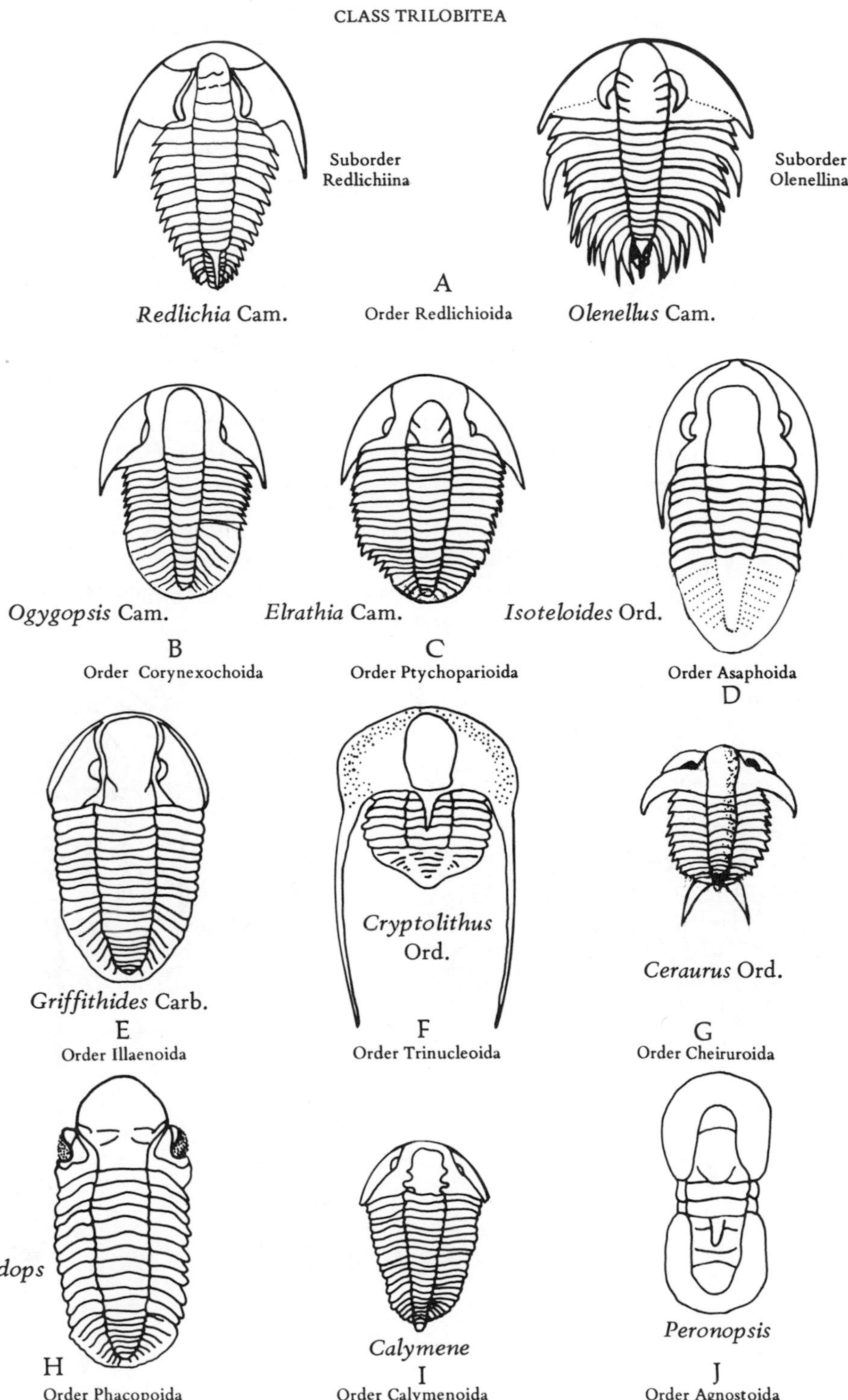

FIGURE 7.80 Representatives of the Major Trilobite Orders.

Subclass Agnostia—These are tiny trilobites (Fig. 7.80J) lacking eyes and, typically, facial sutures also. The cephalon and pygidium are subequal and the very short thorax consists of only two or three segments. They are Cambrian and Ordovician in age, but most are Cambrian.

Order Agnostoida—These are typical forms as described above.

Subphylum Chelicerazoa—Members of this small subphylum are characterized by a chelate (pincer-like) first pair of appendages, no antennae, modified second appendages called pedipalps, and a two-fold body subdivision into cephalothorax and abdomen. They are both aquatic and terrestrial and range from Cambrian to Recent.

Class Merostomea—The aquatic chelicerates have the remaining cephalothoracic appendages modified as walking legs with jaws and the abdominal appendages bear gills. The tail is an elongate, pointed telson.

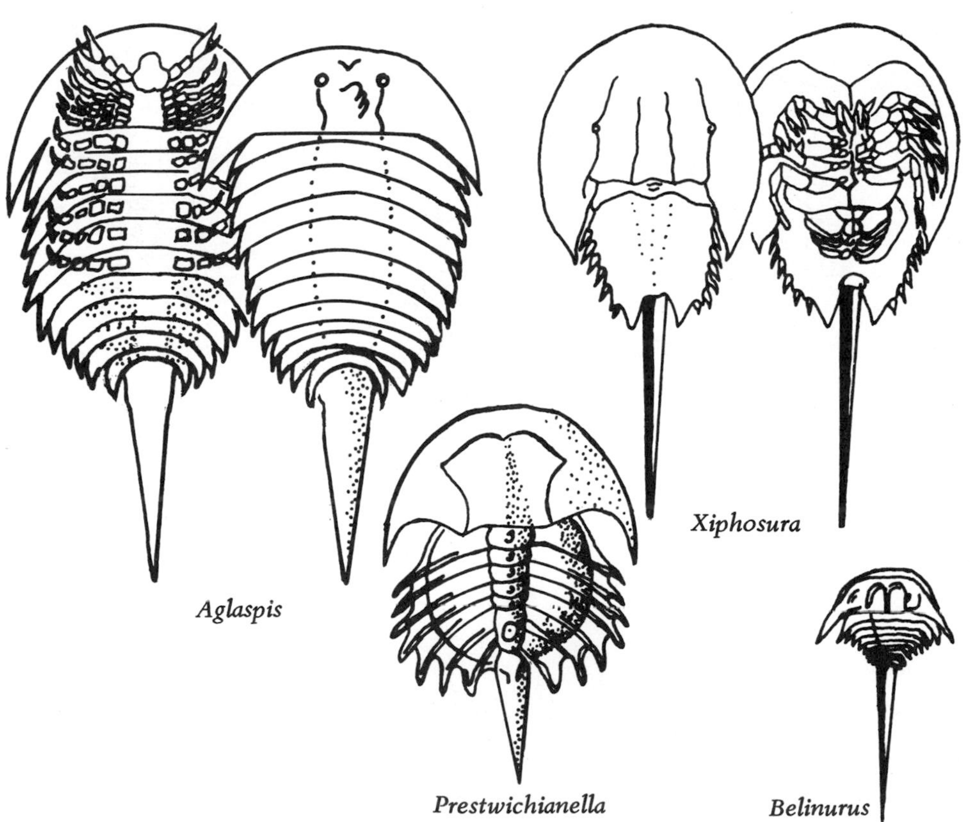

FIGURE 7.81 Fossil Xiphosurian Merostomes.

Subclass Xiphosuria—The horseshoe crabs and extinct aglaspoids (Fig. 7.81) are marine and have a trilobed cephalothorax. They range from Cambrian to Recent.

Subclass Eurypteria—The extinct eurypterans or "giant sea scorpions" (Fig. 7.82) have a very long abdomen, narrower posteriorly than anteriorly, and the last cephalothoracic appendages modified as swimming paddles. They were brackish to fresh water in habitat and lived from Ordovician through Permian, but were only common in the Silurian.

Class Arachnea—The terrestrial chelicerates, including the scorpions, false scorpions, daddy-long-legs, ticks, mites, and spiders (Fig. 7.83), lack abdominal appendages and breathe air by lungs or tracheal tubes. Never abundant in their Silurian to Recent range, they are occasionally common such as in Pennsylvanian coal beds and Oligocene ambers.

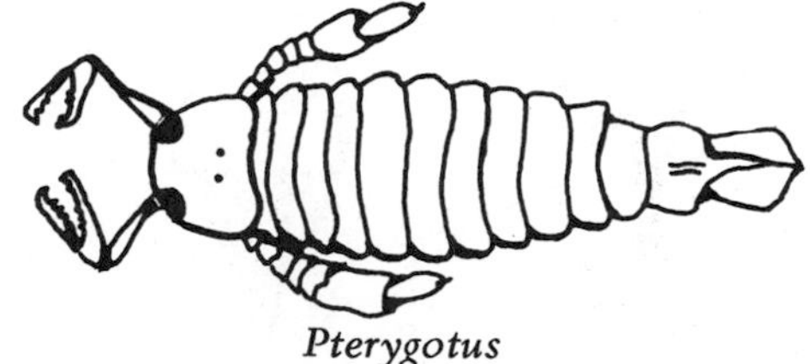

Pterygotus

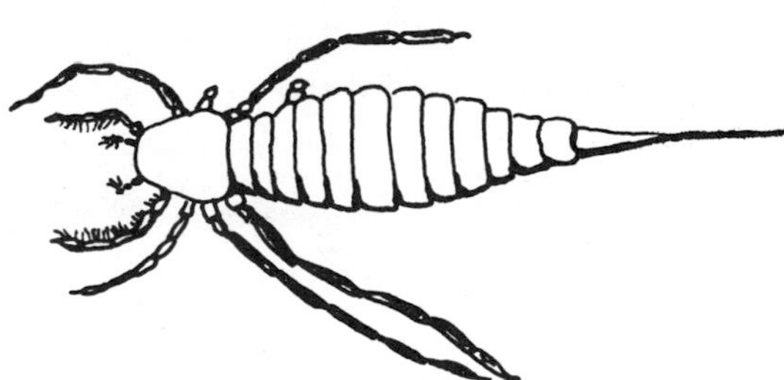

Stylonurus

Carcinosoma

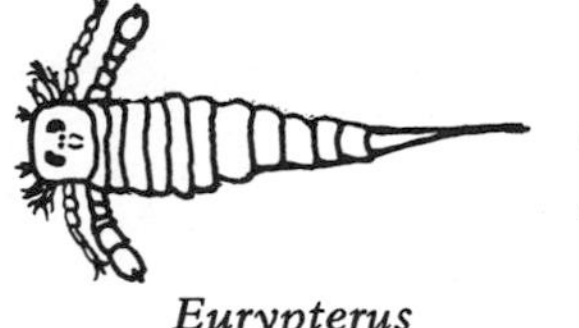

Eurypterus

Eurypterus

FIGURE 7.82 Eurypterans.

Subphylum Mandibulazoa—This is the largest arthropod subphylum and encompasses the crustaceans, millipedes, centipedes, and insects. The first pair of appendages are sensory antennae and the third pair are jaws or mandibles followed by two pair of accessory feeding appendages, the maxillae. The geologic range is Cambrian to Recent.

Class Crustacea—Crustaceans are the primarily aquatic mandibulates. They possess two pairs of antennae and most have a shield or carapace over much of the body. The carapace is mineralized with calcite and/or phosphate in those groups with a significant fossil record. This Cambrian to Recent assemblage is very diverse, but only four groups have left a good fossil record.

Subclass Branchiopodia—This group includes the fairy shrimps, tadpole shrimps, clam shrimps,

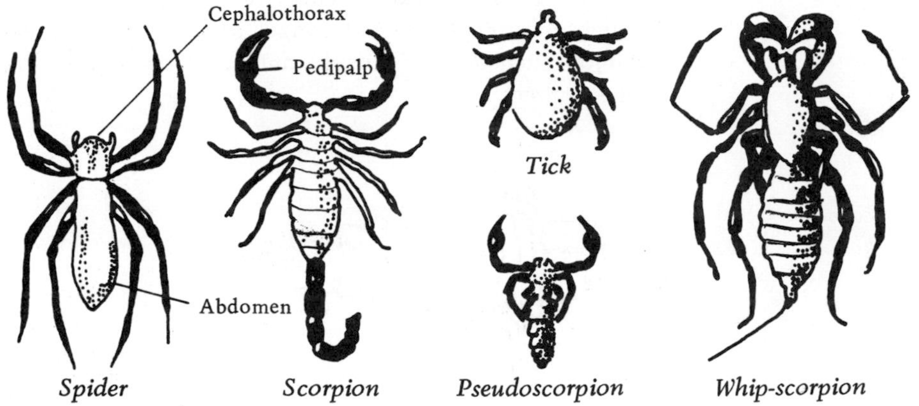

FIGURE 7.83 Arachnean Chelicerates.

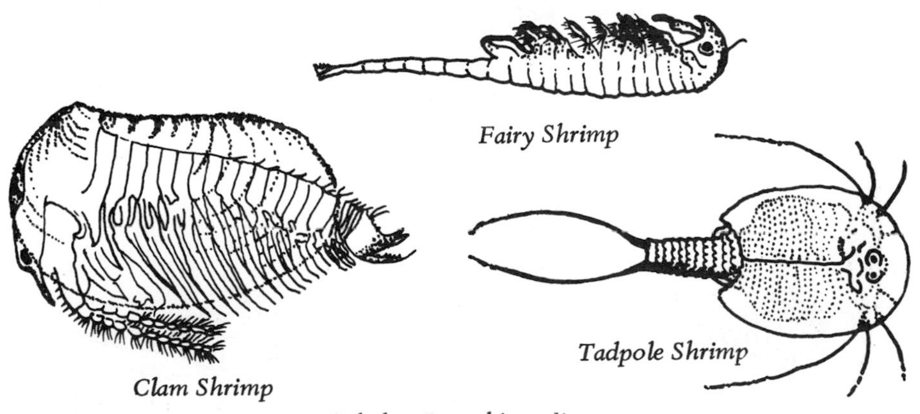

FIGURE 7.84 Branchiopod Crustaceans.

and water fleas (Fig. 7.84). They have large, leaflike gills on the trunk appendages. The clam shrimps have small, bivalved, clam-like carapaces (Fig. 7.85) with growth lines, a feature lacking in other arthropods which molt their shells. They live in brackish to fresh water and range from the Devonian to Recent, though only common in times of abundant non-marine depository such as the Carboniferous and Triassic.

Subclass Ostracodia—The most significant crustaceans to the geologist, the mussel shrimps, have a highly calcified bivalved carapace (Fig. 7.86). The carapace shape and ornamentation are very diverse and there are no growth lines (Fig. 7.87). They are both marine and fresh

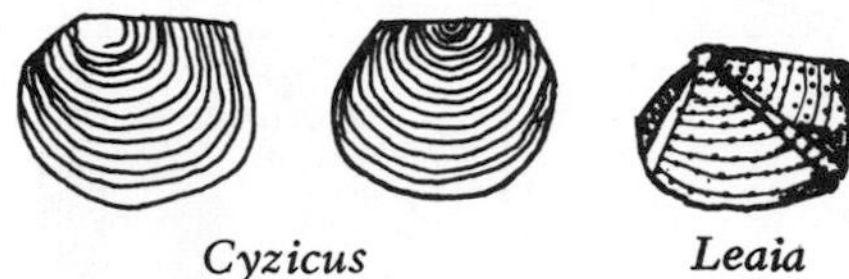

FIGURE 7.85 Fossil Clam Shrimp.

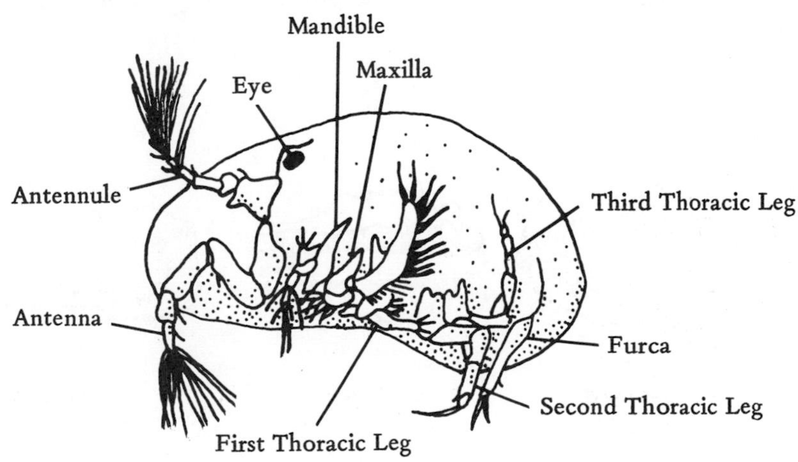

FIGURE 7.86 Ostracod Anatomy.

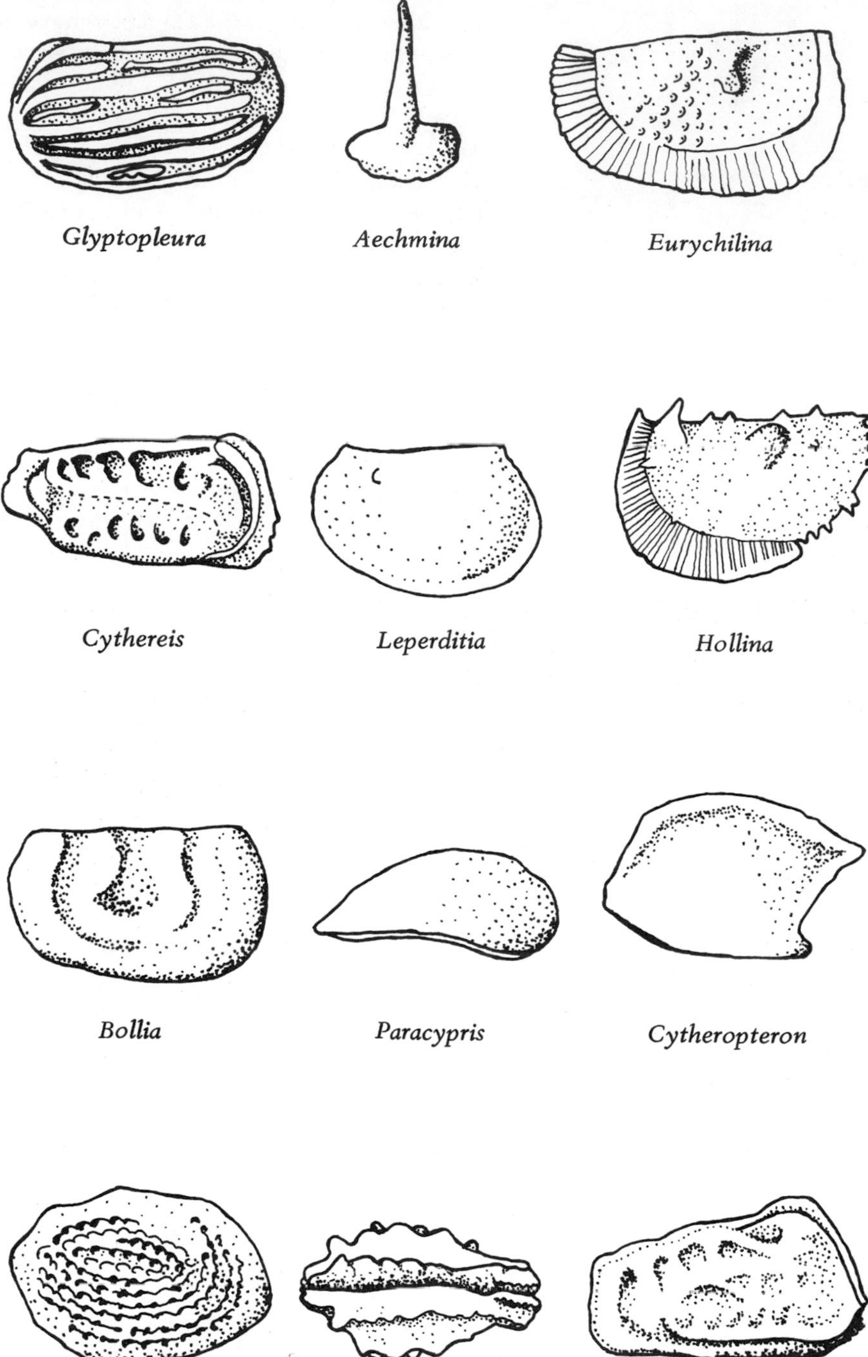

FIGURE 7.87 Fossil Ostracod Carapaces.

water. They range from Cambrian to Recent and have been abundant from the Ordovician onward.

Subclass Cirripedia—The barnacles (Fig. 7.88) have a shrimp-like body concealed in a cluster of overlapping calcite plates which are not shed and which are fixed to the substrate. They are commonly shaped like a tiny volcano or have a stalk. They range from Silurian to Recent and are particularly large and common in the Neogene.

Subclass Malacostracia—There are the large, familiar crustaceans, the shrimp, lobsters, and crabs, and the smaller pill bugs and sand fleas (Fig. 7.89). Distinguished by its segmentation pattern, the group as a whole does not have highly mineralized shells which makes whole fossils uncommon. Fragments of appendages and carapaces are common, particularly in younger

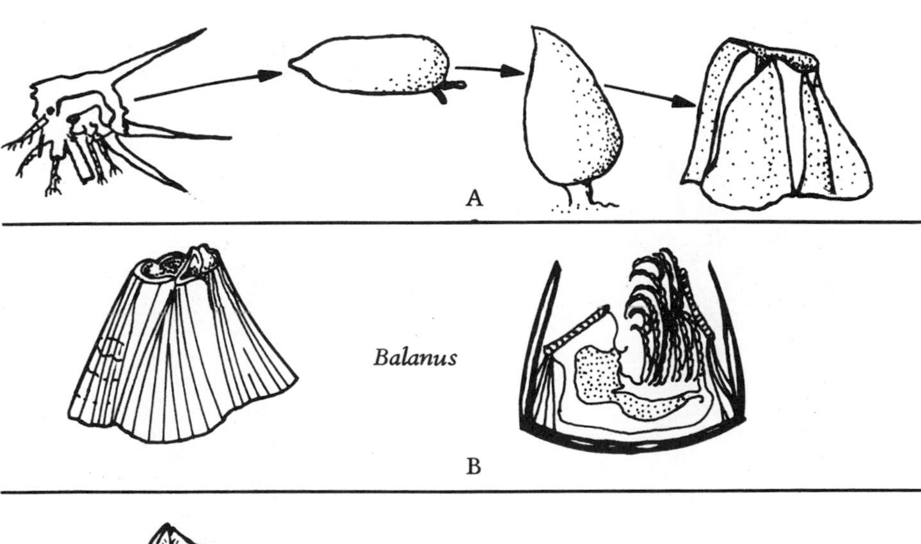

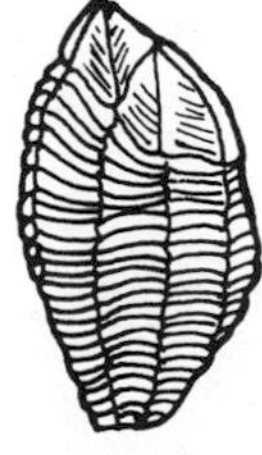

FIGURE 7.88 Subclass Cirripedia. (A) Development of a barnacle. (B) The common genus *Balanus* with cut-away showing soft parts. (C) Fossil barnacle carapaces.

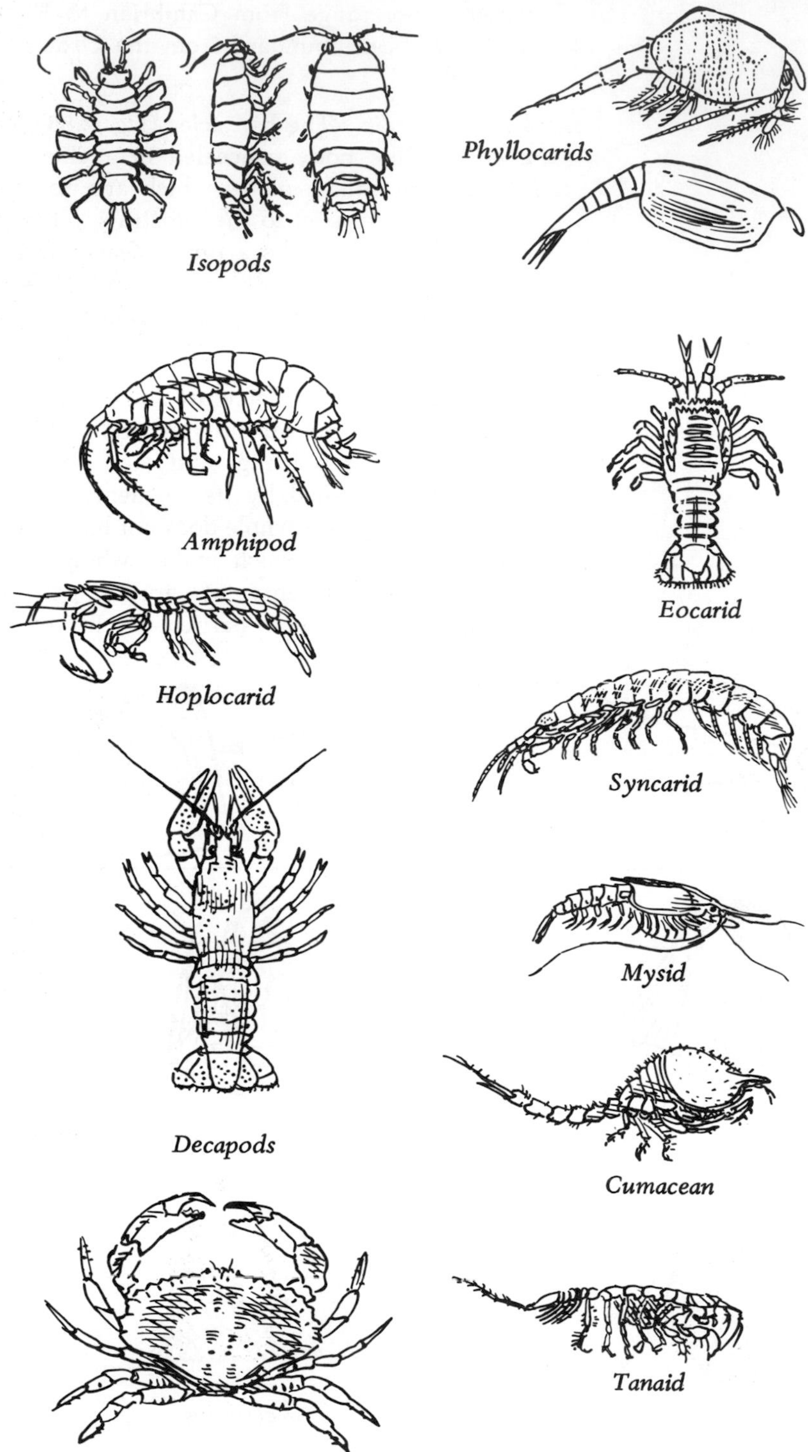

FIGURE 7.89 **Major Groups of Malacostracan Crustaceans** with a Fossil Record.

rocks. They are both marine and fresh water and range from Cambrian to Recent.

Class Insectea—The insects (Fig. 7.90) are by far the largest living group of arthropods and, although they are also the most numerous fossil arthropods, the fossil record displays only a tiny fraction of the whole group. Insects have a single pair of antennae, no abdominal appendages, three pairs of walking legs on the thorax, and typically two pairs of wings on the thorax. It is because of

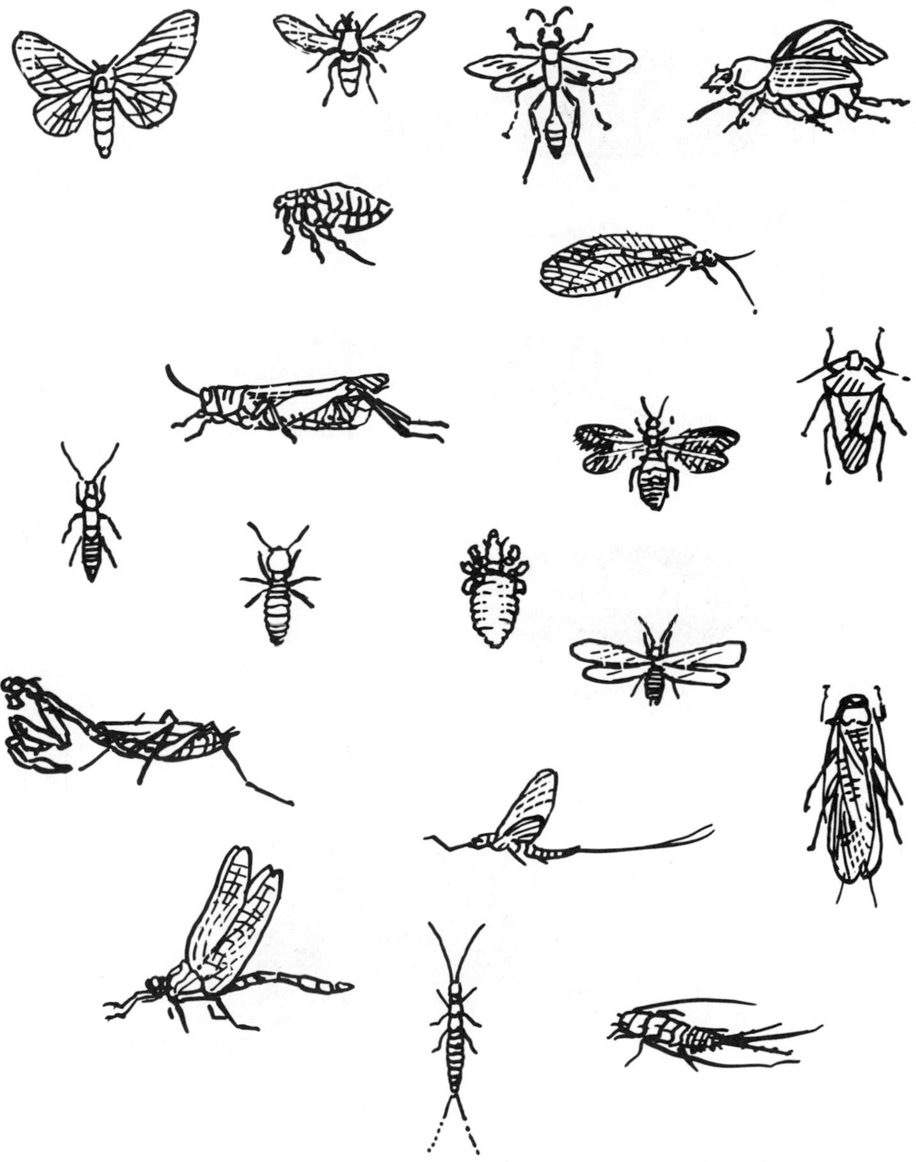

FIGURE 7.90 Insects of the Major Living Orders.

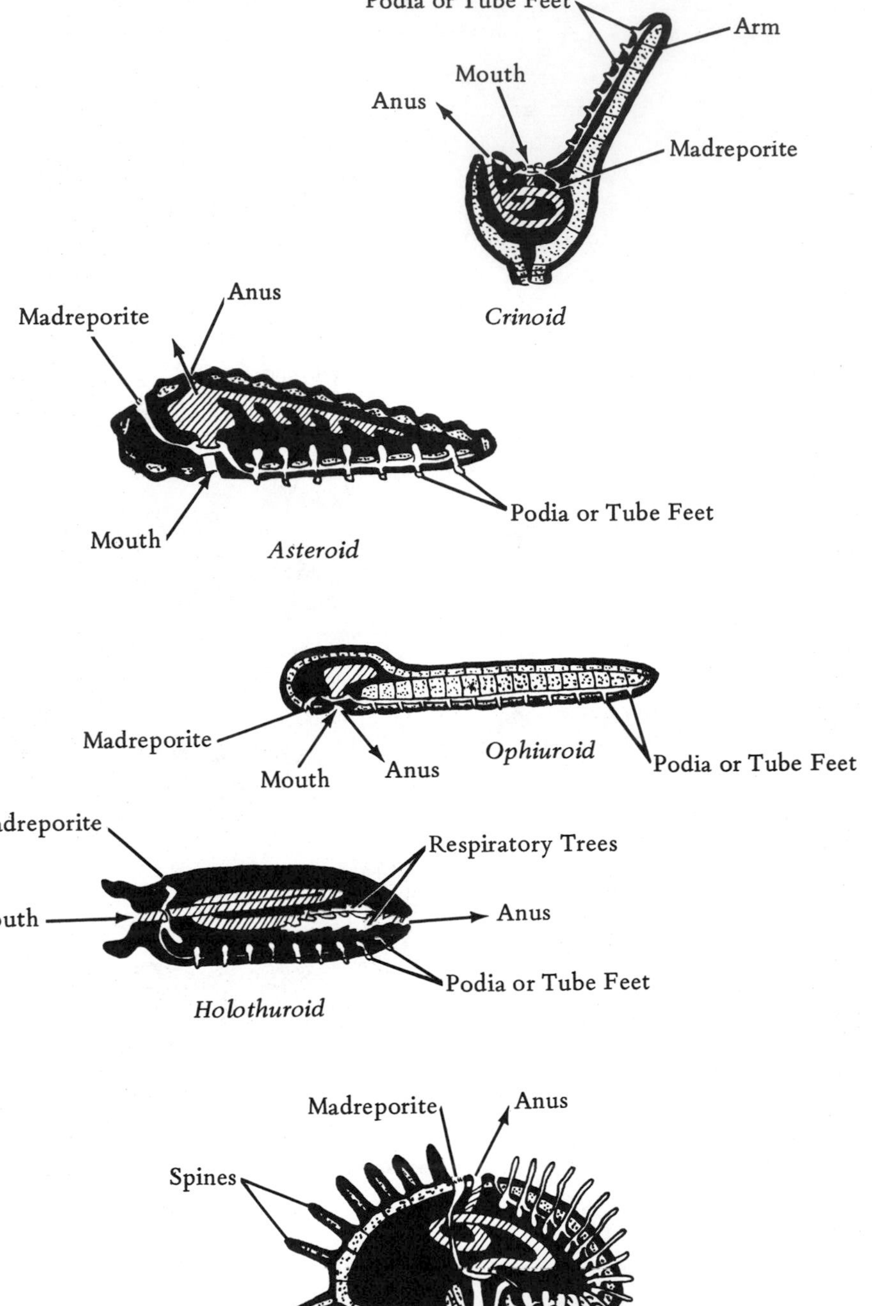

FIGURE 7.91 Internal Anatomy of Major Echinoderm Groups. Note water vascular system.

the aerial habitats and consequently light and poorly mineralized skeleton that the fossil record is chiefly limited to fortuitous circumstances such as swamp, lake, and amber beds. Insects range from Devonian to Recent, with the first flying types known in the Carboniferous. Peaks occur in the Carboniferous, Permian, and Oligocene.

Phylum Echinodermata—Echinoderms are unique among the more complex animals in having a radial symmetry. This symmetry is pentamerous or fivefold and imperfect. They also have a unique series of internal canals called the water vascular system (Fig. 7.91) which operates a series of tube feet used in locomotion, feeding, and respiration. The canals which radiate outward define regions called ambulacra; intervening areas are the interambulacra. The body of nearly all echinoderms possesses an internal calcite test which has a unique meshwork microstructure (Fig. 7.92). All are marine and most are benthic. The fossil record goes back to the Cambrian (possibly Eocambrian) and embraces a wealth of extinct forms surpassing any other phylum. Many are rare such as the chiefly bilateral haplozoans and the homalozoans or *carpoids* (Fig. 7.93) which may link echinoderms and fish.

Subphylum Crinozoa—These are stalked echinoderms with arm-like appendages for gathering microscopic food (Fig. 7.94). They range from Cambrian to Recent but are most abundant in the Paleozoic.

Class Cystoidea—The eocrinoids and cystoids (Fig. 7.95) have the plates of their bodies or thecae perforated by pores or slits. Thecal plates are irregularly arranged in many, but some have a nearly regular pattern. The ambulacra of most are attached to the top and sides of the theca and bear slender food-gathering brachioles. The cystoid-eocrinoid group ranges from Cambrian through Devonian and is occasionally common in the Ordovician and Silurian.

Class Blastoidea—Blastoids (Fig. 7.96) resemble a rose bud with their few, regularly arranged plates. The prominent ambulacra bear brachioles as in cystoids, but blastoids have a unique series of folds (hydrospires) beneath the

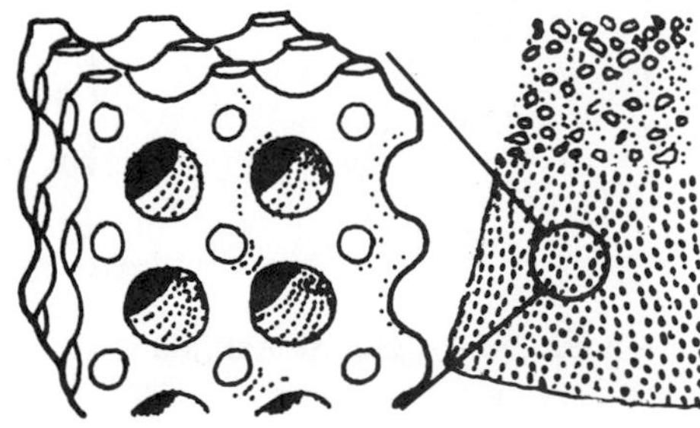

FIGURE 7.92 Echinoderm Skeleton Microstructure.

ambs which are connected to pores or slits along the sides of the ambs and larger openings at the top. Blastoids range from Silurian through Permian, but reach their maximum in the Mississippian.

Class Crinoidea—Called sea-lilies or feather stars, the crinoids (Fig. 7.97) are the only modern survivors of the subphylum. Their body or calyx has regularly arranged plates lacking pores or silts. The ambulacra are not attached, but extend upward and are called arms. The delicate structures attached to them are now called pinnules. The group ranges from Cambrian to Recent, but was most abundant in the Paleozoic, particularly the Mississippian (Fig. 18.2). Paleozoic types were shallow water benthos, but later forms gradually became deep water or pelagic.

Subphylum Asterozoa—The common sea stars, brittle stars, and nearly extinct somasteroids (Fig. 7.98) have star-shaped bodies (Fig. 7.94) and are free-moving. Their skeletons are composed of many tiny

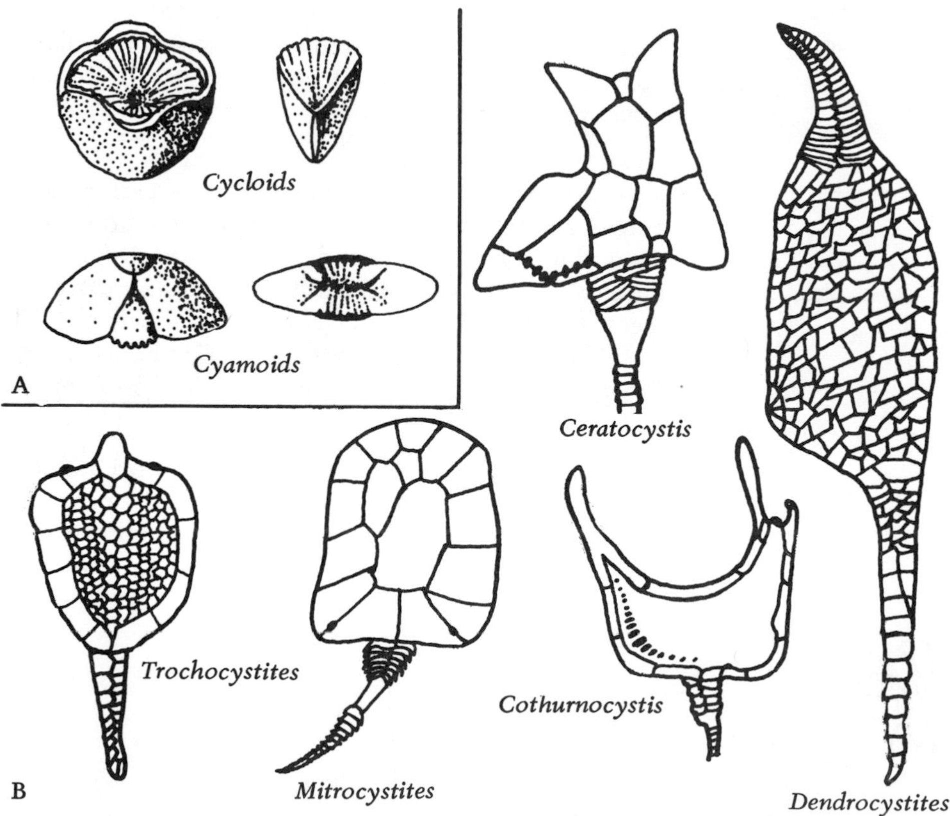

FIGURE 7.93 Extinct Echinoderms of Uncertain Relationships. (A) Haplozoans, (B) Carpoids or Homalozoans.

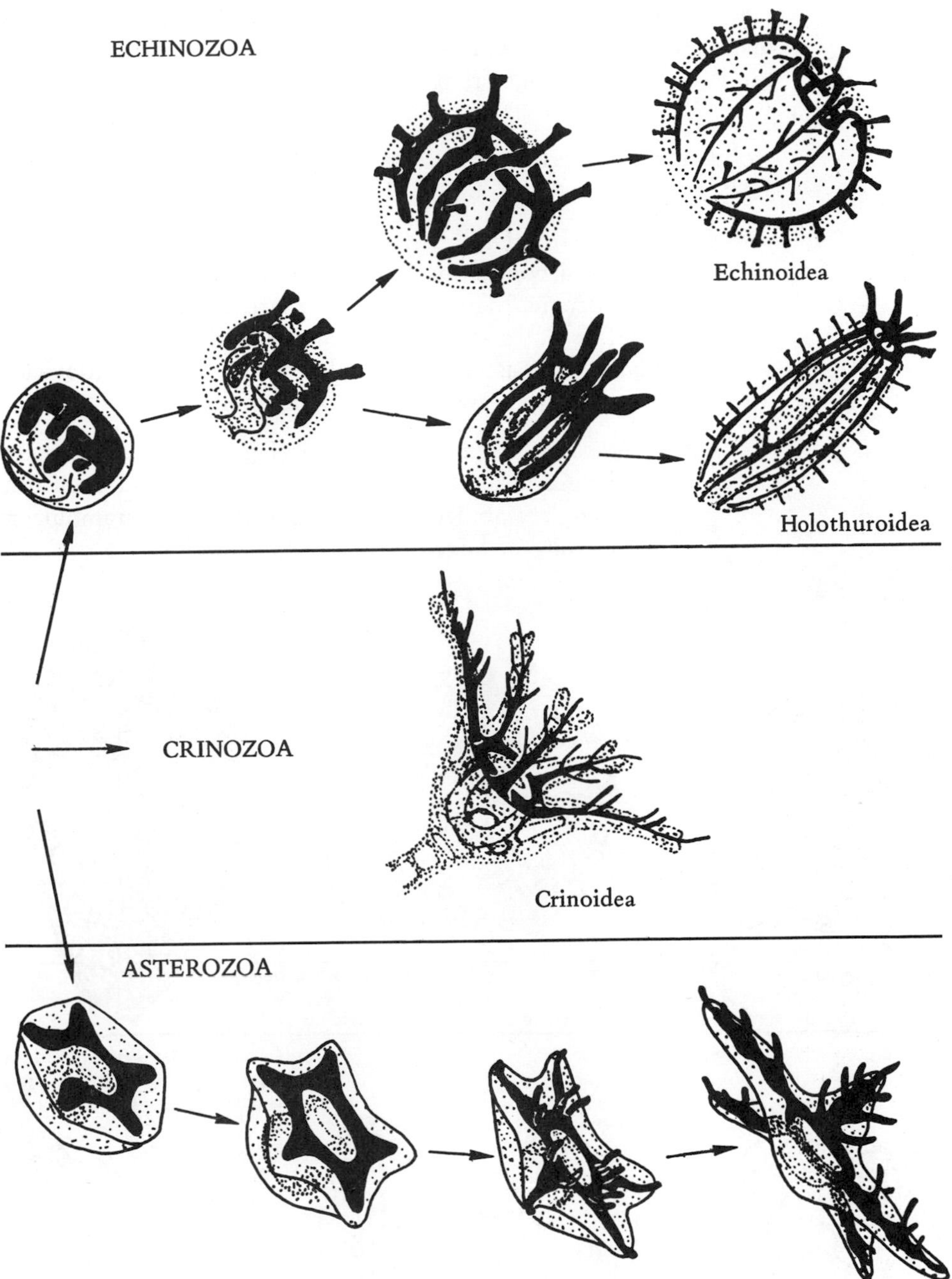

FIGURE 7.94 Growth Gradients in Echinoderms. These are used to define the major subphyla. (From *Treatise on Invertebrate Paleontology*, courtesy of the Geological Society of America and The University of Kansas.)

plates not rigidly joined together. As a result, the test usually disintegrates on death and whole starfish are rare fossils. All three kinds of starfish range from Ordovician to Recent.

Subphylum Echinozoa—This group encompasses the fundamentally globular echinoderms (Fig. 7.94) which are chiefly free-living. They range from Cambrian to Recent.

Class Edrioasteroidea—An extinct group (Fig. 7.99) of Cambrian-Carboniferous age, these are the only attached echinozoans. The 5 ambulaca are set in a mosaic of smaller plates and are typically ringed by a set of larger plates. Edrioasteroids are only common in the Ordovician when many grew on brachiopod shells.

Class Helicoplacoidea—This group (Fig. 7.100) is restricted to the Cambrian. The body is elongate and fusiform. The single ambulacrum and numerous interambulacra are spirally arranged on the expansible body.

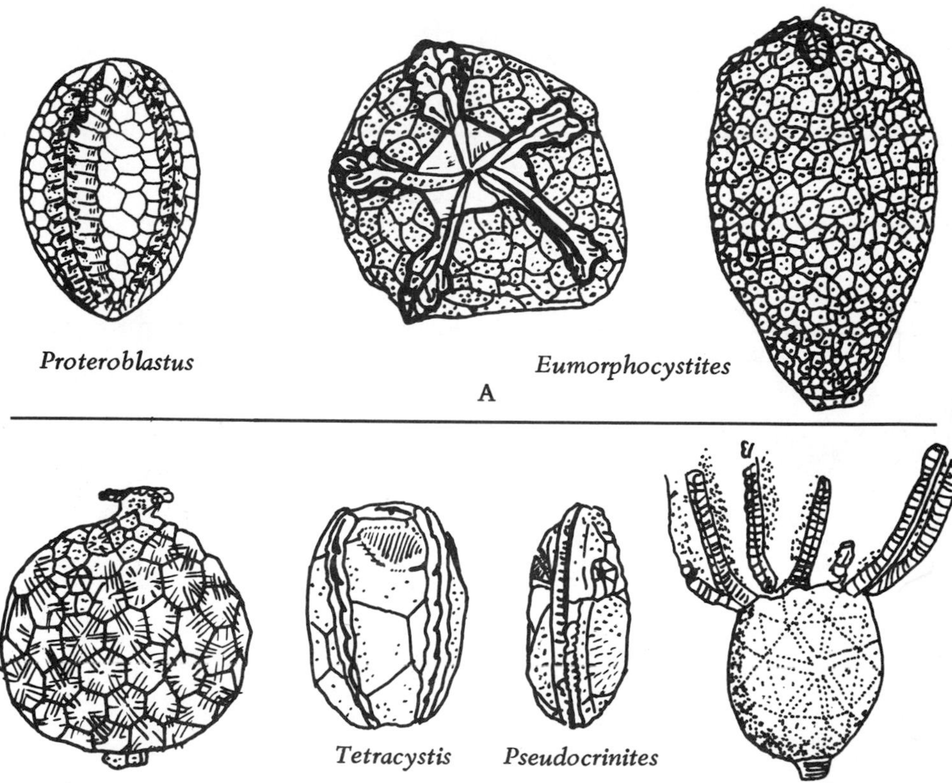

FIGURE 7.95 Class Cystoidea. (A) Diploporoid and (B) Rhombiferoid Cystoids.

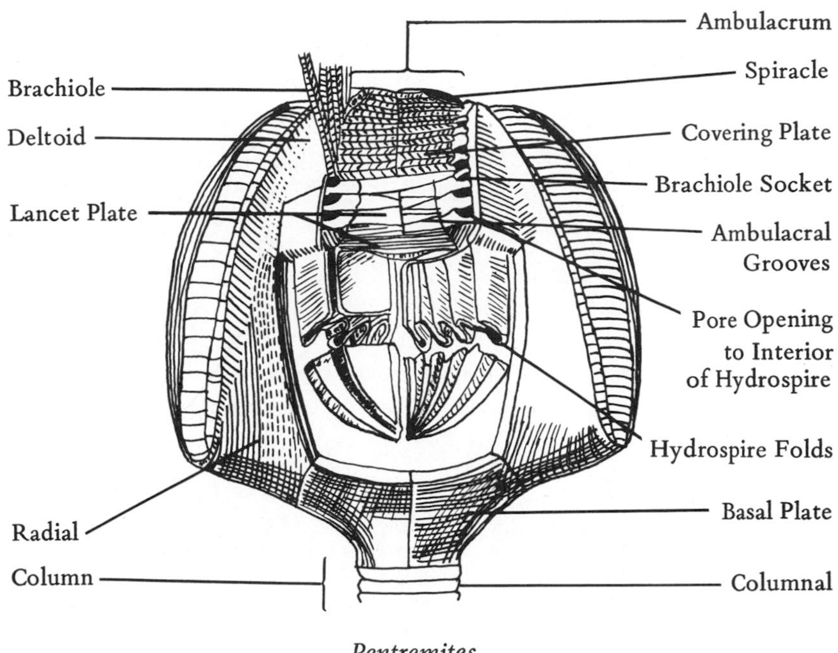

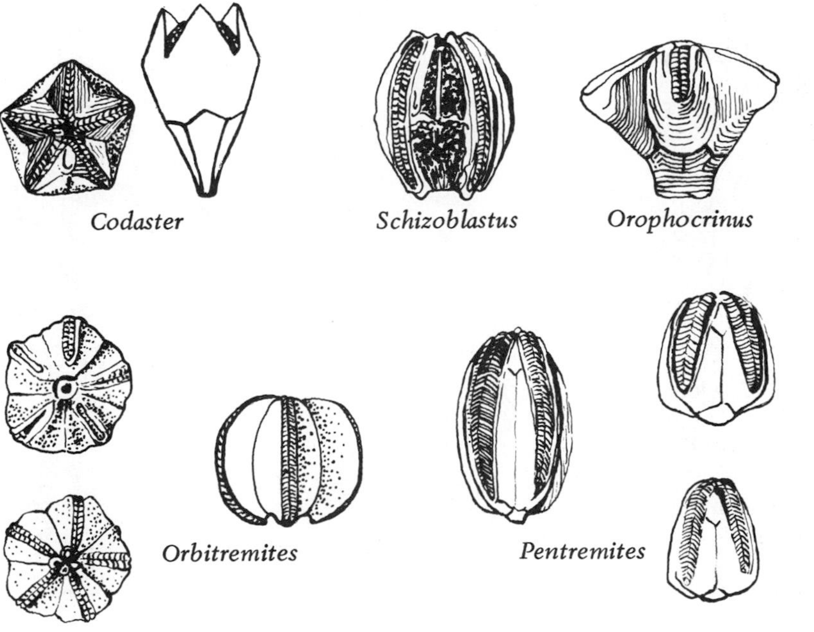

FIGURE 7.96 Class Blastoidea. General morphology and variation in thecal shapes.

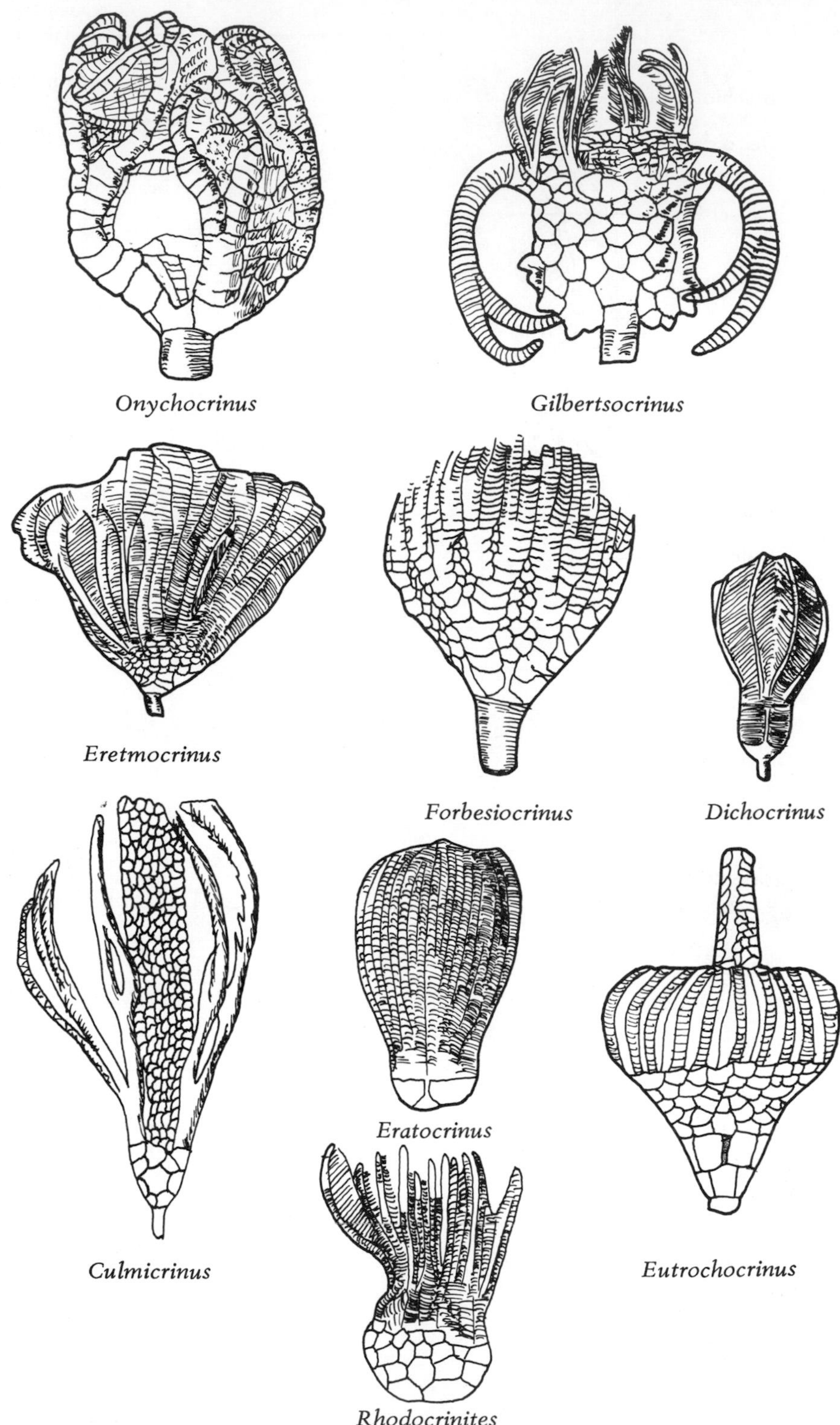

FIGURE 7.97 Fossil Crinoids.

Class Holothuroidea—The sausage-shaped sea cucumbers (Fig. 7.101) have large anterior tube feet modified to tentacles and a skeleton reduced to a calcareous ring in the throat area and isolated plates or ossicles. Only the ossicles are occasionally common fossils.

Class Echinoidea—This is by far the most significant fossil group in the subphylum as it includes the familiar sea urchins, sand dollars, and heart urchins (Fig. 7.102). The plate arrangement is regular with 5 meridional ambs and 5 meridional interambs each consisting of two rows of plates in most echinoids. The mouth is ventral, and opposite it is the apical system which encircles the anus in regular forms and does not in irregulars where the anus can be anywhere in the posterior interambulacrum. Spines set on tubercles cover the test. The spherical regulars or *sea urchins* chiefly inhabit rocks where they use their teeth to eat algae. This environment is not generally favorable for fossilization. Some primitive irregulars, the *holectypoids,* have a sea-urchin-like shape. Most irregulars are bilateral burrowing forms such as the flattened

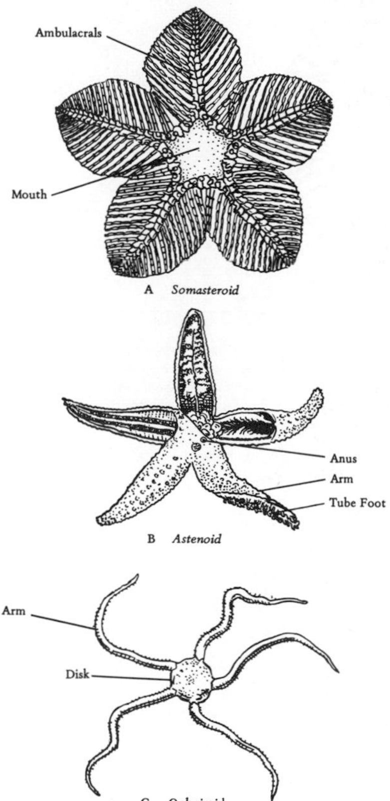

FIGURE 7.98 Subphylum Asteroza. (A) Somasteroid, (B) Asteroid, (C) Ophiuroid.

sand dollars so common in the Cenozoic; the *cassiduloids* with their peculiar floscelle around the mouth and peak of abundance in the Paleogene; and the distinctive *heart urchins*, common from the Cretaceous onward.

Phylum Hemichordata—Living members of this phylum (Fig. 7.103) have bilaterally symmetrical bodies divided into three regions, proboscis, collar, and trunk. The trunk bears branchial or gill slits and there is a small stiffened rod along the back at the anterior end. The acorn worms have left no known fossils, but the deep water, colonial pterobranchs have an irregular or branching scleroprotein coenoecium with a peculiar half-ring or fusellar structure. A few fossils extending back to Cambrian are known, but this group is important chiefly because the structure is virtually identical to that of the extinct graptolites which mystified scientists for so long.

Class Graptolithea—Graptolites (Figs. 7.104, 7.105) are colonial organisms, each of which inhabited an individual cup or theca arranged on a branch or stipe. A collection of stipes is called a rhabdosome and in one major group, the *dendroid* graptolites of

Edrioaster

Carneyella

Cincinnatidiscus

FIGURE 7.99 Class Edrioasteroidea.

Cambrian through Mississippian age, it is attached to the bottom in most types. The most significant fossil group is the *graptoloid* graptolite assemblage in which the rhabdosome was attached to a float or a floating object. These widespread pelagic organisms are limited to Ordovician, Silurian, and Lower Devonian rocks where they are very important for dating purposes. Graptolites are most common in black shales formed in areas where their organic remains settled into a decay-free and scavenger-free environment.

Phylum Conodontophora—For years geologists and biologists have puzzled over tiny tooth-like or plate-like, chiefly Paleozoic, marine fossils called conodonts (Fig. 7.106). Their composition is phosphatic which indicates some relationship with vertebrates, possibly fish, but no comparable parts exist in any living vertebrate. Furthermore, their layered structure shows that growth occurred only on the outer surfaces indicating they were not teeth. Their occurrence in all marine rock types indicates a pelagic creature and rare assemblages show the creature was bilateral and possessed several types of conodonts (Fig. 7.107). Despite the uncertainty of their origin, conodonts are abundant fossils throughout their Cambrian through Triassic range and have become very important fossils for dating Middle to Upper Paleozoic rocks. Recently, what appears to be distillation fossils of the actual animal have been found. The creature is small and resembles a worm with fins. The conodonts are interior structures, apparently part of the digestive system.

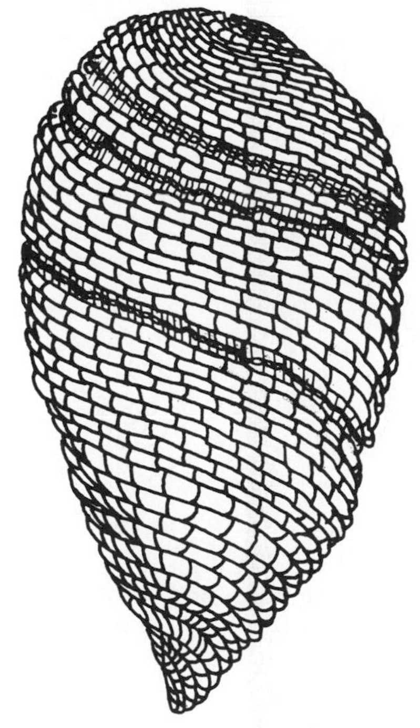

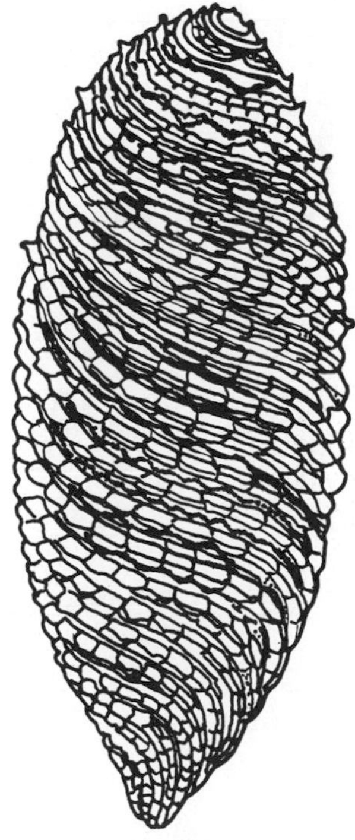

FIGURE 7.100 Class Helicoplacoidea.

Phylum Vertebrata—Most textbooks lump the vertebrates, sea squirts, and amphioxans into a Phylum Chordata based on the fact that all, at some time in their lives, possess a dorsal nerve cord, gill slits, and a dorsal stiffening rod of some type. The last two groups (Fig. 7.108) are very different from vertebrates as adults and have left few fossils so from a paleontologic viewpoint, the vertebrates can be best considered a separate phylum. The vertebrates, except the jawless fish, possess a dorsal supporting structure which is composed of calcium phosphate (either bone or cartilage) called the vertebral column. They all have a phosphatic brain case, hence, their other common name, Craniata. Other bones typically present are the thoracic ribs, axes of appendages, and dermal bones. All bones have a meshwork construction much like that of echinoderms. Most have teeth of hard enamel. The fossil record goes back to the Ordovician or possibly the Cambrian, but is not as good as that of the invertebrates. All vertebrates have numerous parts and are easily disarticulated. Furthermore, nearly half are

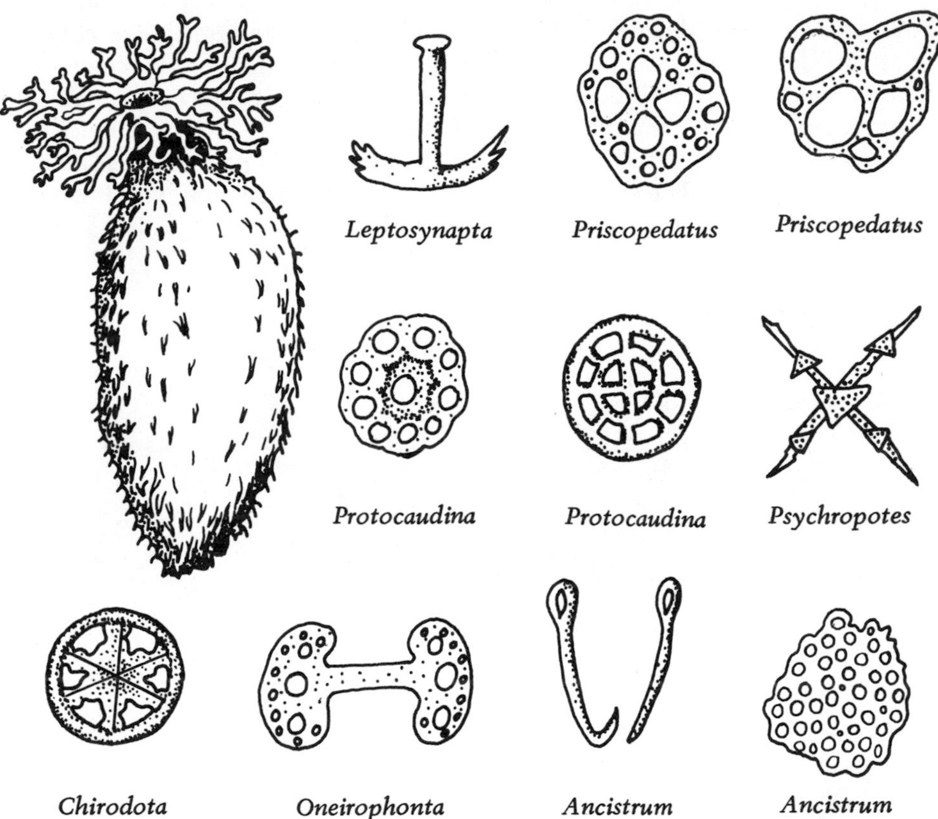

FIGURE 7.101 Class Holothuroidea. A living form and fossil sclerites.

FIGURE 7.102 Class Echinoidea. The major groups and their structural features.

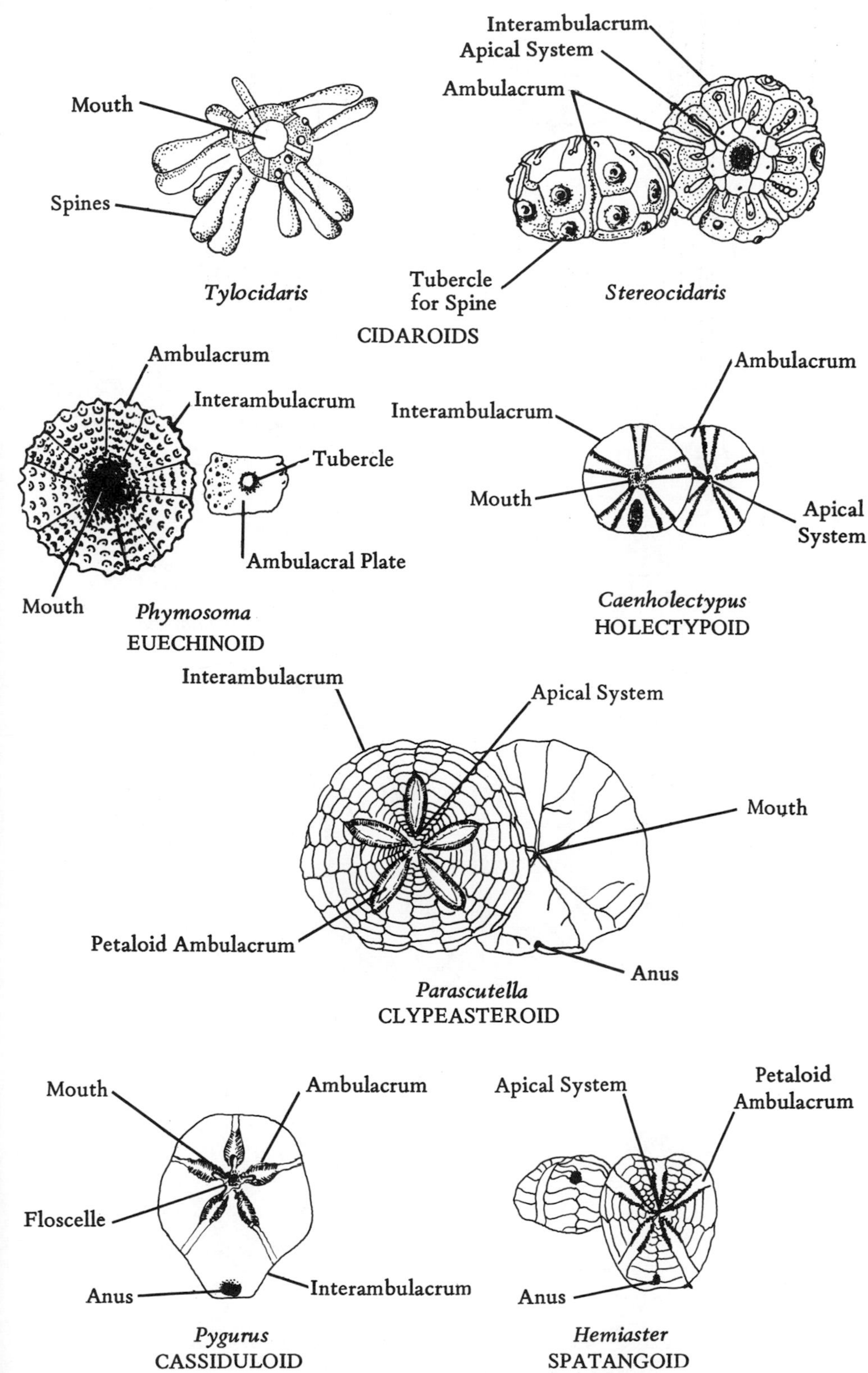
Mouth
Spines
Tylocidaris
Interambulacrum
Apical System
Ambulacrum
Tubercle
for Spine
Stereocidaris
CIDAROIDS
Ambulacrum
Interambulacrum
Tubercle
Ambulacral Plate
Mouth
Phymosoma
EUECHINOID
Ambulacrum
Interambulacrum
Mouth
Apical
System
Caenholectypus
HOLECTYPOID
Interambulacrum
Apical System
Mouth
Petaloid Ambulacrum
Anus
Parascutella
CLYPEASTEROID
Mouth
Ambulacrum
Floscelle
Anus
Interambulacrum
Pygurus
CASSIDULOID
Apical System
Petaloid
Ambulacrum
Anus
Hemiaster
SPATANGOID

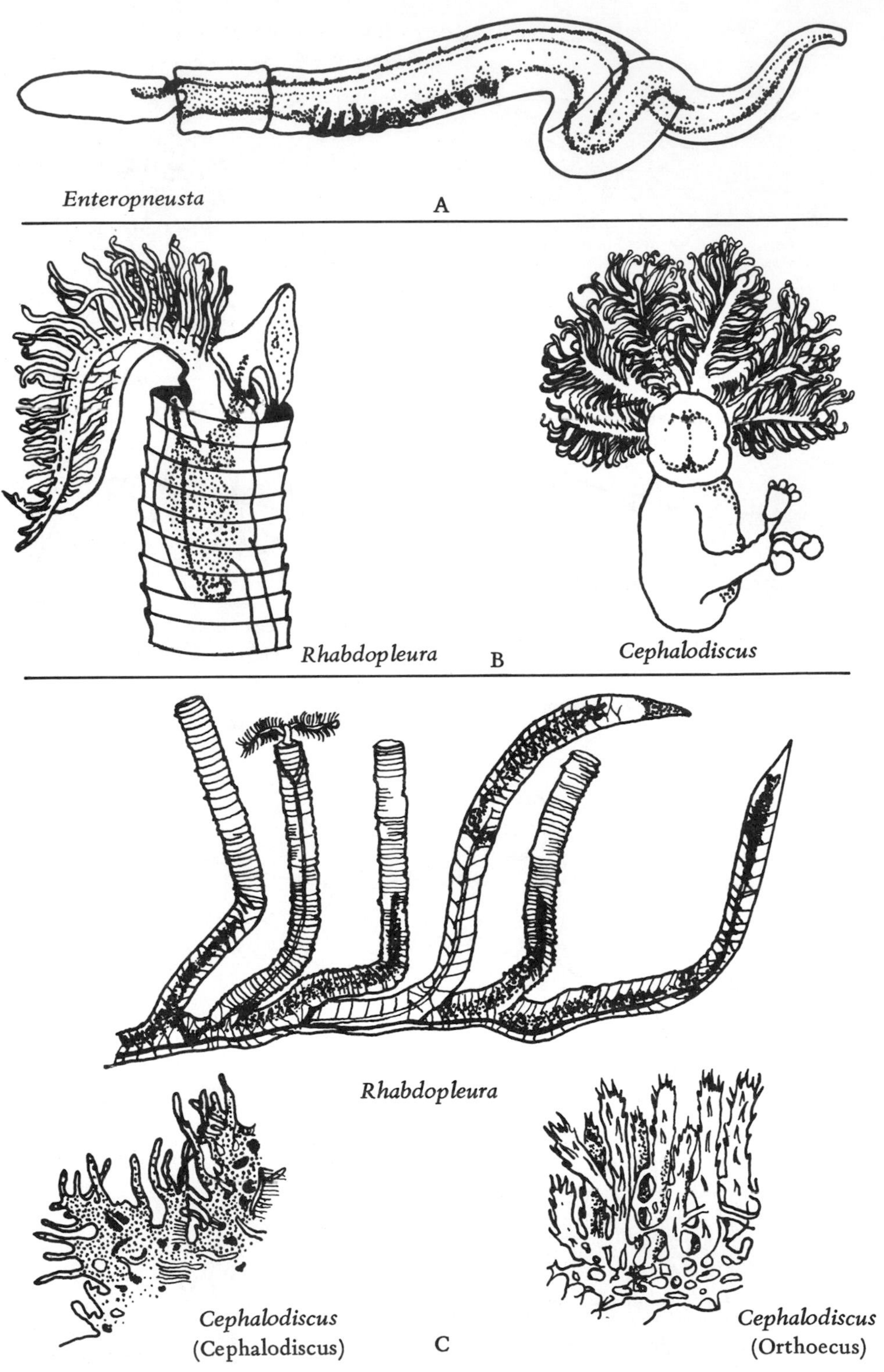

FIGURE 7.103 Living Hemichordates. (A) Acorn worm, (B) Pterobranchs, (C) Coenoecia of pterobranchs.

terrestrial, a poor environment for fossilization. Nevertheless, from the Late Silurian onward vertebrates are not uncommon fossils. Vertebrates are marine, fresh water, terrestrial, and aerial.

Subphylum Piscezoa—The fish are characterized by their typical torpedo-like shape with fins or flaps for steering and an undulating body for swimming. Most respire by gills as all are aquatic.

Class Agnathea—The jawless fish are represented by the living lampreys and hagfish (Fig. 7.109) which have no hard parts and have left few fossils. In the Ordovician through Devonian, however, numerous jawless fish with large bony head shields occurred. These *ostracoderms* have tiny mouths for mud-grubbing, no true fins for stabilization, and large gill areas. Ostracoderms lived in fresh-water, brackish, and marine environments.

Class Placodermea—This is a heterogeneous collection of primitive, armored, jawed fish (Fig. 7.109). The jaws evolved from an anterior gill arch and were not propped securely against the skull (Fig. 7.110). Paired fins of unusual constructions were present. Nearly all placoderms

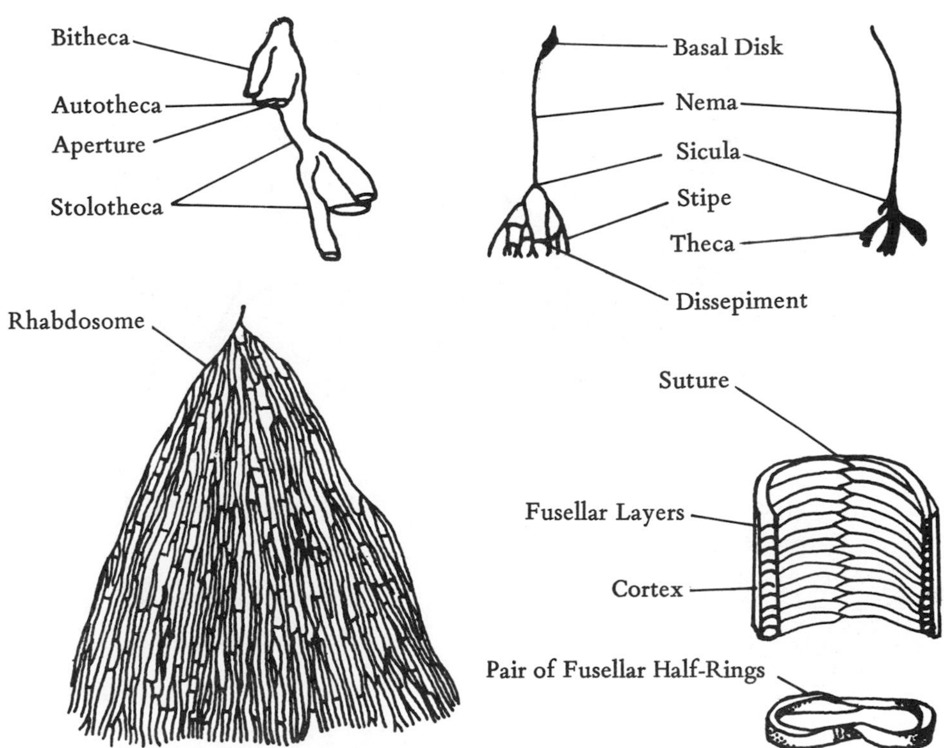

FIGURE 7.104 Morphology of Dendroid Graptolites.

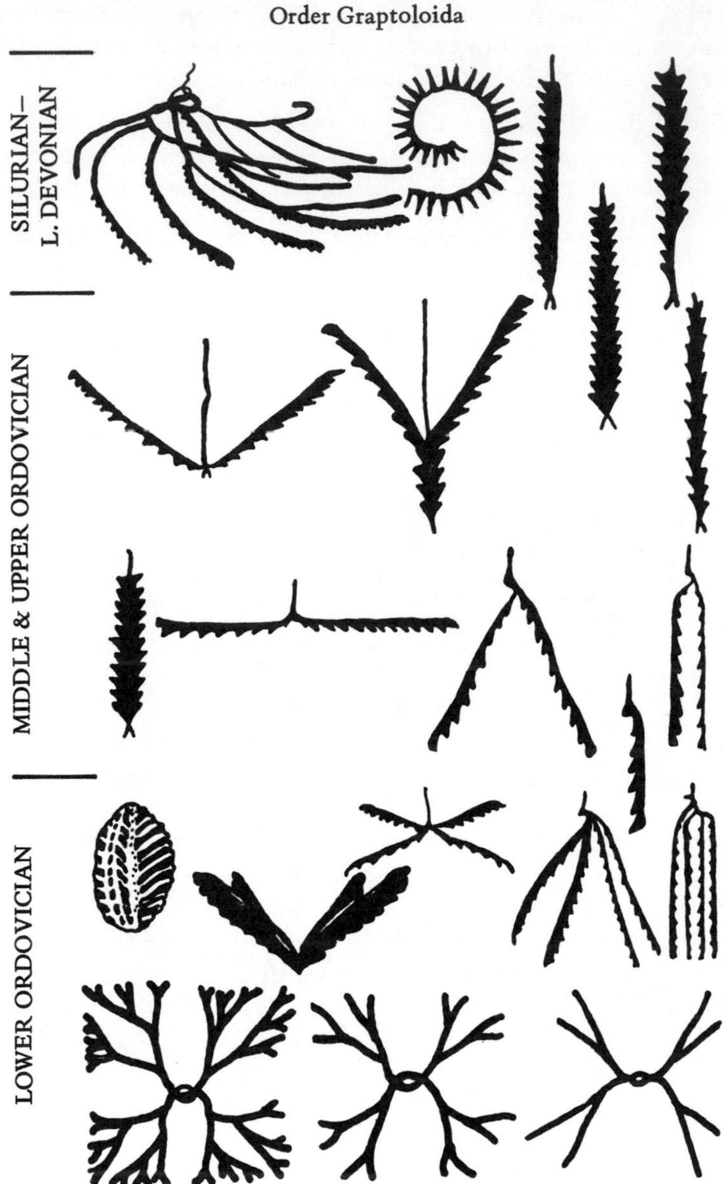

FIGURE 7.105 Variation Among Rhabdosomes of Graptoloid Graptolites.

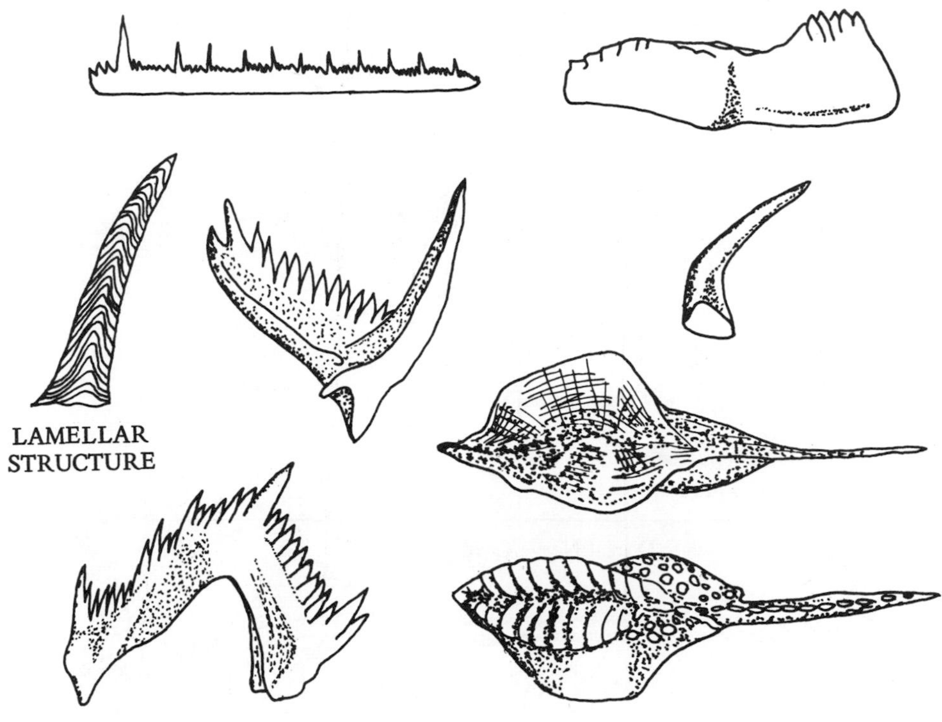

FIGURE 7.106 Conodonts.

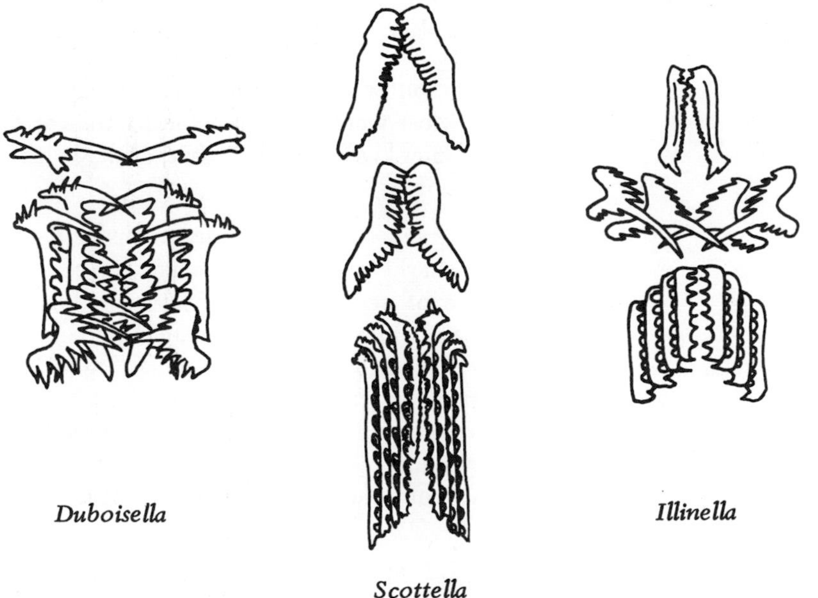

FIGURE 7.107 Conodont Assemblages.

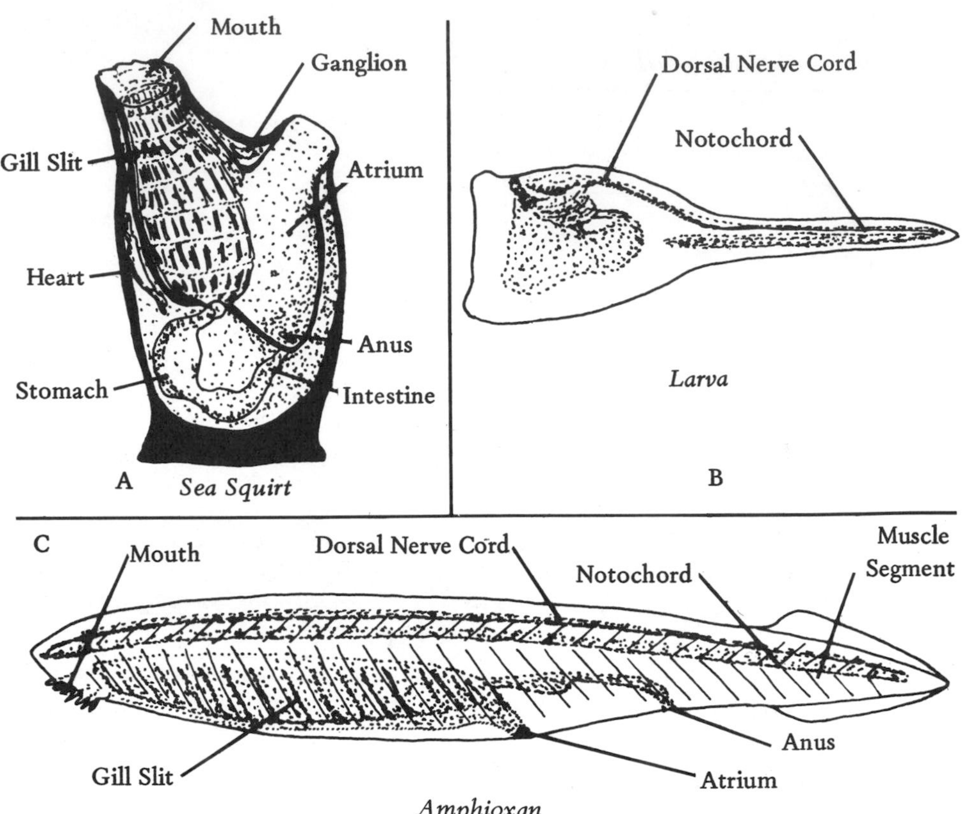

FIGURE 7.108 Sea Squirt (A) Adult, (B) Larval, and (C) Adult **Amphioxan**. These animals are often classified with the vertebrates because of embryological and larval similarities.

are Devonian, but a few Silurian and Mississipian forms are known. Most were marine including the carnivorous arthodires. Some were fresh water such as the mud-grubbing antiarchs.

Class Chondrichthyea—The cartilaginous fish (Fig. 7.109) include the sharks and the skates and rays; the ratfish appear to be of separate origin. Jaws are present and propped against the skull by the next gill arch whose gill slit is reduced to a hole or spiracle (Fig. 7.110). The gills all open separately to the exterior. The internal skeleton is of cartilage and rarely preserved, but the hard teeth and spines (Fig. 7.111) are common fossils, particularly from the Carboniferous onward. Most are marine.

Class Acanthodea—This group (Fig. 7.109) appears to be ancestral to the bony fish. Its members possess a flap covering the gills (the operculum). They have one external opening as in bony fish. In both groups the jaw is propped firmly against the skull and there is no spiracle. Both groups also have similar scales. Acanthodians or spiny sharks have their fins supported by single large spines unlike those of bony fish, however. They were chiefly fresh water and range from Silurian through Permian with a Devonian peak.

Class Osteichthyea—The bony fish (Fig. 7.109) have an internal bony skeleton, a scaly covering, paired fins supported by bony shoulder and hip girdles, jaws propped firmly against the skull with no spiracle present, and an operculum. They are both marine and fresh water and range from Devonian to Recent with a Cretaceous to Recent peak. Bony fish are divided into two subclasses (Fig. 7.112).

Subclass Actinopterygia—The ray-finned fish have slender, parallel rays of dermal bone supporting their paired fins, lack internal nostrils, possess a swim bladder (early types probably had lungs), the vertebrae enter the upper tail lobe, and there is one dorsal fin. They have been far and away the most significant bony fish group after the Devonian.

Subclass Sarcopterygia—The lobe-finned fish have skeletal bones supporting their paired fins which also bear a large and fleshy fringe, possess internal nostrils and lungs, typically have the vertebrae passing between the tail lobes, and there are two dorsal fins. Common only in the Devonian, and rare after the Paleozoic, the lobe-finned fish are notable as the ancestors of the tetrapods which their skeletal structure, particularly the fins (Fig. 7.113), skull, and lungs, foreshadow.

Subphylum Tetrapodazoa—These are basically four-limbed, terrestrial, air-breathing vertebrates. Some have secondarily returned to the sea, but still breathe with lungs; others have taken up flight and the front limbs have evolved into wings.

Class Amphibea—Amphibians, the living frogs, toads, and salamanders (Fig. 7.114), and a host of extinct types, may not be a natural group. The limbs are weak in most, indicating that their habitat is still largely aquatic. All but the earliest forms from the Late Devonian have lost the fish-like tail. The eggs are small, shell-less, and lack protective membranes which necessitates their

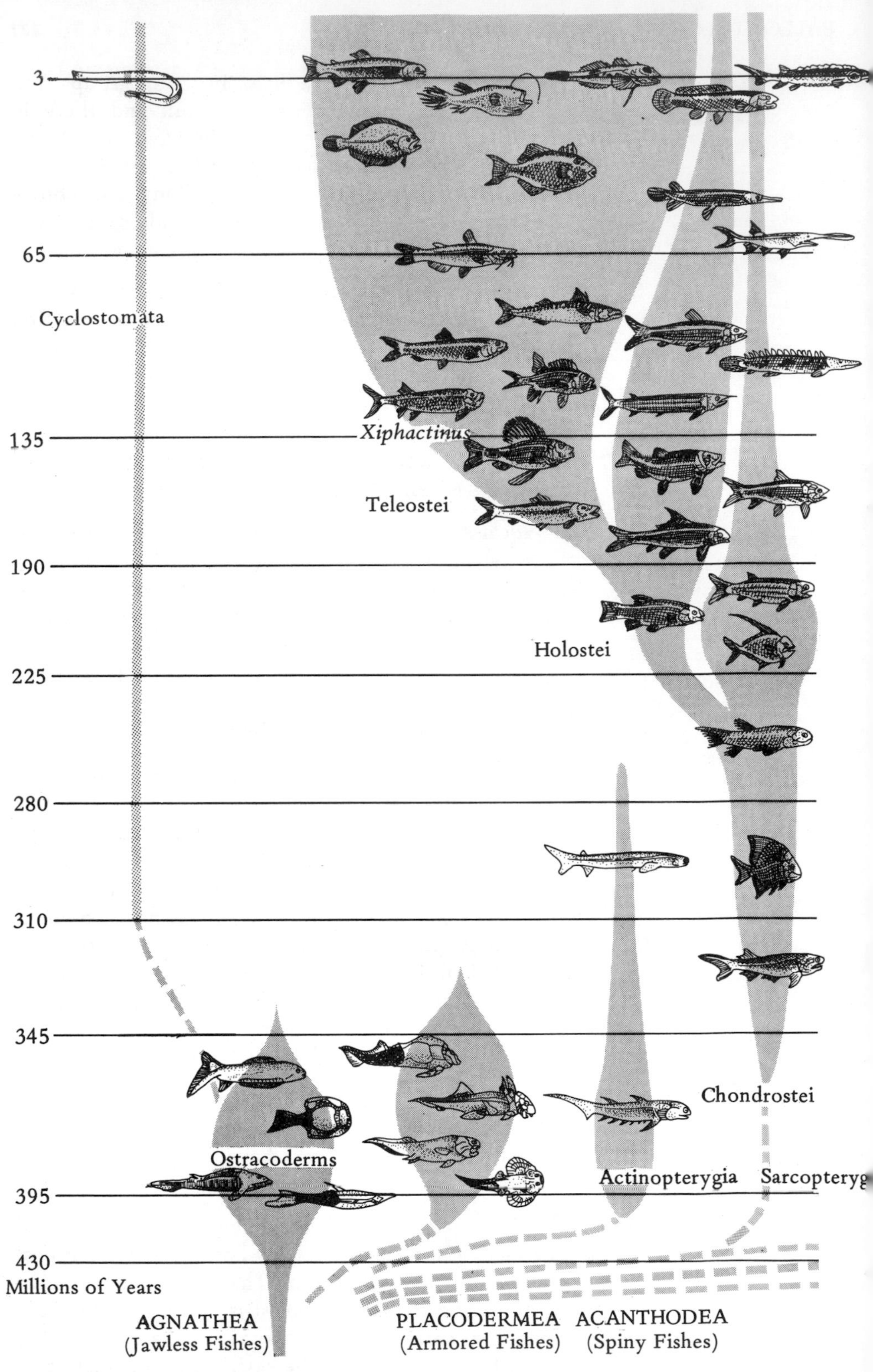

FIGURE 7.109 Subphylum Piscezoa.

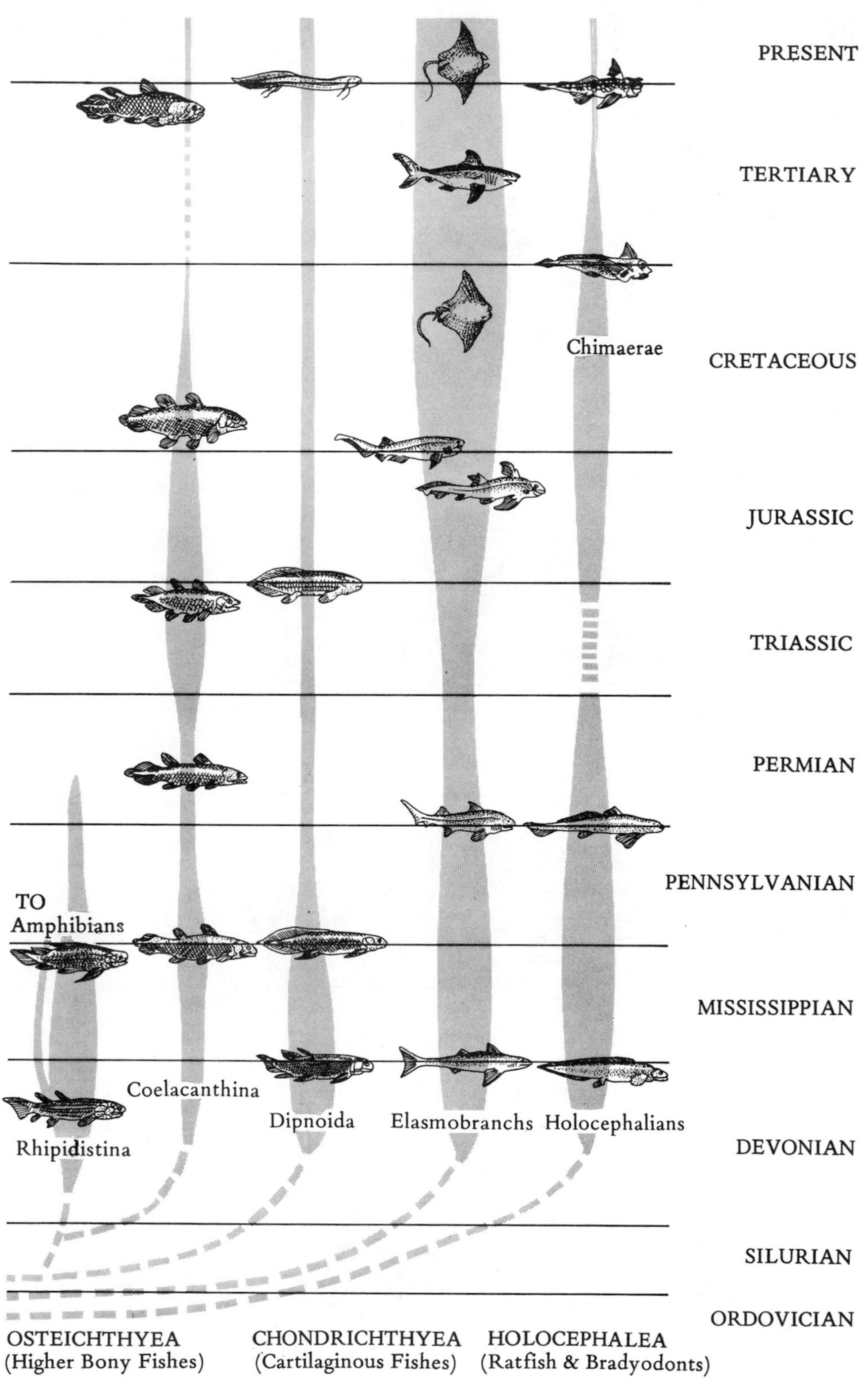

Evolution of the major fish groups.

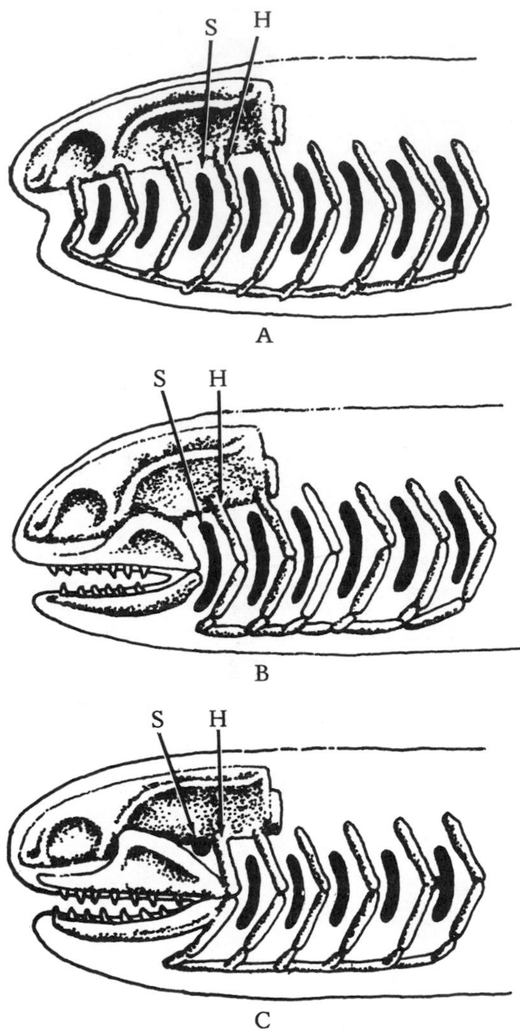

FIGURE 7.110 Evolution of the Fish Jaw. (A) Agnath, (B) Placoderm, (C) Chondrichthyean

TEETH AND DERMAL PLATES

Shark Tooth

Shark Tooth

Dermal Plate

Shark Tooth

FIGURE 7.111 Shark Fossils.

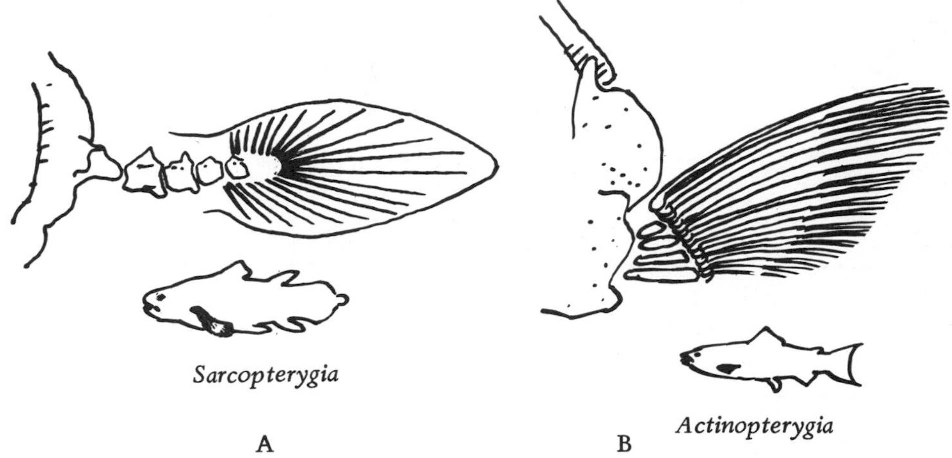

FIGURE 7.112 (A) Lobe-Fin and (B) Ray-Fin Bony Fish.

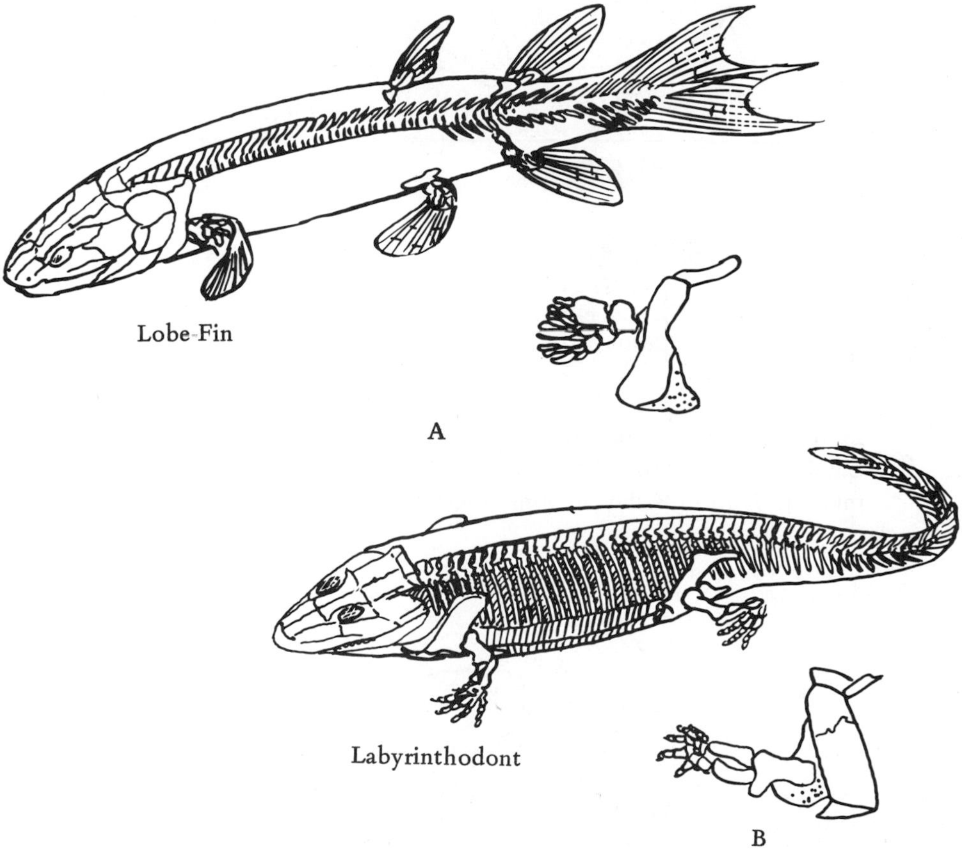

Lobe-Fin

A

Labyrinthodont

B

FIGURE 7.113 Similarities Between Devonian Lobe-Finned Fish and Amphibians. (A) Lobe-fin fish and its shoulder and forelimb, (B) Early amphibian and its shoulder and forelimb.

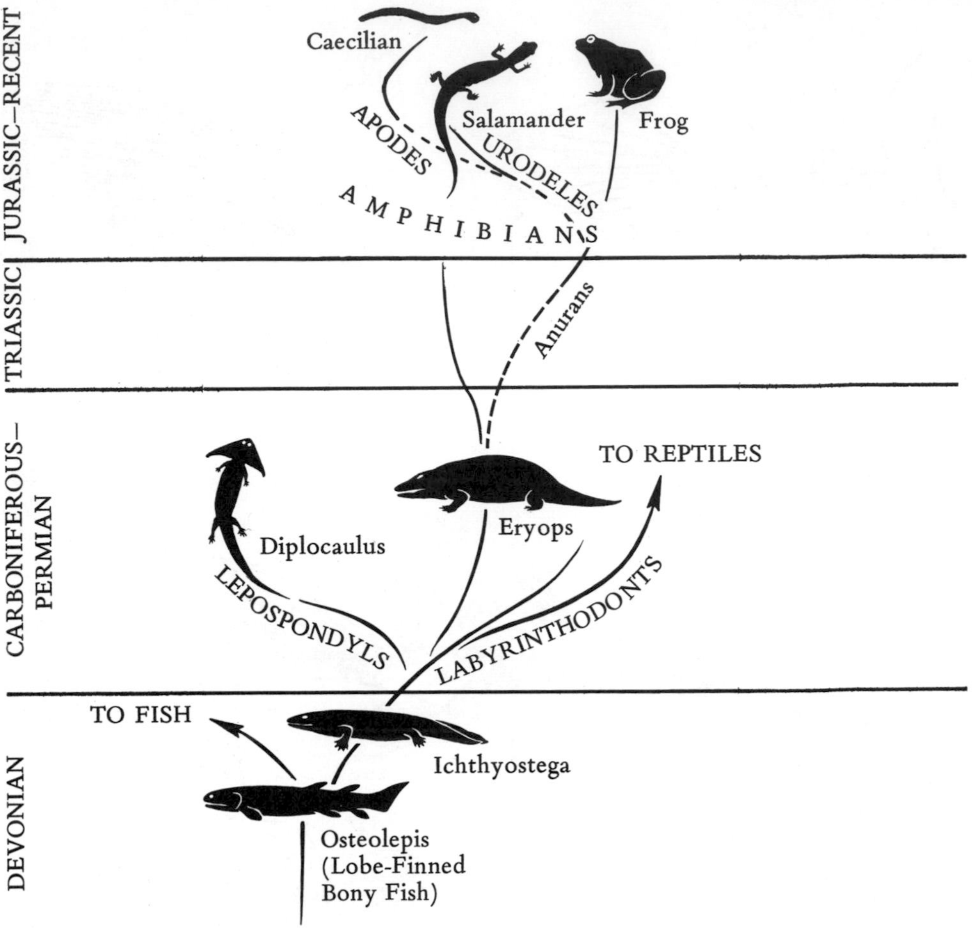

JURASSIC–RECENT
TRIASSIC
CARBONIFEROUS–PERMIAN
DEVONIAN
Caecilian
Salamander
Frog
APODES
URODELES
AMPHIBIANS
Anurans
TO REPTILES
Eryops
Diplocaulus
LEPOSPONDYLS
LABYRINTHODONTS
TO FISH
Ichthyostega
Osteolepis
(Lobe-Finned
Bony Fish)

FIGURE 7.114 Evolution of Major Amphibian Groups. (From Edwin H. Colbert, *Evolution of the Vertebrates,* © 1969, by permission of John Wiley & Sons, Inc.)

Cavity of Ammion Sac
Embryo
Cavity of Allantois Sac
Albumen
Yolk

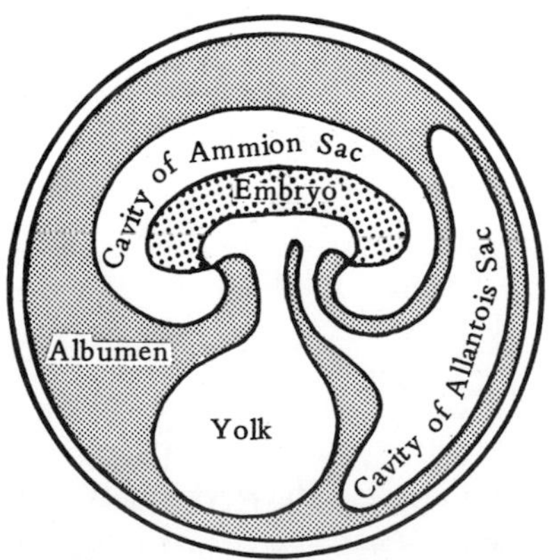

FIGURE 7.115 Schematic Drawing of Amniotic Egg.

being laid in water. The young hatch at an immature stage in which they are aquatic, fish-like, and breathe with gills. Only later do they metamorphose to the air breathing, four-limbed adults. Amphibians, like fish, are "cold-blooded," that is they cannot regulate their own temperature internally, but take the temperature of their environment. They are smooth-skinned as the skin is an important accessory respiratory structure. Amphibians are fresh water and terrestrial. The most common articulated amphibian skeletons are the large *labyrinthodonts* or stegocephelians which resemble huge, flat-headed salamanders with fish-like teeth. They range in age from Devonian through Triassic but peak in the Carboniferous and Permian. The small aquatic salamander or snake-like *lepospondyls* also are occasionally common in the Upper Paleozoic. Remains of the modern types of amphibians, the *lissamphibians,* are largely fragmentary.

Class Reptilea—With the evolution of the reptiles, the vertebrates finally succeeded in escaping the water. Reptiles evolved the amniotic egg with a protective shell, food, and respiratory and excretory membranes (Fig. 7.115). This egg can be laid on land this allowing the young to develop to a more advanced stage before hatching. The limbs are stronger than those of amphibians, but the reptiles cannot regulate their own temperature internally. The skin of reptiles is covered with bony scutes ("scales"). Reptiles are basically terrestrial, but some have returned to the aquatic environment, both marine and fresh water, and some have developed the ability to fly. Reptiles range in age from Carboniferous to Recent, but the Mesozoic Era was the true age of reptiles. Reptiles were also common in the Permian, but became drastically reduced after the Mesozoic. The major reptile groups (Fig. 7.116) are: the *cotylosaurs* (Carboniferous-Triassic) or stem reptiles, common in the Late Paleozoic; the *turtles* (Triassic-Recent) with their unique shell, many of which are aquatic; the *ichthyosaurs* (Triassic-Cretaceous), fish or dolphin-like reptiles common in Triassic and Jurassic; the *sauropterygians* (Triassic-Cretaceous), sea-serpent-like beasts including the amphibious Triassic nothosaurs and the plesiosaurs which reached a Cretaceous peak; the *lizards* (Triassic-Recent) including the marine Cretaceous mosasaurs; the *snakes* (Cretaceous-Recent); the *thecodonts* (Permian-Triassic) including dinosaur ancestors and crocodile-like phytosaurs; the *crocodiles* (Triassic-Re-

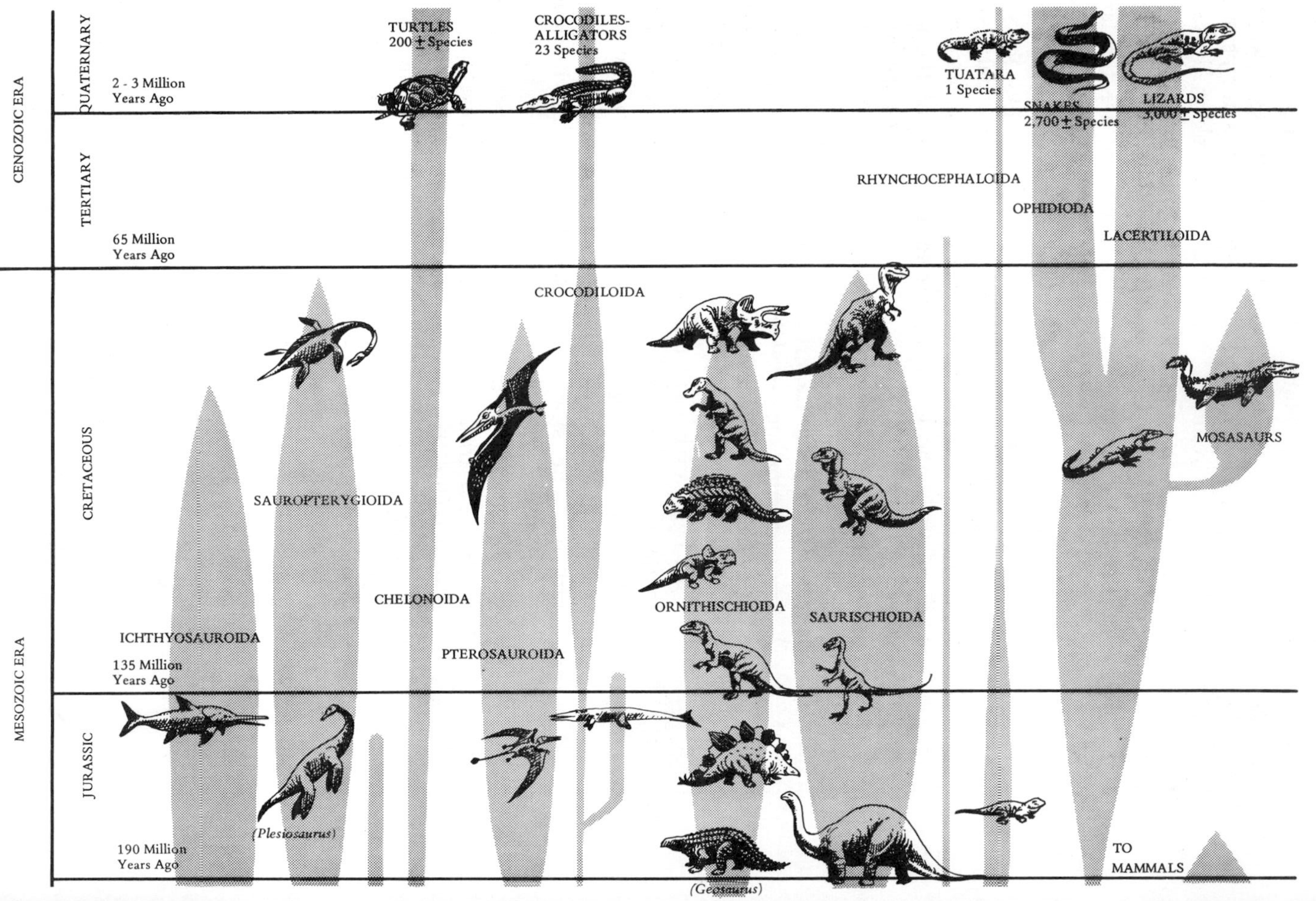
CENOZOIC ERA
QUATERNARY
2 - 3 Million Years Ago
TURTLES 200 ± Species
CROCODILES-ALLIGATORS 23 Species
TUATARA 1 Species
SNAKES 2,700 ± Species
LIZARDS 3,000 ± Species
TERTIARY
RHYNCHOCEPHALOIDA
OPHIDIODA
LACERTILOIDA
65 Million Years Ago
CROCODILOIDA
CRETACEOUS
MOSASAURS
SAUROPTERYGIOIDA
CHELONOIDA
ORNITHISCHIOIDA
SAURISCHIOIDA
ICHTHYOSAUROIDA
PTEROSAUROIDA
MESOZOIC ERA
135 Million Years Ago
JURASSIC
(Plesiosaurus)
190 Million Years Ago
TO MAMMALS
(Geosaurus)

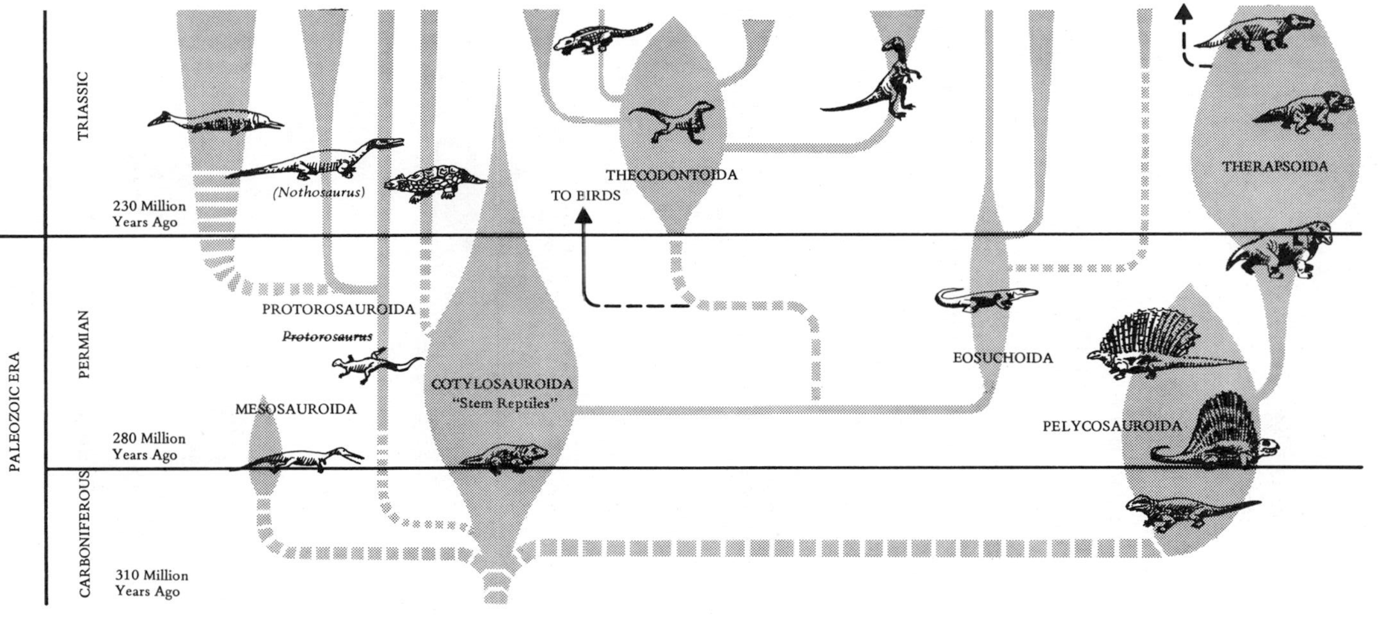

FIGURE 7.116 Evolution of the Reptiles.

cent); the *pterosaurs* or flying reptiles; the *dinosaurs* (Triassic-Cretaceous), two orders of large Mesozoic reptiles; and the mammal-like reptiles (Carboniferous-Jurassic) which evolved into the mammals and also include the large, Late Paleozoic, fin-backed reptiles.

Class Avea—The birds (Fig. 7.117) evolved from the reptiles and a transitional Jurassic form is known. Birds are structured for flight with front limbs modified to wings, the body and wings covered with feathers and the tail made of them, hollow and light bones, the ability to internally control their own temperatures, and large eyes. The Jurassic form had only feathers and, thus, was probably warm-blooded. Birds retain the reptilian egg, but all except the Jurassic genus and a Cretaceous form have lost the teeth. Some large birds have secondarily returned to the ground. Birds are only locally common in Cenozoic rocks.

FIGURE 7.117 Class Avea. Some major adaptive types among the birds. (From Edwin H. Colbert, *Evolution of the Vertebrates*, © 1969, by permission of John Wiley & Sons, Inc.)

Class Mammalea—The mammals evolved from the reptiles in the Late Triassic and the transition is well-documented by fossils. Mammals are viviparous, that is, the shell-less fertilized egg is maintained in the body by the mother and internally nourished until it attains an advanced state. After birth, the young are nursed by the mother's mammary glands insuring a long period

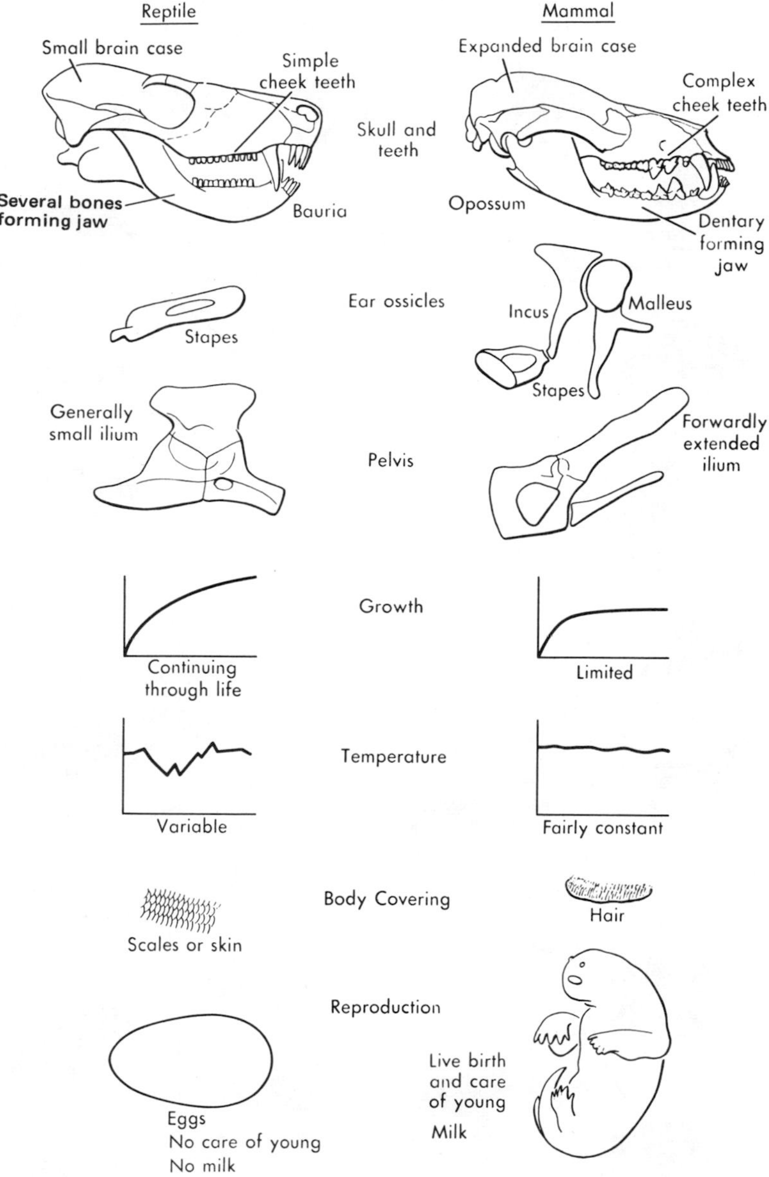

FIGURE 7.118 Some Differences Between Reptiles and Mammals. (From Edwin H. Colbert, *Evolution of the Vertebrates*, © 1969, by permission of John Wiley & Sons, Inc.)

of parental care. Mammals are typically covered with hair and are "warm-blooded." They have large brains, a single bone in the lower jaw, and highly differentiated teeth (Fig. 7.118). The resistant teeth are the most common fossils. Mammals are the major terrestrial vertebrates of the Cenozoic. Some of the more important fossil groups (Fig. 7.119) are: the pouched *marsupials* (Cretaceous-Recent); the ancestral placental *insectivores* (Cretaceous-Recent) including the modern shrews and

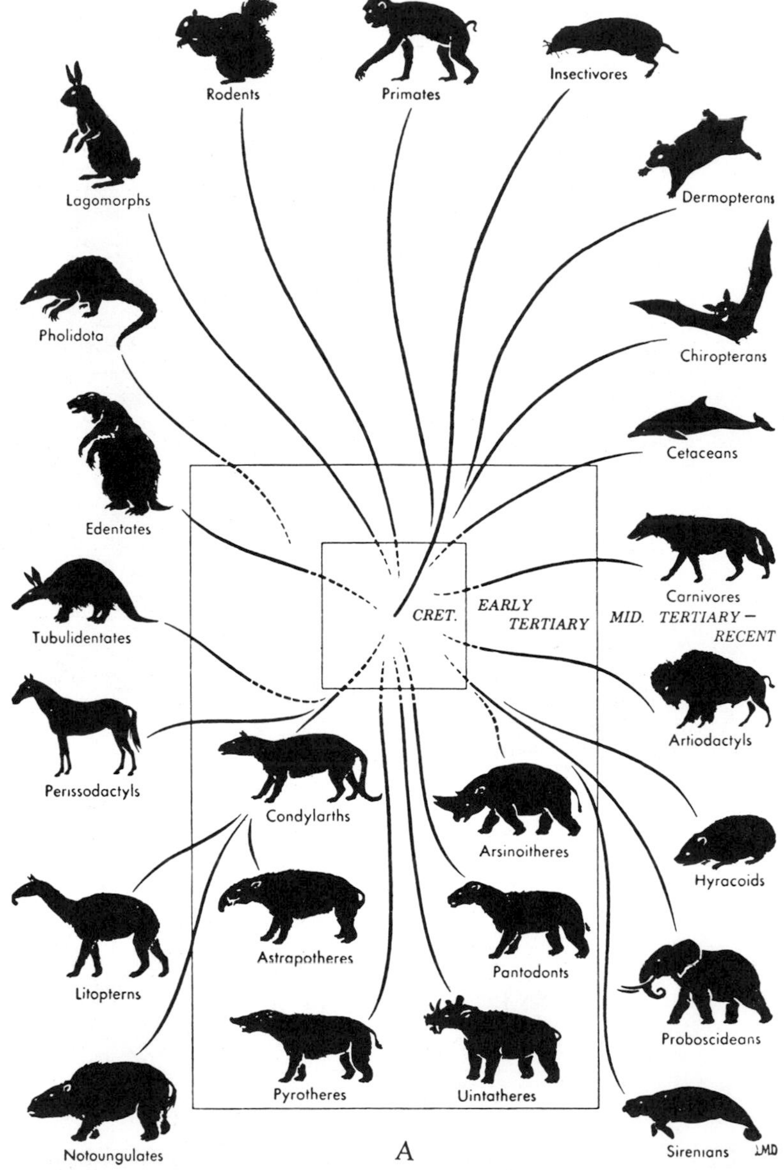

FIGURE 7.119 Class Mammalea. (A) Placental Mammals, (B) Marsupials. (From Edwin H. Colbert, *Evolution of the Vertebrates*, © 1969, by permission of John Wiley & Sons, Inc.)

moles; the flying *bats* (Eocene-Recent); the *primates* (Cretaceous-Recent) which include lemurs, monkeys, apes and men; the *edentates* (Paleocene-Recent) including ground sloths and armadillos; the *rodents* (Paleocene-Recent), the largest order of mammals which includes squirrels, guinea pigs, and mice; the *lagomorphs* (Paleocene-Recent) or rabbits; the *condylarths* (Cretaceous-Oligocene), the ancestors of carnivores and the hoofed mammals; the *cetaceans* (Eocene-Recent), the marine mammals or whales; the carnivorous *creodonts* (Cretaceous-Oligocene) and *carnivores* (Paleocene-Recent); the bulky *amblypods* (Paleocene-Oligocene) including the uintatheres; the *proboscideans* (Eocene-Recent) including mastodons and elephants; the *perissodactyls* (Paleocene-Recent) or odd-toed hoofed mammals including horses, titanotheres, and rhinos; and the *artiodactyls* (Eocene-Recent) or even-toed hoofed mammals including the pigs, oreodonts, camels, deer, antelopes, and cattle.

Kingdom Plantae

Plants, in the restricted sense used here, include those organisms that have a nucleus, are multicellular with a high degree of differentiation, are able to synthesize their own food from inorganic substances by the process of photosynthesis, form their gametes or sex cells in multicellular structures, and form embryos. So restricted, plants consist of two major groups or subkingdoms,

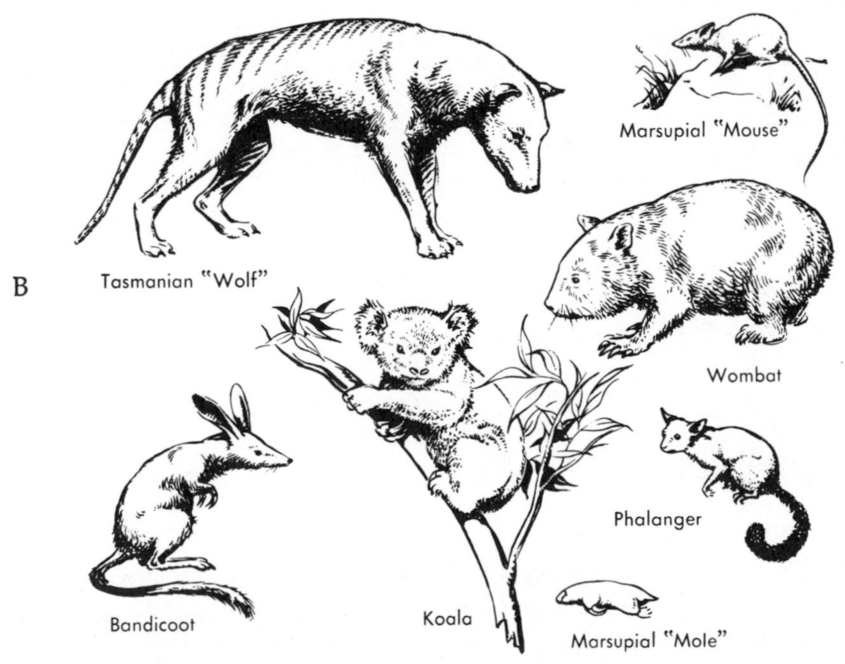

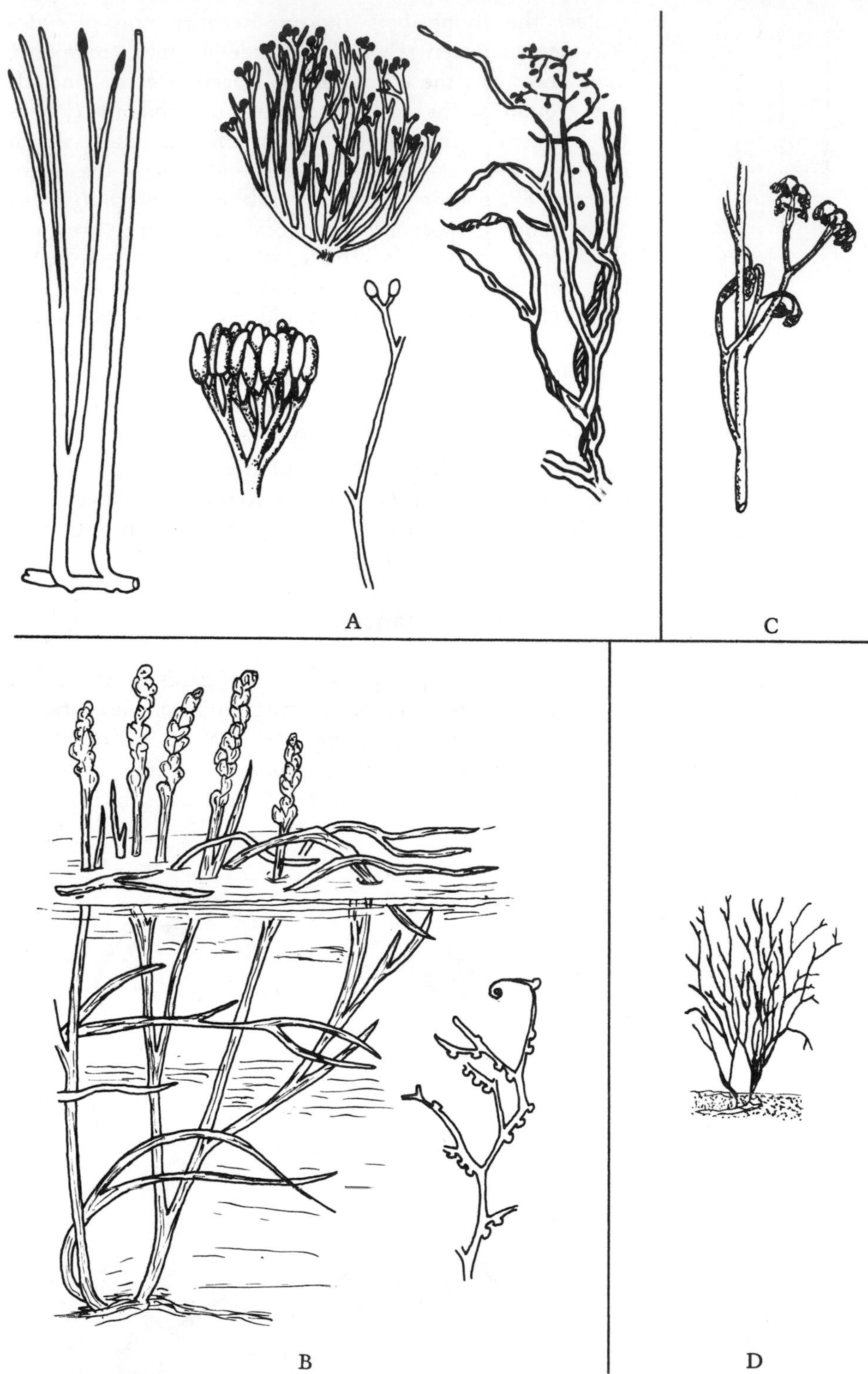

FIGURE 7.120 Psilophytes. (A) *Rhynia*-type, (B) *Zosterophyllum*-type, (C) *Trimerophyton*-type, (D) Living *Psilotum*. (After Banks, 1970.)

the *Bryophyta* comprising the delicate mosses, liverworts, and hornworts which lack a system of conducting tubes; and the *Tracheophyta* which includes all other plants which possess these vessels or tracheids.

Subkingdom Tracheophyta

Phylum Psilophyta—The psilophytes or whisk "ferns" (Fig. 7.120) have no differentiated leaves and roots, just upright stems with chlorophyll and underground supporting and absorptive stems. Branching is of the simple two-fold or dichotomous type and the asexually produced spores are borne in terminal cases or lateral ones aggregated on a spike. They range from Silurian to Recent, but only two genera (Fig. 7.120) apparently belonging to this group survive. The greatest diversity was in the Devonian when several main lineages evolved.

Phylum Lycopodophyta—Lycopods, like the remaining plant groups, have true leaves and roots developed. The leaves are small and spirally arranged, a distinctive feature of this group (Fig. 7.121). Branching is dichotomous or it is monopodial in which one stem becomes dominant. The spore cases are borne on modified leaves which may be aggregated into a cone. The group ranges from Devonian to Recent but is only represented today by 5 small genera of ground "pines" or club "mosses" and quillworts (Fig. 7.122). In the late Paleozoic, particularly the Pennsylvanian, large 100-foot-high scale or seal trees (Fig. 7.123) were a major floral element. The scars or leaf bases on these large trunks resemble scales or the imprint of ancient ring seals, hence, the name.

FIGURE 7.121 Fossil Lycopod Leaves and Leaf Scars.

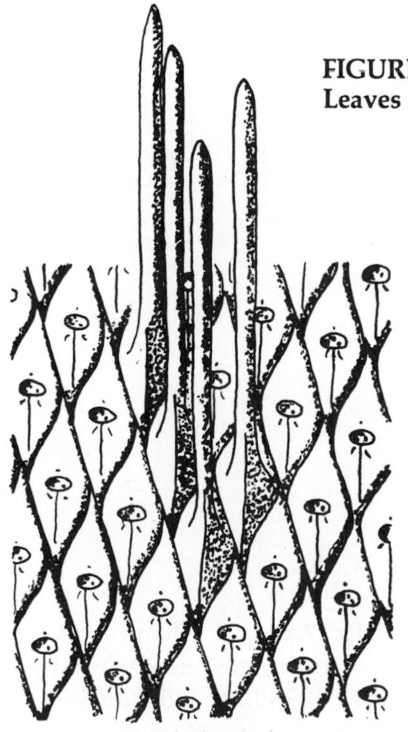

Lepidodendron

FIGURE 7.122 Living Herbaceous Lycopod *Selaginella.*

Phylum Arthrophyta—The articulates or sphenopsids have stems and branches which are jointed. The side branches and their small leaves occur in whorls only at the nodes (Fig. 7.124). The spore cases are terminally arrayed on short lateral branches which are typically aggregated into cones. The single living genus (Fig. 7.125), the small living horsetail or scouring "rush," is the last survivor of this group. Articulates reached the size of large trees in the Pennsylvanian coal swamps. Jointed stems and clustered leaves of these *Calamites* (Fig. 7.126) are common fossils. Articulates range from Devonian to Recent.

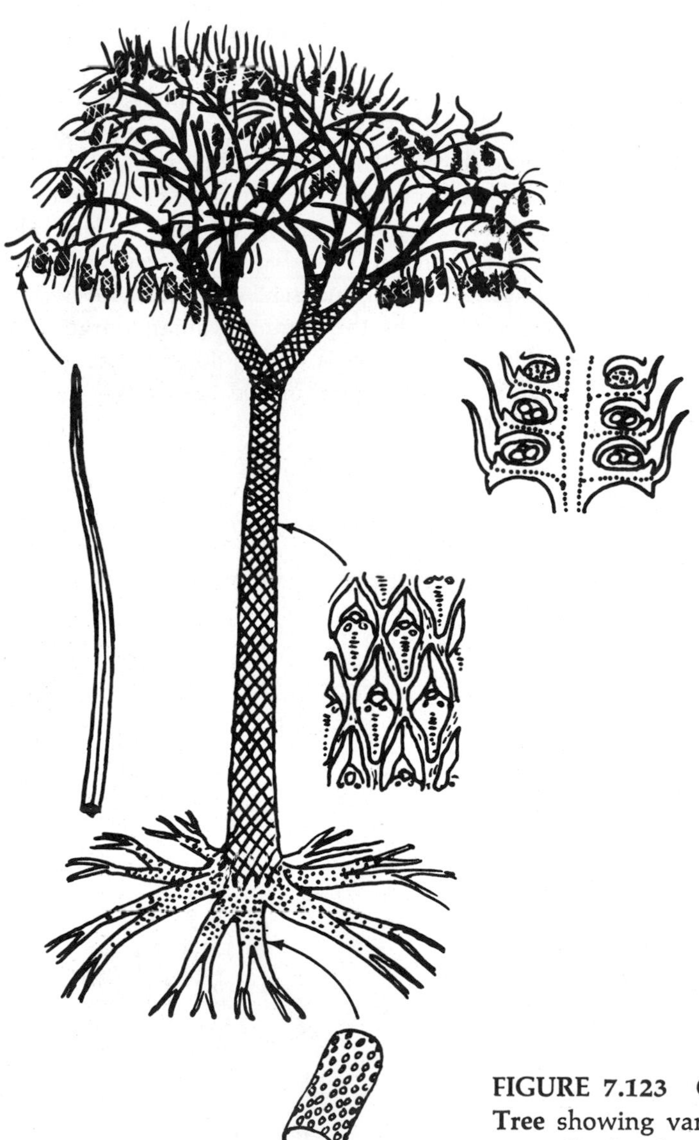

FIGURE 7.123 Carboniferous Scale Tree showing various detached parts typically fossilized individually.

FIGURE 7.124 Arthrophyte Jointing. Fossil Carboniferous foliage.

FIGURE 7.125 Living *Equisetum*.

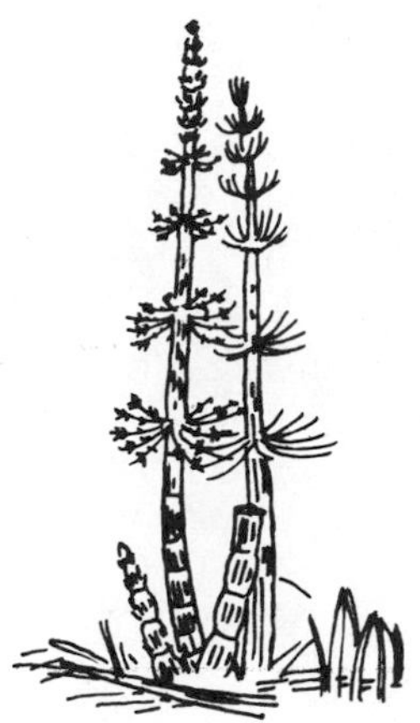

FIGURE 7.126 Carboniferous *Calamites*.

FIGURE 7.127 Ferns. (A) Living fern, (B) Fossil fern foliage.

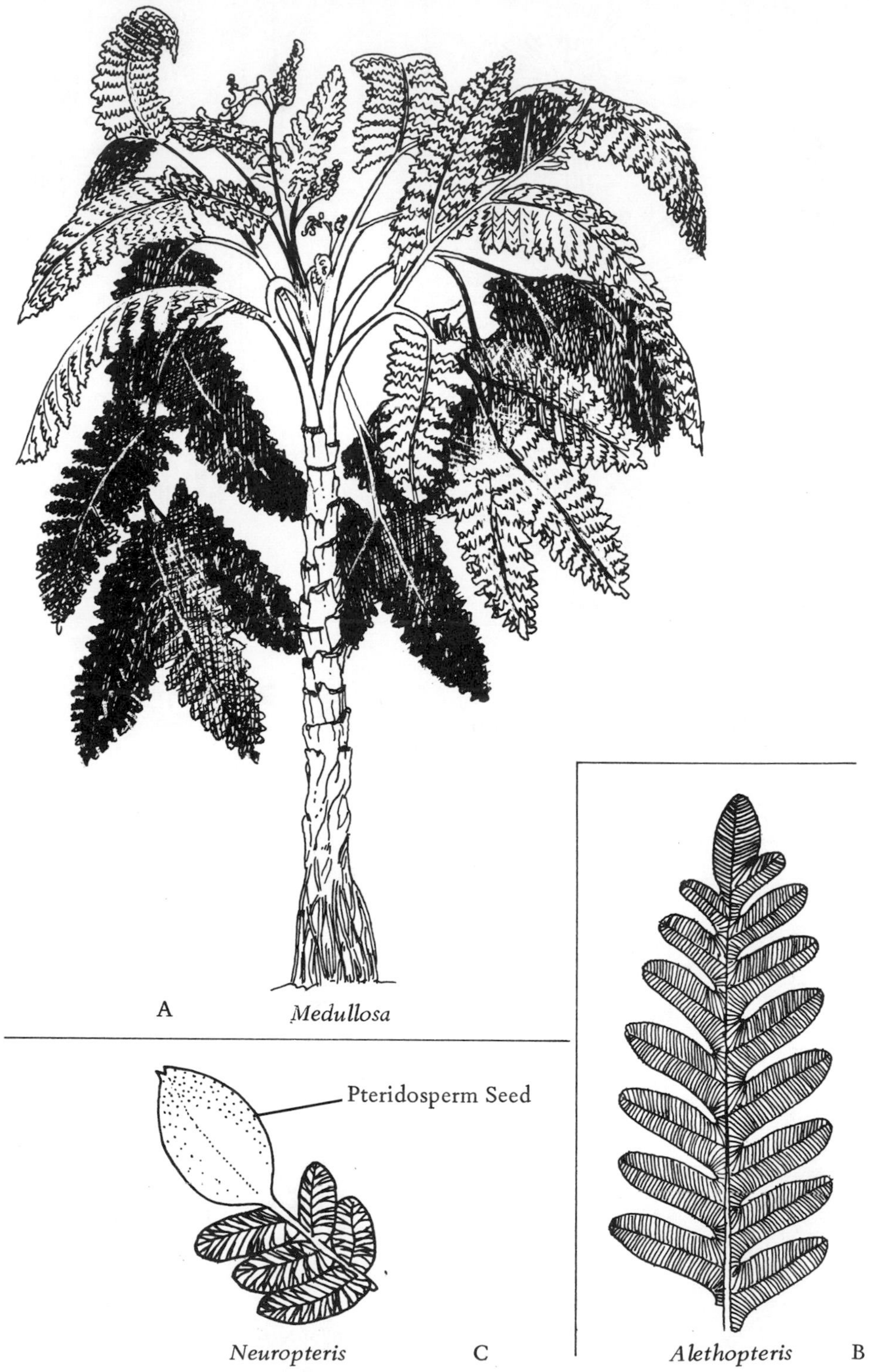

FIGURE 7.128 Pteridosperms or Seed Ferns. (A) Reconstruction of an entire plant, (B) Foliage of another genus, (C) Pteridosperm leaf with seed.

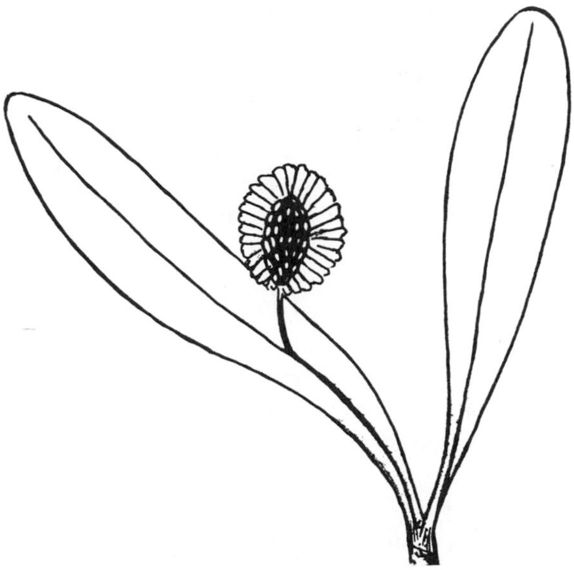

FIGURE 7.129 *Glossopteris.* Foliage and reproductive structure.

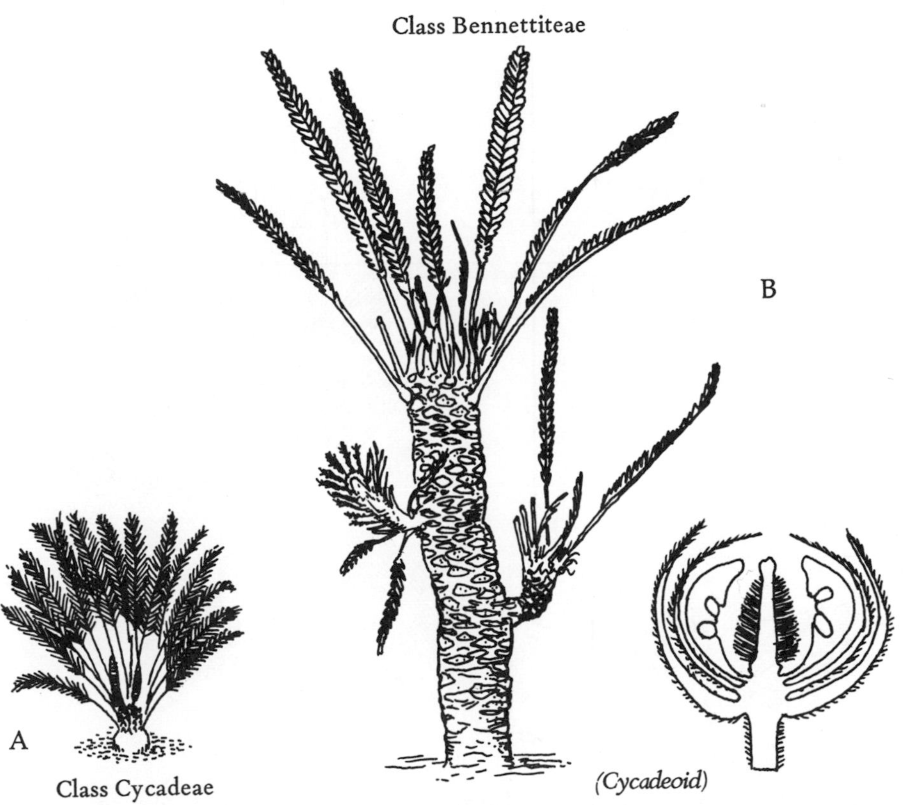

FIGURE 7.130 Cycadophytes. (A) Living cycad, (B) Extinct cycadeoid.

Phylum Filicophyta—The ferns (Fig. 7.127) are the second most diverse group of living plants. The large frondose or pinnate (feathery-looking) leaves with their numerous subdivisions are typically the major part of the body, though some fossil and recent types have attained tree size. The spores are typically borne on the underside of leaves. Ferns range in age from Devonian to Recent and are particularly common in the Mesozoic. Many of the fern-like leaves of the next group were formerly confused with true ferns.

Phylum Pteridospermatophyta—The seed "ferns" (Fig. 7.128) were shrubs or low trees with fern-like foliage that bore seeds or sexually produced reproductive bodies on their leaves. They range in age from Devonian through Cretaceous, but have a Pennsylvanian peak. The peculiar Permian *Glossopteris* (Fig. 7.129) common on the southern continents, was probably a seed fern.

Phylum Cycadophyta—This group encompasses the extinct cycadeoids and the living cycads or sago "palms" (Fig. 7.130). These plants were of the dimensions of shrubs or small trees and bore large coarse fern-like leaves. Seeds were borne on reproductive structures which are specialized leaves

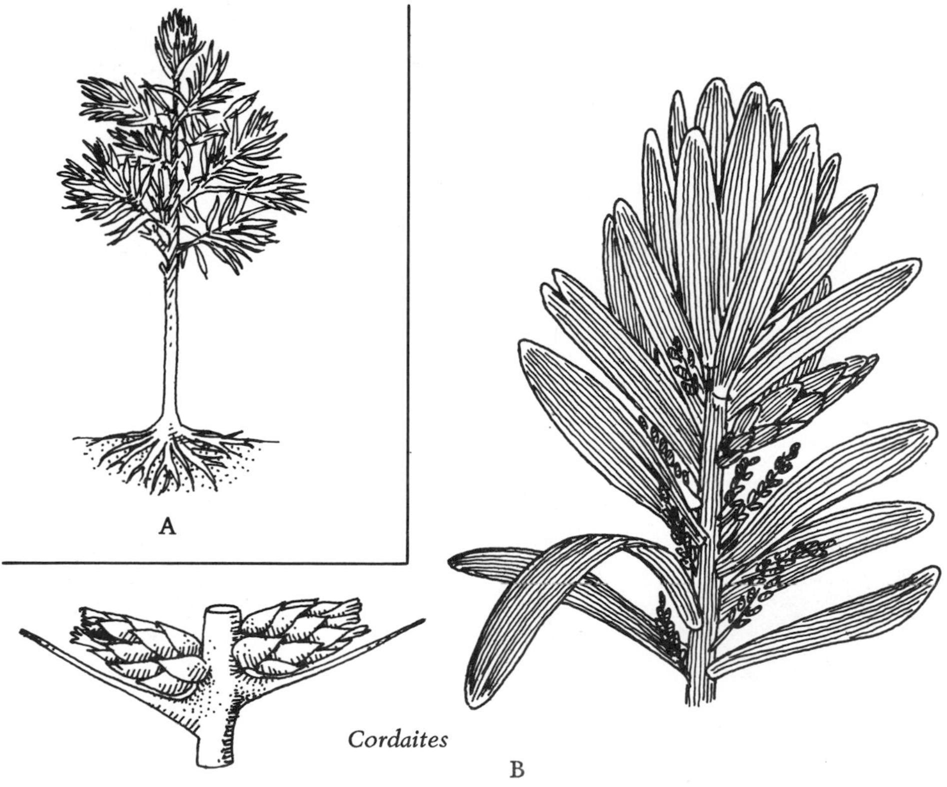

FIGURE 7.131 Cordaitean Conifers. (A) Restoration of the tree, (B) Foliage with cones.

and resemble flowers or cones (Fig. 7.130). The group ranges from Carboniferous to Recent, but was dominant in the Mesozoic Era. Only a few genera of true cycads survive in the tropics and subtropics today.

Phylum Coniferophyta—This group includes the extinct cordaites (Fig. 7.131) and the surviving true conifers (Fig. 7.132). The stem is well-developed as almost all are large trees, some reaching nearly 400 feet in height. The leaves are simple and lack a stalk or petiole (Fig. 7.132). *Cordaites* leaves were strap-shaped; more modern conifers have needle-like leaves. Seeds occur in reproductive structures called cones which are modified branches. Coniferophytes range from Carboniferous to Recent with cordaites dominant in the Pennsylvanian and the true conifers in the Permian and Mesozoic. Conifers are of low diversity today, but cover vast areas in harsh climates.

Phylum Gingkophyta—The surviving gingko or maiden-hair tree is the sole remaining species of this group of trees which evolved in the Carboniferous and was abundant in the Mesozoic. The gingko has a distinctive, dichotomously branched, petiolate leaf (Fig. 7.133). The seeds are borne in small cones which are modified branches similar to conifers.

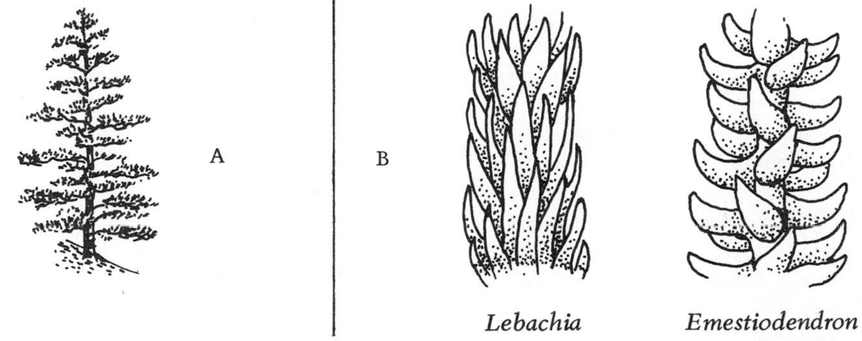

FIGURE 7.132 True Conifers. (A) Living pine tree, (B) Permian conifer foliage.

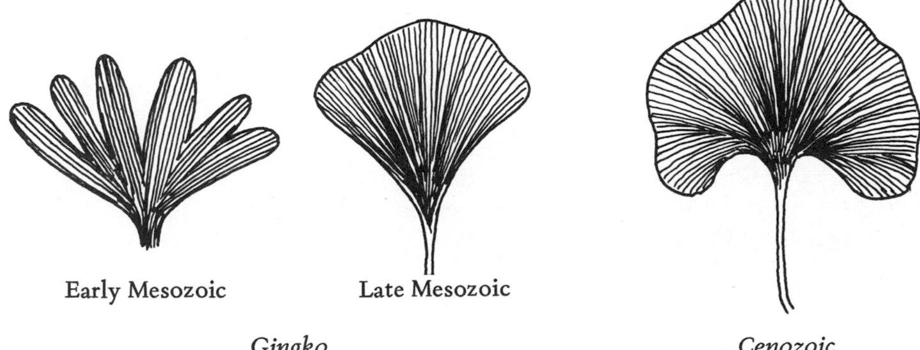

FIGURE 7.133 Gingko Foliage.

Phylum Anthophyta—The angiosperms or flowering plants are by far the largest group of plants in the world today (Fig. 7.134). They range from huge trees such as the eucalyptus to the grasses which are all leaf; they include all garden flowers, vegetables, and shrubs, and many common trees such as oak, maple, and sycamore. They are characterized by the possession of flowers (Fig. 7.134), showy devices to lure insects which aid the reproductive cycle, and fruits, tasty structures housing the seeds. The edible fruits are eaten by mammals and birds who move about and by the time the seeds leave their digestive systems, they, have been dispersed. Angiosperm leaves are typically petiolate. Flowering plants range from the Cretaceous, or possibly the Jurassic or Triassic, to Recent. They became dominant during the Cretaceous Period.

Questions

1. What are the major prerequisites for fossilization?
2. Why do you suppose trilobites have a better fossil record than spiders?
3. What are four common organic skeletal materials?
4. What are the three major inorganic skeletal compounds? Which is most widespread?
5. What is the difference between perminearalization and replacement? Between a mold and a cast?
6. What is the value of trace fossils?
7. What is a phylogenetic classification?
8. What are the distinguishing characteristics of the four kingdoms of organisms?
9. Which groups of protistans are the major oceanic producers of the latter portion of geologic time?
10. How does a foram test differ from a small snail shell? From a small cephalopod shell?
11. What features of archaeocyaths suggest sponge affinities?
12. How can tetracorals and hexacorals be distinguished?
13. What are the differences between stromatolites and stromatoporoids?
14. How can one distinguish bryozoans and colonial corals?
15. How can brachiopod shells and clam shells be differentiated?
16. How can one distinguish snail shells from cephalopod shells?
17. How does one differentiate trilobites and xiphosuroid merostomes?
18. How can ostracods be distinguished from branchiopods? From small clams?
19. What are the distinguishing features of echinoderms?
20. Why are graptolites assigned to the hemichordates?
21. What are the distinguishing traits of vertebrates?
22. What are the major differences between reptiles and mammals?
23. How are ferns and seed ferns differentiated?

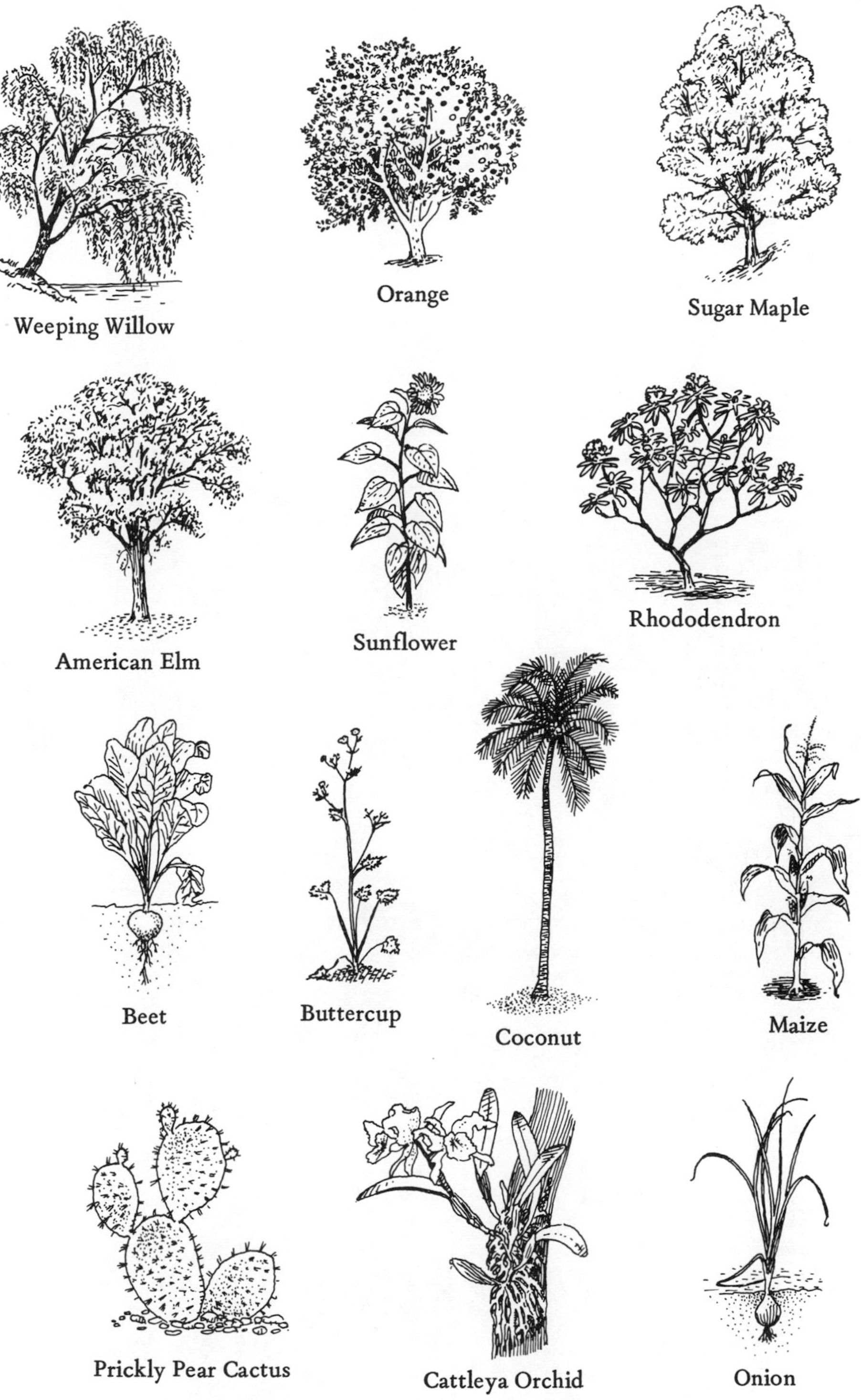

FIGURE 7.134 Angiosperms. A representative sampling of the diversity in this large group.

8

The Stable Heartland:

Sedimentation in Epeiric Seas

Most early geologists, who extrapolated from what they saw in their local area to the world, were not concerned with differentiating large-scale patterns of sedimentation. As knowledge expanded, however, it became apparent that there were several broad categories of regions, each characterized by particular types of sedimentary environments and, hence, rocks. Analysis of how these patterns form and the time relationships of the rock units deposited within them has only begun. As we saw earlier, many early North American geologists did not carefully study modern analogs in interpreting ancient environments.

Geologists have recognized at least five of these broad types of regions or environments. These are the terrestrial regions, the continental shelves, the ocean basins, the geosynclines, and the epeiric seas. Each of these types of regions has a characteristic effect on the patterns in which its sediments are deposited and, hence, profoundly affects the process of correlation discussed in Chapter 10.

The terrestrial environments (fluvial, desert, lacustrine, glacial) and continental shelf (littoral, sublittoral) environments are the best known today because they are the most accessible. The patterns of sedimentation in these environments are treated in physical geology texts in some detail. In historical geology, we are interested in the time context of the rock record, in particular, the correspondence or lack of correspondence of time and rock units. Both the terrestrial and continental shelf depositional records are characterized by rapid lateral and vertical shifts in facies. Thus, there has been less inclination on the part of geologists to confuse time and rock units because the latter typically have little lateral persistence. Still, as we shall see in studying the other main environments, Wernerian concepts die hard. Some geologists continue to have the tendency to assume, for example, that the youngest glacial deposit in Ontario is the same age as one in Ohio despite the fact that they realize it takes time for ice to advance and retreat.

The ocean basins, geosynclines, and epeiric seas are also discussed in physical geology texts, but the full implications of the difference between rock and time units is rarely, if ever, developed. The patterns of sedimentation in the ocean basins will be discussed in Chapter 11; the patterns in geosynclines, in Chapter 9.

Epeiric Seas

In this chapter we will explore the most widespread pattern whose rock record is accessible on the continents, that of the *epeiric, platform,* or *epicontinental* seas. These are the widespread, shallow invasions of the seas that have covered most of the stable continental regions (continental platforms or cratons) of the earth at various times in the past (Fig. 8.1). Their dimensions are measured in thousands of miles and the rock record is very uniform geographically and relatively thin. The exact causes of the advances and retreats of these great seas is still being explored, but their existence has been obvious for a long time. Recently, several geologists have suggested that epeiric seas develop during times of continental fragmentation when the volume of the ocean basins is reduced (see Chapter 11).

One of the great problems in evaluating the epeiric sea environment is apparent by a look at a map of the earth's surface today (Fig. 8.2). With

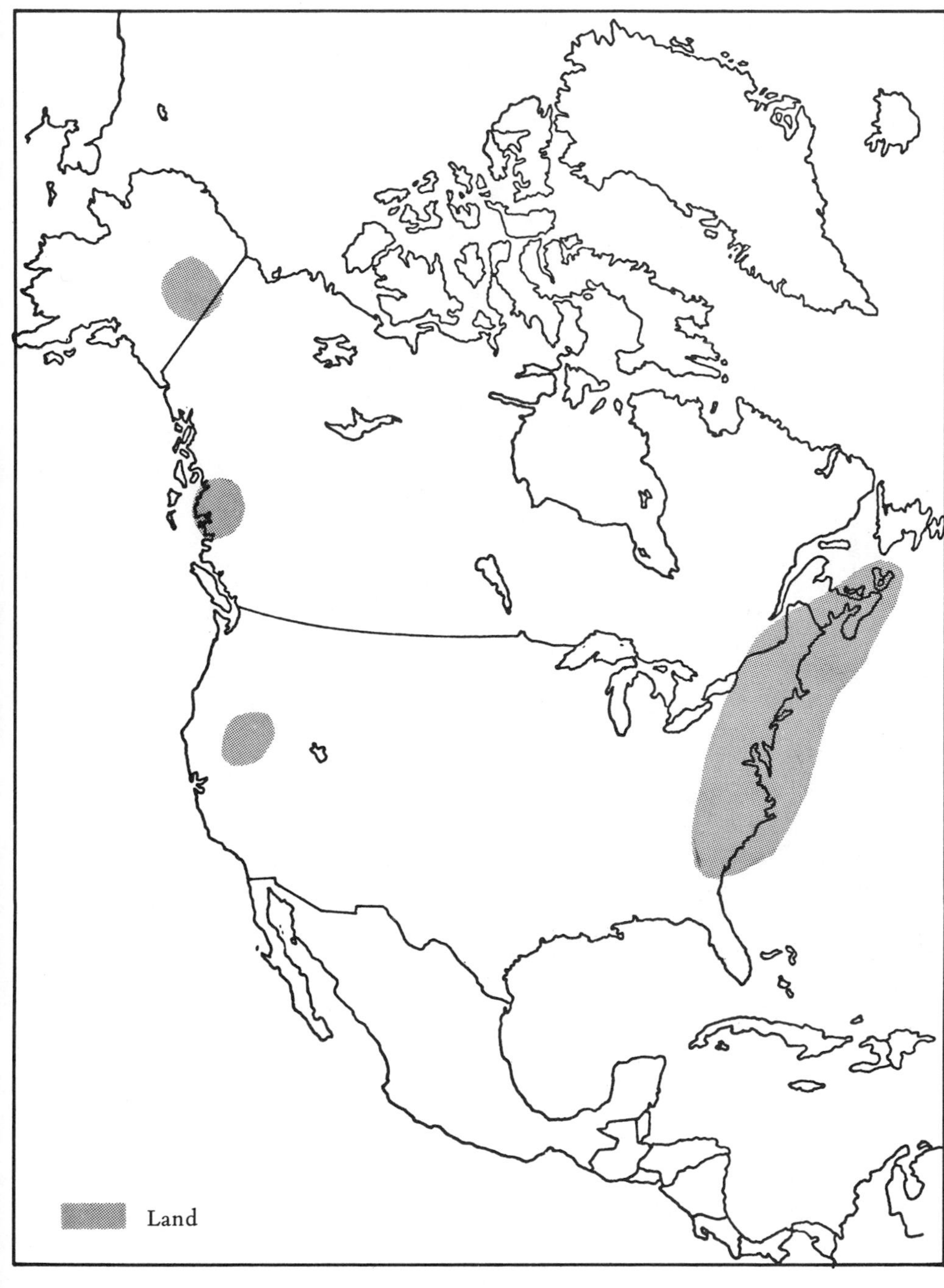

A

FIGURE 8.1 Inferred North American Paleography of (A) Late Ordovician and (B) Middle Cretaceous Showing Epeiric Seas. The maps do not show plate tectonic effects, but only areas of present continental masses which were inundated.

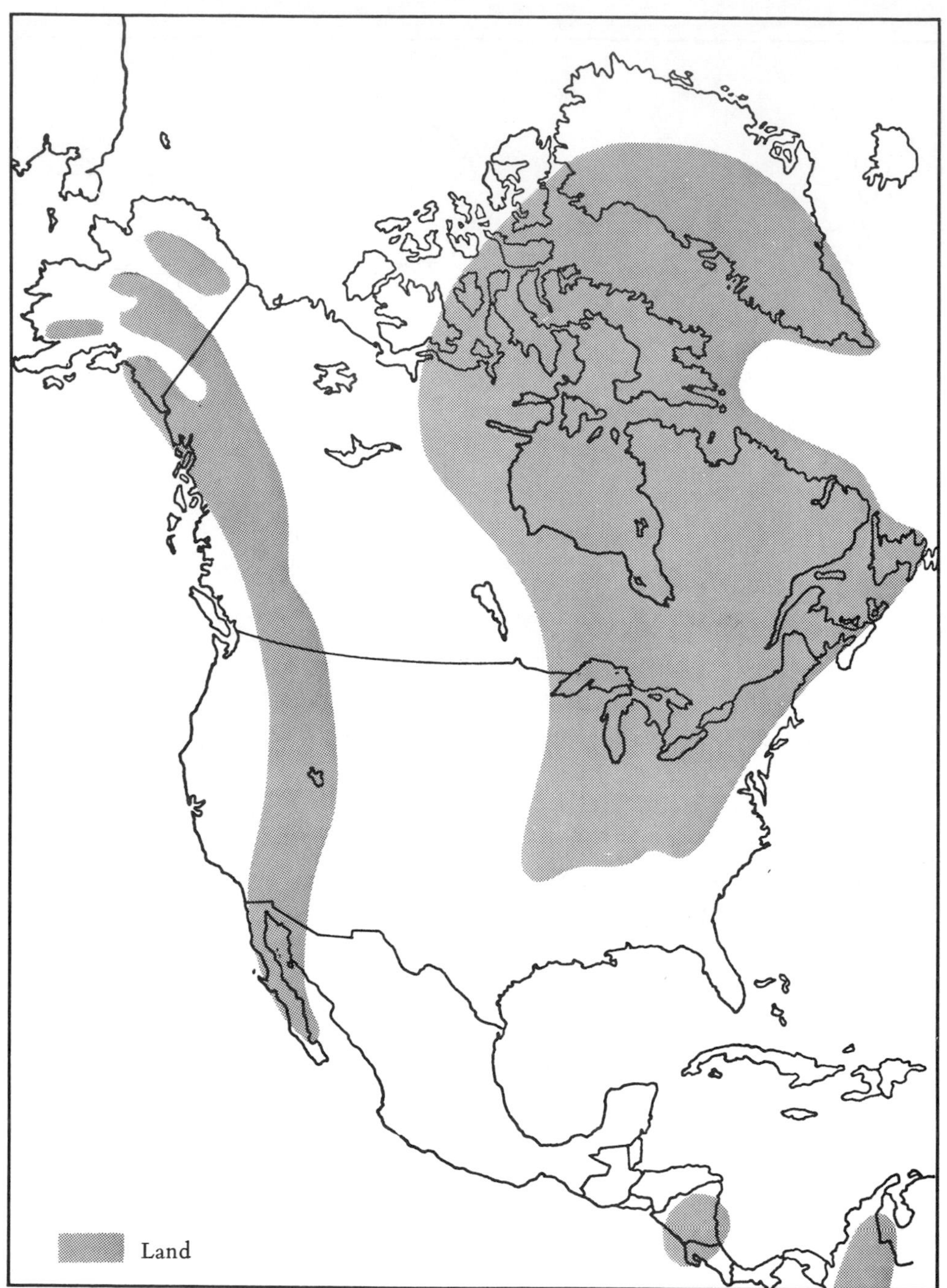

B

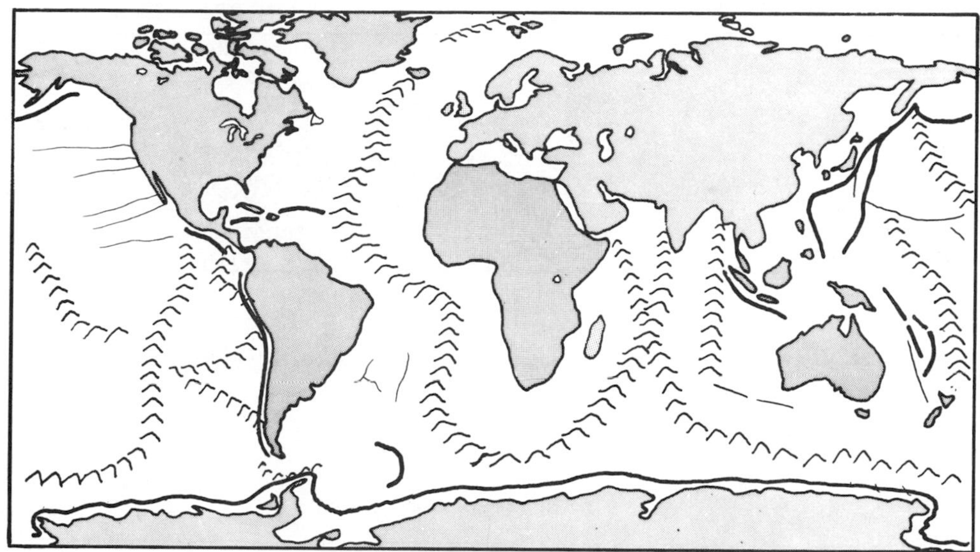

FIGURE 8.2 World Map Emphasizing Sea Floor Topography.
Note small extent of present epeiric seas.

the exception of Hudson Bay in North America, and, perhaps, the Baltic Sea in Europe, there are no widespread marine incursions of this sort on the continents. In addition, these two areas are in cold regions today, away from most of the areas where geology grew up and also differing from the past epeiric seas whose fossils and rocks testify to warmer climates. Hence, our usage of uniformitarianism is restricted; we cannot observe typical examples of these environments today. We must, therefore, apply known physical and chemical laws to the conditions we can deduce from the rock record. This is a type of uniformitarianism in reverse, but the principle of uniformitarianism is so thoroughly validated now, that such a procedure is acceptable.

We know from gross paleontological dating of the rock record of transgressing epeiric seas that the deposits are older near the continental margins than they are toward the center (Fig. 8.8). In other words, the seas transgressed from the ocean basins inward over the continent over a period of time. Hence, the continental surface must have been inclined toward the margins, but very gently because the average slope of their undeformed deposits is typically less than .1 to .5 foot per mile. The type of fossils and rocks indicates a very shallow water depth over immense areas. An important result of these conditions is that the great currents characteristic of today's oceans, both the basins and the shelves, could not exist because the energy would be almost immediately dissipated through bottom friction. (There is much evidence of such agitation in the deposits of epeiric seas.) Furthermore, the normal twice daily rise and fall of the tides could not have occurred, just as it does not today in isolated seas such as the Mediterranean (which is not epeiric) and Florida Bay (which is only a marginal shallow water shelf).

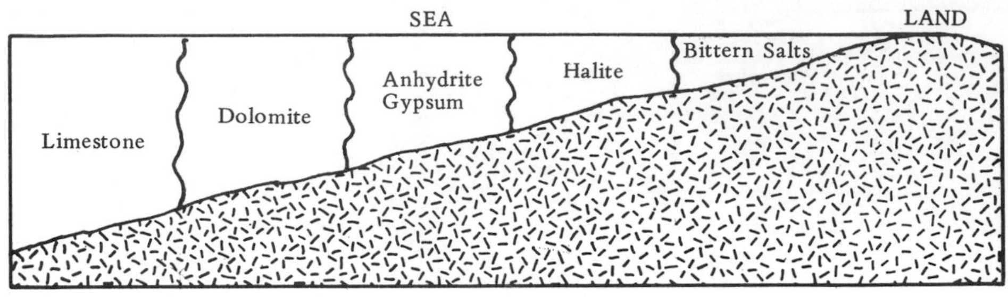

SEA
LAND
Limestone
Dolomite
Anhydrite
Gypsum
Halite
Bittern Salts
A

FIGURE 8.3A The Basic Sequence of Chemical Precipitates in an Epeiric Sea.

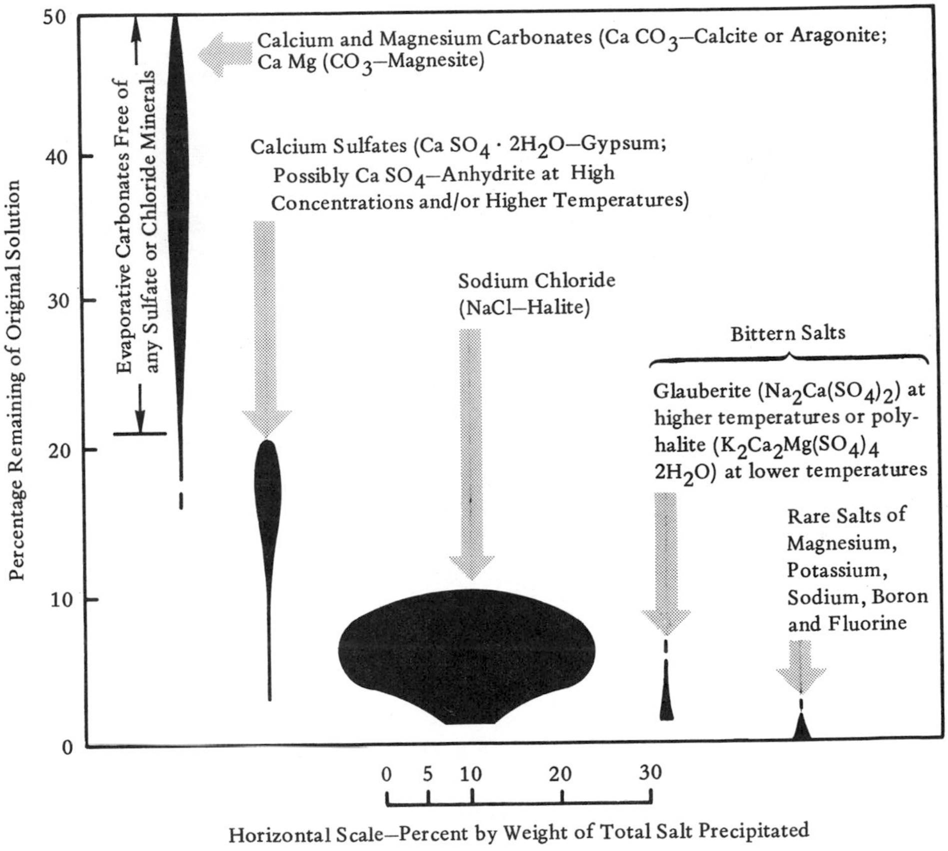

Percentage Remaining of Original Solution
50
40
30
20
10
0
Evaporative Carbonates Free of any Sulfate or Chloride Minerals
Calcium and Magnesium Carbonates (Ca CO3—Calcite or Aragonite; Ca Mg (CO3—Magnesite)
Calcium Sulfates (Ca SO4 · 2H2O—Gypsum; Possibly Ca SO4—Anhydrite at High Concentrations and/or Higher Temperatures)
Sodium Chloride (NaCl—Halite)
Bittern Salts
Glauberite (Na2Ca(SO4)2) at higher temperatures or polyhalite (K2Ca2Mg(SO4)4 2H2O) at lower temperatures
Rare Salts of Magnesium, Potassium, Sodium, Boron and Fluorine
0 5 10 20 30
Horizontal Scale—Percent by Weight of Total Salt Precipitated
B

FIGURE 8.3B Theoretical Evaporite Mineral Precipitation Series.

Chemical Deposits

Because of the lack of the freshening effect of normal tides and currents, epeiric seas would develop a salinity gradient with higher salinities the further toward the interior of the continent the sea progressed (except near the shore where the freshening effects of river water could be effective). The development of this salinity gradient would form a series of facies, based on the degree of salinity, in broad bands parallel to the shore (Fig. 8.3). Hence, the conditions of sedimentation for chemical precipitates would not be uniform over the entire sea and no chemical sedimentary rock unit, such as a thousand-mile wide limestone, could possibly be of the same age throughout, but must instead be older marginally than interiorly. This conclusion is, of course, in direct conflict with the ideas of Ulrich and others, and with the practices of many geologists, particularly as embodied in several standard textbooks and laboratory manuals.

The sequence of epeiric sea chemical precipitates (Fig. 8.3), from least restricted conditions at the margins to most restricted conditions toward the center, is *limestone, dolomite* (not the type produced by later alteration of

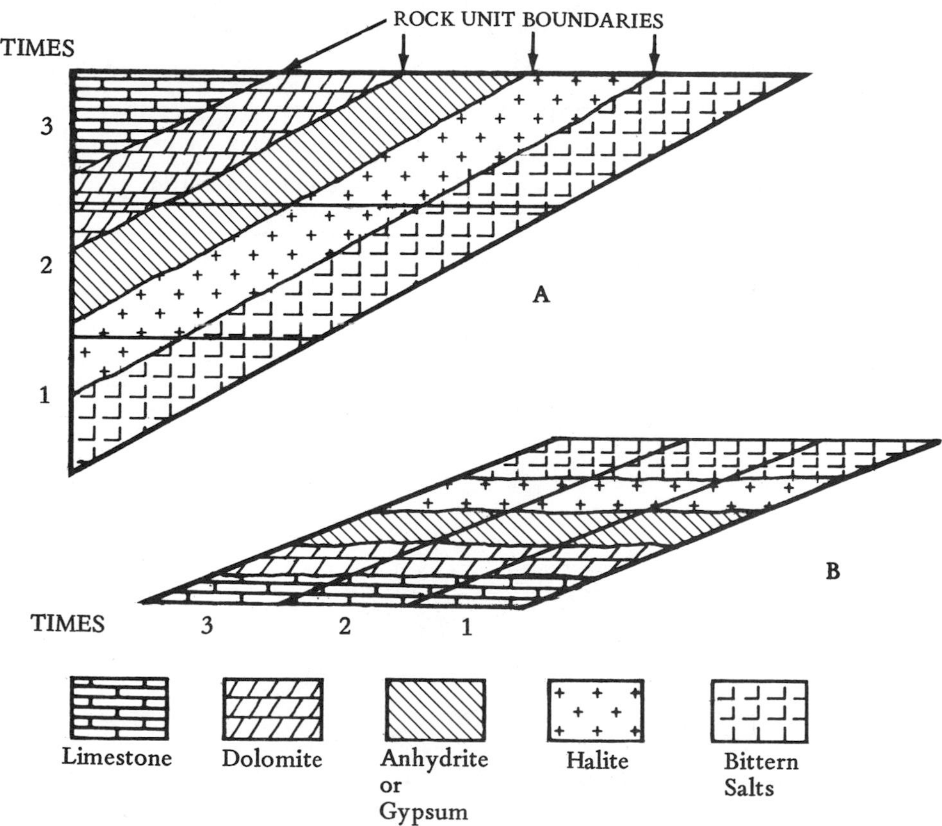

FIGURE 8.4 Patterns of (A) Transgression and (B) Regression in Chemical Precipitate Deposits of Epeiric Seas. Vertical dimension grossly exaggerated.

limestone), *anhydrite* or *gypsum, halite,* and the *bittern* potassium and magnesium *salts.* (The limestones, lying nearest the active open-sea conditions, will show a progressive variation from high to low energy types as one goes inward.) These deposits result from the physical-chemical conditions inherent in epeiric seas and the sea will always attempt to impress this pattern on its sedimentary record. If there is any appreciable influx of clastics, however, the chemical precipitates will be so completely swamped out that they will compose only an insignificant fraction of the rocks. In a strictly chemical sequence of a transgressing sea, the deposits of more open seas overlie those of more restricted conditions and the reverse occurs in regressing seas (Fig. 8.4). You should also note from this discussion, that the presence of evaporite salts does not require an actual physical barrier to set up restricted conditions. The restriction inherently occurs in epeiric seas because of the lack of normal tides and currents.

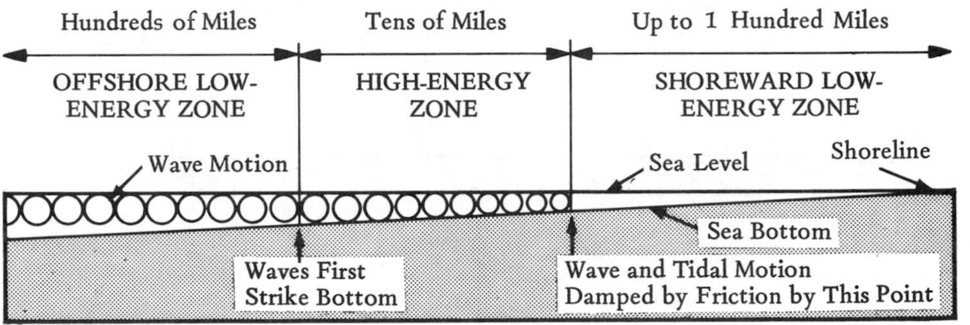

FIGURE 8.5 Energy Zones in Epeiric Seas.

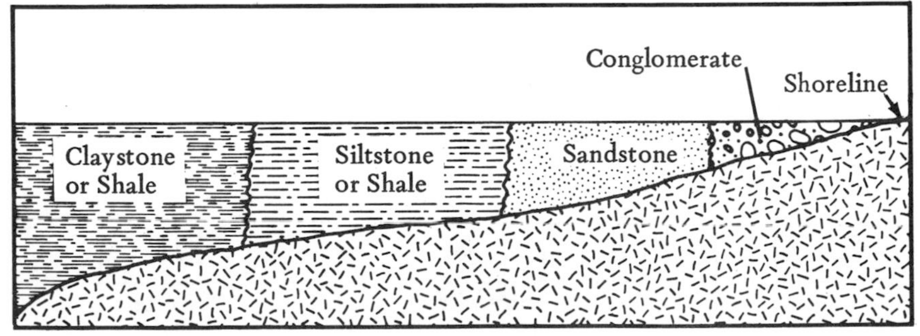

FIGURE 8.6 The Basic Sequences of Clastic Sedimentary Rocks Formed in an Epeiric Sea. This model assumes particles of all sizes are available.

Clastic Deposits

If one considers the clastic deposits of epeiric seas, the effects of the restricted condition are also apparent. Wave action will be restricted to narrow bands where local, wind-caused wave systems develop and feel bottom (Fig. 8.5). The energy is rapidly dissipated and the transportative effects are restricted in scope. Therefore, a widespread clastic rock unit, such as sandstone, which reflects a certain energy environment, cannot be the same age throughout its geographic extent. It too must be older on the cratonal margins than near the center.

The greatest amounts of energy release will occur in the shallower water where the waves feel bottom and there will be correspondingly lesser amounts as one moves outward from the shore toward the margins of the epeiric sea. The basic sequence of clastic rocks (Fig. 8.6), from inshore to offshore, is, therefore, pebbles, sand, silt, and clay, providing all sizes of particles are being provided by rivers. If the particle sizes delivered do not include coarse fractions, there will be a reduction of types at this end of the sequence. In a transgressing sea, then, finer clastics will come to overlie coarser ones and the reverse occurs in a regressing sea (Fig. 8.7).

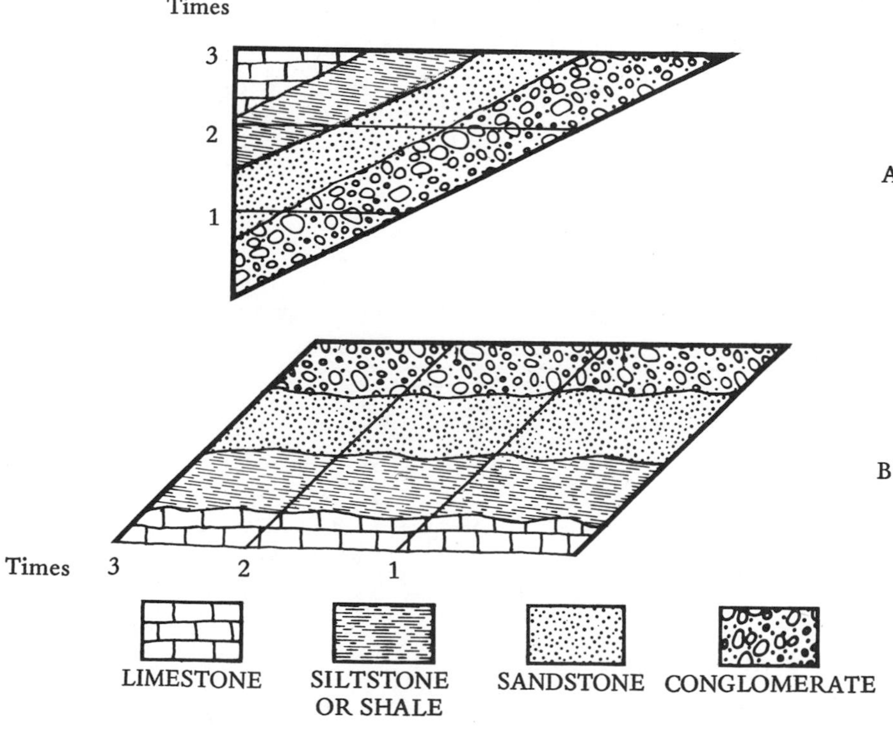

FIGURE 8.7 Pattern of (A) Transgression and (B) Regression in Clastic Deposits of Epeiric Seas. Vertical dimension grossly exaggerated.

Geologists and Epeiric Sea Deposits

Actual rock sequences can be composed of either pure chemical precipitate sequences, "pure" clastic sequences in which the precipitates are swamped out and occur only as impurities, and various mixtures of the two depending on local conditions. Sedimentary rock units such as formations are always of varying ages throughout their geographic distribution (Fig. 8.8). One can determine lateral facies equivalents even at a single locality by noting the overlying and underlying units as noted by *Walther* and discussed in an earlier chapter. Keep in mind that actual sedimentation patterns are rarely as simple as indicated in these generalized diagrams. Local topographic irregularities in the sea bottom typically produce a more complicated pattern when analyzed in detail.

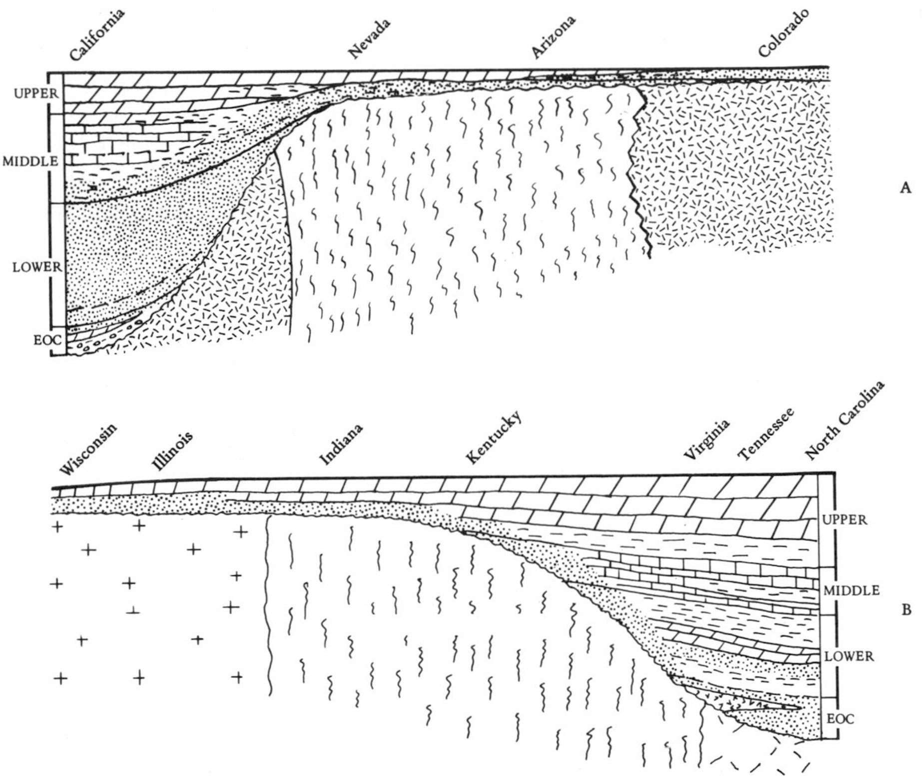

FIGURE 8.8 Restored Cross-Sections of Eocambrian and Cambrian Strata of (A) Western and (B) Eastern North America. Note the time-transgressive nature of all formations.

Finally, we can determine whether the present edge of the marine deposits of an epeiric sea actually represents its original shoreline by checking to see if facies bands parallel the present edge of the deposits or are truncated by it (Fig. 8.9). In the former case, we are dealing with the original margin. In the latter case, the present edge is an erosional feature and we have to try to estimate the original extent of the sea by studying the facies pattern and looking for erosionally isolated remnants. Several series of paleogeographic maps of North America have been prepared by workers who did not accept facies and used the present edge of deposits to mark ancient shorelines. These often-reproduced maps are of diminished value in modern historical geology (Fig 8.10).

Geologists working with the thin, widespread epeiric sea deposits on the continents have often found it convenient to consider a sedimentary rock unit of uniform age throughout, despite the evidence cited above that it cannot be. Even the legends of most geologic maps still portray rock and time unit boundaries as parallel. Very serious consequences result from this practice however.

In Fig. 8.11, we have an advancing epeiric sea transgressing over a continent. At time 1, a band of sand forms nearest shore, a band of mud further out, and finally, beyond the reach of the clastics, a band of lime forms. At time 2, the sea has advanced further inland and so have the facies bands. There is a band of nearshore sand, more offshore mud, and most offshore lime, only now each of these sediment types is a younger age. At time 3, the sea has moved still further inland and the facies bands have shifted correspondingly. A series of maps showing the position of the shoreline and the distribution of sediment types at times 1, 2, and 3 is given on the top of the blocks in Fig. 8.11.

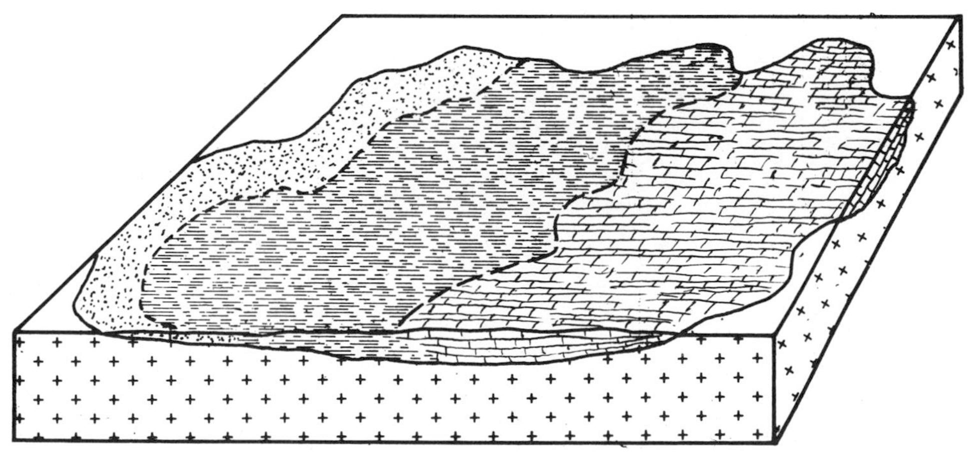

FIGURE 8.9 Determination If the Present Limits of a Rock Unit Mark Its Original Boundaries or Are Erosional. Western margin is probably depositional as facies bands parallel it. Northern margin is erosional as facies bands are truncated.

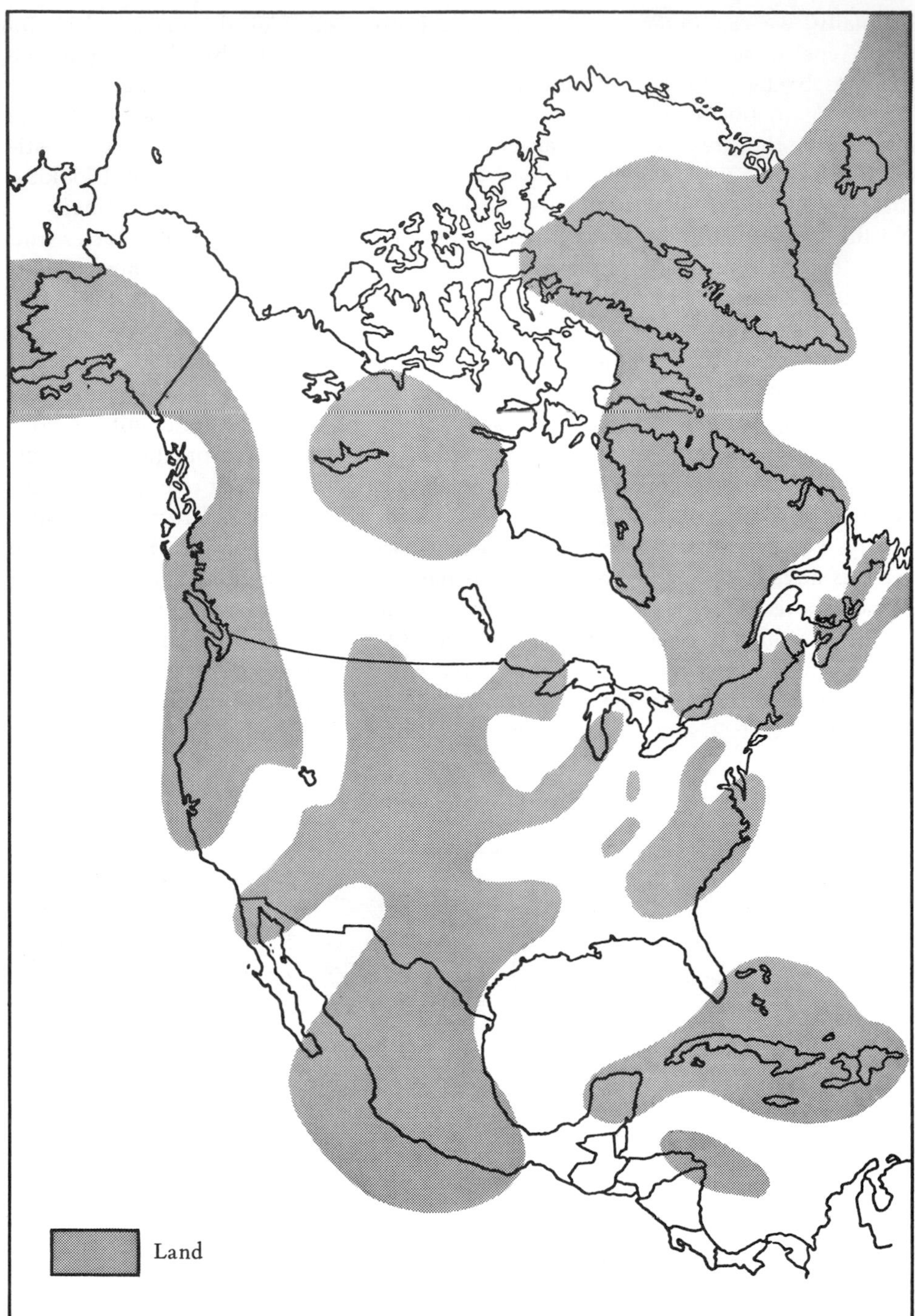

A

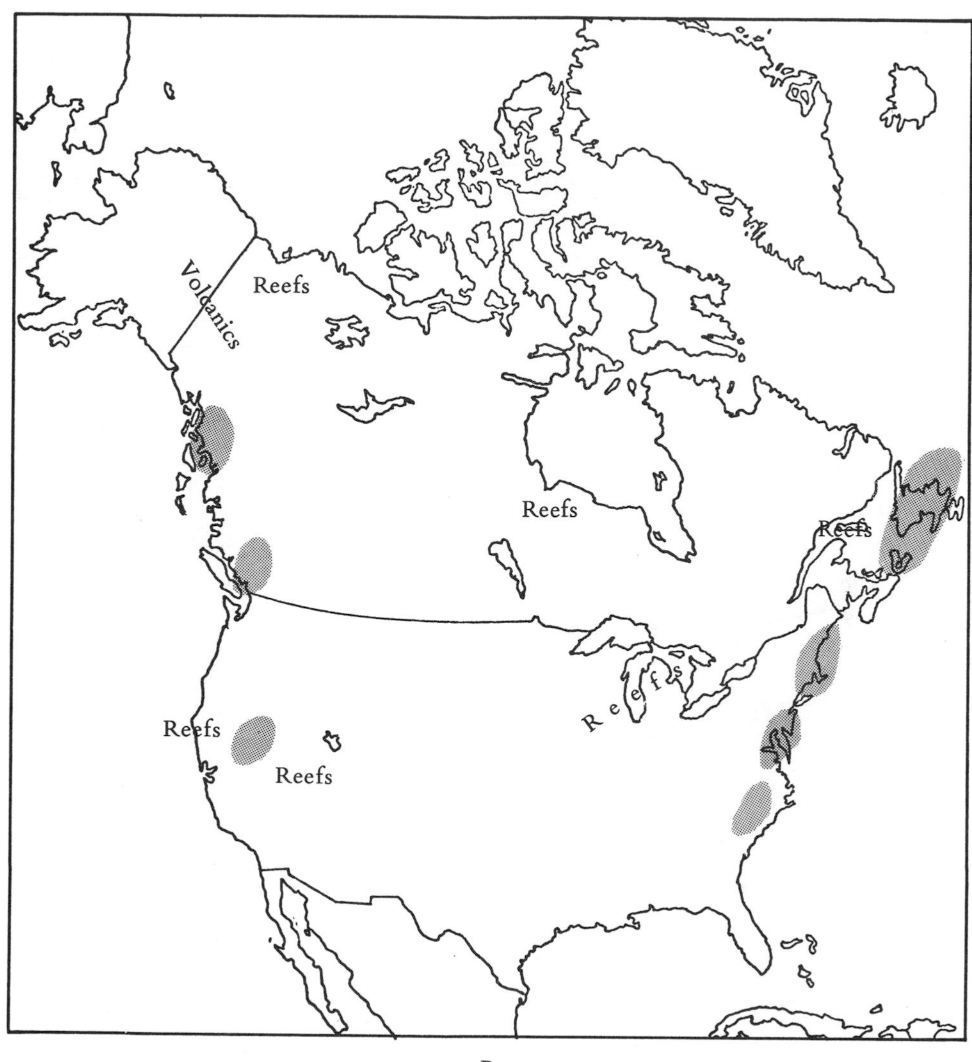

B

FIGURE 8.10 Two Interpretations of Middle Silurian Paleography. (A) Utilizing present limits of marine deposits as shore lines, (B) Utilizing knowledge of facies as explained in Figure 8.9.

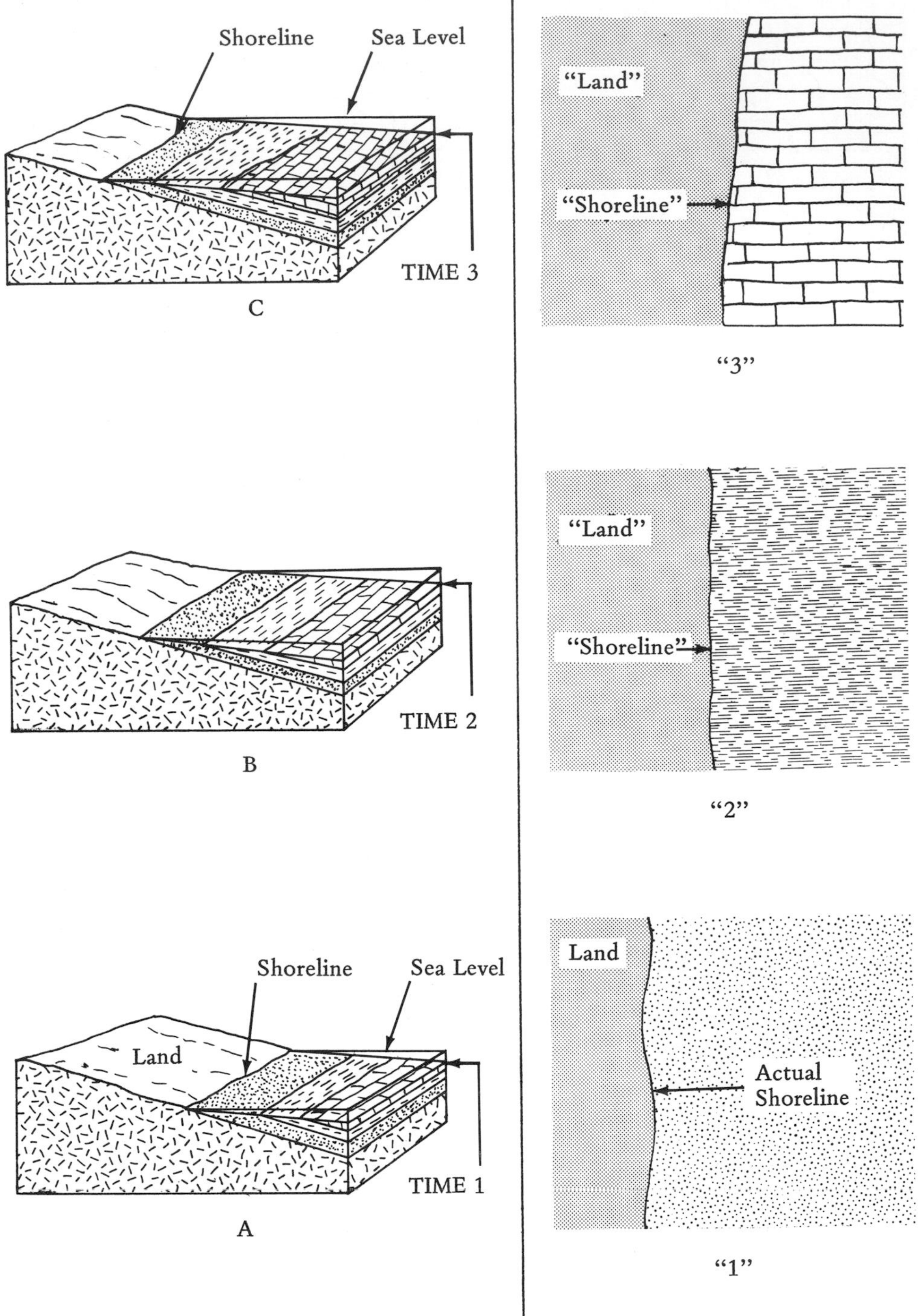

FIGURE 8.11 Errors Resulting From Confusing Lithology with Age. (A–B–C) Deposits of a transgressing sea. Times 1–2–3 are maps of deposits forming at those times. Times "1", "2", and "3" are maps made by substituting rock- for time-correlations.

If one assumes, however, as many geologists do, that each rock type is of uniform age throughout, note what happens. If the sand, which is lowest superpositionally and, hence, "oldest," is assumed to be the same age throughout, the shoreline is far inland at the time of its formation and the only deposit is sand. If the overlying mud, the next younger unit, is assumed to be the same age everywhere, the shoreline has now shifted seaward and only mud occurs offshore. Finally, if the limestone, the "youngest" unit, is considered the same age throughout its extent, the shoreline has now shifted even more seaward and, again, there is only one type of sediment forming. The series of maps, labeled times "1," "2," and "3" in Fig. 8.11, show the consequences of our error. Instead of a transgressing sea, we have a regressing sea. The geologic history of the area has been reversed and the ancient geography has been inverted! And, naturally, one will never find the lateral facies of equivalents of the resulting sandstone, mudstone, and limestone.

What we have just done is to demonstrate again that lithology and time are not synonymous, at least not in the sedimentary deposits of the restricted epeiric sea environment. Let us now proceed to analyze the "non-restricted" record of the geosynclines and ocean basins.

Questions

1. Why is it difficult to determine what conditions were like in ancient epeiric seas?
2. What is the basic sequence of precipitate deposits in epeiric seas?
3. Why is it unnecessary to have an actual physical barrier in epeiric seas to produce hypersalinity?
4. What is the basic sequence of detrital deposits in epeiric seas having a supply of particles of all sizes?
5. What happens to geologic history if lithologic units of epeiric seas are considered time units?

9

The Unstable Margins:

The Development of Geosynclines

Historical Development of Geosynclinal Theory

As early as 1857, the famous American geologist James Hall noted that the marine rocks found on the continents today occur in two different patterns (Fig. 9.1). One is the thin, widespread, uniform pattern of rocks deposited in epeiric seas that was discussed in the last chapter. These sequences are not only thin, but, particularly toward the continent interior, are full of major gaps, which reflect the alternate transgression and regression of the epicontinental seas. At the continental margins, where the major mountain systems such as the Appalachians and western Cordillera occur today, however, are found very thick, deformed sequences of marine strata in more or less linear or arcuate belts. These rocks represent much more geologic time than the platform rocks (Fig. 9.2). Sedimentation on the continental side of the belt occurred in shallow waters as the fossils and rock types indicate. Hence, the trough must have subsided greatly to receive a several-miles-thick accumulation of sediments, but at any one time, the water was shallow. On the oceanic side, rock types are different and generally have been more difficult to decipher because of the more intense deformation there. In 1873, *James Dana* applied the term geosynclinal, later changed to *geosyncline,* to unstable linear belts of thick sediments which have been deformed into mountains. The notion grew that a geosyncline was a subsiding region predestined to undergo mountain-building deformation. Eventually, any thick sequence of strata was called a geosyncline, whether or not it had been deformed. Once again geologists saw a pattern in one area and inferred it was universal.

In Europe, the basic concept of a geosyncline underwent some revision because in that area the major geosynclines occurred between continents rather than marginal to them (Fig. 9.3). For example, the great Tethyean Geosyncline, which spawned the Alps and other Mediterranean-area mountains, lay between Africa and Europe; and the Uralian Geosyncline, parent of the present Ural Mountains, lay between Asia and Europe. The Alpine area, whose deformation was much

NORTH AMERICAN CRATION APPALACHIAN GEOSYNCLINE

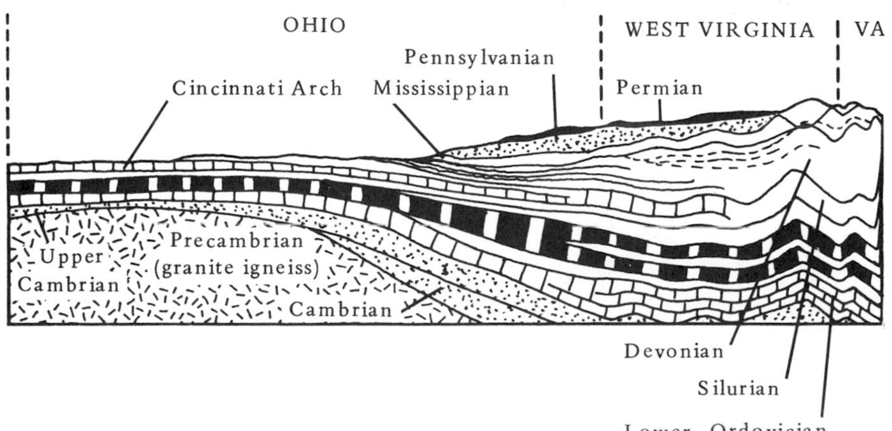

FIGURE 9.1 Cross-section of Eastern North America showing contrast between cratonal and geosynclinal sedimentary sequences.

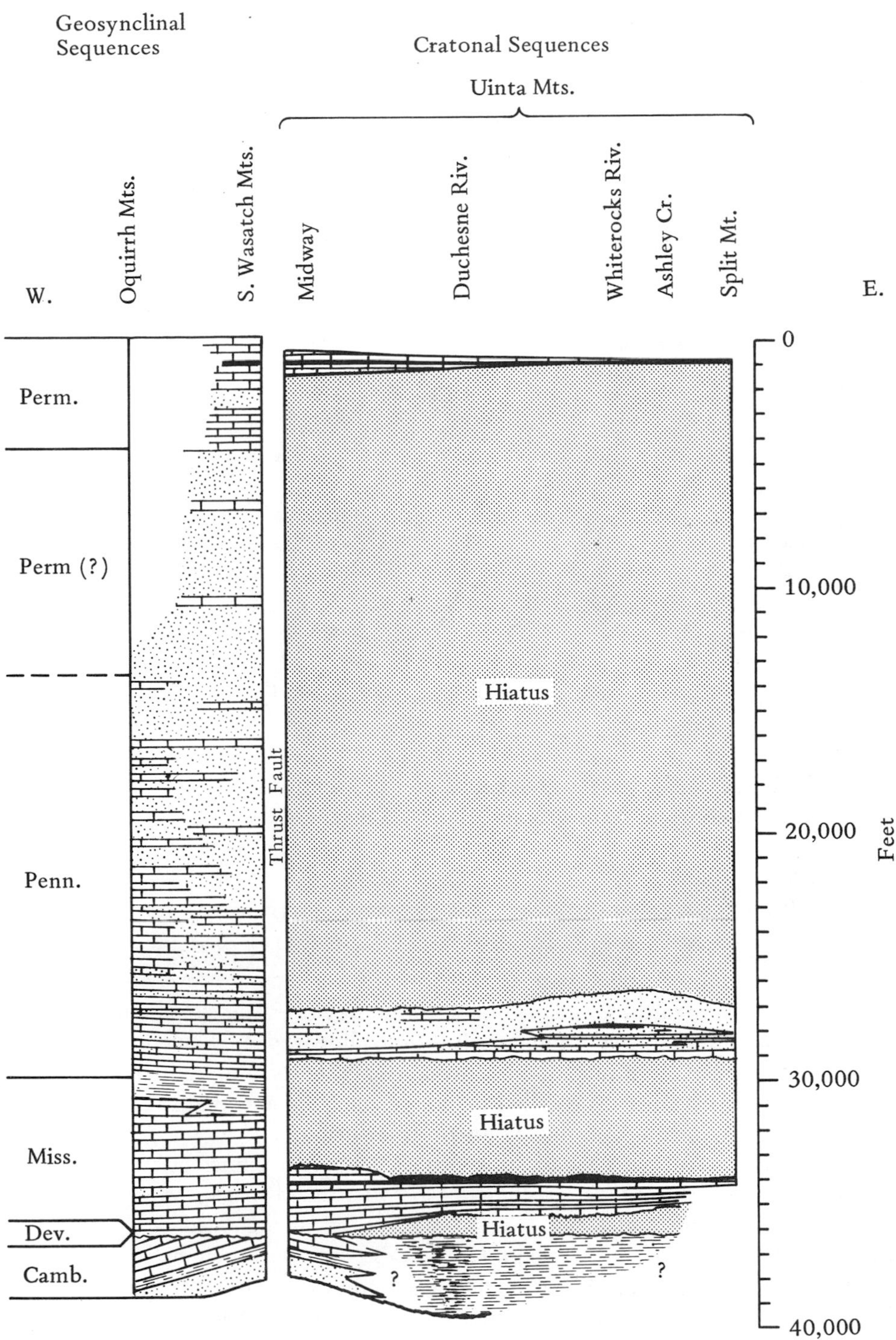

FIGURE 9.2 Diagram Illustrating the Differences in Stratigraphic Columns from Geosynclinal and Cratonal Regions.

more recent than that of American geosynclines and, therefore, has a more complete record, became a standard for study. *Emile Haug* and others developed a complex set of subdivisions of geosynclines into ridges and troughs.

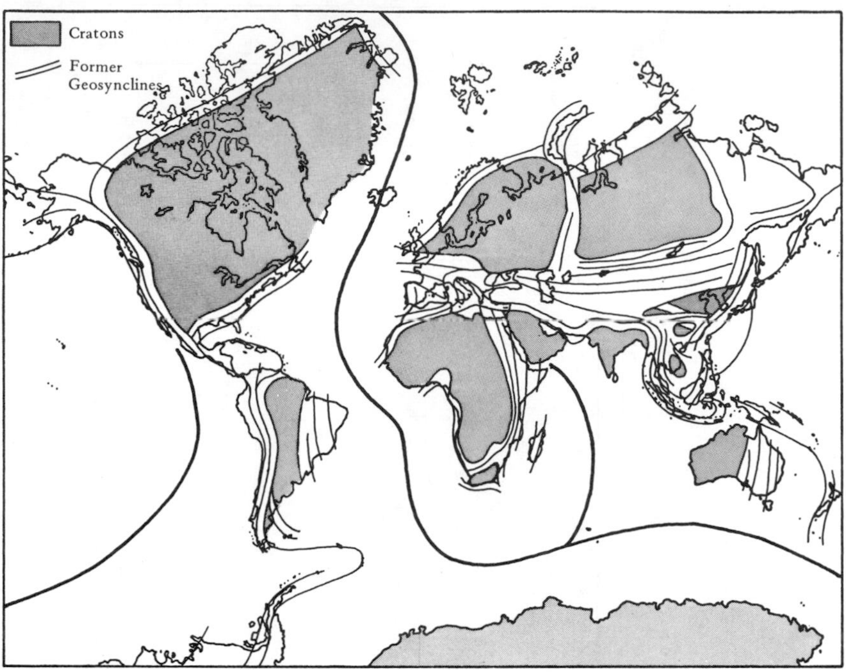

FIGURE 9.3 Present Distribution of Cratons and Former Geosynclines of Late Precambrian to Present Age.

The Kay-Stille Classification

In this century, more detailed studies of geosynclines led to the recognition of many types. Finally, in mid-century, the American geologist Marshall Kay made an attempt to standardize geosyncline terminology. He suggested that geosynclines corresponding to the original definition of linear, deeply subsiding troughs located marginal to continental platforms (whether or not they lay between two continental platforms) be called *orthogeosynclines* and that areas of thicker than average sedimentary sequence on the platforms or their margins be called *parageosynclines*. For the various subtypes of parageosynclines he coined several tongue-twisting names. Most geologists have continued to use *geosyncline* only in the original sense, but Kay's formal subdivision of orthogeosynclines, based on the earlier work of the German *Hans Stille*, has been retained. These subdivisions (Fig. 9.4) are the *miogeosyncline* on the continental platform side and the *eugeosyncline* on the oceanic side. (If the geosyncline was between continents, there was a mirror-image set of both.)

FIGURE 9.4 Classical Interpretation of Geosynclinal History.

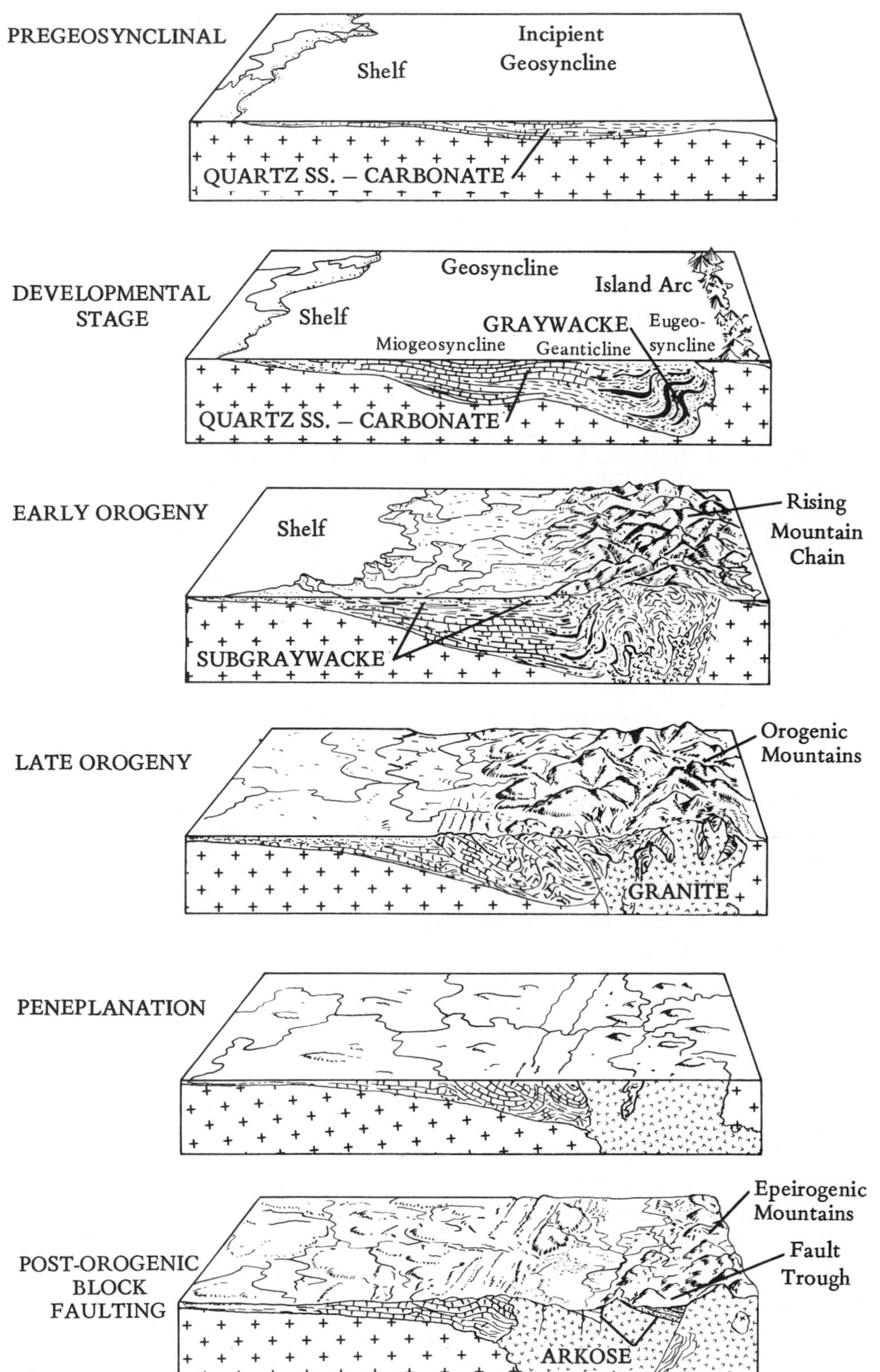
PREGEOSYNCLINAL
Shelf
Incipient Geosyncline
QUARTZ SS. – CARBONATE
DEVELOPMENTAL STAGE
Geosyncline
Island Arc
Shelf
GRAYWACKE
Eugeo-syncline
Miogeosyncline
Geanticline
QUARTZ SS. – CARBONATE
EARLY OROGENY
Shelf
Rising Mountain Chain
SUBGRAYWACKE
LATE OROGENY
Orogenic Mountains
GRANITE
PENEPLANATION
POST-OROGENIC BLOCK FAULTING
Epeirogenic Mountains
Fault Trough
ARKOSE

Miogeosynclines are characterized by thinner sequences of sediments than eugeosynclines and the sediment types belong to the quartz sandstone-limestone associations characteristic of the adjacent epeiric seas. Fossils are also those of shallow-water, shelf organisms. The deformation of miogeosynclines is much less intense than eugeosynclines and progressively dies out toward the craton. The deformation consists chiefly of folds and thrust-faults with little metamorphism and no intrusions. After the compressive mountain-building or *orogenic* stage, tension produces normal block faulting with some terrestrial sediments accumulating in the down-dropped basins between ranges. The miogeosyncline is separated from the eugeosyncline by a ridge or *geanticline.*

Eugeosynclines are characterized by much thicker rock sequences and sediment types of the deeper-water graywacke association. Fossils are uncommon and consist of pelagic organisms only. Volcanic rocks are prominent in the sequence and are derived from a volcanic island arc on the outer side. Deformation begins earlier than in miogeosynclines and not only consists of folding and thrusting, but is followed by the intrusion of coarse-grained granitic rocks and accompanying metamorphism. The post-orogenic block-faulting stage is usually characterized by the eruption of much lava from fissures and the deposition of thicker terrestrial sequences than in miogeosynclines. Not all geosynclines precisely fit all of these definitions, but in their search for universality, geologists were eager to support a unifying hypothesis such as this so-called "geosynclinal theory."

Geosynclines and Plate Tectonics

As outlined in Chapter 11, the new discoveries in the ocean basins, which have revived recognition of continental drift, have drastically revised our notions of how mountain-building (orogeny) proceeds. As this recognition has occurred, geologists have again become aware that no two geosynclines are really alike and that the "geosynclinal theory" is not the unifying concept it was once thought to be. In its place, new unifying theories of continental and oceanic evolution have arisen.

This matter will be more thoroughly discussed in Chapter 11, but for the moment it will suffice to say that the earth's crust is chiefly composed of a few large plates, each consisting of continents and/or ocean basins. The shapes and sizes of these plates continually change as they shift about and collide with and/or slide by one another. These collisions and side-swipes produce mountain-building.

Very thick accumulations of sediment can occur anywhere there is great subsidence. In some instances, such as where the edge of one plate descends beneath another, mountain-building is a part of the process (Figs. 9.6, 9.8, 9.10, 9.11). In other instances, such as the accumulation of land-derived sediments at the base of a continental slope on the foundering edges of separating continents (Fig. 9.5A) or as clastic wedges of material on the continent derived from rising mountains (Fig. 9.5B), it does not follow except if, at some time in the distant future, for unrelated reasons, the region becomes a plate boundary. Thus, if we restrict the term *geosyncline* to the regions of plate contact where subsidence and mountain-building occur, whether it be where oceanic and continental plate edges abut or where two continental plate edges abut, the term has some usefulness. But there are many variations in these zones of contact and it is not possible to list a single sedimentation pattern common to all.

Three basic types of plate collisions/sideswipes that produce mountain belts have been recognized. They are occurring today and ancient geosynclines that fit these patterns have been recognized. These types are: a) a trench-continent collision, b) an island arc-continent collision, and c) a continent-continent collision.

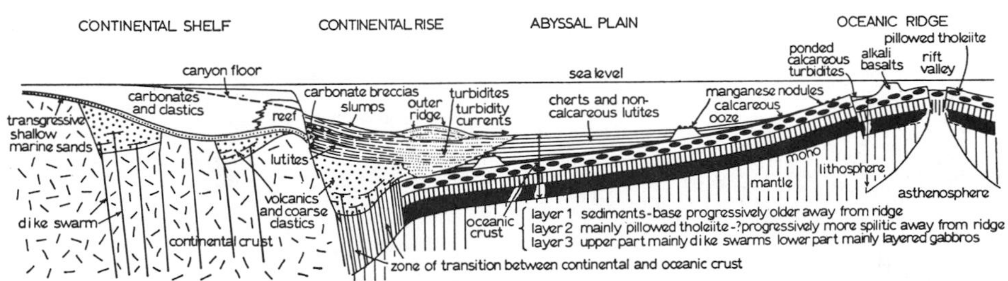

A

FIGURE 9.5A Thick Sedimentary Accumulation on Foundering Continental Edge. The present western Atlantic Ocean off eastern North America closely resembles this idealized diagram. (From Dewey and Bird, *Journal of Geophysical Research,* vol. 75, no. 14, May 10, 1970. By permission of authors and AGU.)

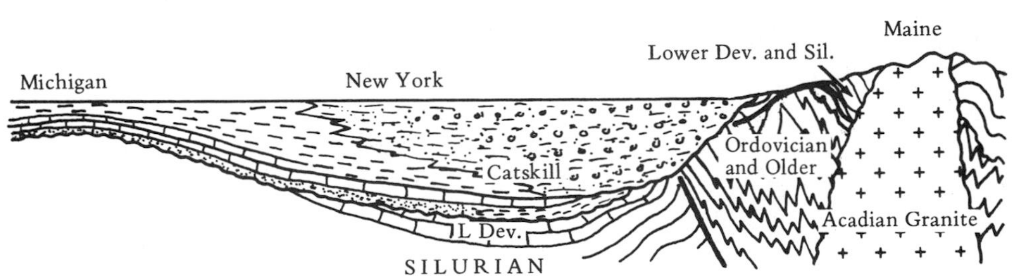

B

FIGURE 9.5B A Clastic Wedge, a thick accumulation of sediments adjacent to a mountainous region.

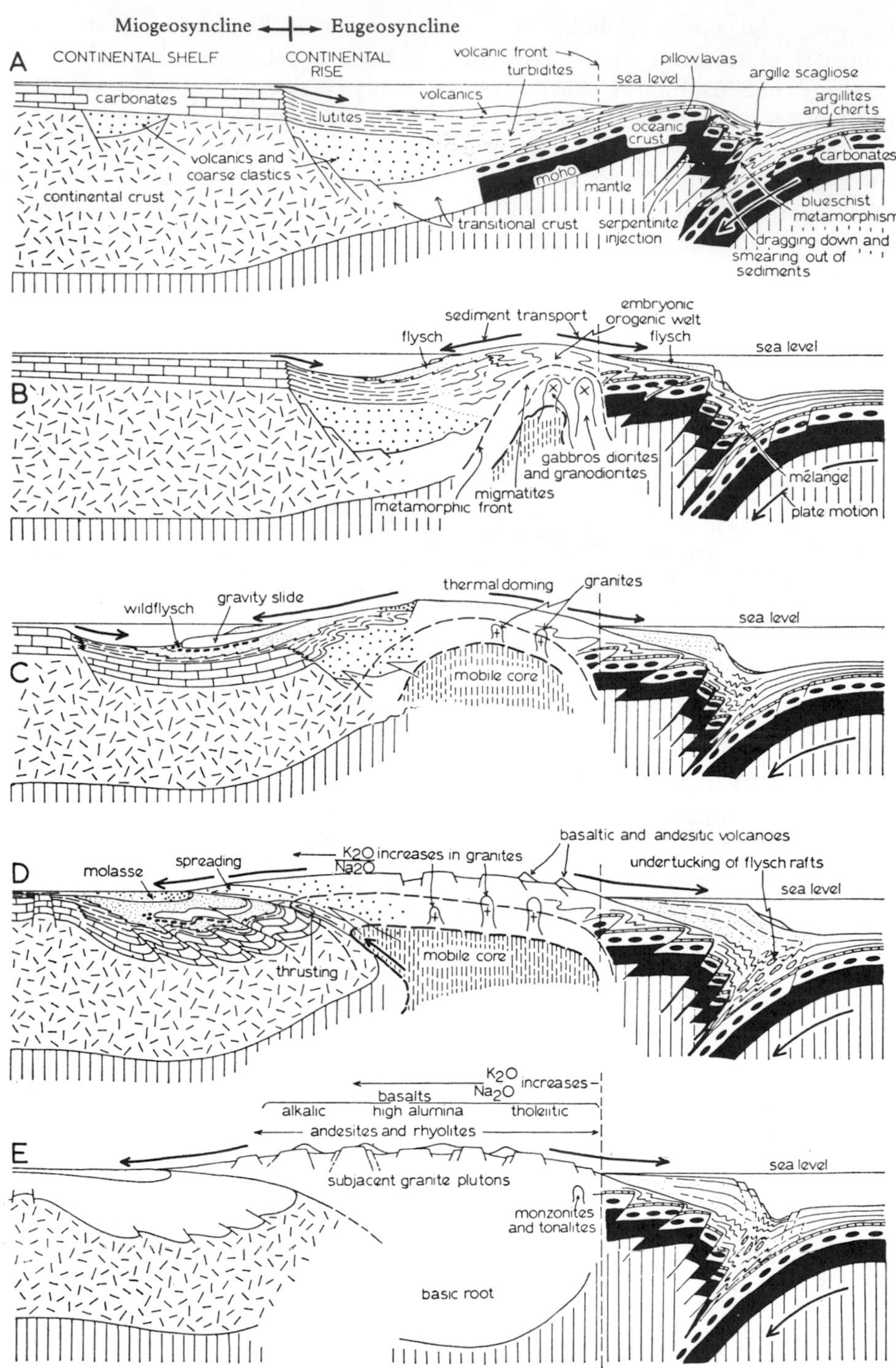

FIGURE 9.6 Trench-Continent Collision. (From Dewey and Bird, *Journal of Geophysical Research,* vol. 75, no. 14, May 10, 1970. By permission of authors and AGU.)

Trench-Continent Collisions

When two plates are in contact and the margin of the advancing one is an oceanic trench and the margin of the other, a continent, the results are as indicated in Figure 9.6. In this instance, oceanic crust is being thrust under a continental margin. (Because continental margins form by the rupturing and foundering of an original continent, the continental margin is like that shown in Figure 9.5A.) This process is occurring today where the eastern Pacific is being thrust under western South America, where the western Pacific is being thrust under eastern Asia with its island arcs, and where the eastern Mediterranean is being thrust under eastern Europe and Asia Minor (Fig. 9.7). The Cordilleran geosyncline of western North America of the Paleozoic and Mesozoic is but one of many excellent examples in the geologic record.

Where the oceanic edge of a plate is descending beneath or "underthrusting" the edge of a continent, a great melange of deposits forms in a trench. Jumbled, land-derived, but trench-deposited, turbidity current sediments (flysch) and pieces of offshore oceanic crust, including both pelagic sediments and volcanics, are piled together, not in the time sequence of their deposition, but in the sequence in which they come into contact at the colliding or side-swiping margins. As a result, no one rock type can be traced very far, so there has been little tendency to assign similar ages to rocks on the basis of similar lithologies.

The whole unit, such as the Franciscan Formation of California, has, however, been assigned the same age by many geologists. A few Jurassic fossils and the supposed superpositional location of the Franciscan beneath the Cretaceous, gave rise to a Jurassic date for the deposit as a whole. The picture was com-

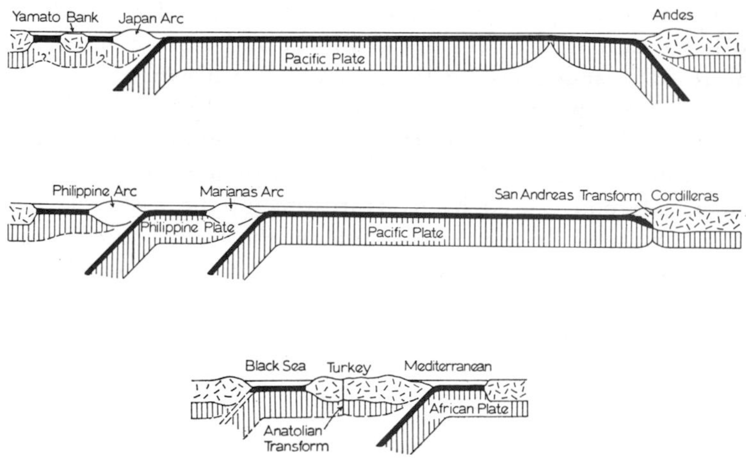

FIGURE 9.7 Modern Trench-Continent Collisions. (From Dewey and Bird, *Journal of Geophysical Research,* vol. 75, no. 14, May 10, 1970. By permission of authors and AGU.)

plicated when several upper Cretaceous fossils were found in widely scattered places. The deposit then became considered the eugeosynclinal facies of the western geosyncline. Now that its mode of origin is understood, the jumbled interrelationships of its components become more intelligible, and its time of formation appears to range up into the Paleogene. The upper pieces are superpositionally the youngest to be incorporated into the Franciscan, but the original time of formation of incorporated rocks is variable and unrelated to the formation of the Franciscan. The fossils only tell us the age of the incorporated blocks, but not the time of formation of the Franciscan, except that they set a maximum age. The age of the Franciscan must be determined by cross-cutting relationships with adjacent strata (unless someone could somehow recognize those fossils which settled into the trench at the time a particular part of the rock unit was being formed).

The chiefly oceanic material descending in the trench is metamorphosed by high pressure and relatively low temperature conditions producing the blueschist metamorphic facies (glaucophane, lawsonite, jadeite mineral assemblages). When this material reaches depths of over 100 kilometers, the effects of temperature become dominant and magmas are generated producing basic volcanic rocks and basic to intermediate plutonic rocks. This high temperature deformation produces a rising and expanding dome or orogenic welt which grows toward the continent by the pressure from the approaching oceanic trench and its own isostatic rise. This dome may take the form of an island arc as in east Asia or a volcanic mountain chain such as the Andes. The spreading mobile core produces a series of gravity-slide thrust faults, flysch-type clastic wedges, and chaotic melanges (wildflysch) which fill any remaining sea between the orogenic welt and the continent. As the trench moves and approaches the continent, the continental rise sediments and the continental crust itself are melted producing acidic (granitic) plutons and extrusive rocks. Metamorphosed sheets of hard rock are now thrust over the continent and erosion from these sheets produces terrestrial sandstones (molasse). As the welt expands, it fractures by tension producing normal faults and a basin-range topography. Note that the thrust faulting of Cordilleran-type orogenic belts occurs in two distinct phases, one early and caused by underthrusting of the ocean floor beneath the continent, and the other, late and caused by overthrusting of the rocks of the orogenic welt over the continent. In Kay's terminology, the miogeosyncline is the area from the edge of the continental shelf inward to the limit of deformation and contains the preorogenic quartz sandstone-limestone shelf association covered by the overthrust flysch-molasse plates, while the eugeosyncline encompasses all the clastic, oceanic, and volcanic rocks involved in deformation around the trench and seaward of the continental margin.

Island Arc-Continent Collision

When two plates are in contact and the margin of the advancing one is a continent and the other, an island arc, the results are portrayed in Figure 9.8. This process is occurring today where southeast Asia is closely approaching the northern Philippine Arc and it occurred in the Miocene during the mountain building that produced New Guinea when the Australian continent collided with the Bismarck Arc. The volcanic arc consists of a flysch wedge and a trench melange (Fig. 9.9), while the continental margin contains its typical assemblage of sediments shown in Figure 9.5A. The ocean basin caught in the impending crunch (such as the South China Sea today) fills with flysch. But when the continent reaches the trench, it is too light to be thrust under the arc. As a

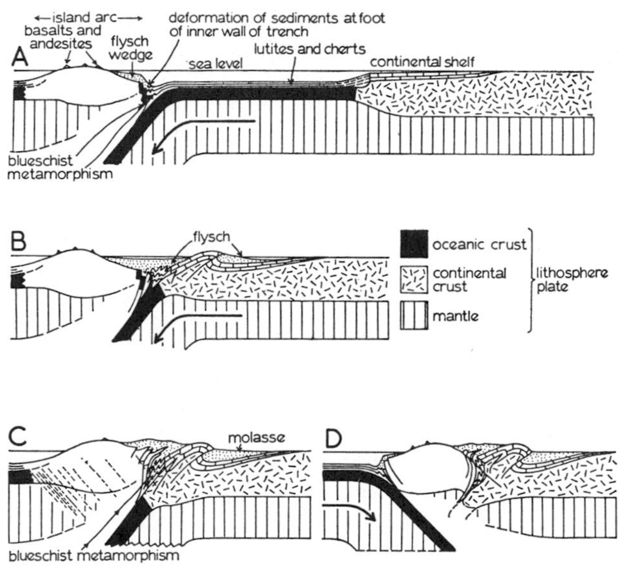

FIGURE 9.8 Continent-Island Arc Collision. (From Dewey and Bird, *Journal of Geophysical Research*, vol. 75, no. 14, May 10, 1970. By permission of authors and AGU.)

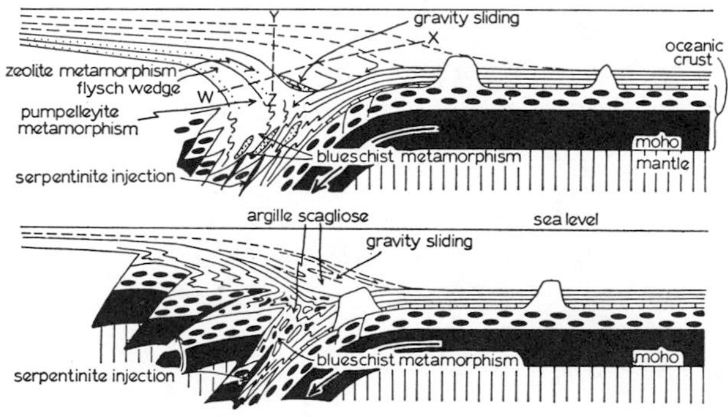

FIGURE 9.9 Trench Development. (From Dewey and Bird, *Journal of Geophysical Research*, vol. 75, no. 14, May 10, 1970. By permission of authors and AGU.)

result, the arc rocks are thrust over the continental margin and molasse from these thrust sheets forms on the interior side of the continent. If the colliding pressure continues, the oceanic crust behind the island arc may buckle forming a new trench to take up the crustal material. This new underthrust, however, will have the opposite inclination of the initial trench. A variation of this pattern occurs when there is no island arc and a continent collides with oceanic crust along an intraoceanic trench (Fig. 9.10). In this case oceanic crust may actually be thrust over continental crust. This has and is happening along the boundary between the Australian and Pacific plates adjacent to New Zealand.

Continent-Continent Collision

Finally, when the two edges of colliding plates are both continental, the results are those seen in Figure 9.11. Such processes have occurred and are occurring in the continuing collision of India and Asia. Before the continental margins approach, the situation on one margin is much like that of the trench-continent or continent-arc collisions discussed above. Oceanic crust and sediment may be squeezed out of the trough and thrust toward the approaching continent. As the oncoming continent approaches the trench, it too may be thrust-faulted. Because both colliding edges are light and will not sink appreciably, the splintered edges will be buoyed up to great heights by the great double thickness of continental crust. A broad zone of deformation will occur on both continents adjacent to the line of their suture.

The Ages of Rocks in Geosynclines

Much mountain building is infinitely more complex than any of these three generalized cases because mixtures of two or three of them can occur in the historical sequence of any one geosyncline. All of these types of geosynclines are, however, unstable zones of the earth's crust at or near plate boundaries which become deformed into mountains in the geologic sense of folding, thrusting, intrusion, and vulcanism. The process, however, is not instantaneous and its effects usually proceed in waves of deformation and orogenically produced sediments proceeding in one or both directions from the zone of collision. Thus, the sediments issuing from the mountains are of different ages as one crosses the geosyncline.

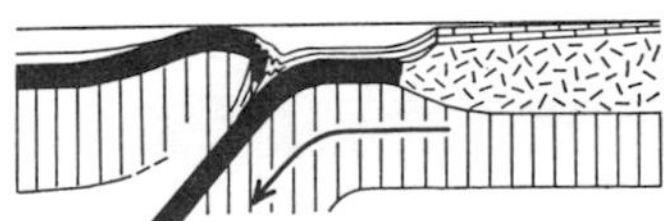

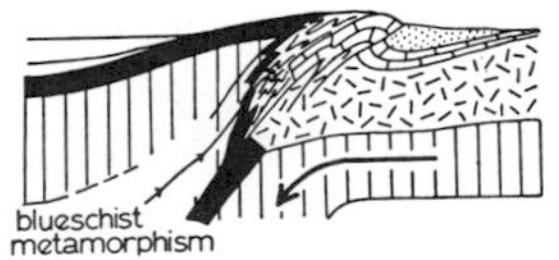

FIGURE 9.10 Continent-Ocean Basin Collision. (From Dewey and Bird, *Journal of Geophysical Research,* vol. 75, no. 14, May 10, 1970. By permission of authors and AGU.)

For example, the famous *flysch* deposits, which consist of land-derived clastics (graywackes, shales) in rhythmic alternation and which form from turbidity currents, are of varying ages as can be seen in Figures 9.6, 9.8, and 9.11. They are typically older toward the orogenic welt and younger away from it, and older on the oceanic side of a geosyncline than on the continental side. These deposits, which reflect orogeny, typically span many periods or even a whole era. To assume they are the same age throughout because they have the same lithology would cause a gross misinterpretation of earth history. Yet this has been done and only discovered when fossils of vastly different ages have been found in different areas of flysch deposits in one geosyncline. Likewise, after the orogenic phase of geosynclinal evolution passes and the time of isostatic adjustment and normal faulting (which often produces greater relief) occurs, another characteristic rock type is formed. This is the *molasse*, a very thick sequence of coarse detrital sediments, chiefly homogeneous calcareous sandstones, which collects in the basins. Molasse is typically a terrestrial deposit but may be marine in basins of very rapid subsidence. It also begins to form earlier near the orogenic welt where the mountains first develop, and proceeds toward the continent. It too, then, is diachronous or of different ages in different places.

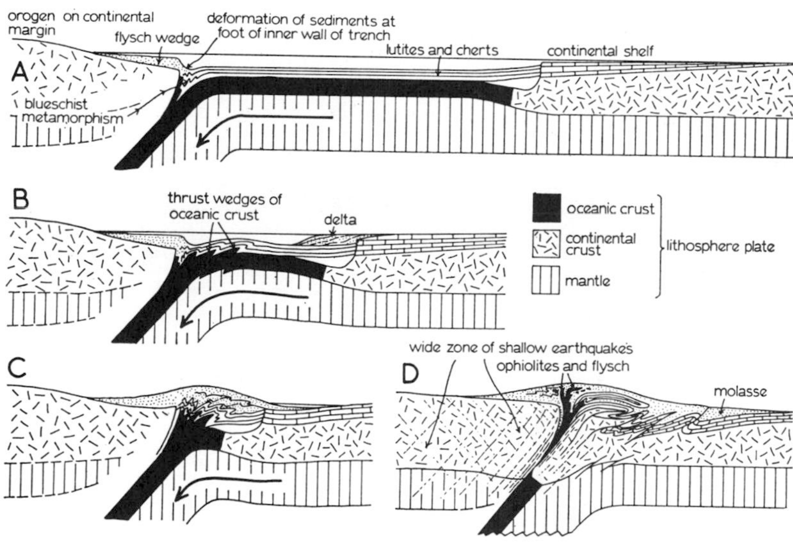

FIGURE 9.11 Continent-Continent Collision. (From Dewey and Bird, *Journal of Geophysical Research*, vol. 75, no. 14, May 10, 1970. By permission of authors and AGU.)

In summary, geosynclinal sedimentation is characterized by rapid lateral shifts in rock types and their wide horizontal migration with time. It is even more obvious in geosynclines than in epeiric seas that rock units, such as formations, are not units of time. It is perhaps because of this that much of the significant work on facies and the lack of correspondence of time and rock terms occurred in areas of former geosynclines.

Questions

1. What has been the traditional concept of a geosyncline?
2. In this model, what was the difference between a miogeosyncline and a eugeosyncline?
3. How has the concept of a geosyncline been altered by the recent discovery of plate tectonics?
4. Describe the results of a trench-continent collision. An island arc-continent collision. A continent-continent collision.
5. Why are flysch and molasse diachronous?
6. Why is it difficult to date the time of formation of a trench melange?

10

The Maelstrom:

Correlation

Since the early chapters of this book, we have been talking about dating rocks and determining whether or not two bodies of rock were deposited at the same time. We have seen that this cannot be done on the basis of lithology. We have also seen that this is usually done with fossils in sedimentary sequences and by radiometric means in igneous and metamorphic terrains. This process is called *correlation* by most geologists. Unfortunately, this term, like so many in geology, has been used in other senses. For example, many geologists also use the word *correlation* for the process of determining whether two bodies of rock defined lithologically are or were once part of the same lithologic body of rock. For example, they want to determine if a certain sandstone in one area originally connected with one in another area, and, hence, they are really the same lithostratigraphic unit (Fig. 10.1). Because so many geologists are still Wernerian in their outlook, they see little difference between the two processes.

In the preceding two chapters, we saw that time equivalency and rock equivalency are not the same thing and that disastrous consequences can result for historical geology if one operates on the assumption that they are the same. Nevertheless, the word *correlation* has been so widely used for both processes that it seems unwise to restrict its usage to only one of them. Therefore, an adjective must be applied to the word *correlation* to assure that there is no confusion. Accordingly, *time-correlation* refers to the process of determining whether two rock bodies were deposited at the same time, and *rock-correlation* to the process of determining whether two rock bodies are really part of the same lithologic rock body. Only in those instances where a rock body results from a single interval of deposition such as an ash fall, a lava flow, or a turbidity current deposit do the two correspond.

Rock-Correlation

There are four major methods of *rock* or *lithologic-correlation*. The most desirable method is that of tracing the units into one another as this proves unequivocally that the two are one (Fig. 10.2). This technique, called the *physical continuity* method, is only applicable, however, in regions of sparse vegetation and soil that lack great structural disturbance. Where vegetation and soil cover intervening exposures and/or faults terminate rock units, the geologist must resort to more indirect methods.

If the two rock bodies have absolute or near *lithologic identity*, that is their mineralogy, texture, primary structures, thickness, etc., correspond, the geologist can be fairly secure in rock-correlating the two (Fig. 10.3) even if he cannot trace them into one another. A problem arises, however, if there are two or more rock units at one or both of the localities that have lithologic identity.

The next guideline to be utilized by the geologist is the *position* of the unit he is trying to correlate *in* the *stratigraphic sequence.* (Fig. 10.3) For example, if he is trying to decide which of two sandstones at one outcrop he wishes to correlate with a sandstone at another outcrop, he examines their positions in the geologic column. If one sandstone is underlain by shale and overlain by limestone and the other is underlain by coal and overlain by salt, he correlates the first one with the sandstone at the other outcrop which is also underlain by shale and overlain by limestone. Even this method encounters difficulties when one is dealing with rocks that occur in repetitious sequences such as Pennsyl-

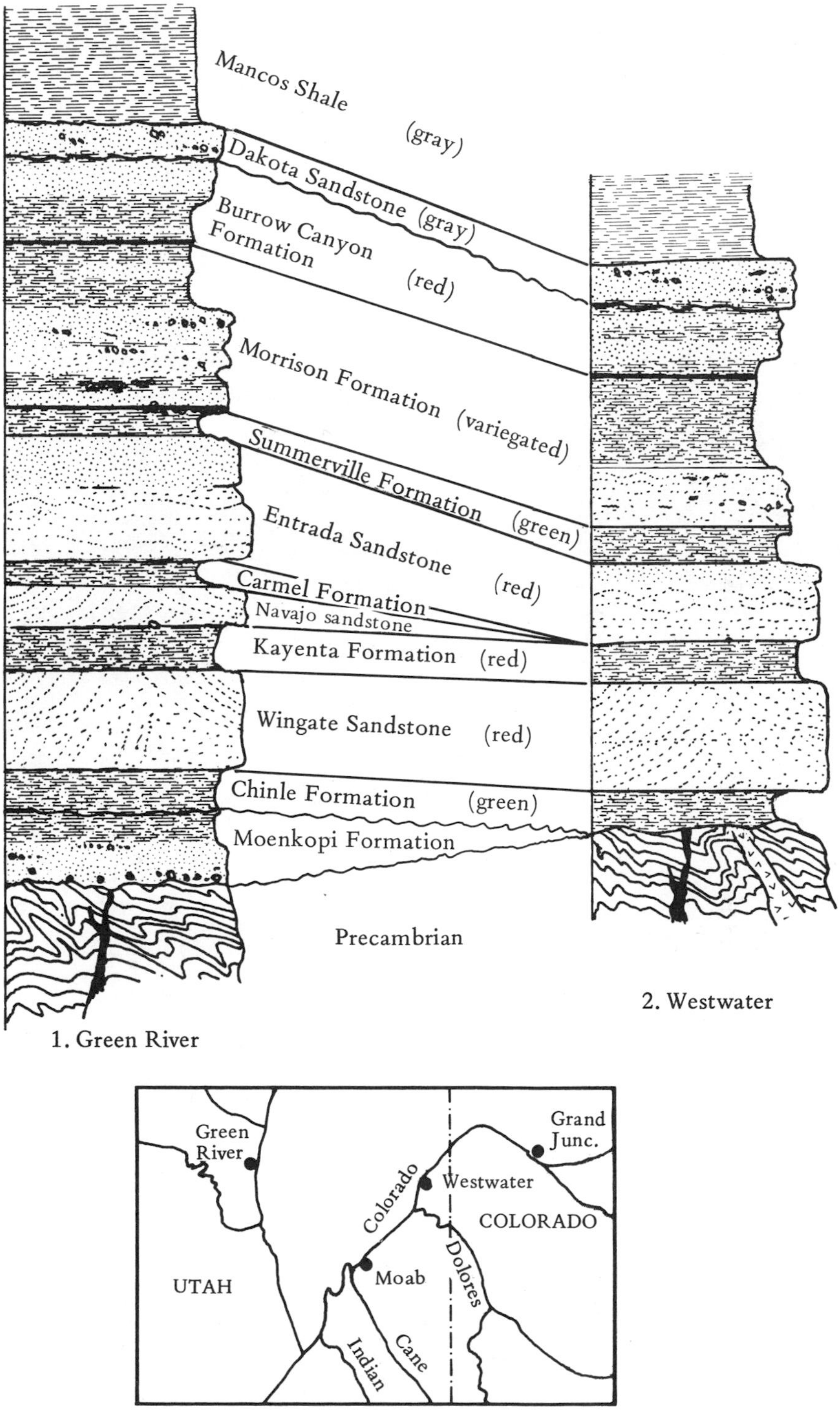

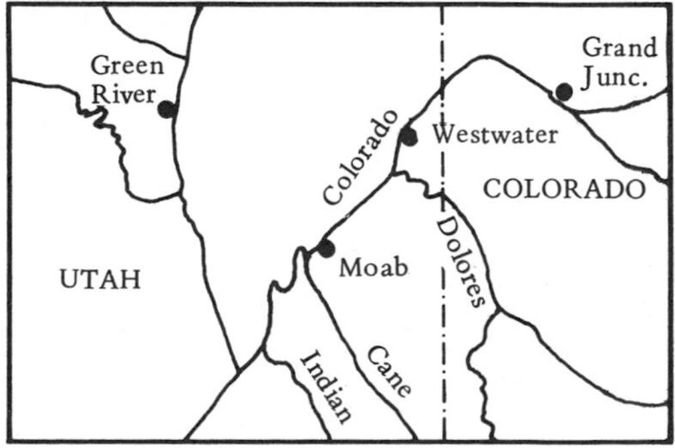

FIGURE 10.1 Rock-Correlation. Strata are inferred to have once been continuous between the two localities.

vanian Mid-Continent cyclothems (Fig. 10.4) and Cretaceous Great Valley turbidites (Fig. 10.5). In some of these instances it is virtually impossible to rock correlate individual units. They have often been time-correlated on the basis of fossils. This, of course, only proves that they cannot be rock correlated unless they are very near one another and, hence, probably still part of the same rock unit before its time lines cross over into another rock unit.

FIGURE 10.2 Physical Continuity Method of Rock-Correlation. Mesozoic strata in the Canyonlands area of Utah can be traced continuously along barren mountainsides and canyonsides. (Courtesy of William M. Mintz.)

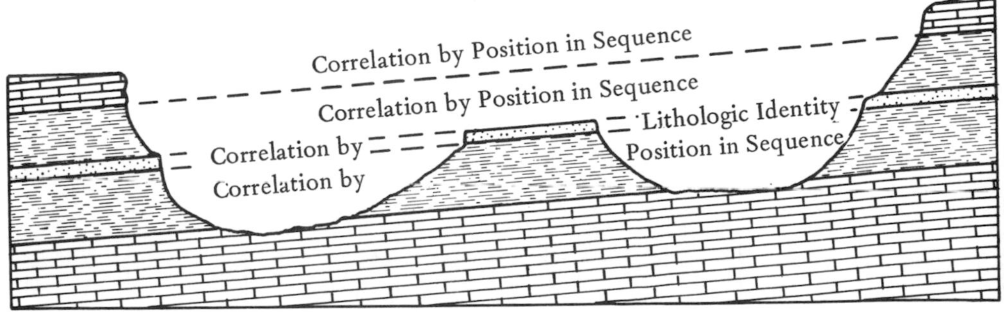

FIGURE 10.3 Diagram Illustrating Various Surface Methods of Rock-Correlation.

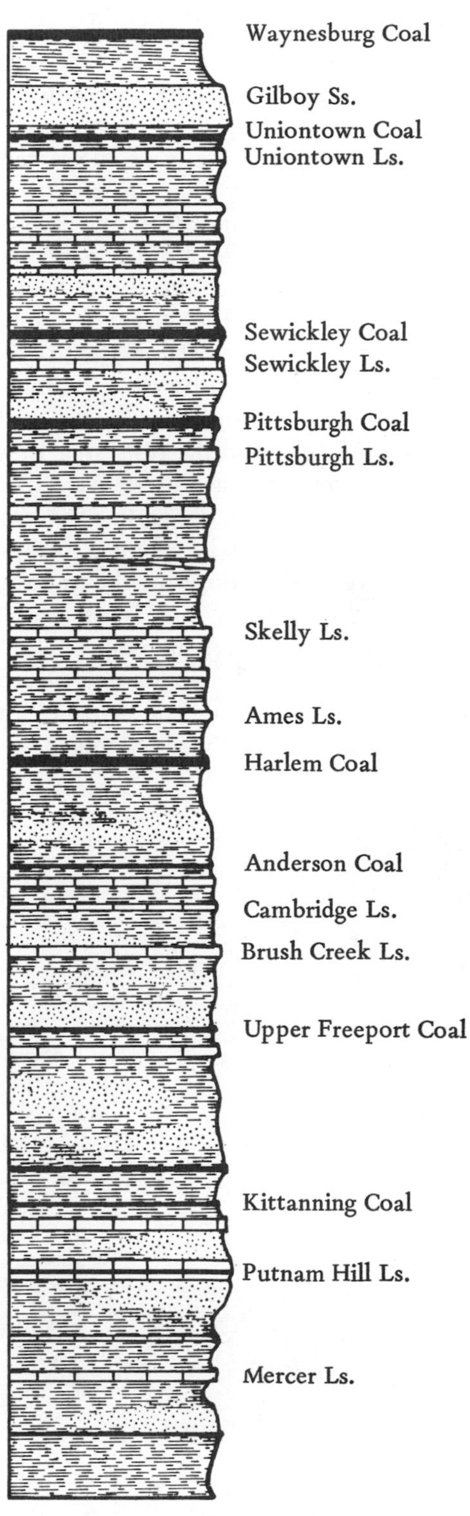

FIGURE 10.4 Cyclothems. Position in a sequence and lithologic identity are of little help in rock-correlating these strata.

The subsurface geologist, who works with the data gathered from drilling wells, has developed a battery of tools for rock correlation that parallel those of his surface counterpart. The only exception of course, is physical continuity which cannot be used between well holes. If actual cores of sediment are brought to the surface, a geologist can rock-correlate by the lithologic-identity or position-in-a-sequence method. The taking of rock cores is a long and expensive process, however, and is only occasionally done. Therefore, the subsurface geologist uses indirect methods of gathering data about rock properties so he can apply the identity or position-in-sequence methods. These indirect methods fall under the general heading of *instrumental well logging*.

A very commonly used method is that of *electric logging*. A set of electrodes is lowered into a well by electric cable and they measure both the electromotive force produced in the rocks by the contact of their fluids with the drilling fluids in the drill hole, and the resistance they offer to an artificial electromotive force. The measurements are carried by the cable to the surface and recorded as a continuous function of depth. The result is a set of curves (Fig. 10.6) whose shape changes with lithology. Other varieties of logs are used which measure the *natural radioactivity* of the various rock types, the effects of *neutron bombardment* on various rock types (Fig. 10.6), and the velocity of propagation of *sonic waves* in various lithologies (Fig. 10.7). These are all methods of measuring rock properties at a distance and must be interpreted by geologists for lithologic identity and position in a sequence for purposes of rock correlation. (Do not confuse radioactivity well-logging, a method of rock correlation, with radiometric

FIGURE 10.5 Flysch Turbidites, Pedro Point near Pacifica, California. (Courtesy of Mary Hill, California State Division of Mines.)

dating, a method of time correlation. Both depend on natural radioactivity, but are entirely different techniques for entirely different purposes.)

Unconformities of all varieties can also be used for rock-correlations, but, like rock units, they vary in age from place to place. For example (Fig. 8.8), the uppermost Precambrian-lowermost Cambrian interval is an unconformity over most of the North America craton, but on the margins it embraces only the Eocambrian and lowest Cambrian, while in the center of the continent, the lower two-thirds of the Cambrian is missing. The basal sandstone lying on this unconformity can be rock-correlated across the area, but it is not the same age everywhere, being Eocambrian or Lower Cambrian marginally and Upper Cambrian centrally.

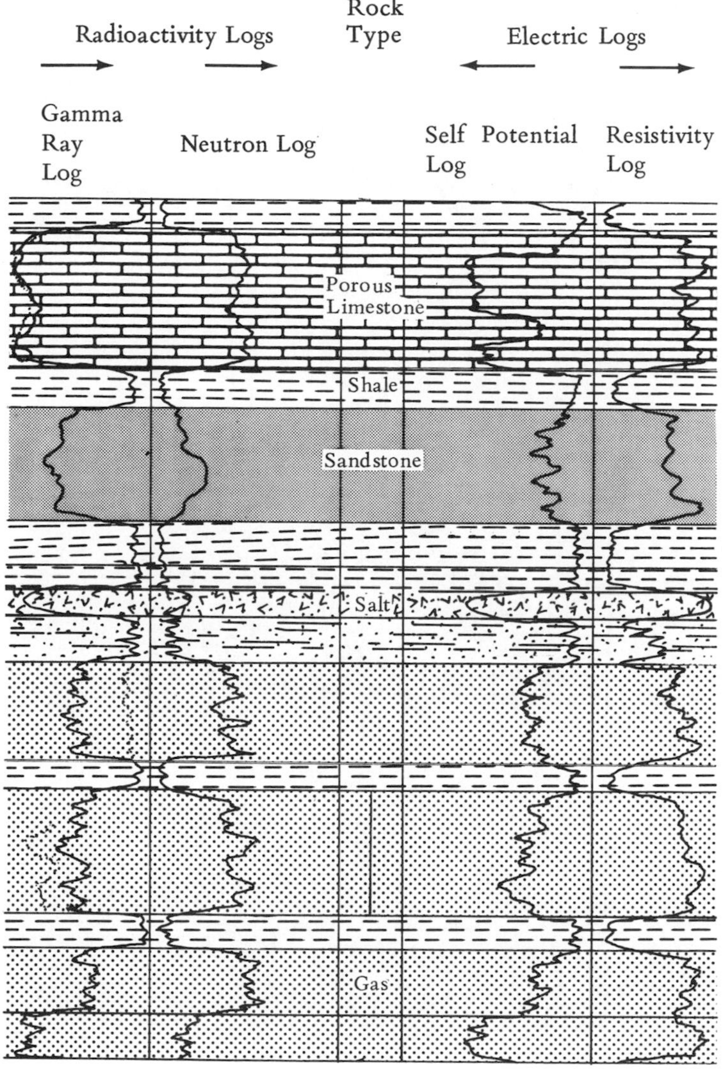

FIGURE 10.6 Electric and Radioactivity Logs.

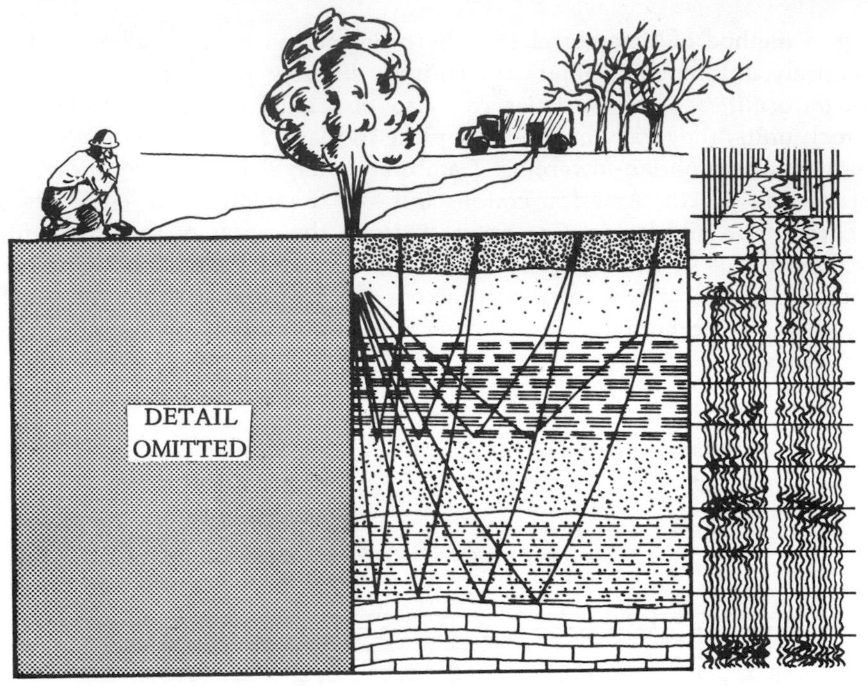

FIGURE 10.7 Sonic Logging.

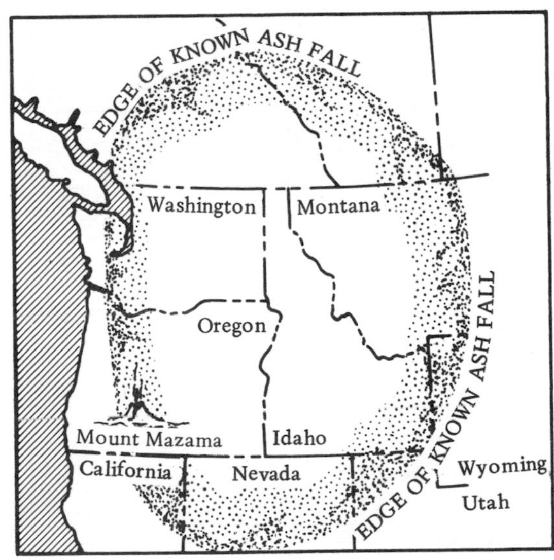

FIGURE 10.8 Formation of a Recent Key Bed for Time-Correlation. The Mount Mazama (Crater Lake rim) eruption of 6,000 years ago scattered a wide-spread distinctive ash over the West.

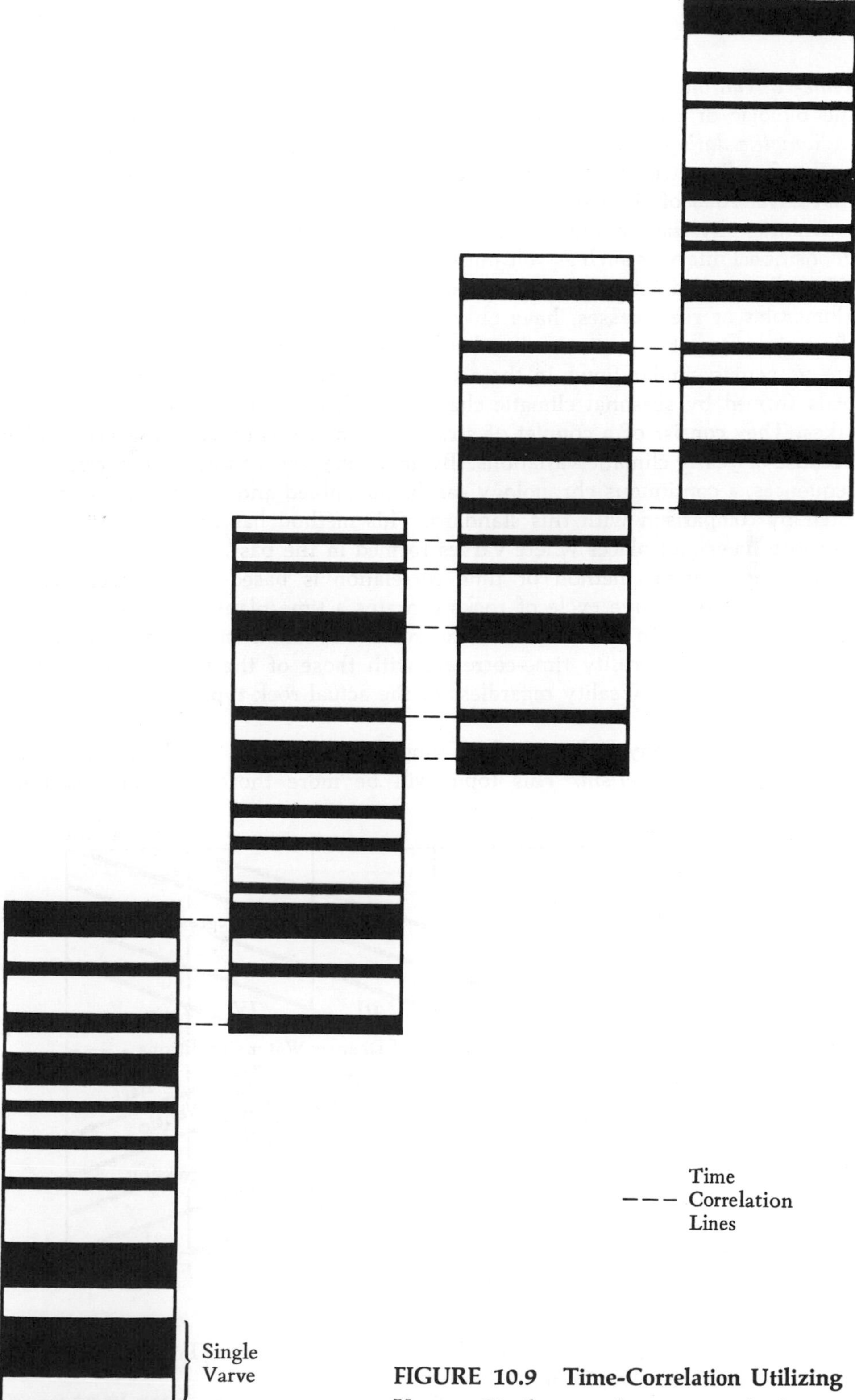

FIGURE 10.9 Time-Correlation Utilizing Varves. Similar couplets are matched.

Time-Correlation

Time-correlation methods fall into two categories, the physical methods and the biologic or paleontologic methods. The major *physical* method is that of *radiometric dating* discussed in Chapter 6. There, we pointed to a major drawback of radiometric dating, that it is usually not applicable to the sediments that cover 75% of the earth's surface.

Another physical method is that of *key* or *marker beds* which result from single depositional intervals such as volcanic eruptions (Fig. 10.8) and turbidity currents. (Not all so-called "key beds" have time significance. Many, such as thin conglomerates or reef masses, have only environmental significance.) Recognizing the uniqueness of a particular key bed is often a difficulty, particularly if there are numerous similar flows in the section. *Varves* (Fig. 10.9) are types of key beds formed by seasonal climatic changes, such as commonly occur in glacial lakes. They consist of a couplet of strata and no two couplets are exactly alike because of yearly climatic variations. By matching the couplets of overlapping sequences, a continuous chronology can be assembled and other couplets correlated by comparison with this standard. This method has limited applicability to those times and places where varves formed in the past.

Another physical method of time correlation is based on the fact that a transgressive-regressive cycle of rocks contains a time plane, the *time of maximum inundation*. In other words, the rocks deposited in the most offshore conditions at one locality time-correlate with those of the most offshore conditions at any other locality regardless of the actual rock types or water depths. (Fig. 10.10).

Finally, a new physical method of time correlation has developed from the study of *paleomagnetism*. This topic will be more thoroughly discussed in

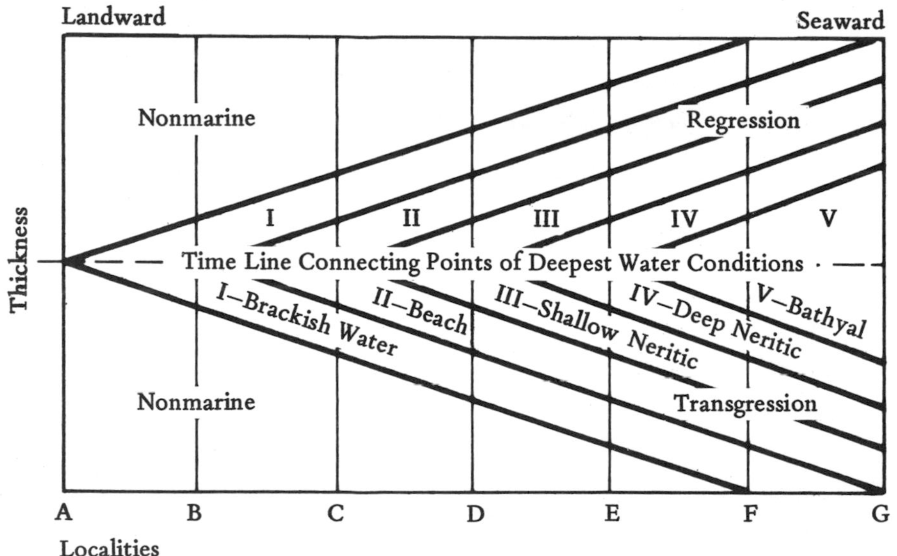

FIGURE 10.10 Time-Correlation by Position in a Transgressive-Regressive Cycle.

Chapter 11 which covers continental drift. Here it will suffice to note that magnetic iron oxide particles in lavas or sediments align their polarity with the earth's magnetic field at the time they crystallize or settle. In older rocks, this alignment of particles differs significantly from that of the present because the continents have been moving with respect to the poles. If one assumes

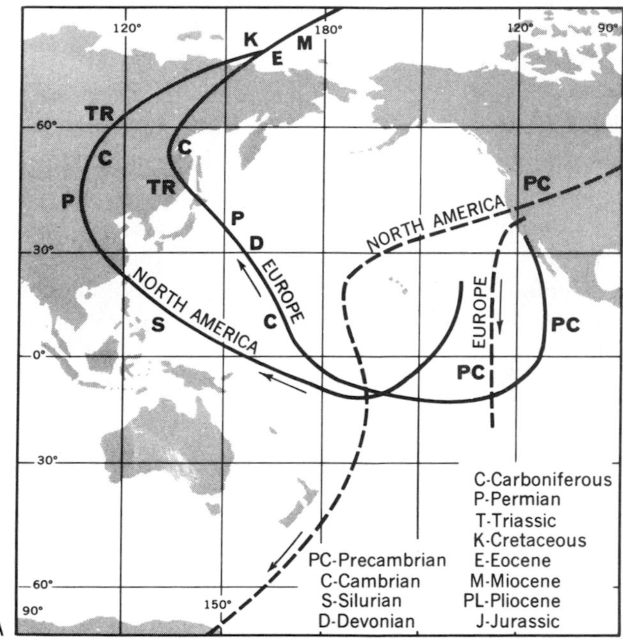

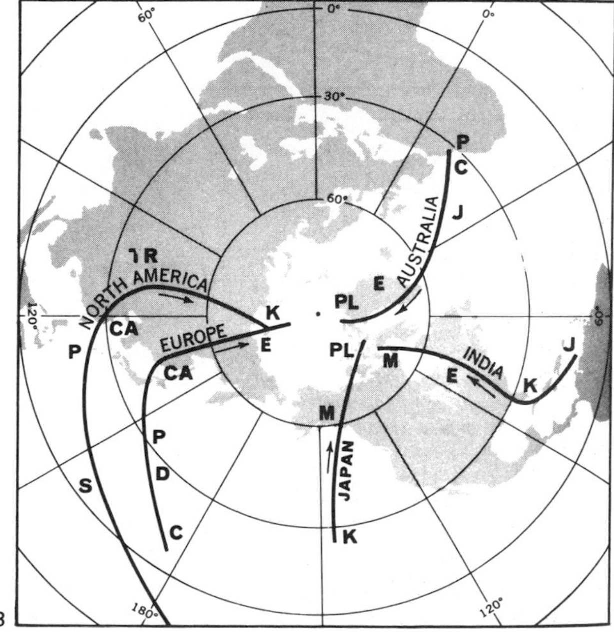

FIGURE 10.11 Polar-Wandering Curves. (From Kay and Colbert, *Stratigraphy and Life History,* © 1965, by permission of John Wiley & Sons, Inc.)

a particular continent was fixed and then determines the successive positions of the pole relative to that continent, a "polar-wandering" curve can be prepared (Fig. 10.11). Then one can correlate an unknown rock sample by comparing the position of the pole determined from its remnant magnetism with the curve. For example, if the pole position for an unknown sample fell on the Ordovician part of the curve, it formed in that period.

It has also been found that the earth's poles have reversed themselves many times in the geologic past and that this phenomenon is also recorded in rocks containing magnetic minerals (Fig. 10.12). Unfortunately, one normal or one reversed time span is like any other (except for its length), so simply finding that a rock has reverse or normal magnetism doesn't help too much with correlation. If, however, one is fairly certain by other means of time correlation of the approximate time interval under consideration, the polarity can be useful. For example, if one already knew the stage in which he was working, but one zone was a time of normal polarity and the other of reversed polarity, the magnetism could identify the zone if there were no diagnostic fossils. Magnetic reversals can also be used to date strips of igneous rocks on the ocean floors as we shall see in the next chapter.

Biologic Methods of Time-Correlation

The biological methods of time-correlation are based fundamentally on the principle of faunal succession with its underpinning of evolution. In other words, like fossils means like ages. We have seen the difficulties that facies introduce, particularly at lower levels like stages and zones, though these are less of a problem with systems and erathems. No paleontologist would be likely to confuse Devonian fossils from any environment with Jurassic fossils regardless of what environment they lived in. However, it may be very difficult to correlate sublittoral and abyssal faunas of the same Devonian stage. We have also seen how this problem can be alleviated by the presence of wide-ranging species of fossils such as pelagic forms or by locating the lateral intertonguing of facies. In the early days of geology, workers such as the Englishman Thomas Huxley, suggested that like fossils only meant like environments and not necessarily like ages. He called this concept homotaxis. We now know that because of evolution, the likelihood that a whole fauna or flora can exist for any recognizable length of time without evolutionary changes in at least one lineage is impossible. Hence, the concept of non-synchronous homotaxis of faunas (or floras) need not concern us in time correlation. Individual species, however, may exist at different times in different places as discussed in the next paragraph.

Index Fossils

There are several widely used biologic methods of time correlation. A very popular technique is the so-called *index fossil* method. After considerable collecting, it was discovered that some lineages of fossil organisms evolved rapidly so that the vertical stratigraphic range of any species is short, spread widely so that the horizontal geographic range of any species is wide, and was common

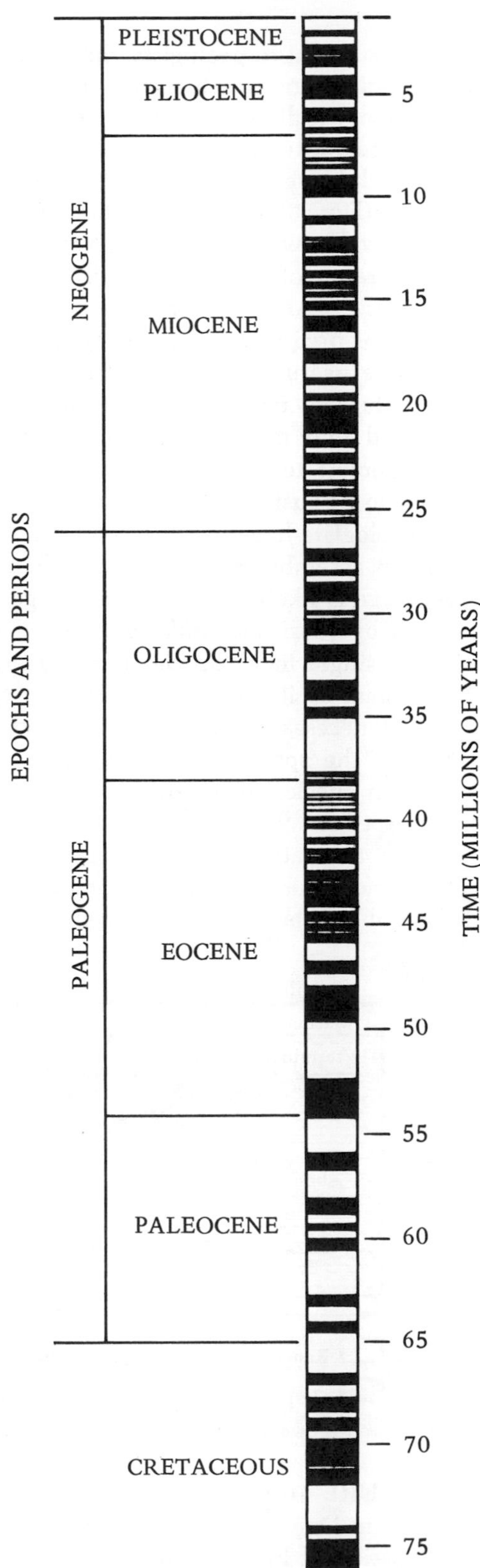

FIGURE 10.12 Reversals of the Earth's Magnetic Field Since Late Cretaceous Times. Black bars are times of normal polarity; white bars, times of reversed polarity. (Knowledge as of 1968.)

enough to be useful to geologists. Species of these lineages have become known as *index fossils.* One problem that always arises with index fossils is that their ranges are empirically determined; they cannot be known in advance. Therefore, there is no guarantee that an index fossil didn't survive later in some restricted area than it did elsewhere (Fig. 10.13). For example the screw-shaped bryozoan *Archimedes* (Fig. 7.53) was long thought to be an index of the Mississippian, but it is now known from Pennsylvanian and Permian rocks in a few places. So the time range of any index fossil is always subject to modification based on new discoveries. Another supposition of the index fossil method is that only a few types of organisms (Fig. 10.14) evolved rapidly enough and spread widely enough to be useful as index fossils. This double condition of restriction is a limiting feature to be sure, but the development of new techniques and tools has indicated that many groups of organisms, such as the pelagic radiolaria, formerly thought to be too evolutionarily static for correlation purposes have turned out to be just as rapidly evolving as the widely used organisms.

Other geologists have reasoned that if a single species is not a perfect correlation tool, an assemblage of organisms should provide better resolution because an assemblage is likely to have a greater geographic range than a single species. Correlation by *fossil assemblages* is based on the similarities of whole faunas rather than single index species (Fig. 10.15). The improvement, however, may not be as great as it first appears. Finding ten species in common between two units is not necessarily ten times as good a correlation as one based on one species if all the species are controlled by the same environmental conditions. Also, the ten species may belong to lineages that did not evolve as rapidly as one index species. Nevertheless, it appears that the usage of more species does improve the accuracy of time-correlation because the more that are used, the less the chance that the one changing lineage has been overlooked or that the whole assemblage has changed its geographic range.

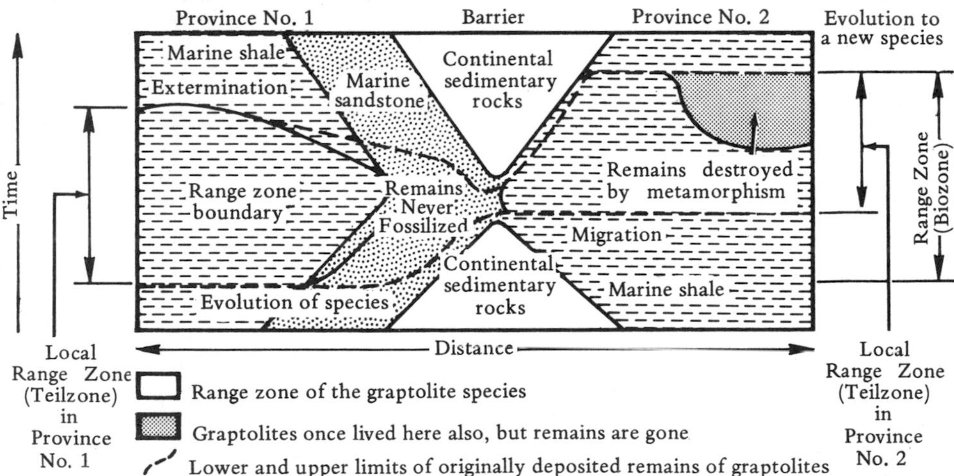

FIGURE 10.13 Major Factors Affecting the Range of a Species. (From Don L. Eicher, *Geologic Time,* © 1968, reprinted by permission of Prentice-Hall, Inc., Englewood Cliffs, N.J.)

Concurrent-Range Zones

A very popular type of fossil assemblage time correlation is that of utilizing *concurrent-range zones* or Oppelian zones (Fig. 5.1). As discussed earlier, these are based on the overlapping stratigraphic ranges of species and are characterized by unique assemblages that do not occur above or below in the geologic record. A characteristic fossil, common in the zone but not necessarily limited to it or found throughout it, is selected as the name giver and is known as a *guide fossil.*

Keep in mind that any range is empirically determined and that, hence, no range, even one that is used to establish a zonal boundary, can be known with absolute certainty. Formal time-stratigraphic zones or chronozones of the concurrent-range variety are not established until after exhaustive collecting from a wide region in an attempt to overcome the problem, but it can and still does

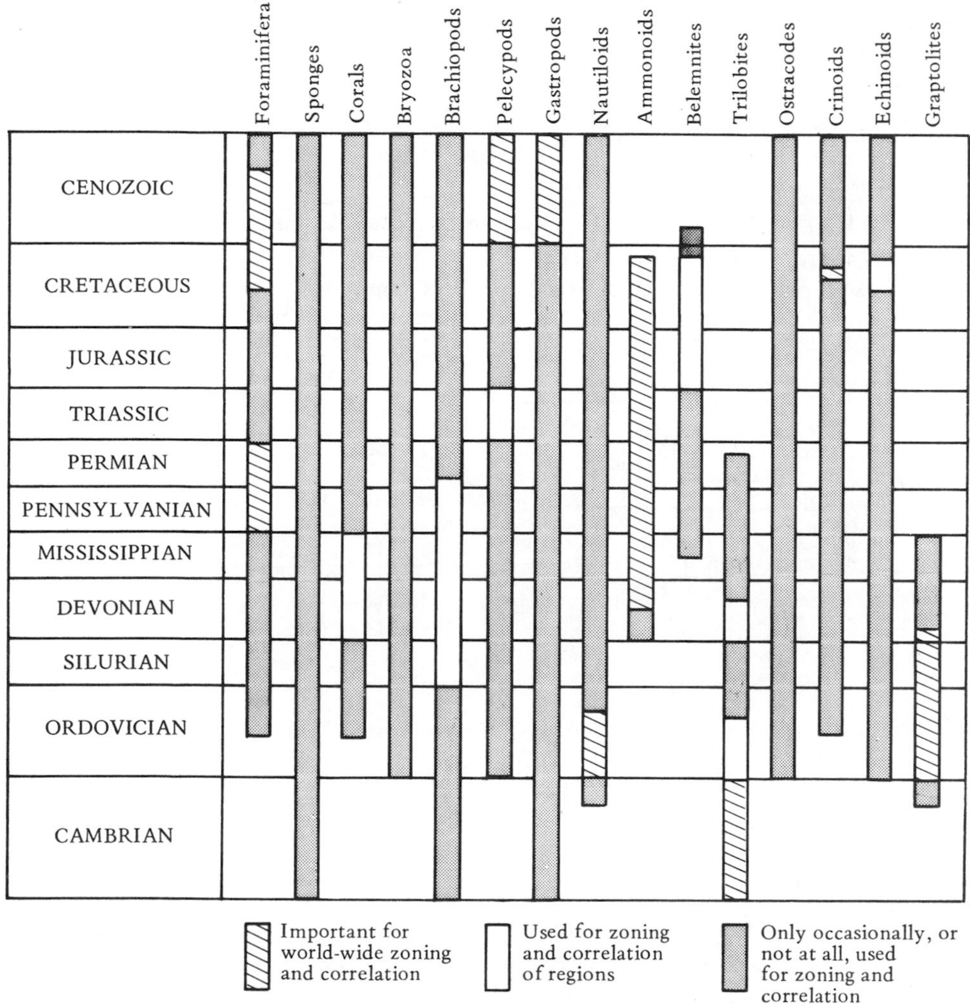

FIGURE 10.14 Major Fossil Animal Groups Containing Index Fossils. (From Don L. Eicher, *Geologic Time* © 1968, reprinted by permission of Prentice-Hall, Inc., Englewood Cliffs, N.J.)

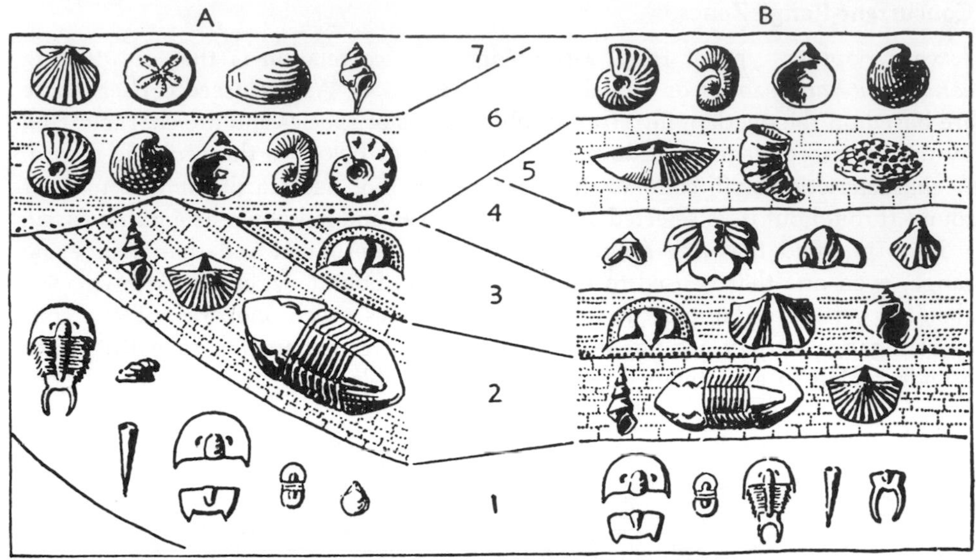

FIGURE 10.15 Time-Correlation by Fossil Assemblages. (From Moore, Lalicker, and Fischer, *Invertebrate Fossils,* © 1952, used with permission of McGraw-Hill Book Company.)

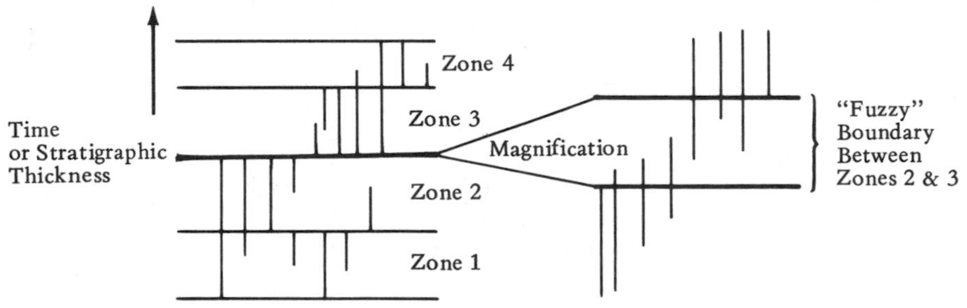

FIGURE 10.16 Fuzzy Zonal Boundaries. Detailed collecting usually reveals that coincident endings and beginnings of species do not occur.

FIGURE 10.17 Allan Shaw's Method of Time-Correlation Based on Ranges of Fossil Species. (A) Two sections with ranges of fossils indicated, (B) Correlation line of sections in (A), (C) Dog-legs in correlation line produced by cessation of deposition in one section. (From Don L. Eicher, *Geologic Time* © 1968, reprinted by permission of Prentice-Hall, Inc., Englewood Cliffs, N.J.)

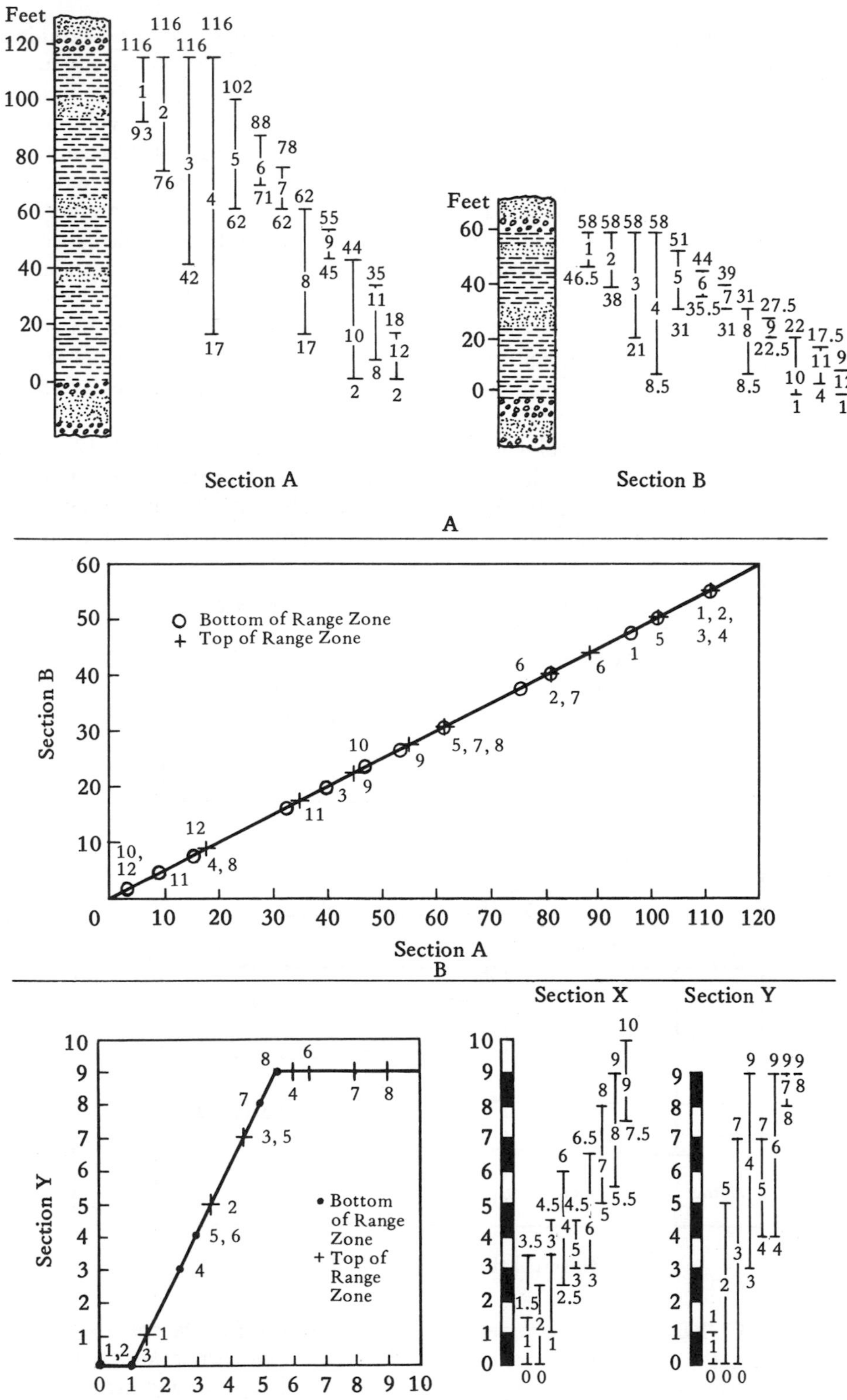
Feet
Section A
Section B
A
Bottom of Range Zone
Top of Range Zone
Section B
Section A
B
Section X
Section Y
Bottom of Range Zone
Top of Range Zone
Section Y
Section X
C

occur. As a result, a zonal boundary established on the basis of the beginning of two species and the ending of two other species eventually becomes a fuzzy zone as slight extensions upward and downward are found (Fig. 10.16). Indeed, the process is inevitable for there is nothing systematic about the evolution of species by natural selection that could cause the simultaneous appearances or disappearances of several species at once except in those instances where two species cannot survive without each other. If one form were absolutely dependent on another form as its food, then, of course, the two could die out at once, but there is no reason why two species of forams, clams, or ammonites should die out simultaneously. And there is no good reason why several species of these groups should all arise at the same time either. Evolution is based on random mutations occurring continuously, most of which are not selected for by the environment. Hence, coincident beginnings and endings of ranges are most unlikely.

Yet they occur in the geologic record at any one locality (Fig. 2.17). This simply indicates that when such coincidences occur, they represent environmental shifts in the region or gaps in the record there. This is still further evidence that nonsynchronous homotaxis does not exist. Zones can be used then for

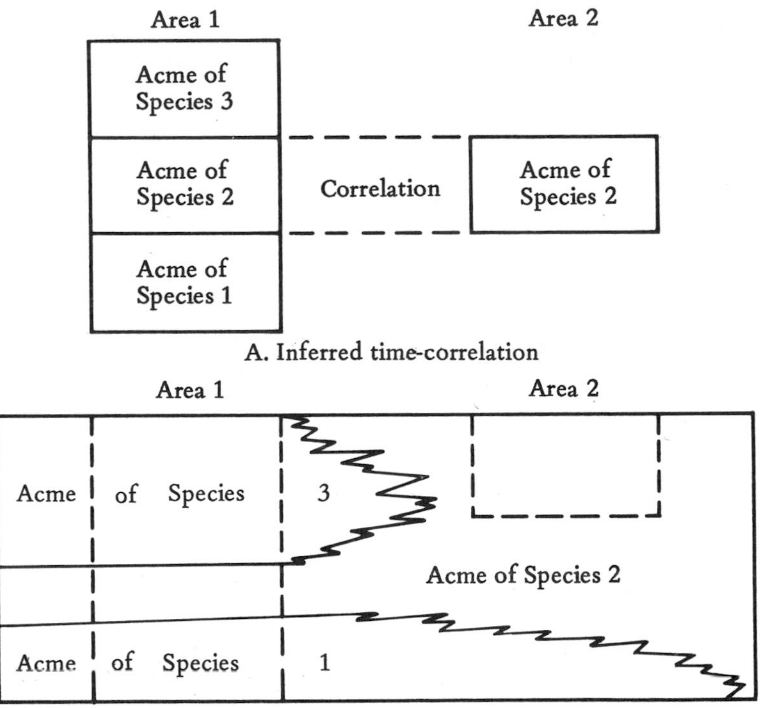

FIGURE 10.18 Miscorrelation Using Epiboles (Strata containing the acme of a species). (A) Inferred time-correlation, (B) Actual time relationships before erosion. The preserved strata in area 2 equal the youngest strata in area 1, not the middle ones.

correlation, but only down to a certain level of refinement. Still more detailed time correlations are becoming desirable and it has been suggested that statistical methods based on the ranges of individual species be used.

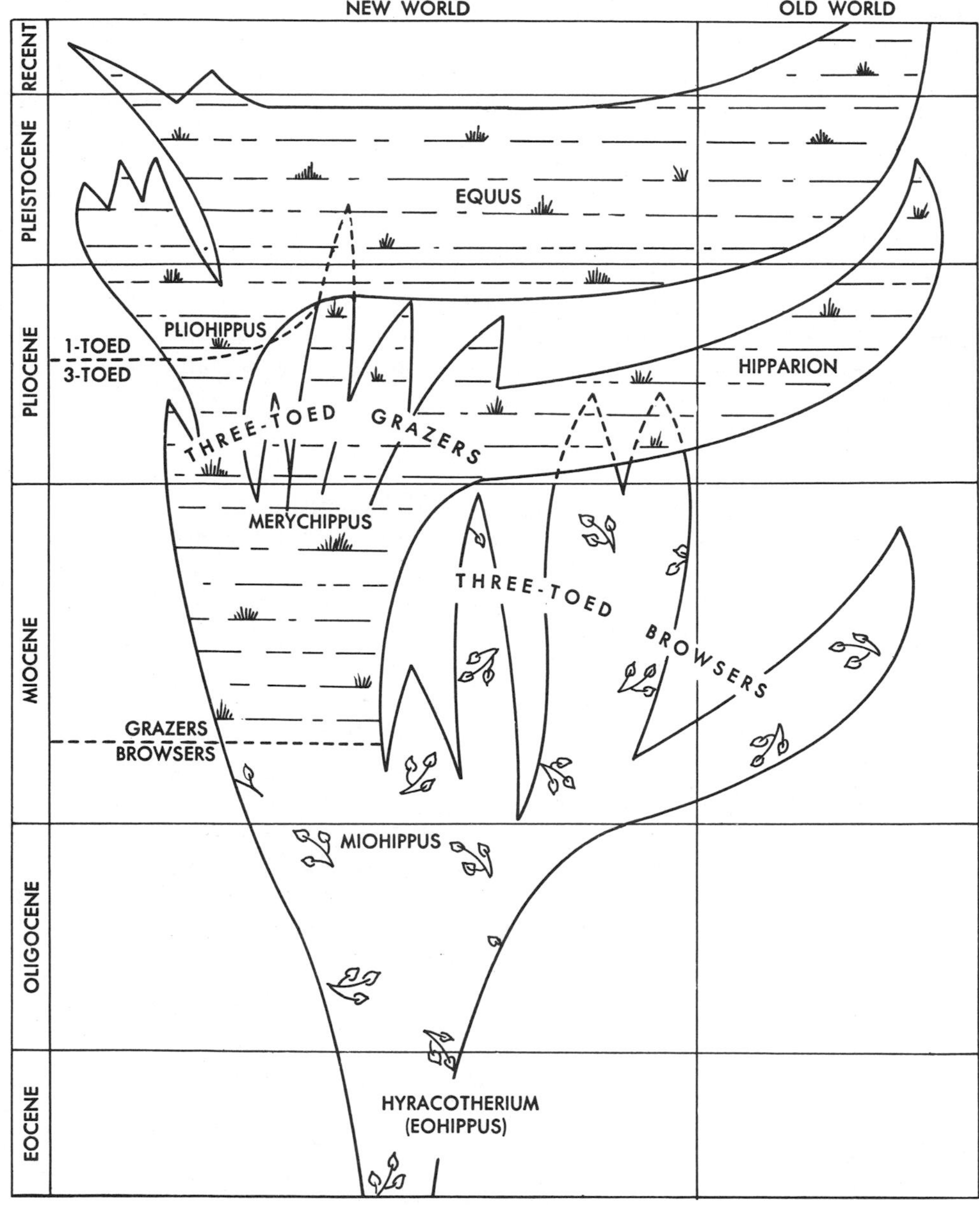

FIGURE 10.19 Evolution of the Horse Family showing descendant genera living at the same time as their ancestral genera. (From Thomas H. Clark and Colin W. Stearn, *Geological Evolution of North America,* Second Edition, Copyright © 1968, The Ronald Press Company, N.Y.)

Statistical Methods

For example, Alan Shaw has devised a method based on graphing the ranges of fossils in two geologic sections (Fig. 10.17). If the two sections correlate, there must be a correlation line on the graph because for every point in one column there must be a time-equivalent point on the other. Where deposition ceased or a gap exists in one section, while sedimentation continued in the other, a dog-leg occurs as the number of feet increases on one axis and stays the same on the other. Of course, ranges will not precisely coincide from section to section so a scattering of points rather than an exact line will typically occur. The corresponding line can be visually approximated or statistically determined. The method can be expanded to correlate one section after another until a composite standard is built up showing the total stratigraphic distribution of all species in the geographic area. Eventually, if tie points can be established between regions, a standard can be constructed for the whole world. Many geologists object to statistical methods of correlation arguing that most fossils are facies fossils and only a few forms, such as pelagios, can be used for correlation.

Acmes

From time to time in the past, geologists have attempted to correlate strata that contain the *maximum abundance* or *acme* of a species. The maximum abundance of a species of fossil is chiefly ecologically controlled, however. It may represent a locally favorable environment or simply a mechanical aggregate of shells. There is no reason why a species should necessarily have a peak of abundance, or if it does, why it should be recorded by layers of abundant specimens. But even if a species did reach a peak and it is recorded, there would be no way to distinguish an abundance of fossils produced this way from one produced ecologically and, hence, the concept of correlating on the basis of the acme of a species is unworkable. One could assign the same age to strata containing an abundance of specimens of the same species and they could actually have been formed at different times as a particularly favorable environment shifted about (Fig. 10.18). About the only time value of such an occurrence of fossils is simply that of contributing to the range of the species.

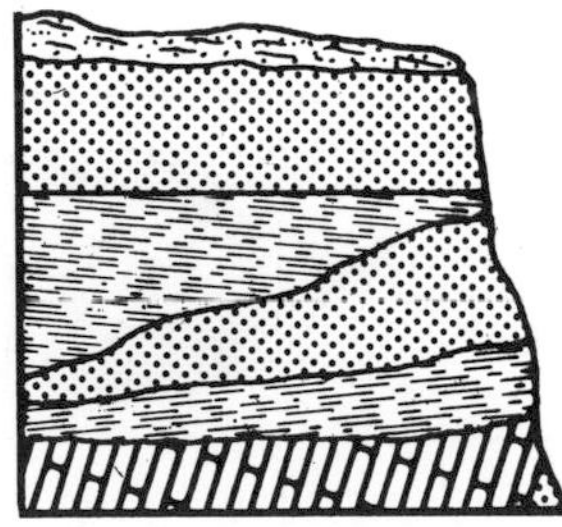

RECENT ________ Living Species 100 Percent
PLEISTOCENE _____ Living Species 90 Percent
PLIOCENE ________ Living Species 33-50 Percent

MIOCENE ________ Living Species 18 Percent
EOCENE _________ Living Species 3½ Percent
CRETACEOUS______ No Living Species

Percentage Method—With Invertebrates

FIGURE 10.20 Lyell's Percentage of Living Species Method of Time-Correlation.

Reconnaissance Methods

There are two other widely used biologic means of time correlation important in the reconnaissance of new regions, but they lack the precision and reliability of the index fossil, fossil assemblage, or range of fossil species methods described above. These three methods should be used in the detailed studies that follow.

The *stage of evolution* method is based upon the thorough knowledge of the history of an evolutionary lineage such as that of the horse (Fig. 4.18). In such a well-established lineage, the age of every step or evolutionary transformation in the series is known. To correlate rocks of unknown age, a geologist would compare a fossil from them with the stage of evolution reached by the group in the rest of the world. For example, a fossil horse tooth from a new locality in Nevada could immediately date the strata containing it because the structure of horse teeth for every stage of the Cenozoic from the Upper Paleocene onward is known. The method is not infallible, however, because there is no advance guarantee that an ancestral species did not live on at the same time as its descendants or even outlive them. For example (Fig. 10.19), the horse genus *Hipparion* lived on in the Old World at the same time and later than another horse genus *Pliohippus* which evolved after it from the same ancestor. (One could imagine the difficulties an early geologist might have had if he only knew the European *Hipparion* and the American *Pliohippus*. He could get the whole evolutionary lineage upside down!) The method also has limited application because few lineages are as well known as that of the horse family though the broad outlines of the evolution of many groups is known.

Finally, we come to Lyell's *percentage of living species* method of time correlation. Lyell, you may recall, established the Cenozoic epochs on the basis of the percentage of surviving species of mollusks they contain (Fig. 10.20). He eliminated the problem of the subjectivity of species by utilizing the identifications of only one man, the French paleontologist Deshayes. Lyell's percentages, 3½, 18, 33–50, 90, are discontinuous and, hence, indicate that there was missing time in his sections. We saw above that such discontinuities cannot exist biologically and, hence, must have a geologic cause. Furthermore, there is no guarantee that mollusks in environments other than western Europe and other types of organisms all evolved at the same rate. Thus Lyell's percentages serve as useful approximations for time-correlating rocks at the series level, but for detailed correlation, more refined methods are needed.

As noted in Chapter 6, there may be a new biologic method of reconnaissance time-correlation coming into use. This involves comparing the numbers of daily *growth rings* on a fossil with a geophysically determined scale of days in the year for each period of geologic time.

Rock-Correlation Versus Time-Correlation

Before we leave the topic of correlation, it is necessary to emphasize once again that rock-correlation and time-correlation are not the same and, therefore, cannot be substituted for one another. We cannot substitute rock-correlations

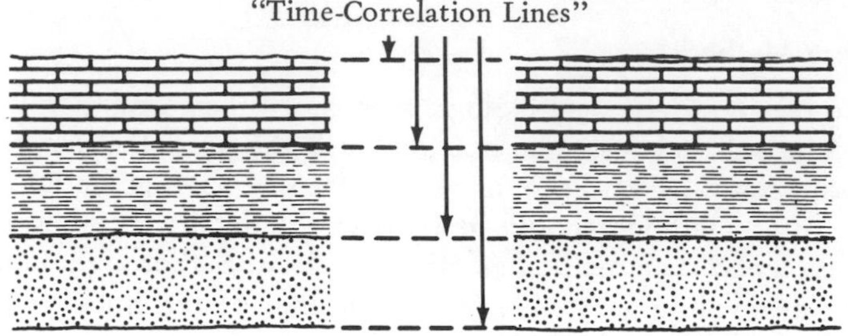

A Incorrect Technique

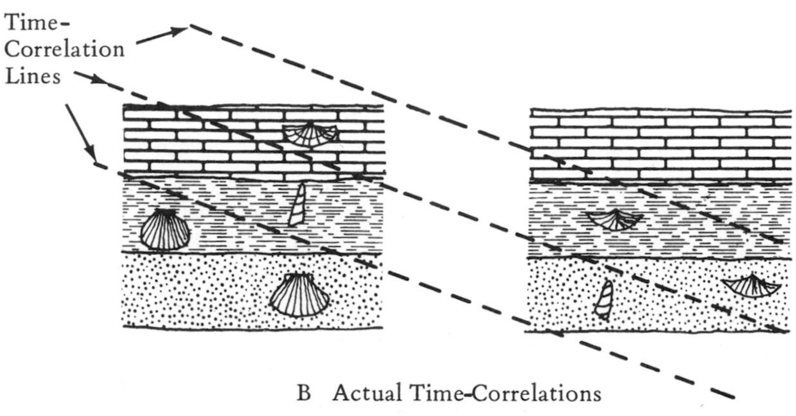

B Actual Time-Correlations

"Time-Correlation Lines"

Unconformity

C Incorrect Technique

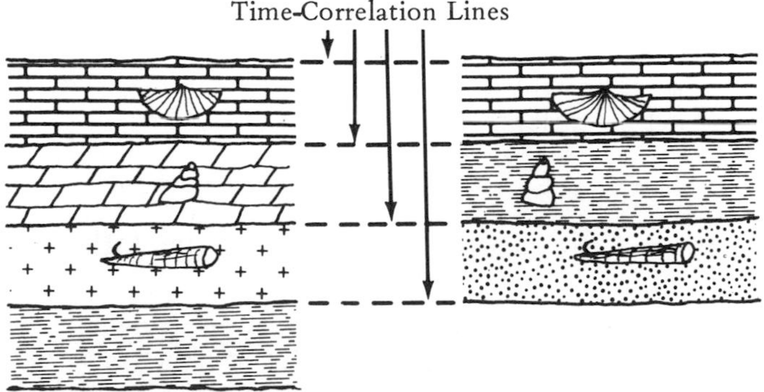

D Actual Time-Correlations

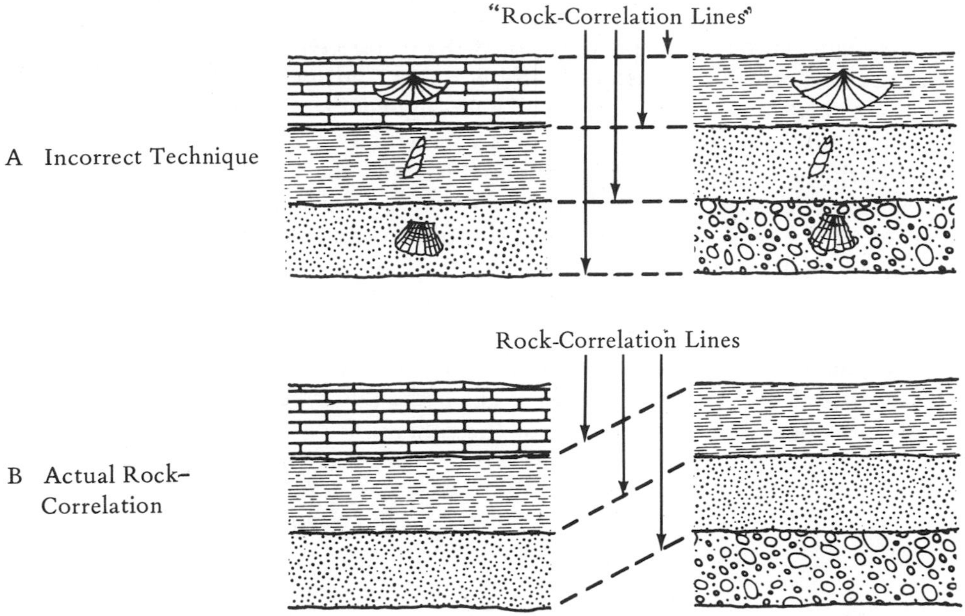

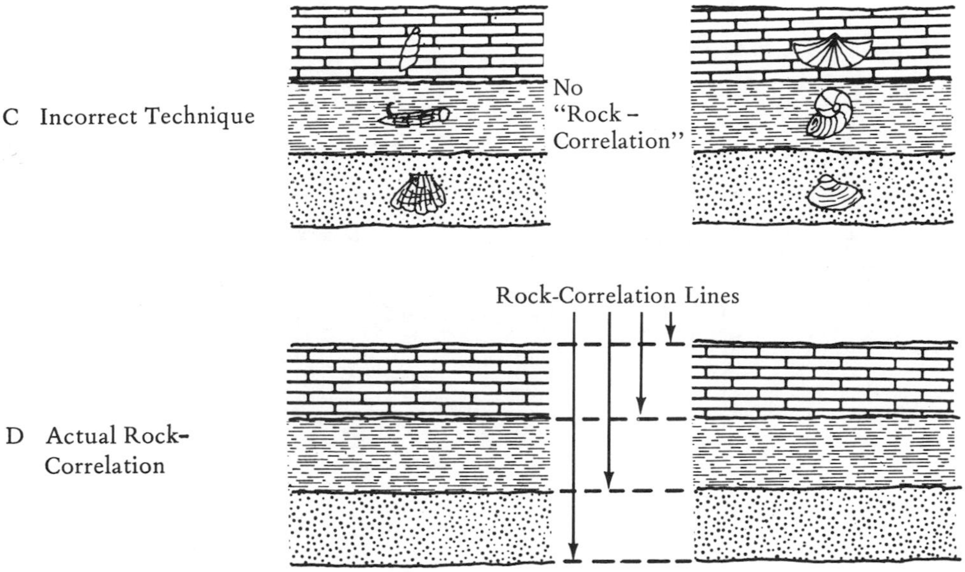

FIGURE 10.22 Diagram Illustrating Errors that result when time-correlations are substituted for rock-correlations. (A, C) Incorrect technique, (B, D) Actual time-correlations.

FIGURE 10.21 Diagram Illustrating Errors that result when rock-correlations are substituted for time-correlations. (A, C) Incorrect technique, (B, D) Actual time-correlations.

for time-correlations because if we do, rocks of different lithologies that formed at the same time will be assigned different ages and rocks of the same lithology which formed at different times will be assigned the same age (Fig. 10.21). If we substitute time-correlations for rock-correlations, rocks with the same fossils but differing lithologies will be considered parts of the same formation and rocks of the same lithology but different fossils will be considered separate formations (Fig. 10.22). A particularly troublesome problem can arise with bodies of rock of identical lithology and fossils (Fig. 10.23). The identical fossils indicate contemporaneity and time-correlation, but they do not prove lithologic continuity and, hence, rock-correlation. They may be parts of entirely different formations forming at the same time as indicated in the diagram on the right in Fig. 10.23. Hence, it is always important to distinguish the two types of correlation and not to confuse them or substitute one for the other.

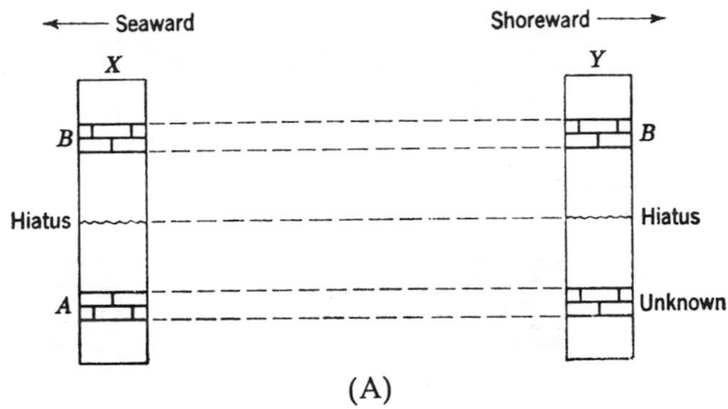

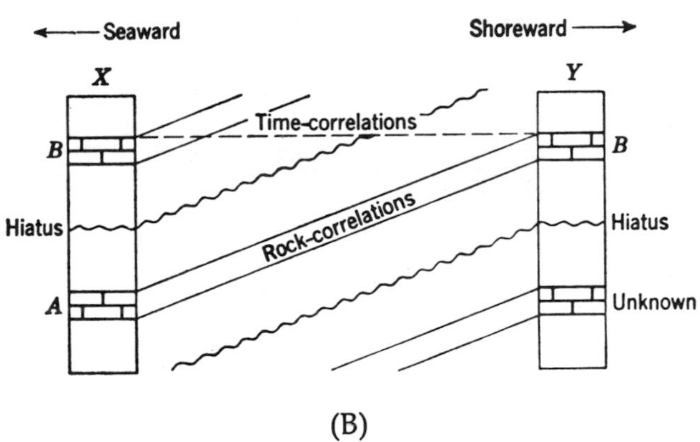

FIGURE 10.23 Difference Between Rock- and Time-Correlations. (A) Assuming rock- and time-correlations are the same, (B) Actual rock- and time-correlations. (From Alan B. Shaw, *Time in Stratigraphy*, © 1964, used with permission of McGraw-Hill Book Company.)

Questions

1. What is the difference between rock-correlation and time-correlation?
2. What are three major techniques of surface rock-correlation?
3. What is the purpose of instrumental well-logging?
4. What are three major physical methods of time-correlation?
5. What is an index fossil?
6. Why is non-synchronous homotaxis not considered an obstacle to time-correlation with fossils?
7. How does the use of fossil assemblages in time-correlation improve correlations?
8. Why is the use of acmes in time-correlation considered of little value?
9. What groups of fossils would be most amenable to the stage of evolution method of time-correlation?
10. What major errors result when time-correlations are substituted for rock-correlations? When rock-correlations are substituted for time-correlations?

11

Footloose Continents:

Continental Drift and Treadmill Ocean Basins

Historical Development of Concept

As early as 1620, Francis Bacon suggested that Europe and Africa had once been joined with North and South America because of the remarkable similar outlines of their Atlantic coasts (Fig. 8.2). In 1668, *P. Placet* stated for the same reason that before the deluge, the Western Hemisphere continents had been joined to the Old World. Buffon, about 1750, suggested that Europe and North America had once been joined because of the similarities of their present fauna and flora. He also suggested continental and polar dislocations to explain anomalous climatic changes in the past. In 1858, Antonio *Snider* noted the great similarity of the Carboniferous coal age plants on both sides of the North Atlantic. Reasoning that it was most unlikely that the spores of these plants could have blown across the Atlantic Ocean, Snider proposed that all of the continents had been joined together in the past. Near the end of the 1800s, the Austrian geologist *Edward Suess* noted that much of the geologic record of the Southern Hemisphere continents of South America, Africa, India, and Australia was nearly identical. He thought that they were all united at once in the past forming a great supercontinent which he called *Gondwanaland* after a group of rocks which defined a geological province of India. (Knowledge of Antarctica, gained at a later date, led to its inclusion in Gondwanaland also.)

The first comprehensive theories about the drifting of continents were independently proposed around 1910 by the American geologists *F. B. Taylor* and *H. H. Baker* and the German meteorologist *Alfred Wegener* (Fig. 11.1). Wegener's work was particularly thorough and detailed and most geologists today would call him the father of the continental drift concept. Wegener spent many years matching the geological and paleontological similarities on the two sides of the Atlantic. He felt that all of the continents had been united in the late Paleozoic into the vast supercontinent *Pangaea* (Fig. 11.2). (Others noted that

FIGURE 11.1 Alfred Wegener (1880–1930).

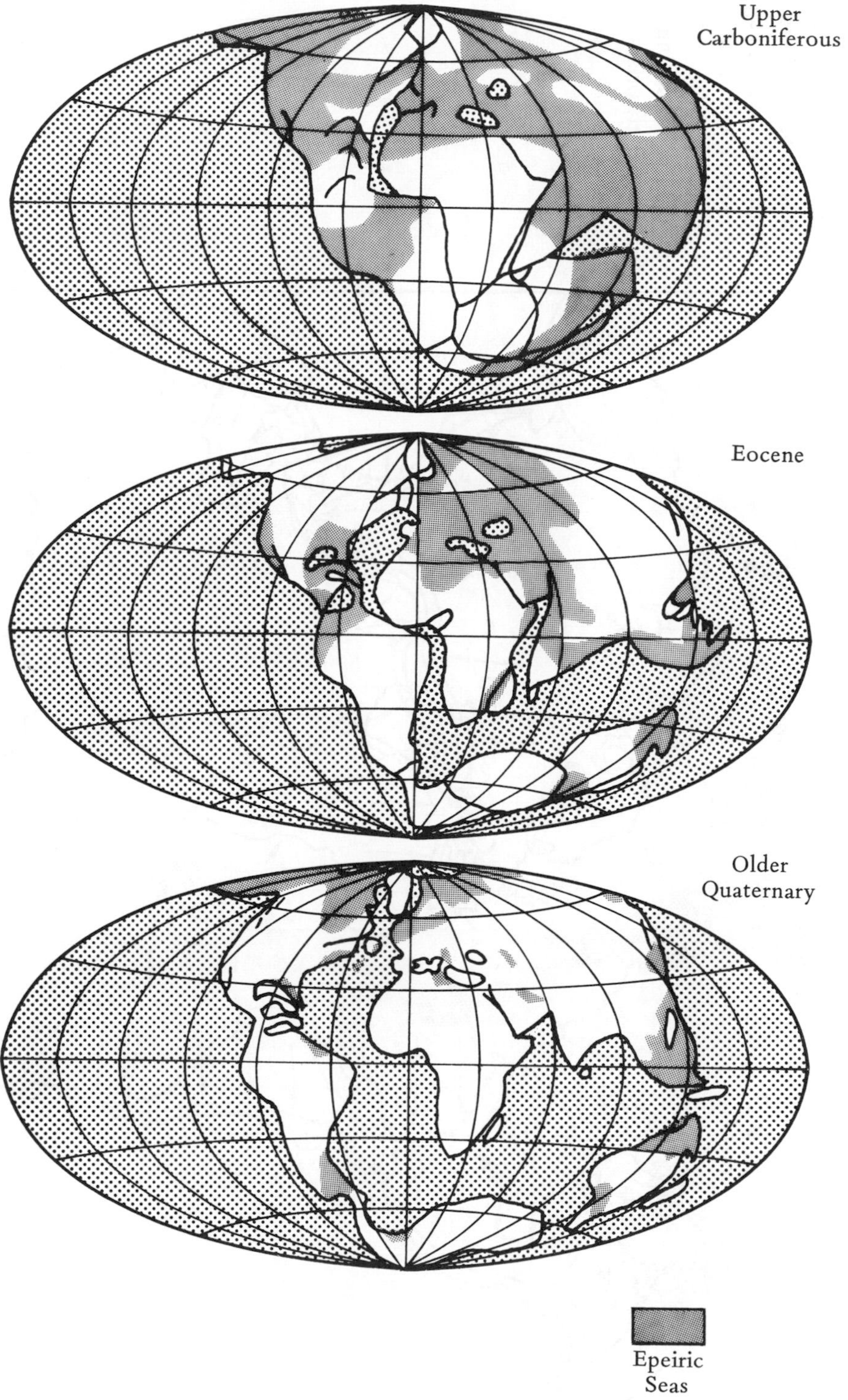

FIGURE 11.2 Wegener's Reconstructions of Pangaea and Its Fragmentation.

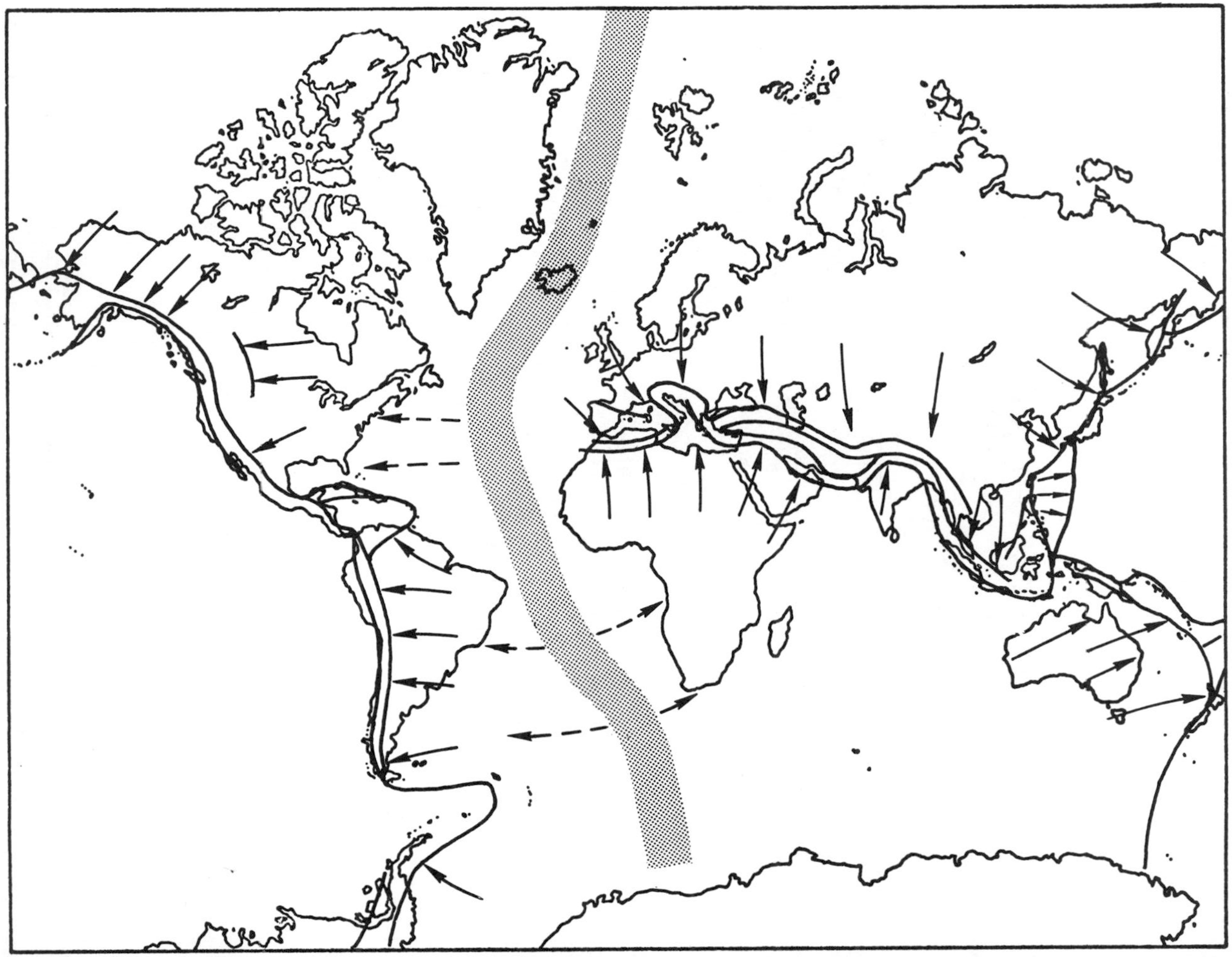

FIGURE 11.3 Mountain-Building as a Result of Continental Drift. The leading edges of moving continents buckled to form mountains according to early drift theorists.

connections between the northern and southern continental masses were not complete and preferred *Laurasia* and Gondwanaland respectively.) Wegener suggested that the continents broke up in the Mesozoic and have since drifted to their present positions. He proposed rotational and tidal forces as the driving mechanisms of continental drift and suggested that the long puzzling origin of mountain-building could also be explained by his theories. As the continents plowed their way through the oceans, he reasoned that their leading edges would be crumpled forming geosyclines and mountain belts (Fig. 11.3). The South African geologist *Alexander du Toit* continued and expanded Wegener's work in the 1930s.

Early Evidence

Some of the lines of evidence gathered by these men are the similarity of the geologic rock units and structures of continents on the opposite sides of the oceans (particularly on the Southern Hemisphere continents), the paleontologic similarities of now widely separated landmasses, and the anomalous climatic bands in the past if one assumed static continents.

They noted that not only do the edges of the sundered continents fit like the torn edges of pieces of newspaper, but the geology matches on the two sides just as the lines of type match when the two halves of the torn newspaper are put together. Particularly striking was the geologic sequence of the Southern continents in the late Paleozoic and early Mesozoic (Fig. 11.4). In all these continents, a Permocarboniferous series of glacial deposits is followed by a thick sequence of Permotriassic terrestrial beds including coal. The sequence is capped by a thick Jurassic sequence of basaltic lava flows erupted from fissures. Geological structures such as the folded Cape Mountains of Africa and the Sierra de la Ventana of Argentina also line up if one puts the continents together along their matching coasts (Fig. 11.5).

The fossils found in these rocks are also strikingly similar. A Permotriassic group of plants with large tongue-shaped leaves, *Glossopteris* and *Gangamopteris* being the most prominent (Fig. 11.6), have been found only on all the southern continents. If one assumes a static position for the Southern continents, it would be necessary to conclude either that the *Glossopteris* flora migrated between the Southern continents via the more closely connected Northern continents or that the seeds of these plants blew across thousands of miles of open ocean. The absence of the *Glossopteris* flora in the north (a few elements of the flora occur along the southern edge of the northern continents in the Mesozoic) refutes the first suggestion and the second seems most unlikely, particularly in view of the later discoveries of *Glossopteris* in very isolated Antarctica. Also, the tiny fresh-water reptile *Mesosaurus* (Fig. 11.7), which is found only in a short interval around the Carboniferous and Permian boundary, occurs in South America and Africa and nowhere else. It seems most unlikely that the *Mesosaurus* could have crossed the present Atlantic in light of its ecologic occurrence and it is not found in the northern continents even where the proper facies occur. Other fossil organisms with disjunct distributions have also been cited.

Finally, the climatic evidence seemed to require a different position for the continents in the late Paleozoic than they now have. The Permocarboniferous

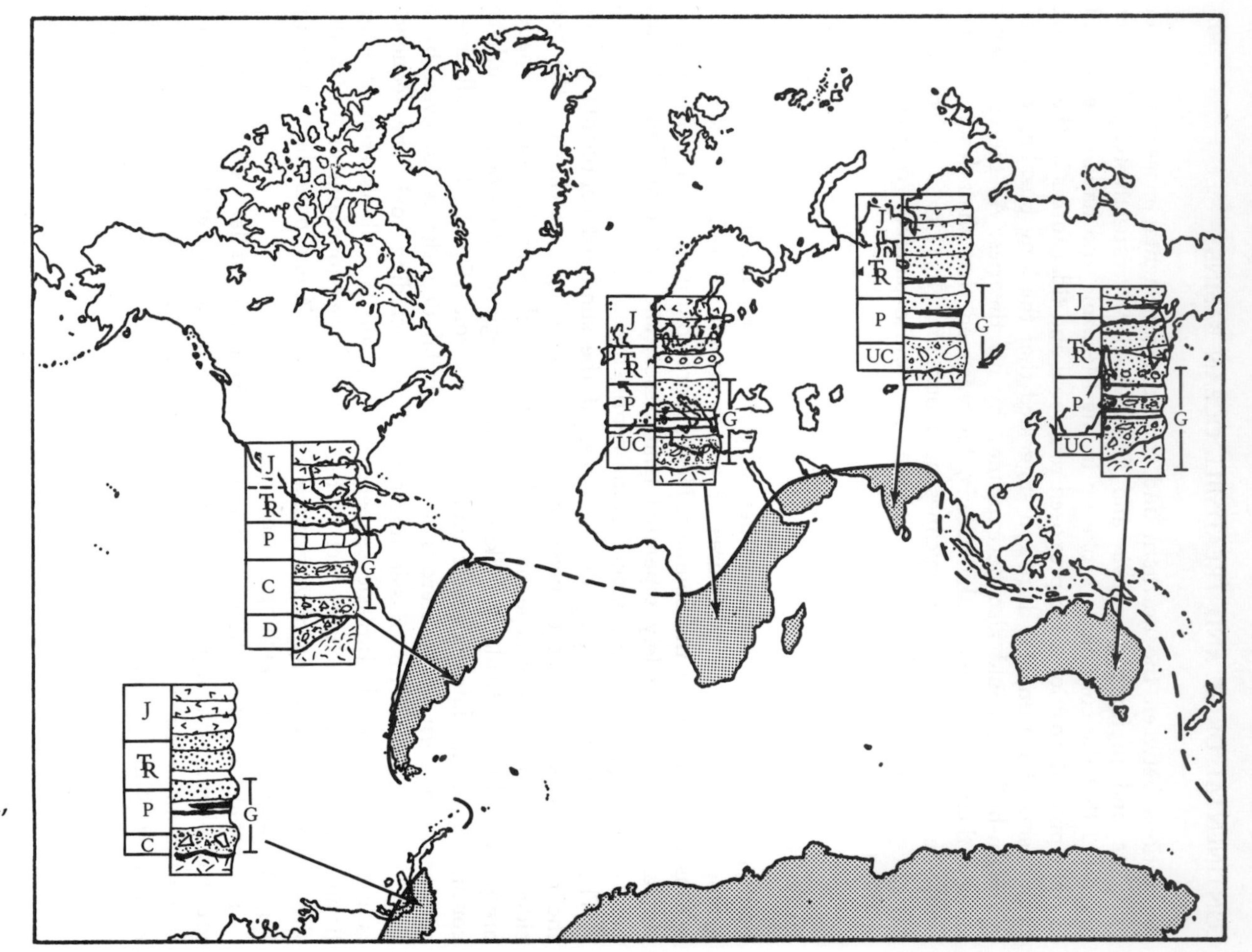

FIGURE 11.4 Similarities of the Geologic Records of the Gondwana Continents. The Carboniferous-Jurassic records of Africa, Antarctica, Australia, India, and South America are virtually identical despite their present wide separation.

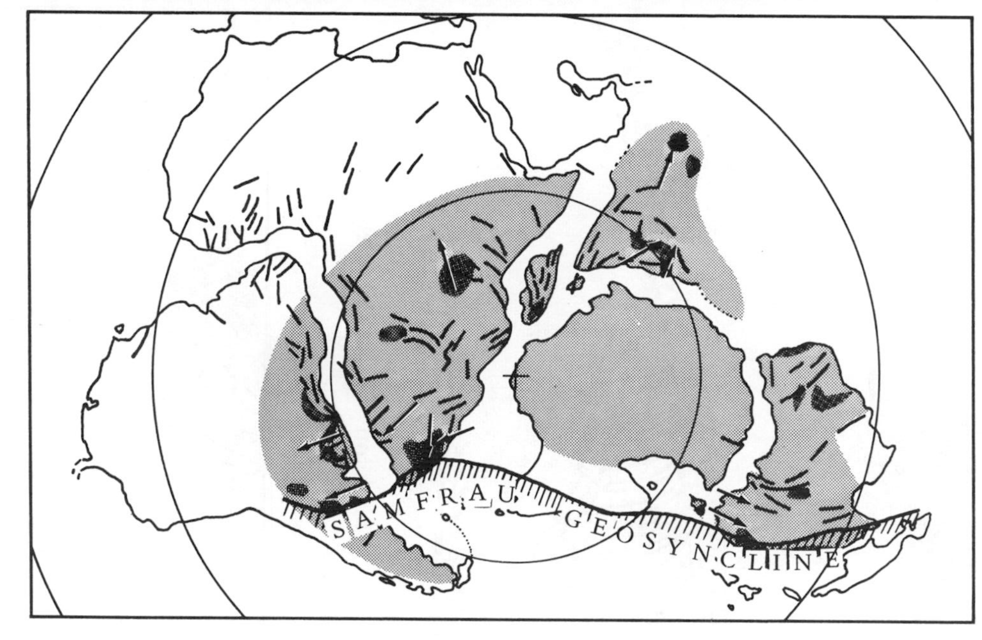

FIGURE 11.5 Reconstruction of Gondwanaland in the Late Paleozoic, showing mountain belt alignments, glacial deposits, and ice flow directions. (After DuTоit, 1937)

FIGURE 11.6 *Glossopteris* **Flora.** Fossil leaves of *Glossopteris* and *Gangamopteris.*

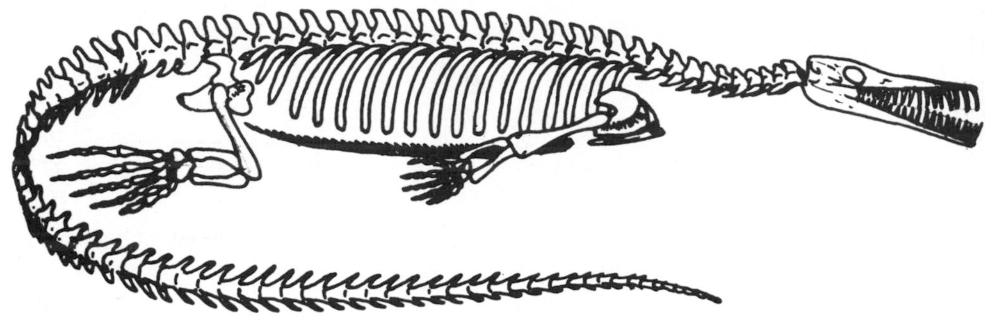

FIGURE 11.7 *Mesosaurus.* Skeleton of fresh-water reptile known only in South America and Africa.

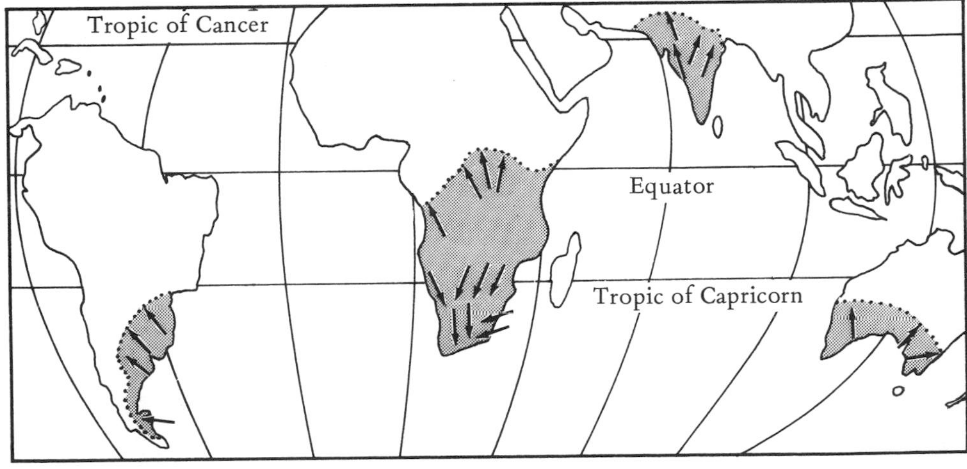

FIGURE 11.8 Present Distribution and Flow Directions of Late Paleozoic Continental Glaciers and Glacial Deposits. (Compare with Figure 11.5.)

glaciations (Fig. 11.8) not only occurred in areas such as Africa and India far removed from the present poles, but the lines of ice flow are presently away from the equator and toward the poles, the exact opposite of what would be expected if the continents were situated as they now are. In addition, the lines of flow indicate that the ice flowed from the sea onto the land in many areas, also the reverse of what one would expect. Clearly, these phenomena could only be explained by assuming the continents were once joined so that the ice was flowing across a single land area. The distribution of evaporites, redbeds and tropical vegetation at various times in the past also suggested that the continents were in different climatic zones than at present (Fig. 11.9).

Wegener's ideas on continental drift were largely rejected by Northern Hemisphere geologists, though they continued to thrive in the Southern continents because of the work of Du Toit and others and the more striking evidence there. The reason for the Northern Hemisphere rejection was that the rotational and tidal forces proposed by Wegener were too small to move continents and the physicists could not imagine any larger forces in the earth. In the 1930s and 1940s there were few northern hemisphere geologists who would support continental drift.

Convection Currents

Two who did were the Dutchman *F. A. Veining Meinesz* and the Scotsman *Arthur Holmes.* These men were geophysicists and they noted that gravity measurements around the oceanic trenches indicated that material there was being held down in opposition to its tendency to attain isostatic equilibrium (Fig. 11.10). It was as if someone beneath a water surface were pulling a cork down against its natural tendency to bob up to a certain level consistent with its height and low density. By this time, geologists were aware of the fact that continental crust is lighter and thicker than oceanic crust (Fig. 11.11). The continents are somewhat like corks floating in a sea of denser material. Veining Meinesz and Holmes hypothesized that the earth's internal heat generates convection currents in a liquid mantle (Fig. 11.12). Hot material rises to the surface, moves outward carrying the lighter continents with it, cools and descends at the trenches. Here where the cooler material is sinking and its drag effect produces a trench, a piece of the earth's crust is being held down against its own level-seeking tendencies. Trenches are present-day geosynclines in this theory.

Once the convection stops in an area, the trench with its great thickness of land-derived light sediment is no longer held down and bobs up to form mountain ranges. Despite the fact that the convection current theory unified many observations in geology with a coherent explanation, it was ridiculed in most geology texts even through the 1950s.

Modern Concepts

After World War II, however, several startling discoveries were made by geologists studying the earth's *magnetism* and its *ocean basins.* These discoveries completely reversed the previous trend and won over the scientific community in the 1960s just as evolution did a century ago. Some geologists still refuse to accept dynamic continents much as their counterparts in the 1860s denied a dynamic biology.

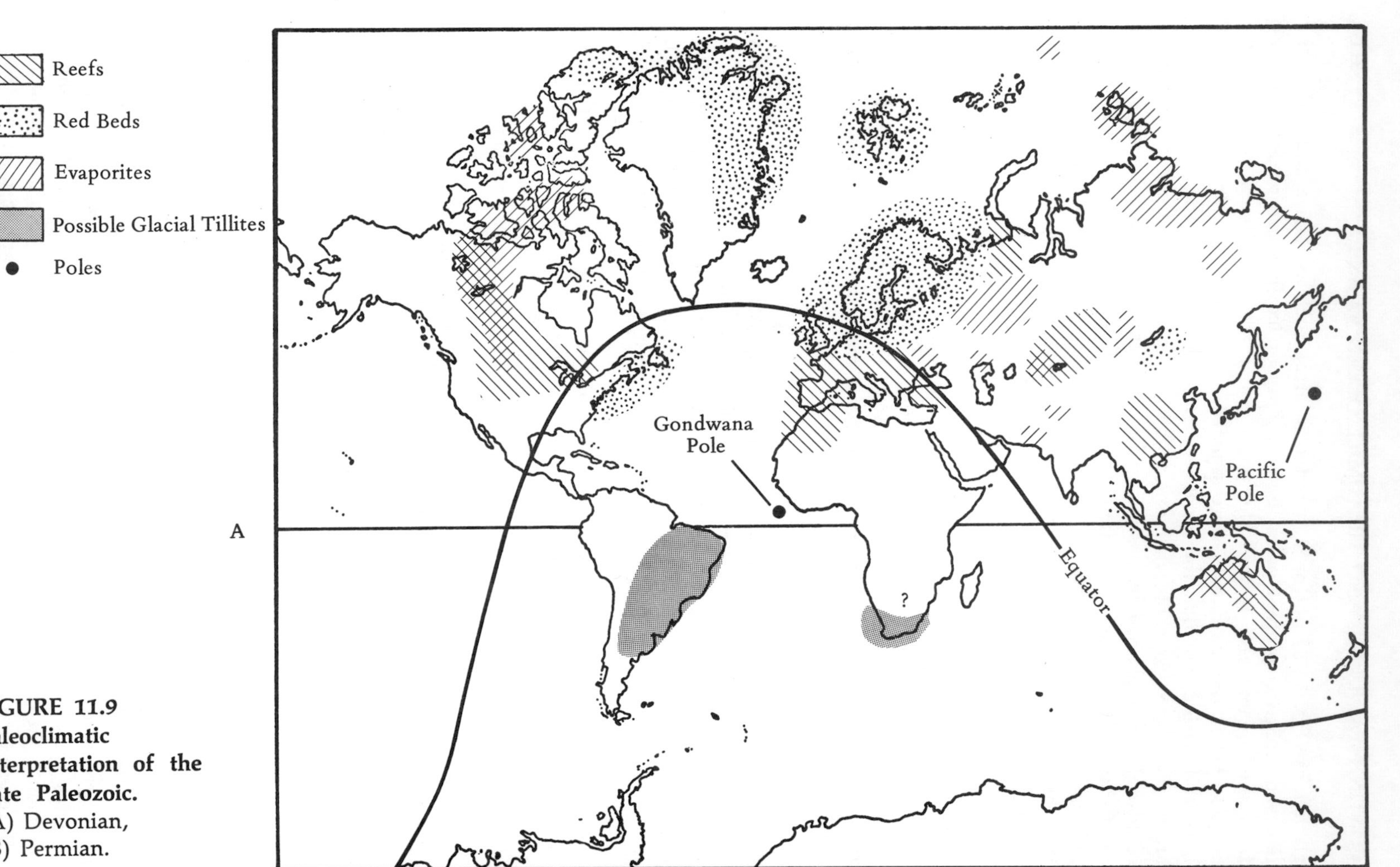

FIGURE 11.9 Paleoclimatic Interpretation of the Late Paleozoic. (A) Devonian, (B) Permian.

B

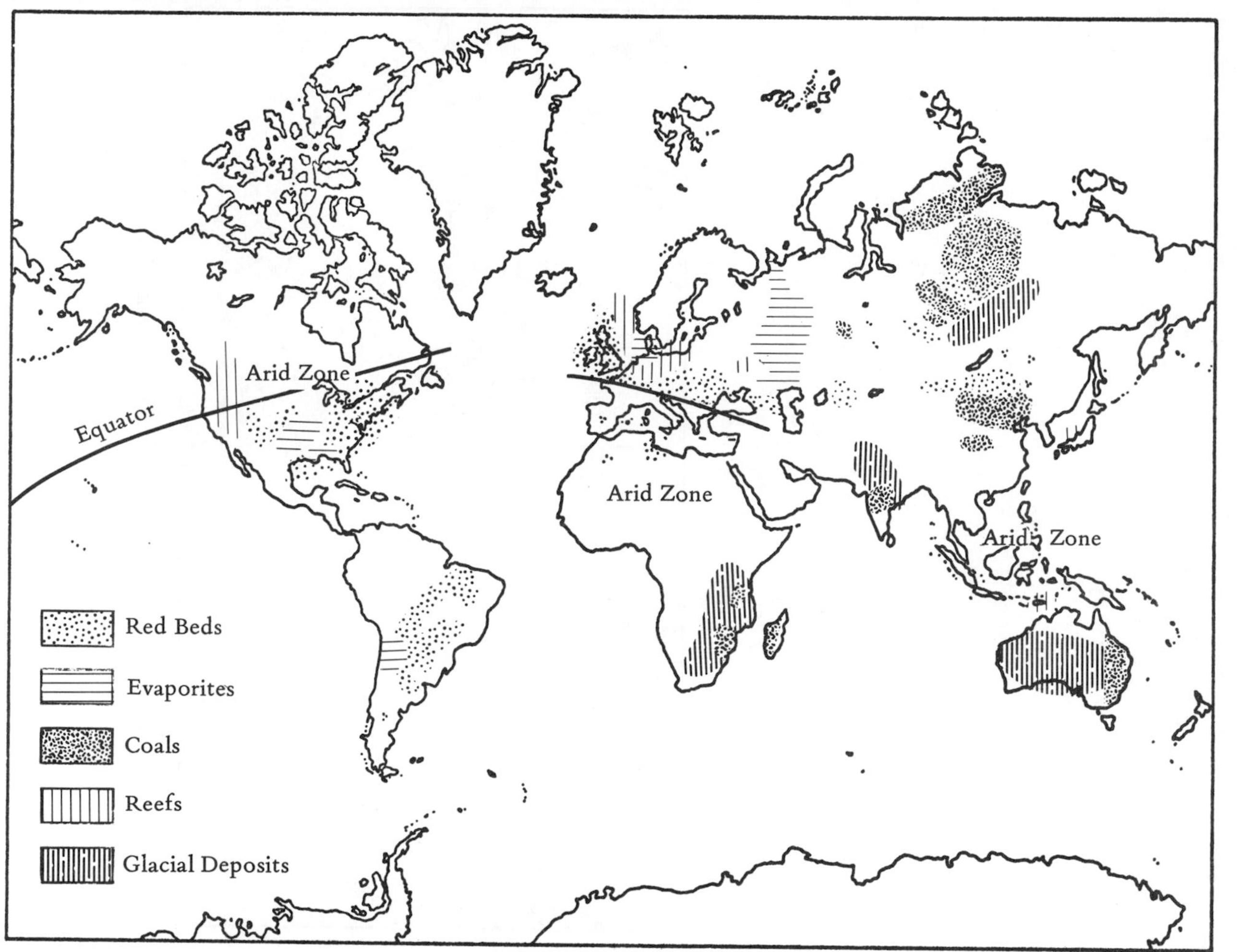

Arid Zone
Equator
Arid Zone
Arid Zone
Red Beds
Evaporites
Coals
Reefs
Glacial Deposits

Paleomagnetism

We noted in Chapter 9 that magnetic iron oxide particles in a cooling lava flow or a settling sediment align themselves with the earth's magnetic field. The direction of alignment and inclination from the horizontal indicate the position of the magnetic pole. (A magnetic needle points straight down at the magnetic poles, is horizontal at the equator, and occupies intermediate angles of inclination in between.) Likewise, in the geologic past, magnetic particles lined up in the same manner and were "frozen" in the earth's crust. Thus, geophysicists have a means of locating the earth's past magnetic poles. They found, surprisingly, that the evidence indicated that poles were in different positions, relative to a continent, at various times in the geologic past. There were two alternatives: either the poles or the continents had moved. To test these hypotheses, they sampled other continents. If the continents were stationary and the poles had moved, then the paleomagnetically determined pole positions for

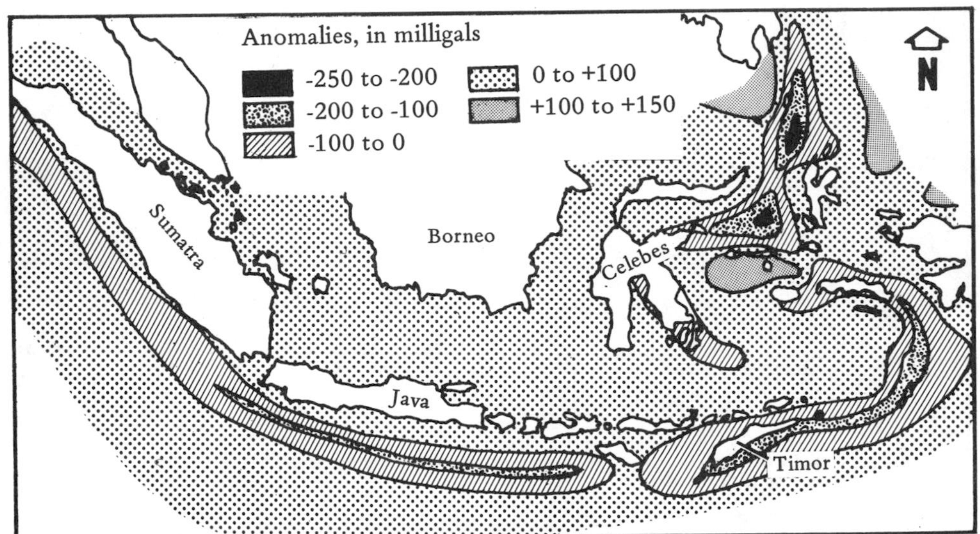

FIGURE 11.10 Gravity Anomalies and Oceanic Trenches.

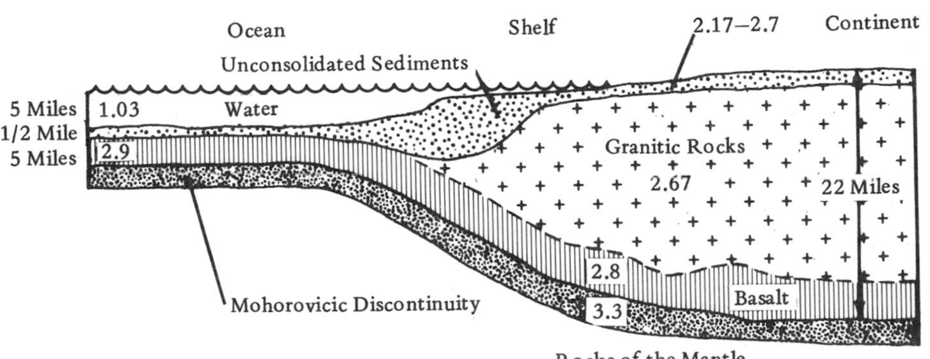

FIGURE 11.11 Cross-Section of Earth's Crust and Upper Mantle showing relative densities and thicknesses of constituents.

a particular time should be the same for all continents. If, on the other hand, the continents had moved and the poles remained stationary, the pole positions for a particular time would differ among the continents. The latter case turned out to be the reality (Fig 10.11).

The line connecting the pole positions relative to a particular continent for various times in the geologic past is still called a *"polar wandering" curve* in reference to the original supposition that the first hypothesis was true. Its form is the same for all continents, but its position is not. The polar wandering curves converge toward the present poles as the continents moved to their present positions (Fig. 10.11).

Unfortunately, all paleomagnetism can do is tell us how far from the pole a continent was situated, in other words, its latitude. It cannot tell where around the world at that latitude a continent was situated, in other words, its longitude. One pole, the present North Pole (but not always so, because the poles have reversed) has been in the Pacific area much of the time, and is often called the *Pacific Pole.* The other, which has been in or near the Gondwana continents, has been named the *Gondwana Pole.* Keep in mind that there are two assumptions inherent in paleomagnetic reconstructions. One is that the earth's magnetic field has always been dipolar, that is with two opposite poles, and that the magnetic and rotational poles have always been very close. The common shape of all the polar wandering curves supports the first supposition, the match of the geologically determined climatic zones with the geophysically determined latitudes supports the second.

Oceanography

Even more significant discoveries came from oceanographers during this interval. Computer fits of the continents at their actual submarine edges proved more striking than at the shorelines (Fig. 11.13). Geologists had assumed that the

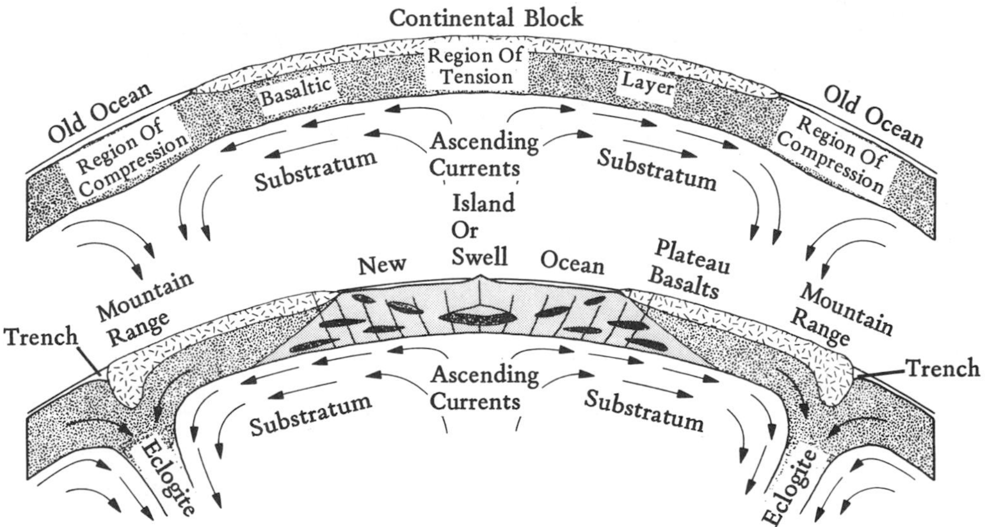

FIGURE 11.12 Convection-Current Hypothesis proposed by Arthur Holmes.

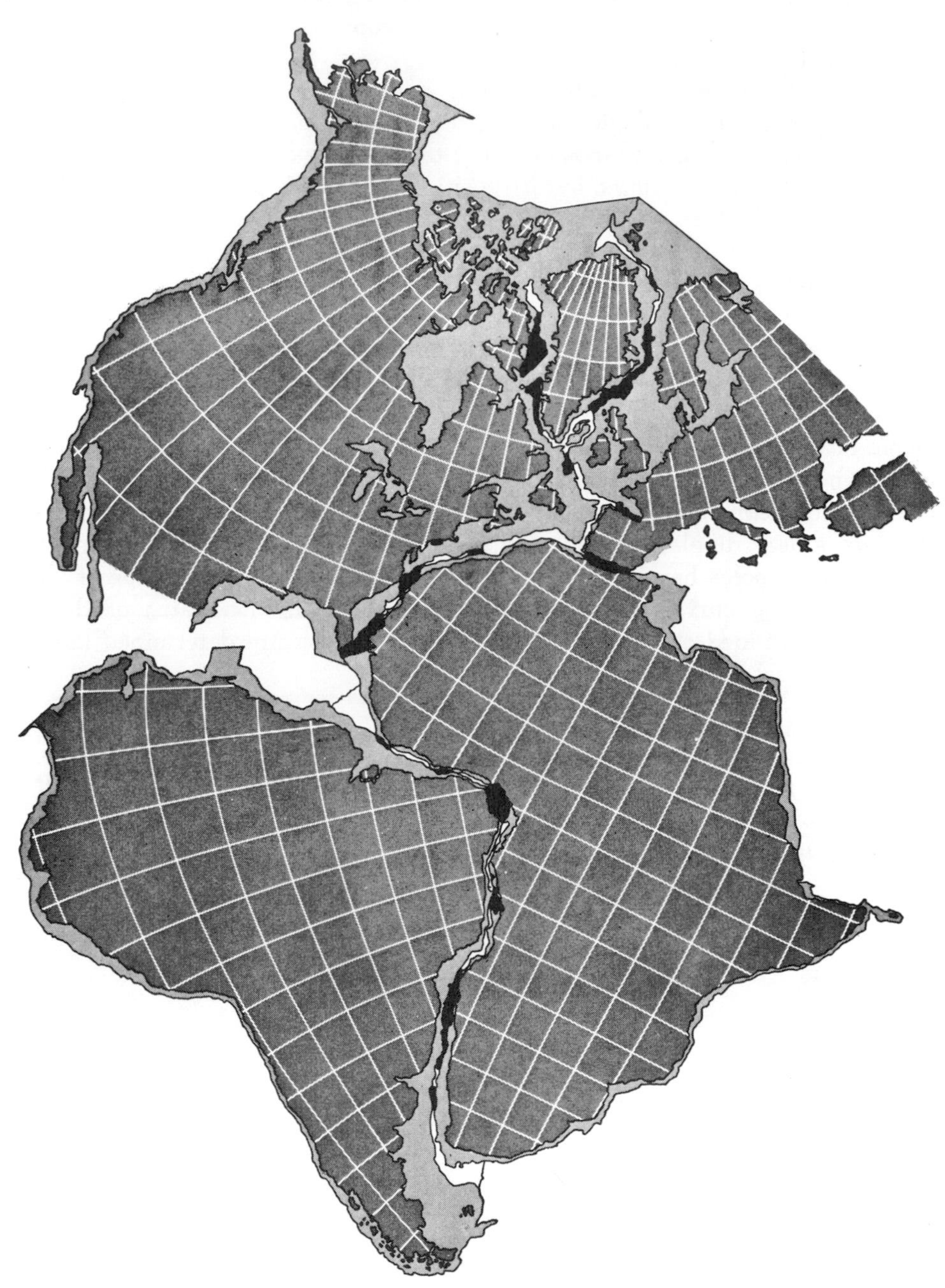

A

FIGURE 11.13 Computer Fits of Sundered Continents. ((A) Across the Atlantic Ocean, (B) Across the Indian Ocean. (A) present ocean basins have been in existence since the beginning of geologic time and hence, have been receiving a steady fall of sediments. (Oceans had been in existence even in the Precambrian as indicated by marine rocks on the continents.) Unlike the continents with their numerous unconformities, the ocean basins should contain a complete geologic record. Surprisingly, however, not only are there unconformities in the oceanic sections (Fig. 11.14), but no rocks older than Jurassic have been encountered. The oceanic sediments were found to contain great quantities of calcareous debris at depths where shells falling from above today would dissolve (Fig. 11.15). In addition, the rocks regularly are older toward the margins of the ocean basins (Fig. 11.16).

The topography of the ocean floor is not the predicted flat surface from side to side, but is crossed by a world-girdling system of ridges, often called mid-ocean ridges (Fig. 11.17) after the centrally located Atlantic one. The ridges, whose continuity was first noted by *Maurice Ewing* and *Bruce Heezen,* are generally cracked down the middle by a rift zone characterized by earthquakes, vulcanism, and high heat flow (Fig. 11.18). The age of volcanic islands increases outward from the mid-ocean ridges (Fig. 11.19). The ridges go ashore at various places such as the Rift Valleys of Africa and the head of the Gulf of California. These are active zones of tension also and the same types of activity occur as on the mid-ocean ridges (Fig. 11.18). The ocean-margin trenches not only have

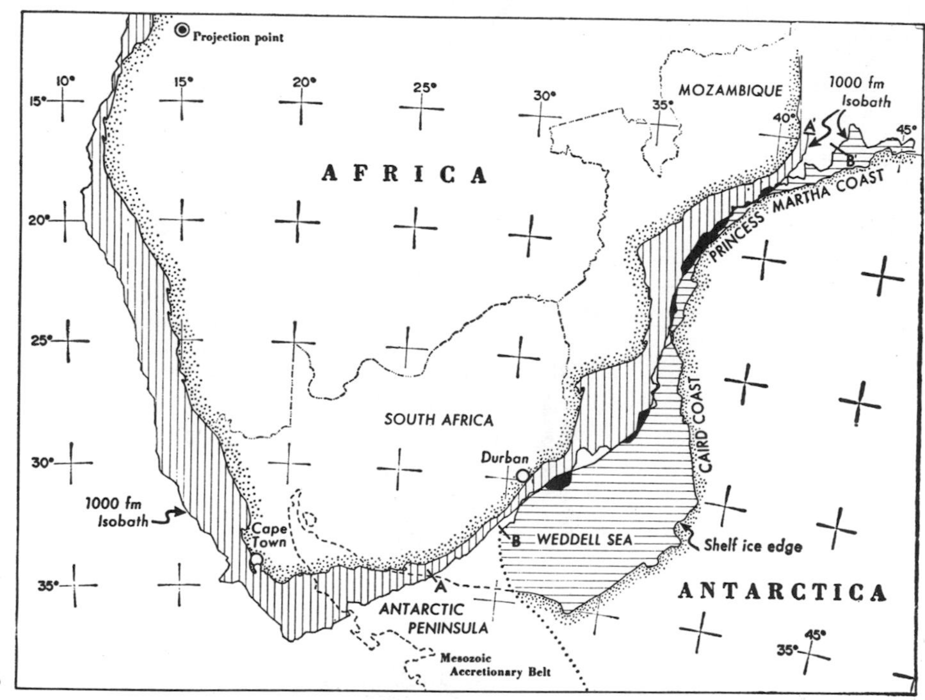

from *The Confirmation of Continental Drift* by Patrick M. Hurley, Copyright © 1968 by Scientific American, Inc. All rights reserved. (B) from Dietz and Sproll, *Science*, vol. 167, p. 1613, March, 1970, Copyright © 1970 by AAAS, by permission of the authors and AAAS.)

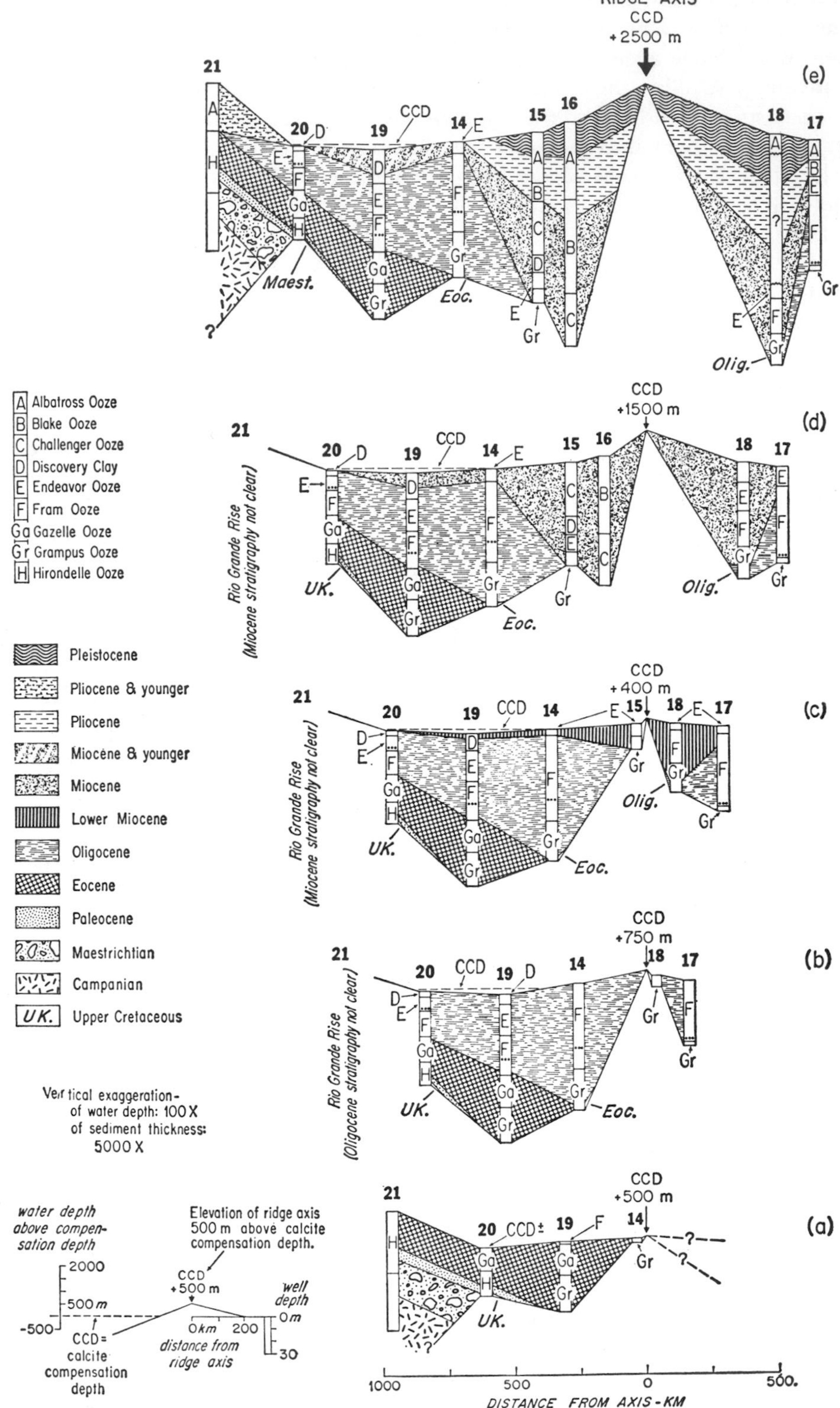

RIDGE AXIS
CCD
+2500 m
(e)
21
20
19
14
15
16
18
17
CCD
Maest.
Eoc.
Olig.
CCD
+1500 m
(d)
Rio Grande Rise
(Miocene stratigraphy not clear)
UK.
CCD
+400 m
(c)
CCD
+750 m
(b)
Rio Grande Rise
(Oligocene stratigraphy not clear)
CCD
+500 m
(a)
CCD±
A Albatross Ooze
B Blake Ooze
C Challenger Ooze
D Discovery Clay
E Endeavor Ooze
F Fram Ooze
Ga Gazelle Ooze
Gr Grampus Ooze
H Hirondelle Ooze
Pleistocene
Pliocene & younger
Pliocene
Miocene & younger
Miocene
Lower Miocene
Oligocene
Eocene
Paleocene
Maestrichtian
Campanian
UK. Upper Cretaceous
Vertical exaggeration-
of water depth: 100 X
of sediment thickness:
5000 X
water depth
above compensation depth
2000
500 m
-500
CCD =
calcite
compensation
depth
Elevation of ridge axis
500 m above calcite
compensation depth.
CCD
+500 m
0 km
200
distance from
ridge axis
well
depth
0 m
30
1000
500
0
500.
DISTANCE FROM AXIS - KM

FIGURE 11.14 History of the South Atlantic Ocean Basin since the Late Cretaceous. Note unconformities in present-day geologic column at top. Lower diagrams represent inferred history. (From Maxwell, et al. *Science*, vol. 168, p. 1057, May, 1970, Copyright © 1970 by AAAS, by permission of the authors and AAAS.)

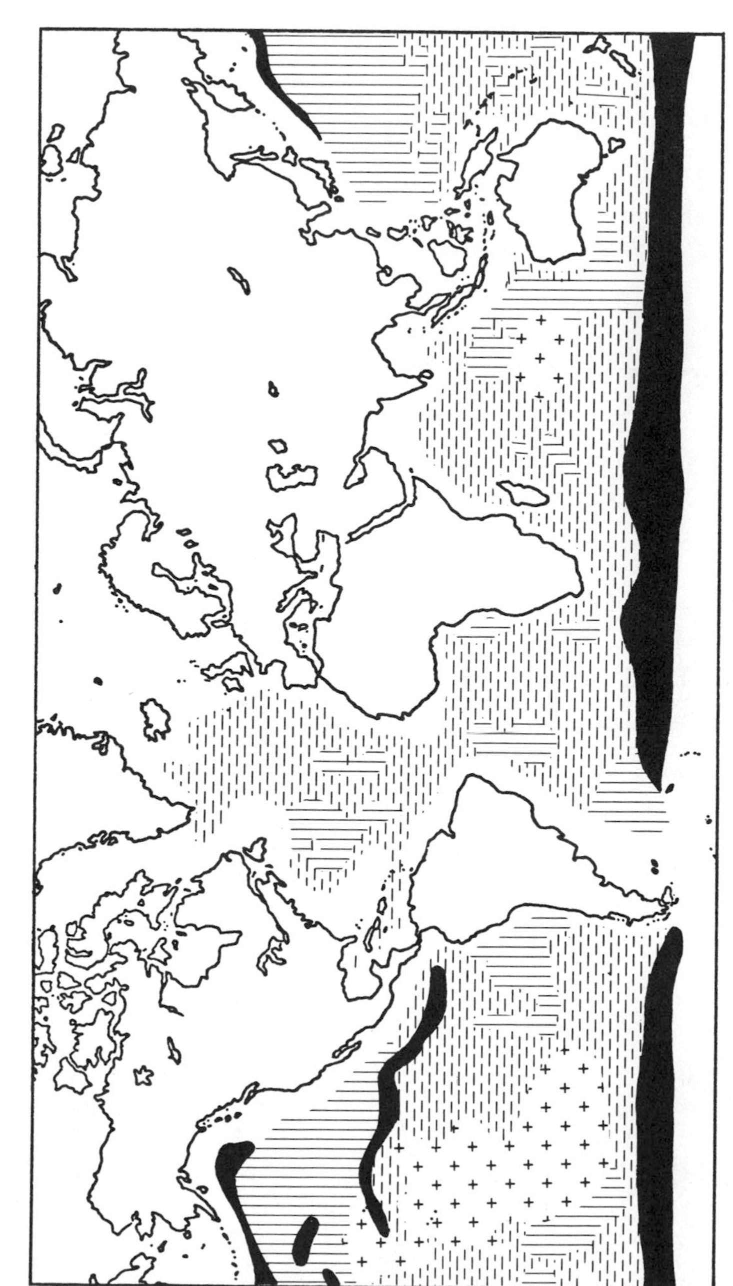

FIGURE 11.15 Distribution of Major Ocean Basin Sediments. Note decrease in calcareous sediments away from ridge crests.

prominent gravity anomalies, but also are active volcanic and seismic regions with the earthquakes occurring along a plane (*Benioff zone*) sloping downward and inward toward the continent (Fig. 11.20).

Another notable discovery was that in the oceans, the intensity of the earth's magnetic field varied. There was a series of parallel bands of alternating intensity (Fig. 11.21). It was discovered that the pattern of these bands was symmetrical about the mid-ocean ridges and the width of each paired set of symmetrical bands was proportional to the duration of the periods of normal or reversed polarity found on the continents.

Finally, the mid-ocean ridges and other ocean basin features are typically offset by large faults, but when the direction of movement of earthquakes on these faults was determined, it was the reverse of the direction of offset (Fig. 11.22).

All of these observations have been synthesized in the past ten years into a coherent theory of the constant evolution of the earth's crust called *plate tectonics* or *global tectonics.*

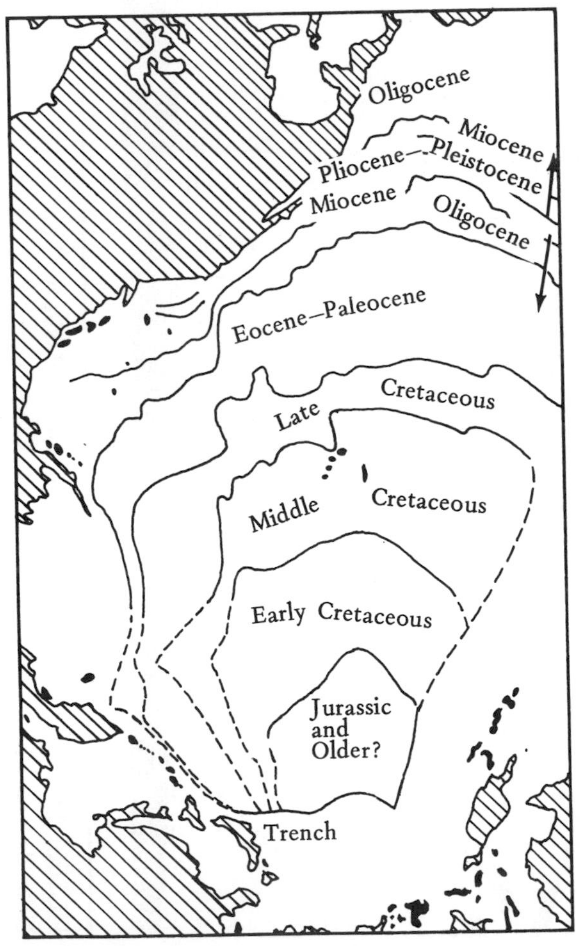

FIGURE 11.16 Ages of Oldest Sediments in Various Parts of North Pacific Basin. (From Fischer, et al, *Science*, vol. 168, p. 1211, June, 1970, Copyright © 1970 by AAAS, by permission of the authors and AAAS.)

FIGURE 11.17 Major Modern Oceanic Ridges, Fracture Zones, and Trenches.

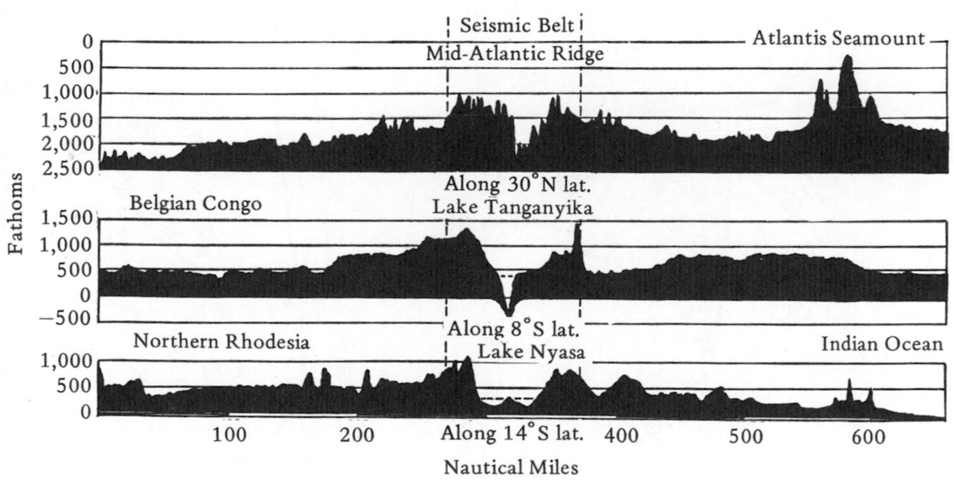

FIGURE 11.18 Profiles Across an Oceanic Ridge and Analogs on the Continents. (From Bruce C. Heezen, *The Deep Sea Floor*, © 1962, by the Academic Press, N.Y. Used by permission of the author and publisher.)

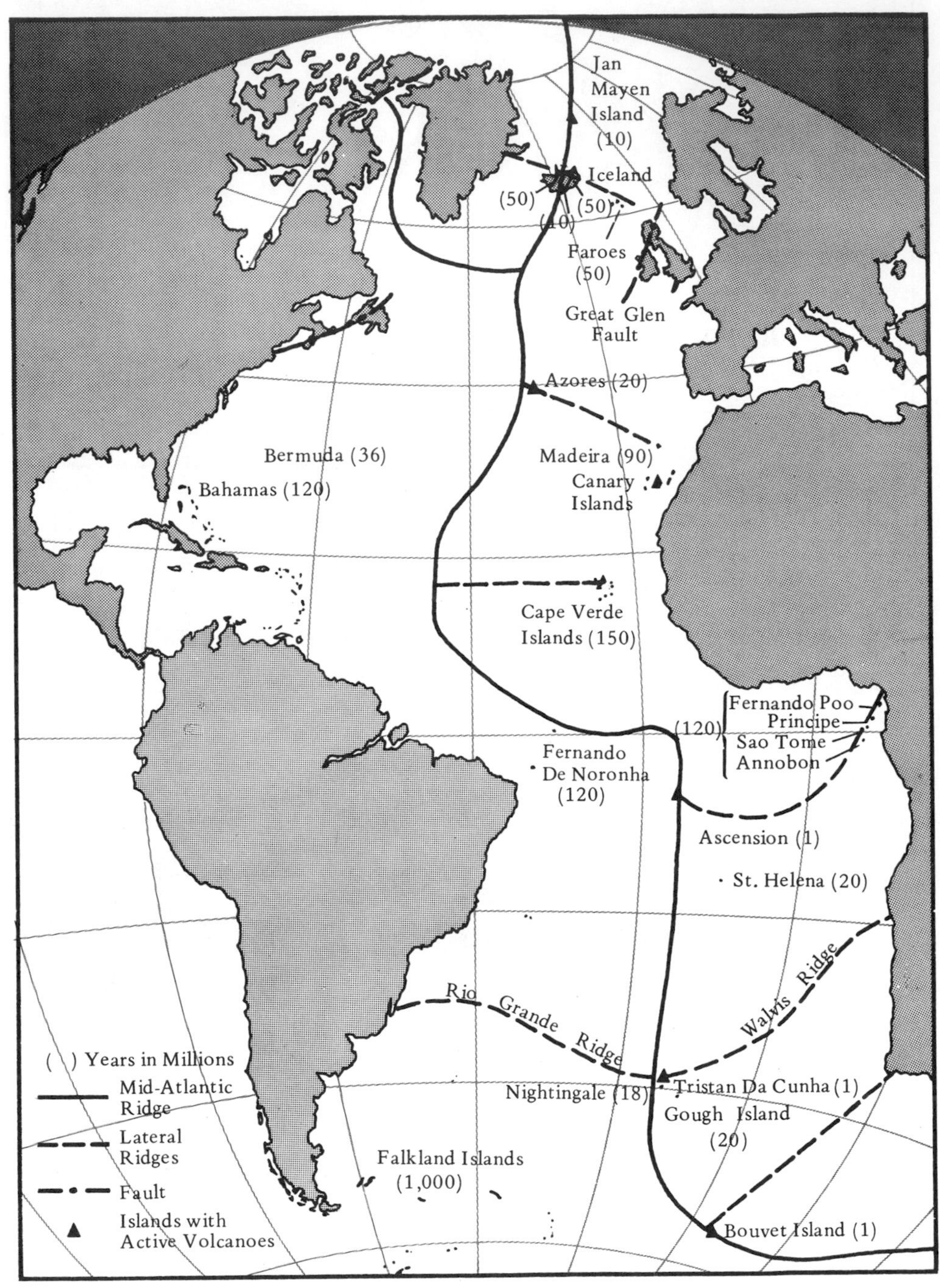

FIGURE 11.19 The Age of Volcanic Islands in the Atlantic. (After Wilson.)

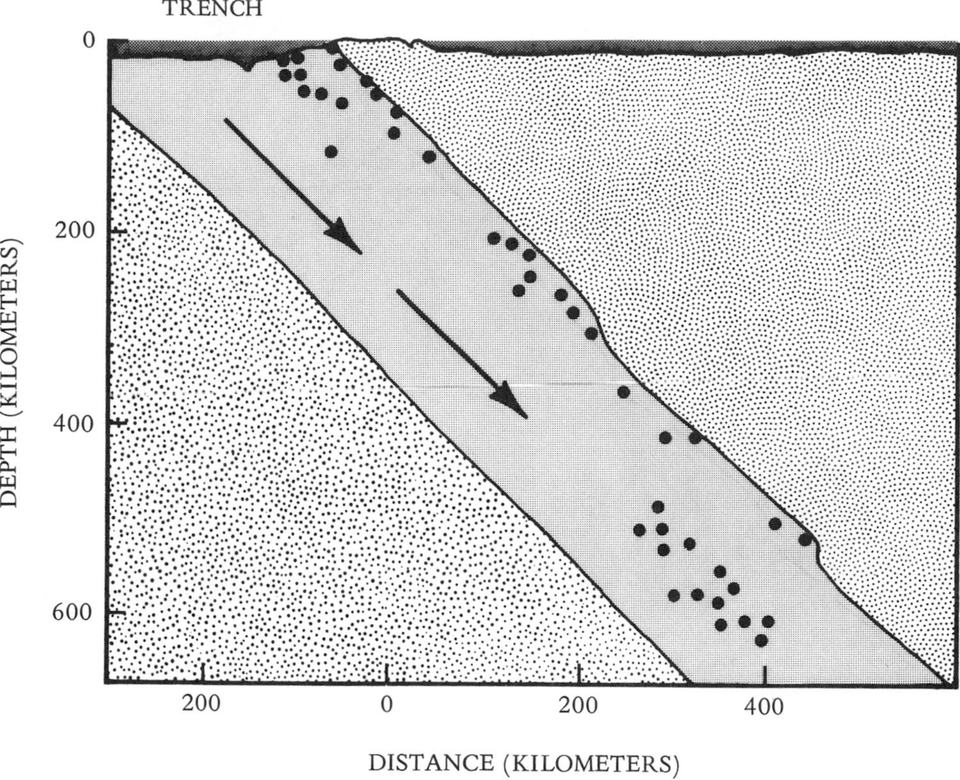

FIGURE 11.20 Benioff Zone Beneath a Pacific Trench.

Plate Tectonics

According to this theory, new crustal material is constantly being generated at the ridge centers where the crust is thin (Fig. 11.23). Lava wells up from below, cools below the so-called Curie point, and the magnetism of the earth's field at the time is frozen in the rock. The constant rising of magma into the tensional rift zones of the ridges is responsible for their vulcanism. This material gradually moves outward from the ridge in both directions. As it does, new magma wells up from below and solidifies, also taking on the magnetism prevailing at the time of its formation. If the earth's two poles have reversed, the rock acquires magnetism in this new direction. This explains, as *F. J. Vine* and others have shown, the striped pattern of high and low magnetic intensities symmetrical about the ridges (Fig. 11.24). Where the strip is magnetized in the direction of earth's present field, the effect is additive and strong magnetic intensity results. Where the strip is magnetized in the opposite direction, it subtracts from the present magnetic intensity leaving a low value. The various strips can be matched with the normal and reversed periods of magnetization dated on the continents (Fig. 11.25). (Those older than Late Cenozoic are dated by

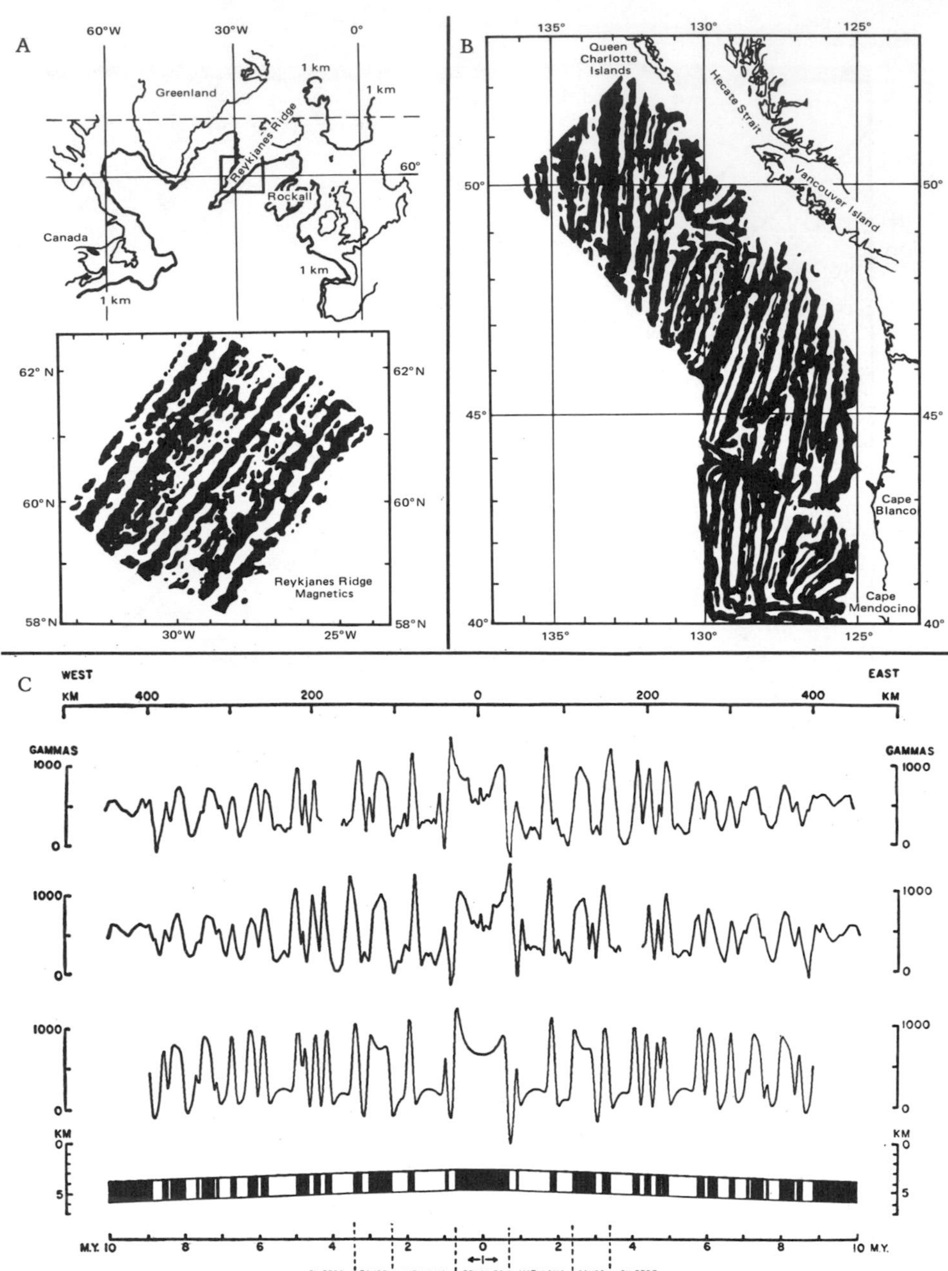

FIGURE 11.21 Bands of Alternating Magnetic Intensity in the Ocean Basin Crust. (A) Map pattern in North Atlantic, (B) Map pattern in North Pacific, (C) Profiles of anomalies across ridge in South Pacific.((A) from Pittman and Heirtzler, *Science*, vol. 154, Dec., 1966, p. 1170, Copyright © 1966 by AAAS. (B) from Raff and Mason, *GSA Bulletin*, vol. 72, August 1961, p. 1268, (C) from Vine, *Science*, 1966, vol. 154, Dec. 1966, p. 1407, Copyright © 1966 by AAAS. All figures used by permission of authors and societies.)

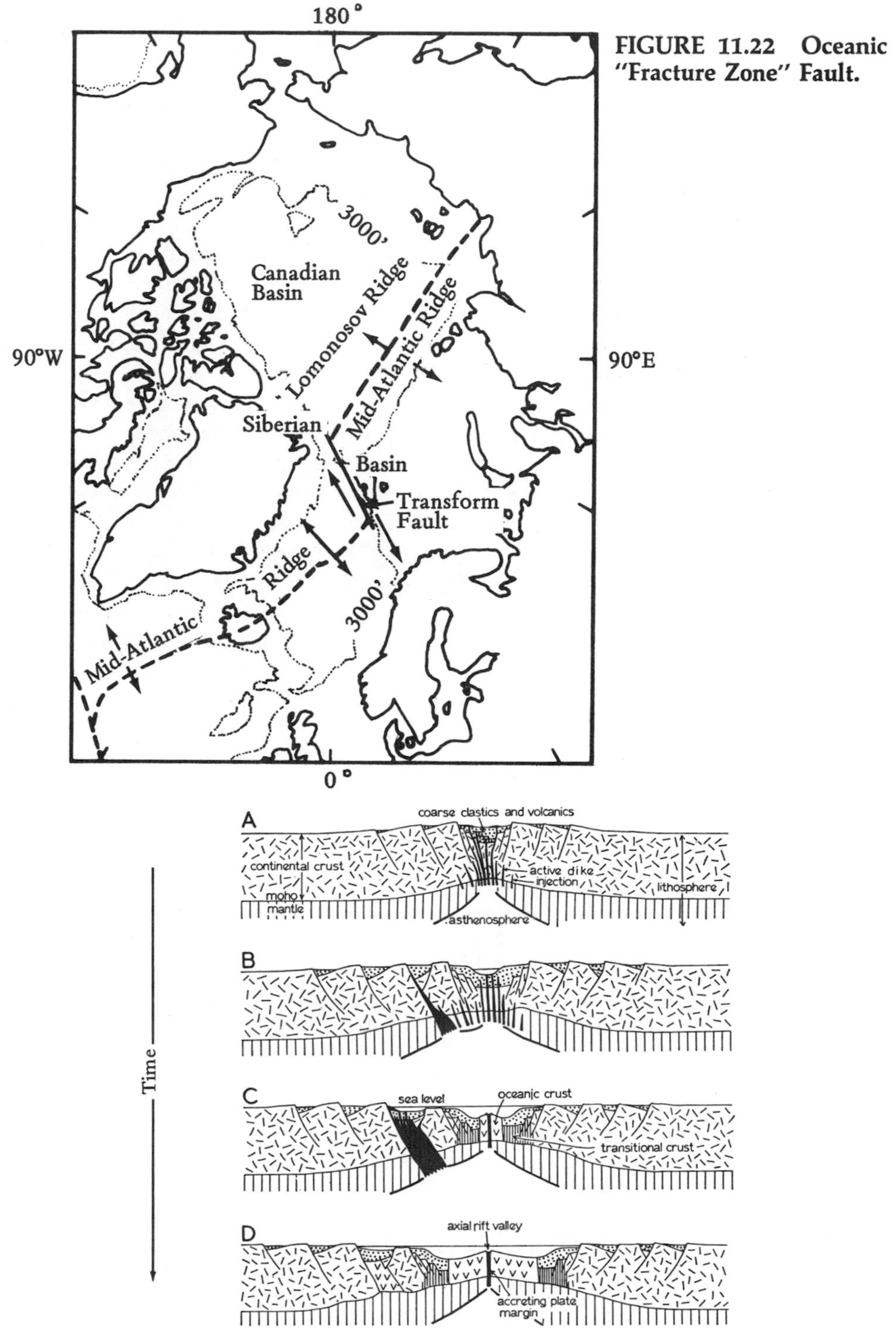

FIGURE 11.22 Oceanic "Fracture Zone" Fault.

FIGURE 11.23 Development of an Oceanic Ridge by the Rupturing of a Continent. (From Dewey and Bird, *GSA Bulletin*, vol. 75, May, 1970. Used by permission of authors and G.S.A.)

extrapolation assuming a constant spreading rate, but radiometric tie-points indicate the assumption is generally valid back into the Cretaceous.) Thus the age of these strips of the ocean floor can be dated simply by towing a magnetometer behind a ship (Fig. 11.26).

The sediment falling on the new material near the ridges forms in relatively shallow water. Calcium carbonate shells of pelagic organisms (Fig. 11.27), such as those of coccolithophores, globigerine foraminifera, and small snails called pteropods, accumulate in great numbers. When these move out onto the abyssal floor, they enter deeper waters in which the calcareous shells no longer accumulate, though siliceous ones such as those of radiolaria and diatoms will if abundant in the surface waters. As a result, the calcium carbonate is slowly dissolved, leaving behind the insoluble residue called brown or red clay that characterizes the more marginal zones of the ocean basins (Fig. 11.15).

The spreading of the sea floor explains why both volcanic and sedimentary rocks are older away from the ridges and why volcanic islands change the composition of their lavas as they migrate across the ocean floors tapping magmas of different compositions. It also explains the formation of guyots, slow-growing volcanic seamounts with tops planed off as they moved outward and, hence, downward away from the ridges (Fig. 11.28). Finally, sea-floor spreading accounts for the thinness of the crust beneath mid-ocean rift zones.

Rifts can open either in ocean basins or under continents. In the latter case, a continent is torn asunder and a new ocean basin gradually develops in between (Fig. 11.23). This is now occurring in Africa (Arabia has already been

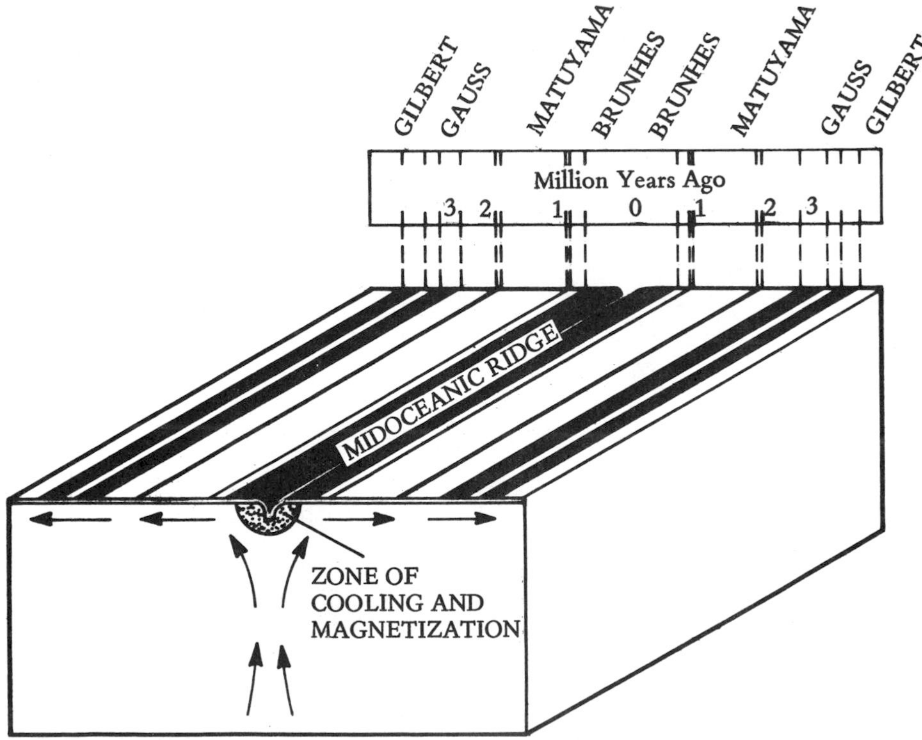

FIGURE 11.24 Formation of Magnetic Anomaly Patterns.

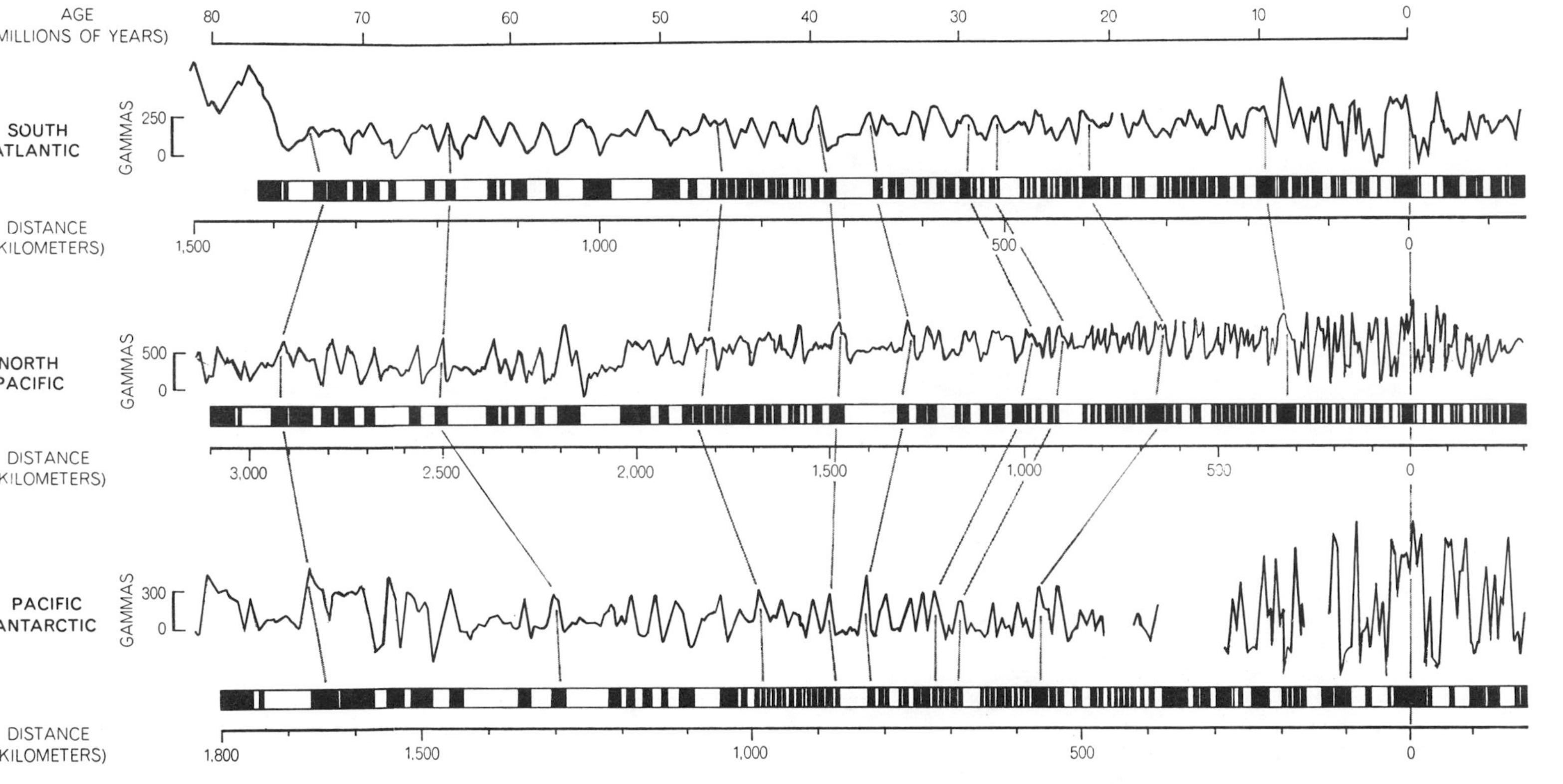

FIGURE 11.25 Dating Oceanic Crust by the Magnetic Anomaly Method. (From *Sea-Floor Spreading* by J. R. Heirtzler,)

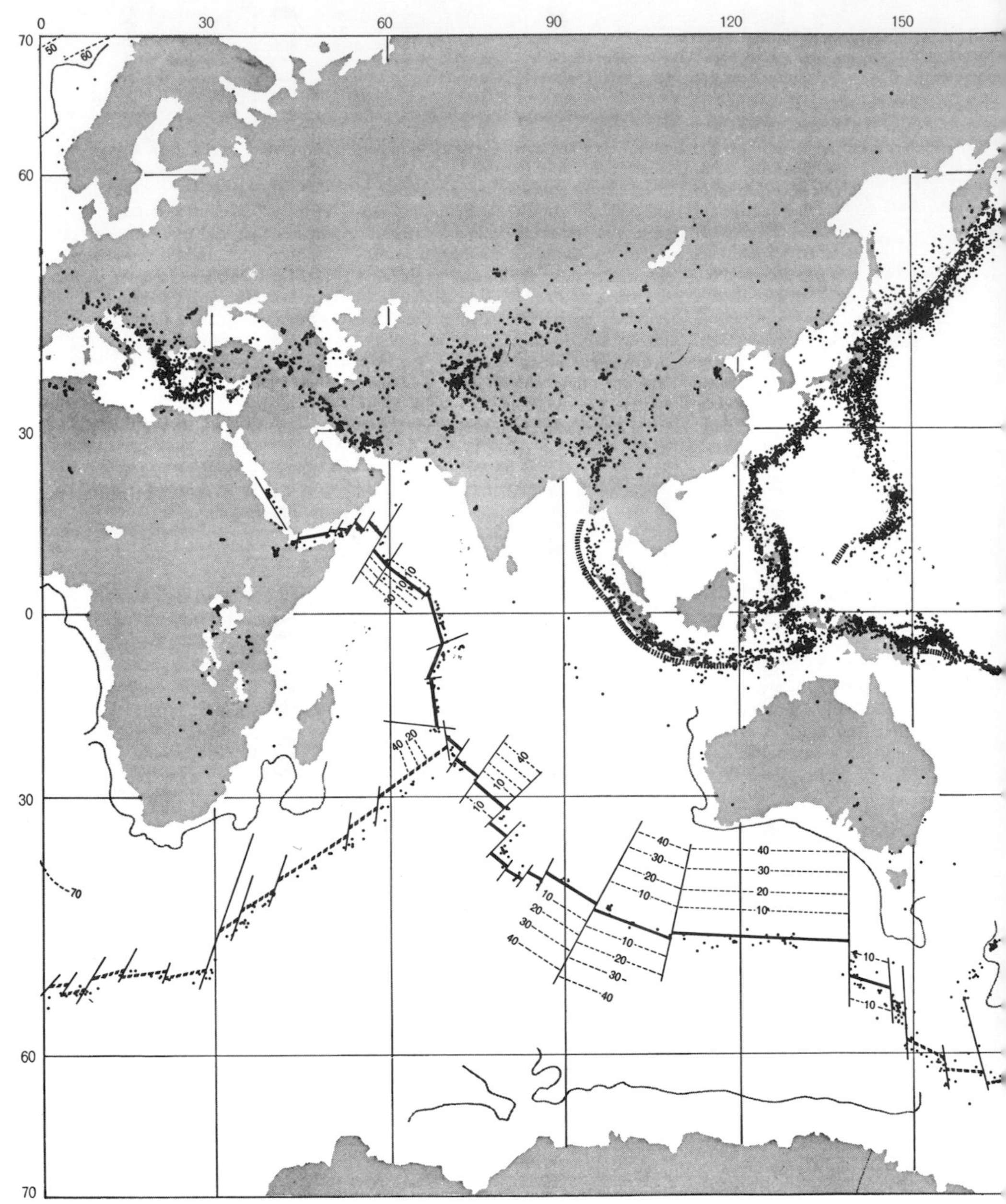

FIGURE 11.26 The Age of the Ocean Crust Determined by the Magnetic Anomaly Method. (From *Sea-Floor Spreading* by

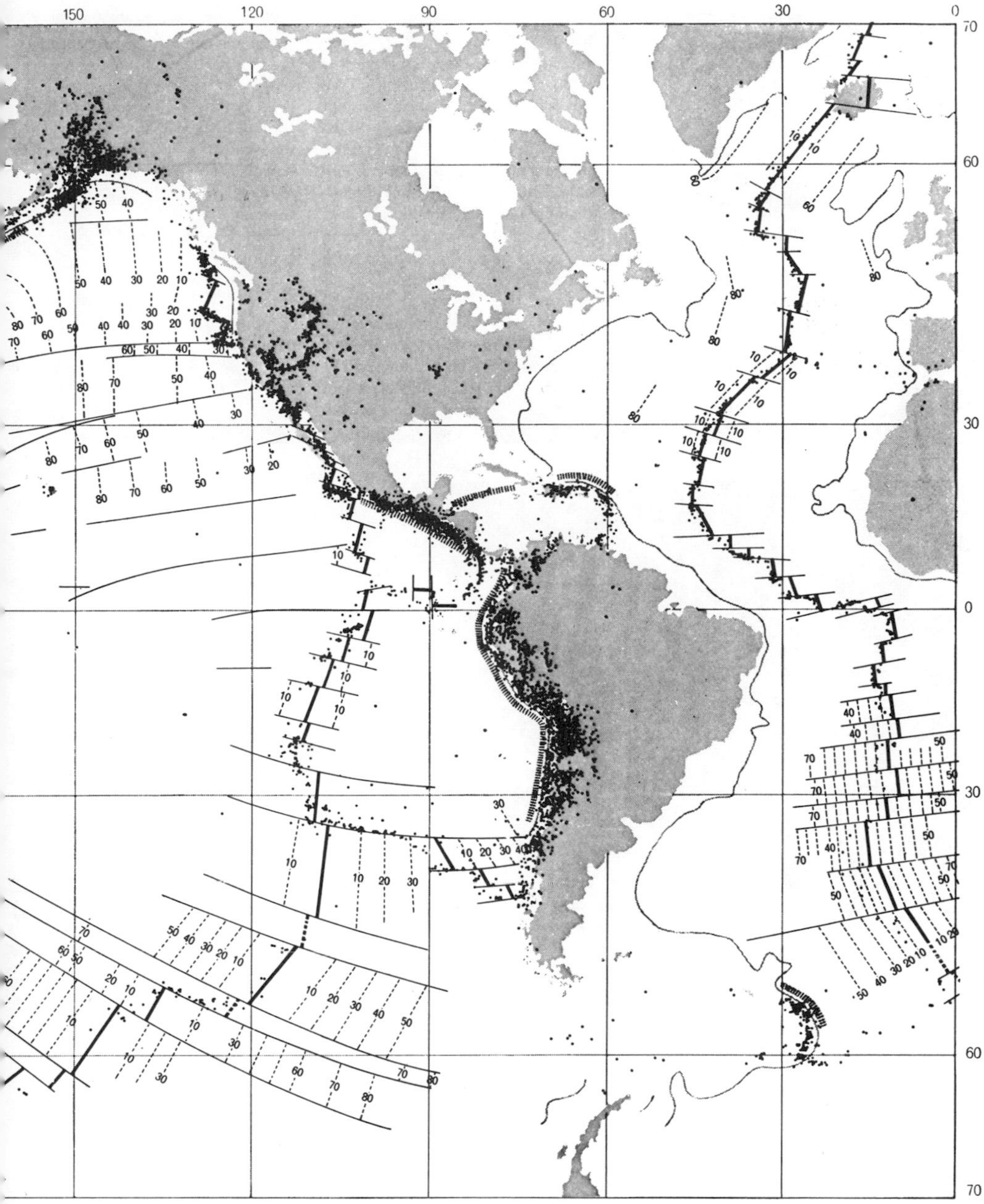
150
120
90
60
30
0
70
60
30
0
30
60
70

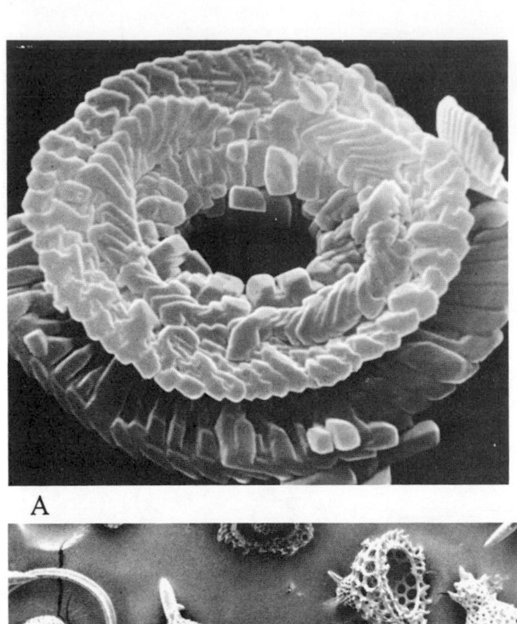

A

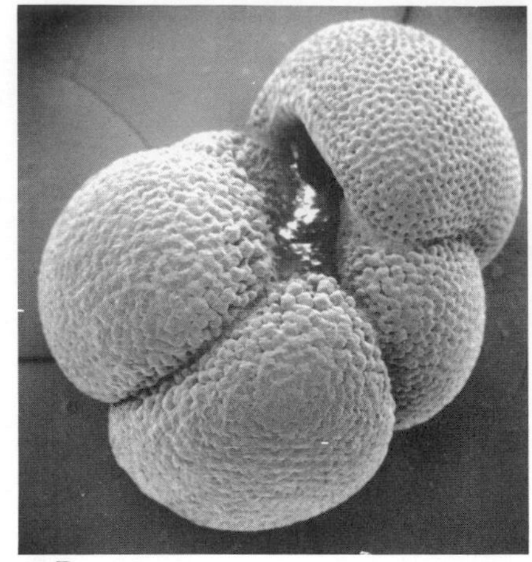

B

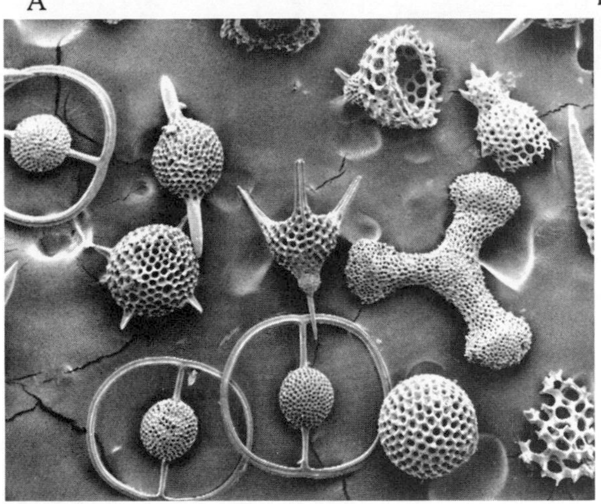

C

FIGURE 11.27 Major Ooze Formers: (A) Coccoliths, (B) Pelagic Forams, (C) Radiolarians. (Courtesy of Ron Parker and James I. Jones, Florida State University.)

torn off creating the Red Sea) and occurred in the past when Gondwanaland ruptured forming the Atlantic and Indian Oceans. Each block of the earth's crust (and some of the upper mantle) moving outward from a ridge is called a *plate,* whether it consists only of an ocean basin or an ocean basin and a continent. All the plates together compose the *lithosphere.* The very viscous fluid material on which they move has been named the *asthenosphere* or the rheosphere.

Two reasons have been advanced to account for plate motion. Either plates are being pushed outward from the ridges by the rising magma probably generated by a convection current, or they are being dragged into the trenches by the sinking, cooling, heavier margin of the plate. The "push" hypothesis makes it difficult to account for the nearly vertical tension cracks in the mid-ocean rift zones (Fig. 11.23). The "pull" hypothesis accounts for them and the conditions in the trenches. In a trench, a denser oceanic plate dives under the margin of the other plate (Fig. 11.29). This creates the inclined Benioff zone

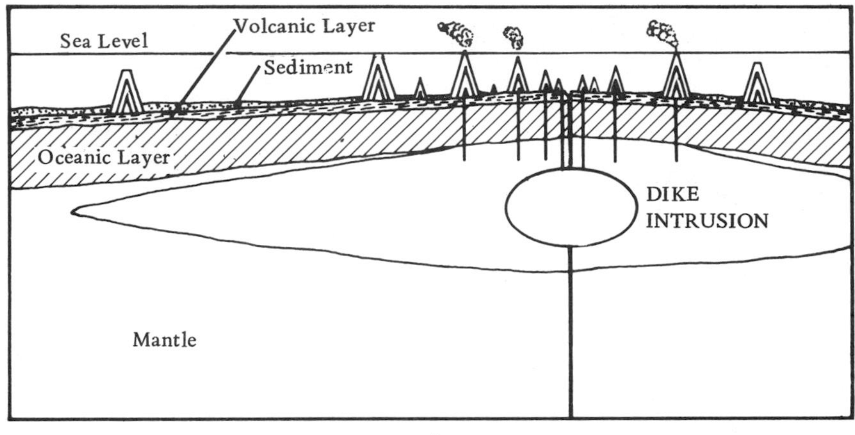

A

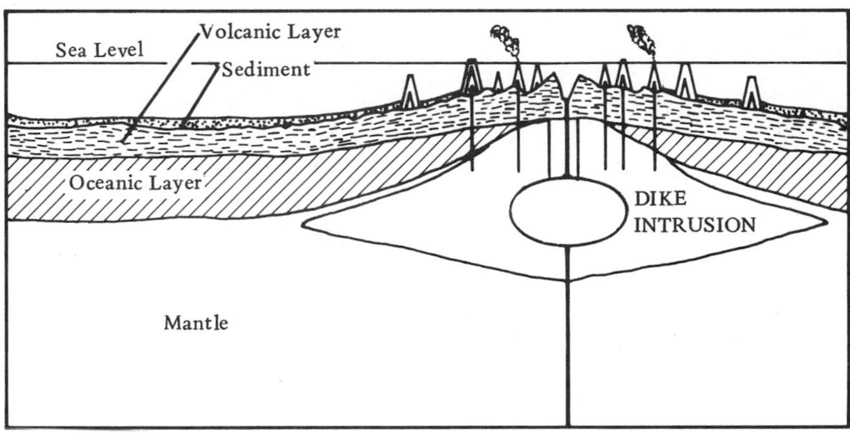

B

FIGURE 11.28 The Formation of Guyots on (A) Fast- and (B) Slow-spreading Ridges. (A) represents the current Pacific Ocean and (B) the Atlantic. (After Menard.)

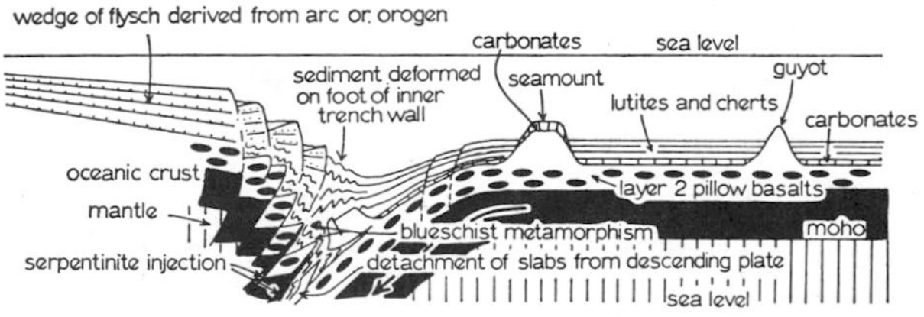

FIGURE 11.29 Cross-Section of a Modern Trench. (From Dewey and Bird, *Journal of Geophysical Research,* vol. 75, no. 14, May 10, 1970. Used by permission of authors and AGU.)

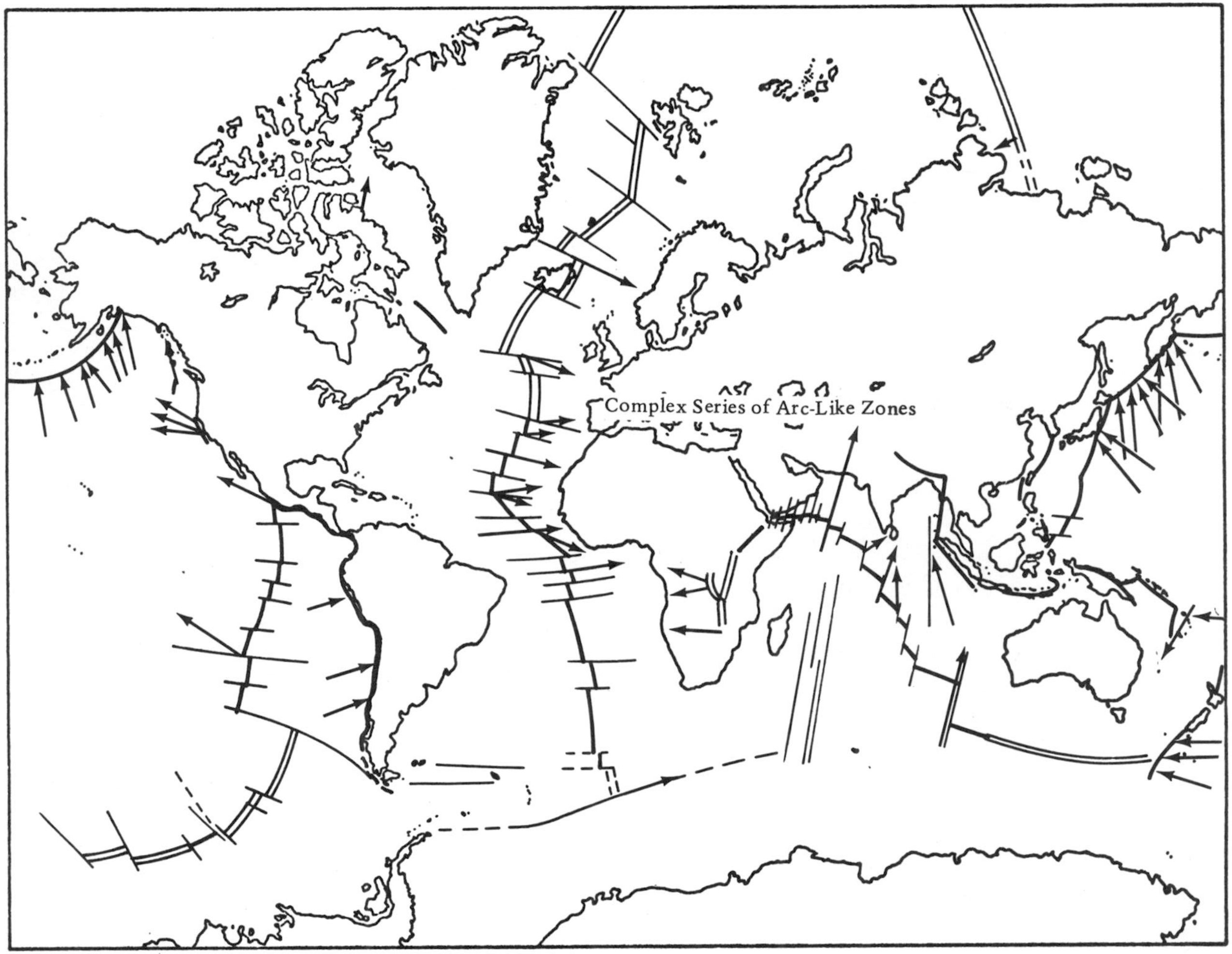

Complex Series of Arc-Like Zones

FIGURE 11.30 Present-Day Patterns of Ridges and Trenches Showing Relative Plate Motions. (From Bryan Isacks, et al, *Jour. Geophys. Res.*, vol. 73, p. 5861, 1968. Used by permission of authors and A.G.U.) (After Wilson.)

of deep focus earthquakes. As the plate margin sinks, two factors come into play: the increasing pressure which tends to increase the density of the sinking edge or *subduction zone* and the increasing temperature which tends to decrease its density. The pressure effect occurs more rapidly than that of the temperature, thus, increasing the sinking. This pulls the plate away from the ridge and toward the trenches. The rapid descent of the cold crust into the high pressure and temperature zone also metamorphoses and melts the sinking plate edge, as we saw in Chapter 9.

One apparent difficulty with the pull hypothesis is that there are no trenches between some of the earth's ridges. For example (Fig. 11.30), there is no trench between the Mid-Indian and Mid-Atlantic Ridges (in fact, new ridges are forming beneath the African Rift Valleys). In this instance, the ridges themselves must be moving apart to provide space for the constantly generated new material. The slack must be taken up somewhere, apparently in the Pacific and Indian Ocean trenches. Regardless of whether the surface plate is pushed or pulled there must be a compensating reverse flow of mantle material to complete the cycle. The type and depth of this flow are still matters of major debate among geologists.

The side boundaries of plates are termed *megashears* which are essentially gigantic strike-slip faults. They provide guidelines to the relative motions of plates in the past.

The absolute motion of plates is very difficult to determine, but one line of evidence seems reliable for the more recent plate movements. Hot spots or plumes apparently may arise from a deep mantle source and remain fixed in position along a ridge. As new crust is generated adjacent to the plume, basaltic volcanic outpourings occur and the trace of these flows leaves a *thread ridge* or *nematath* extending from the site of the hotspot outward across the newly-formed plates on either side of it (Fig. 11.31). The Walvis-Rio Grande Ridges of the South Atlantic and the Chagos-Laccadive Ridge of the Indian Ocean appear to be nemataths and thus allow the absolute positioning of South America, Africa, and India since the Jurassic. By coupling these data with that from the magnetic anomalies and megashears, the positions of the other continents can be determined with some accuracy (Figs. 19.2, 19.3).

The details of the structure and composition of consuming or colliding and accreting or separating plate margins were discussed and illustrated in Chapter 9.

As noted above, several plate boundaries are zones where the involved plates slide past one another rather than colliding edge-on. As they do, both marginally and internally, a series of great transverse *fracture zones* appear which offset the structural grain of a plate (Fig. 11.17). Along a fracture zone across a ridge, the two offset parts of a ridge continue to form new material at their centers which spreads outward from both (Fig. 11.32). The material on each side of a fracture zone moves from the ridge on its side in the direction of the ridge on the opposite side. This produces the anomalous pattern of ridges offset in one sense with the motion of present earthquakes in the opposite sense. These peculiar faults are now called *transform faults* by *J. Tuzo Wilson.* They are best known in the ocean basins where they offset most ridges, but also occur on land. For example, the San Andreas Fault connects the ridge in the Gulf of California, Mexico with the Gorda Rise off Northern California.

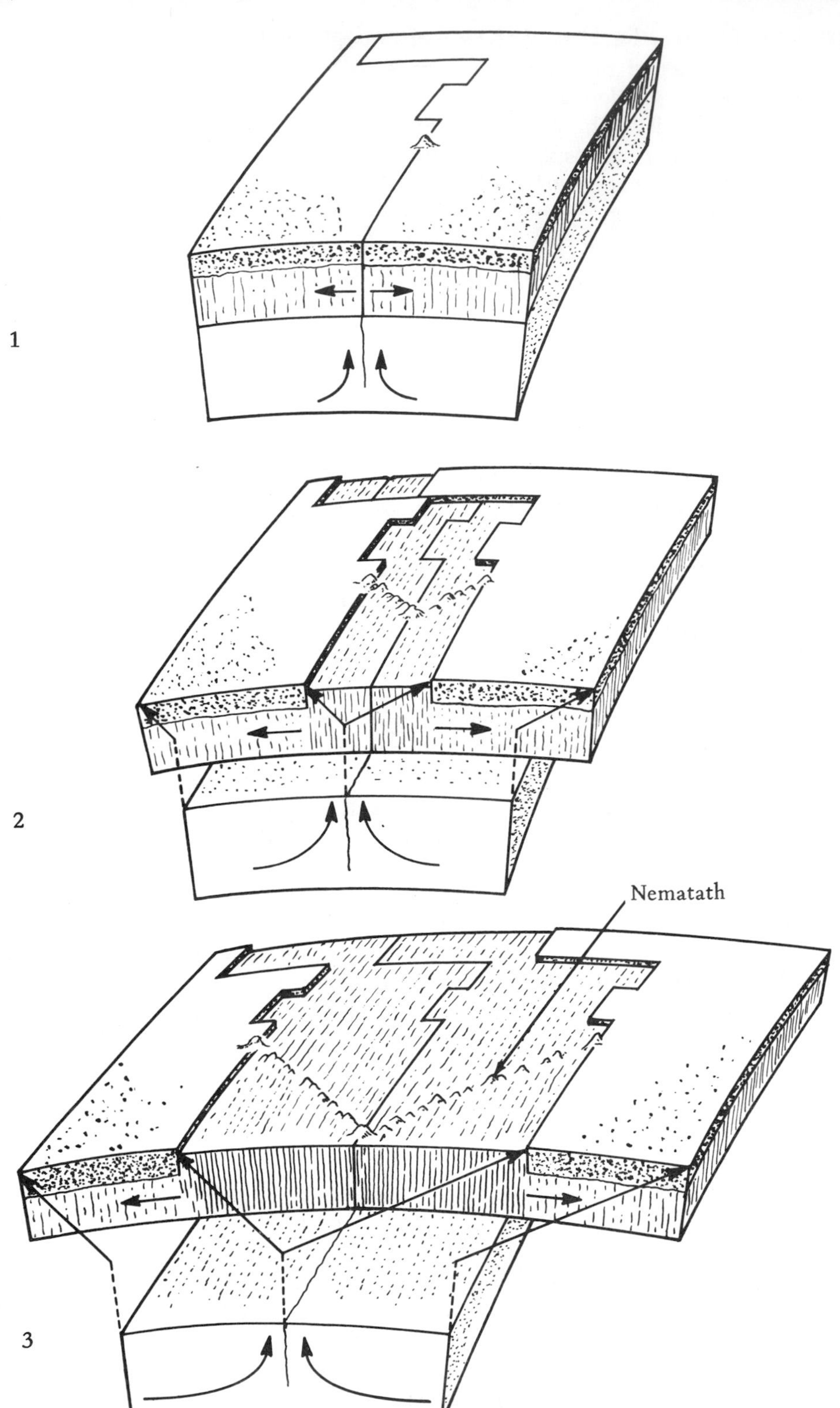

FIGURE 11.31 Formation of Nemataths or Thread Ridges. (After Dietz and Holden.)

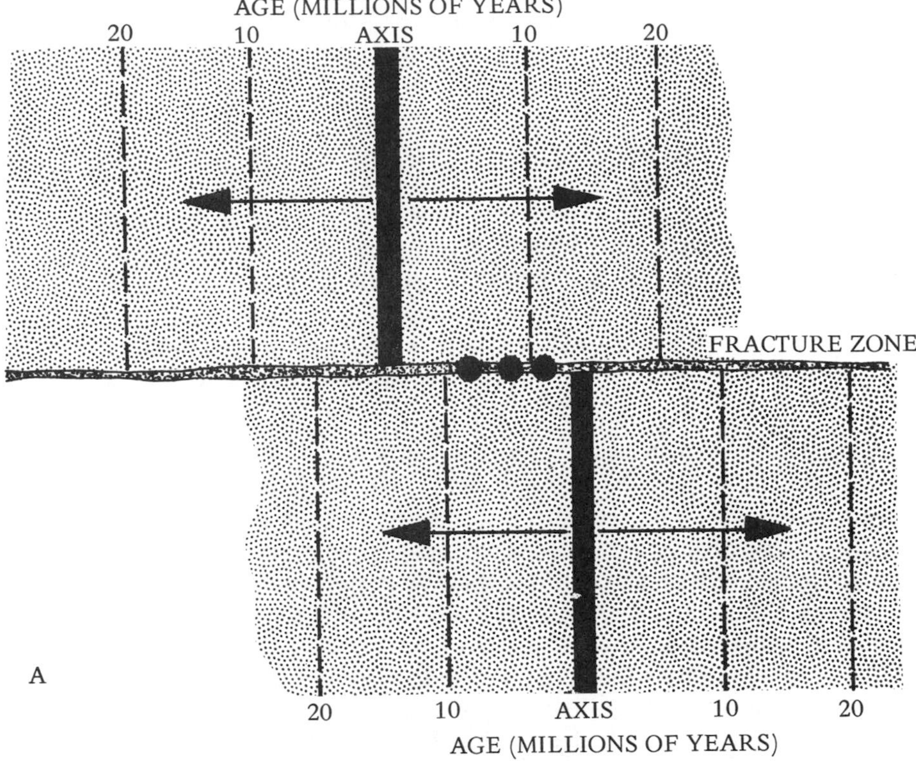

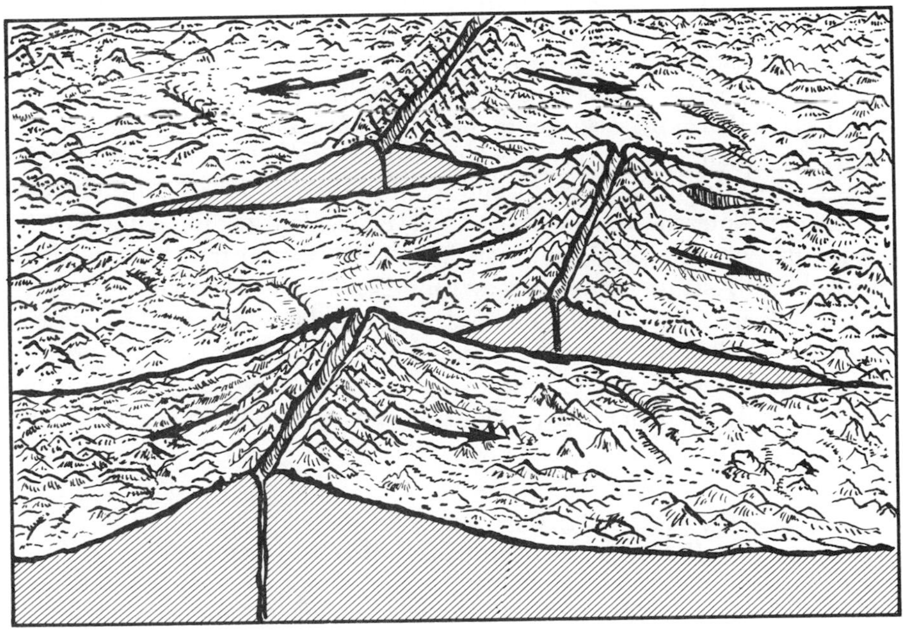

FIGURE 11.32 Formation of Transform Faults. (A) Map view, (B) Oblique view of ocean basin floor.

Thus we can see that the older names of "continental drift" and "sea-floor spreading" are not inclusive enough terms for the new global tectonics. It is rigid plates with continents and/or ocean basins on them which are moving, not just "floating" continents or "conveyor-belt" ocean basins. This is why the name "plate tectonics" has been widely adopted.

The Plates Today

There are eight major and several minor plates of the earth's crust today (Fig. 11.33). The major plates are as follows: 1) *North American,* including the Atlantic Ocean west of the Mid-Atlantic Ridge and north of Central America, and all of North America except Baja California and Southwest California (Alaska is a question mark at present, but may be part of the Eurasian plate); 2) *South American,* including the Atlantic Ocean west of the Mid-Atlantic Ridge, south of Central America, and north of the South Sandwich Trench; and all of South America except the extreme north; 3) *Eurasian,* including the Atlantic Ocean east of the Mid-Atlantic Ridge and north of the Mediterranean Ridge and Eurasia north of the Near Eastern and Himalayan Ranges (eastern Siberia is a question mark at present, but may be part of the North American Plate); 4) *African,* including the Atlantic Ocean east of the Mid-Atlantic Ridge, south of the Mediterranean area and north of the Atlantic-Indian Ridge; the Indian Ocean west of the Mid-Indian Ridge; and Africa; 5) *Indian* including the Indian Ocean east of the Mid-Indian Ridge, north of the Indian-Pacific Ridge, and west of the Ninety East Ridge; and India; 6) *Australian* including the Indian Ocean east of the Ninety East Ridge; the Pacific Ocean west of the New Zealand Ridge and Kermadec-Tonga Trenches, north of the Indian-Pacific Ridge, and south of the southwest Pacific Trenches; and Australia; 7) *Pacific* including the Pacific Ocean west of East Pacific Rise, east of the western Pacific trenches, south of the Aleutian Trench and north of the Indian-Pacific Ridge; 8) *Antarctic* including the Pacific Ocean south of the Indian-Pacific and Juan Fernandez Ridges, the Atlantic Ocean south of the South Sandwich Trench and Atlantic-Indian Ridges, the Indian Ocean south of the Atlantic-Indian and Indian-Pacific Ridges, and Antarctica. Somewhat lesser plates include the *Southeast Asian* between the western Pacific and northern Indian Ocean trenches and the Himalayan-central Asian mountain chains; the *East Pacific* between the East Pacific Rise, the Peru-Chile Trench, the Galapagos Ridge, and the Juan Fernandez Ridge; the *Caribbean* Plate; the *Arabian* Plate, the *Philippine Sea* Plate; and some smaller ones off western North and Central America. Not all boundaries have been certainly identified yet, particularly in the Arctic Sea–northern Asia region.

Plate Tectonics and Historical Geology

One of the clearest implications of the new global tectonics is that the framework of historical geology will require some drastic alterations. Nearly all textbooks of historical geology until now have been written with the underlying theme of continent and ocean basin stability. A few recent ones include the

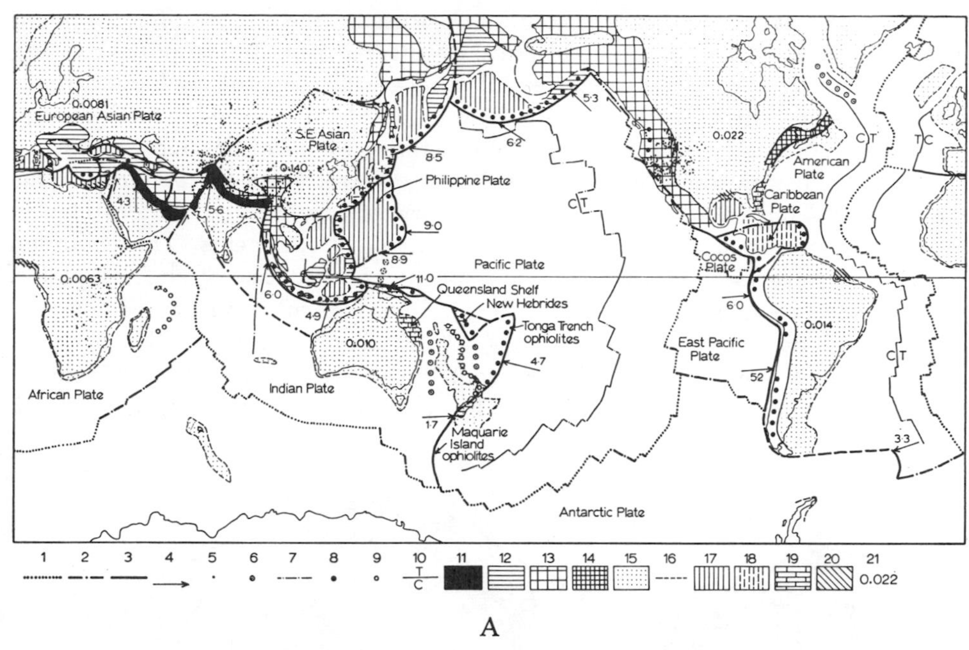

A

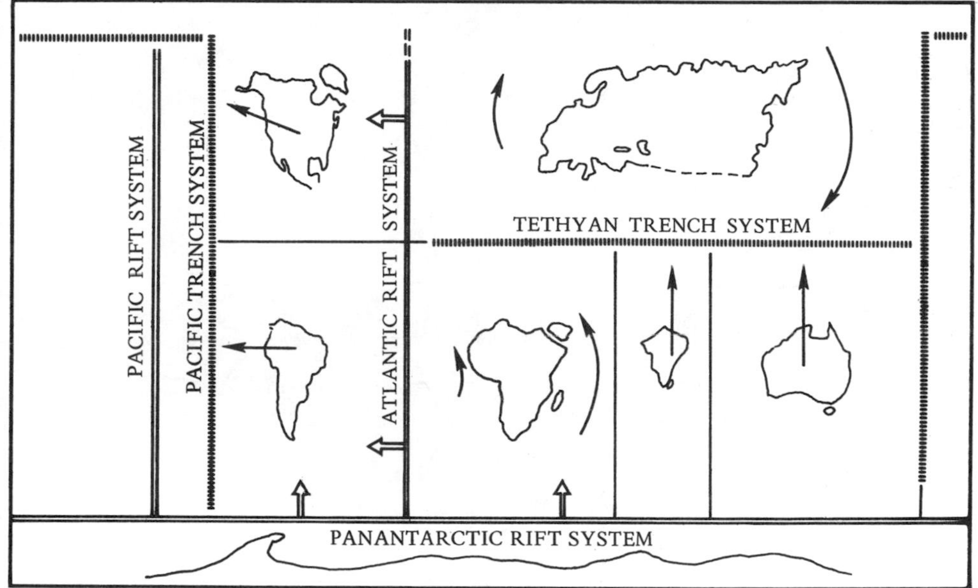

B

FIGURE 11.33 Lithosphere Plates Today. (A) Plates, margins and relation to geologic structures, (B) Diagrammatic representation of plates, margins, and movements. (A) From Dewey and Bird, *G.S.A. Bulletin*, vol. 75, May, 1970. (B) From Deitz and Holden, *Jour. Geophys. Res.*, vol. 75, p. 4954, 1970. Used by permission of authors and societies.

evidence for plate tectonics, but continue discussing the history of a continent as if it occurred in isolation. The physical events, particularly mountain-building and flooding by epeiric seas, and the biological events, particularly biogeographical dispersion with its effects on evolution, can clearly be understood only in the context of plate movements and interactions. Many of the most-cherished concepts in historical geology have had to be discarded and the pattern for the future is still only dimly seen. Physical history will have to be reinterpreted in light of Gondwanaland, and Laurasia and previous continents. The evolution of life will have to be viewed in the perspective of a mosaic of shifting plates.

Already geologists have matched Precambrian zones of similar radiometric age dates across the South Atlantic (Fig. 11.34). The boundary in South America is exactly where it was predicted from the African data. Zones of identical lithologies, ages, and structure have been matched between eastern South

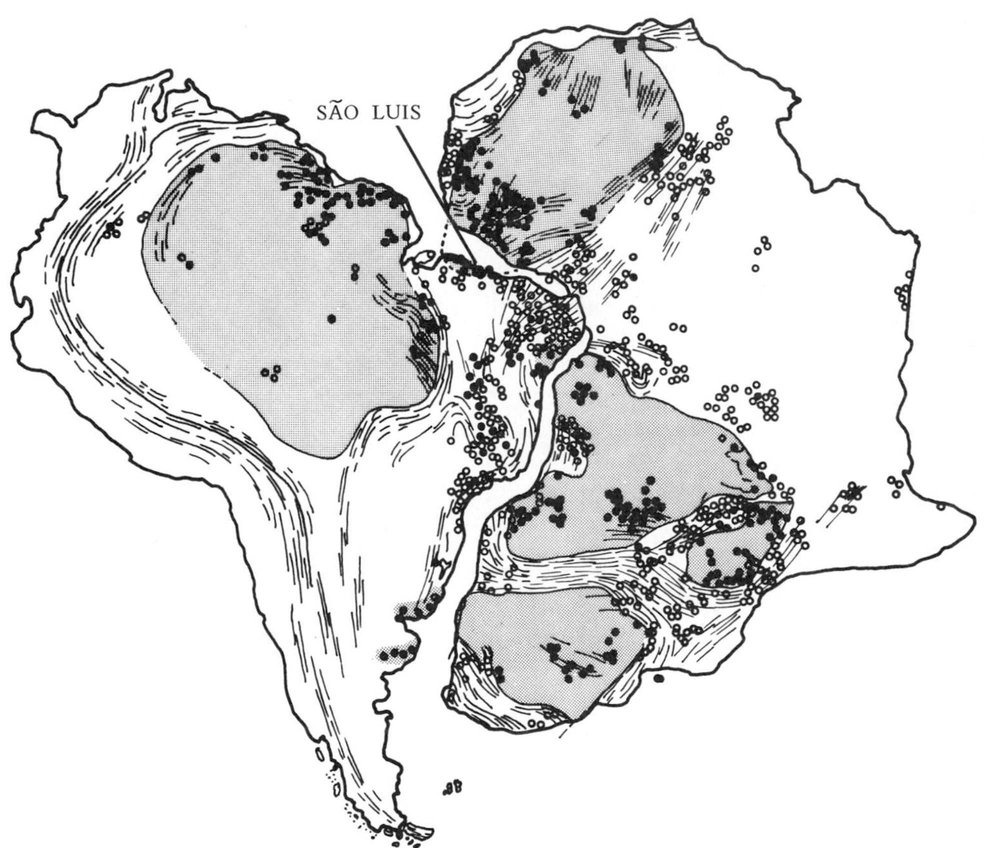

FIGURE 11.34 Match of Cratons and Belts of Deformation. Dark areas are cratons; shaded areas are former geosynclines. Solid dots are rocks 2,000,000,000 years of age; open dots are younger rocks. (Dates determined radiometrically.) (From *The Confirmation of Continental Drift* by Patrick M. Hurley, Copyright © 1968 by Scientific American, Inc. All rights reserved.)

America and west Africa, between India and western Australia, between Australia and Antarctica (Figs. 11.35A-C, 11.36). Structures, lithologies, and fossils match between Newfoundland and the British Isles if North America and Europe are reunited by computer fits of shorelines (Fig. 11.37). The sequence of events in the current stage of spreading beginning with the Mesozoic have been roughly dated and assigned proper places in both the relative and radiometric time scales (Fig. 11.38). Fossil amphibians and reptiles have been found in Antarctica, a continent to which they could never have migrated with its present isolated geographic situation. Plate tectonics may also explain the epeiric sea inundations of the continents. The parting of plates along a ridge produces a shallower ocean basin than that being consumed by an advancing edge of the split continent (Fig. 11.39). This decrease in oceanic volume may have been at least partially responsible for flooding the continents.

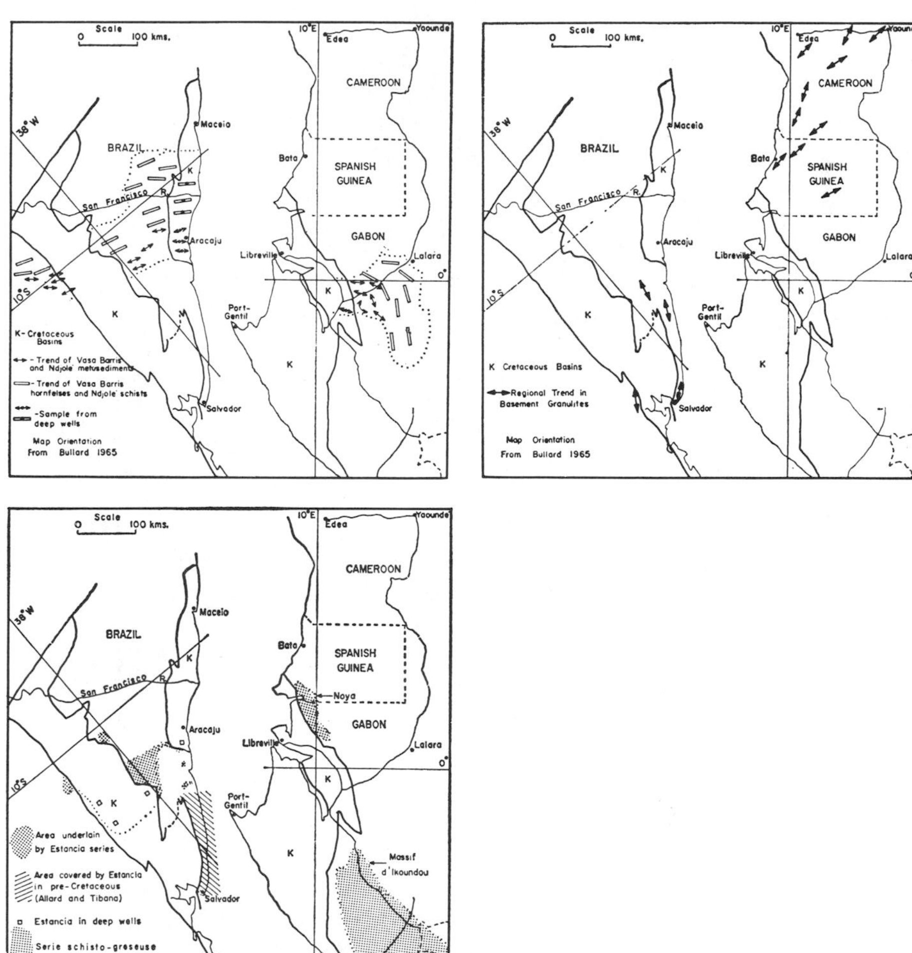

FIGURE 11.35A Match of Precambrian Trends Between South America and Africa. (All figures from Allard and Hurst, *Science*, vol. 163, February, 1969. Copyright © 1969 by AAAS. Used by permission of authors and AAAS.)

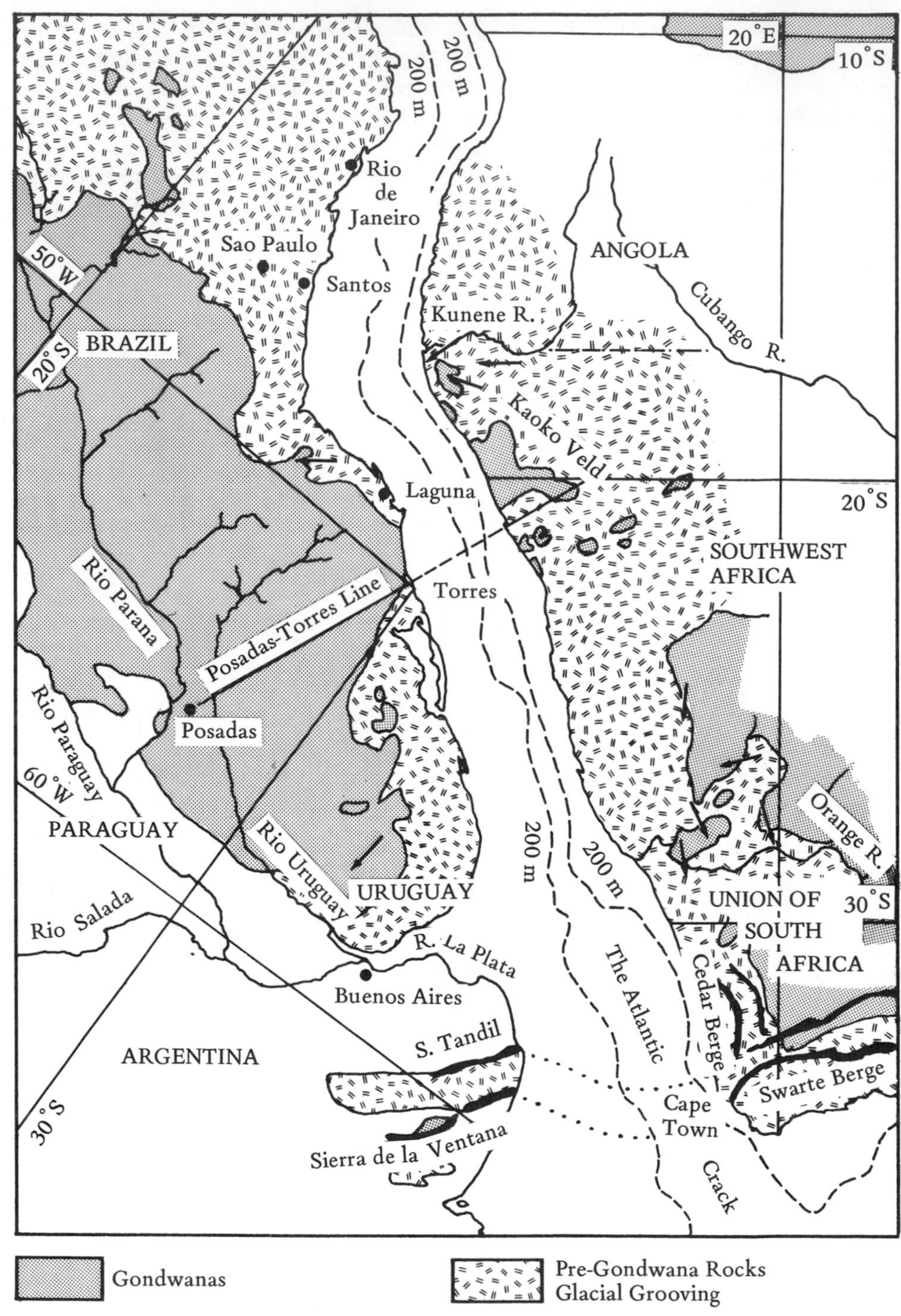

FIGURE 11.35B Match of Late Paleozoic Geology Between South America and Africa. (From *Historical Geology* by A. O. Woodford, W. H. Freeman and Co., Copyright © 1965.)

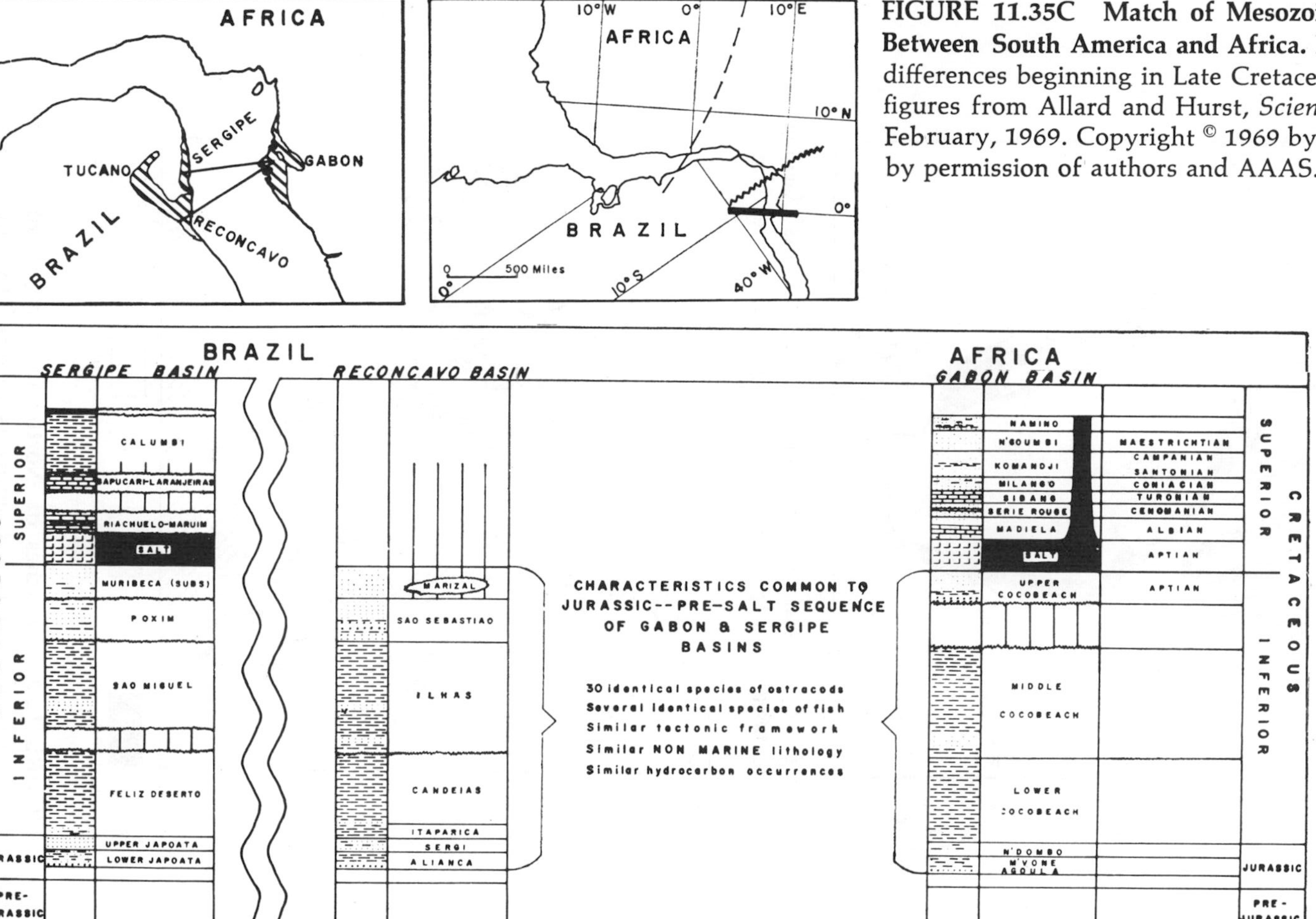

FIGURE 11.35C Match of Mesozoic Geology Between South America and Africa. Note differences beginning in Late Cretaceous. (All figures from Allard and Hurst, *Science*, vol. 163, February, 1969. Copyright © 1969 by AAAS. Used by permission of authors and AAAS.)

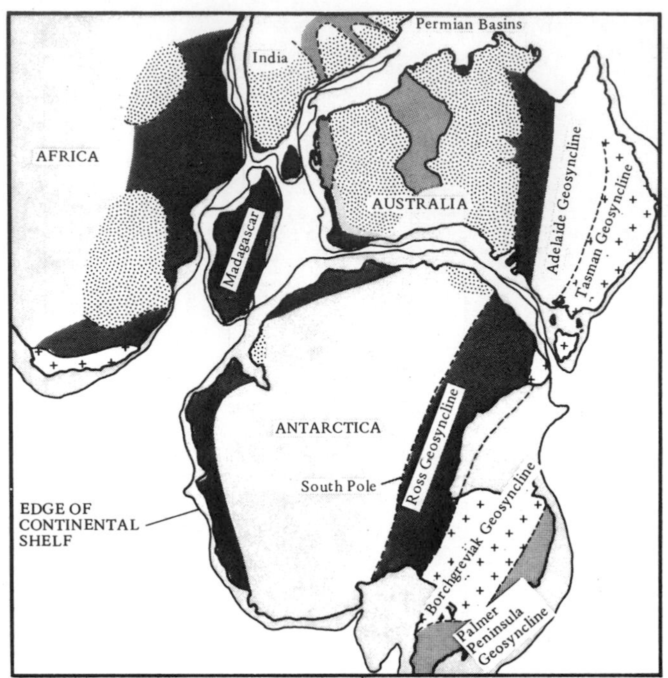

FIGURE 11.36 Matches of Gondwana Continents' Geology. Continents restored to Late Paleozoic positions. (After Hurley.)

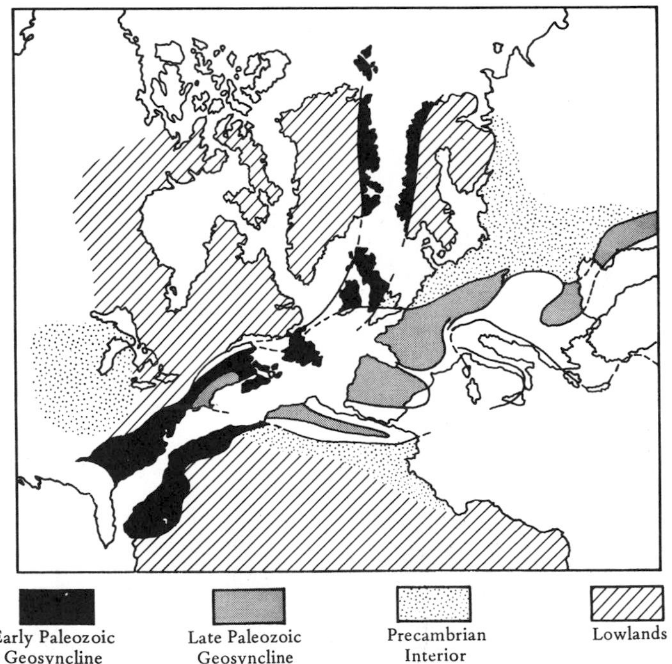

FIGURE 11.37 Matches of North American, European, and North African Geology. Continents restored to Late Paleozoic positions. (After Hurley.)

AGE (MILLIONS OF YEARS)		GONDWANALAND	LAURASIA
Quaternary	0		
Pliocene			Opening of Gulf of California
Miocene			Change in Spreading Direction, Northeast Pacific
			Birth of Iceland
		Beginning of Spreading Near Galapagos Islands	
Oligocene		Opening of Gulf of Aden	
Eocene		Separation of Australia from Antarctica	Opening of Bay of Biscay
Paleocene			Separation of Greenland from Norway
		Separation of New Zealand from Antarctica	Change in Spreading Direction, Northeast Pacific
	100		
Cretaceous		Separation of Africa from India, Australia, New Zealand and Antarctica	
Jurassic		Separation of South America from Africa	
	200		
Triassic			
Permian			Separation of North America from Eurasia
	300		
Pennsylvanian			
Mississippian			

FIGURE 11.38 Events Dated by Use of Magnetic Reversal Time Scale. (After Heirtzler.)

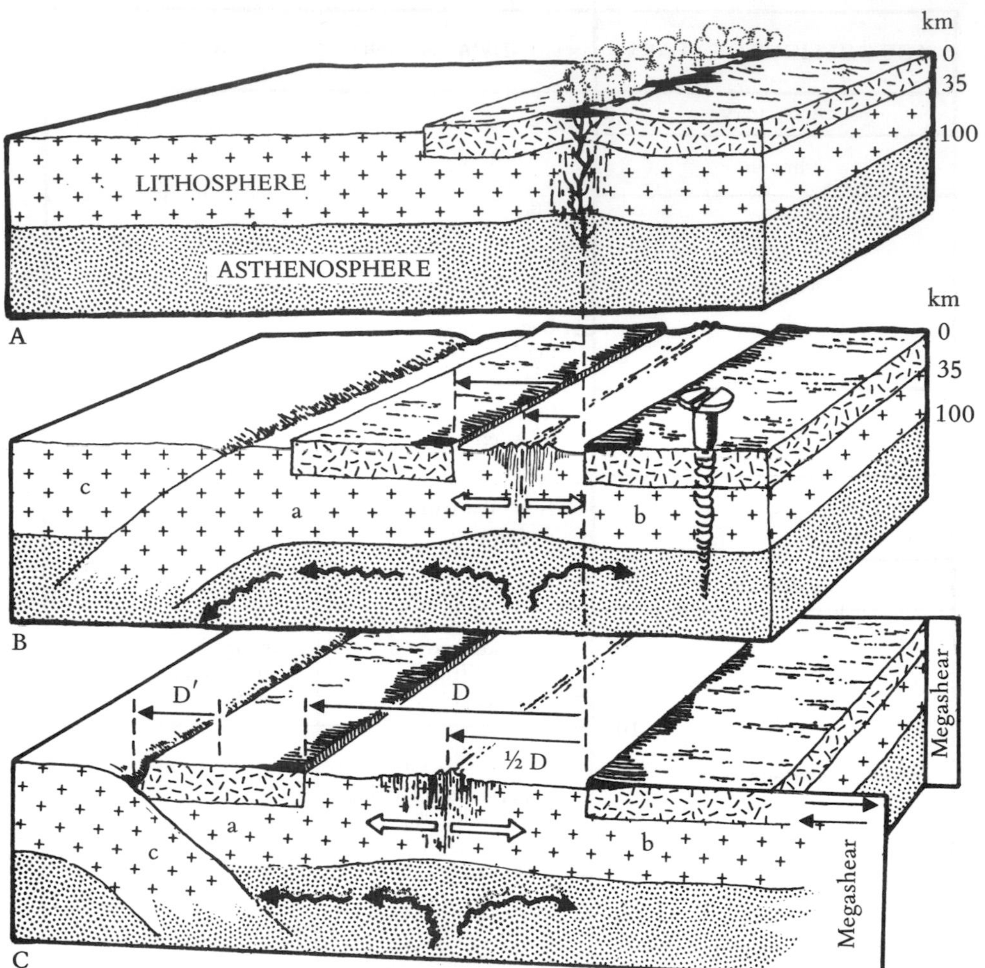

FIGURE 11.39 Diagram of Plate Tectonics illustrating attendant movement of ridge and decrease in ocean basin volume. (From Dietz and Holden, *Jour. Geophys. Res.*, vol. 75, p. 4942, 1970. Used by permission of authors and A.G.U.)

Events before the breakup of Gondwanaland and Laurasia are only now beginning to be unraveled in the light of the new plate tectonics. The similarity of structures and lithologic associations in rocks both older and younger than the Permian clearly indicates that plate motion was significant before these supercontinents formed, but the details are much less clear (Fig. 11.40). In the remainder of this book we will be studying the actual events in earth history as they unfold. We will attempt to interpret these events in the new framework, recognizing the predictive value of plate tectonics now that this theory has been validated by continuous empirical testing. There are still many significant details not yet understood and a meshing of the history read from the rocks by the historical geologist with the new discoveries of geophysicists and oceanographers will help us to understand them.

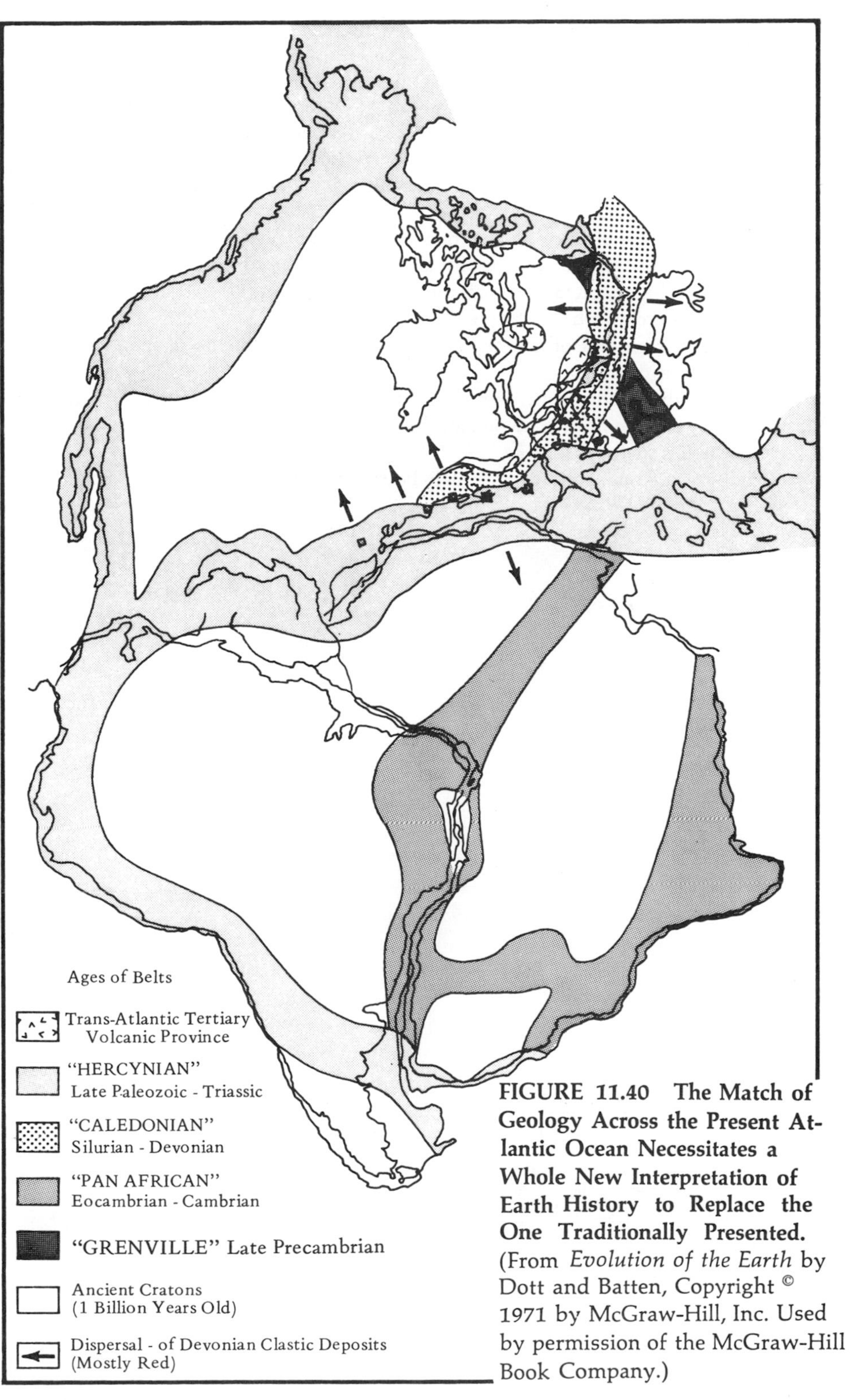

FIGURE 11.40 The Match of Geology Across the Present Atlantic Ocean Necessitates a Whole New Interpretation of Earth History to Replace the One Traditionally Presented. (From *Evolution of the Earth* by Dott and Batten, Copyright © 1971 by McGraw-Hill, Inc. Used by permission of the McGraw-Hill Book Company.)

Questions

1. What promped Edward Suess to postulate the existence of Gondwanaland?
2. What man is usually considered the father of continental drift ideas?
3. Describe the Late Paleozoic-Mesozoic rock sequence of the Gondwana continents.
4. What is the significance of the *Glossopteris* flora?
5. Give at least one example of paleoclimatologic evidence that supports continental dislocations.
6. What is the significance of negative gravity anomalies?
7. How does paleomagnetism support the idea that the continents have moved?
8. What are the characteristics of the oceanic ridge systems that suggest they are places where new crust is formed?
9. How have Vine and others explained the alternating bands of magnetic intensity in the ocean floors?
10. What are the strong points and the weak points of the pull theory of plate motion?
11. What is the significance of nemataths or thread ridges?
12. What is a transform fault?
13. What are the major plates and their boundaries today?

PART 2

Earth History

12

"In the Beginning":

The Origin of the Earth and Life

MAN HAS SPECULATED ABOUT THE ORIGIN OF THE EARTH AND LIVING THINGS AT LEAST since the dawn of recorded history. All of the world's religions have creation myths of some sort, either original or borrowed from their neighbors. The Hebrew creation myth, much of which is borrowed from other Near Eastern religions and which has become part of the Christian heritage of Western man, speaks of a world and its inhabitants supernaturally created in six days. By the 1600s and 1700s, however, men were beginning to think in terms of natural processes and longer time spans. In the 1700s, the German philosopher *Emmanuel Kant* speculated that the sun, the earth, and the other planets had condensed from a hot nebular cloud, while the French naturalist *Buffon* suggested that a comet had collided with the sun and spattered out the planets. These two ideas have come down to us in this century as the nebular and encounter hypotheses respectively.

Nebular Hypothesis

The *nebular hypothesis,* as scientifically promulgated by the French mathematician *P. S. Laplace* in 1820, envisioned a great spherical cloud of hot gases that slowly rotated (Fig. 12.1). As it rotated, it continually speeded up by the law of conservation of momentum and flattened out producing a thin disk, the ancestral planets, and a still somewhat inflated center, the ancestral sun. When the outward centrifugal force at the outer rim of the disk exceeded the gravitational attraction of the massive protosun, an outer ring of material detached and the inner mass shrunk. This process repeated itself several times until a ring ancestral to each of the modern planets had formed. The material in each ring was gradually swept up into a planet, though some remains detached today as meteorites. Each ring, in turn, went through the same process of differentiation producing satellites. The planets and satellites, being much smaller than the sun, lost their heat more rapidly and developed cold outer crusts. (The continued cooling and shrinking of this originally hot earth was a popular hypothesis to explain mountain-building even into this century.) The major problem with the nebular hypothesis was that it indicated the sun should be rotating at a very rapid rate in order to have spun off the planets. Yet 99% of the speed of rotation (angular momentum) in the solar system occurs in the rapidly spinning planets and only 1% in the slowly turning sun.

Encounter Hypothesis

The *encounter,* or *second body, hypothesis,* given a scientific basis by the geologist *T. C. Chamberlin* and the astronomer *F. R. Moulton* in the early 1900s, dealt with this problem readily. In this hypothesis, another passing star contributed greatly to the tidal and eruptive forces of the sun causing great clouds of gas to be thrown out of the sun's interior into large spirals (Fig. 12.2A). This material was given a great angular momentum and quickly cooled into solids and liquids which accumulated into planets by collisions. In this hypothesis then, the planets originally began as cold bodies, in contrast to the nebular hypothesis. Calculations soon indicated the sun had essentially no tides and

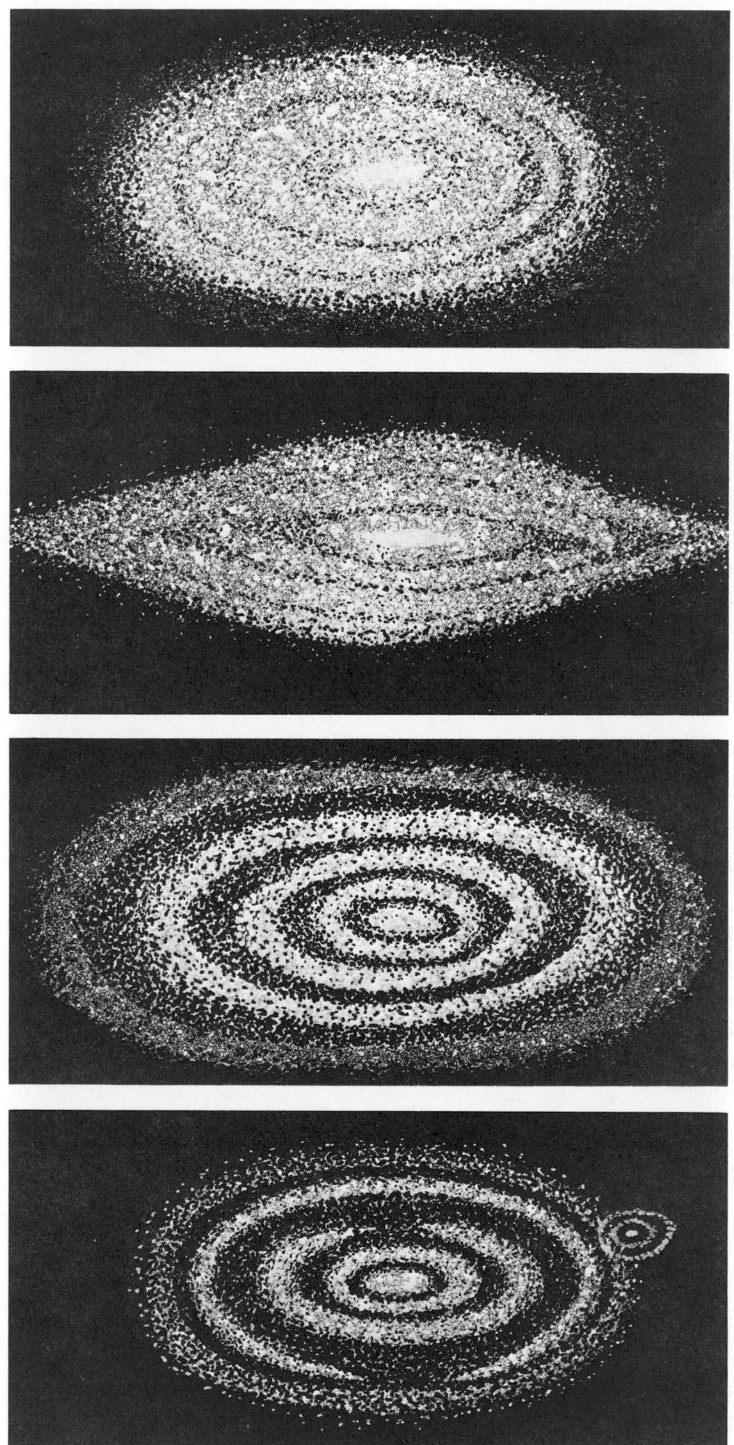

FIGURE 12.1 The Nebular Hypothesis of Solar System Origin.

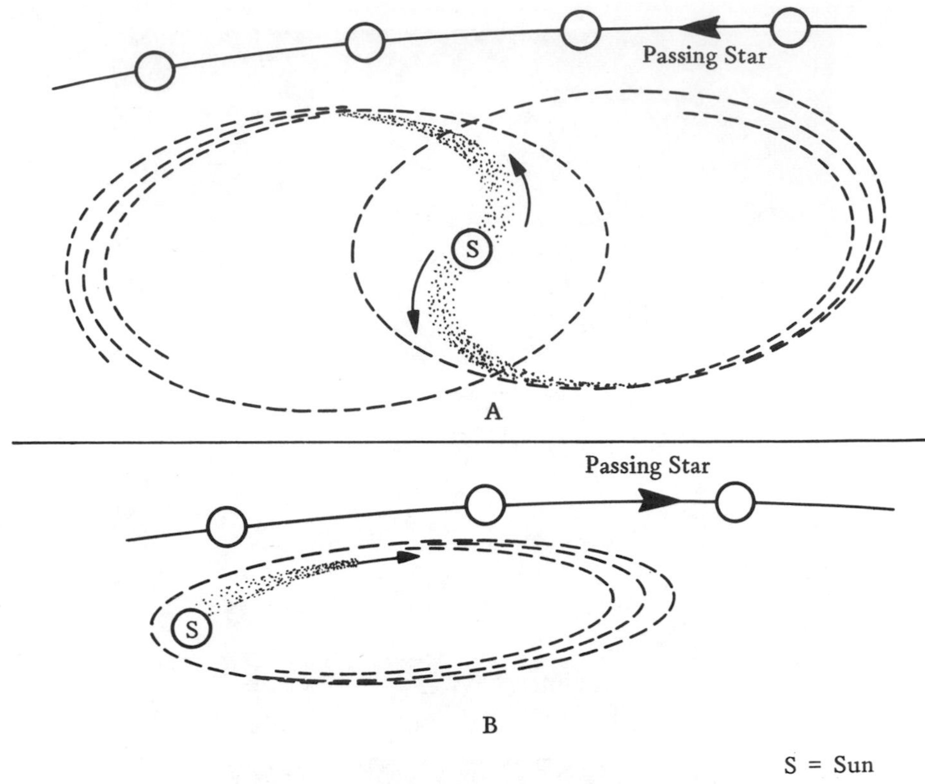

FIGURE 12.2 Encounter or Second-Body Hypothesis. (A) Original Chamberlin-Moulton Version (1900). (B) Modified version of Jeans and Jeffreys (1925).

its eruptions were surface features only. The hypothesis was modified to assume a near collision with the passing star which drew a long hot filament out of the sun in a very eccentric orbit (Fig. 12.2B). The internal resistance of the cloud eventually produced today's nearly circular orbits. The major difficulties with the encounter hypothesis were that calculations indicated no filament pulled out of the sun this way would ever have enough gravitational cohesion to collect into a planet. Furthermore, the likelihood of any two stars colliding was so statistically remote as to be virtually impossible. Finally, the hypothesis could not explain how the satellites formed.

Modern Theory

The modern theory of the earth's origin, as expounded by *Fred Whipple, C. F. von Weizsacker*, and *Gerard Kuiper*, borrows the strong points of both older ideas, but goes further by viewing the earth and solar system in their setting of the galaxy and the universe. The origin of the universe is still widely debated and opinions fluctuate between the *big-bang hypothesis* of an initial explosion of all matter which is now flying outward from that center and a *continuous-*

FIGURE 12.3 Cold Dust Clouds. Horsehead Nebula in Orion. (Courtesy Hale Observatories.)

creation hypothesis in which the processes of star formation and destruction are continuous everywhere. Either hypothesis is compatible with the major points of present ideas on solar system formation and both are probably partly right.

Astronomers claim we can see all the stages in the formation of a star and its solar system going on before our eyes in space today. In certain regions of space, interstellar dust particles, composed of frozen non-luminous hydrogen, carbon dioxide, methane, and ammonia are aggregated together, probably by the pressure of light and other electromagnetic radiation. These *cold diffuse clouds,* probably the initial stage of star formation, can be recognized because they shut off the light from stars behind them (Fig. 12.3). The next stage is represented by small, dark, round nebulae called *globules,* which apparently evolved from the diffuse clouds when the forces of gravitational attraction became effective (Fig. 12.4). Finally, there are *clouds* of *luminous hot gases* and *dust particles* with bright *stars* scattered in them (Fig. 12.5). These probably evolved from the globules when the amount of aggregated material was large enough for its gravitational attraction to produce the high temperatures and pressures necessary to start the nuclear reactions. These reactions, chiefly the fusion of hydrogen to form helium, are the basis of the energy of stars. At this stage a star becomes luminous and visible by its own emitted radiation. Stars beyond our own solar system are too distant to visually determine whether

FIGURE 12.4 Globules. Rosette Nebula in Monoceras. (Courtesy Hale Observatories.)

FIGURE 12.5 Luminous Clouds with Stars. Great Nebula in Orion. (Courtesy Hale Observatories.)

they have planets, so the process of planetary formation is based on inference about the collection process of dust clouds into stars.

As now envisioned (Fig. 12.6), a cold nebular cloud that will become a solar system condenses to form a globule. Rotation, a prevalent physical property in the universe, begins producing an oblate form. Rotation speeds up, again by the law of the conservation of momentum, and turbulence, another prevalent physical property, develops, forming eddies (Fig. 12.7). The more rapid central spinning is dissipated by communication outward from eddy to eddy. As the cloud continues condensing, a balance between outward centrifugal force and the inward pull of gravity is reached and cold bodies start to collect in rings around the center. These protoplanets form by collecting debris in their vicinities, including many frozen gases which make them considerably larger than the modern planets. The scars of this sweeping process remain today as great craters on planets with no atmospheres (Fig. 12.8). Many uncollected particles still remain as meteorites. Finally, the gravitational self-compression of the protosun produces the critical temperature and pressure necessary to start the nuclear reactions. A great flood of radiant solar energy, the solar wind, fills the solar system, vaporizing and driving off most of the volatile constituents of the planets. Each ring of cold particles undergoes its own differentiation in the process of planet formation producing satellites. The sun's rotation can be further slowed by magnetic braking action. Ions ejected from the sun encounter magnetic lines of force generated by the sun's magnetic field (Fig. 12.9). These propel the ions away from the sun at high speeds, thus carrying off a relatively large amount of the sun's angular momentum for an infinitesimal loss of mass.

Early Development of Earth

Radiometric dating of meteorites and, more recently, of material on the moon's surface, indicates that these events occurred between 4.5 and 5 billion years ago. Extrapolating backwards the ratios of the various isotopes of lead to a time when there was no radiogenic lead also yields these dates. The oldest known earth rocks are dated at about 3.5 to 3.8 billion years. This leaves a long interval of the earth's history, over a billion years, unrepresented by direct evidence on the earth's surface today. Most of our theorizing about this interval comes from seismic studies of the earth, the composition of the other planets and their atmospheres, and physical-chemical studies in the laboratory.

The Earth's Interior

The passage of earthquake-generated waves through the earth's interior has disclosed much about its composition and behavior (Fig. 12.10). These waves behave differently depending on the density and rigidity of the material through which they pass. Seismic studies have demonstrated that the earth's interior is stratified into layers of differing density. There is a very dense core, a less dense mantle, and a light surface crust (Fig. 12.11).

Clearly the original particles were collected at random so that some form of segregation has occurred with the heavy material sinking inward to form

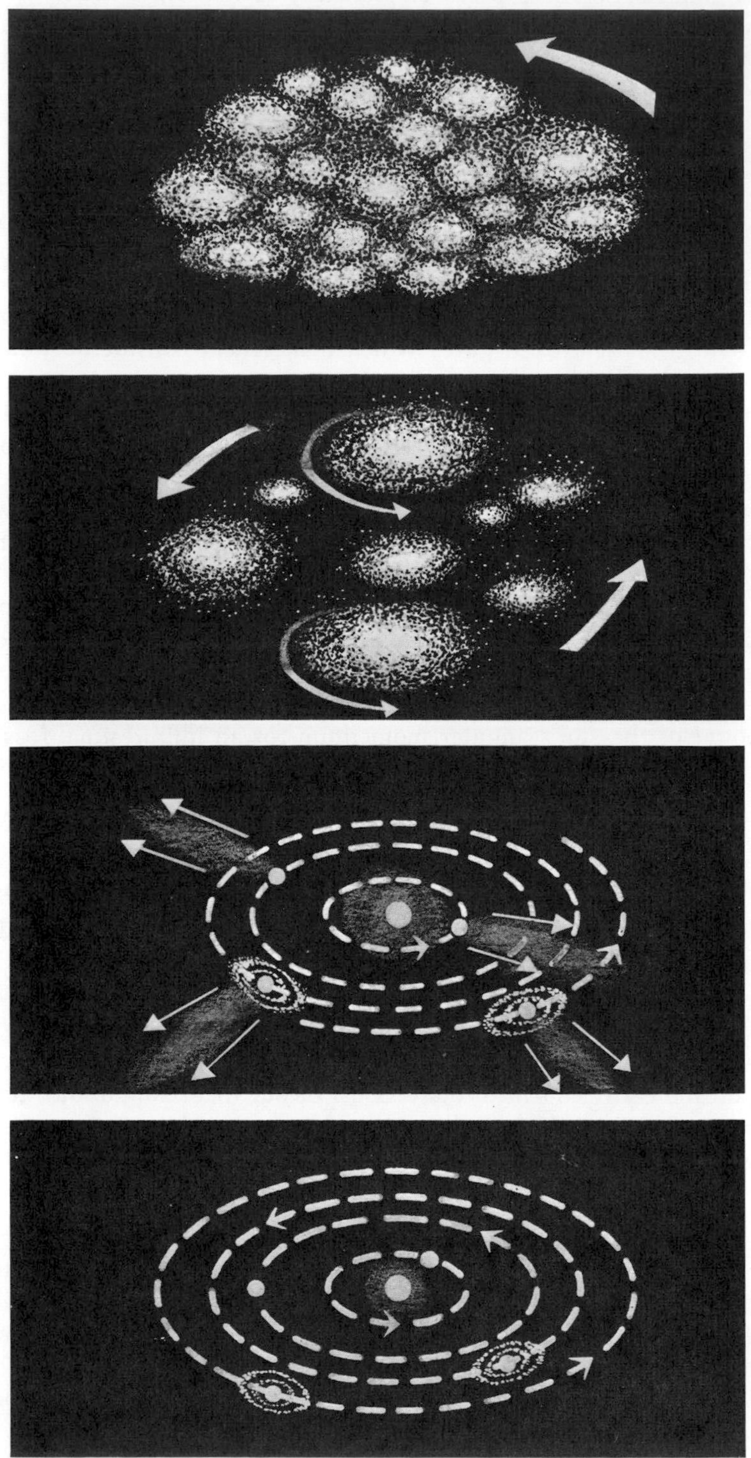

FIGURE 12.6 Modern Protoplanet Theory of Solar System Formation.

FIGURE 12.7 Von Weizsacker's Concept of Turbulent Eddies in Solar Nebulae.

the core and the lighter material rising to the surface to form the crust. The outer core behaves as if it were a liquid and the mantle material can also flow plastically under great pressure as demonstrated in the lab. This suggests some form of convection currents caused the differentiation. Heat from gravitational compression, from decaying radioactive isotopes trapped in the interior, and/or from tidal friction generated by the moon being much nearer the earth could have been the driving force or forces. We are uncertain as yet whether the earth's interior ever completely melted or whether the heating is still increasing, decreasing, or about steady. The segregation of the earth's interior does suggest, however, that it has been largely molten. Some geophysicists dispute this contention, however. They reason that diffusion in the solid state was more likely because if the earth were largely molten, it would have lost most of the volatiles it now possesses. The discovery of plate tectonics on the earth's surface also supports the existence of convection currents, though these may only be surface features unrelated to conditions deep in the earth's interior. The speed and timing of the segregation of the earth's interior is another much debated question. We shall explore the evidence from the rock record in the next chapter.

The composition of the earth's interior is presently unknown, but certain ultra-basic (low silica) igneous rocks, such as peridotite and serpentine, brought up from great depths along plate margins suggest the upper mantle may be similar in composition. Eclogite, a high-density rock of the same composition as basalt, occurs in these same settings and may indicate that chemical differences are not the only ones between crust and mantle. The core is very dense and generates the earth's magnetic field (Fig. 12.12) which suggest it is composed of iron and related elements. (The so-called "transition metals" of the chemist's periodic table are the only elements with the correct density characteristics which are common enough to compose the core.) Certain meteorites, the iron-nickel type, have this composition, but if they are material left over from the origin of the solar system, they could not have formed in the interior of a planet by segregation. On the other hand, they may have resulted from the disruption of a planet or satellite. Mathematical calculations indicate that any protoplanet which began to form in the present asteroid belt between Jupiter and Mars would probably be disrupted before undergoing a very long history.

As the original earth, lacking an atmosphere and a magnetic field and swept by the solar wind, underwent heating, segregation, and core formation, a magnetic field developed which shielded its surface from the solar wind.

The Earth's Atmosphere and Oceans

The presence of a primeval atmosphere and ocean at this stage has been much debated. Astronomers, judging from the abundance of various gases present in the universe and particularly in the atmospheres of the outer planets which are less affected by the solar wind, have postulated an original residual atmosphere of methane, ammonia, water, and hydrogen (Fig. 12.13). Geochemists, on the other hand, have noted that the inert or noble gases, xenon, krypton, and neon, are much depleted on the earth relative to the rest of the universe. The loss of these gases indicates that the original earth must have been swept clean of an atmosphere by the solar wind. Geochemists have also shown that the rock record of the early Precambrian indicates that the presence of free methane or ammonia was most unlikely. Most geologists have now come around to this way of thinking and envision a primeval earth without an atmosphere and, hence, without oceans.

They reason that the atmosphere must have outgassed from the earth's interior, both from frozen gases originally trapped within and from their chemical contribution to compounds. These include clay minerals and other hydrated silicates which contain water, non-biogenic organic compounds which contain carbon, potassium-bearing minerals with ammonium ions which contain nitrogen, and silicates which contain some chlorine ions as substitutes for oxygen. These compounds occur in carbonaceous chondrite meteorites thus supporting this derivation. The volatiles that accumulated from outgassing would be chiefly carbon dioxide and monoxide, water vapor, nitrogen, and hydrogen (Fig. 12.14). These compounds could also result if the original atmosphere had contained methane, ammonia, and water (Fig. 12.15). The action of sunlight dissociates water releasing oxygen which reacts with methane to produce carbon dioxide and with ammonia to produce nitrogen. Because the volume of these gasses on earth today vastly exceeds the amount that could have resulted from the

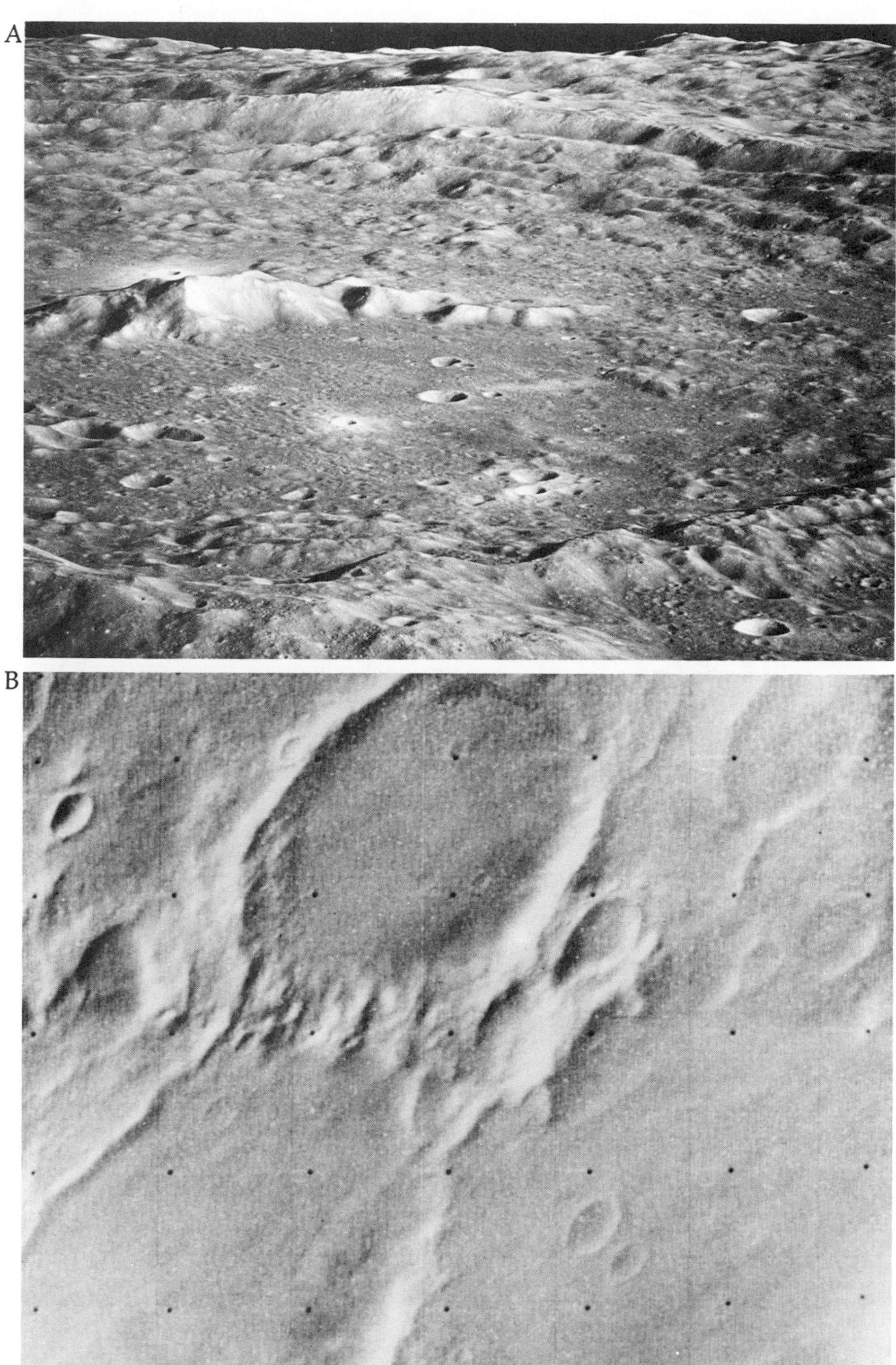

FIGURE 12.8 (A) Moon and (B) Mars Craters. (Photographs courtesy of N.A.S.A.)

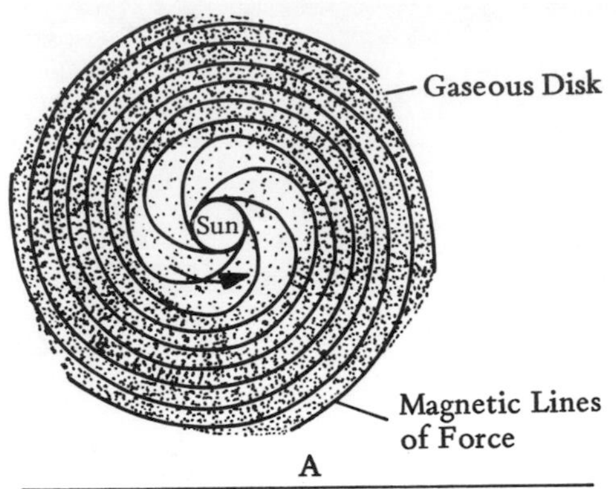

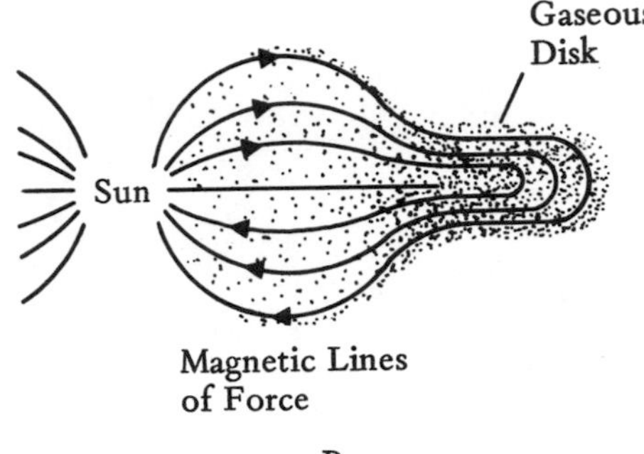

FIGURE 12.9 Magnetic Braking Action. Particles ejected along magnetic lines of force carry away much momentum but little mass. (A) Polar view, (B) Equatorial view.

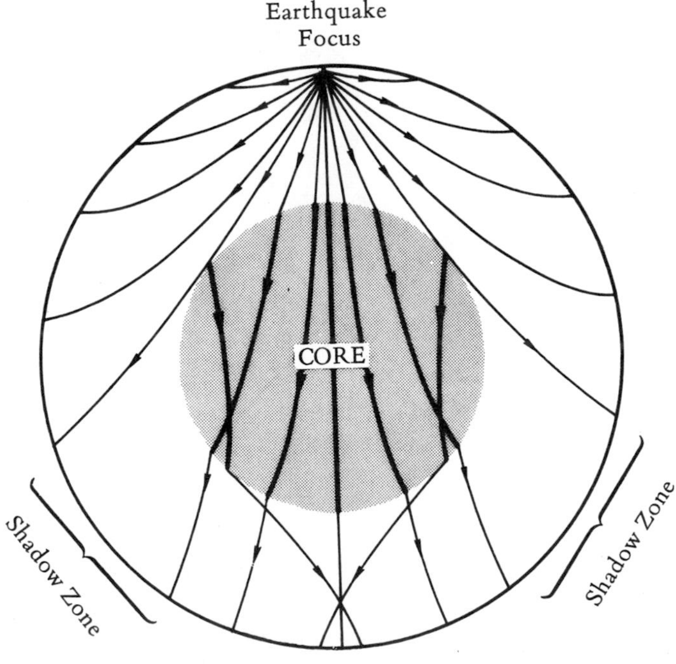

FIGURE 12.10 Earthquake Wave Behavior Gives Information About the Earth's Interior. Note bending of waves by core.

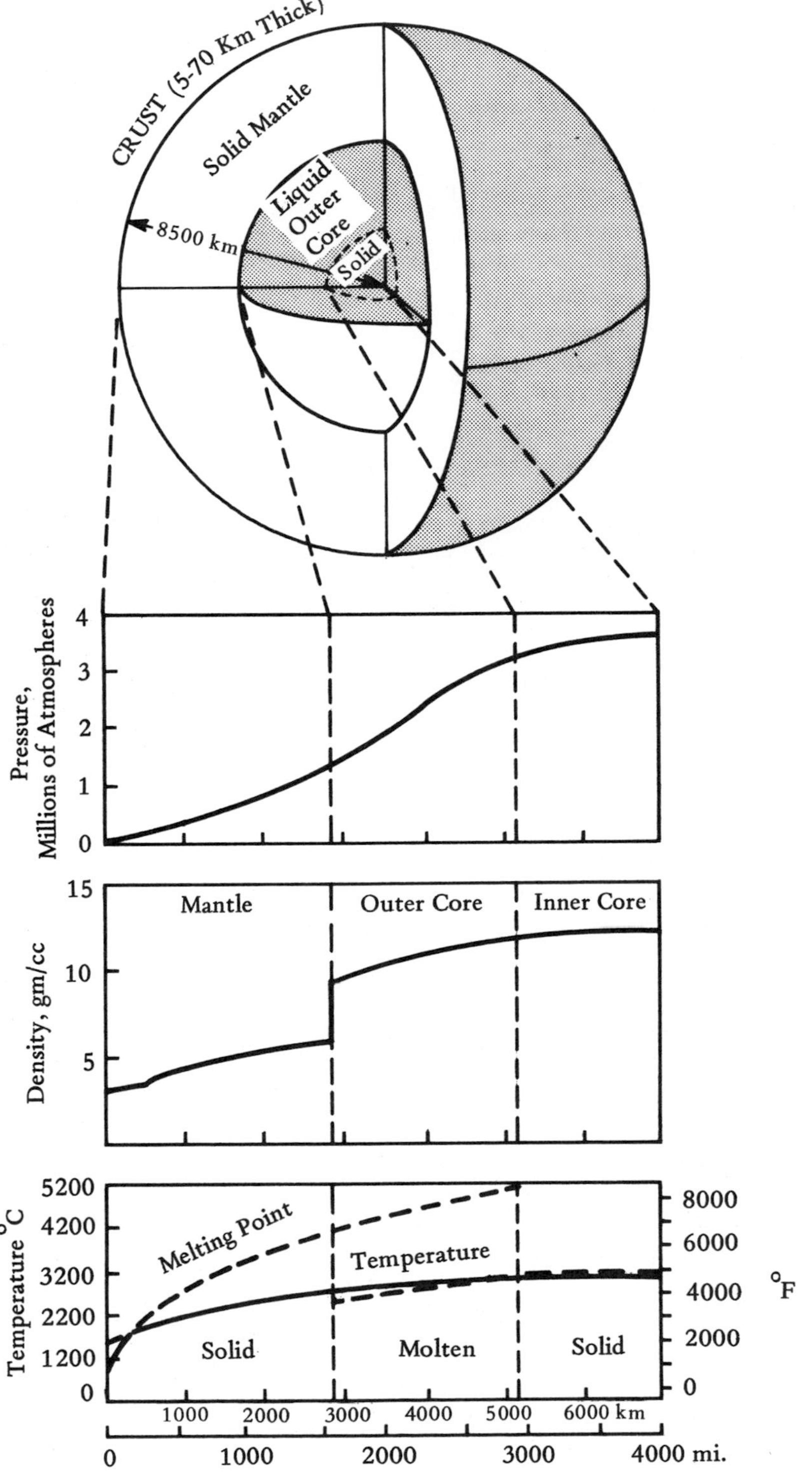

FIGURE 12.11 The Structure of the Earth's Interior.

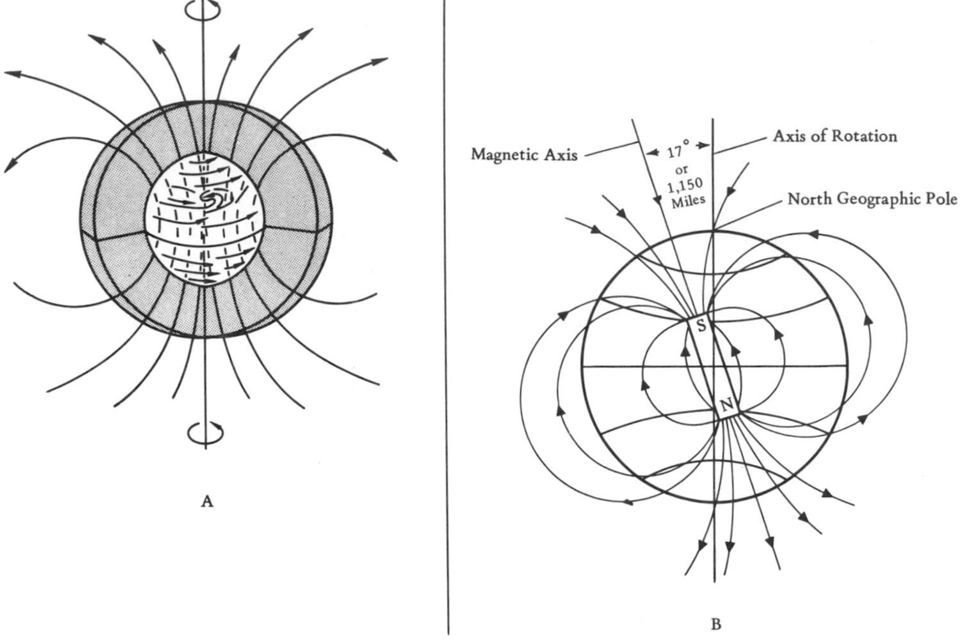

FIGURE 12.12 The Earth's Magnetic Field. (A) Generation of the Field—flow of charged particles in core induces stronger field maintained by earth's rotation. (B) Relationship to the Rotational Axis.

weathering of rocks at the earth's surface, they are called *"excess volatiles."* Oxygen was not present in the earth's early atmosphere as evidenced by the presence of many elements such as iron and uranium in their reduced states, a condition that does not occur in rocks of mid-Precambrian and younger ages.

Geochemists debate whether the outgassing was a relatively rapid event that occurred in the earth's early history or has been a slow continuous process. The lack of a rock record for the first billion years and the impossibility of distinguishing in present-day volcanic emanations the volatiles derived from groundwater and those coming from outgassing have made a final decision impossible at present. The low boiling point of the volatiles found in carbonaceous chondrite meteorites indicates an early quick release when the earth heated up. On the other hand, even if only 1% of the present volcanic emanations actually represent outgassing, they could produce the present atmosphere and oceans. The widespread occurrence of marine rocks in Precambrian strata of 3.5 billion years age and younger indicates that the amount of ocean water present then was probably as great as now and argues for the first alternative.

The Origin of Life

Thus, at some time before 3½ billion years ago there was probably an earth with an atmosphere of CO_2, CO, H_2O, N_2, H_2, oceans of water, and sedimentation processes occurring. The oldest rocks, though metamorphosed, clearly

Earth's present atmosphere			Volcanoes, average	Geysers and fumaroles	Meteorites, average	Jupiter's atmosphere
Gas	Percent	Stability				
Major:						
Nitrogen	(78%)	Stable	Water vapor (73%)	Water vapor (99%)	Carbon dioxide	Methane
Oxygen	(21%)	Unstable (reacts with Fe and C)	Carbon dioxide (12%)	Hydrogen	Carbon monoxide	Ammonia
Argon	(0.9%)	Stable	Sulphur dioxide	Methane	Hydrogen	Hydrogen
Carbon dioxide	(0.03%)	Unstable (reacts with silicates)	Nitrogen	Hydrochloric vapor	Nitrogen	Helium
Water vapor	(variable)	Unstable	Sulphur trioxide	Hydrofluoric vapor	Sulphur dioxide	Neon
Minor: (traces only)			Carbon monoxide	Carbon dioxide	Methane*	
Neon		Stable	Hydrogen	Hydrogen sulfide	Nitrous oxide*	
Helium		Stable	Argon	Ammonia	Carbon disulfide*	
Krypton		Stable	Chlorine	Argon	Benzene*	
Xenon		Stable		Nitrogen	Toluene*	
Hydrogen		Unstable (reacts with oxygen to form water)		Carbon monoxide	Naphthalene*	
					Anthracene*	

*Only present in Carbonaceous chondrites.

FIGURE 12.13 Comparison of the Earth's Atmosphere with That of an Outer Planet and gases present in meteorites, volcanic emanations, and geysers.

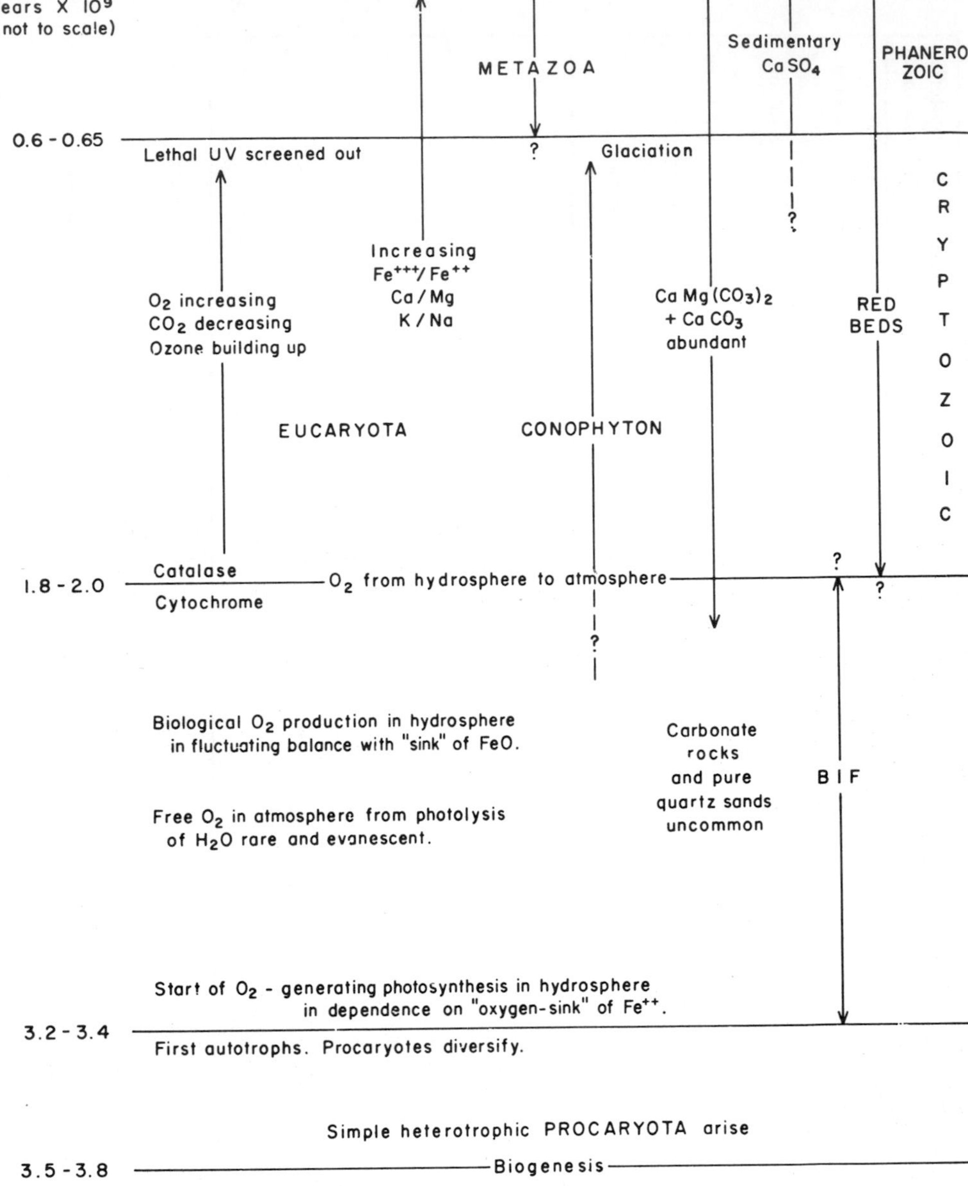

FIGURE 12.14 Evolution of the Earth's Early Atmosphere and Life. (From Preston E. Cloud, Jr., in Drake, *Evolution and Environment*, by permission of Yale University Press.)

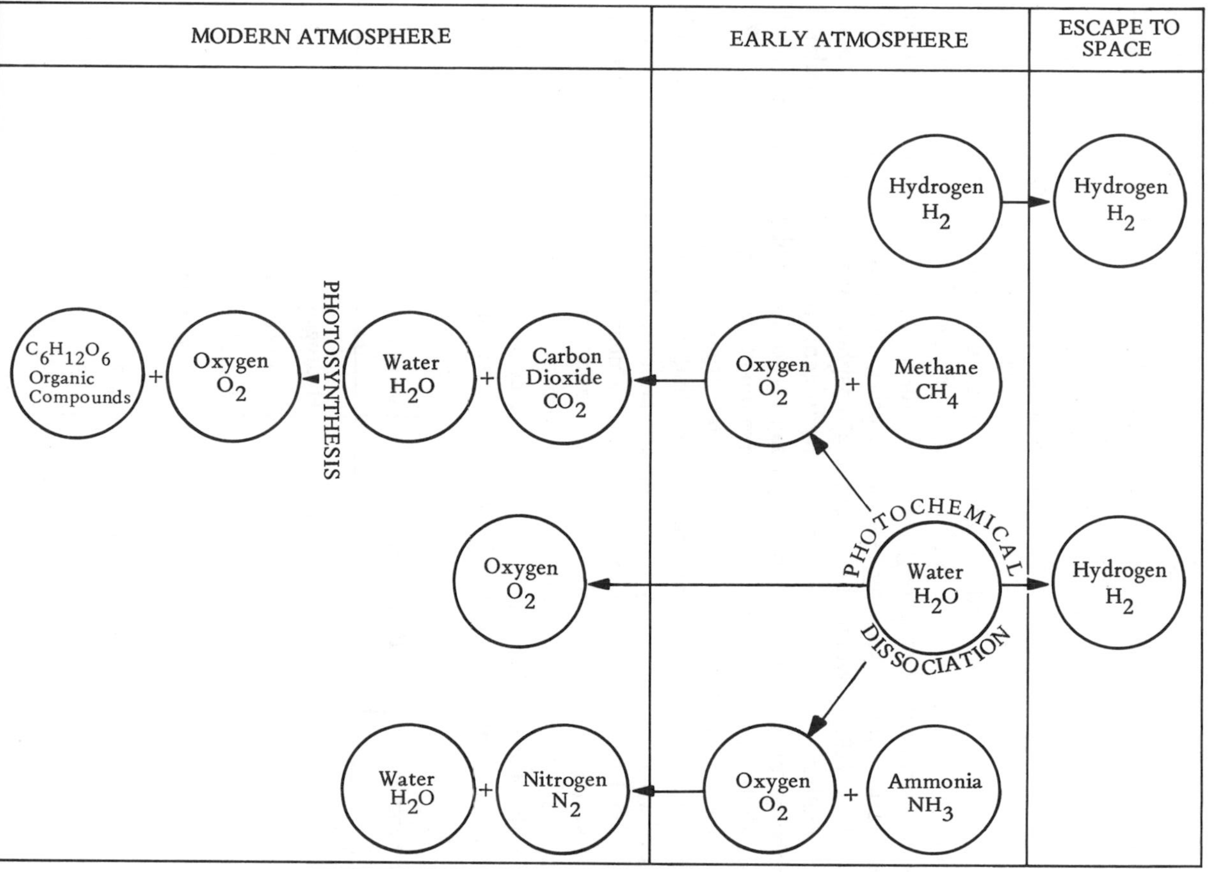

FIGURE 12.15 **Possible Derivation of CO_2, O_2, and N_2 from Original Atmosphere of CH_4, H_2, NH_3, and H_2O.**

show the uniformitarian nature of the sedimentary processes that formed them. (As we shall see in the next chapter, however, Precambrian rocks do have peculiarities not found in Phaenerozoic strata.) These oldest known rocks, the Swaziland and Fig Tree Groups of South Africa, also contain tiny bodies which appear to be the remains of bacteria and blue-green algae, the simplest of living things. The fact that the oldest rocks contain fossils indicates that the fossil record cannot directly help us determine the steps in the origin of life. We must turn again to the biochemist and geochemist to suggest how life might have originated given the physical conditions described above.

Even the simplest modern type of organism consists of complex organic compounds such as carbohydrates, lipids, adenosine phosphates, proteins, and nucleic acids in conjunction with water, the universal solvent (Fig. 12.16). Our original environment consisted of simple compounds such as carbon dioxide and monoxide, water, nitrogen, and hydrogen (or possibly methane, ammonia, water, and hydrogen). It is a long way from these to the complexly organized, energy-capturing and expending, self-duplicating thing we call life. Or is it?

In the days (1953) when the original atmosphere was believed to consist chiefly of methane, ammonia, water, and hydrogen, *Miller* and *Urey* circulated such a mixture over an electric spark, analagous to ultraviolet radiation (which could penetrate the then oxygenless environment) or lightning (Fig. 12.17). They obtained in this experiment the complex organic substances called amino acids, the building blocks of proteins. Since that date, using various possible mixtures of volatiles and energy sources available on the early earth, chemists have succeeded in synthesizing in the laboratory nearly every major constituent of living systems. For example, by heating and cooling various amino acid mixtures, *Fox* was able to aggregate them into protein-like chains. *Oro* found the nucleotide-base building blocks of nucleic acids in one of his experiments. *Bar-Nun* found that shock waves from thunderclaps are very efficient producers

Compounds	Functions	Composition
Water	Universal Solvent	Hydrogen, Oxygen
Carbohydrates	Energy Source	Hydrogen, Oxygen, Carbon
Fats	Energy Storage	Hydrogen, Oxygen, Carbon
Adenosine Phosphates (ADP, ATP)	Energy Transfer	Hydrogen, Oxygen, Carbon, Nitrogen, Phosphorus
Proteins	Structural; Facilitation of Chemical Reactions	Hydrogen, Oxygen, Carbon, Nitrogen, Phosphorus, Sulfur (Organized Into 20 Amino Acid "Building Blocks")
Nucleic Acids (DNA, RNA)	Patterns for Protein Synthesis	Hydrogen, Oxygen, Carbon, Nitrogen, Phosphorus (Organized Into 5 Nucleotide "Building Blocks")

FIGURE 12.16 Principal Chemical Constituents of Life.

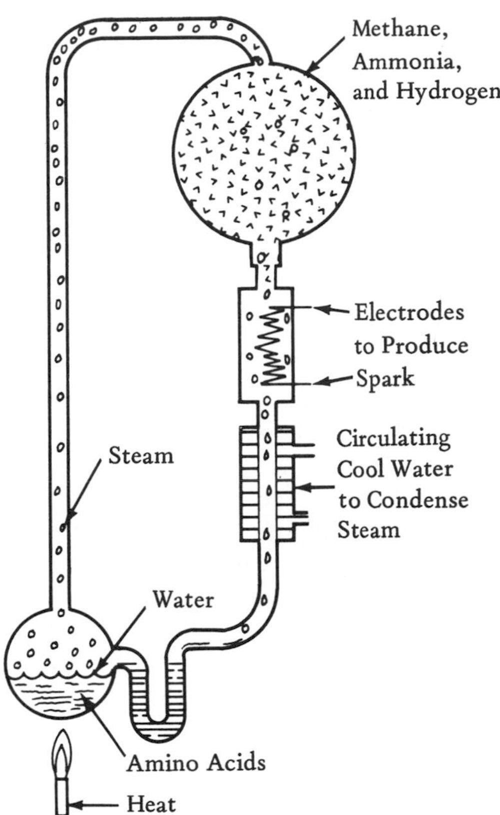

FIGURE 12.17 The Miller-Urey Experiment.

of amino acids. And, in a very significant breakthrough made in 1968, *Ponnamperuma* and *Hodgson* synthesized chlorophyll, the compound responsible for photosynthesis. Other workers found that even in ordinary mixtures of simple carbon compounds, small amounts of the complex ones are present because those chemical reactions, like all others, are equilibrium reactions. That is, every reaction goes both ways even if only a small amount in one direction under natural conditions. So, by an early date in earth history after degassing, the oceans could develop into organic soups of complex molecules. There were no scavengers nor any oxygen to destroy the organic compounds and if they sank below the level of ultraviolet penetration they were safe from dissociation.

Complex organic compounds by themselves are not organisms, however. The chance fusion of all these compounds into anything even as simple, by biologic standards, as a bacterium is clearly impossible. But chemists began to notice other interesting things about their non-biogenically produced complex compounds. *Oparin* found that the mutual attraction among these large molecules produced oriented clumps called *coacervates*. In other words, much of the order thought to be found only in living things is derived from the structure of molecules themselves and actually occurs in the non-living world. In fact, many of the steps leading to life could probably occur anywhere in space where the right materials and conditions exist. It was also found that some of the compounds in the soup had greater affinity for the simple building blocks than others and, hence, became more numerous, a sort of incipient natural selection. Furthermore, some of the complex organic compounds are *catalysts,* that is they speed up reactions that without them would proceed very slowly.

This would include not only the self-catalysis just described, but also the step-wise process of continually aggregating more and more complex structures until membranes and other cellular structures resulted producing the first cell. Even some simple inorganic compounds found in the water inefficiently catalyze organic reactions that today are speeded by more efficient organic compounds, the protein catalysts called *enzymes.* Many of the steps in this chain of catalytic reactions leading from the organic building blocks in the soup to the simplest cells have now been performed in the lab including the synthesis of the hereditary material itself, *DNA* or deoxyribonucleic acid. Because the transition from the non-living to the living is a continuous, evolutionary one, it is hard to decide on a definition of life to be used in deciding if life has been created. The creation of the self-duplicating DNA molecule, however, is certainly on or near the boundary by most definitions.

The chemical processes and structure in all living things are so similar that with the possible exceptions of some groups of bacteria, it seems certain that all life had a common origin. Thus, one type of molecular aggregate in the ancient oceans or soup must have been a superior competitor for building blocks and more efficient in self-duplication than the others, and eliminated them. A survivor from this age of subsistence on simple building blocks may be *Kakabekia* (Fig. 14.2), a tiny umbrella-shaped creature requiring only ammonia as an energy source and known from fossils of two billion years of age.

Fermentation

In the absence of oxygen, the main energy-releasing compound used by cells today, the earliest organisms probably utilized the process of fermentation to release energy. Fermentation, still utilized today by some small organisms such as yeast, is the process of splitting organic compounds such as sugars to release energy, carbon dioxide, and alcohol (Fig. 12.18). Clearly an organism employing fermentation as an energy source depends on an external supply of organic molecules and is heterotrophic or animal-like.

Photosynthesis

One of the early creatures either incorporated into self or acquired as a parasite the compound chlorophyll which, we have already seen, can be generated abiotically. Chlorophyll allows its possessor to take the carbon dioxide produced by fermentation, water, and solar energy and make its own organic compounds, such as sugar, without depending on an external supply. Oxygen is liberated

Fermentation: $C_6H_{12}O_6 \rightarrow 2CO_2 + 2C_2H_5OH + \text{energy}$

Photosynthesis: $6CO_2 + 6H_2O + \text{sunlight} \rightarrow C_6H_{12}O_6 + 6O_2$

Respiration: $C_6H_{12}O_6 + 6O_2 \rightarrow 6CO_2 + 6H_2O + \text{energy}$

FIGURE 12.18 Major Chemical Equations Important to the Evolution of Life.

as a by-product in this process (Fig. 12.18). *Photosynthesis* had appeared. (A few bacteria utilizing a slightly different compound and process can also manufacture their own food along with producing oxygen.) These first simple autotrophs could then make their own organic molecules by photosynthesis and derive energy from them by fermentation.

The free oxygen produced by photosynthesis is a poison to life. Today it is removed by complex organic compounds in the cell which reduce it to nonpoisonous forms. Before these evolved, however, some other device was necessary to get rid of the oxygen. The seas at that time must have had large quantities of unreduced or ferrous iron derived from the weathering of rocks because there was no free oxygen in the atmosphere or oceans to oxidize it. The oxygen generated by photosynthesis readily combined with this soluble ferrous iron and oxidized it to insoluble oxides and hydroxides which form precipitates. Here then is the explanation for the large marine iron precipitate deposits of the Precambrian rocks older than two billion years (Fig. 12.19). Such rocks do not occur later in the earth's history because once oxygen became abundant after the spread of photosynthesis, it completed the precipitation of the ferrous iron in the sea and then began to collect in the atmosphere. There it oxidized the ferrous iron before it reached the sea producing terrestrial redbeds which first appeared around 2 billion years ago as the marine iron precipitation ceased.

FIGURE 12.19 Banded Iron Formation, Michigan. (Courtesy Michigan Dept. of Natural Resources.)

Further evidence for the biologic origin of the iron is the fact that the iron occurs in alternating bands with silica, a chemical association that could not occur without organic intervention.

Respiration

If life had remained dependent on inefficient fermentation reactions to obtain energy, the process of photosynthesis would have proceeded much more slowly in supplying the earth with oxygen. However, once the oxygen-reducing compounds evolve in cells, such cells could survive in the presence of free oxygen and eventually use it for a more efficient metabolic process, *respiration* (Fig. 12.18). In cellular respiration, sugars and oxygen combine to produce energy, water, and carbon dioxide. All those by-products are harmless to life; in fact, carbon dioxide is a necessary raw material in photosynthesis. The complementary nature of the photosynthesis-respiration reactions coupled with the much greater efficiency of respiration over fermentation led to the tremendous expansion of life. With it came the gradual evolution of oxygen throughout the oceans and atmosphere. This screened out ultraviolet light, thus removing one of the energy sources which may have started the evolution of living things, but which was also detrimental to near-surface or terrestrial life. Free oxygen also meant the end of the non-oxidizing conditions necessary for the beginning of life and has precluded any additional origins of life.

Organisms then evolved which could obtain their organic compounds by devouring others and utilizing the increasingly abundant oxygen to release energy from these compounds. Animal-like heterotrophs had thus appeared. By mid-Precambrian times then, the earth had formed, evolved its present atmospheric and oceanic constituents (although the percentages of gases were probably not the same as at present), and was the scene for the unfolding drama of the continuing evolution of life.

Questions

1. How does Laplace's nebular hypothesis differ from the modern theory of the origin of the earth?
2. What have been the major problems with the encounter hypothesis?
3. What is the nature of the earth's interior as disclosed by earthquake waves?
4. What direct evidence do we have about the nature and composition of the mantle?
5. Why do geochemists believe the earth's atmosphere outgassed instead of being a remnant of the protoplanet stage?
6. What are the major compounds found in living things?
7. Give some examples of the non-biogenic generation of complex organic compounds.
8. What is fermentation and why is it a relatively unimportant process today?
9. Compare and contrast photosynthesis and respiration.

13

The First Four Billion Years:

The Precambrian

THE PRECAMBRIAN OR THE CRYPTOZOIC EON, THAT IMMENSELY LONG INTERVAL OF geologic time preceding the appearance of abundant megafossils, has not been subdivided into a series of units of worldwide application such as those established for later geologic time. Because it covers approximately 8/9 of the earth's entire time of existence, from over 4.5 to .6 or .7 billion years ago, it may seem incongruous to devote only two and a half chapters in a textbook of earth history to this interval. But because there has been no worldwide tool for correlating Precambrian rocks and, hence, for dividing them into universally recognizable units, we simply do not have the detailed knowledge of these strata that fossils give us in later geologic time. The best we can do is assign dates of occurrence to events in terms billions and hundreds of millions of years ago as determined from radiometric dating techniques. But there is still no way to obtain the fine subdivisions possible with fossils. New tools, such as the scanning electron microscope, are enabling paleontologists to uncover new structural details in Precambrian microfossils. Perhaps there will be enough diversity and rapid evolution discovered to provide subdivisions in the future.

In the days when geologists believed that the earth's history had been a series of quiet intervals of sediment accumulation separated by short periods of worldwide mountain-building, they unravelled the local Precambrian history of their regions and extrapolated it to the world. As a result, one sees, in older textbooks, a series of "eras" and "periods" in the Precambrian time scale. The eras called *Archeozoic* or *Archean* and *Proterozoic* or *Algonkian* in North America were based on the supposition that all the granites, gneisses, and associated highly deformed and metamorphosed sediments were very old, and that the slightly deformed sedimentary rocks were younger. Once again the illusion that lithology is a guide to age was shattered.

The advent of radiometric dating showed that rocks of both types occurred throughout the Precambrian from the oldest 3.5 billion-year-old rocks down to the beginning of the Paleozoic Era. As a result it became obvious that the original usage of these terms was for rock and not time-rock units. Some geologists have now switched usages, however, and call all old rocks, regardless of their lithology, Archean, and all younger Precambrian rocks, Proterozoic.

The incongruity of the switch of definitions is pointed up dramatically in the Grand Canyon. At the bottom of the Canyon are highly deformed, intensely metamorphosed, intruded, and very "old-looking" rocks originally assigned to the Archean (Fig. 13.1). Radiometric dating, however, has disclosed that they were deformed only 1.3 billion years ago, relatively late in the Precambrian. In the new usage of "Proterozoic" as applying to rocks deformed less than 2.4–2.5 billion years ago, these Grand Canyon rocks are Middle Proterozoic. In recent years then, there has been more of a tendency to recognize that most names applied to Precambrian subdivisions have a lithostratigraphic and not a chronostratigraphic basis. Geologists in each area of Precambrian exposures have described the various episodes of sedimentation and deformation, assigning them local era and period names and determining their radiometric ages.

Shields

Precambrian igneous, metamorphic, and sedimentary rocks are found well-exposed on all continents today. In some areas, the Precambrian rocks are

FIGURE 13.1 Grand Canyon Precambrian. Lowest cliff is composed of dark Vishnu Schist, a highly deformed and metamorphosed group of rocks. Cliff above this on right side of photograph (not present on left side) is composed of younger Precambrian Grand Canyon Group, a series of more normal sedimentary rocks. (Courtesy of William M. Mintz.)

widely exposed at the surface over broad, low areas called *shields*. The major shields of the earth today (Fig. 13.2) are the *Canadian* in North America, the *Amazonian* in South America, the *Baltic* in Europe, the *Ethiopian* encompassing most of Africa and Arabia, the *Angaran* in Siberia, the *Indian*, the *Chinese*, the *Australian*, and the *Antarctic*. Several small shield areas are also known. In the days when continents were assumed to be static, a great deal of genetic significance was attached to shields. They were thought to be stable continental nuclei or cratons around which geosynclines developed in a series of concentric rings, each successively plastered to the shield by mountain-building.

In North America, where the Phanerozoic geosynclines ring the shield, this concept has been widespread (Fig. 13.3). It has found little support elsewhere,

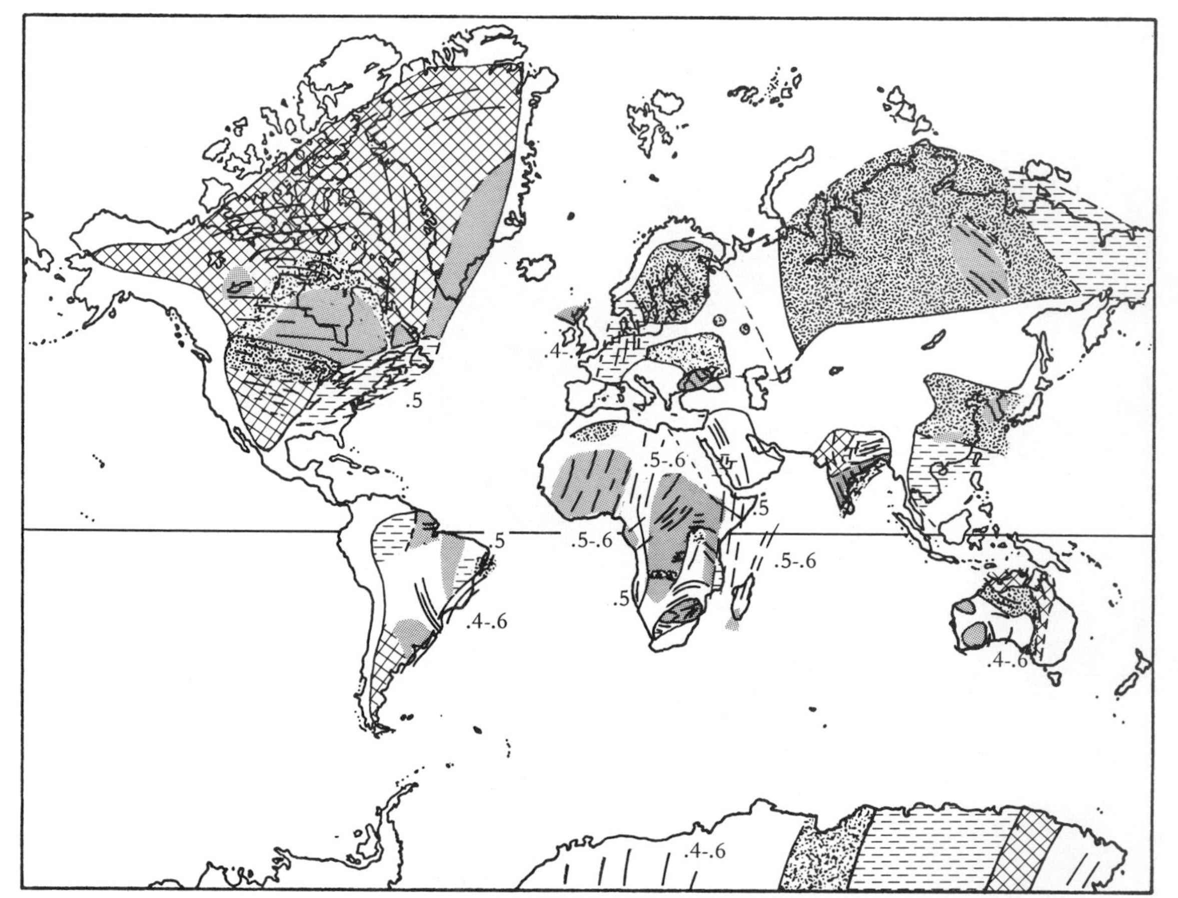

FIGURE 13.2 Major Shield Areas of the Earth and Their Ages of Deformation.

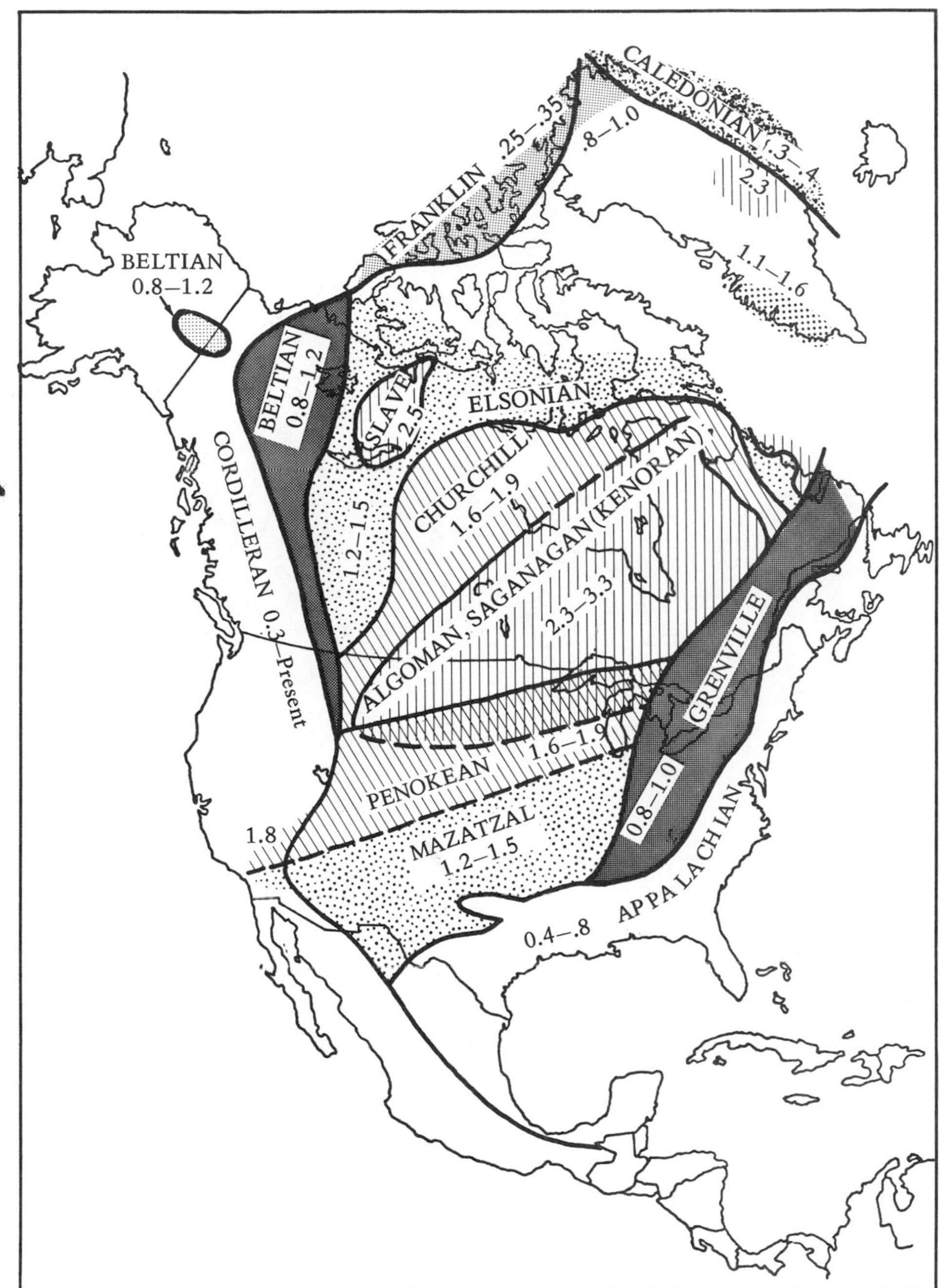

FIGURE 13.3 Isotopic Age Provinces of North America.

however, because Eurasia and the southern continents do not show any hint of such a concentric pattern (Fig. 9.3). For example, a Paleozoic geosyncline ran down the middle of Eurasia and most margins of Africa are Precambrian. Even a glance at the radiometric dates of the North American shield (Fig. 13.4) shows that there are three or four belts of very old Precambrian rocks and

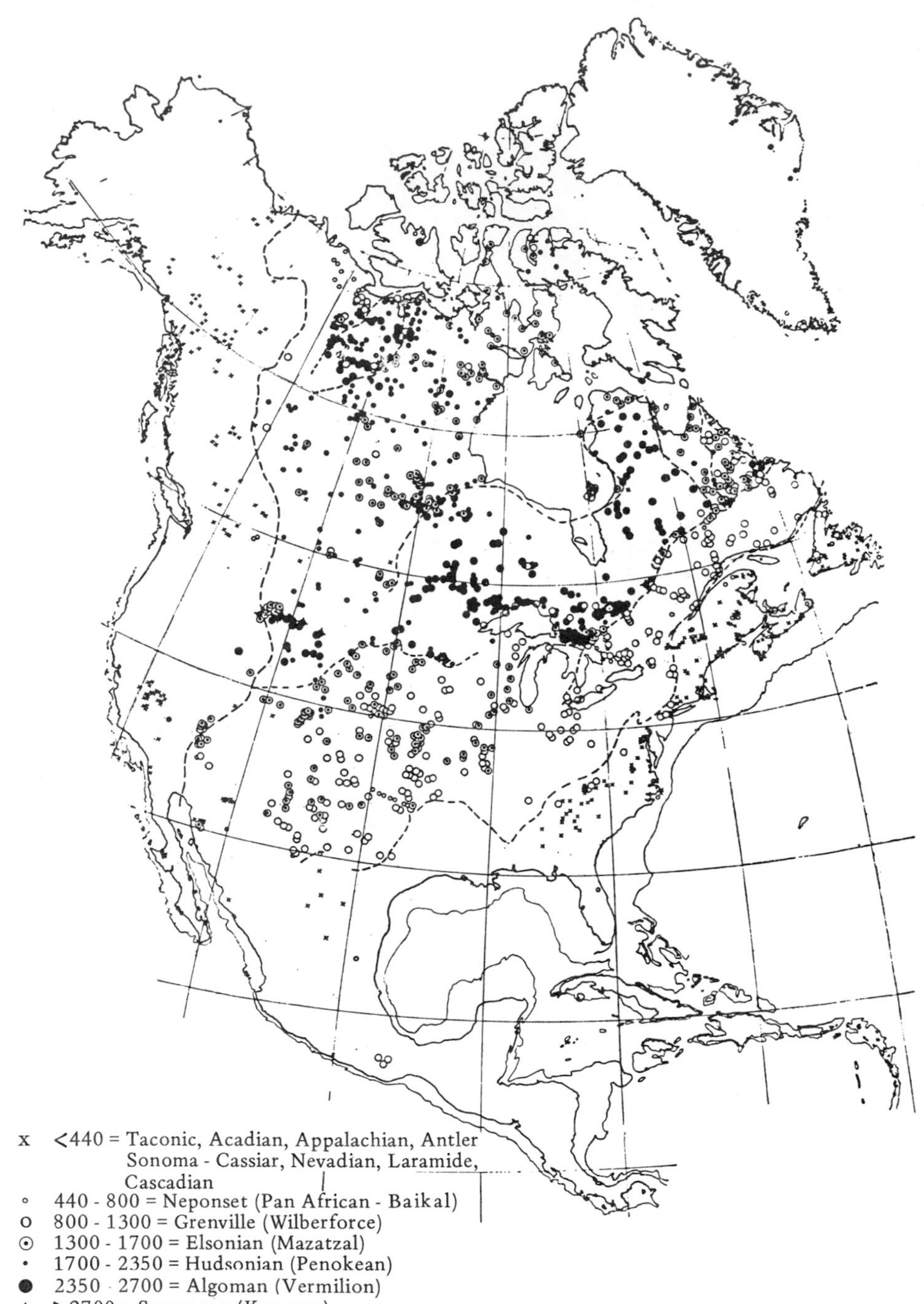

FIGURE 13.4—Isotopic Age Dates for North America. (From Hurley and Rand, *Science*, vol. 164, June, 1969, Copyright © AAAS. Used by permission of authors and AAAS.)

although one is fairly central, the others are not. It now appears that the significance of shields is simply that they are blocks of exposed Precambrian rocks which have not been significantly deformed or destroyed since the Precambrian by plate collisions.

Many have been sundered apart by rifting, however. For example, eastern South America, Africa, India, western Australia, and eastern Antarctica are all part of the same shield that has been broken into fragments. The matching boundaries of Precambrian deformed belts can be found on opposite sides of the Atlantic in eastern South America and western Africa (Fig. 11.34). Whatever controls the positions of the margins of the plates, it appears effectively random as far as the position of continents on the lithosphere is concerned. So, sooner or later, even the existing shields will be near plate margins and be deformed again. This has occurred to many areas of shields in the past for Precambrian rocks occur in the cores of many of the earth's great mountain ranges including the Appalachians and western Cordillera of North America.

Precambrian Rocks

The most abundant Precambrian rock types are granites and gneisses (Fig. 13.5). This is not surprising considering the vast length of time these rocks have been in existence which greatly increased their chances of finding themselves at plate margins and being melted and metamorphosed. The volume of the earliest Precambrian ($>$ 2.3 billion years) granites is so great, however, that some geologists believe they represent the relatively rapid density differentiation of the earth with light materials rising to the surface. The sediments of this age are oceanic types everywhere they have been found. This may indicate that the original crust was oceanic and then large-scale differentiation within the earth sent most of the light continental material bobbing to the surface at this time. Another possibility is that capture of the moon by the earth at this time could have produced tidal forccs bringing up the granite. If these hypotheses are true, granites of middle Precambrian age and younger simply represent reworked sialic material from this one original event. Certain rocks apparently derived from the mantle, such as layered ultramafics (Fe and Mg rich rocks) and anorthsites (rocks made entirely of Ca-plagioclase feldspars) occur in quantity only in the Precambrian. This may indicate that the crust was still thin and in the process of forming.

The Precambrian rocks found in areas deformed since the Precambrian, that is Phanerozoic mountain belts, have been deformed several times so their history is very difficult to decipher. In the shields, however, where Precambrian rocks have thus far escaped Phanerozoic mountain-building, the history is easier to read. Granites and the metamorphic rock gneiss characteristically form only at former plate margins in the cores of mountain ranges in the Phanerozoic. With the possible exception of the earliest granites, we can be fairly certain, therefore, that we are dealing in large part with the roots of old mountain ranges that formed by plate collisions in what are now the relatively flat shields. These intrusive-metamorphic zones are highly mineralized and are responsible for the great metallic ore deposits found in most shields.

FIGURE 13.5 Precambrian Granite. Grand Teton Mountains, Wyoming. (Courtesy of William M. Mintz.)

In scattered pods in the shields are occasional remnants of once-overlying sedimentary sequences (Fig. 13.6), now metamorphosed to varying degrees. (These occur within the granites of all ages.) We are fortunate in having even this much left of the sedimentary cover, considering the more than half billion intervening years during which erosion has occurred. This erosion has not only stripped away most of them, but has also cut down into the very roots of the Precambrian mountains as evidenced by the abundant migmatites (Fig. 13.7), rocks in which igneous and metamorphosed sedimentary materials are interlayered.

Not all surviving Precambrian rocks were eugeosynclinal in origin, however. The Middle Precambrian Huronian Group of the southern Canadian shield (Fig. 13.8) is a shallow sea sequence. The later Precambrian Keewenawan Group of the Canadian Shield (Fig. 13.9) is a great sequence of tilted lava flows and shallow sea sediments, probably representing a rift zone in the area.

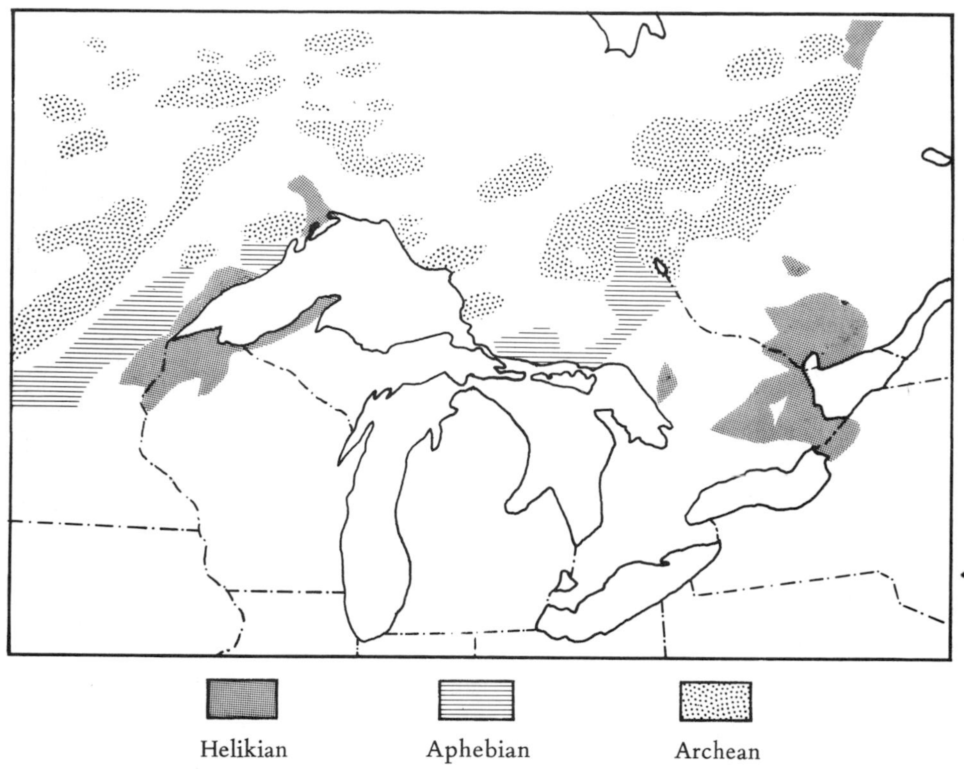

FIGURE 13.6 Map Showing Pods of Sedimentary Rocks in Southern Canadian Shield.

FIGURE 13.7 Migmatite in the Canadian Shield. (Courtesy of E. R. Brooks, C.S.C.H.)

A B

C

FIGURE 13.8 Huronian Sediments (A) and (B) Animikian rocks at Kakabeka Falls, Ontario, courtesy of Ontario Dept. of Tourism and Information. (C) Ripple-marked quartzite near Sault Ste. Marie, Ontario, from *Evolution of the Earth* by Dott and Batten. Copyright © 1971 by McGraw-Hill, Inc. Used with permission of McGraw-Hill Book Company.)

The Lipalian Interval

It was formerly believed by North American geologists that a great unconformity separated Precambrian and Cambrian rocks all over the world. This reasoning was based on their ideas of the universality of mountain-building and regressing seas at various intervals in the past. In North America, except

A

B

C

D

FIGURE 13.9 Keewenawan Lavas and Sediments. (A) Mott Island, Isle Royale National Park, (B) Monument Rock, Isle Royale National Park, (C) Lake of the Clouds, Porcupine Mountains State Park, (D) Presque Isle River Falls, Michigan. ((A-B) Courtesy of National Park Service, (C-D) Courtesy of Michigan Department of Natural Resources.)

for the very margins of the the craton, there are no latest Precambrian-Eocambrian marine rocks (Fig. 13.10). Hence, there is a craton-wide unconformity in this interval and geologists assumed it was worldwide.

The notion could also explain why there were no megafossils except algae in Precambrian rocks and abundant animal megafossils in the Cambrian rocks. The organisms had evolved in the deformed marginal geosynclines or the ocean basins where they were not well enough preserved or accessible for examination. Then, when the seas reinvaded the craton, they brought with them this influx of creatures whose ancestors were unknown.

This gap, assumed to be worldwide, was called the *Lipalian Interval.* We now know it is a regional unconformity non-existent in the geosynclines but encompassing increasing amounts of late Precambrian-Eocambrian time and up to two-thirds of the Cambrian as one moves toward the center of the craton. On other continents, however, a sequence of shallow marine sediments continues right through this interval across the Precambrian-Eocambrian and Eocambrian-Cambrian boundaries. These rocks are called *Riphean* in Russia, *Sparagmitian* in Scandinavia, *Sinian* in Asia, and *Late Proterozoic* in Australia and Antarctica. Eocambrian strata are also known in many former geosynclines outside of North America.

Banded Iron Formations

The rock record of the Precambrian contains several interesting phenomena. One is the widespread occurrence in rocks over two billion years old of *marine iron precipitate deposits.* In North America, these have served as the base of the iron and steel industry. As noted in the last chapter, these deposits were most probably of biologic origin, precipitated by primitive photosynthesizers as a device for disposing of the poisonous by-product oxygen. In the interlayed bands of silica, numerous types of fossil micro-organisms (Fig. 14.2) occur offering further evidence that their metabolism caused the precipitates. In rocks younger than 2 billion years, the banded iron formations no longer occur, but, instead, terrestrial redbeds colored by iron oxides are common. This change probably reflects the evolution of oxygen reducers by primitive organisms, thus allowing them not only to survive in the presence of oxygen, but eventually to utilize it in respiration. Oxygen was then produced in large amounts, which precipitated the remaining ferrous iron in the sea, and accumulated in the ocean and, eventually, the atmosphere. There it could oxidize ferrous iron before it reached the sea, producing the iron-oxide stains and coatings on terrestrial clastics.

Precambrian Glaciations

Another notable rock type of the Precambrian is *tillite,* or lithified glacial till, a poorly sorted mixture of clastics of all sizes from boulders to clay. The Lower Proterozoic rocks (about 2.2 billion years old) of the Canadian Shield contain apparent tillites (Fig. 13.11). Whatever the exact cause or causes of glaciation, a topic we will discuss in Chapter 20, it appears certain that continental glaciers

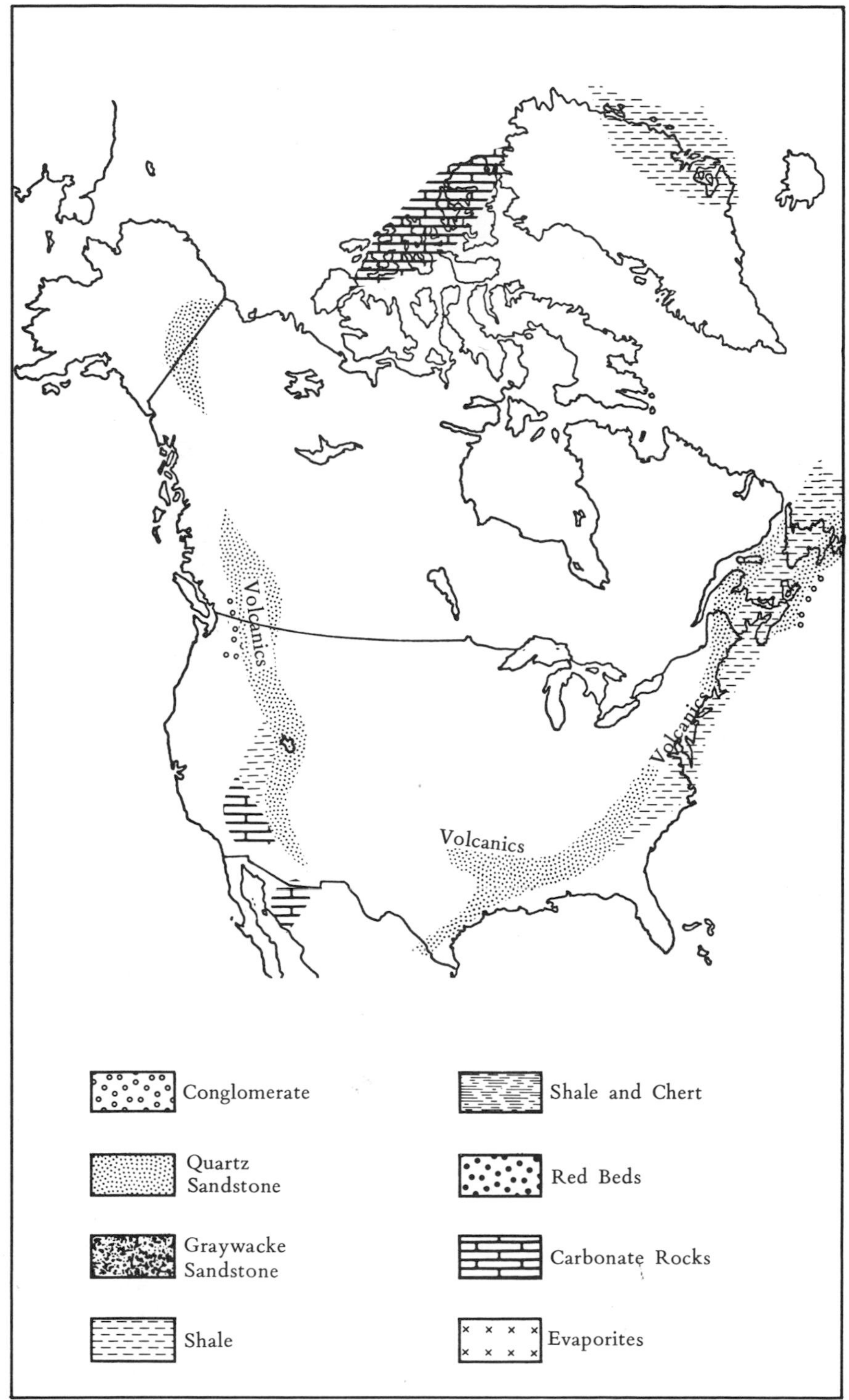

FIGURE 13.10 Present Distribution of Eocambrian Sedimentary Rocks in North America.

FIGURE 13.11 Gowganda Tillite. Huronian Group. (Courtesy Geological Survey of Canada, Ottawa.)

originate in polar or subpolar areas. Thus, we have evidence that at this time in the Precambrian, these regions were near a pole, probably the Pacific Pole. Some apparent tillites of this age also occur in South Africa indicating it was near the Gondwana Pole.

Much more extensive apparent tillites, many associated with striated, polished pavements, occur in the Upper Precambrian-Eocambrian rocks of the Gondwana continents, Soviet Asia, Scandinavia, China, Scotland, Greenland, Newfoundland, Yukon, and Utah. The paleomagnetic positions determined for the interval indicate, however, that the equator ran through Australia and central Asia very near some of these deposits. It is possible that some of the so-called tillites are fluvial deposits which are very much like glacial tills but lack a striated pavement. Some, however, do have striated pavements and yet were apparently near the equator (Fig. 13.12). Evidence from the fossil record (see Chapter 14) and geophysics indicates that the moon may have been much closer to earth than now and perhaps it triggered large submarine gravity slides which produced the till-like lithology and pavements.

If the deposits are indeed glacial, they indicate the most latitudinally widespread glaciation in earth history. It has been suggested that the onset of glacial conditions was triggered by a decrease in the amount of atmospheric carbon dioxide, a gas which helps the earth retain solar heat by trapping it in the atmosphere. This is called the greenhouse effect. The reduction in CO_2 could have been caused by a great increase in its complement oxygen. Oxygen was apparently then released in large quantities by phytoplankton colonizing the upper levels of the oceans because the atmosphere had sufficiently developed to shut off harmful ultraviolet rays. If less CO_2 means less heat retained at

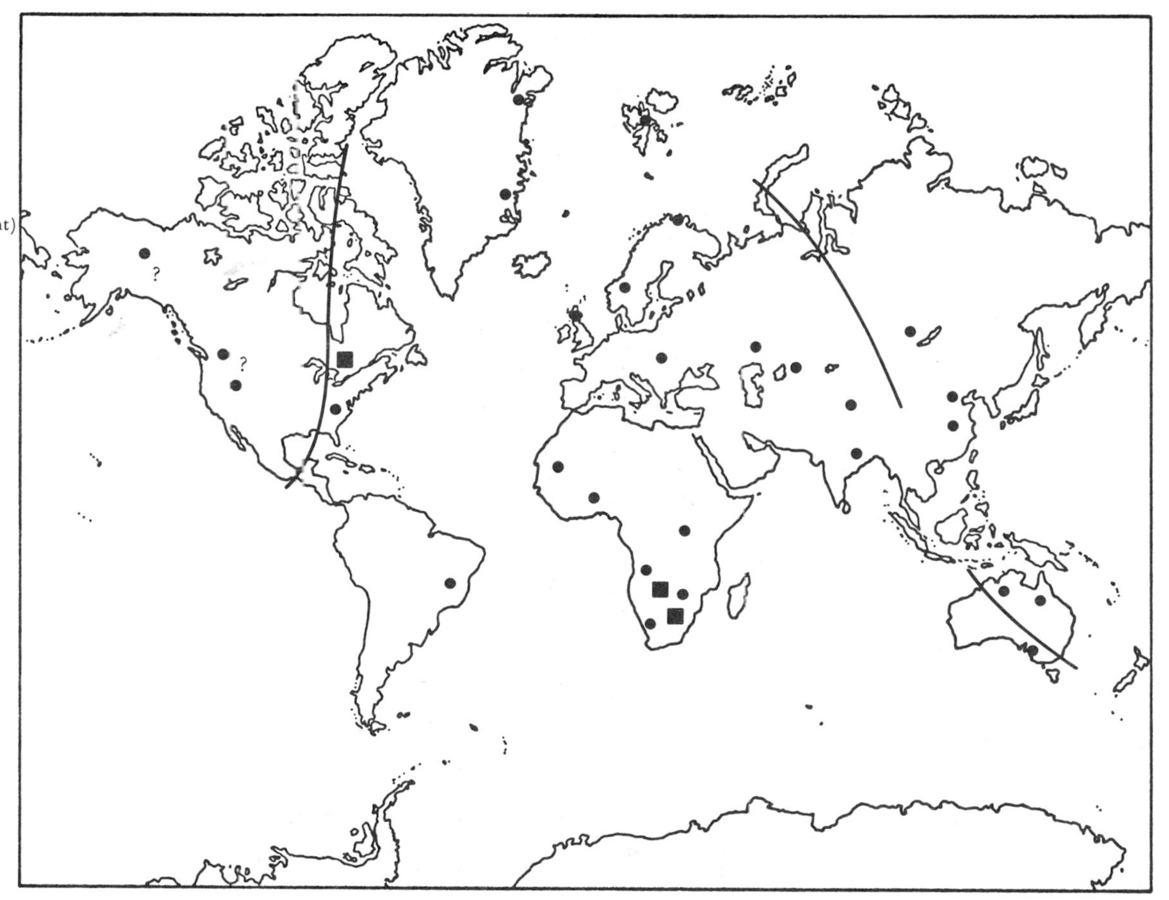

FIGURE 13.12 Distribution of Supposed Precambrian and Eocambrian Tillites.

the earth's surface, this may well have started the peculiar Late Precambrian-Eocambrian glaciation. It seems unlikely that one could invoke the same argument to explain earlier or later glaciations.

Lack of Carbonates and Sulfates

Notable by their *absence* or *rarity* in the early Precambrian are *carbonates* and *sulfates,* rock types that abound in later geologic time. The lack of shell-forming marine organisms and efficient photosynthesizers (photosynthesis removes CO_2 from solution thus precipitating $CaCO_3$) in the early Precambrian probably accounts for the lack of carbonates. Sulfates are absent in the entire Precambrian, probably because the presence of the sulfate ion requires abundant oxygen, not then available, to oxidize reduced sulfur compounds.

Precambrian Physical History

It would serve little purpose in an introductory text such as this to give a detailed region by region Precambrian history for the entire earth. This is not only because it is poorly known and correlated in detail, but also because it consists of a series of depositional and orogenic episodes varying from area to area that chiefly record normal sedimentary, igneous, and metamorphic events and processes. Geologists are only now beginning to unravel the more intriguing side of the history, the patterns of shifting plates and, hence, continents and ocean basins. Utilizing the age of deformation of Precambrian mountain belts (Figs 13.13A-F), they should be able to locate some plate boundaries for various times in the Precambrian. And, by combining this with paleomagnetic data, geologists should be able to approximately position the continents at those various times. Therefore, some general pattern of Precambrian plates will be forthcoming, but unfortunately very little has yet been done in this regard.

The Canadian Shield

The Precambrian history of the present North American craton deserves some special comment because it is probably the best known and the most familiar to readers of this book. The dates of deformation cluster around five periods, 3.3, 2.4–2.5, 1.6–1.7, 1.3–1.4 and .9–1 billion years ago (Fig. 13.14). The two oldest orogenies, called the *Saganagan, Kenoran* or *Laurentian,* and the *Algoman* or *Vermilion,* respectively, deformed a broad east-west trending geosynclinal belt north of the Great Lakes called the *Superior Province* (Fig. 13.15). The deformed rocks (Fig. 13.16) are assigned to an Archean or Archaeozoic interval, sometimes divided into lower or Keewatin and upper or Timiskaming subdivisions. (Keewatin rocks, like all other very old sediments, appear to be oceanic in origin and consist chiefly of metamorphosed lavas and turbidites.) These mountain-building episodes also affected the Beartooth and Bighorn Mountain areas of Montana and Wyoming, the Slave Province north of Great Slave Lake, and the eastern Nain Province of coastal Labrador. These may have

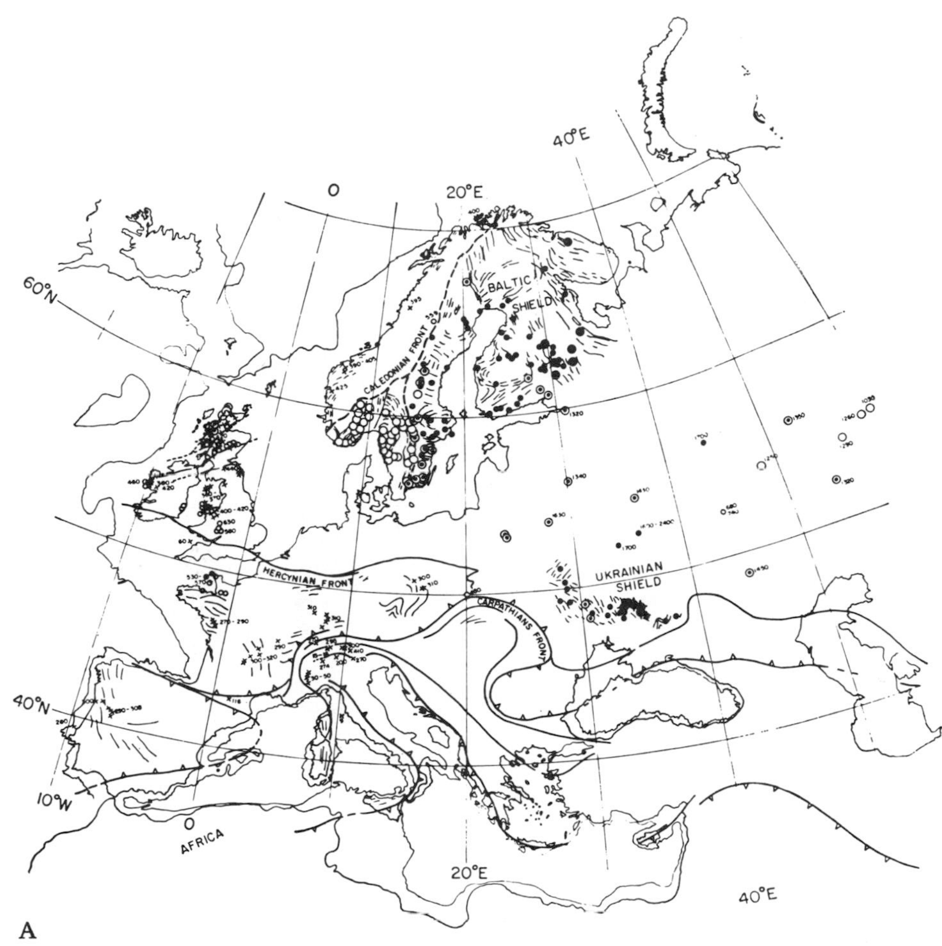

FIGURE 13.13A European Isotopic Dates. ((A-F) From Hurley and Rand, *Science,* vol. 164, 13 June, 1969, Copyright © by AAAS. By permission of authors and AAAS.)

been separate plate edges or they may have originally been continuous across the intervening regions which now record younger orogenies, the radioactive "clocks" having been reset.

The younger cycles of sedimentation are assigned to the Proterozoic interval. The next major disturbance, the *Hudsonian, Penokean,* or *Killarneyan,* chiefly affected a broad geosynclinal band around much of the Superior Province, the *Churchill-Ungava Province.* Most of this province lies to the north and west of the Superior Province. Its southern part, whose connection with the main body is now buried under epeiric sea deposits, is often called the Southern Province. The Hudsonian orogeny culminated about 1.6–1.7 billion years ago. The deformed rocks (Fig. 13.8) are assigned to a subdivision of time called the Early Proterozoic *Aphebian* "Era."

The *Elsonian* or *Mazatzal* orogeny, often lumped with the Hudsonian as a continuation of it, affected the *Central Province,* a belt extending from the

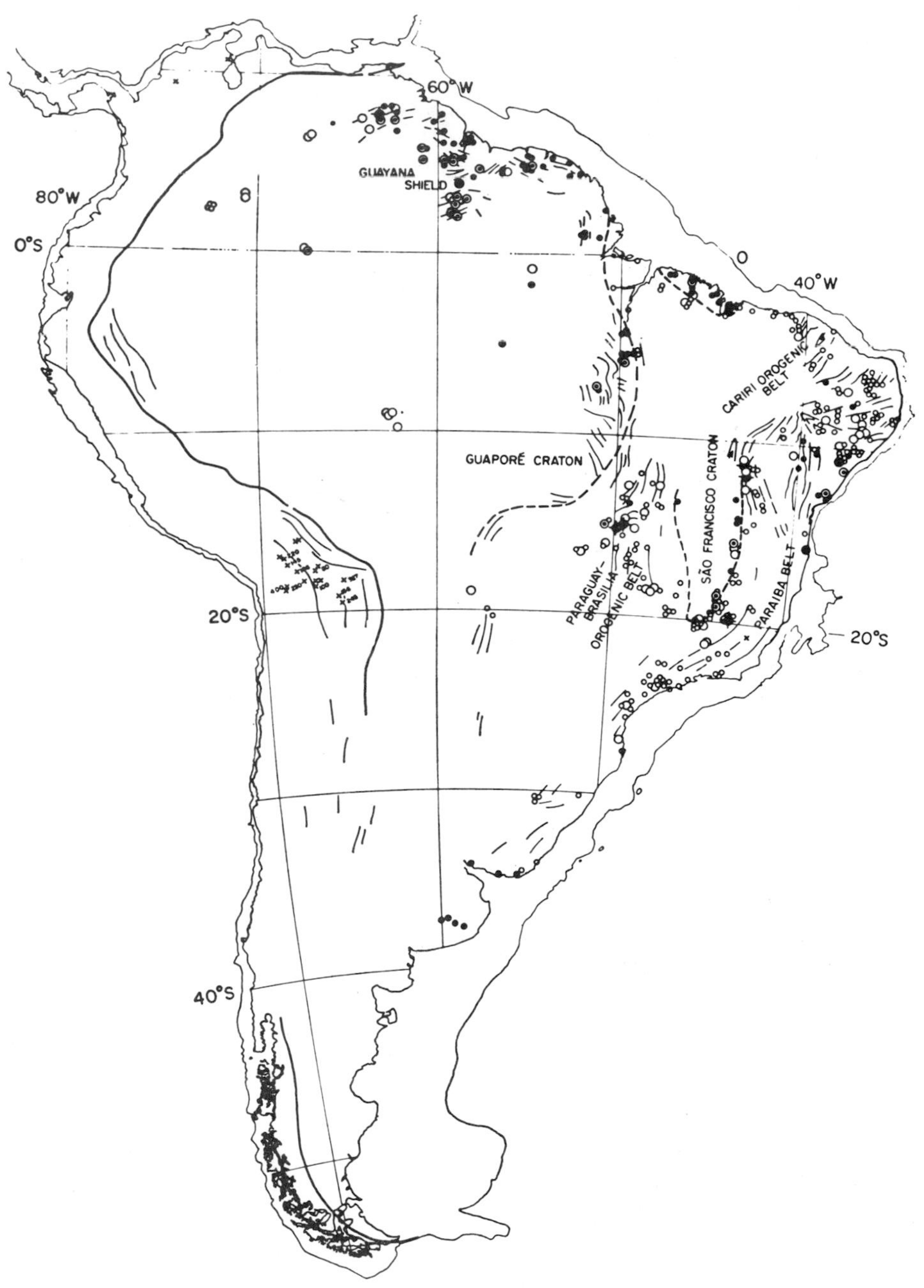

B

FIGURE 13.13B Isotopic Age Dates for South America.

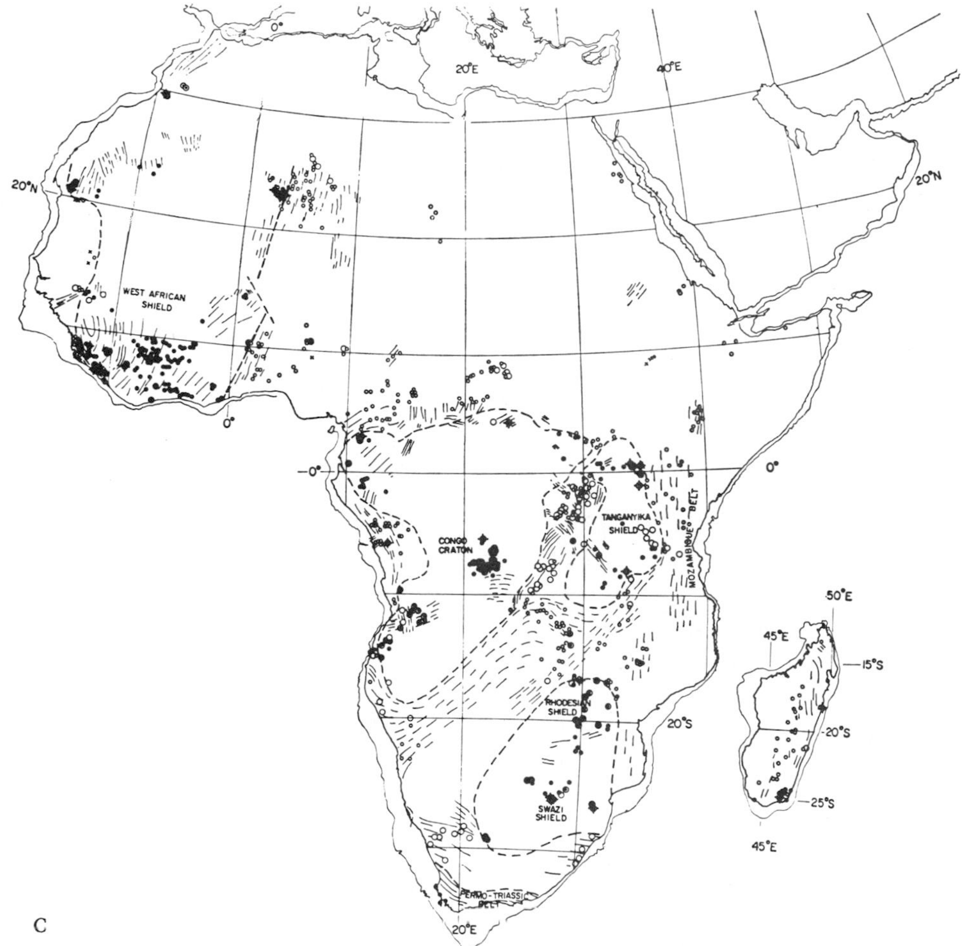

FIGURE 13.13C Isotopic Age Dates for Africa.

present southern edge of the Precambrian shield outcrops across the midcontinental United States to Arizona (Figs. 13.1, 13.15, 21.30) and adjacent states. The western Nain Province and some of the Churchill Province were also affected by this deformation which occurred 1.3–1.4 billion years ago.

The final orogeny affecting what is now the Canadian Shield was the *Grenville* or *Wilberforce* orogeny which occurred about 900 million–1 billion years ago and deformed the areas of the present Appalachian Mountains and southeastern Canadian Shield. Rocks deformed by the orogeny, which include abundant carbonates (Fig. 13.17), are assigned to the *Helikian* "Era." In the northern Rocky Mountains are strata called the Belt Supergroup (Fig. 13.18) which yield isotopic dates of this interval also. Though in high mountains today, they are still essentially flat-lying strata in contrast to the deformed shield rocks.

The last segment of Precambrian time in North America, represented only in the sediments and metamorphics of the Appalachian and Cordilleran Geo-

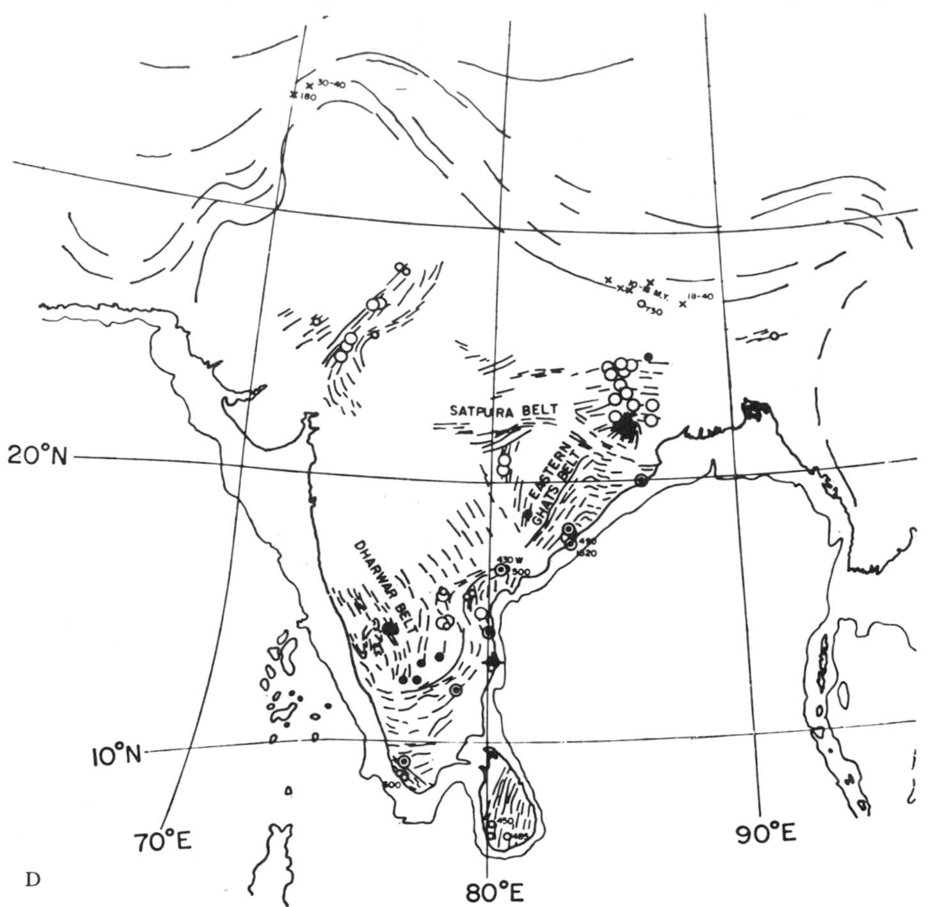

FIGURE 13.13D Isotopic Age Dates for India.

synclines deformed after Precambrian, is assigned to the Late Proterozoic or *Hadrynian "Era."*

Other Cratons

The other areas of Precambrian rocks on earth also display sequences of sedimentation and orogeny, but chiefly of differing dates from those in North America (Fig. 13.13). These also record the shifting about and collisions of plates, but worldwide synthesis is still not possible. It is perhaps appropriate to end a chapter on the Precambrian on this note of uncertainty because so many of the statements about this vast time interval are in the realm of hypothesis and weak theory. There are many more unanswered than answered questions about these enigmatic first four billion years of earth history.

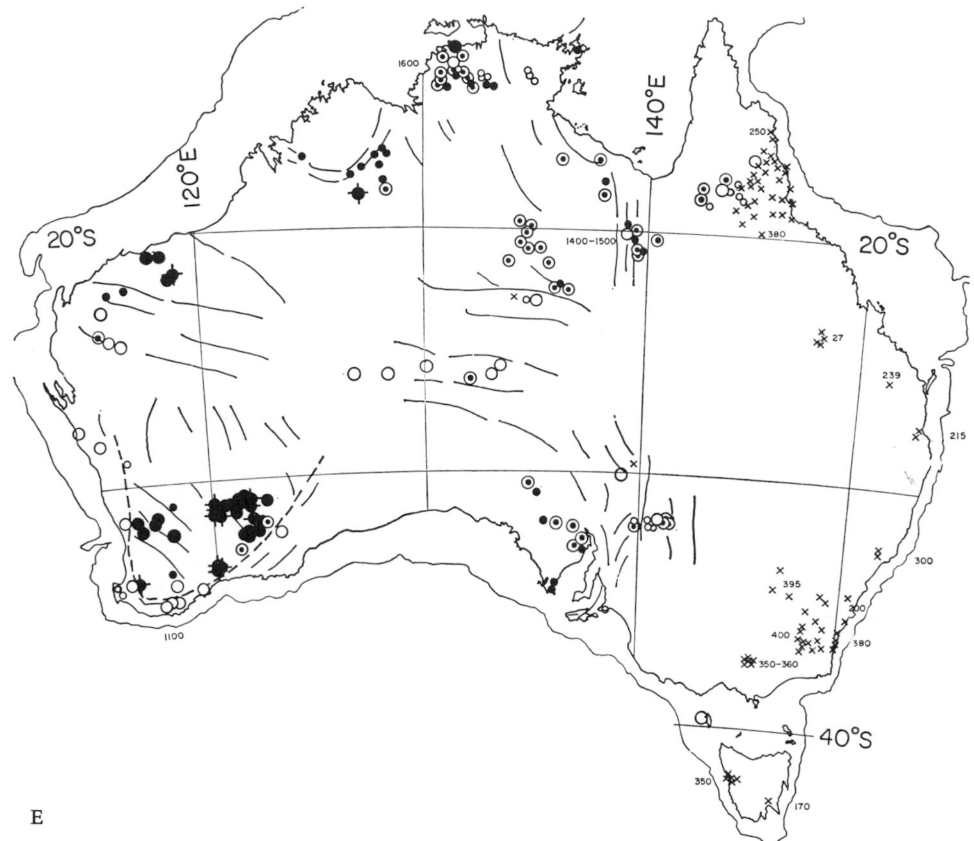

E

FIGURE 13.13E Isotopic Age Dates for Australia.

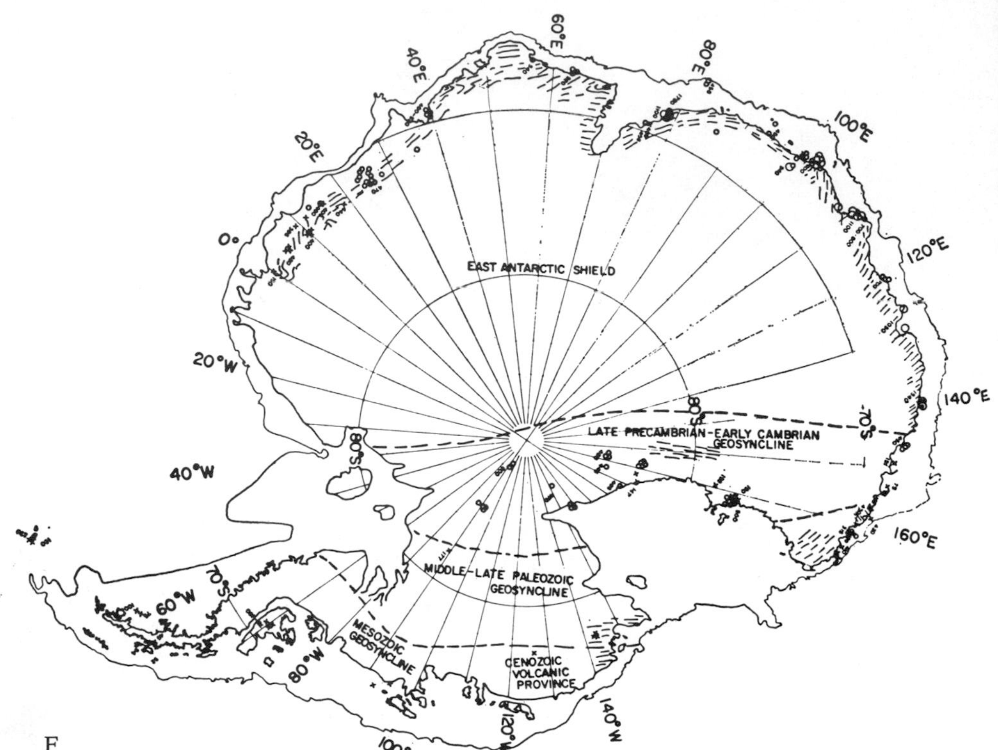

FIGURE 13.13F Isotopic Age Dates for Antarctica.

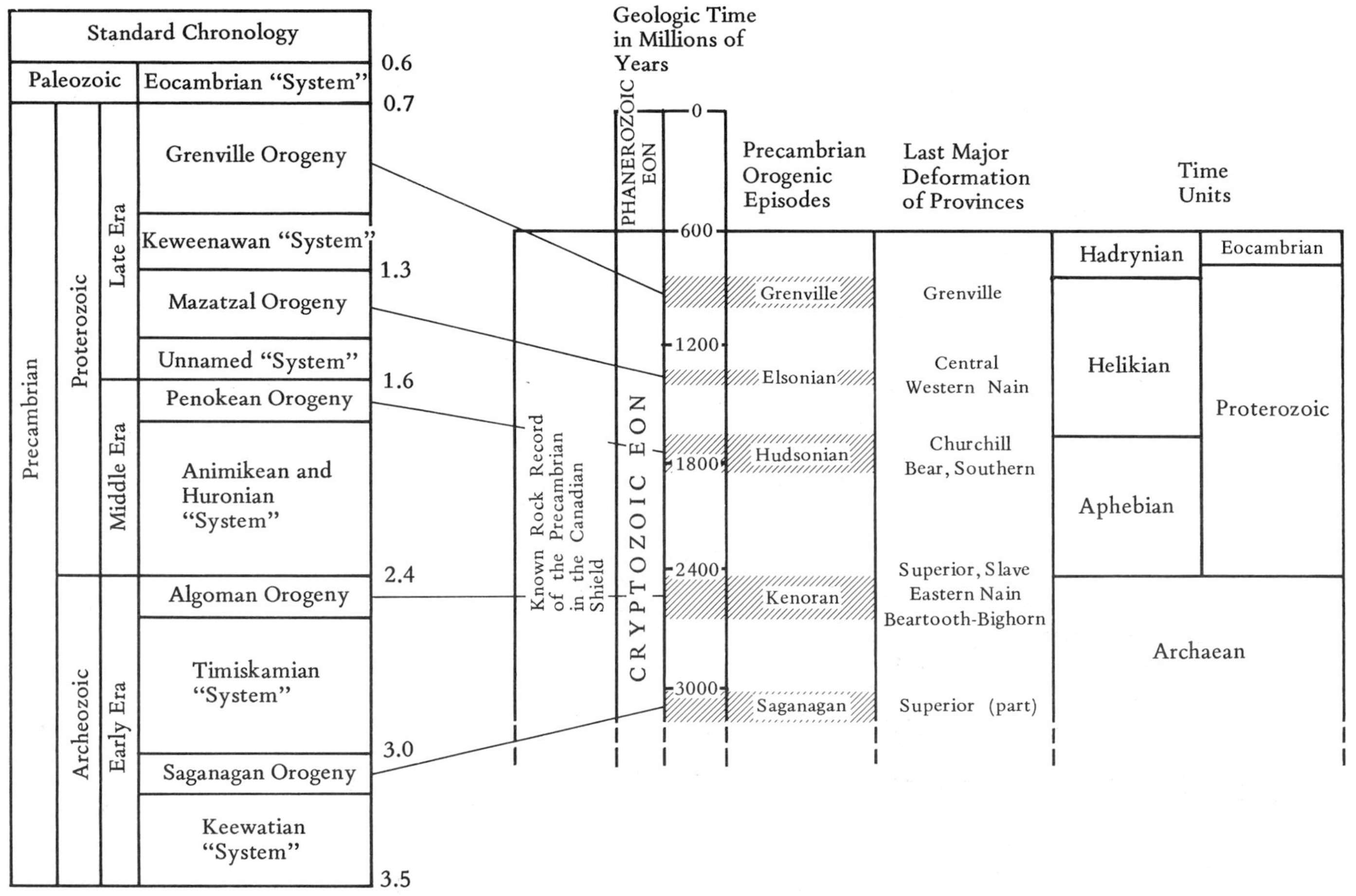

FIGURE 13.14 Precambrian History of Canadian Shield.

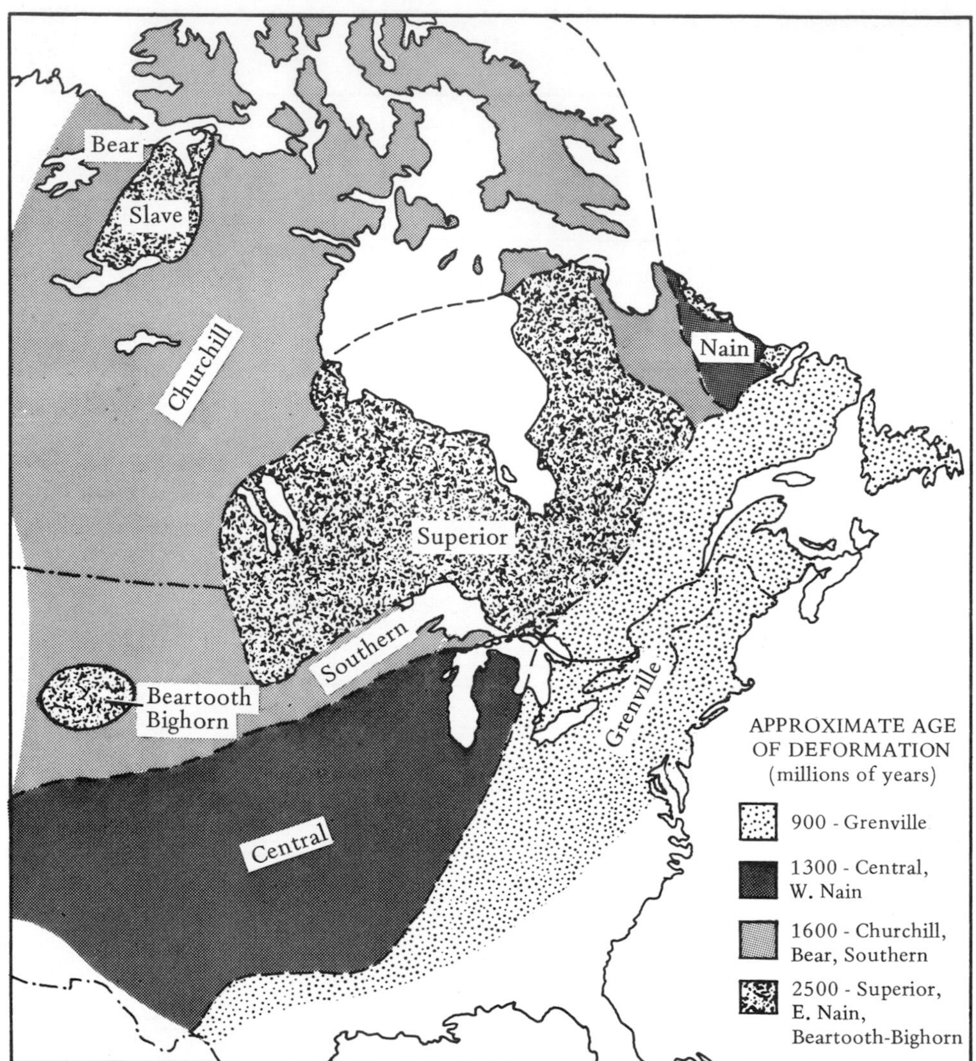

FIGURE 13.15 Precambrian Structural Provinces of North America.

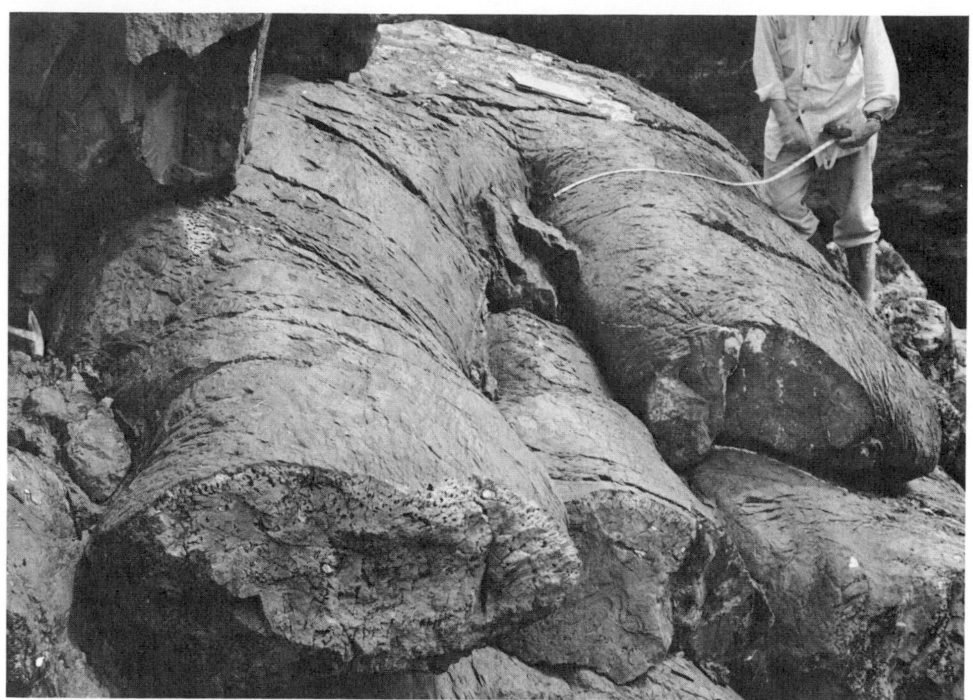

FIGURE 13.16 Keewatin Volcanic Rocks. (Courtesy Geological Survey of Canada, Ottawa.)

FIGURE 13.17 Grenville Limestone Rocks, Bancroft area, Ontario. (Courtesy Geological Survey of Canada, Ottawa.)

FIGURE 13.18 Belt Rocks, Glacier National Park, Montana. (Courtesy William M. Mintz.)

Questions

1. Explain the differences between the original definitions of Archeozoic and Proterozoic and their modern usages.
2. What are the major shields of the earth today and what is their significance?
3. What are the major Precambrian rock types and what may they signify?
4. What do geochemists believe is the major significance of the banded iron formations?
5. What is the evidence for Precambrian and Eocambrian glaciations?
6. How can the presence of Eocambrian tillites near the Equator be explained?
7. How do geochemists account for the rarity or absence of carbonates and sulfates in the Precambrian?
8. What are the Precambrian provinces of North America and when were they last deformed?

14

The Great Enigma:

Precambrian Versus Cambrian Life

The Oldest Fossils

We have noted earlier that the oldest known rocks, the approximately 3.5 billion-year-old *Swaziland Group* of South Africa, contain tiny spherical bodies resembling blue-green algae (Phylum Cyanophyta). The somewhat younger 3.1 billion-year-old *Fig Tree Group* which overlies it contains microscopic bodies (Fig. 14.1) which look like the remains of both *bacteria* and *blue-green algae.* In the 2.7–3.0 billion-year-old *Soudan Iron Formation* of Minnesota, similar bodies occur. Biologically produced organic compounds also occur in these rock units supporting the biologic origin of the spheres. Thus, at the beginning of the rock record, procaryotic, single-celled monerans had already evolved. The monerans even today are more resistant to ultraviolet radiation than other organisms which indicates that when life arose, these harmful rays reached the earth's surface. These fossils all occur in bands of cryptocrystalline quartz or chert, originally gelatinous silica, probably precipitated by the organisms to act as an ultraviolet shield. The bands of chert alternate with the silicic iron oxides of the iron formation which probably represent the disposal of harmful oxygen by primitive photosynthesizers. Iron probably also served as a non-biologic catalyst in the intracellular reactions that took place in the earlier cells which had not yet evolved their own enzymes.

The Gunflint Chert Biota

The oldest unequivocal remains of a diversity of microorganisms occur in the *Gunflint Chert* of the Canadian Shield. These abundant and structurally preserved fossils (Fig. 14.2) are approximately 2 billion years in age and include not only unquestioned bacteria, blue-green algae, and the ammonia-consuming *Kakabekia,* but some things that resemble green algae and fungus-like organisms and others of unknown origin. Some of these forms of uncertain relationships are probably algal cysts and are assigned to the unnatural group of *acritarchs* pending determination of their affinities.

Stromatolites

At about this same time, 2 billion years ago, the oldest stromatolites appeared. Stromatolites (Fig. 14.3) are the concentrically layered mounds of calcium carbonate constructed by blue-green algal colonies as a by-product of their metabolism and trapped by their mucilage. The height of their colonies is indicative of the thickness of the intertidal zone in which this type of algae live (Fig. 14.4). The Precambrian colonies are much higher than later ones, a fact which supports the idea that the moon was much closer to the earth then. The calcareous layers of stromatolites serve as an effective ultraviolet shield today, enabling the cyanophytes to live in very shallow marine waters. (The living forms also occur in fresh waters and on land.) Up until this time, organisms were undoubtedly virtually all pelagic, having originated in the organic soup of the oceans and occurring in the upper illuminated, photic zone where the energy to generate the first life and later support photosynthesis was available. Only where this narrow band of life in the oceans intersected the bottom

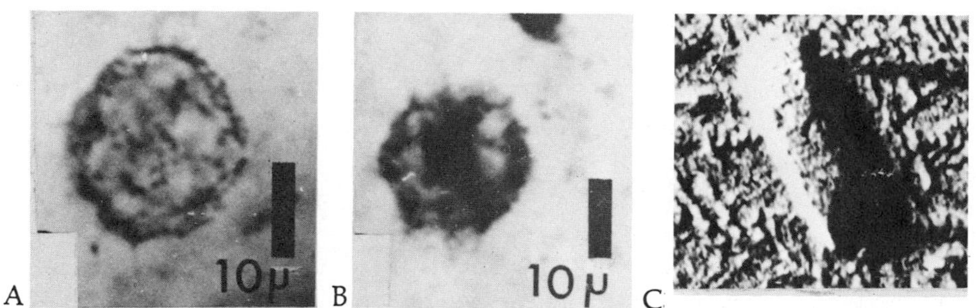

FIGURE 14.1 Fig Tree Fossils. (A-B) *Huronia,* an algal-like form, (C) *Eobacterium,* a bacterium. (Courtesy of E. S. Barghoorn, Harvard University.)

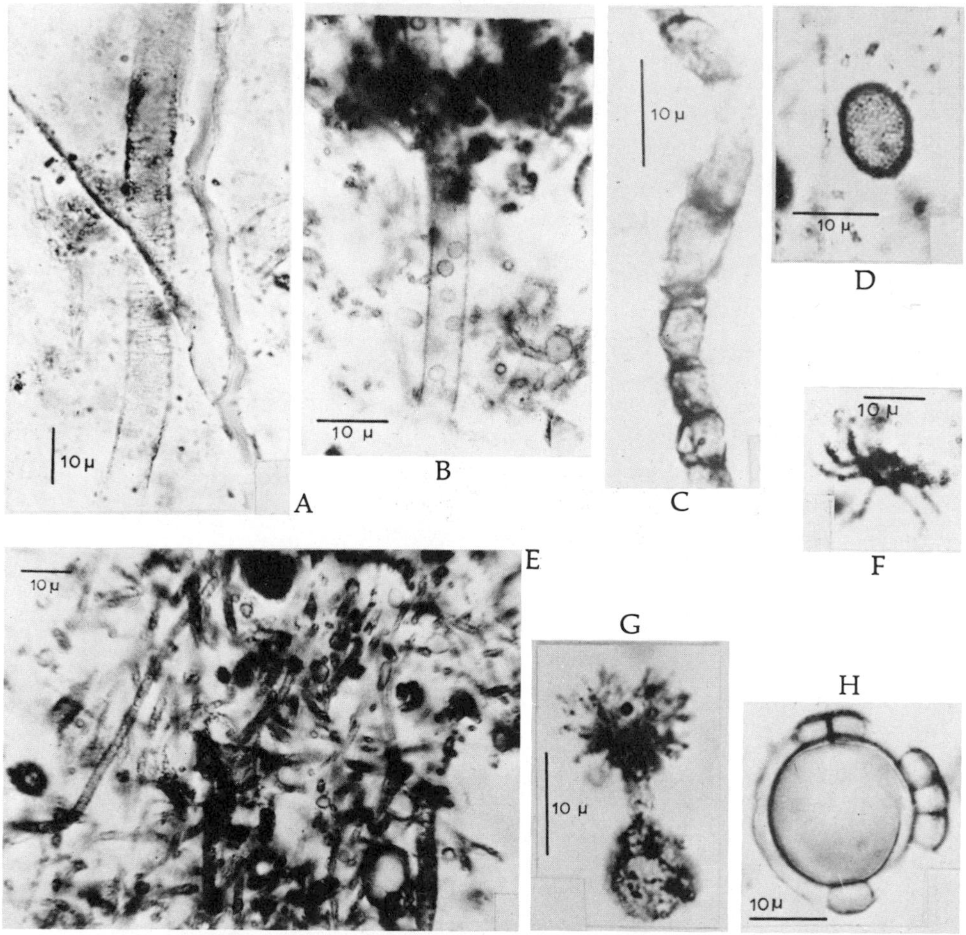

FIGURE 14.2 Gunflint Fossils. (A-C) Blue-green algae: *Animikia, Entosphaeroides* and *Gunflintia,* (D) *Huroniospora,* an algal spore, (E) *Gunflintia* and *Hurionospora,* (F) *Euastrion,* a bacterium, and enigmatic forms, (G) *Kakabekia,* (H) *Eosphaera.* (Courtesy of E. S. Barghoorn, Harvard University.)

FIGURE 14.3 Precambrian Stromatolites. (Courtesy Geological Survey of Canada, Ottawa.)

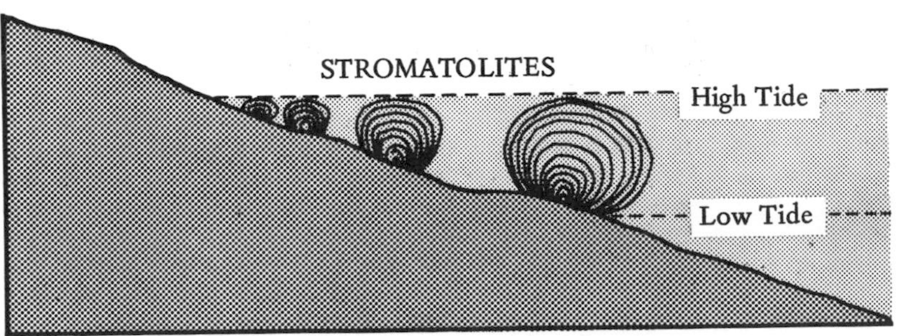

FIGURE 14.4 Schematic Profile Showing Relation of Domal Stromatolites to Intertidal Zone.

at their margins, in the sublittoral zone, could benthic organisms develop. Some workers have felt that the location of this narrow band was the thermocline, the zone of rapid temperature change that occurs in the upper levels of the ocean. Water in this zone may have had the proper density to support these floating organisms and been deep enough to be protected from ultraviolet radiation. At any rate, by 2 billion years ago, oxygen from the layer containing life must have been abundant enough to diffuse widely throughout the oceans for stromatolites and limestones gradually become abundant. They occur in profusion in Africa, Australia, Canada, China, Scandinavia, the Soviet Union, and the United States. Oxygen was also entering the atmosphere at this time as the banded iron formations cease and terrestrial redbeds begin to appear. As free oxygen (O_2) invaded the atmosphere, ultraviolet radiation converted it to ozone (O_3) which began to shield the earth's surface from the deleterious ultraviolet rays.

The Bitter Springs Biota

These conditions made it possible for the evolution of the more efficient and complex eucaryotic cell of protistans. In the one billion-year-old *Bitter Springs Formation* of Australia not only do we find the usual bacteria, blue-green algae, and organisms of uncertain affinities, but structurally preserved *green algae* (Phylum Chlorophyta) displaying nuclei and even nuclear division (Fig. 14.5). From this point onward in the Precambrian, numerous microfossil localities with similar organisms and biologically produced organic compounds occur. These are known in China, France, Michigan, Montana, Poland, Scotland, Sweden, the Soviet Union, and elsewhere.

Summary of Precambrian Fossils

The Precambrian interval is paleontologically characterized then by pelagic microfossils of monerans first and monerans and one-celled protistans later, and by benthic stromatolites produced by blue-green algae in its latter half. Search as they might, geologists have never been able to uncover any unquestioned megafossils of the larger, multicellular animals or metazoans in Precambrian rocks. Various markings and organic residues interpreted as jellyfish, brachiopods, worms, and arthropods were reported from time to time in Precambrian rocks, but all have been rejected either because they are of non-biologic origin, are not Precambrian in age, or were misidentified.

Cambrian Fossils

The Cambrian System stands in marked contrast to this dearth of fossils. At or near the base of the Cambrian, sponges, archaeocyathans, brachiopods, primitive mollusks of several extinct classes, trilobites, helicoplacoids, eocrinoids, and edrioasteroids occur (Fig. 14.6). Worm tubes, perhaps of annelids, also are common. Before the Cambrian was over, the coelenterates (hydroids, conularoids), snails, clams, cephalopods, chelicerate and mandibulate arthropods, crinoids, graptolites, conodonts, and some less important groups appeared. This remarkable change from pelagic microfossils and benthic stromatolites to a host of shell-bearing benthic metazoans is one of the great mysteries of geology.

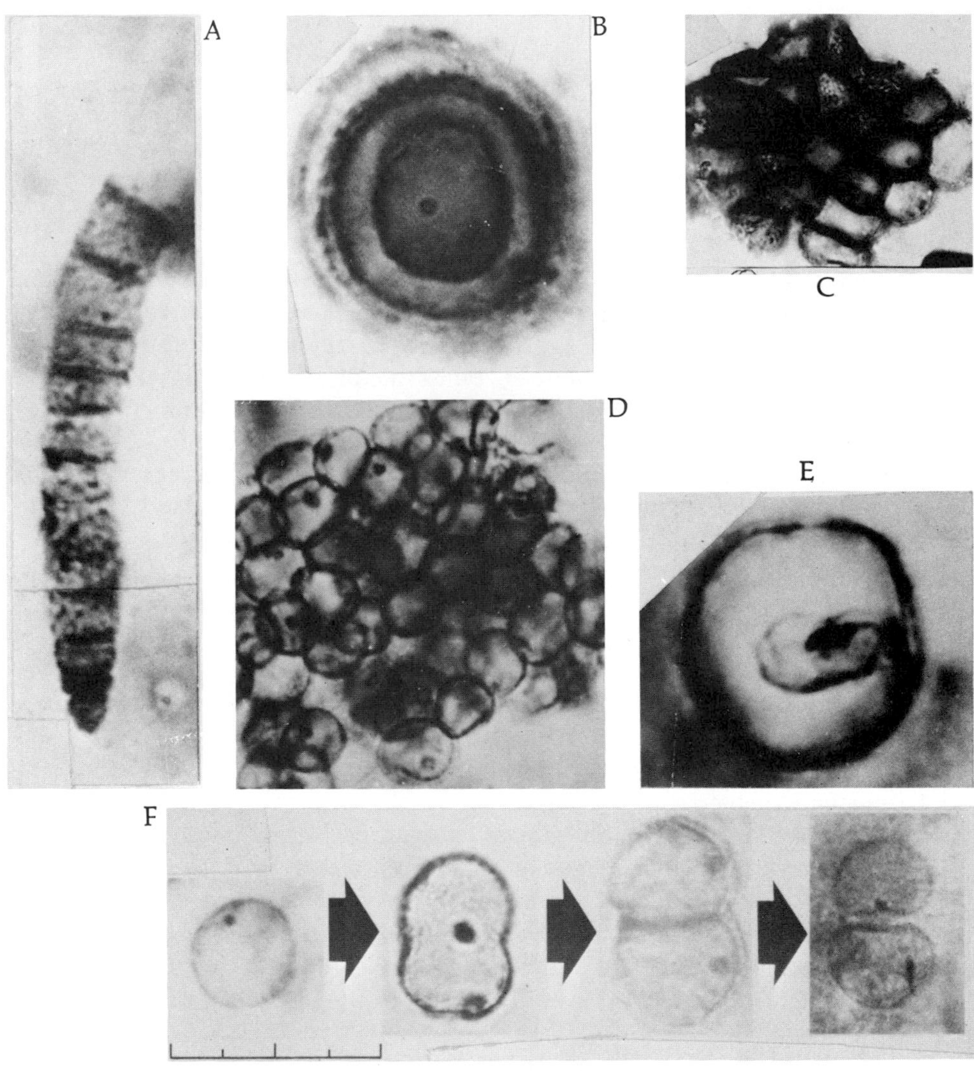

FIGURE 14.5 Bitter Springs Fossils. (A, C, D) Blue-green algae: *Cephalophytarion, Myxococcoides*, and *Palaeonacysts*, (B) ?Dinoflagellate *Gloeodiniopsis*, and (E, F) Green algae *Caryosphaeroides, Glenobotrydion*. (Courtesy of J. W. Schopf, U.C.L.A.)

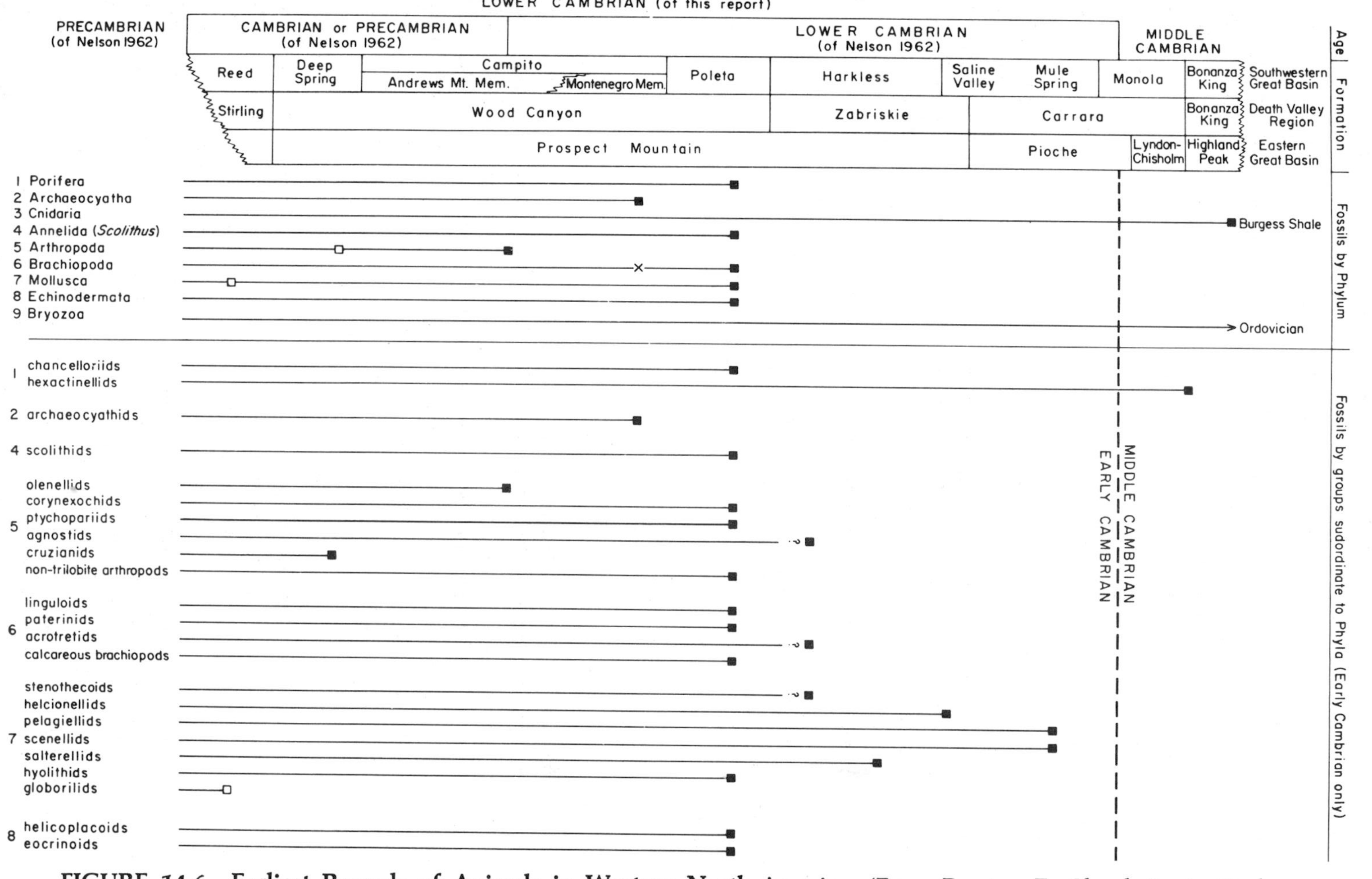

FIGURE 14.6 Earliest Records of Animals in Western North America. (From Preston E. Cloud, Jr. in Drake, *Evolution and Environment*, by permission of Yale University Press.)

Causes of the Difference

Many reasons have been advanced down through the years to explain this phenomenon. Some workers suggested that the metamorphism and igneous activity abundantly shown in Precambrian rocks had destroyed any traces of fossils that might have been preserved. But other geologists pointed to the virtually unaltered sedimentary rocks of Montana's Belt Group (Fig. 13.18) and Arizona's Grand Canyon Group (Fig. 13.2). Both contain stromatolites and no other megafossils. Other workers suggested that there was a lack of calcium carbonate in Precambrian seas or that the oceans were acidic. The presence of stromatolites and abundant Grenville carbonates refutes that notion. Some suggested life originated on land and that most Precambrian rocks were marine and thus could not record the event. As noted in Chapter 12, it appears that the physical-chemical conditions for the evolution of life as we know it could only have existed in the sea, however. In addition, Precambrian terrestrial redbeds are common in many areas. Still other geologists suggested that Precambrian fossils had either been overlooked or were so different from modern forms that we wouldn't easily recognize them. After several centuries of searching, it now appears that the first supposition is unlikely and the second, if true, would create two similar problems instead of only one. Finally geologists fell back on the idea of the Lipalian interval discussed in the last chapter. In this hypothesis, benthic metazoans originated in the ocean basins and/or geosynclines because the cratons were emergent. The discovery that the Lipalian interval was not worldwide and a careful search of geosynclinal rocks undercut this idea, too.

Eocambrian Fossils

Finally, in the last two decades, unequivocal impressions of metazoans have been found in the youngest Precambrian rocks. The largest and best-preserved fauna is known from Australia in rocks of so-called Adelaidian age. This fauna (Fig. 14.7), called the *Ediacara* after the Ediacara Hills where it was discovered, consists of coelenterates including scyphozoan and possible hydrozoan jellyfish and octocoral-like forms, segmented worms including some with head shields that may be transitional to trilobites, and a number of enigmatic animals that may have been conularoids, annelids, arthropods, or edrioasteroid echinoderms. Trace fossils of several varieties also occur. Evidently the creatures lacked hard parts as they are all preserved by molds and casts. Similar faunas of less diversity and poorer preservation have been found in England, South Africa, the Soviet Union, and in the Great Basin of the United States (Fig. 14.8). The age of these fossils has not yet been determined because they are not bracketed by datable intrusives or metamorphic rocks. In Australia and England, they are clearly beneath and, hence, older than rocks containing typical Cambrian fossils such as archaeocyathans and trilobites. The lowest trilobites are usually poorly calcified and beneath them are tracks probably left by trilobites but no actual remains occur suggesting that mineralized shells had not yet evolved. The base of the Cambrian has long been defined by the appearance of actual trilobite remains (redlichioids) and this boundary has been

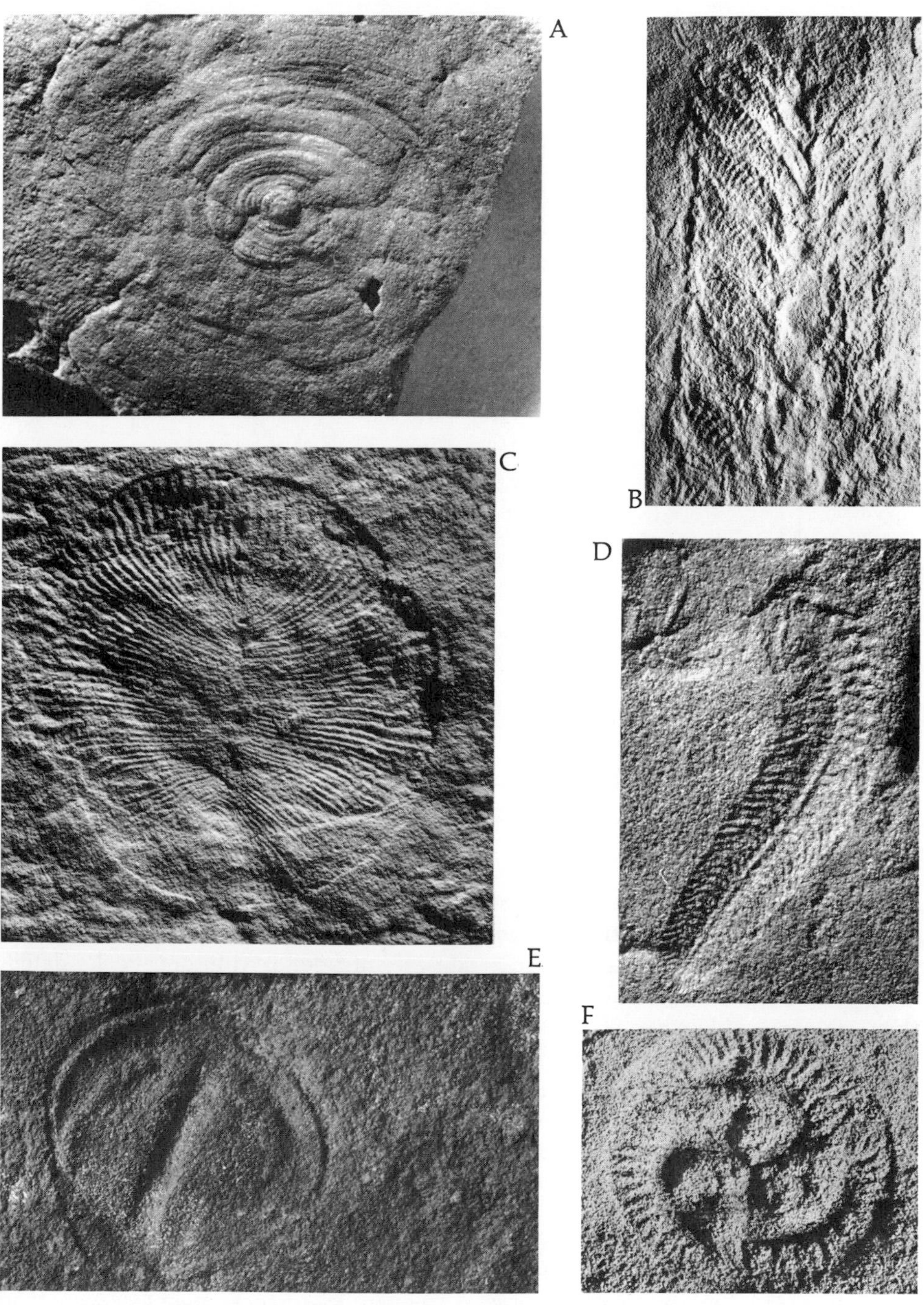

FIGURE 14.7 Ediacara Eocambrian Fauna. (A) *Cyclomedusa*, a jellyfish, (B) *Rangea*, an octocoral, (C) *Dickinsonia*, a jellyfish or a worm, (D) *Spriggina*, a worm, (E) *Parvancorina*, an arthropod, (F) *Tribrachidium*, possibly an echinoderm. (Photographs courtesy of M. F. Glaessner.)

FIGURE 14.8 Geographic Distribution of the Ediacara Fauna in the Old World.

roughly dated by radiometric means at 600 million years ago. The Ediacara fauna is not much below this boundary and thus is almost certainly not older than 700 million years. The English strata are intruded by plutonic rocks dated at between 574 and 684 million years old and the Soviet localities are older than 590–670 million-year-old glauconites found in the same sequences. These dates also support a 600–700 million-year age for the Ediacara fauna.

The chronostratigraphic assignment of the Ediacara fauna has been much disputed. The earliest workers placed it in the very basal Cambrian because it contains metazoans, the appearance of which marks the beginning of that period. *Glaessner,* the man who has studied the fossils and their occurrence in most detail, thinks they are Late Precambrian. They are certainly pre-Cambrian in the sense that they antedate the trilobite remains used to mark the basal Cambrian. On the other hand, they fall more clearly within the literal definition of the Phanerozoic as opposed to the Cryptozoic Eon because they contain megafossil animals. Rocks of latest Precambrian age are called *Eocambrian* in parts of Europe and many geologists feel that a separate period and system with this name should be established for the Ediacara Fauna and its correlatives. The period is usually assigned to the Paleozoic Era. This gives an Eocambrian-Cambrian boundary, but still does not explain the abrupt appearance of metazoans in the rock record.

It also creates a problem in terminology because now there is a pre-Cambrian part of the Paleozoic Era. Strict interpretation of the derivation of a word means that now the word "Precambrian" is no longer available for the vast interval before the Eocambrian, and that Prepaleozoic must be substituted for the venerable term. But Prepaleozoic is still Precambrian and the latter word is so ingrained in geological usage that it will probably never be replaced. Nor is there any strict reason why it should. If biologists changed the name of a phylum every time they added or subtracted a species from it, chaos would result. In the interest of nomenclature stability, this book will continue to use Precambrian for pre-Eocambrian time and rocks, while recognizing an Eocambrian Period and System assigned to the Paleozoic.

New Theories

This still does not resolve the issue of why the metazoans occur so late and relatively suddenly in geologic history. Their late occurrence has been correlated with the attainment of sufficient levels of atmospheric oxygen required for the metabolism of such large and complex organic entities. *Berkner* and *Marshall* and others have suggested that not until the amount of atmospheric oxygen reached 1–3% of its present level could metazoans evolve. This could occur relatively rapidly once the level of atmospheric oxygen had built up to this point because when it did, the harmful ultraviolet radiation would be screened out and organisms could invade much more of the upper levels of the sea. This would cause a dramatic increase in atmospheric oxygen as the photosynthesizers rapidly filled the new space by increasing in numbers. The sudden increase in oxygen not only could have caused the evolution of metazoans, but probably triggered a corresponding decrease in carbon dioxide, because the two gases tend to be mutually exclusive. As stated earlier, this decrease in CO_2 may have caused a decrease in temperature leading to the ice age that occurred around the Precambrian-Eocambrian boundary. Thus, the physical and biological histories of the earth could be inextricably linked again as they are throughout geologic history. One worker's summary of the above theories is shown in Fig. 14.9.

Weyl has recently revived another plausible explanation for the rapid appearance of the metazoans by giving it a new twist. He reasoned that most early organisms were pelagic because life originated in the organic soup, photosynthesizers had to remain near the surface to receive sunlight (but not at it because of ultraviolet radiation), and their density held them in the thermocline or zone of rapid temperature change. There were no predators, with the possible exception of protozoans, so shells were not necessary for protection. Furthermore, a shell would cause an organism to sink and this would be lethal when the oxygen was still concentrated in the thermocline and above because it was not yet generated in enough abundance to diffuse throughout the oceans. Thus, throughout the Precambrian, most life was shell-less and pelagic. Tiny floating microorganisms predominated except where the thermocline intersected the bottom in a relatively narrow band around the oceans where the early stromatolite-forming algae lived. Only when oxygen had been produced in enough abundance and diffused downward through the thermocline and throughout the water masses of the oceans could an extensive benthonic fauna develop. According to Weyl, this occurred at the beginning of the Eocambrian when benthos first appear. A shell was now a favorable adaptation to keep the

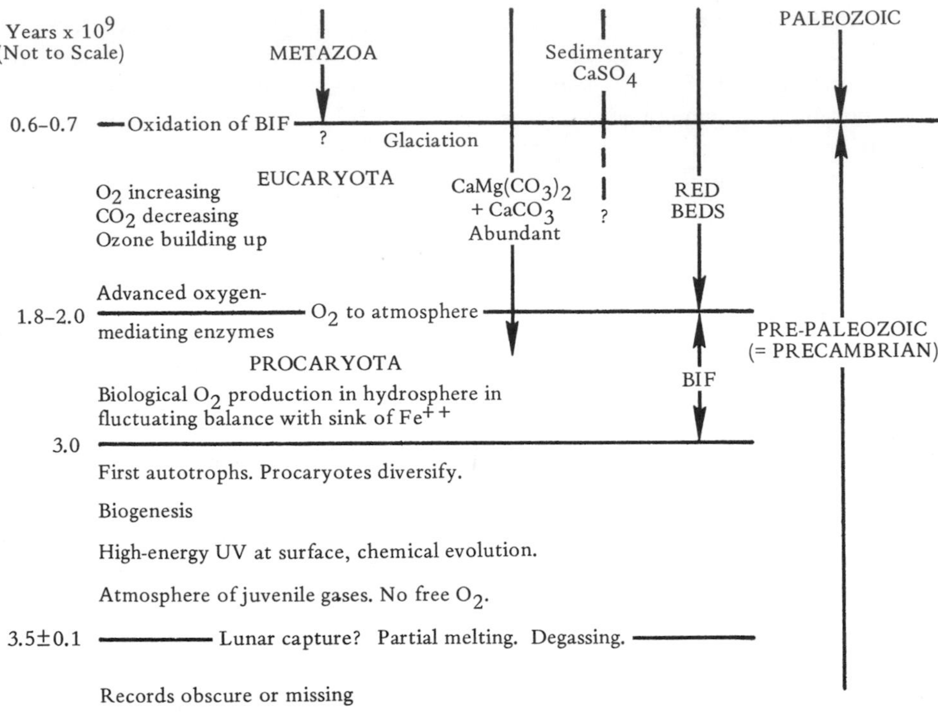

FIGURE 14.9 One Theory of the Origin and Interrelationships of the Major Groups of Organisms. (From Preston E. Cloud, Jr. in Drake, *Evolution and Environment*, by permission of Yale University Press.)

animal on the bottom and several lineages independently evolved the ability to mineralize shells and skeletons by or during the Cambrian. Large size was no longer the hindrance it is to floaters, so multicellular organisms could evolve.

Weyl's hypothesis is supported by the fact that there are no metazoan trails or tracks known before the Eocambrian. If metazoans had been around earlier without shells they would certainly have left trails or tracks, but none are known. Yet they appear in some abundance in the Eocambrian indicating that previous to that time organisms were probably pelagic, hence, tiny and shell-less as postulated by Weyl. However, the presence of abundant and widespread blue-green algae, limestone and redbeds in the latter half of the Precambrian rock record argues that oxygen was generated in some abundance considerably earlier than the Eocambrian. Perhaps, however, it diffused upward into the upper levels of the water and into the atmosphere before it could diffuse downward to any appreciable degree.

For whatever reason caused their apparently sudden appearance in the record, multicellular organisms spread widely in the Eocambrian and by and during the Cambrian had evolved into nearly all of today's major phyla and several probably totally extinct ones (Fig. 14.10). The Cambrian fauna, even

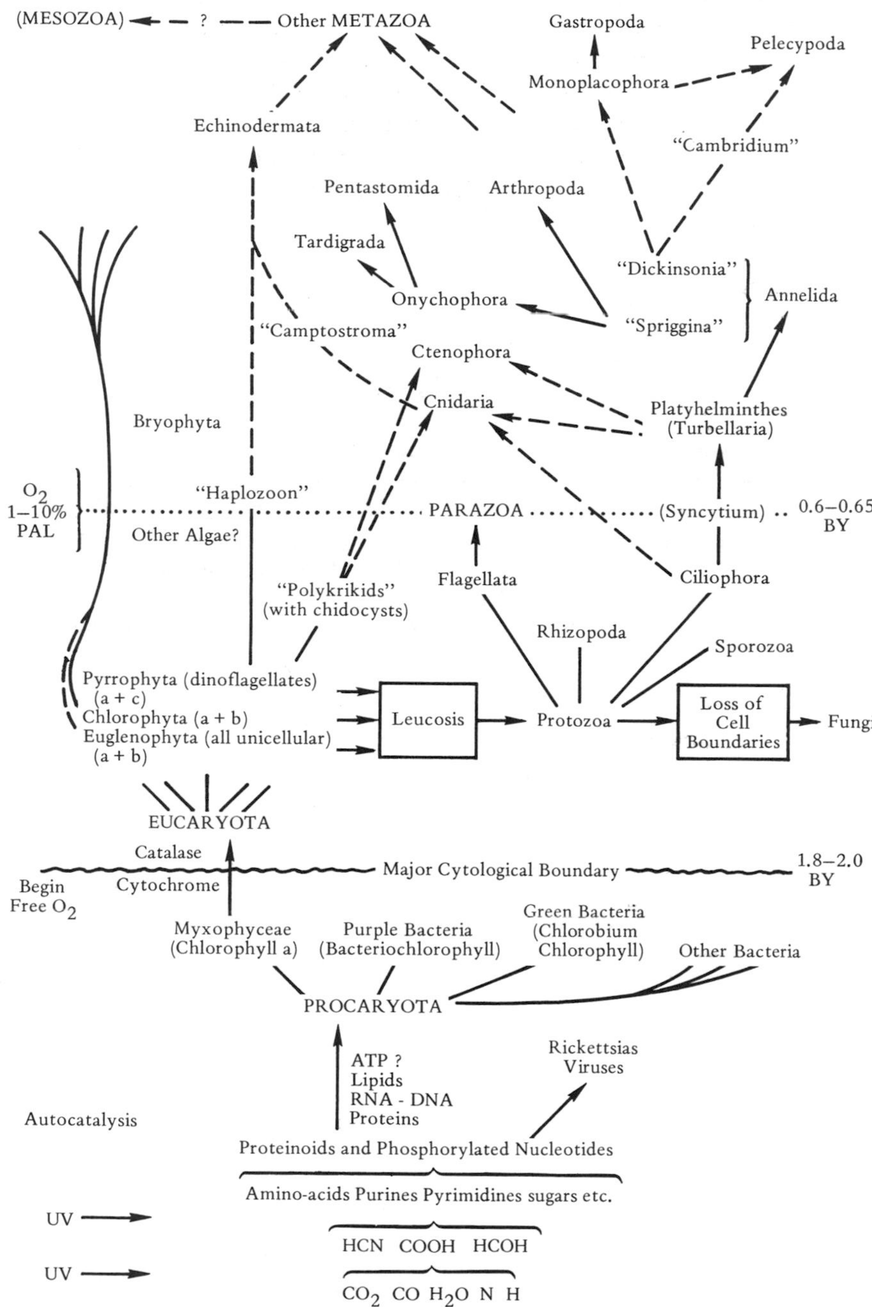

FIGURE 14.10 Early Evolution of the Earth, Life, and Atmosphere. (From Cloud, *Science*, vol. 160, 17 May, 1968, Copyright 1968 by AAAS. Used by permission of author and AAAS.)

though it contains all modern phyla with hard parts except possibly the vertebrates (although fish-like teeth have been reported from the Upper Cambrian of Wyoming), is largely composed of unfamiliar and extinct orders and even classes and phyla (Fig. 14.6). These include many groups of arthropods, echinoderms, probable mollusks, and the archaeocyatha. Most of these primitive forms disappeared or diminished appreciably by or during the course of the Ordovician. Thus, the Cambrian appears more like the close of the great initial radiation of metazoans than its beginning. Cambrian life will be discussed in detail in Chapter 17.

Questions

1. What are the oldest known evidences of life on earth?
2. What is the significance of the Gunflint Chert biota?
3. What are stromatolites and when do they appear in the geologic record?
4. What is the oldest unequivocal evidence of eucaryotic organisms?
5. Contrast the typical Precambrian biota with that of the Cambrian.
6. What is the significance of the Ediacara fauna?
7. What is the Berkner-Marshall theory of the origin of multicellular animals?

15

The Transition Mountains (Ubergangsgebirge):

Cambrian, Ordovician, and Silurian

THE THREE PERIODS OF THE EARLY PALEOZOIC THAT SUCCEED THE EOCAMBRIAN, THE *Cambrian, Ordovician,* and *Silurian,* encompass approximately 200 million years of earth history from about 600 million to 400 million years ago. All three systems have their type areas in Wales, in or marginal to the Caledonian Geosyncline (Fig. 2.19). The geologic history of individual regions in this interval is moderately well known, but geologists are just now starting to interpret this history in the larger framework of the new global tectonics using paleomagnetism, computer fits of shorelines, and more conventional geologic tools such as the matching of lithologic and biologic trends.

Plate Positions and Motions

The evidence suggests that *Gondwanaland,* composed of present-day South America, Africa, India, Antarctica, and Australia was in existence at this time and probably somewhat earlier as well. Late Precambrian-Early Paleozoic geosynclinal bands match between Brazil and west Africa (Fig. 11.34); between eastern South America, South Africa, west Antarctica and east Australia (Figs. 11.36, 15.1); and between east Africa, India, east Antarctica, and west Australia (Fig. 15.1). A great mountain-building event, the *Pan-African Orogeny,* occurred between Middle Eocambrian and Middle Ordovician time. It deformed

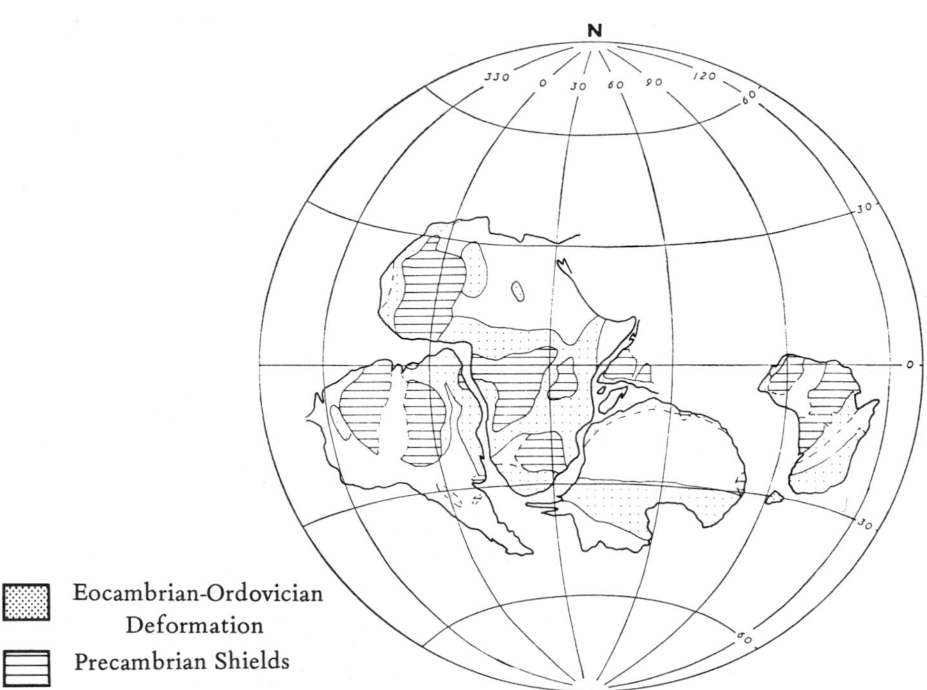

FIGURE 15.1 Reconstruction of Gondwanaland in the Paleozoic. Note zones of Early Paleozoic (Eocambrian-Ordovician) deformations indicating formation of the supercontinent at that time. (From McElhinny and Luck, *Science,* vol. 152, 15 May, 1970, Copyright © by AAAS. Used by permission of authors and AAAS.)

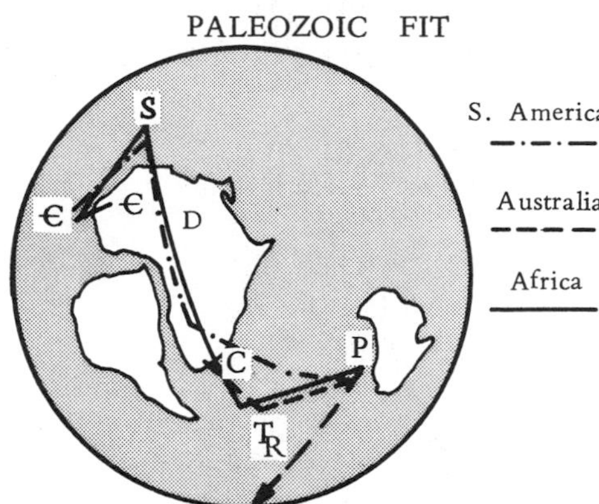

FIGURE 15.2 Partial Reconstruction of Gondwanaland in Paleozoic and Early Mesozoic showing Overlapping Polar-Wandering Curves.

areas between all the pre-existing Gondwana continents indicating their collision in this interval. The polar wandering curves for the Early Paleozoic interval are superimposed if the present southern continents are hypothetically reunited into Gondwanaland (Fig. 15.2). Much of Gondwanaland was in the South or Gondwana polar region during the Cambrian-Silurian interval as indicated by the paleomagnetic data (Fig. 15.3). The pole "migrated" southeasterly from northwest to south Africa during this time. (Actually, of course, Gondwanaland was moving to the northwest.) The present-day continent of Australia was at the northern edge of Gondwanaland and on the equator (Fig. 15.4).

The positions of the present Northern Hemisphere continental masses are less well known. Paleomagnetic and paleoclimatic evidence suggests that North America was turned about 90° from its present position so that the equator passed through the continent in approximately the present north-south direction (Figs. 15.5, 15.6). The continent gradually rotated counter-clockwise through the Early Paleozoic. The rock record indicates a proto-Atlantic Ocean opened in the Eocambrian and Cambrian (Fig. 15.13), an event which probably triggered the extensive Cambrian epeiric seas of North America. In the Ordovician and Silurian (and continuing through the Devonian), this ocean basin closed as the European craton approached North America and Laurasia began to form. The zone occupied by this compressed ocean basin and crumpled continental margins was the *Appalachian* (including the Ouachita belt)-*Caledonian Geosyncline* (Fig. 15.7), a long mobile belt that was split into its current eastern North American and northwest European parts by the Mesozoic rifting that produced the modern Atlantic Ocean. The Paleozoic zone of collision clearly did not parallel the modern shores as part of the ancient geosyncline and both cratons are now found on both sides of the Atlantic. Paleomagnetic evidence (Fig. 15.8) indicates that the European and Siberian cratons were not in proximity in the Early Paleozoic. The Uralian Geosyncline that formed between them as they approached in the Late Paleozoic did not exist earlier in the era. Instead, there was an ocean basin between them with an offshore island arc or arcs to the present east of the European craton and a trench-subduction zone

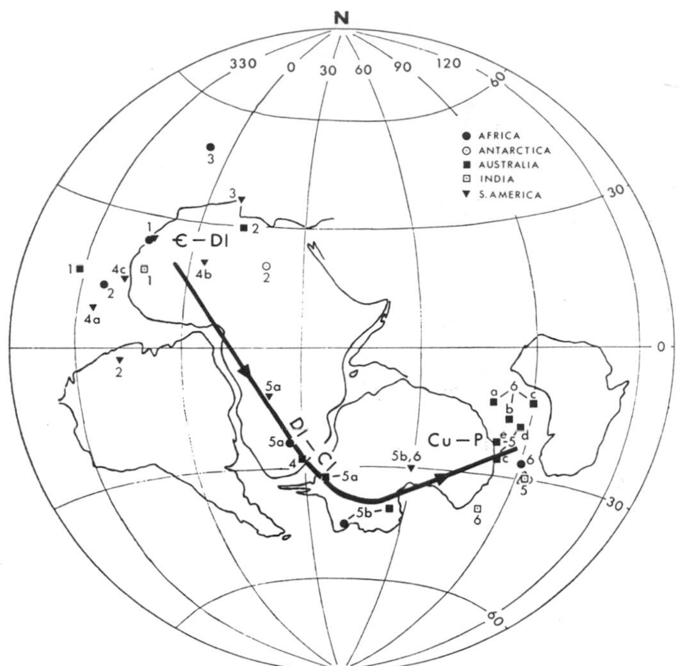

FIGURE 15.3 Reconstruction of Gondwanaland by Superposition of Polar Wandering Curves. Polar positions for various periods indicated. (From McElhinny and Luck, *Science*, vol. 152, 15 May, 1970, Copyright © by AAAS. Used by permission of authors and AAAS.)

complex bounded the present western margin of the Siberian craton. The situation on the European side probably resembled that of the modern northern Philippine Trench (Fig. 15.9) while that on the Asian side resembled the current Pacific side of South America (Fig. 4.7). Fragments of both these complexes were brought together in the Late Paleozoic Ural belt. The relationship of the three northern continents in the present North Pole region in the Early Paleozoic is uncertain, but a proto-Arctic Ocean was probably present because the Late Paleozoic Franklin Geosyncline contains ocean basin sediments (in addition to shelf facies) of Cambrian-Silurian age. This ocean narrowed as the Middle Paleozoic approached, however, for faunas of North America and Asia show many similarities and there was extensive mountain-building in the Franklin Geosyncline in the Devonian-Mississippian interval.

The question of the relationship between Gondwanaland and the present northern continents is not answerable with certainty. The present southern margin of Europe was bounded by the *Hercynian Geosyncline* in the Paleozoic. A plate (or plates) was apparently colliding with Europe from the present southerly direction. The margins of this plate, probably the one containing Gondwanaland, were apparently oceanic with the possible exception of the area of present northwest Africa-northern South America where rocks and fossils similar to those of Europe are found. This does not prove, however,

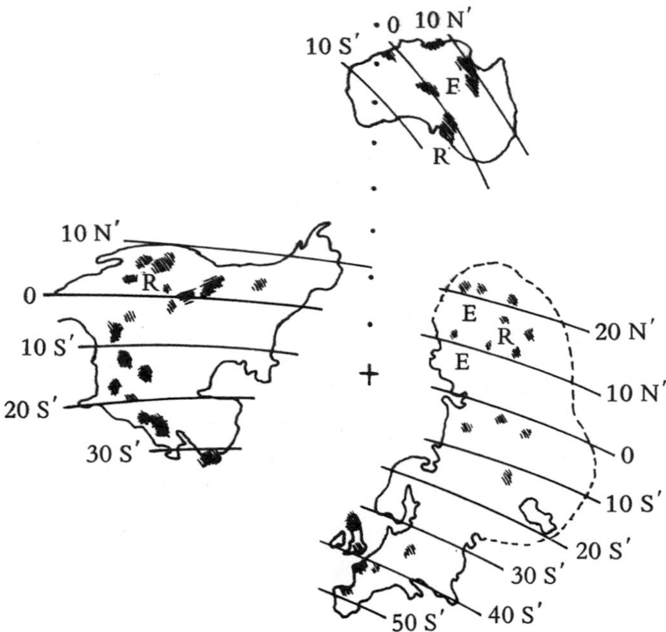

FIGURE 15.4 Latitudes for North America, Eurasia, and Australia in the Cambrian. (Figs. 15.4–15.6 from Irving, *Paleomagnetism and Its Application to Geological and Geophysical Problems*, Copyright © 1964, by permission of John Wiley & Sons, Inc.)

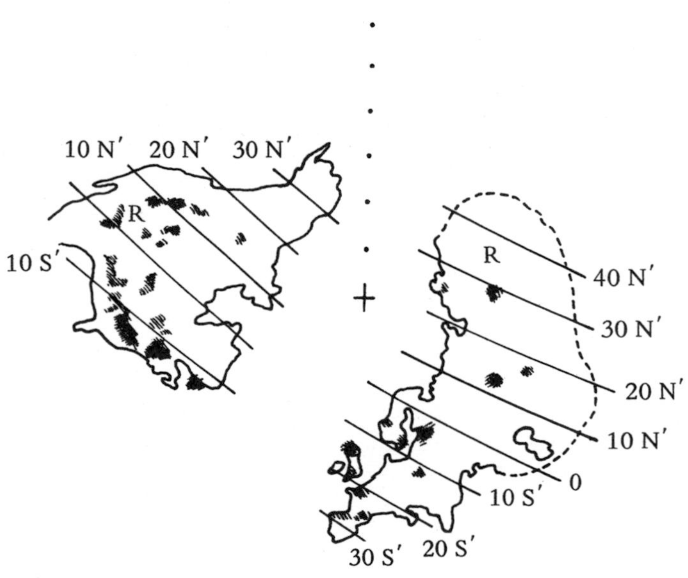

FIGURE 15.5 Latitudes for North America and Eurasia in the Ordovician. (From Irving, see Fig. 15.4)

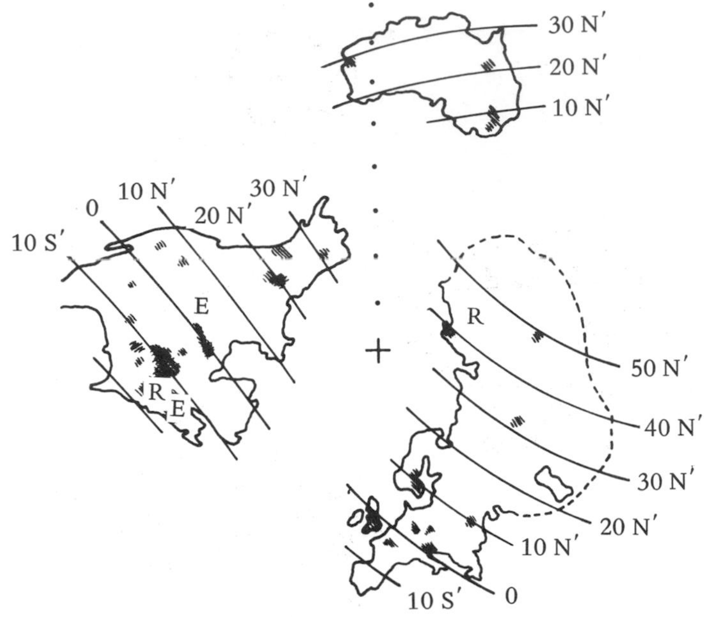

FIGURE 15.6 Latitudes for North America, Eurasia, and Australia in the Silurian. (From Irving, see Fig. 15.4)

that Gondwanaland and Europe were in close proximity in the Early Paleozoic. The strata could have formed elsewhere and been plastered against Africa during the Late Paleozoic collision of Europe, North America, and the African part of Gondwanaland (see Chapter 17). The present southern margin of Asia was bound by the *Mongolian Geosyncline* during the Paleozoic. Hence, a plate (or plates) was also impinging on Asia from the present southerly direction. This may have been the plate containing Gondwanaland also, for the Australian part of the supercontinent was in low latitudes as was Siberia and the two share many faunal similarities (Fig. 15.10).

The Pacific Ocean, probably in much larger form, must have been in existence at this time to take up all the remaining space left on the globe. The *Cordilleran Geosyncline* of western North America (Figs. 15.11, 15.12), the *Andean Geosyncline* of western South America, the *Cape Geosyncline* of South Africa, and the *Adelaide* and *Tasman Geosynclines* of eastern Australia, which continued into Antarctica as the *Ross* and *Borchgrevink* respectively, were all in existence in the Lower Paleozoic (Fig. 9.3). This indicates that a Pacific plate or plates were actively impinging on these areas at the time. The situation was probably similar to that in the eastern South Pacific today (Fig. 9.7). The paleomagnetic evidence indicates that the North or Pacific Pole was in the ancestral central Pacific Ocean in the Cambrian-Silurian interval.

Caledonian-Appalachian Area

Several major physical events occurred in the Cambrian, Ordovician, and Silurian history of the earth. As noted above, the Cambrian was a time during which

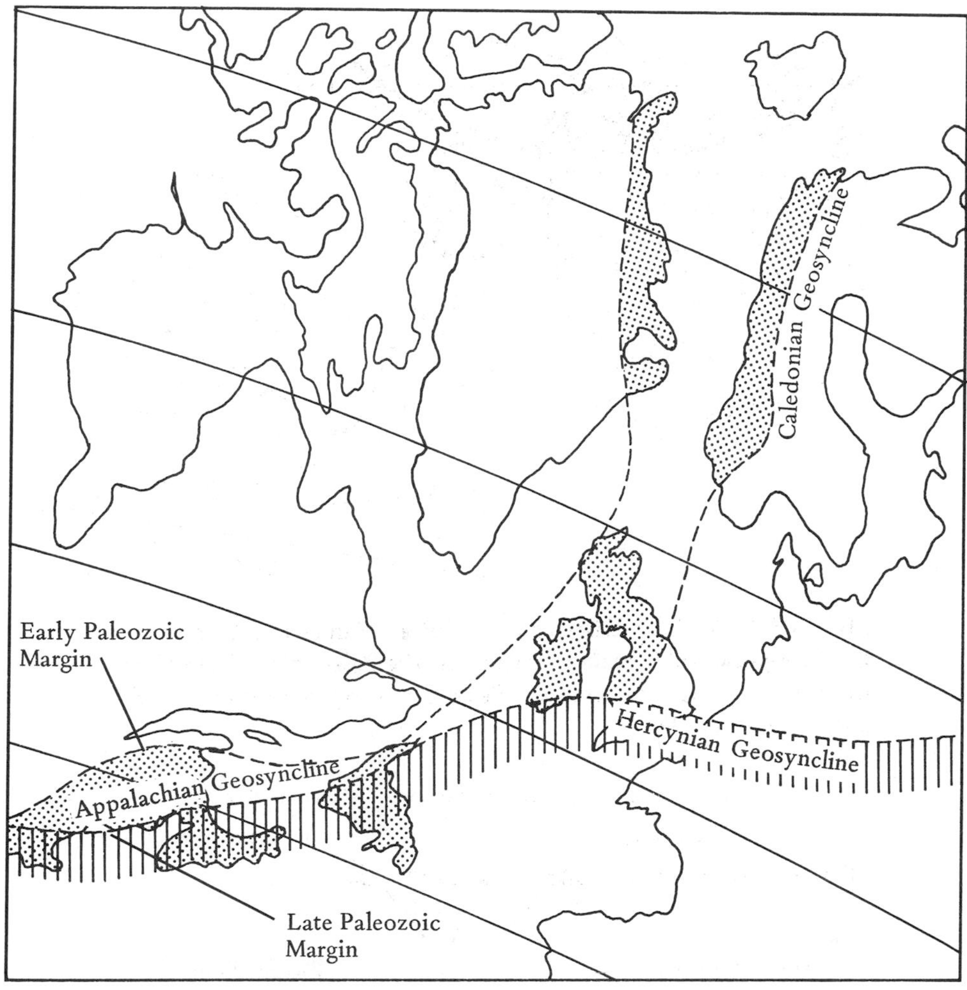

FIGURE 15.7 The Continuity in Structure of the Taconic-East Greenland-Caledonian Mountains and the Appalachian-Hercynian Mountains. Continents reassembled to Paleozoic positions.

a proto-Atlantic Ocean opened (Fig. 15.13), an occurrence which may have triggered the great epeiric sea invasion of North America in that period. The Ordovician-Silurian closing of the proto-Atlantic produced the great Caledonian-Appalachian Geosyncline (Fig. 15.14) which stretched from present-day northern Greenland and northern Norway, down along present east Greenland and the length of Norway, through Great Britain and Ireland, and down the present eastern side of Northern America from Newfoundland through what is currently southeastern United States and on to the present Ouachita Mountains. The extreme northwest Africa of today was also probably part of this belt. This belt was unstable throughout the Ordovician-Silurian interval with constant mountain-building disturbances occurring. The Early Ordovician dis-

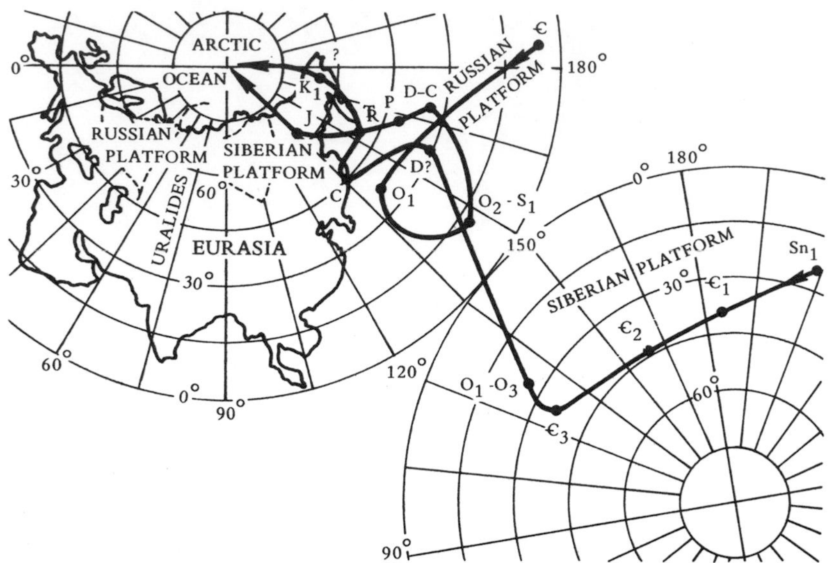

FIGURE 15.8 The Disparity of Polar Wandering Curves for the European and Siberian Cratons in the Paleozoic. (From Hamilton, *G.S.A. Bulletin,* vol. 81, Sept. 1970. By permission of the author and G.S.A.)

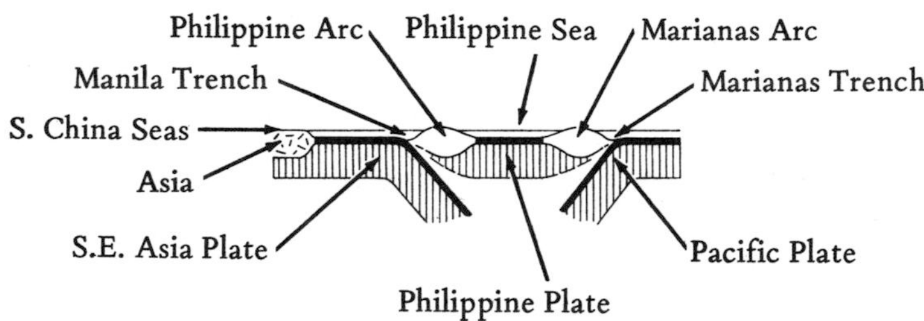

FIGURE 15.9 Cross-Section of the Philippine Plate Region Today. (From Dewey and Bird, *Journal of Geophysical Research,* vol. 75, no. 14, May 10, 1970. By permission of authors and AGU.).

turbances are called the *Penobscot Orogeny,* the Middle-Late Ordovician ones, the *Taconic Orogeny,* and the Late Silurian ones, the *Caledonian Orogeny.* Apparently North America was colliding with the margins of two other plates, both oceanic at this time, but bearing Europe and Africa on them (Fig. 15.13). These disturbances continued into the Late Paleozoic as the African-South American edge of Gondwanaland became a major contributor to the three-way collision. The Silurian-Devonian period of mountain-building produced the most intense and final deformation of the northern Appalachians and the Caledonian Mountains as Europe and North America came into contact.

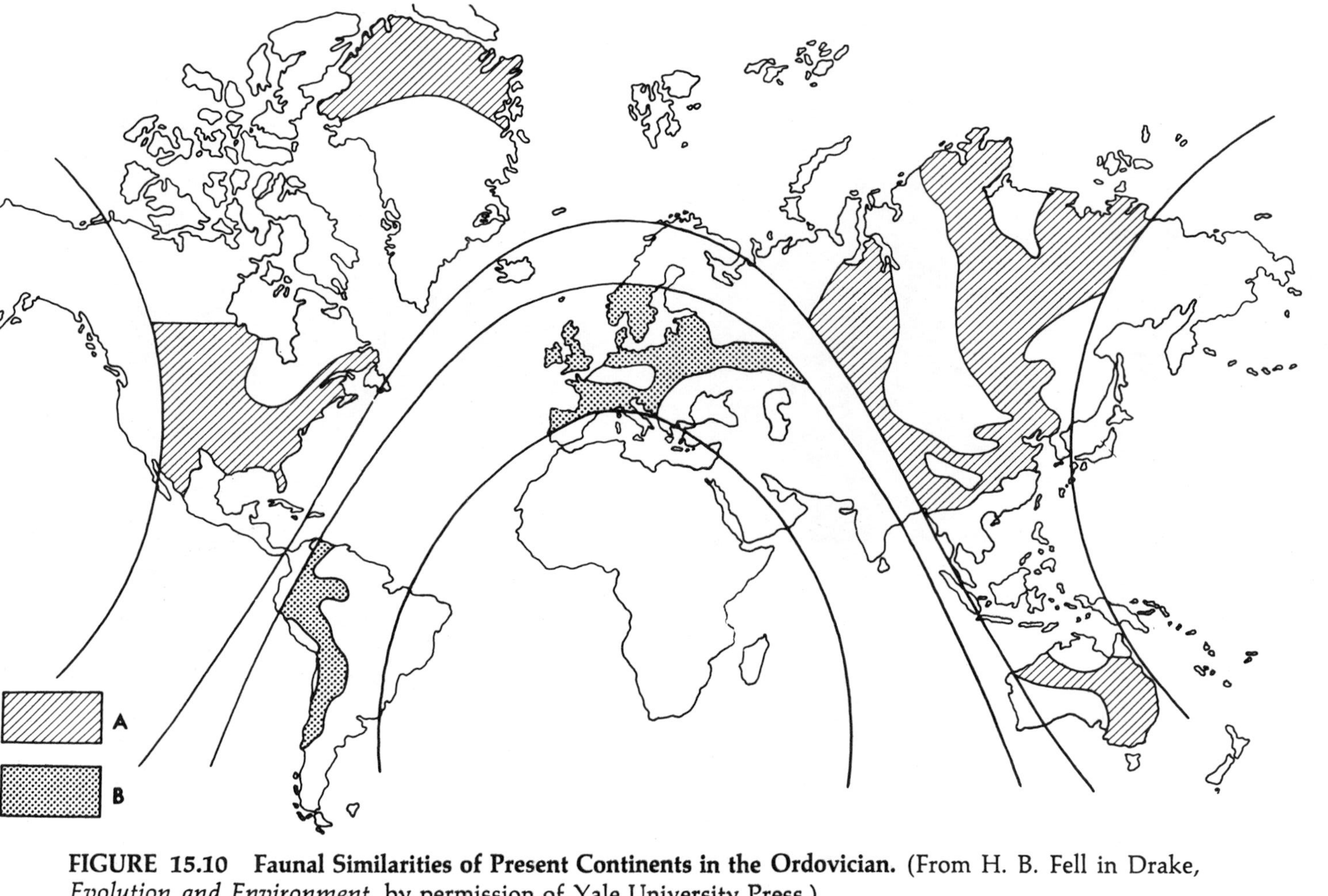

FIGURE 15.10 Faunal Similarities of Present Continents in the Ordovician. (From H. B. Fell in Drake, *Evolution and Environment*, by permission of Yale University Press.)

A

B

FIGURE 15.11 Lower Paleozoic Rocks in Canadian Rockies. (A) Mt. Stephen, Yoho National Park, British Columbia, (B) Mt. Eisenhower, Banff National Park, Alberta. Strata are chiefly Cambrian. (Courtesy British Columbia Dept. of Travel Industry and Alberta Dept. of Industry and Tourism.)

A

B

FIGURE 15.12 Lower Paleozoic Rocks in Canadian Rockies. (A) Moraine Lake, Valley of the Ten Peaks, (B) Lake Louise, Mt. Victoria, Banff National Park, Alberta. Strata are chiefly Cambrian. (Courtesy Alberta Dept. of Industry and Tourism.)

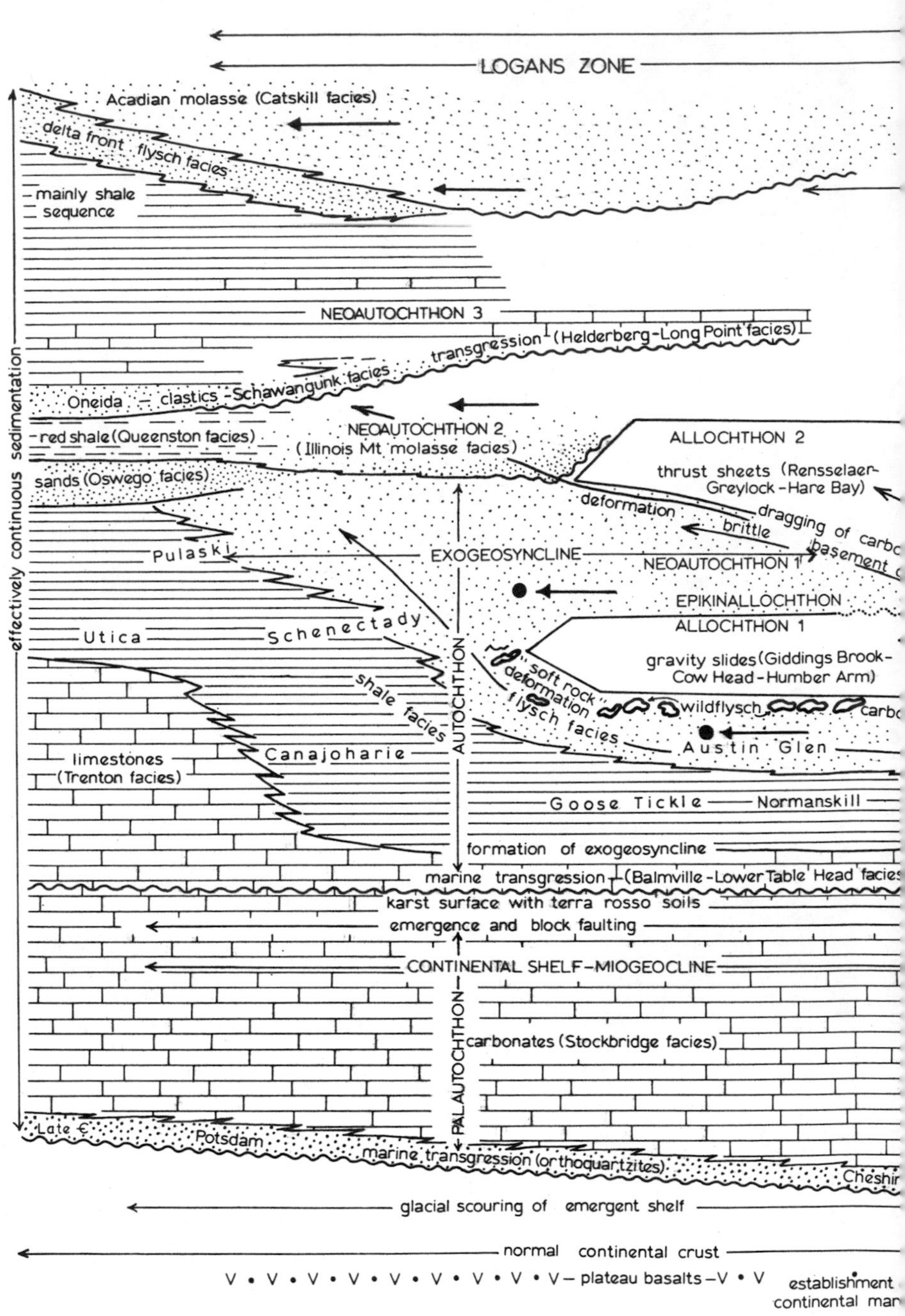

FIGURE 15.13 The Appalachian Geosyncline. (A) Time/space diagram showing

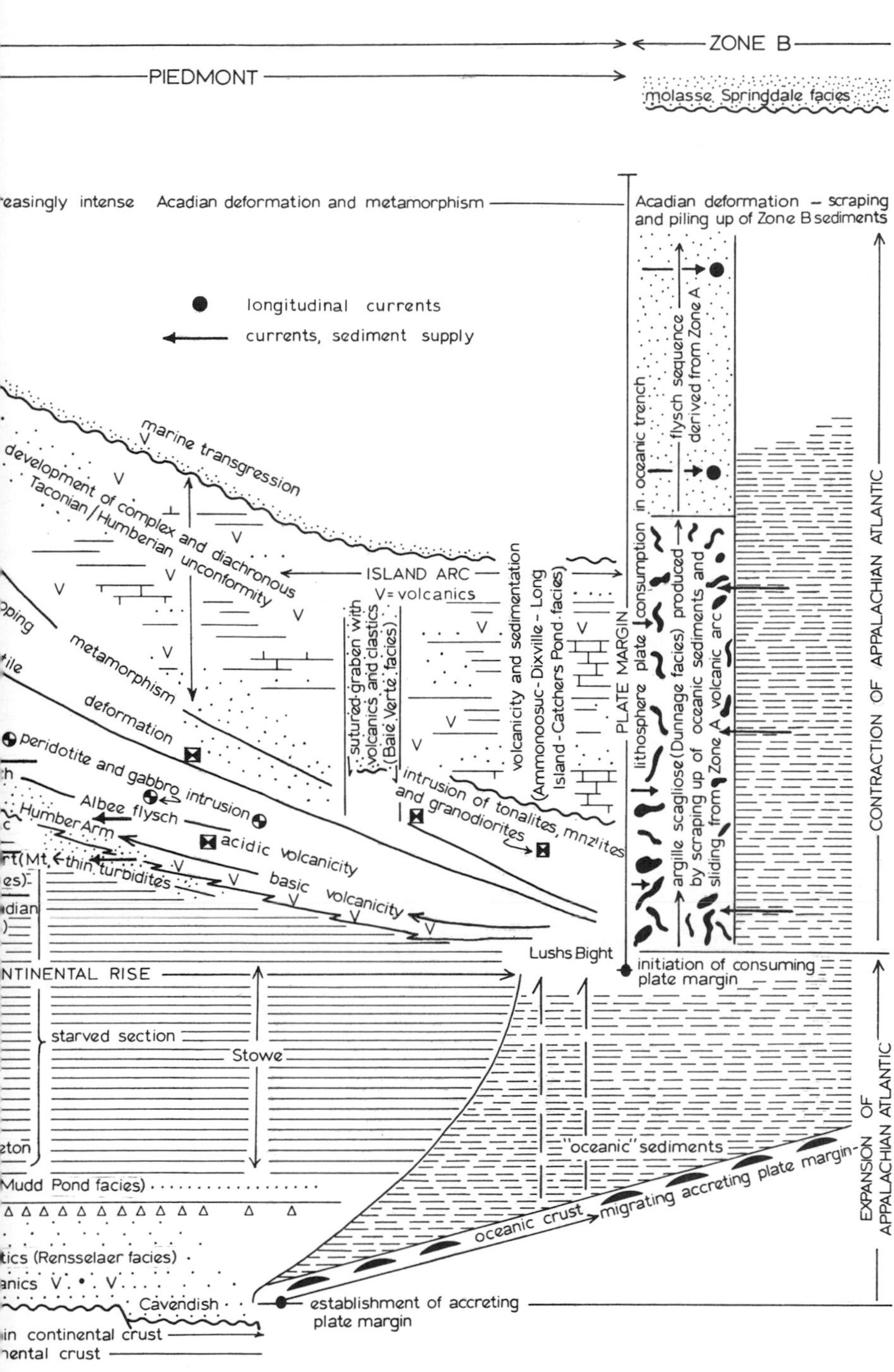

framework of sedimentation, vulcanism, deformation, and metamorphism.

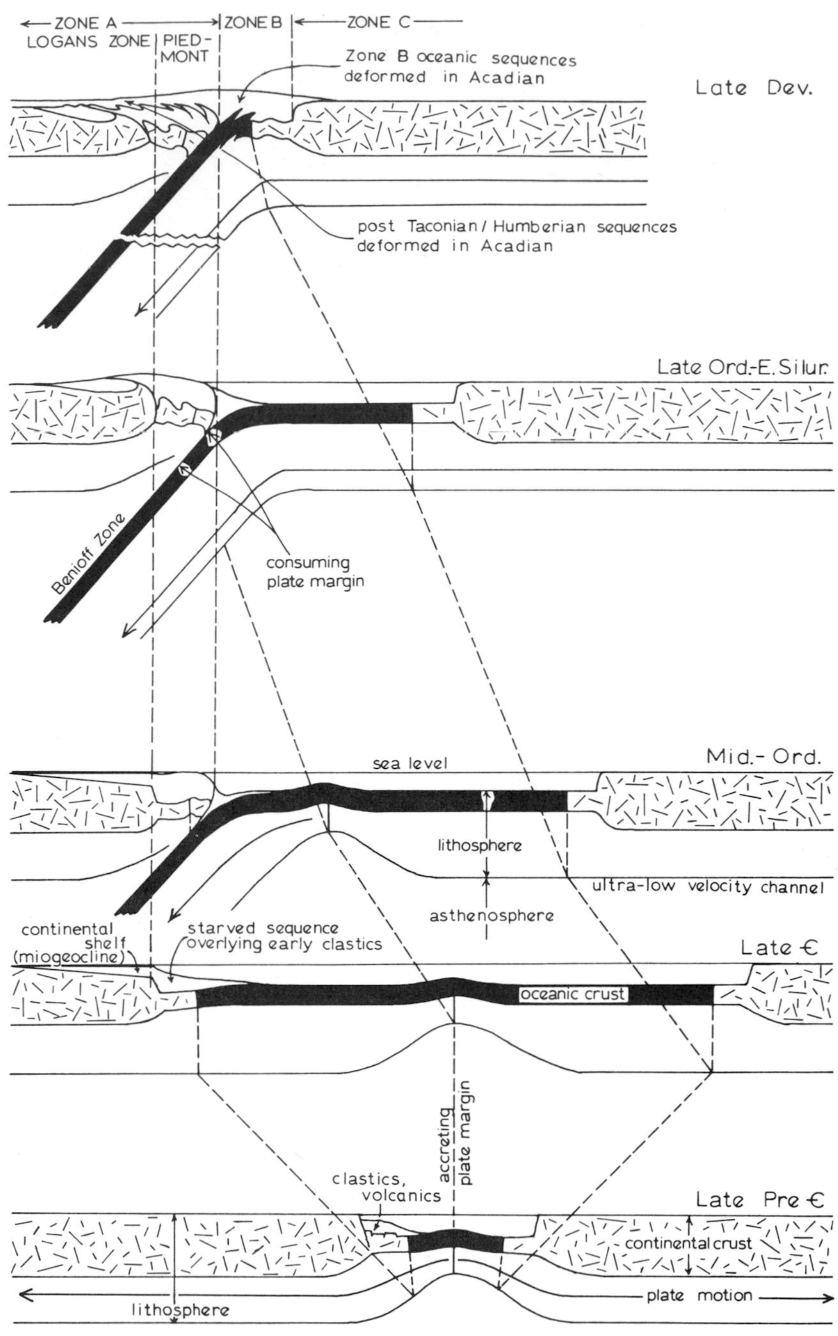

FIGURE 15.13 **The Appalachian Geosyncline.** (B) Suggested plate evolution and orogenies. (From Bird and Dewey, *G.S.A. Bulletin,* vol. 81, April, 1970. By permission of the authors and G.S.A.)

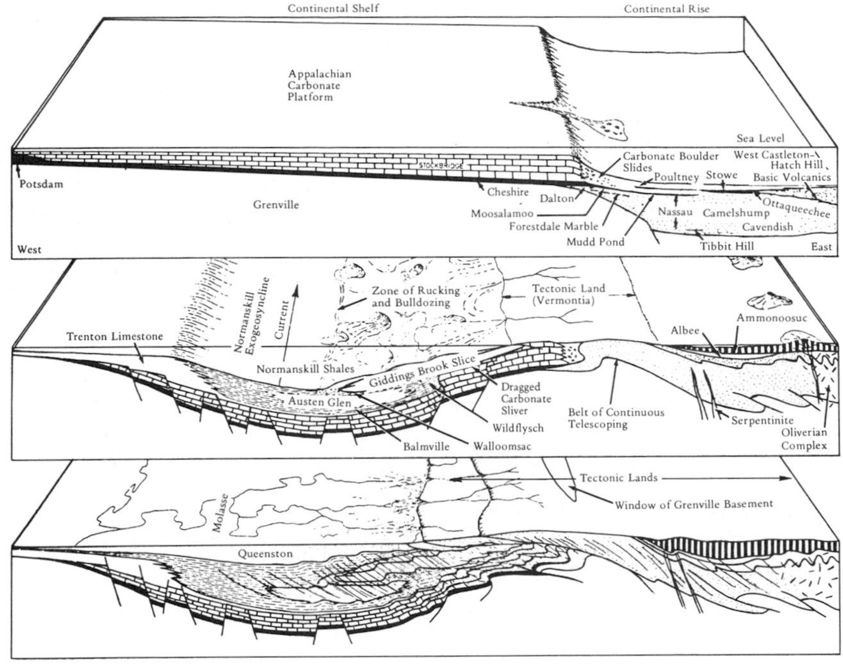

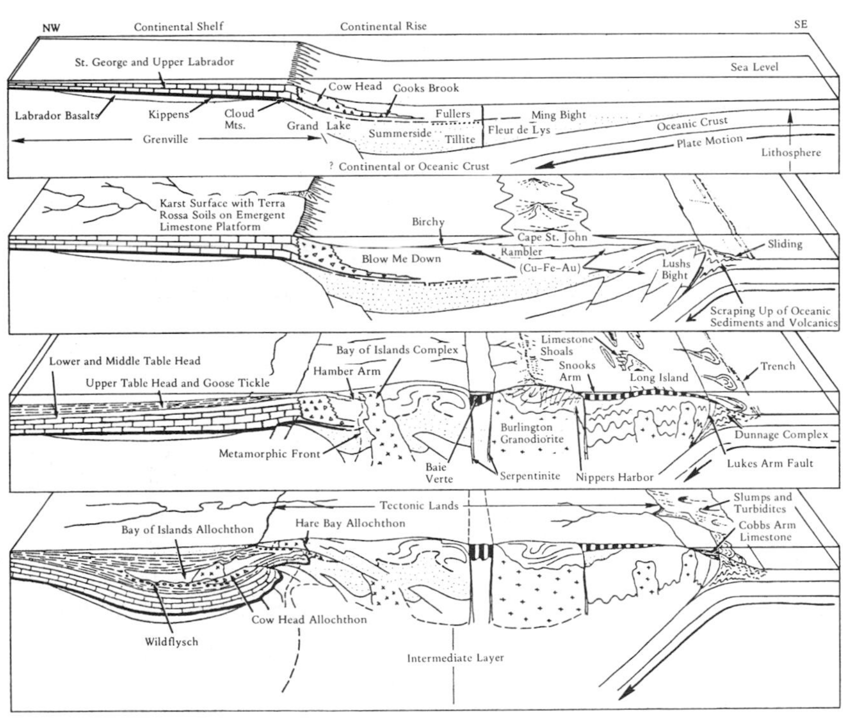

FIGURE 15.14 Evolution of the Northern Appalachian Geosyncline in the Early Paleozoic. (A) New England. (B) Newfoundland. (From Bird and Dewey, *G.S.A. Bulletin*, vol. 81, April, 1970. By permission of the authors and G.S.A.)

The Taconic Orogeny produced a great clastic wedge of terrestrial redbeds or molasse, called the *Queenston-Juniata* (Fig. 15.15), in North America which drove the sea out of the present eastern cratonal area. A large influx of clastics into the epeiric sea occurred even as far west in terms of present geography as Michigan, Ohio, Kentucky, and Tennessee. The largest surviving remnants of the orogeny are the Taconic Mountains of eastern New York (Fig. 15.16), a block of metamorphosed rocks of eugeosynclinal facies thrust westward over strata of miogeosynclinal-cratonic facies.

The Caledonian Orogeny is best recorded in Europe, though its Devonian continuation in the Appalacians, the *Acadian,* has also left a good record. The flood of clastics produced by the rising Caledonian-Acadian Mountains built out a great clastic wedge of terrestrial redbeds in all directions as the colliding edges were both continental by this time. This wedge has been termed the *Old Red Sandstone "Continent"* (Fig. 15.17) in Europe and the *Catskill "Delta"* (Fig. 17.8) in North America. It formed a large band down the present eastern side of the North American craton and covered the northern two-thirds of the present European craton. The clastics from this disturbance resulting from continental collision muddied the waters of the adjacent epeiric seas to a much greater extent than the Taconic Orogeny produced by oceanic-continental collision. They reached as far as present-day Iowa in North America and overran virtually the entire remaining European seas.

North American Epeiric Seas

North America was inundated by two major epeiric seas in the Lower Paleozoic (Fig. 15.18). The *Sauk Sea* began to spread westward from the Appalachian Geosyncline and eastward from the ancestral Cordilleran Geosyncline in the Eocambrian, reached its time of maximum flooding in the Late Cambrian and Early Ordovician, and then withdrew throughout the remainder of the Lower Ordovician. The *Tippecanoe Sea* began to flood the continent in the Middle Ordovician, reached one maximum in the Late Ordovician, then partially with-

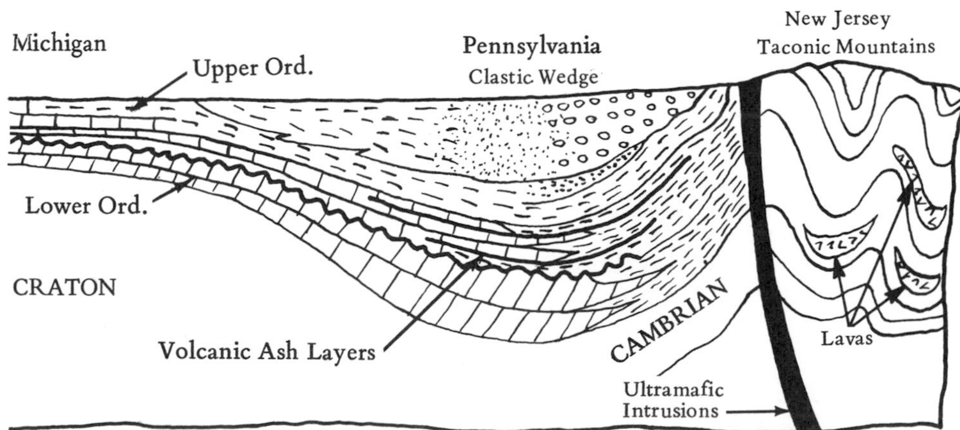

FIGURE 15.15 Restored Cross-Section through Queenston-Juniata Clastic Wedge.

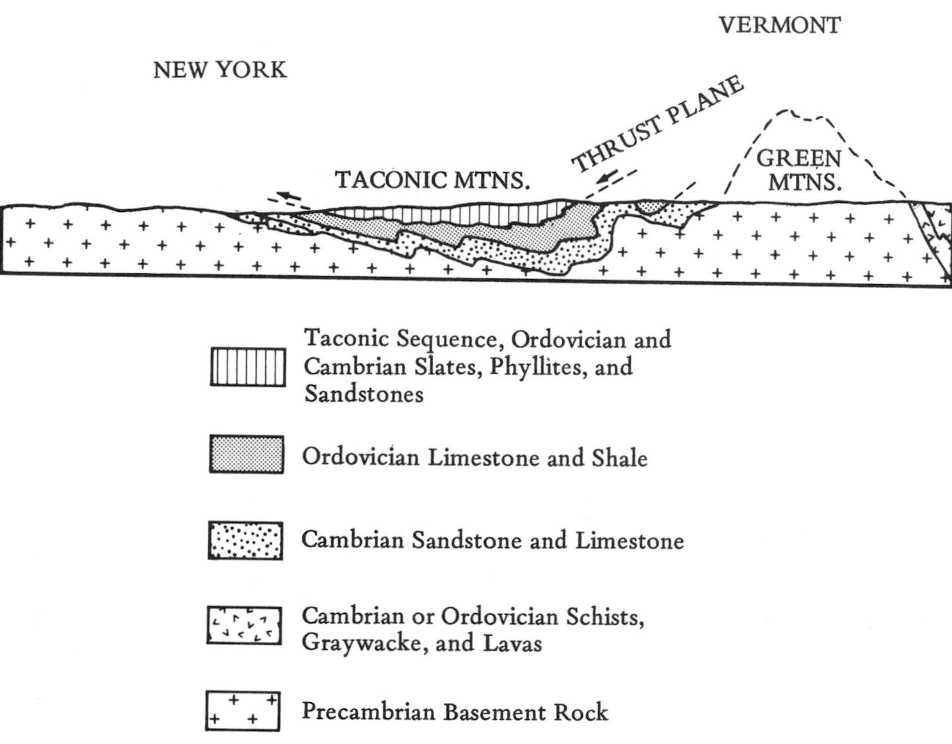

FIGURE 15.16 Cross-Section of Taconic Mountains.

drew in response to the flood of clastics from the Taconic Orogeny, returned to reach another maximum in the Early to Middle Silurian, and finally regressed through the Late Silurian into the Early Devonian. Both sequences are characterized by basal nearshore sandstones (Figs. 15.19–23), derived from the low Canadian Shield, that get progressively younger toward the continental center. These are followed by offshore carbonates (Figs. 15.24–28) that cover huge areas. The carbonates have large admixtures of clastics or even turn to clastics near the eastern margin at the close of the Ordovician as the Taconic Mountains formed. The great Silurian peak of the Tippecanoe Sea is characterized by numerous large reefs, particularly around the Michigan Basin (Fig. 15.27). The regressive phases of the Tippecanoe Sea are characterized by extensive halite beds (Fig. 15.29) in the Great Lakes area.

The North American craton did not subside uniformly to receive sediments during the Lower Paleozoic. Some areas, such as the *Michigan Basin, Illinois Basin, Arbuckle Basin,* and *Williston Basin* (Fig. 15.30) subsided much more rapidly than the remainder and received greater thicknesses of sediment. This does not mean that the water was deeper at any time than in the adjacent areas, in fact the facies indicate it was not, only that the basin subsided more rapidly and the sedimentary piles are thicker. Other regions, such as the *Cincinnati-Nashville Arch,* the *Ozark Dome, Transcontinental Arch, Montana Dome,* and *Peace River-Tathlina Uplift* (Fig. 15.30), did not subside as rapidly as the rest of the craton and have thinner sequences than the remainder. In some cases, such as the Ozark area, the sea did not completely cover a high area of the shield, and islands were left in the epeiric seas.

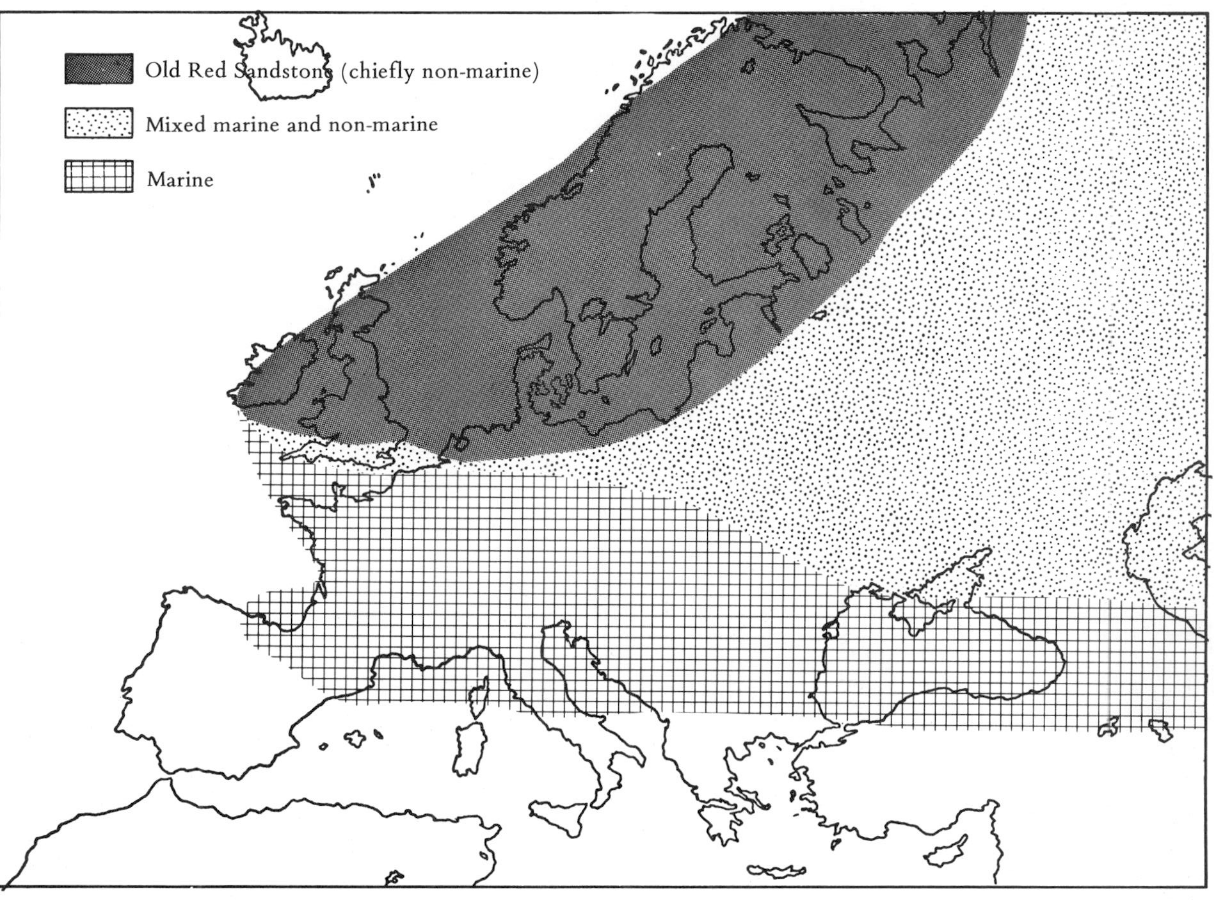

FIGURE 15.17 The Old Red Sandstone "Continent." Effects of plate tectonics not shown.

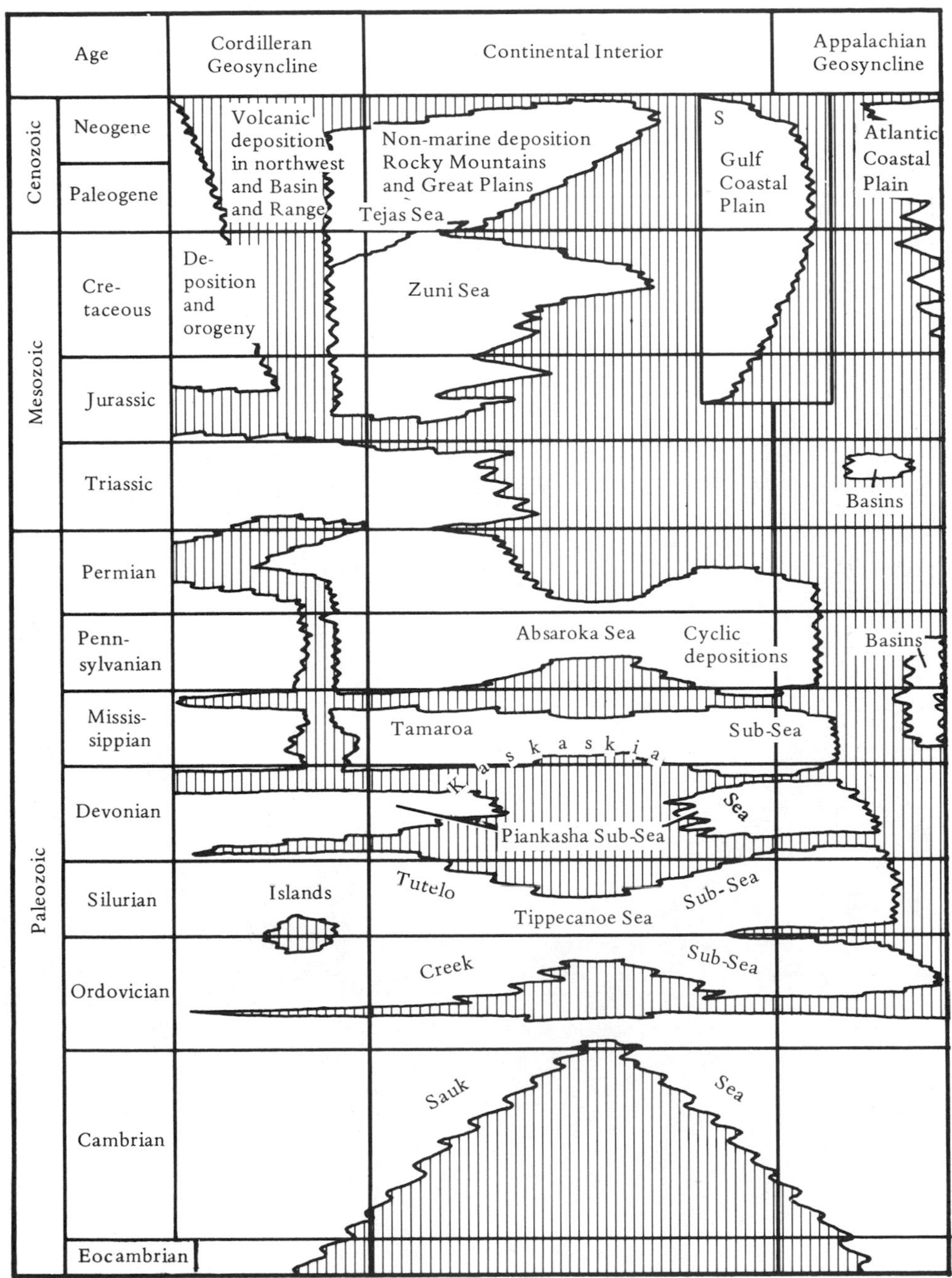

FIGURE 15.18 Major Epeiric Seas of North America. (After Foster.)

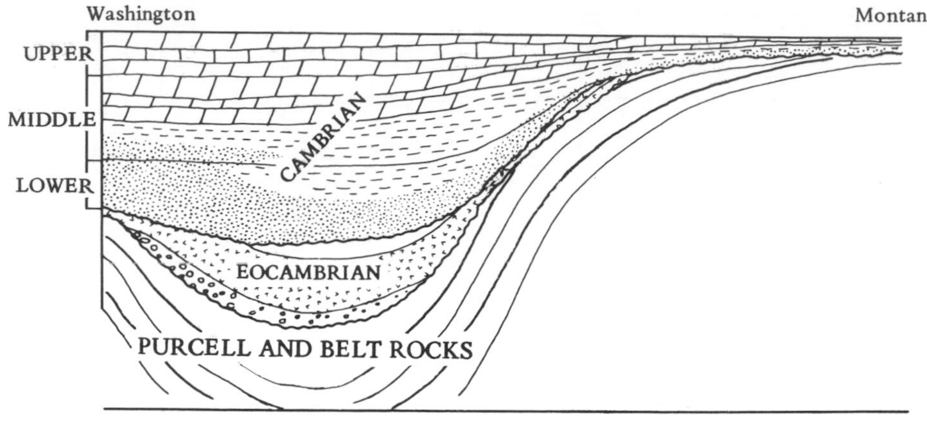

FIGURE 15.19 Sauk Sea Transgressive Deposits of Cambrian Age. Restored cross-section across western North America.

FIGURE 15.20 Glenwood Canyon, Colorado—Upper Cambrian Sandstones. (Courtesy Colorado Dept. of Public Relations.)

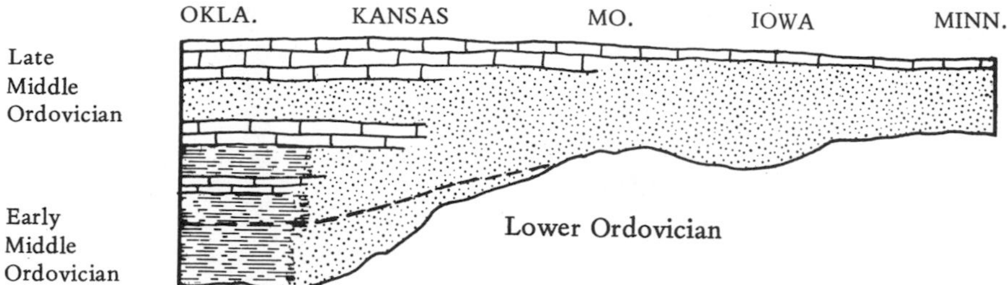

FIGURE 15.21 Tippecanoe Sea Transgressive Deposits of Middle Ordovician Age. Restored cross-section across Midwest.

FIGURE 15.22 Wisconsin Dells—Upper Cambrian Sandstones. (Courtesy of Wisconsin Natural Resources Dept.)

FIGURE 15.23 Ausable Chasm, N.Y.—Upper Cambrian Sandstones. (Courtesy of William M. Mintz.)

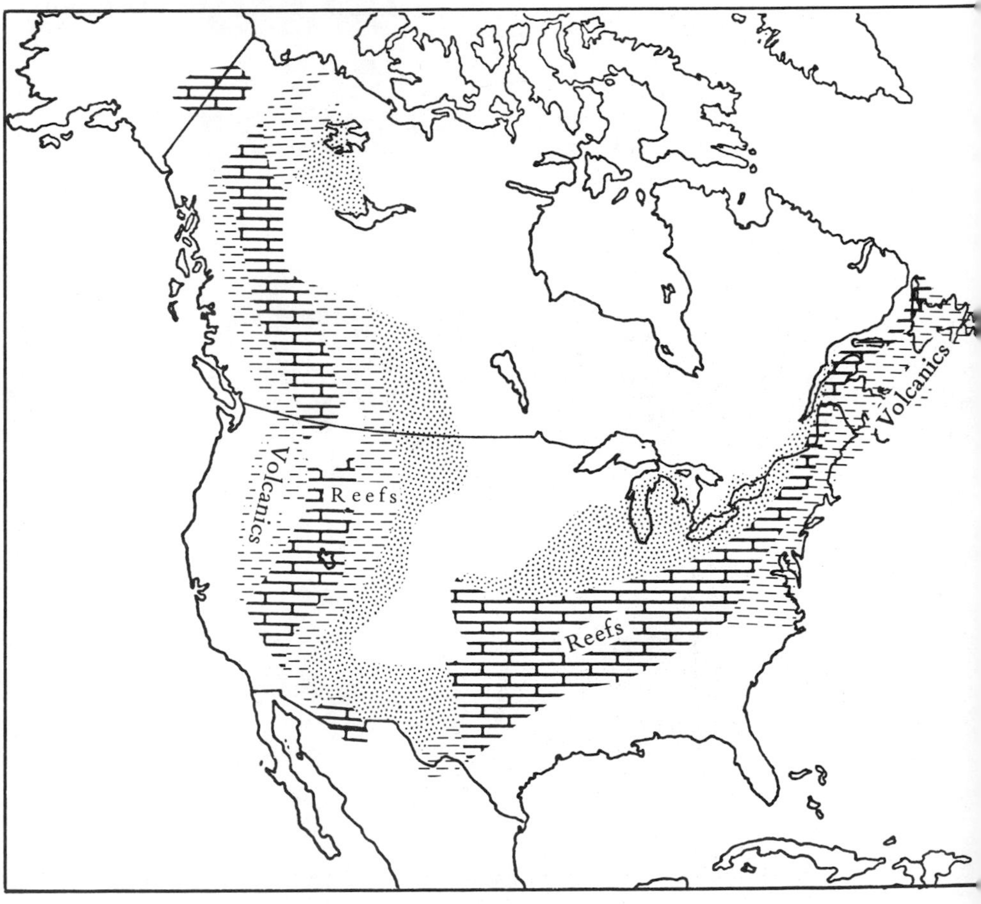

FIGURE 15.24—Upper Cambrian Sedimentary Deposits of Present-Day North America.

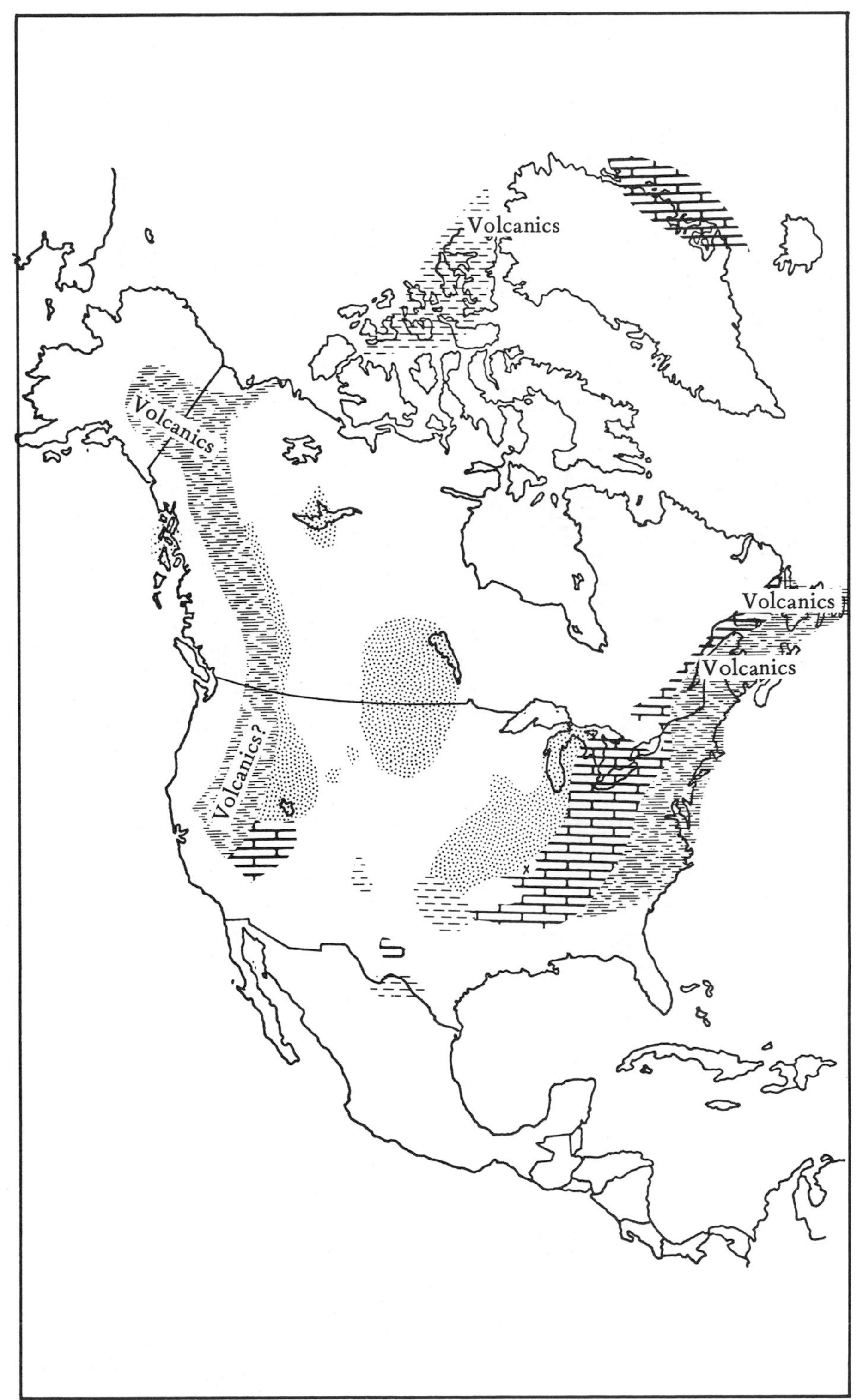

FIGURE 15.25 Middle Ordovician Sedimentary Deposits of Present-Day North America.

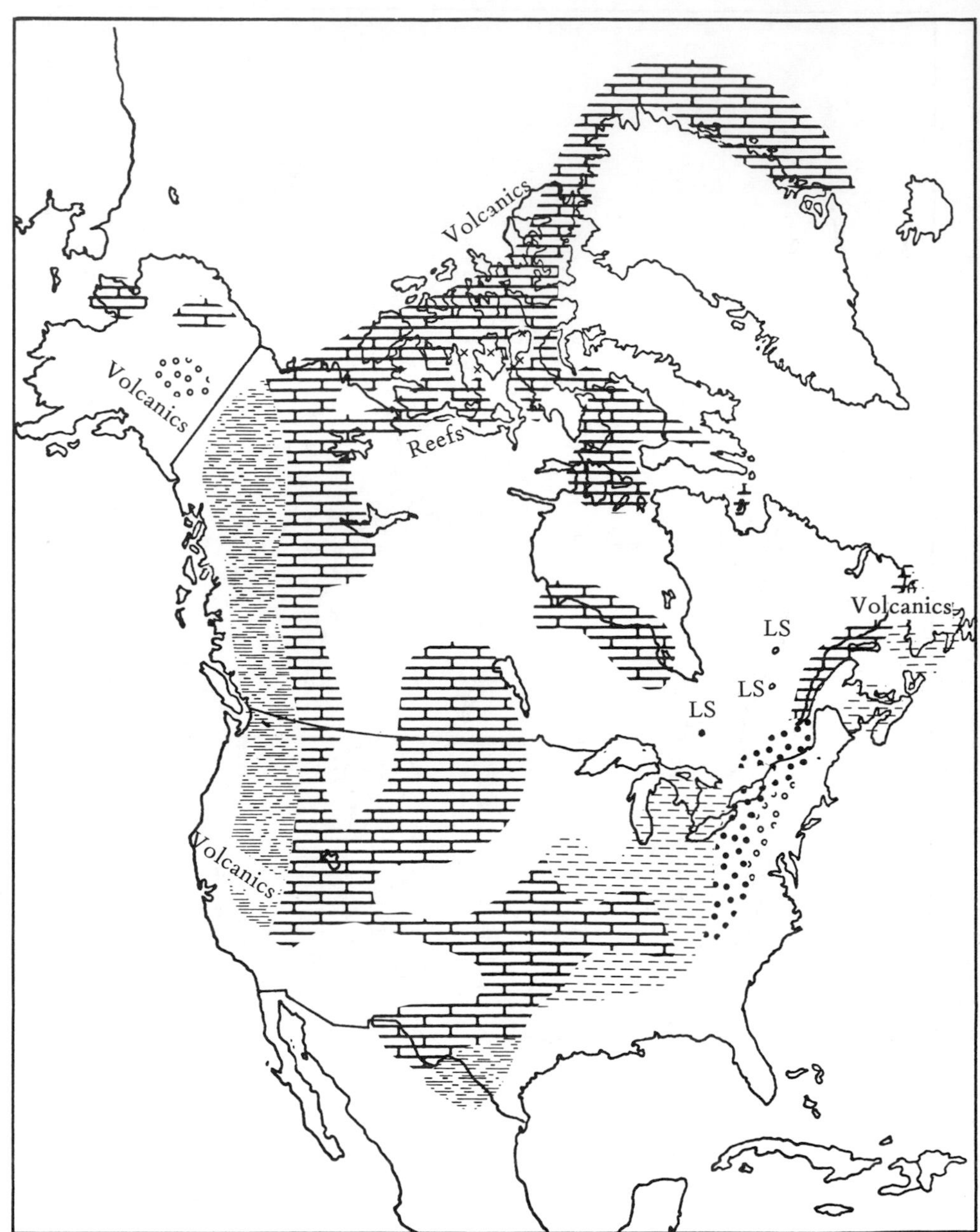

FIGURE 15.26 Late Ordovician Sedimentary Deposits of Present-Day North America.

Events on Other Continents

The *European Craton* (Fig. 9.3), embracing the areas between the present Caledonian, Uralian, and Hercynian mountains, was flooded by epeiric seas during most of the Early Paleozoic. The Lower Paleozoic rocks are now deeply buried across most of the platform in European Russia but crop out in the Baltic Sea

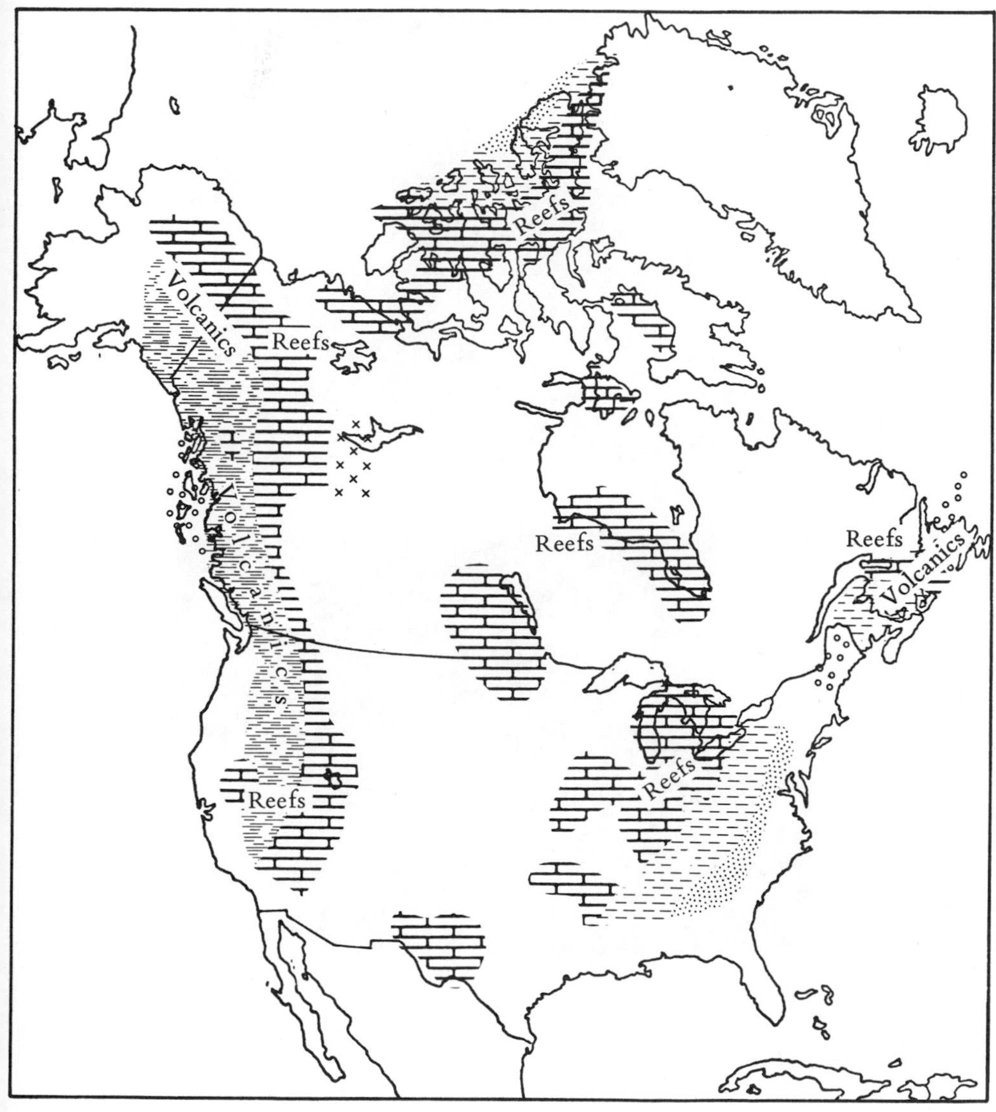

FIGURE 15.27 Middle Silurian Sedimentary Deposits of Present-Day North America.

region. The *Angara Shield,* forming the *Asian craton* (Fig. 15.10), was flooded by epeiric seas throughout the Lower Paleozoic, but not thereafter. The *Gondwanaland Craton* was not extensively flooded in the Cambrian-Silurian interval. Flooding occurred chiefly along the present northern margin including the Amazon River area of present South America (Fig. 15.31), northern Africa, and northern and western Australia (Fig. 15.32).

In Gondwanaland, the *Adelaide Geosyncline* (Figs. 9.3, 11.36), running down what is now eastern Australia and extending across present Antarctica where it is called the *Ross Geosyncline,* developed during the *Eocambrian* and *Early Cambrian.* It was then thoroughly deformed probably by compression from the Pacific plates or the small New Zealand plate which also contains abundant

FIGURE 15.28 (A-B) Niagara Falls, (C) Flowerpot Island, Georgian Bay Islands National Park, Ontario. (A-B) Courtesy N.Y. State Power Authority, (C) Courtesy Ontario Dept. of Tourism and Information.

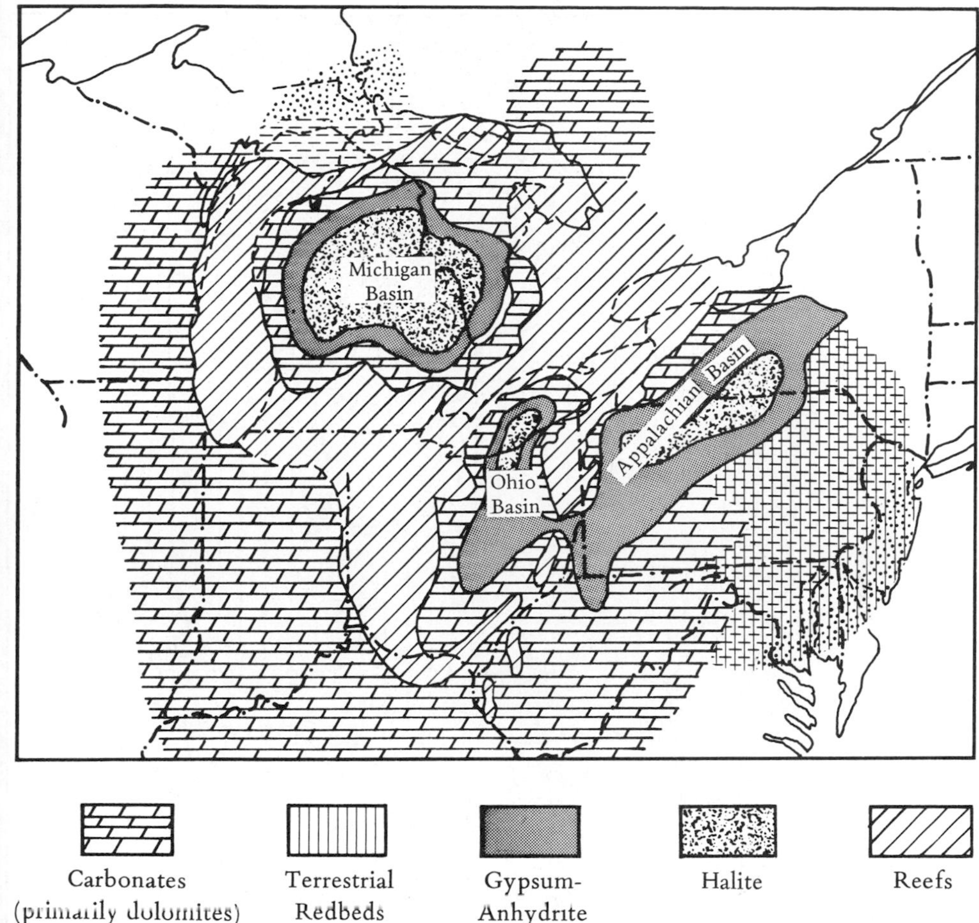

FIGURE 15.29 Late Silurian Evaporite Basins of Great Lakes Region.

Lower Paleozoic geosynclinal rocks. Compression may have ceased for awhile, but resumed shortly because another geosyncline, called the *Tasman* in Australia, developed adjacent to it on the Pacific side. The geosyncline underwent continuous sedimentation and orogeny throughout the remainder of the Paleozoic.

With the South or Gondwana Pole situated in a large continental mass such as Gondwanaland, it is not surprising that continental glaciations occurred on the supercontinent during much of the Paleozoic. In the Early Paleozoic record, extensive evidence of continental glaciers occurs in the Ordovician of present North Africa, the region in which the pole was situated at that time. Features produced by both glacial erosion and deposition (Fig. 15.33) abound over a broad area from Morocco, through Mauretania, Algeria, Niger, to Libya and Chad. Some evidence of glaciation has been uncovered in present-day Brazil and South Africa as well.

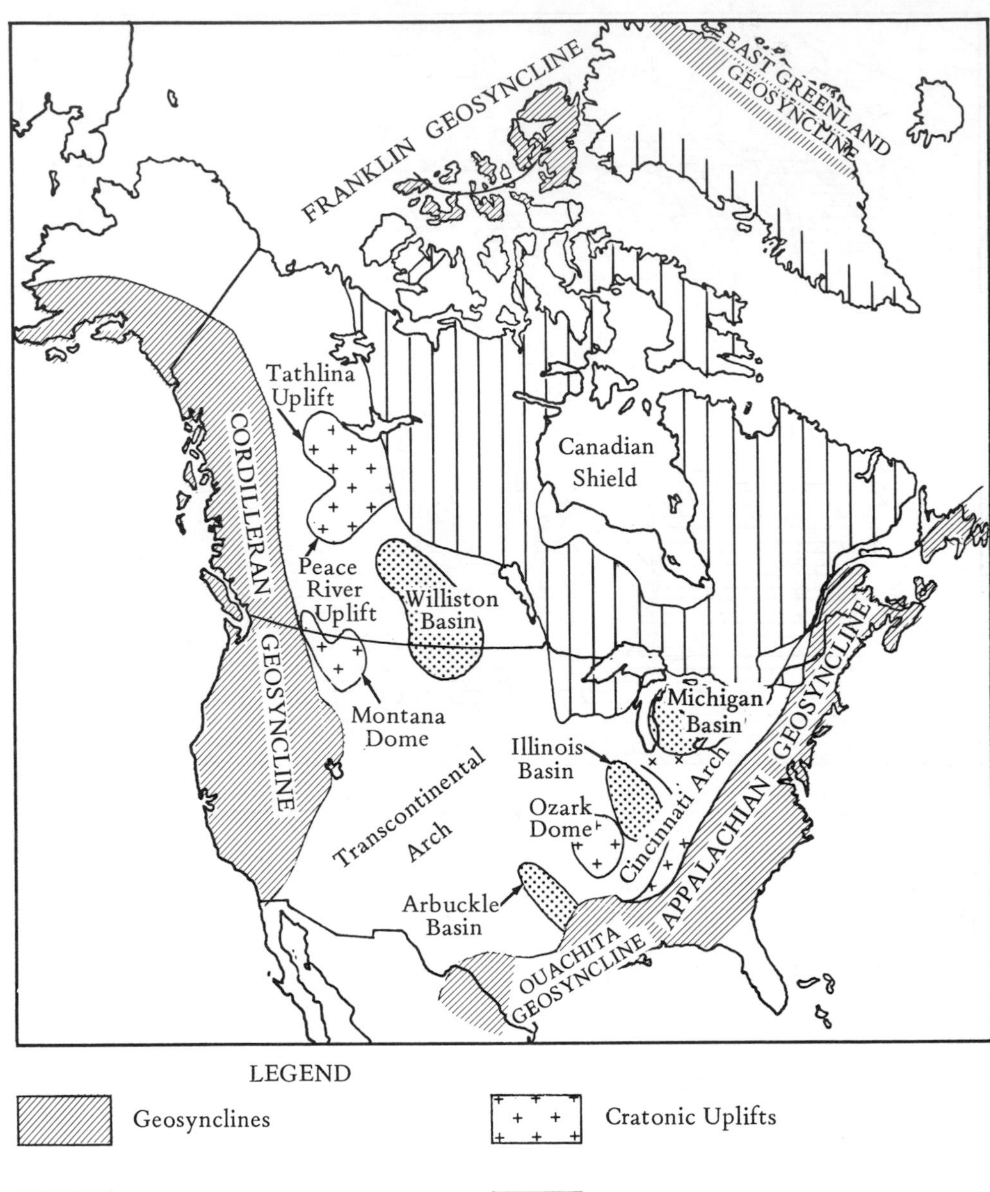

FIGURE 15.30 Domes and Basins of Present-Day North America which Existed in the Early Paleozoic.

FIGURE 15.32 Cambrian, Ordovician, and Silurian Sedimentary Deposits of Present-Day Australia.

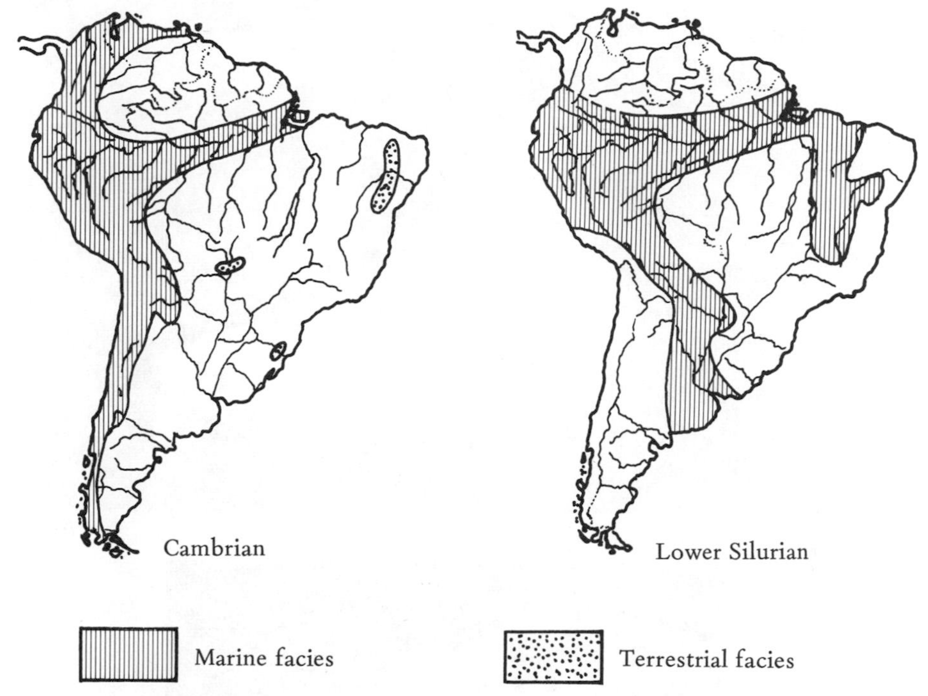

FIGURE 15.31 Cambrian and Early Silurian Sedimentary Deposits of Present-Day South America.

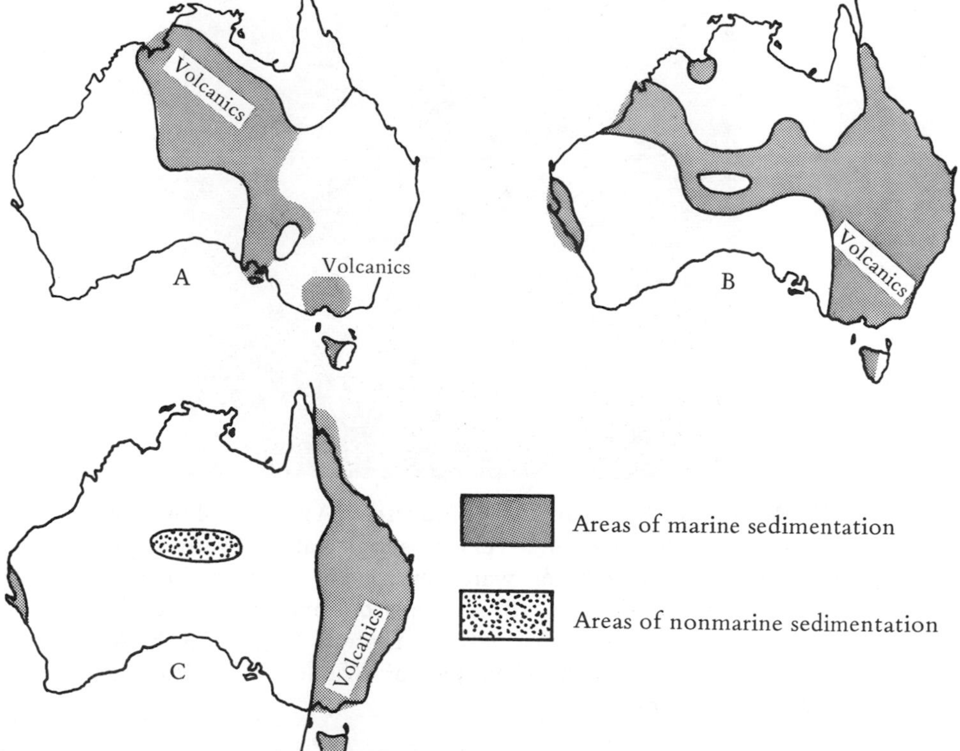

A

B

FIGURE 15.33 Saharan Glacial Features. (A) Horse-shoe pressure scours produced by high-velocity subglacial meltwater. Open ends point in direction of water movement. Eastern Hoggar, Algeria. Upper Ordovician. (B) Glacial scour and grooves rippled by subglacial melting of permafrost. Eastern Hoggar, S .E. Algeria. Upper Ordovician. (Courtesy of Rhodes W. Fairbridge, Columbia University.)

Against and within this background, the life of the Lower Paleozoic evolved. Fossils of the larger marine organisms became abundant for the first time as hard parts evolved. In the Lower Paleozoic epeiric sea deposits of North America, particularly the widespread limestones, the fossils occur in such abundance that they compose the bulk of many rock units. The Early Paleozoic seas were widespread yet not widely separated for most of the continents were approaching. Let us now turn to the fossil record of the Cambrian, Ordovician, and Silurian Periods and examine the effects of this situation.

Questions

1. What is the significance of the Pan-African Orogeny?
2. Where was the Gondwana Pole in the Early Paleozoic?
3. Why is it more difficult to determine plate positions and margins in the Early Paleozoic than in later geologic time?
4. What does the paleomagnetic and paleoclimatic evidence indicate about the position of North America in the Early Paleozoic?
5. What lines of evidence indicate that Europe and Asia were separate continents in the Early Paleozoic?
6. What is the significance of the Appalachian-Caledonian Geosyncline?
7. What was the major mountain-building event in the Appalachian area in the Early Paleozoic and what type of plate collision produced it?
8. What were the major epeiric seas of North America in the Early Paleozoic and when did they reach maximum flood stage?
9. What were the major tectonic elements of the craton in the Paleozoic?
10. What cratons other than North America underwent extensive flooding in the Early Paleozoic?
11. What evidence do we have of the margins of the Pacific Plate or plates in the Early Paleozoic?
12. What evidence is there of continental glaciation in the Early Paleozoic and where does it occur?

16

Abundance in the Sea:

Early Paleozoic Life

LIFE IN THE EARLY PALEOZOIC PERIODS WAS APPARENTLY LIMITED TO THE SEA UNTIL the very end of the Silurian when the first land plants and animals appear in the fossil record. Early Paleozoic marine life is also dominated by the invertebrates, with only scraps of fish preserved until the Late Silurian. The Early Paleozoic is thus best described as the age of marine invertebrates (Figs. 16.1–16.4).

Major Marine Invertebrates

Dominating the invertebrate animal fossils in abundance and diversity were the *corals, bryozoans, brachiopods, trilobites,* and *graptolites.* Brachiopods and trilobites appear right at the base of the Cambrian; in fact, the first appearance of redlichioid trilobites is used to mark that boundary in most areas. Brachiopods continue to flourish throughout the Paleozoic, but trilobites rise to a Late Cambrian–Early Ordovician peak and then decline. Corals do not appear until the beginning of the Middle Ordovician, but diversify rapidly through the Middle Paleozoic. Graptolites appear late in the Cambrian, and diversify greatly in the Ordovician and Silurian.

Corals

The major coral groups of the Paleozoic, the *tabulate* and *tetracorals,* appear in the Middle Ordovician, but the tabulates (Fig. 16.5) dominate the Lower Paleozoic. By the Late Ordovician, tabulates were constructing coral reefs, and in the Silurian, their zenith, they were the major framework builders of the abundant reefs of the period (Fig. 15.27). Such genera as the honeycomb coral, *Favosites,* and the chain coral, *Halysites,* as well as numerous others, occur in great abundance in Silurian rocks. The tetracorals (Fig. 16.6) of the Early Paleozoic were chiefly small, horn or cup corals such as *Streptelasma.*

Bryozoans

The *bryozoans* (Fig. 16.7) appear for certain in the Ordovician, though some possible Cambrian specimens are known. The major Lower Paleozoic groups are the twig-like or massive *trepostomes* and *cystoporoids,* the stony bryozoa. In the Upper Ordovician, their remains are so abundant that they actually compose the bulk of many rock bodies. Common genera include the trepostomes *Hallopora, Homotrypa, Prasopora,* and *Monticulopora,* and the cystoporoid *Constellaria.* These two orders were reduced, but still common, in the Silurian.

Brachiopods

The brachiopods underwent a tremendous adaptive radiation in the Lower Paleozoic. The *inarticulates,* such as *Dicellomus, Lingulepis,* and *Acrothele* (Fig. 16.8), dominated the Cambrian brachiopod fauna, and, although the group was even more diverse in the Ordovician, inarticulates were no longer the major brachiopod group. Attached inarticulates such as *Petrocrania* (Fig. 16.8) were common in the Ordovician. The *articulates* appeared in the Cambrian also and chiefly consisted of *orthoids* such as *Billingsella* and *Nisusia* (Fig. 16.9). In the Ordovician, the articulate brachiopods diversified greatly. The orthoids and *strophomenoids* are the most abundant and widespread, but pentameroids,

FIGURE 16.1 Diorama of Cambrian Marine Life. (Courtesy Field Museum of Natural History.)

FIGURE 16.2 Diorama of Ordovician Marine Life. (Courtesy Field Museum of Natural History.)

FIGURE 16.3 Diorama of Silurian Marine Life. (Courtesy Field Museum of Natural History.)

FIGURE 16.4 Diorama of Silurian Brackish-Water Life. (Courtesy Field Museum of Natural History.)

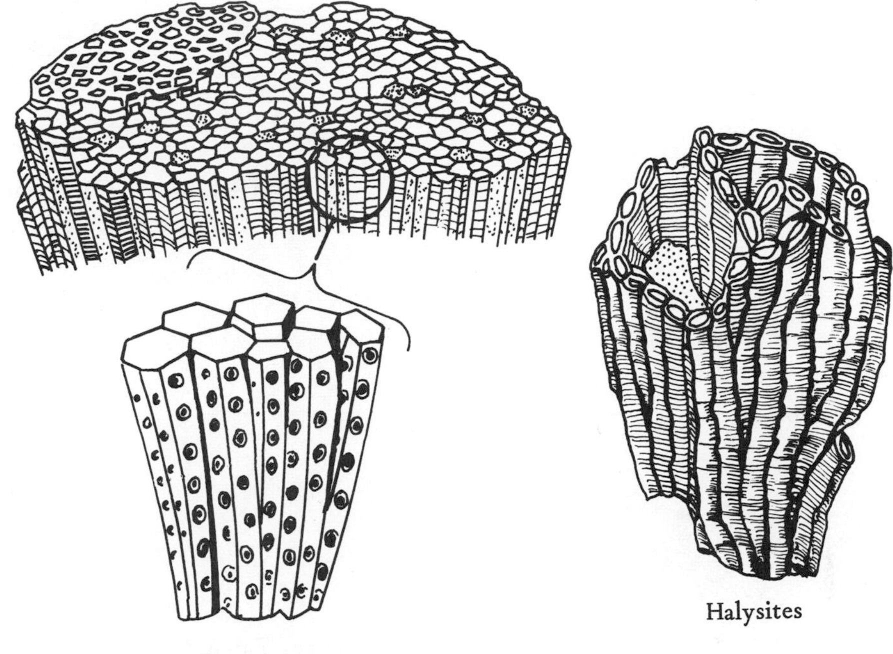

FIGURE 16.5 Lower Paleozoic Tabulate Corals.

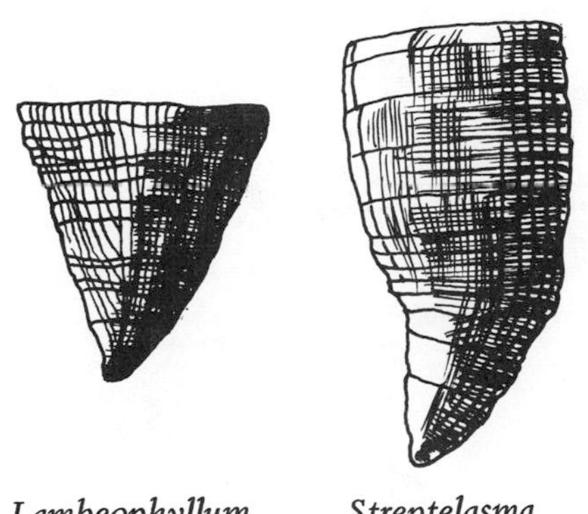

FIGURE 16.6 Lower Paleozoic Tetracorals.

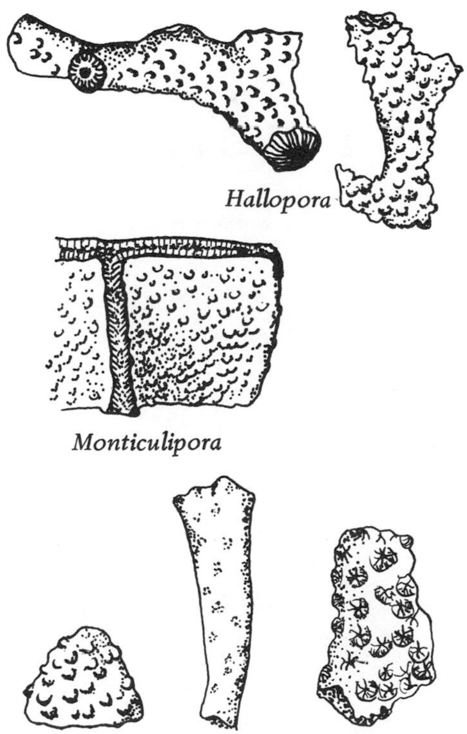

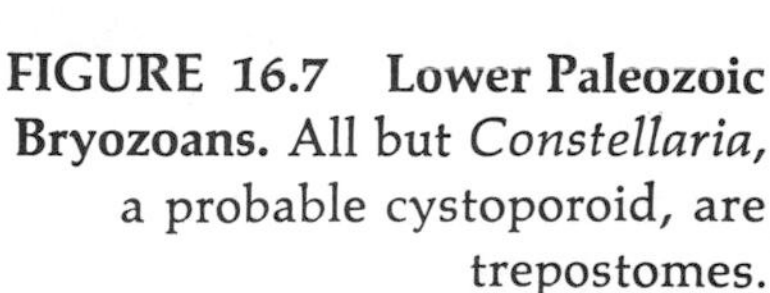

FIGURE 16.7 Lower Paleozoic Bryozoans. All but *Constellaria*, a probable cystoporoid, are trepostomes.

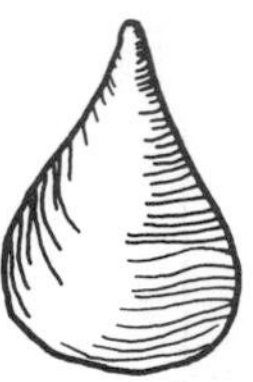

Lingulepis *Dicellomus*

Acrothele

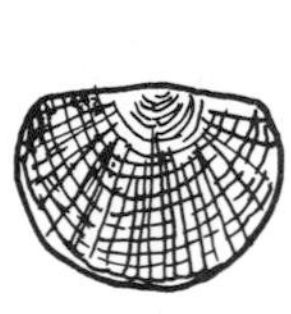

Micromitra *Petrocrania*

FIGURE 16.8 Lower Paleozoic Inarticulate Brachiopods.

Billingsella *Nisusia*

FIGURE 16.9 Cambrian Orthoid Brachiopods.

Resserella *Plaesiomys*

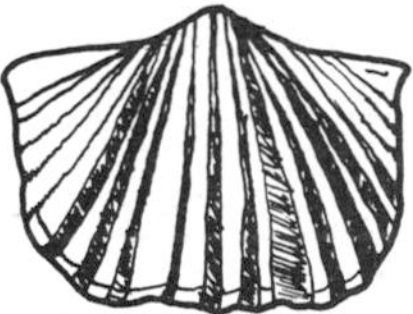

Platystrophia

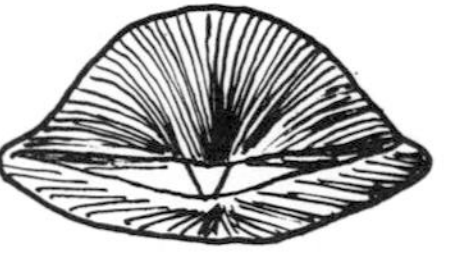

Hebertella

FIGURE 16.10 Ordovician Orthoid Brachiopods.

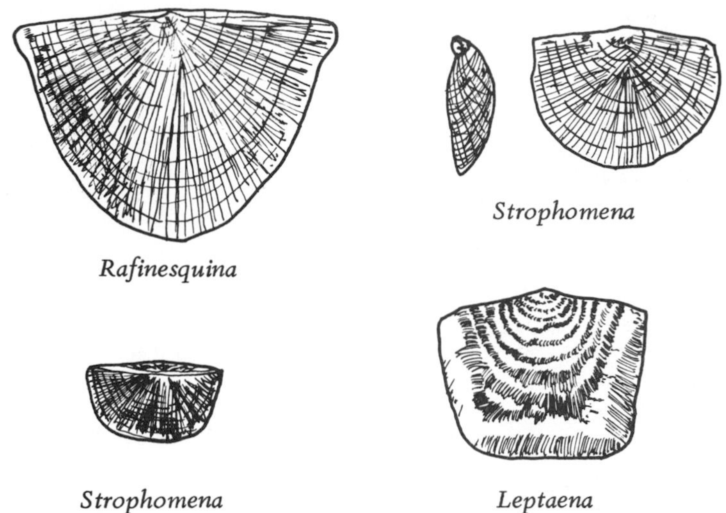

FIGURE 16.11 Ordovician Strophomenoid Brachiopods.

rhychonelloids, and some atrypoids are also common. Characteristic orthoid genera (Fig. 16.10) include *Plaesiomys, Platystrophia, Hebertella,* and *Resserella.* Among the common strophomenoids (Fig. 16.11) are *Sowerbyella, Leptaena, Rafinesquina,* and *Strophomena. Camerella, Rhynchotrema,* and *Zygospira* are common representatives of the pentameroids, rhynchonelloids, and atrypoids, respectively (Fig. 16.12). The Silurian brachiopod fauna is dominated by the *pentameroids* (Fig. 16.13) such as *Pentamerus* and *Conchidium.* Pentameroids are useful fossils to geologists for time-correlating rocks within regions. Rhynchonelloids (Fig. 16.14A) such as *Rhynchotreta* and *Stegerhynchus* are also common Silurian rocks. Some athyridoids (Fig. 16.14B) such as *Whitfieldella,* are occasionally common in the Silurian.

Trilobites

The major trilobites of the Cambrian are the *redlichioids, corynexochoids, ptychoparioids,* and *agnostoids.* The first two orders are restricted to the Cambrian, the ptychoparioids are also common in the Ordovician and the agnostoids are chiefly Cambrian in age. Representative redlichioids (Fig. 16.15) include *Olenellus* and *Paradoxides;* common corynexochoids (Fig. 16.16) include *Ogygopsis* and *Bathyuriscus; Elrathia* (Fig. 16.17) is an abundant ptychoparioid; and *Peronopsis* (Fig. 16.18) is a typical agnostoid. Trilobites are the major fossils for both regional and worldwide time-correlation. In the Ordovician, the ptychoparioids such as *Triarthrus* (Fig. 16.19A) continue to be important, but the *asaphoids* such as *Isotelus* (Fig. 16.19B) and *trinucleoids* such as *Cryptolithus* (Fig. 16.19C) become the major groups. Some *illaenoids* (*Illaenus*), *cheiruroids* (*Ceraurus*), and *calymenoids* (*Flexicalymene*) were also common (Fig. 16.20). Trilobites are very useful for time-correlation within regions in the Ordovician. In the Silurian, the calymenoids (Fig. 16.21) reach a peak with such genera as *Calymene* and *Trimerus* characteristic. Some illaenoids (*Bumastus*) and phacopoids (*Dalmanites*) are also common (Fig. 16.22).

Camerella

A

Zygospira

C

Rhynchotrema

B

FIGURE 16.12 Ordovician Brachiopods of the Orders: (A) *Pentameroida,* (B) *Rhynchonelloida,* (C) *Atrypoida.*

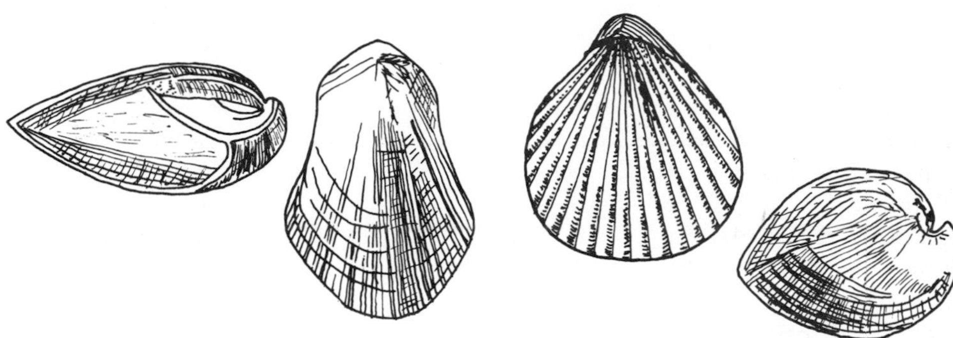

Pentamerus *Conchidium*

FIGURE 16.13 Silurian Pentameroid Brachiopods.

Rhynchotreta

Stegerhynchus

A

Whitfieldella

Meristina

B

FIGURE 16.14 Silurian Brachiopods of the Orders: (A) Rhynchonelloida, (B) *Athyridoida.*

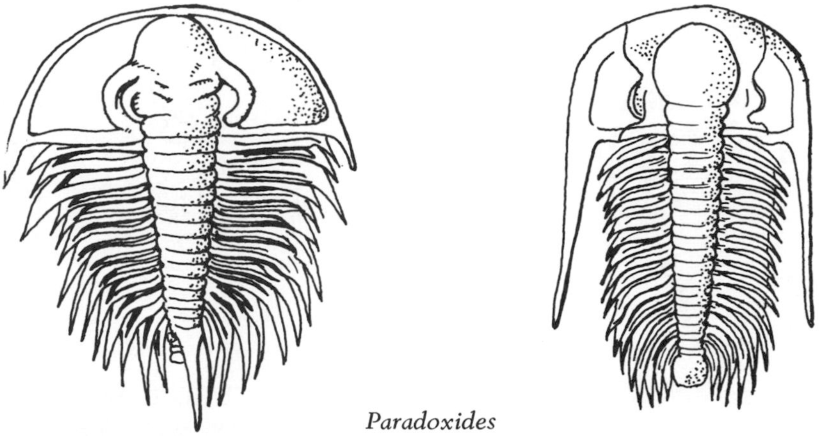

Paradoxides

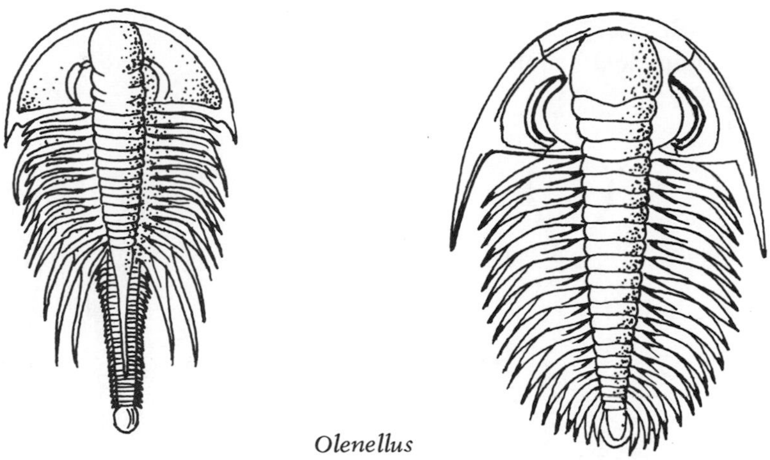

Olenellus

FIGURE 16.15 Cambrian Redlichioid Trilobites.

Ogygopsis

Bathyuriscus

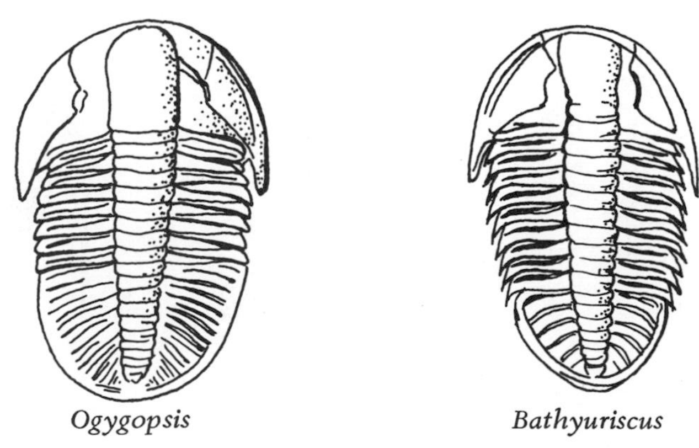

FIGURE 16.16 Cambrian Corynexochoid Trilobites.

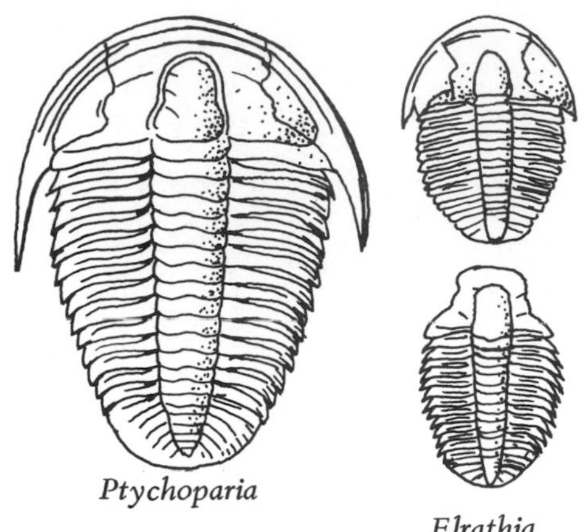

FIGURE 16.17 Cambrian Ptychoparioid Trilobites.

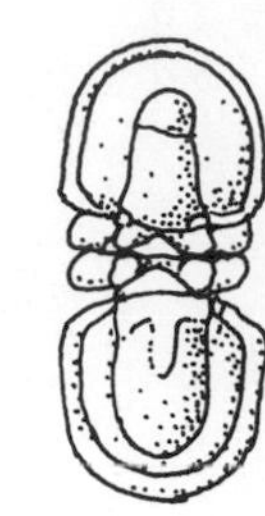

FIGURE 16.18 Cambrian Agnostoid Trilobites.

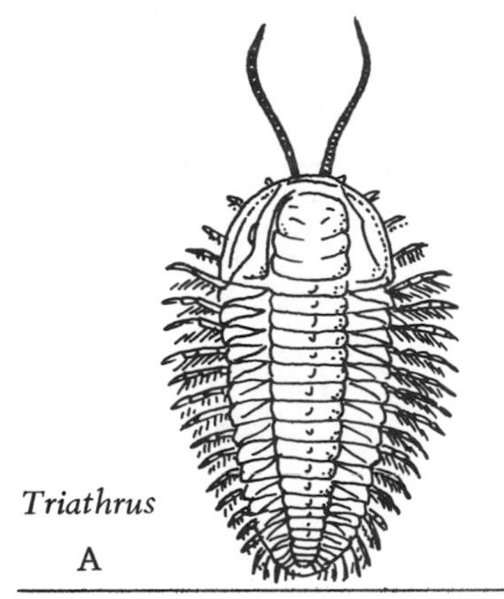

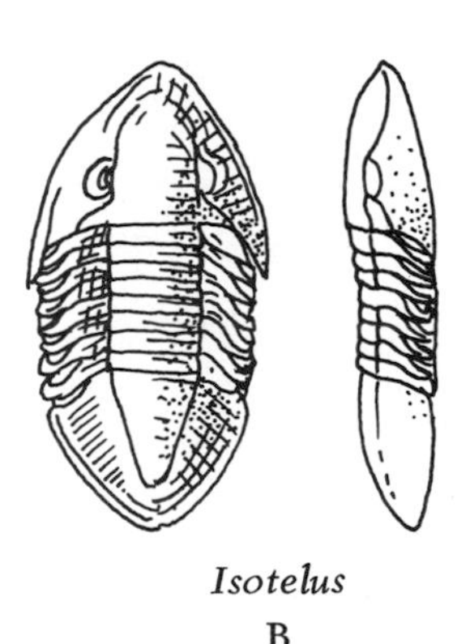

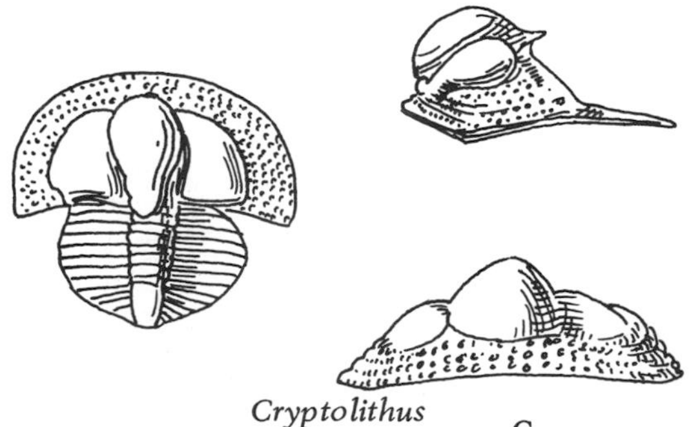

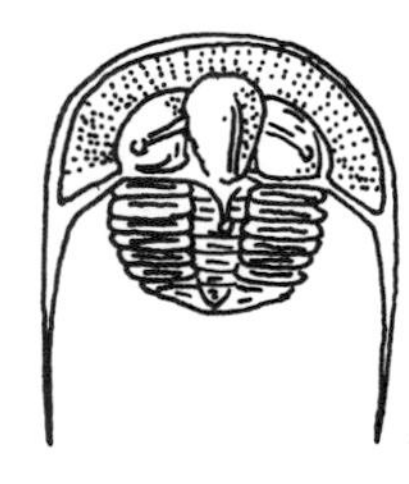

FIGURE 16.19 Ordovician Trilobites of the Orders: (A) Ptychoparioida, (B) Asaphoida, (C) Trinucleoida.

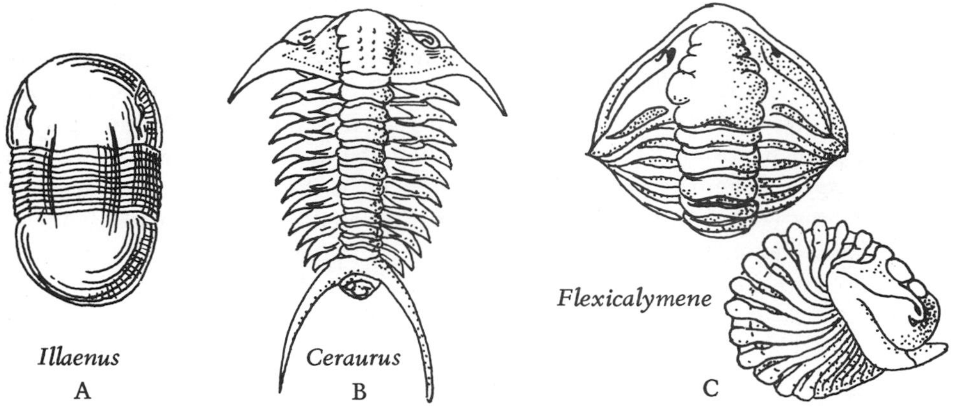

FIGURE 16.20 Ordovician Trilobites of the Orders: (A) Illaenoida, (B) Cheiruroida, (C) Calymenoida.

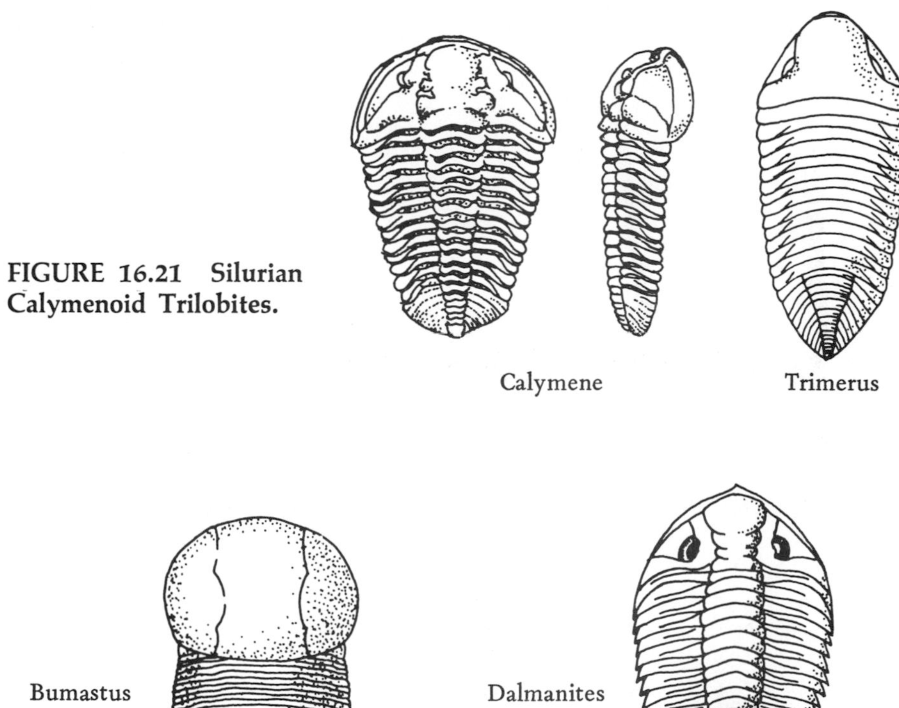

FIGURE 16.21 Silurian Calymenoid Trilobites.

FIGURE 16.22 Silurian Trilobites of the Orders: (A) Illaenoida, (B) Phacopoida.

Graptolites

The graptolites appear at the very end of the Cambrian but are only represented by rare attached dendroid types. In the Lower Ordovician, however, the pelagic *graptoloid* graptolites (Fig. 16.23, 16.25) appear and undergo a large radiation producing such forms as *Tetragraptus, Didymograptus, Climacograptus,* and *Dicellograptus.* There is a reduction in numbers in the Silurian, but the genus *Monograptus* (Fig. 16.24, 16.25) is abundant throughout the period, dying out in the Early Devonian. Graptoloid graptolites are the major fossils used in time-correlation in the Ordovician, Silurian, and lowest Devonian. As planktonic forms, species and genera of this group have wide geographic ranges which is

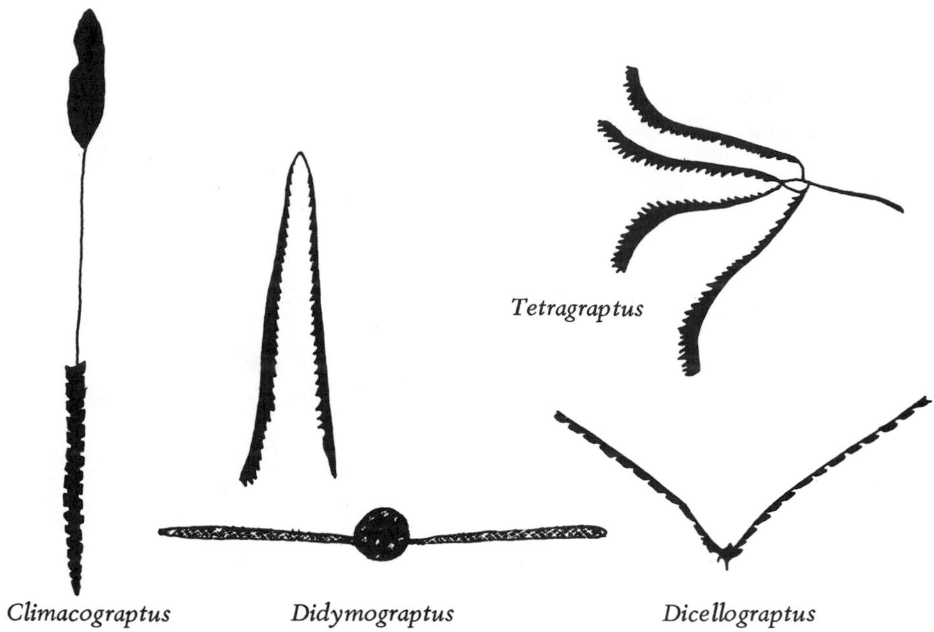

FIGURE 16.23 Ordovician Graptoloid Graptolites.

FIGURE 16.24 Silurian Graptoloid Graptolites.

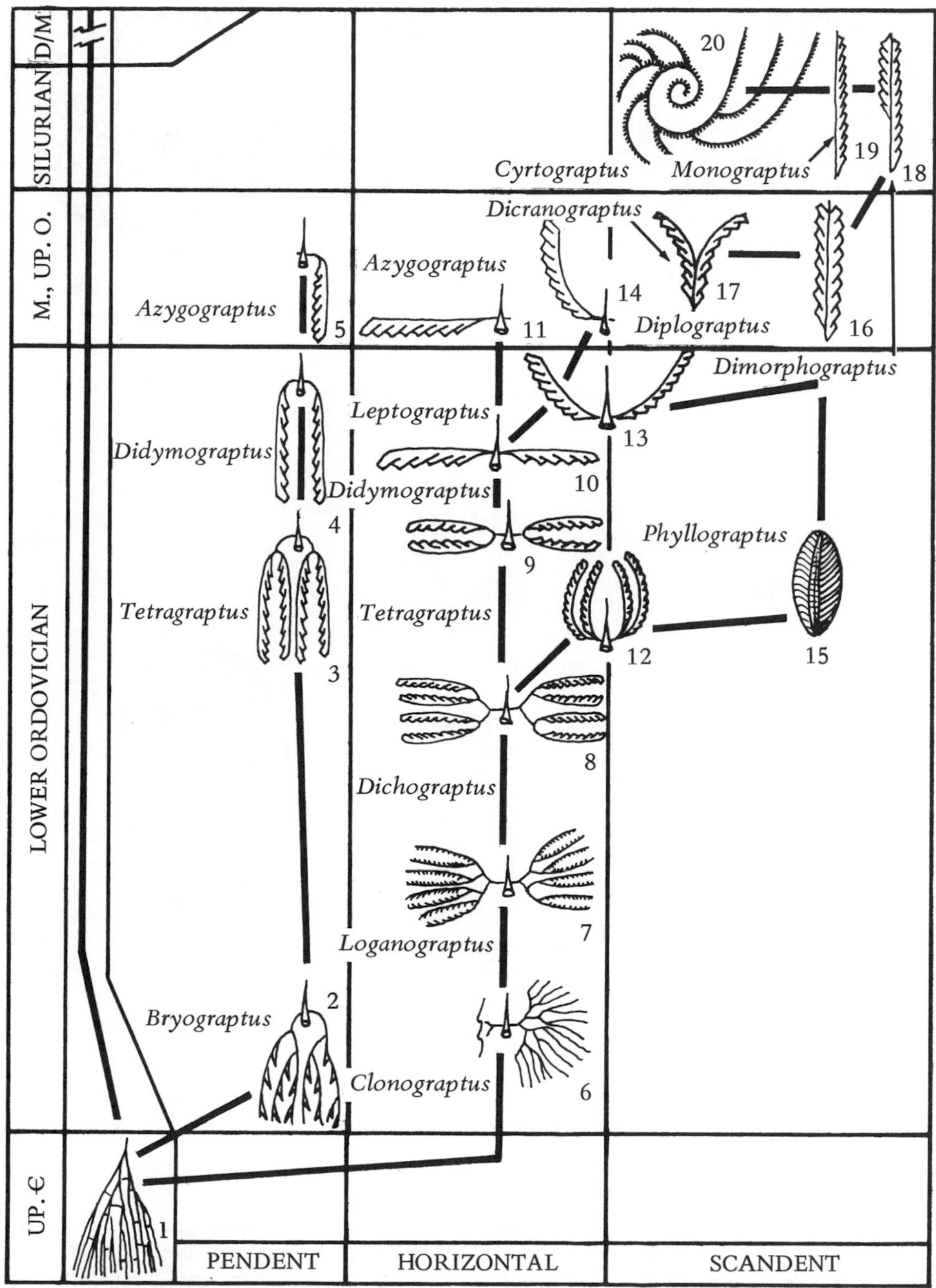

FIGURE 16.25 Evolution of the Graptolites. (From William H. Easton, *Invertebrate Paleontology*, Fig. 4.7 "Evolutionary Patterns of Graptozoans," After Elles, 1922, *Graptolite Faunas of the British Isles*, Proc. Geologists Assoc., and Swinnerton, 1938, *Development & Evolution*, British Assoc. Adv. Sci., Harper and Row, 1960.)

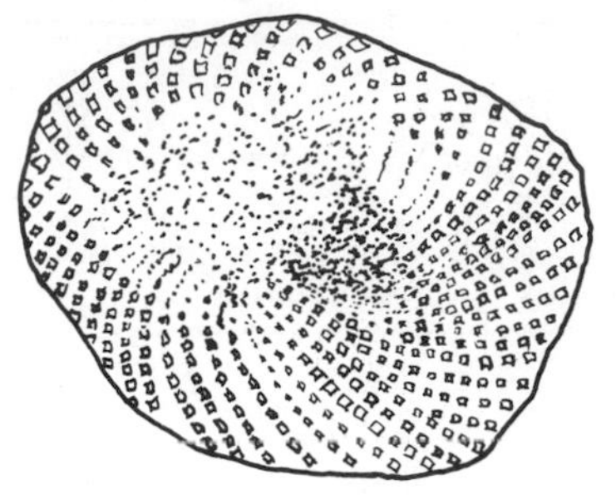

Receptaculites

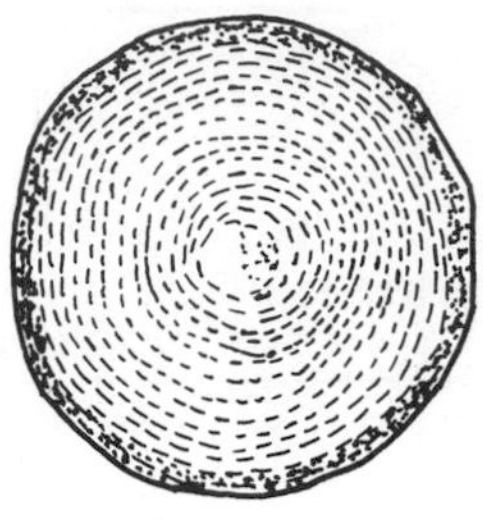

Ischadites

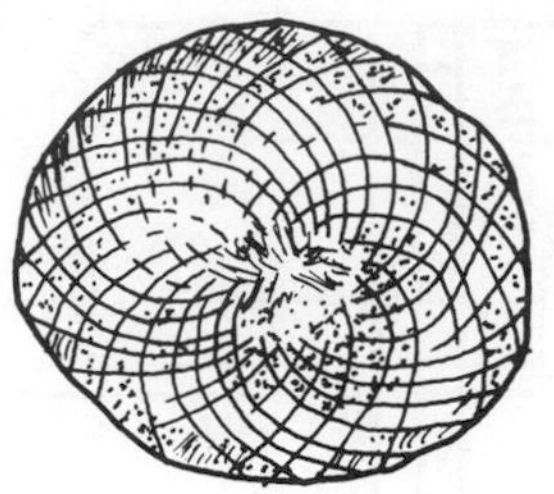

Receptaculites

FIGURE 16.26 Ordovician Receptaculitids.

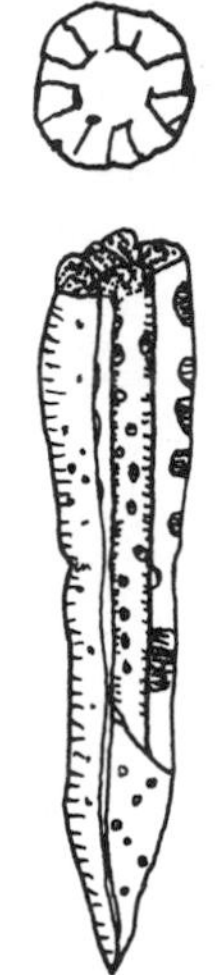

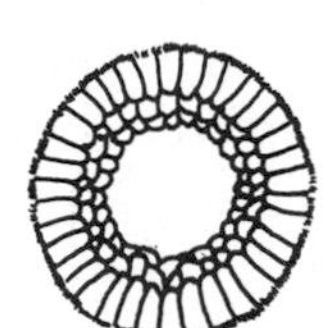

Ethmophyllum

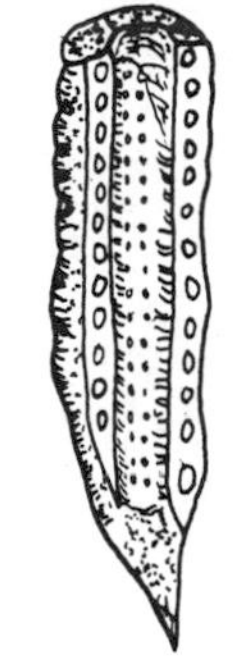

Tercyathus

Ajacicyathus

Nevadacyathus

FIGURE 16.27 Cambrian Archaeocyathans.

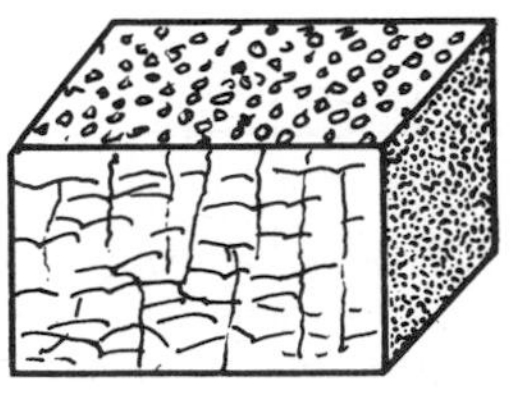

Stramatocerium

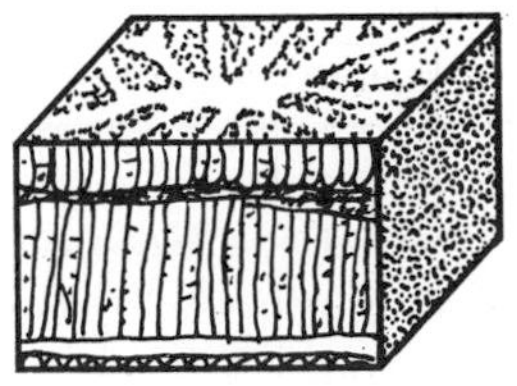

Stromatopora

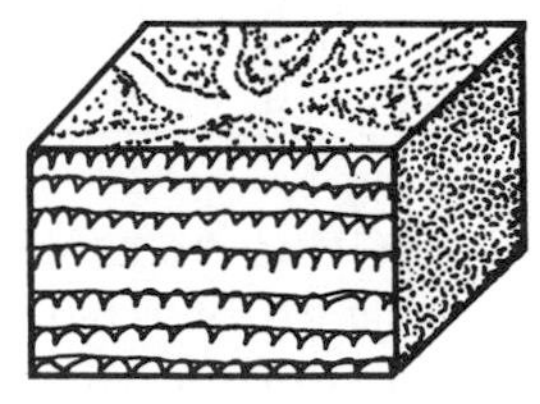

Clathrodictyon

FIGURE 16.28 Silurian Stromatoporoids.

vital for making long-distance time-correlations. Unfortunately, they are chiefly found in black shales formed on toxic bottoms free of scavengers and decay bacteria. Their organic hard parts do not survive as readily on normal marine bottoms that support a wealth of living things which destroy the scleroprotein stipes. Nevertheless, they do occur in sufficient numbers even in these strata to make them the most important group of fossils for long-disance time-correlation in the Lower Paleozoic.

Other Significant Marine Groups

Several other groups of fossil organisms are significant for various shorter intervals within the Cambrian through Silurian interval. These include the receptaculitids, archaeocyathans, stromatoporoids, nautiloids, eurypterans, ostracods, and conodonts. *Receptaculitids* (Fig. 16.26) were large dasycladaceous green algae abundant in the Ordovician. The extinct *archaeocyathans* (Fig. 16.27) are limited to the Cambrian System, in fact they are chiefly Early Cambrian in age. These enigmatic creatures, which may be sponge, coral, or green algae relatives, are the major reef builders of the Cambrian. The *stromatoporoids* (Fig. 16.28) are another extinct group of uncertain affinities, but seem most like hydrozoan coelenterates or sclerosponges. Some questionable Cambrian forms are known, certain stromes are found in the Ordovician, and by the

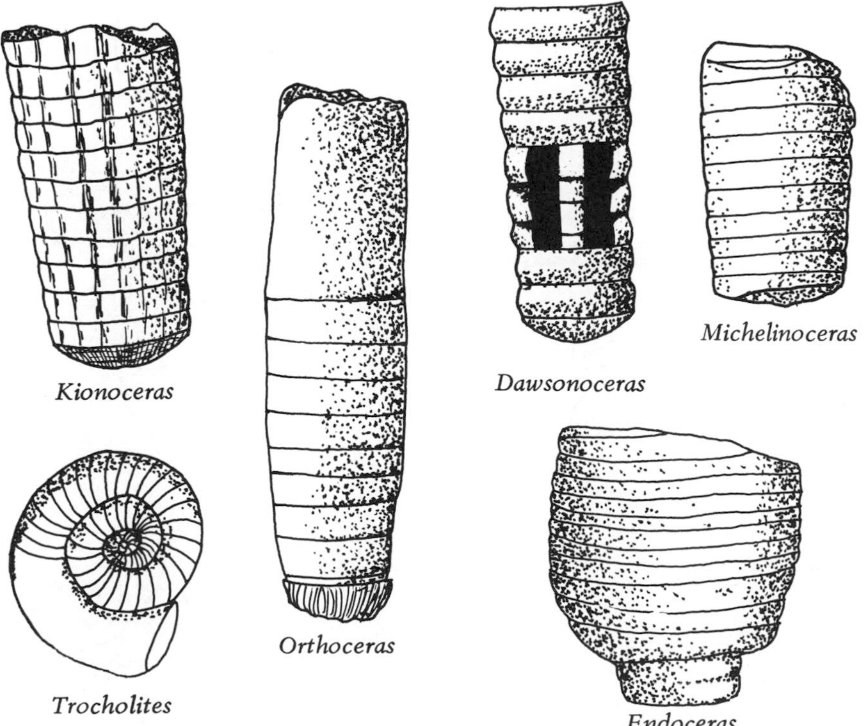

FIGURE 16.29 Lower Paleozoic Nautiloids.

Silurian, the group is a significant faunal element. The encrusting, amorphous stromatoporoids are major cementing agents holding together the framework of Silurian reefs.

The *nautiloids* (Fig. 16.29) appeared in the Late Cambrian and underwent a large adaptive radiation in the Ordovician, the peak of their abundance and diversity. Nautiloids are used for worldwide zoning and time-correlation in the Lower and Middle Ordovician Series. Nautiloids decline in the Silurian, but are still a common faunal element. The *eurypterans*, or sea scorpions, were chiefly

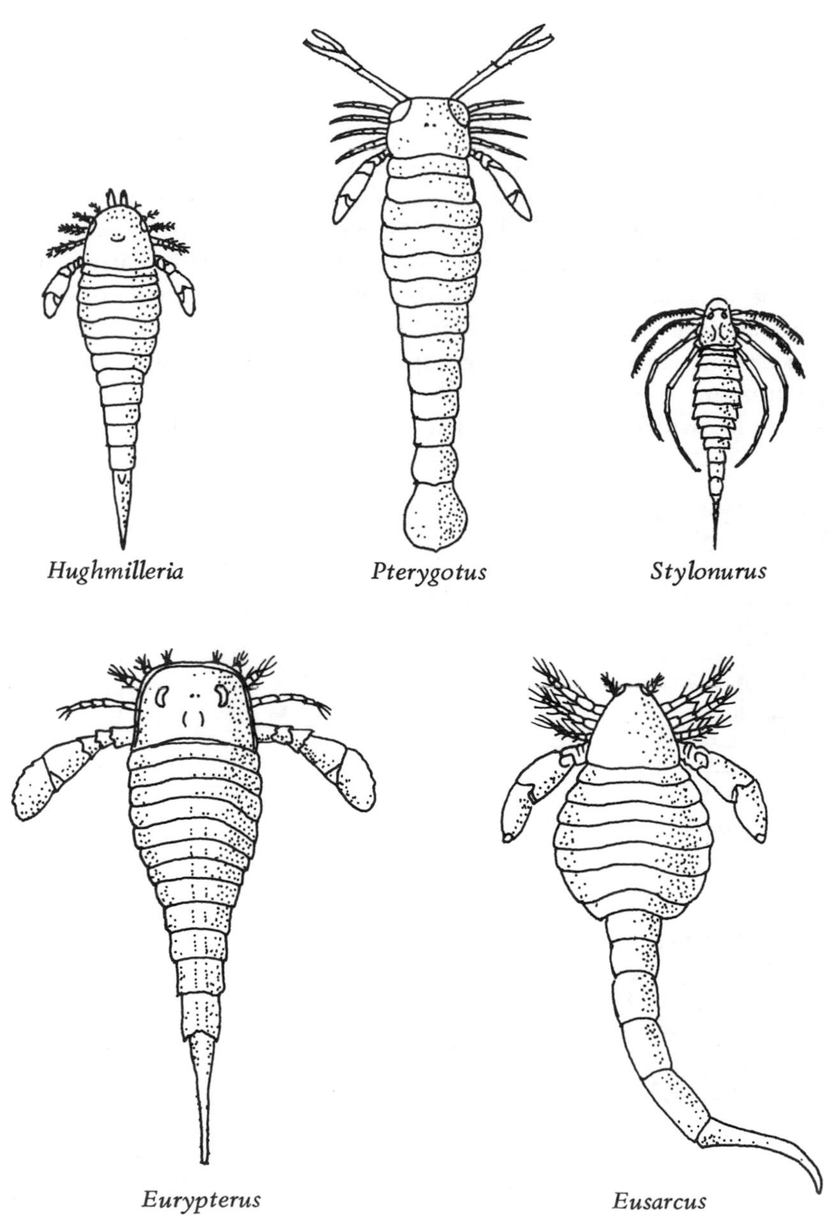

FIGURE 16.30 Silurian Eurypterans.

denizens of brackish waters, an environment not abundantly represented in the fossil record (Fig. 16.30). They range from Ordovician-Permian, but it is only in the widespread brackish water deposits of the Upper Silurian that they are abundantly represented in the fossil record (Fig. 16.4). Most were less than a foot in length, but some may have reached nine feet with their appendages extended in front of them. The tiny crustaceans known as *ostracods,* are rare in the Cambrian, but their bean-shaped, microscopic shells are frequently common in the Ordovician and Silurian (Fig. 16.31). *Leperditia* is a common genus. The *conodonts* (Fig. 16.32) are common from mid-Cambrian times onward, but their peak is in the Upper Paleozoic.

Other lesser groups of fossils reach a peak in the Lower Paleozoic. These are the enigmatic hyolithans and tentaculites, and the cystoid, helicoplacoid, and edrioasteroid echinoderms. The mollusk-like *hyolithans* (Fig. 16.33) reach a Cambrian peak, while the *tentaculites* (Fig. 16.34), which may also be mollusks, are locally common from the Ordovician through the Devonian. *Cystoids,* never abundant, are most common in the Ordovician and Silurian (Fig. 16.35). The spirally plated *helicoplacoids* (Fig. 16.36A) are restricted to the Lower Cambrian, while the long-ranging (Cambrian-Carboniferous) *edrioasteroids* (Fig. 16.36B) reach an Upper Ordovician peak where they are commonly found cemented to brachiopods.

Lesser Groups

The groups just discussed are the major fossil invertebrate elements of the Lower Paleozoic. Note how many belong to extinct orders, classes, and even phyla. The modern groups appeared during this interval, but most were still subsidiary faunal elements and only occasionally were major contributors to the Lower Paleozoic record. The forams and radiolarians were uncommon throughout this interval. The sponges (Fig. 16.37) were common only in the Silurian when various siliceous types such as *Astylospongia* and *Astraeospongium* multiplied. Snails and clams (Fig. 16.38) appeared in the Cambrian, but except for some

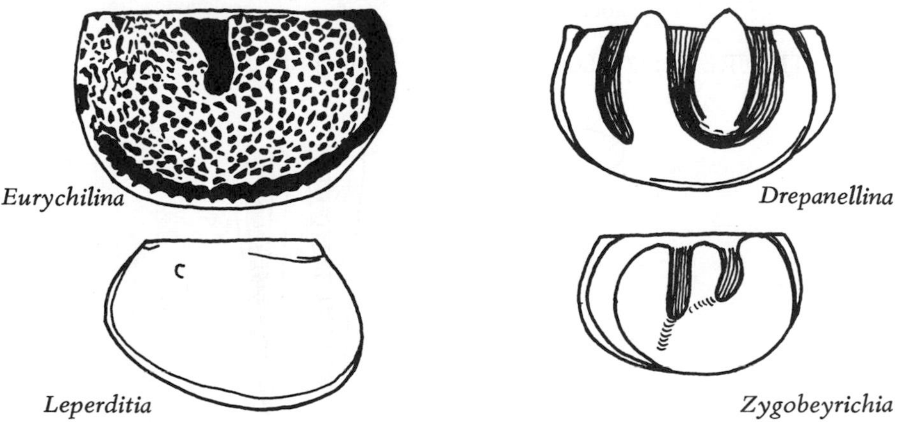

FIGURE 16.31 Ordovician and Silurian Ostracods.

Loxodus
Belodus
Cordylodus
Cyrtoniodus
Dichognathus
Trichognathus
Acodus
Bryantodina
Microelodus
Pteroconus
Oulodus
Phragmodus
Plectospathodus
Paltodus
Distacodus
Oistodus
Drepanodus
Acontiodus
Stereoconus
Polygnathoides
Coleodus
Chirognathus
Erismodus
Neocoleodus
Multioistodus
Leptochirognathus
Cardiodella
Curtognathus
Trucherognathus
Polycaudous
Amorphognathus
Acanthodus
Ambalodus
Clavohamulus
Ulrichodina
Scolopodus

FIGURE 16.32 Lower Paleozoic Conodonts.

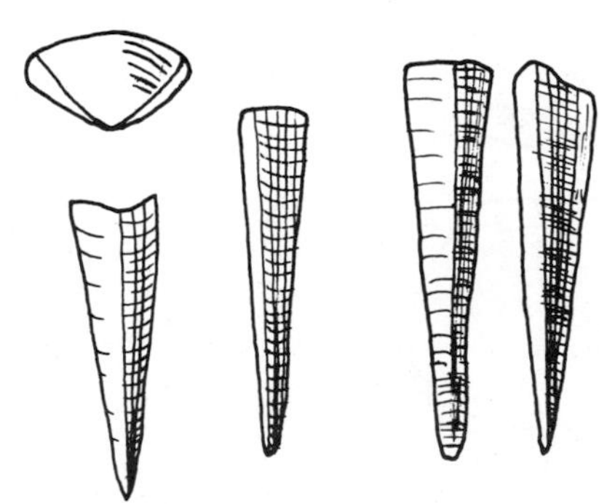

FIGURE 16.33 Lower Paleozoic Hyolithans.

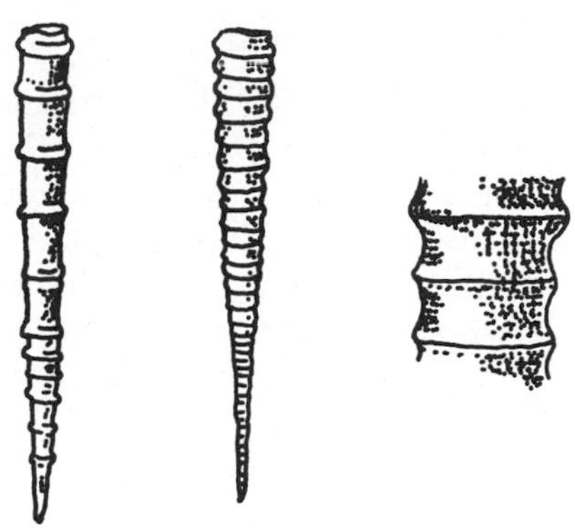

FIGURE 16.34 Lower Paleozoic Tentaculites.

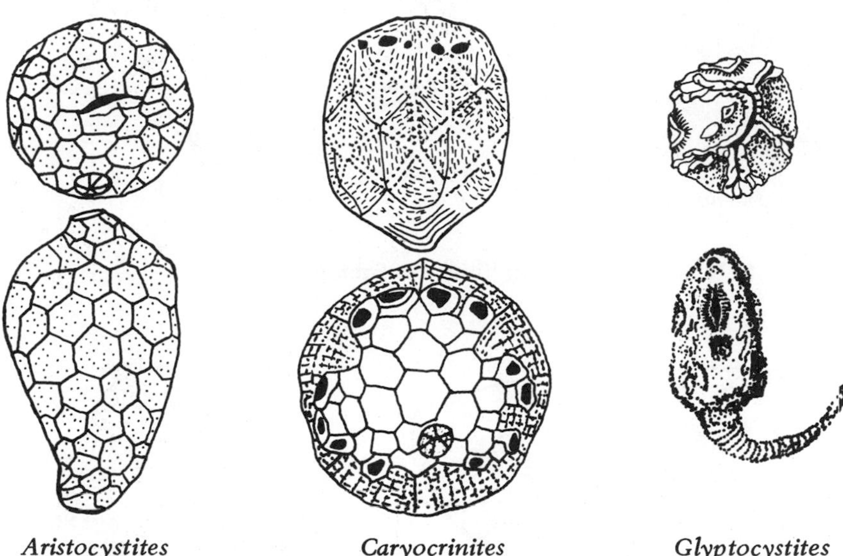

Aristocystites *Caryocrinites* *Glyptocystites*

FIGURE 16.35 Lower Paleozoic Cystoids.

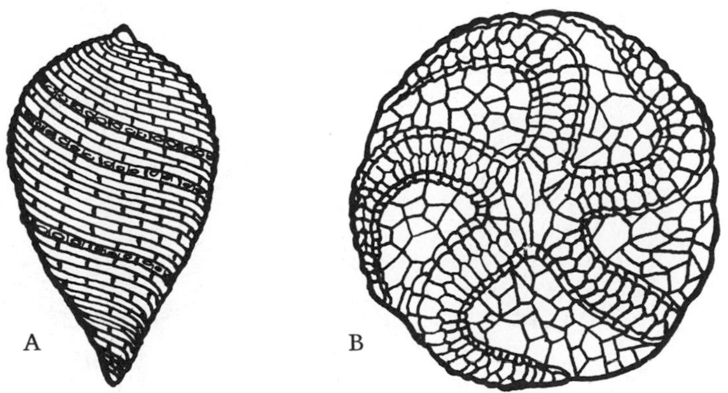

FIGURE 16.36 Lower Paleozoic Echinozoans. (A) Cambrian Helicoplacoid, (B) Ordovician Edrioasteroid.

gastropod genera such as *Maclurites* (Ordovician) and *Cyclonema* (Ordovician–Silurian) and pelecypods such as *Byssonychia* (Ordovician) and *Megalodon* (Silurian), they were not very common. Modern types of crustaceans such as barnacles and malacostracans occur, but are rare. Among the echinoderms, crinoids first appear in the Cambrian and are represented by some common genera (Fig. 16.39) in the Ordovician (*Glyptocrinus*) and Silurian (*Eucalyptocrinites*). Starfish and echinoids appear in the Ordovician but are rare.

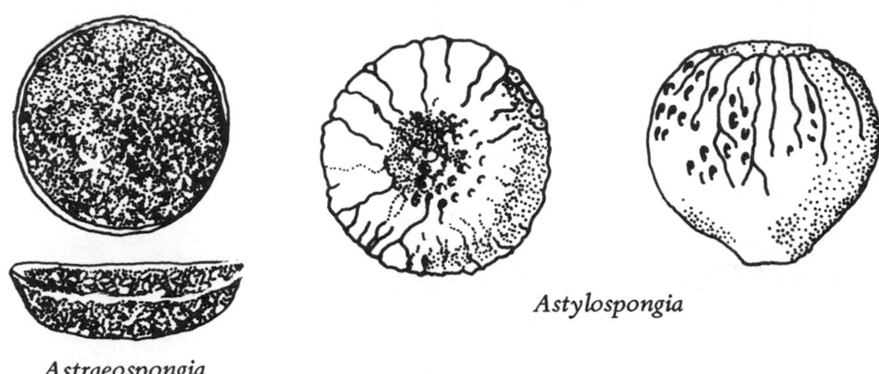

FIGURE 16.37 Silurian Siliceous Sponges.

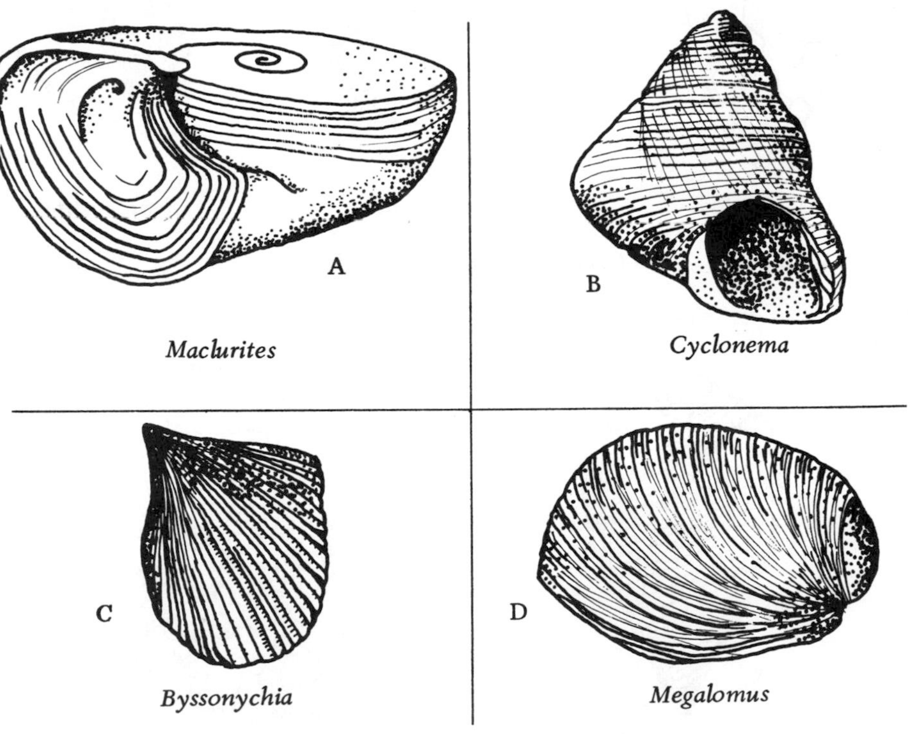

FIGURE 16.38 Lower Paleozoic Mollusks. (A) and (B) Gastropods, (C) and (D) Bivalves.

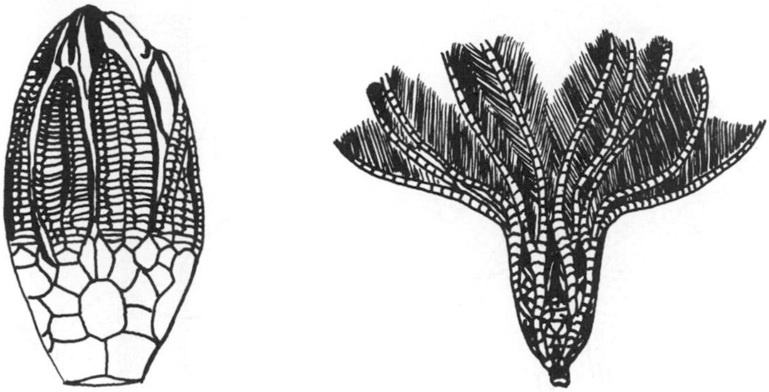

Eucalyptocrinites *Glyptocrinus*

FIGURE 16.39 Lower Paleozoic Crinoids.

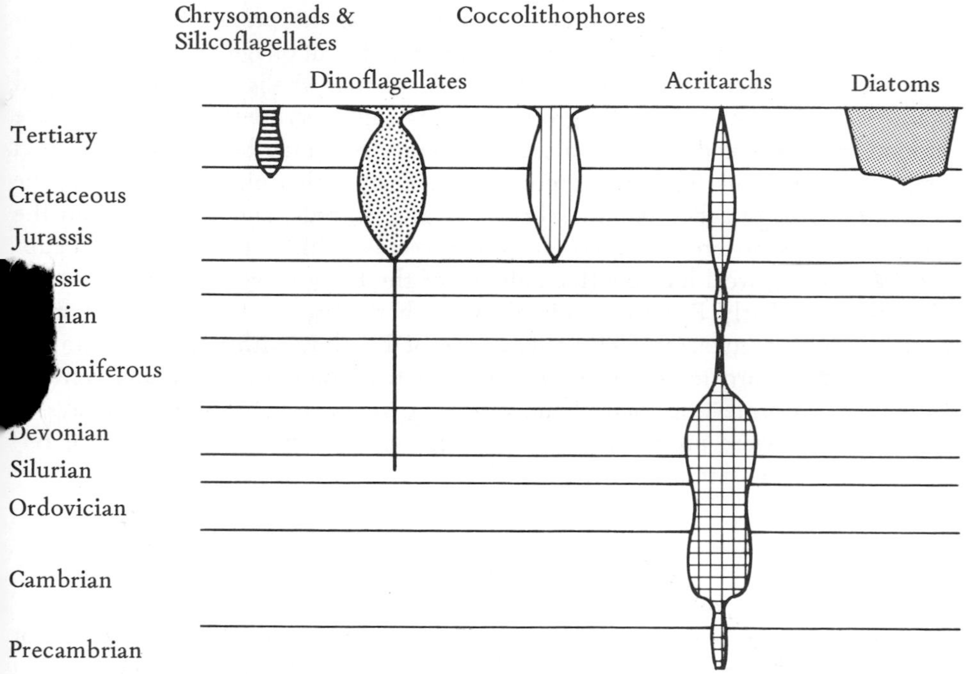

FIGURE 16.40 Geologic Ranges and Diversity of Marine Phytoplankton Groups. (From Raup and Stanley, *Principles of Paleontology*, Copyright © 1971, W. H. Freeman and Co.)

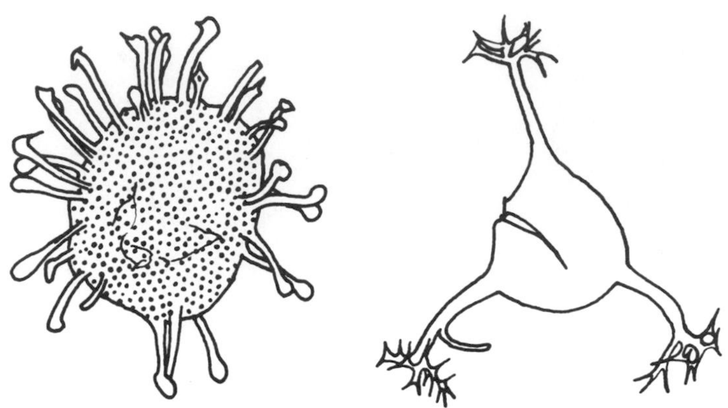

FIGURE 16.41 Lower Paleozoic Acritarchs.

The Marine Ecosystem

It is difficult to reconstruct the food chains of the Early Paleozoic seas. The major producers in the Late Mesozoic and Cenozoic seas, the pelagic phyloplankton groups, are unknown in the Paleozoic except for the dinoflagellates (Fig. 16.40). They may have been present, but lacked shells or had shells of easily destroyed organic materials. An enigmatic group of cysts called *acritarchs* (Fig. 16.41) are common in Lower and Middle Paleozoic strata. They may belong to an extinct group of phytoplankton that supported the Early Paleozoic food chains. The major groups of attached benthic algae, the greens, reds, and possibly even the entirely soft-bodied browns, do occur, but it seems unlikely that they could support the amazing wealth of benthic animals of the Early Paleozoic seas.

Virtually all Early Paleozoic animals were herbivores, scavengers, and m detritus-plankton collectors. The abundance of benthos with hard shells in fossil record is hard to reconcile with this fact. Perhaps the skeletons were major value in protecting their shallow-water owners from the powerful rays the sun. These were probably not yet adequately screened out by the still relatively low oxygen atmosphere.

Biogeography

The most thorough biogeographic study of the Early Paleozoic has been done for the Ordovician (Fig. 16.42). The results agree well with the inferences about plate positions derived from physical criteria. (The positions of the Gondwanaland continents shown in Figure 16.42 do not coincide with the positions deduced from the physical data, but a shift to those positions would not invalidate the conclusions.) Present-day western North America and eastern Asia were in the north temperate zones, while present-day eastern North America, northern Europe, western Asia, western South America, and Australia were in the tropics. Modern southern Europe and northern Africa were in the south temperate zone. In this

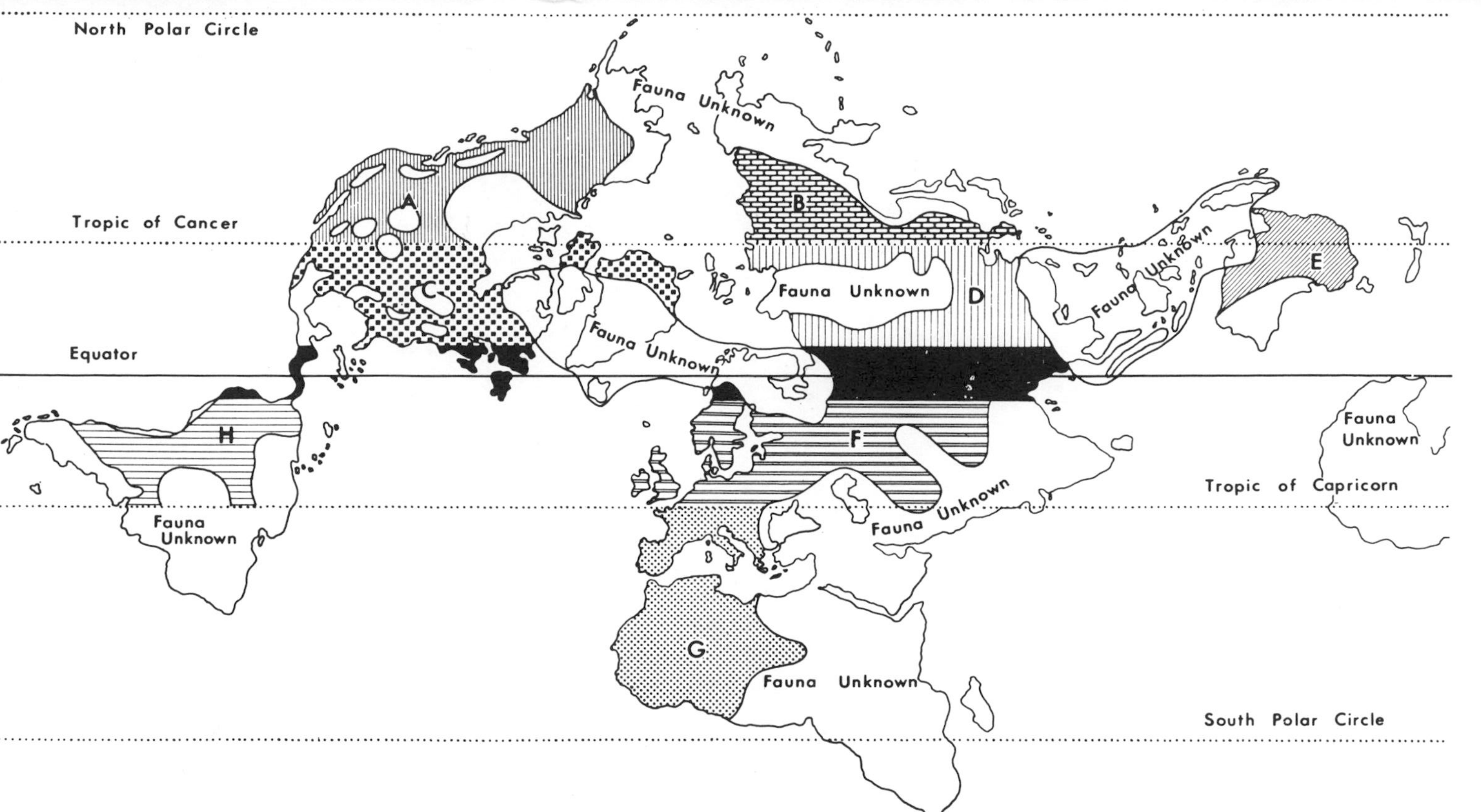

FIGURE 16.42 An Interpretation of Ordovician Biography Based on Distribution, Type, and Affinities of Marine Faunas. (After H. B. Fell in Drake, *Evolution and Environment,* by permission of Yale University Press.)

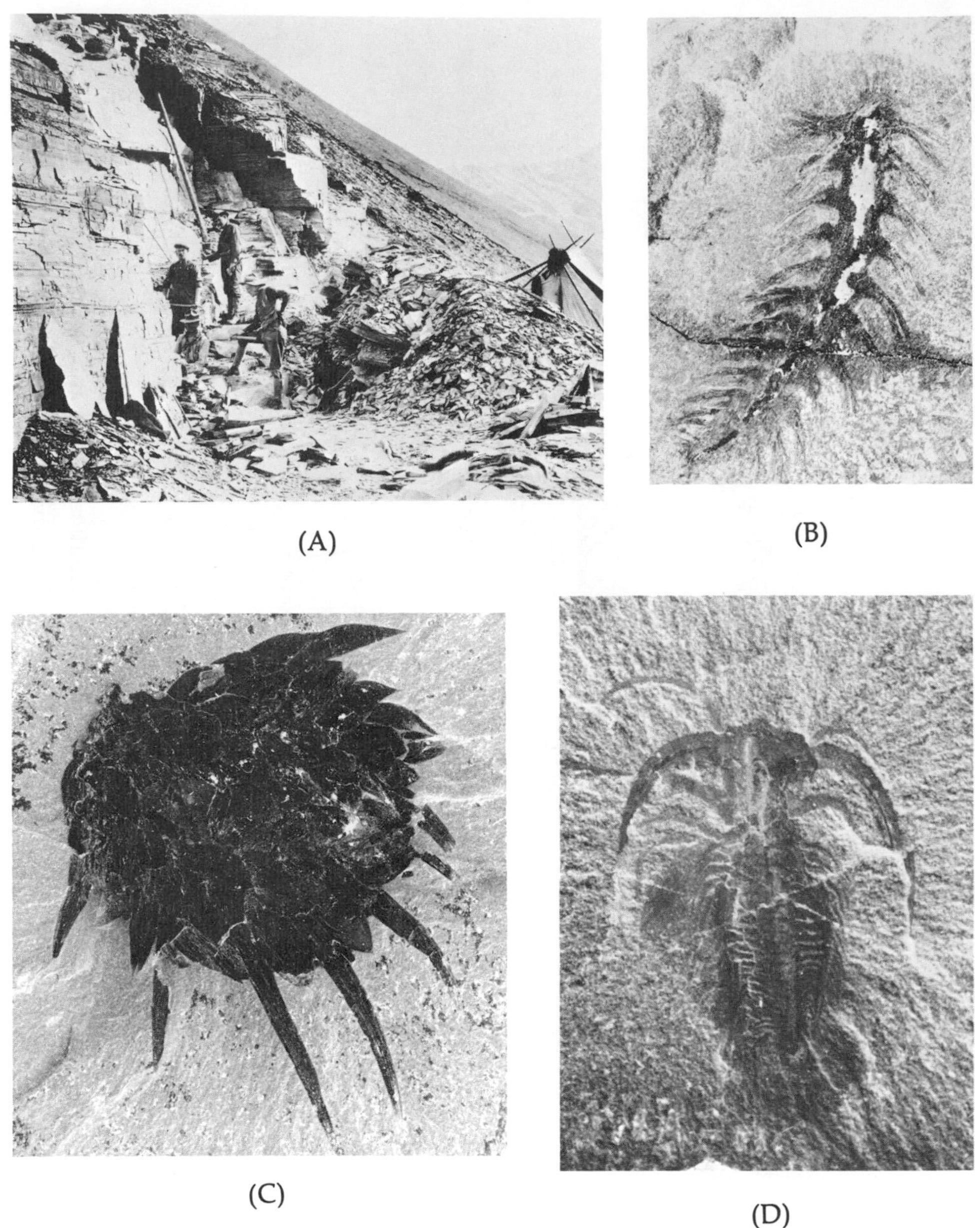

FIGURE 16.43 Burgess Shale Locality and Fossils. (A) The locality high in the Canadian Rockies of British Columbia, (B) *Canadia*, an annelid worm, (C) *Wiwaxia*, an annelid worm, (D) *Marrella*, a trilobitazoan, (E) *Sidneyia*, a chelicerate-like trilobitazoan, (F) *Emeraldella*, a chelicerate-like trilobitazoan, (G) *Naroia*, a chelicerate-like trilobitazoan, (H) *Waptia*, a crustacean-like trilobitazoan, (I) *Olenoides*, a corynexochoid trilobite, (J) *Hymenocaris*, a primitive malacostracan crustacean. (Courtesy of U.S. National Museum.)

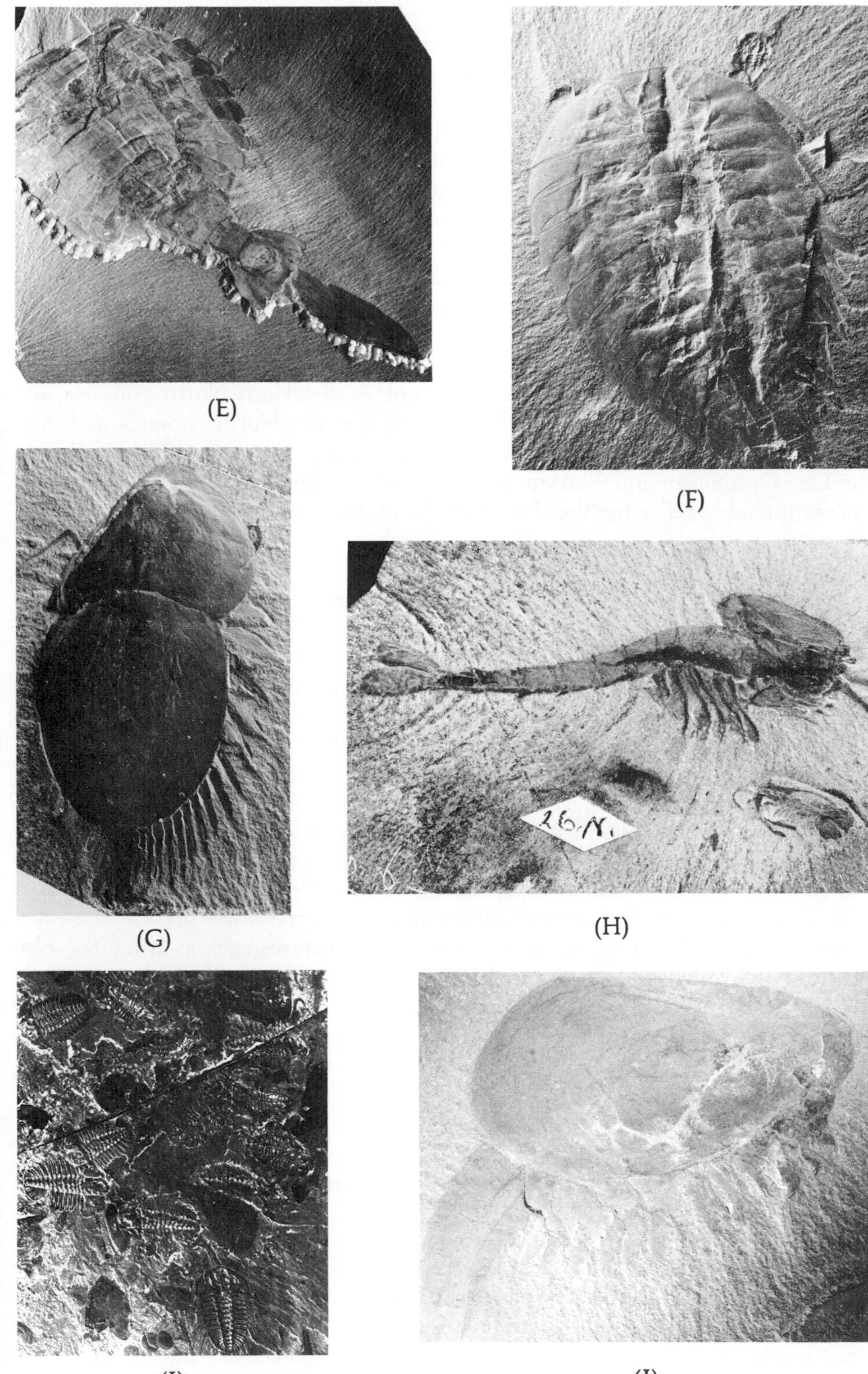

(E) (F) (G) (H) (I) (J)

reconstruction, coral reefs occupy a 70° belt on either side of the Ordovician equator. This is approximately the present width of the coral reef zone and indicates that climatic zones were not appreciably different from today's. The reconstruction also explains faunal similarities between Australia and North America and between Europe and South America because the clockwise currents of the proto-Pacific and the counter-clockwise currents of the proto-Atlantic would favor the transportation of pelagic forms or pelagic larvae in these directions. Coincidently, the data on wind directions deduced from cross-stratification of sandstones indicates that eastern North America lay in the zone of the equator-ward blowing trade winds in the Early Paleozoic. This corresponds exactly with the biogeographic and paleomagnetic reconstructions.

The picture presented for the Ordovician varies only in details for the other Early Paleozoic periods. In the Cambrian, when the proto-Atlantic was still narrow, the faunal similarities between present northeastern North America and Europe were more pronounced. Also, present western North America and Asia must have been closer in the Cambrian because their faunas are quite similar. In the Late Ordovician and Silurian, there was a high degree of cosmopolitanism of faunas probably reflecting the closing of the proto-Atlantic.

The Burgess Shale

No discussion of Early Paleozoic marine invertebrate life would be complete without mention of the famous Burgess Shale localities in the Middle Cambrian of the Canadian Rockies (Fig. 16.43). Known only from a few scattered outcrops high on the west slopes of the Rocky Mountains in Yoho National Park, British Columbia, this black shale unit is noted for its distillation fossils of a cross-section of Cambrian life. The deposit apparently formed in a toxic, hydrogen sulfide-generating environment which excluded bacteria and scavengers. As a result, organisms which crawled or swam into the area were asphyxiated, died, and were buried without being destroyed. Their remains give the most complete glimpse of Lower Paleozoic life that we have because the soft parts are preserved as carbonaceous films. Not only are the soft parts of creatures such as trilobites, typically known only from their exoskeletons, preserved, but also a host of soft-bodied forms otherwise unknown as fossils (Fig. 16.43). These include many worm phyla, including the annelids, and several groups of soft-bodied arthropods transitional in structure between the trilobites and the modern arthropod subphyla.

The Oldest Vertebrates

The *vertebrates* appear in the Lower Paleozoic, but until the Upper Silurian, they are not common. Some apparent tooth and jaw fragments from the Cambrian of Wyoming may be the oldest vertebrates. Other jaws are known in the very Early Ordovician of Missouri and Colorado and then fragments of bony *ostracoderm* head shields or spines occur in several localities of Lower to Middle Ordovician age. These disarticulated remains occur in nearshore deposits which has led to a spirited debate among vertebrate paleontologists about whether fish (and, hence, vertebrates) originated in the sea or in fresh water and were washed into the sea. Silurian fish were apparently rare until the end of the period when the rising Caledonian-Acadian Mountains produced the Old Red Sandstone "Continent." In lake and stream deposits interbedded in these terrestrial redbeds, both complete and partial fish remains are common for the first time. They are chiefly ostracoderms, but some fragments of placoderms and acanthodians also occur.

Invasion of the Land

The invasion of the land also occurred late in the Lower Paleozoic. Although occasional carbonaceous smudges, algal cysts, and supposed wood fragments have been reported from Cambrian, Ordovician, and Silurian rocks, it is not until the very close of the Silurian that unequivocal plant remains occur. These uncommon fossils are members of the simple tracheophyte group, the *psilophytes* (Fig. 16.44). The animals also may have invaded the land at this time. *Scorpions* (Fig. 16.45) are known from Upper Silurian rocks, but it is uncertain if these earliest representatives of the group were fresh water or terrestrial. Recent studies suggest that these ancestral types were probably aquatic, unlike their descendants. Also

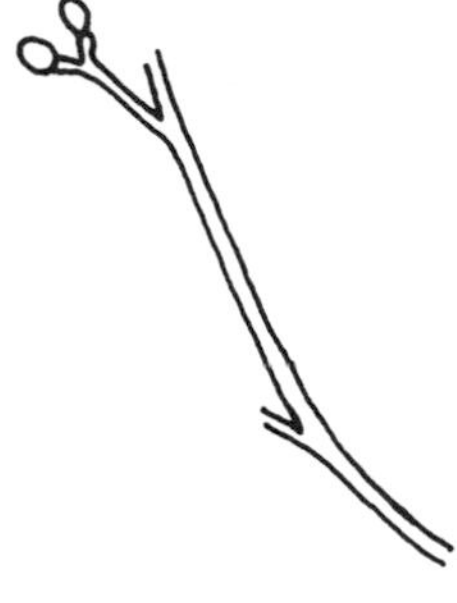

FIGURE 16.44 Silurian Psilophyte.

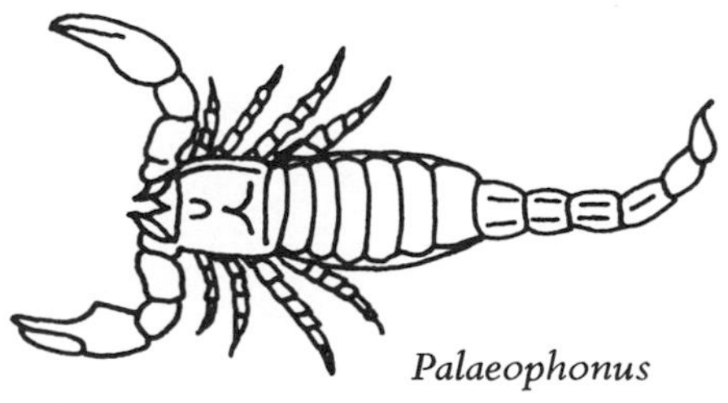

FIGURE 16.45 Silurian Scorpion.

found in Upper Silurian rocks is a member of the archipolypods, a group of arthropods related to the *millipedes*. The millipedes are entirely terrestrial today, but again it is not certain if these early fossils were aquatic or terrestrial because the preservation does not permit identification of breathing structures. Some Lower Paleozoic malacostracan crustaceans may have been fresh water or even partially terrestrial.

Questions

1. Why is the Early Paleozoic called the Age of Marine Invertebrates?
2. What were the major marine invertebrate groups of the Early Paleozoic?
3. Which type of corals dominated Silurian reefs?
4. Which types of bryozoans reached an Ordovician peak?
5. What were the four major brachiopod groups of the Early Paleozoic?
6. How could you distinguish a Cambrian trilobite fauna from an Ordovician one?
7. Which group of fossils is most useful for intercontinental time-correlations in the Ordovician and Silurian? Why?
8. In which Early Paleozoic period did each of the following groups reach a peak: receptaculitids, archaeocyaths, stromatoporoids, nautiloids, eurypterans?
9. Which significant modern groups of marine invertebrates were relatively unimportant in the Early Paleozoic?
10. Why is it difficult to reconstruct Early Paleozoic marine food chains?
11. What is the significance of the Burgess Shale fauna?
12. What evidence occurs of vertebrates and land life in the Early Paleozoic?
13. How well does Early Paleozoic biogeography fit with plate tectonics models?

17

Limestone, Coal, and Redbeds:

Devonian, Carboniferous, and Permian

The three periods of the Late Paleozoic, the Devonian, Carboniferous, and Permian, encompass approximately 175 million years of earth history from about 400 to 225 million years ago. The *Devonian* has its type area in the Hercynian Geosyncline of southwest England; the *Carboniferous,* in central Great Britain on a shelf on the site of the former Caledonian Geosyncline; and the *Permian,* in Russia on the European craton just west of the Uralian Geosyncline (Fig. 2.19). In North America, the Carboniferous is typically subdivided into two periods, the *Mississippian,* named from the Upper Mississippi Valley of Illinois and Iowa, and the *Pennsylvanian,* named from the state of Pennsylvania. Both type areas are on the craton, but the latter is near the edge of the Appalachian Geosyncline. Geologists have a considerably better picture of the global tectonics of the Late Paleozoic than they do of the Early Paleozoic interval.

Gondwanaland

Gondwanaland was still a single entity as indicated by the paleomagnetic, geological, and paleontologic data. However, it had shifted position because the location of the Late Paleozoic Gondwana Pole (Fig. 15.3) changed from present South Africa in the Devonian, across present Antarctica from the Weddell Sea area to the Ross Sea area, and then to Wilkes Land in the Carboniferous, and finally to the vicinity of present southern Australia in the Permian. The situation of the pole in Gondwanaland is reflected by the widespread Carboniferous and Permian glacial deposits and erosional features (Fig. 17.1) of eastern South America, southern Africa, Antarctica, India, and Australia. The succeeding terrestrial redbed sequences with coal (Fig. 17.2) are strikingly similar on all the present southern continents which once composed Gondwanaland. The best known is the Karroo Group of South Africa with its large mammal-like reptile fauna. Nearshore marine rocks of the same lithofacies and with the same fossil assemblages occur in the Permocarboniferous sequence near Buenos Aires, Argentina, in Southwest Africa, the Salt Range of Pakistan and southeastern Australia.

The marine *Eurydesma* fauna (Fig. 17.3) contained in these rocks is only one of the impressive paleontologic links between the now widely separated southern continents. We have already mentioned the small fresh-water reptile *Mesosaurus* (Fig. 18.59) found in the Permocarboniferous of South America and South Africa. Most striking of all in stimulating early thought about continental drift, however, is the *Glossopteris* flora (Fig. 17.4) of Late Carboniferous to Early Triassic age. This apparent seed fern and its attendant plants are the major contributors to the important coal beds of this interval on today's southern continents.

The Pacific Plate(s) with the New Zealand continental fragment must have been moving against Gondwanaland along the present eastern side of Australia. The Tasman Geosyncline, as well as New Zealand, underwent continuous disturbances which culminated in the mid-Carboniferous *Kanimblan Orogeny.* Its effects remain today in the Great Dividing Ranges (Fig. 17.5) of eastern Australia. The Antarctic continuation of the geosyncline, the Borchgrevink, was also affected. In the Permian, terrestrial beds with interbedded coal and tillites succeed this orogenic disturbance. In the present east Pacific area, the Pacific

FIGURE 17.1 Gondwana Glacial Deposits and Erosional Features. (A) Polished pavement on quartzite, friction cracks, South Australia, (B) Striated and faceted stone in tillite, Tasmania. (Courtesy of Warren Hamilton, U.S.G.S. from Hamilton and Krinsley, *G.S.A. Bulletin 78,* June 1967, by permission of authors and G.S.A.)

Plate must have been impinging against South America as the Andean Geosynclinal trough accumulated sediments and underwent deformations throughout much of the Late Paleozoic. The Andean belt of the Late Paleozoic bends eastward through present northern Argentina (Sierra de la Ventana) and the Cape Geosyncline of South Africa, also deformed in the Permo-Triassic interval, and continues into the present Antarctica and eastern Australian belt discussed above. Geologists call this great geosyncline (Fig. 11.5) the *Samfrau* after South America, Africa, and Australia.

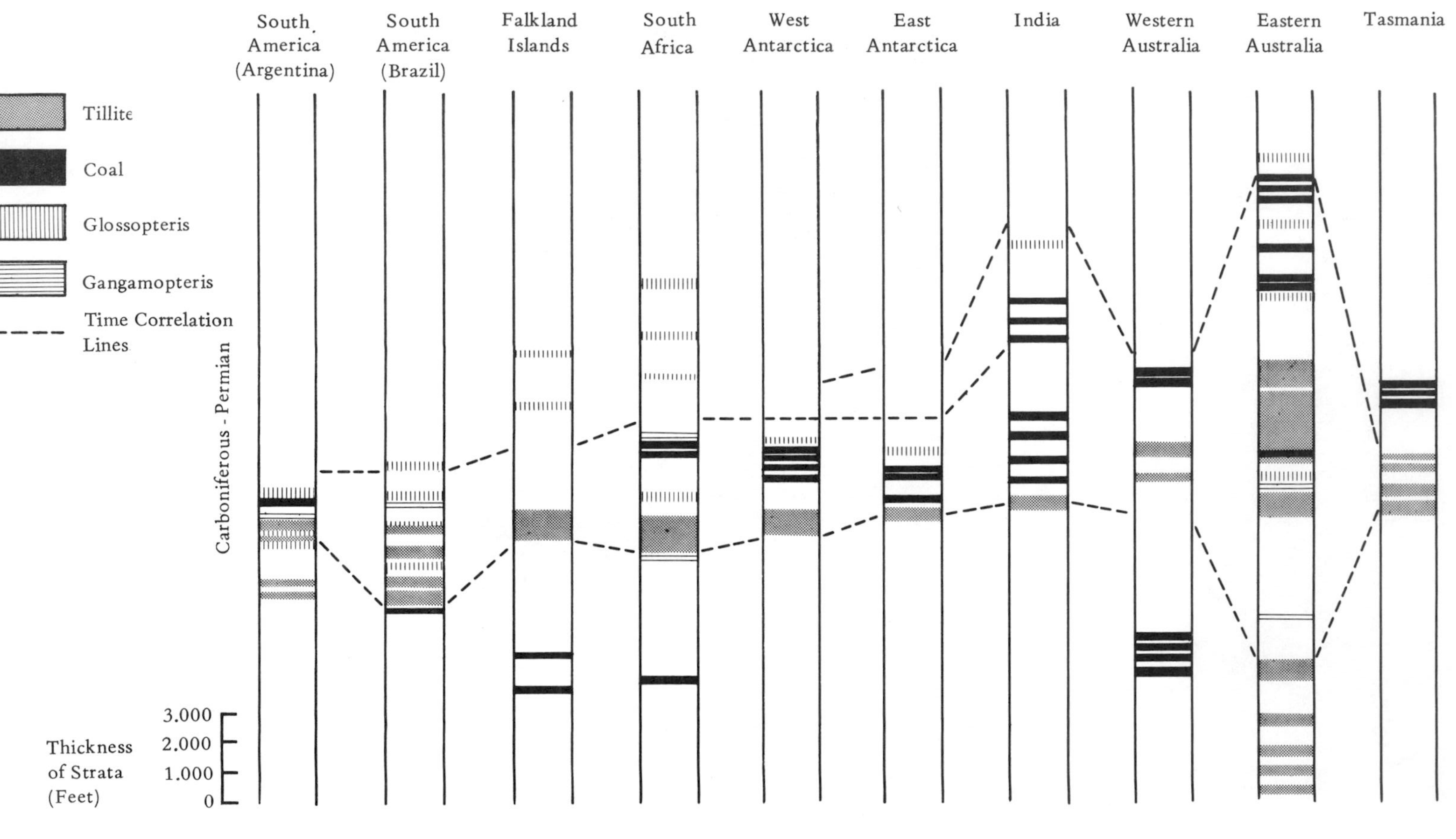

FIGURE 17.2 The Gondwana Succession on the Southern Continents. (A) Generalized cross-sections of all Gondwana Areas.

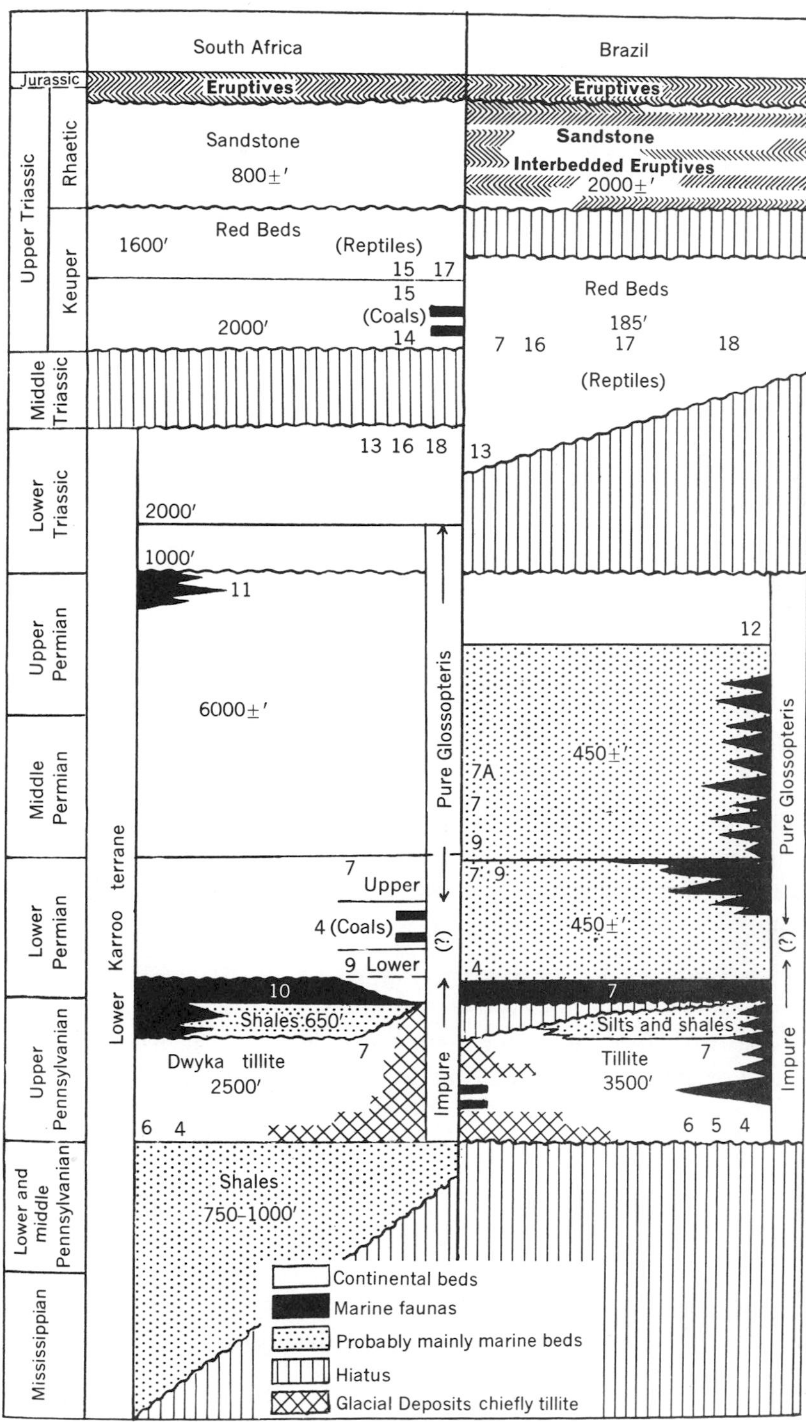

(B) Detailed Gondwana Succession of South Africa and Brazil. (Numbers refer to significant fossils found in both areas.) (B) From Kay and Colbert, *Stratigraphy and Life History*, Copyright © 1965, by permission of John Wiley & Sons, Inc.

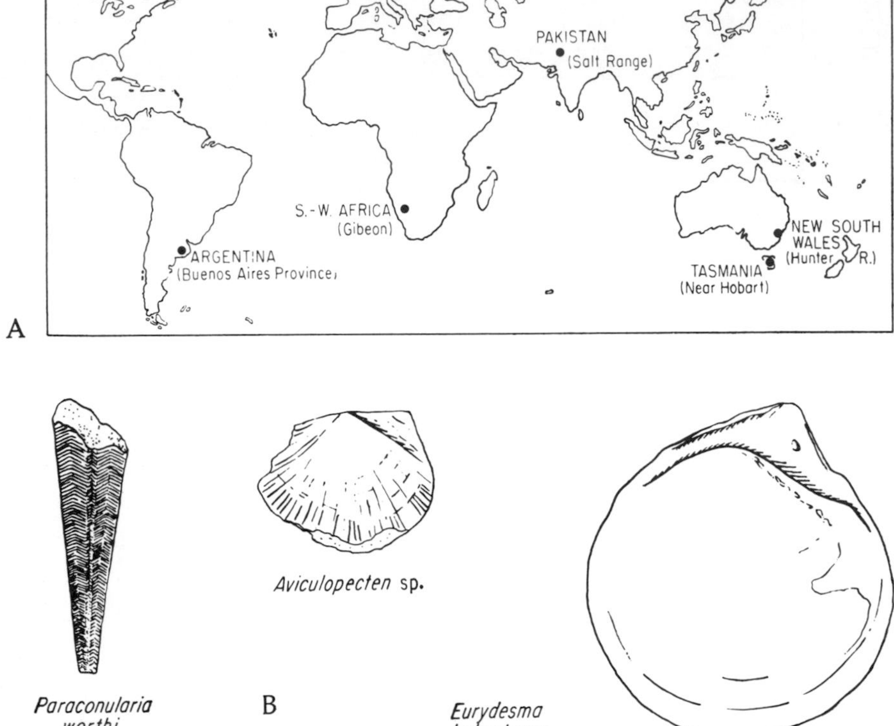

FIGURE 17.3 The Eurydesma Fauna. (A) Geographical Distribution Today. (B) Typical Members. (From Ager, *Principles of Paleoecology,* Copyright © 1963, used with permission of McGraw-Hill Book Co.)

Laurasia

Laurasia also took form as a single super-continent in the Late Paleozoic. The collision of present eastern North America and northwest Europe which had begun in the Late Silurian continued into the Devonian producing the *Acadian Mountains* (Fig. 17.6) as they are called in North America or the *Old Red Sandstone "Continent"* (Figs. 15.17, 17.7) as it is known in Europe. Acadia refers to a region in the Canadian Maritime provinces where the effects are well displayed. From this region to the present south, North America collided with the present northwestern edge of Africa (Fig. 17.8). The *Catskill* clastic wedge (Fig. 17.9) which extends from southeastern New York to Virginia is the main expression of the Acadian Orogeny in the United States. Deformation continued later into the Paleozoic as evidenced by the formation of thick sequences of clastics in the Mississippian. The *Pocono* clastic wedge of Pennsylvania is one of the best known. Several fault-block basins developed in the Canadian Maritime Provinces (Fig. 17.10).

The final phases of orogenic deformation in the Appalachian-Caledonian belt occurred toward the close of the Paleozoic Era. In North America, this major deformation is called the *Appalachian* or *Alleghenian Orogeny* and extends

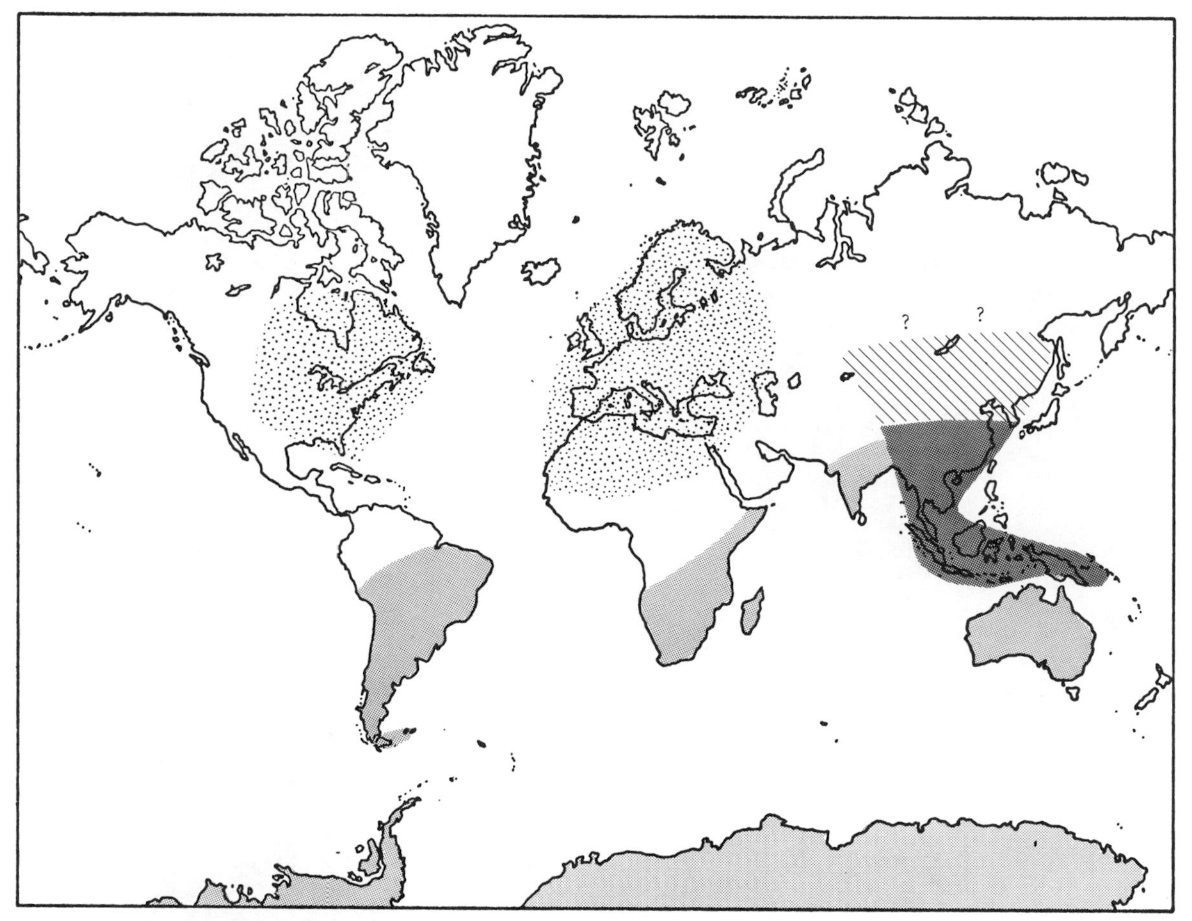

FIGURE 17.4 Present-Day Distribution of Upper Paleozoic Floras.

A

B

FIGURE 17.5 Great Dividing Ranges of Australia. (A) Mt. Kosciusko, highest mountain in Australia, (B) The Three Sisters in the Blue Mountains. (Courtesy Australia Tourist Commission.)

from the Pennsylvanian, through the Permian into the Triassic. Its major expressions today are the folds, thrust faults, and intrusions preserved in our present Appalachian Mountains (Fig. 17.11) and the *Dunkard* clastic wedge which extends westward from them. The causes of the Appalachian-Allegheny Orogeny are unclear. Some authorities consider it to be the final result of the collision of North America, Europe, and the African edge of Gondwanaland, while others believe it represents the initial stages in the formation of the rift that eventually formed the present Atlantic Ocean basin. The fact that the Late

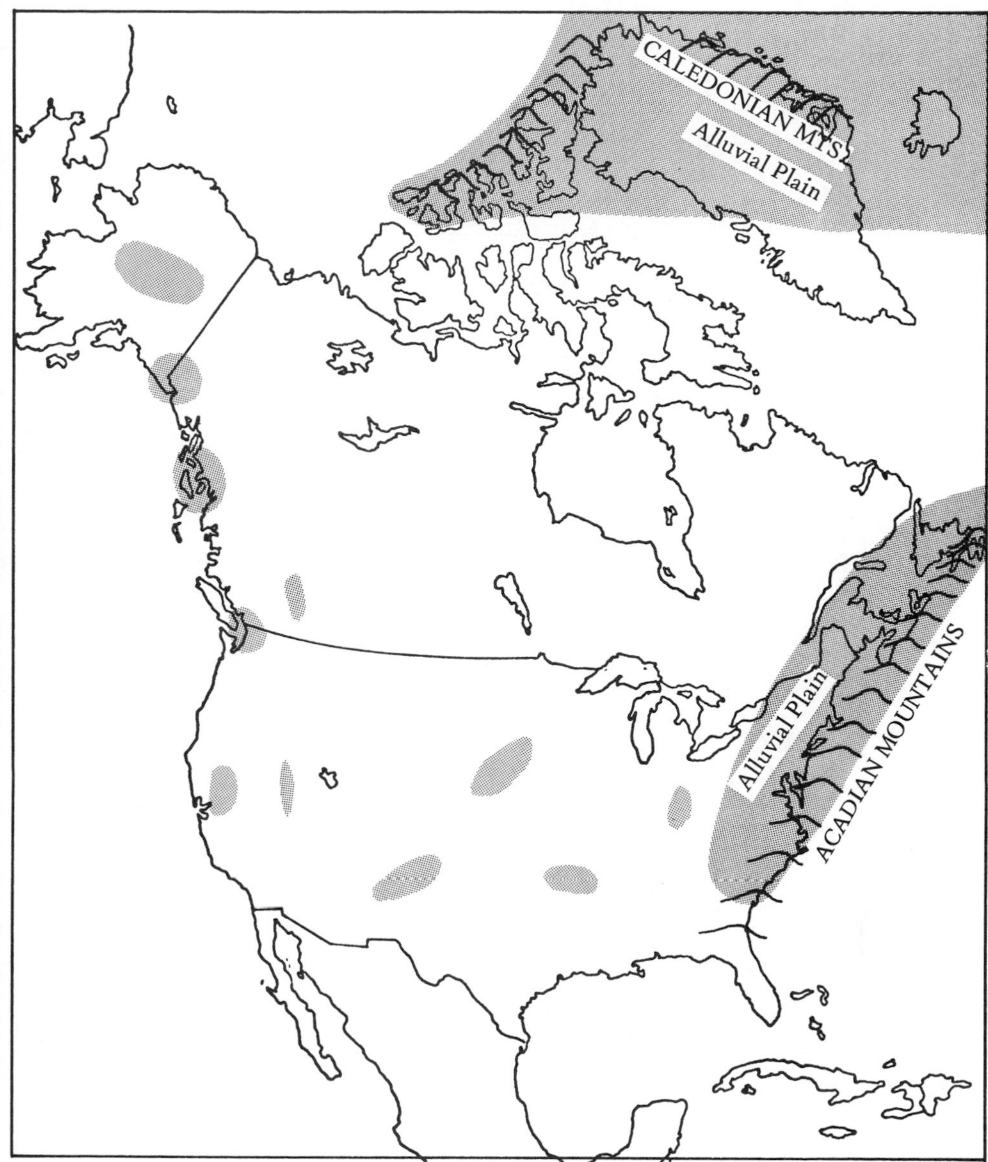

FIGURE 17.6 Present-Day Distribution of Acadian-Caledonian Mountain Belt in North America.

Paleozoic structural trends parallel the present coastlines and cut across older Appalachian structures tends to support the latter contention.

In Europe, the collision of present northwest Africa (Fig. 17.7) produced extensive mountain-building. This episode is named the *Hercynian* (after the Harz Mountains, Germany) or the Variscan *Orogeny*. Culmination was in the Late Carboniferous and Early Permian. Its present remnants (Fig. 17.12A-B) occur in isolated low mountain massifs in southwestern England, northern France and Germany, Bohemia, and the Iberian Peninsula (Spain and Portugal). The disturbance is also recorded in the Alps of southern Europe and the Atlas Moun-

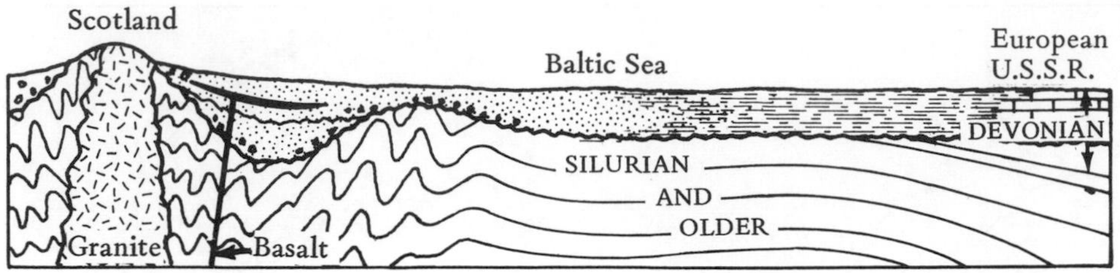

FIGURE 17.7 Cross-Section of Old Red Sandstone "Continent" of Europe.

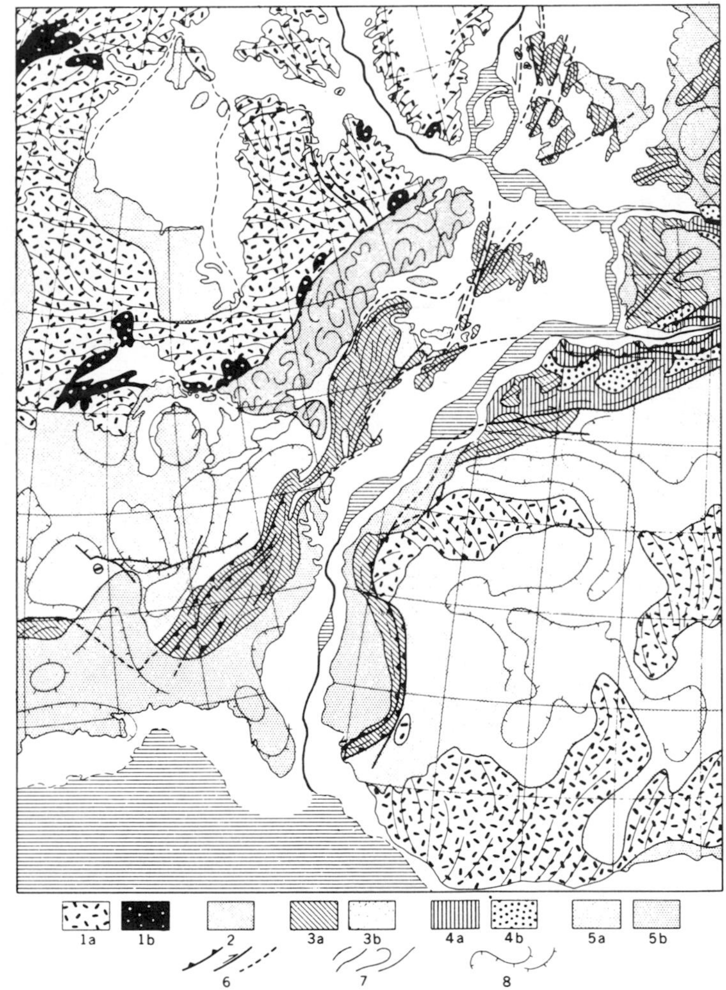

FIGURE 17.8 Relationships of Present-Day North America, Europe, and Africa in the Late Paleozoic. (From: Johnson and Smith, *The Megatectonics of Continents and Oceans.* Copyright © 1970, Rutgers University Press, New Brunswick, N.J.)

WEST EAST

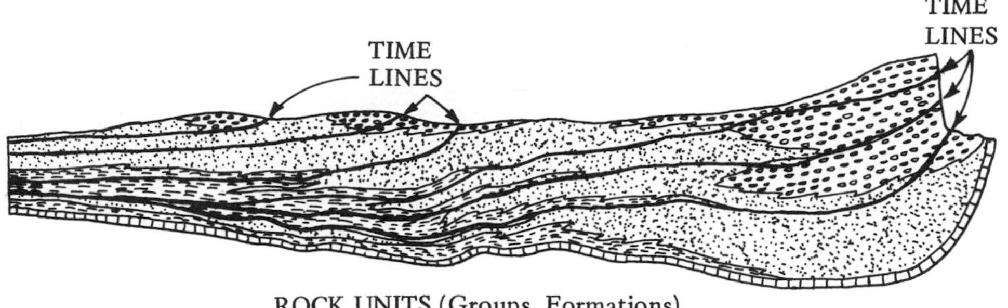

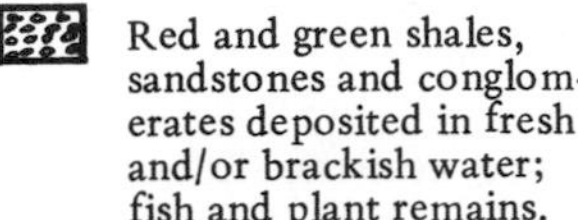
Red and green shales, sandstones and conglomerates deposited in fresh and/or brackish water; fish and plant remains.

Interbedded gray shales, siltstones and sandstones deposited in shallow marine water; abundant and varied fossil invertebrates.

Dark gray and black shales of deeper-water origin; only ammonoids and thin-shelled bivalves common.

Shallow-water marine limestones, often cherty; fossils abundant and varied.

FIGURE 17.9 Cross-Section of Devonian System in Southern New York Showing Catskill Clastic Wedge.

tains of northwest Africa, but the effects are largely obscured by later orogeny.

The collision of the present continental masses of Europe and Asia also occurred in the Late Paleozoic as the *Uralian Orogeny* (Fig. 17.13) resulted from compression of the Uralian Geosyncline between the Baltic and Angara cratons in the Permian. In the Early Devonian, an island arc collided with the present eastern edge of the European craton. This brought a subduction zone to the edge of the craton, but this zone then reversed its Early Paleozoic dip away from the continent to a direction toward and beneath the continent. From Middle Devonian through Early Permian, the situation probably resembled that in eastern Asia today where the Ryukyu arc is being driven against the craton by the Pacific. On the Asian side, a subduction zone dipping beneath the continent had been present since the Eocambrian. Both subduction zones collided as Europe and Asia joined in the Permian.

Southern Laurasia was deformed along its entire southern and eastern margins by the Hercynian Orogeny. As stated earlier, the westernmost part of this belt in Europe resulted from the collision of two continental plate edges (Fig. 17.12). To the present east, through eastern Europe, the Near East and

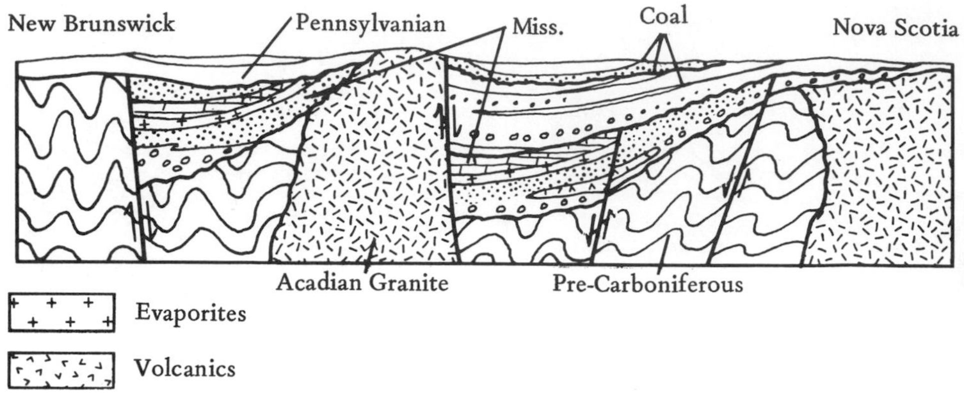

FIGURE 17.10 Late Paleozoic Fault-Block Basins of the Canadian Maritime Provinces.

across southern Asia, however, the colliding margins were continental on the Laurasian side, but oceanic on the Gondwana side. The southeast Asia compression may have resulted from its collision with the East China craton. Along the present eastern side of Asia, an oceanic plate was impinging against the craton.

Deposition continued in the Cordilleran Geosyncline (Fig. 17.14) which underwent Late Paleozoic deformations (Fig. 17.15) called the *Antler Orogeny* (chiefly Mississippian) and the *Sonoma* or *Cassiar Orogeny* (chiefly Permian). The causative compression came from a Pacific Ocean plate(s) to the present southwest. That edge of the North American craton was buckled, chiefly in the Pennsylvanian, into the so-called *Ancestral Rockies* and *Oklahoma Mountains* (Fig. 17.16A-C). Great thicknesses of marine and terrestrial clastics accumulated in the basins between the various ranges produced by this deformation. Later deformed by the disturbances that produced the present western Cordillera, some of the arkoses now stand with steep eastward slopes along the flanks of the Colorado Rockies forming the picturesque Flatirons, Red Rocks, and Garden of the Gods (Fig. 17.17). Some of the marine strata have been subsequently elevated to form high mountains (Fig. 17.18). Completing our circle of Laurasia was a westward continuation of the southern Appalachian belt called the *Ouachita-Marathon Geosyncline* (Fig. 17.19) which underwent a great Mississippian filling followed by the Pennsylvanian-Permian *Ouachita-Marathon Orogeny* (Figs. 17.20, 17.21). This deformation probably was caused by the approaching edge of the South American region of Gondwanaland.

The collision of Gondwanaland with Laurasia in the present South America-Marathon-Ouachita region and the Africa-Appalachian-Hercynian region produced a true Pangaea in the Triassic (Fig. 17.22).

As a final note on orogenies, one must mention the Devonian–Mississippian *Innuitian Orogeny* (Figs. 17.23–24) that deformed the northern margin of present

FIGURE 17.11A The Appalachian-Alleghenian Orogeny. Map showing present distribution of Appalachian Orogenic Features.

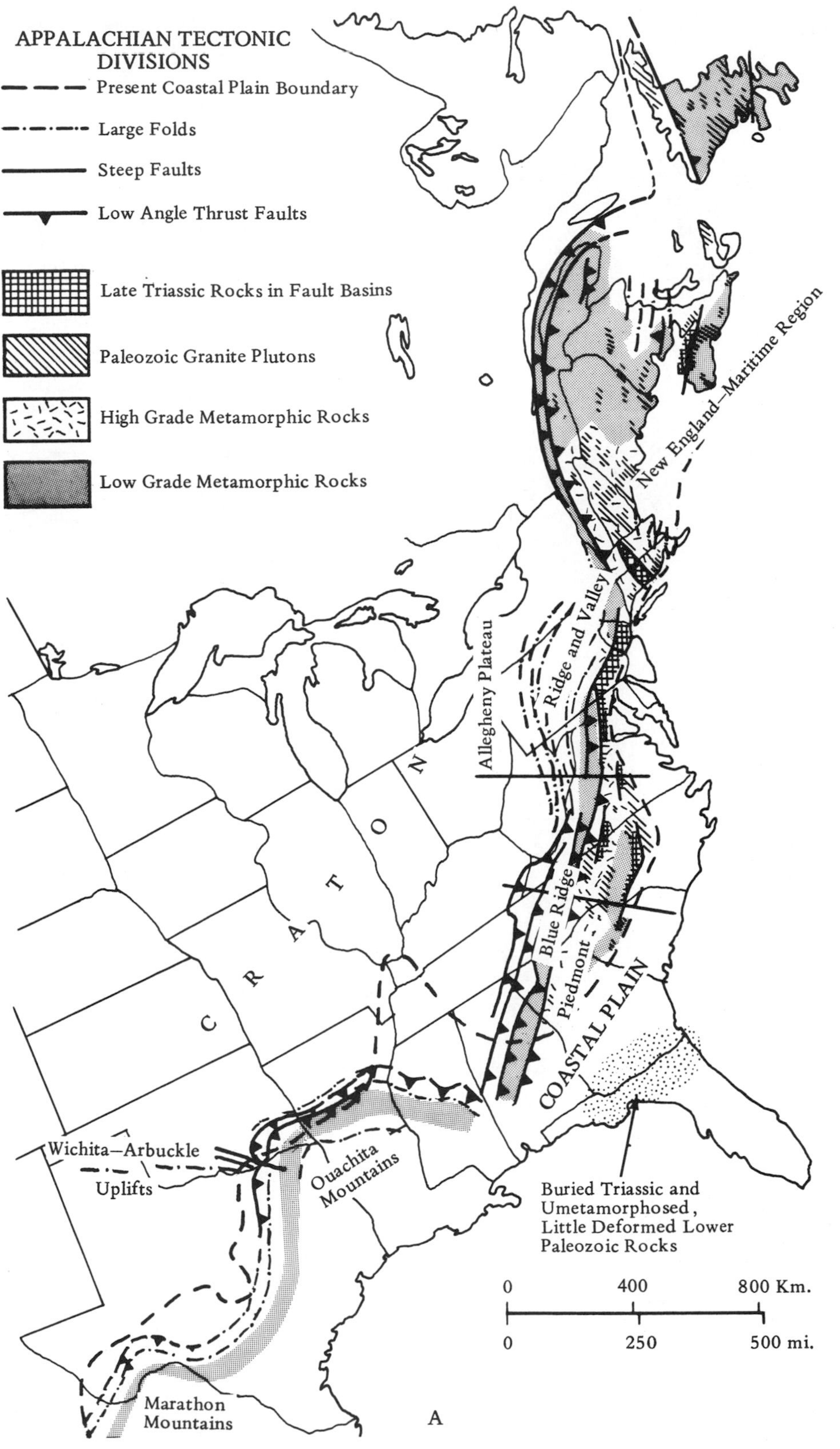
APPALACHIAN TECTONIC DIVISIONS
Present Coastal Plain Boundary
Large Folds
Steep Faults
Low Angle Thrust Faults
Late Triassic Rocks in Fault Basins
Paleozoic Granite Plutons
High Grade Metamorphic Rocks
Low Grade Metamorphic Rocks
New England—Maritime Region
Allegheny Plateau
Ridge and Valley
Blue Ridge
Piedmont
COASTAL PLAIN
CRATON
Wichita—Arbuckle
Uplifts
Ouachita Mountains
Marathon Mountains
Buried Triassic and
Umetamorphosed,
Little Deformed Lower
Paleozoic Rocks
0 400 800 Km.
0 250 500 mi.
A

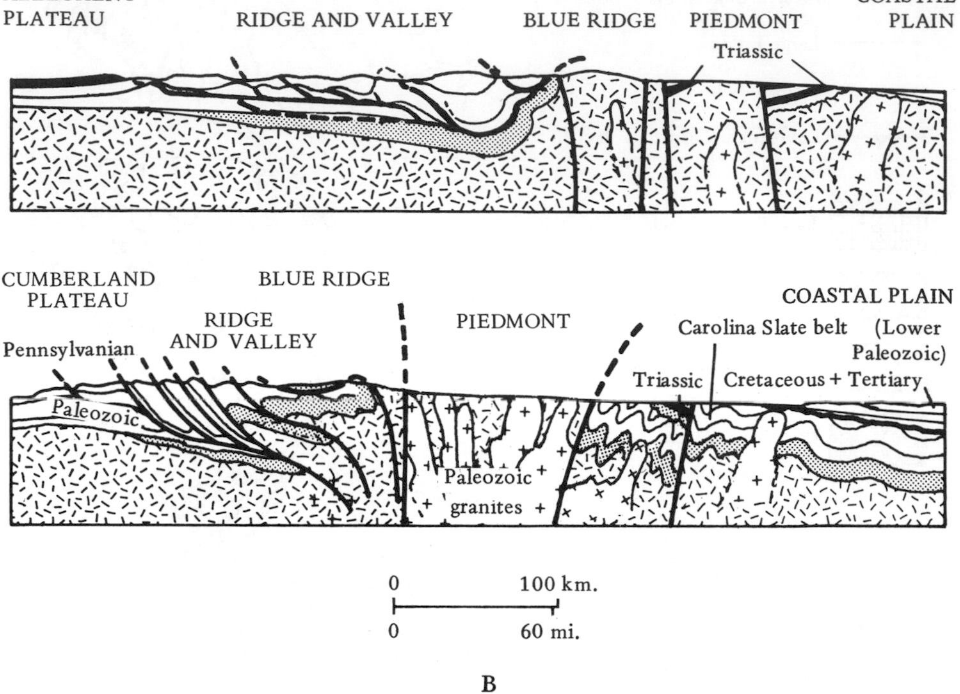

FIGURE 17.11B Cross-sections through present Appalachians along sections indicated in 17.11A.

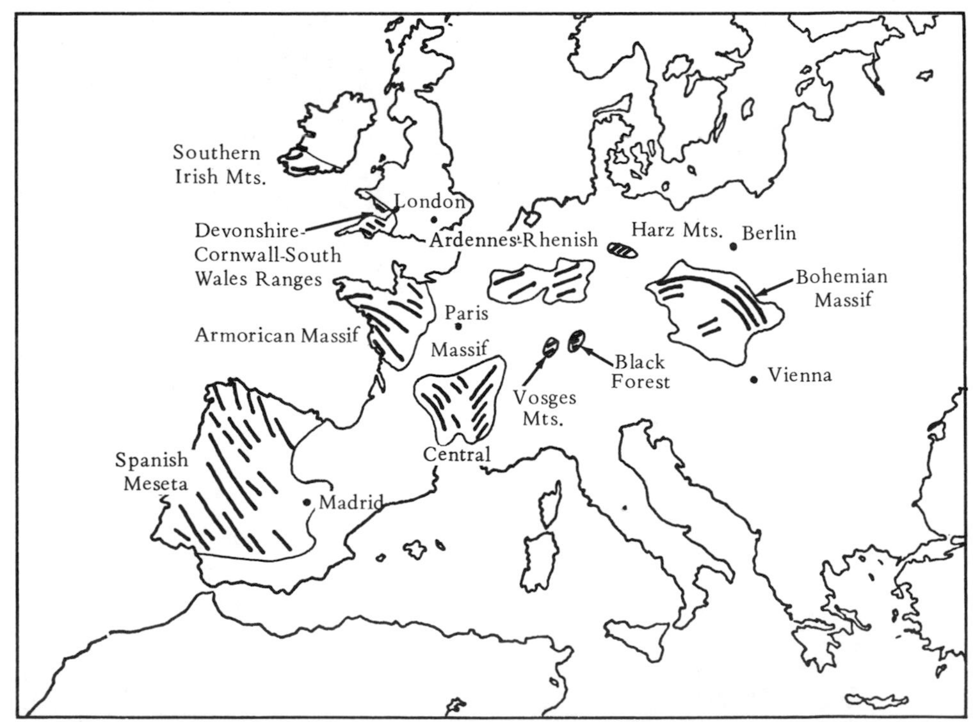

FIGURE 17.12A Present-Day Distribution of Hercynian Fold Belts in Europe.

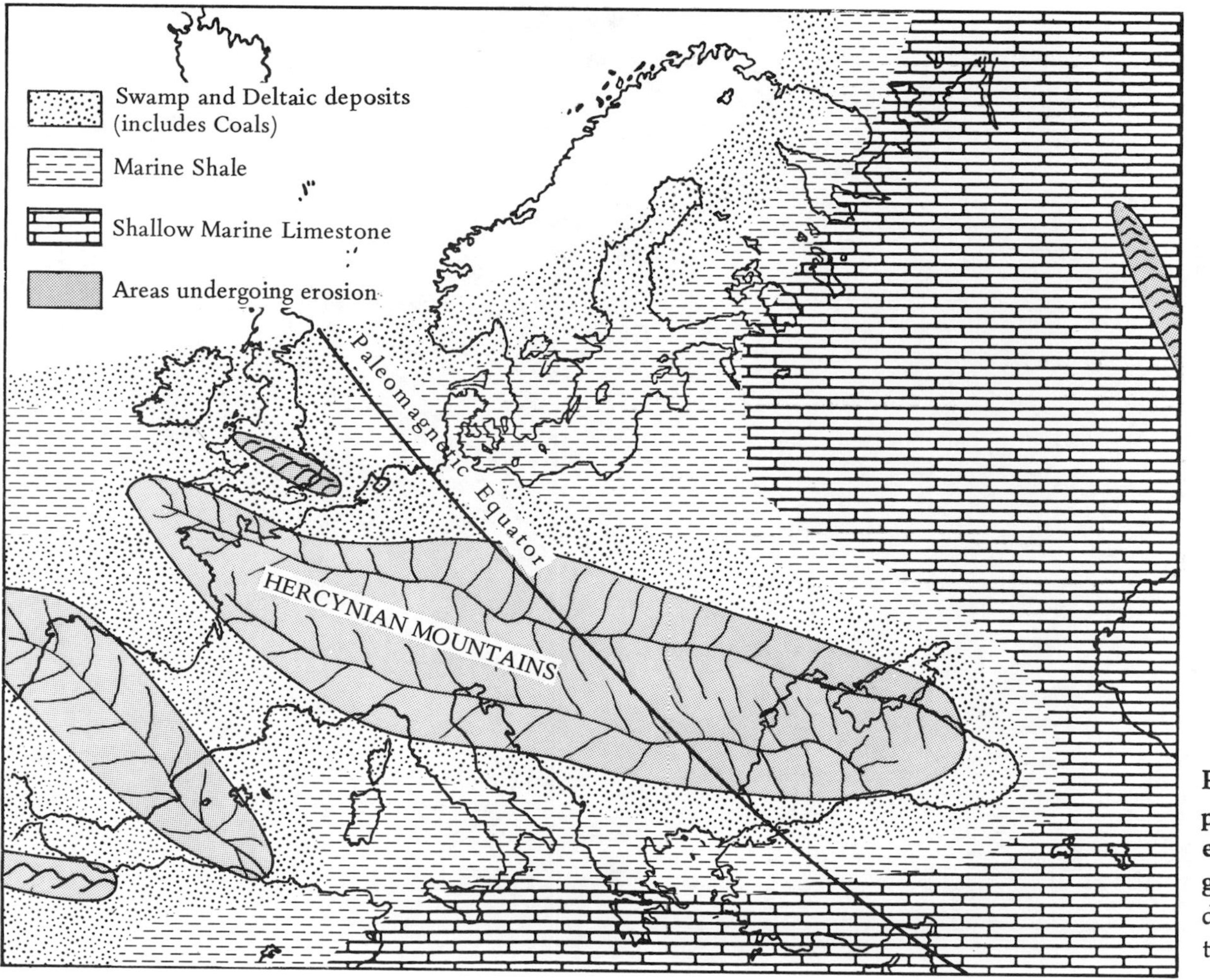

FIGURE 17.12B An Interpretation of Late Carboniferous European Paleogeography. This reconstruction does not take into account the affects of plate tectonics.

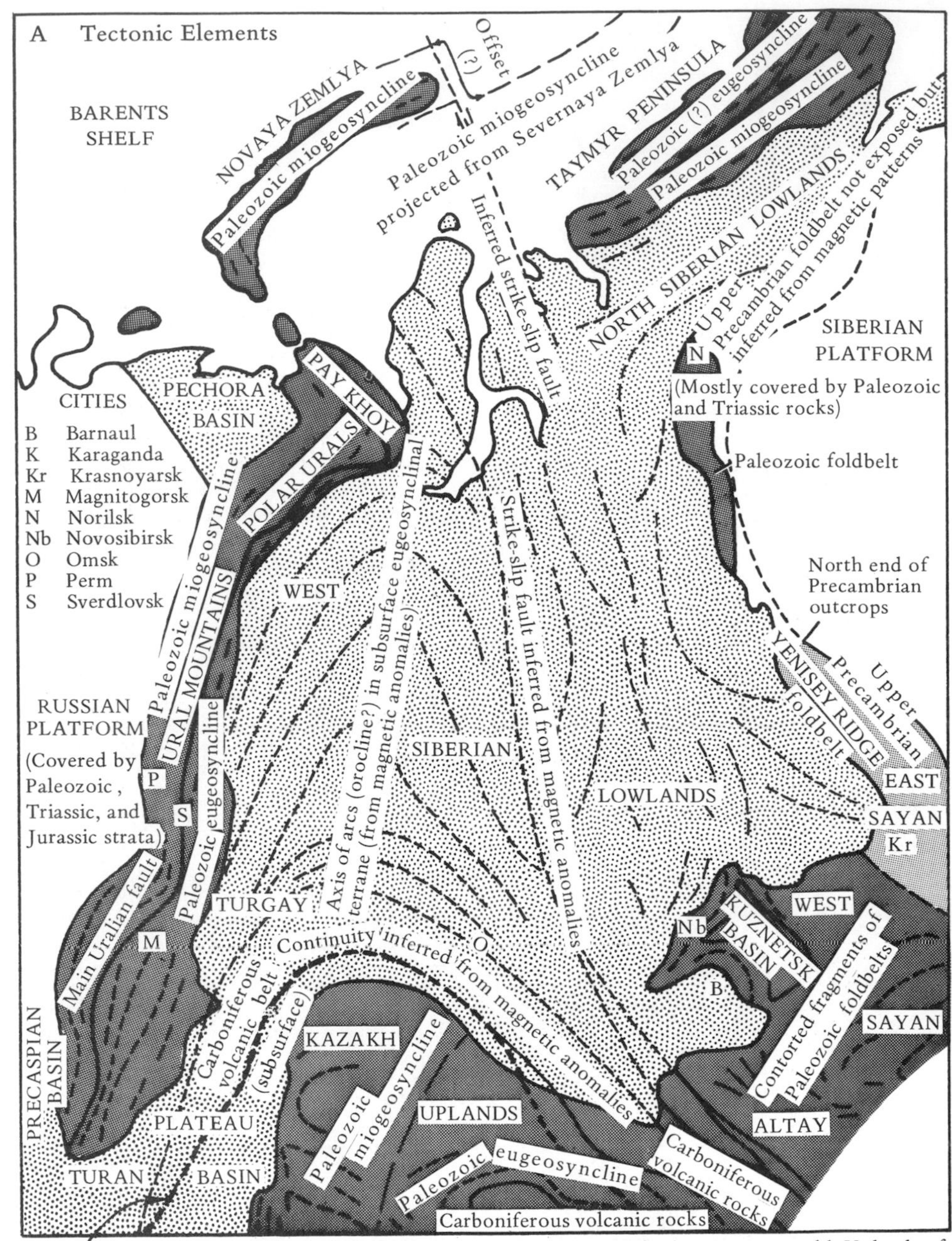

Hypothethetical strike-slip fault drawn to account for apparent duplication in Kazakh Uplands of belts either of Ural Mountains and Turgay Plateau (left lateral) or of southern Siberia (right lateral).

– – – – – – – – Structural trends in exposed foldbelts, and trends of magnetic anomalies inferred to indicate basement structure in sedimentary basins.

Cenozoic, Cretaceous, and Jurassic basins | Paleozoic Foldbelts | Upper Precambrian Foldbelts

FIGURE 17.13 Present-Day Distribution of Uralian Orogenic Features. (From Warren Hamilton, *G.S.A. Bulletin*, vol. 81, Sept. 1970. By permission of author and G.S.A.)

FIGURE 17.14 Cordilleran Geosynclinal Deposits in The Canadian Rockies. Cascade Mountain, Banff National Park, Alberta. Strata are chiefly Mississippian in age. (Courtesy, Alberta Dept. of Industry and Tourism.)

North America, the *Franklin Geosyncline*. This compression resulted from the collision of another plate with the present northern margin of North America. Whether this was an ancestral Arctic Ocean plate, part of Alaska (Fig. 17.25), or a separate small east Siberian continental mass is uncertain at present. Post-orogenic block-faulting and basin-filling succeeded the mountain-building.

North American Epeiric Seas

The North American craton portion of Laurasia was widely flooded in the Devonian. The Early Devonian is characterized by the final phases of the Tippecanoe Sea regression. Beginning at the close of the Early Devonian and culminating in the Middle Mississippian is the major epeiric sea called the *Kaskaskia* (Fig. 15.18). A minor regression occurred in the Late Devonian because of the mountain-building disturbances to the present east. The Devonian epeiric seas were characterized by widespread limestones containing abundant coral reefs (Fig. 17.26). The deposits along the eastern side of the sea consist chiefly of shales (Fig. 17.27) reflecting the rising Acadian Mountains. This

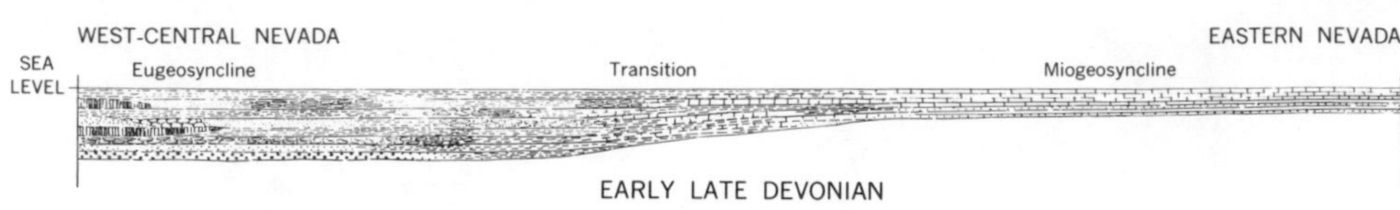

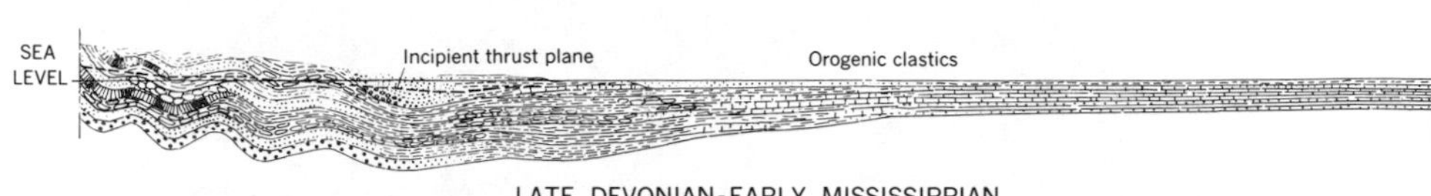

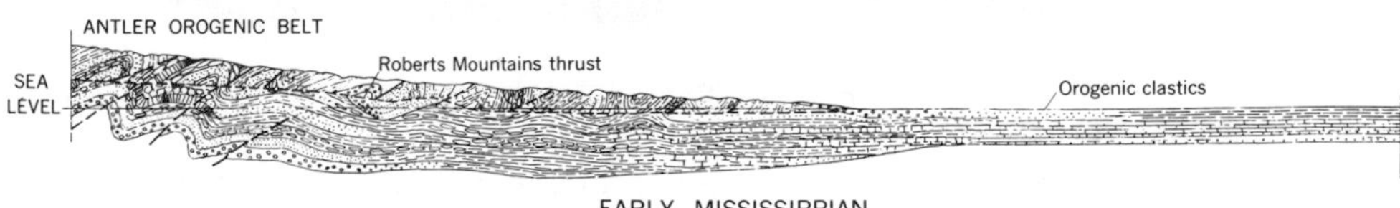

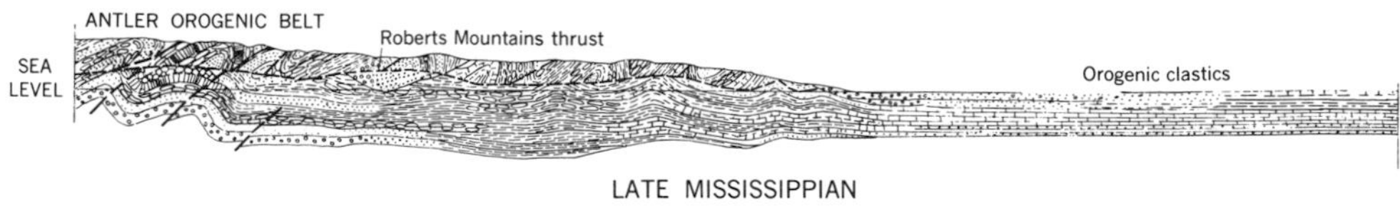

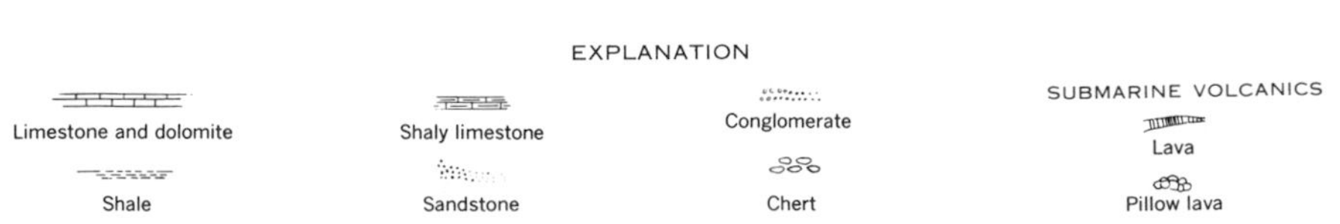

FIGURE 17.15 Cross-Sections of Cordilleran Belt Showing an Interpretation of the Antler and Sonoma-Cassiar Orogenies. (From R. J. Roberts, *U.S.G.S. Professional Paper*, 459A.)

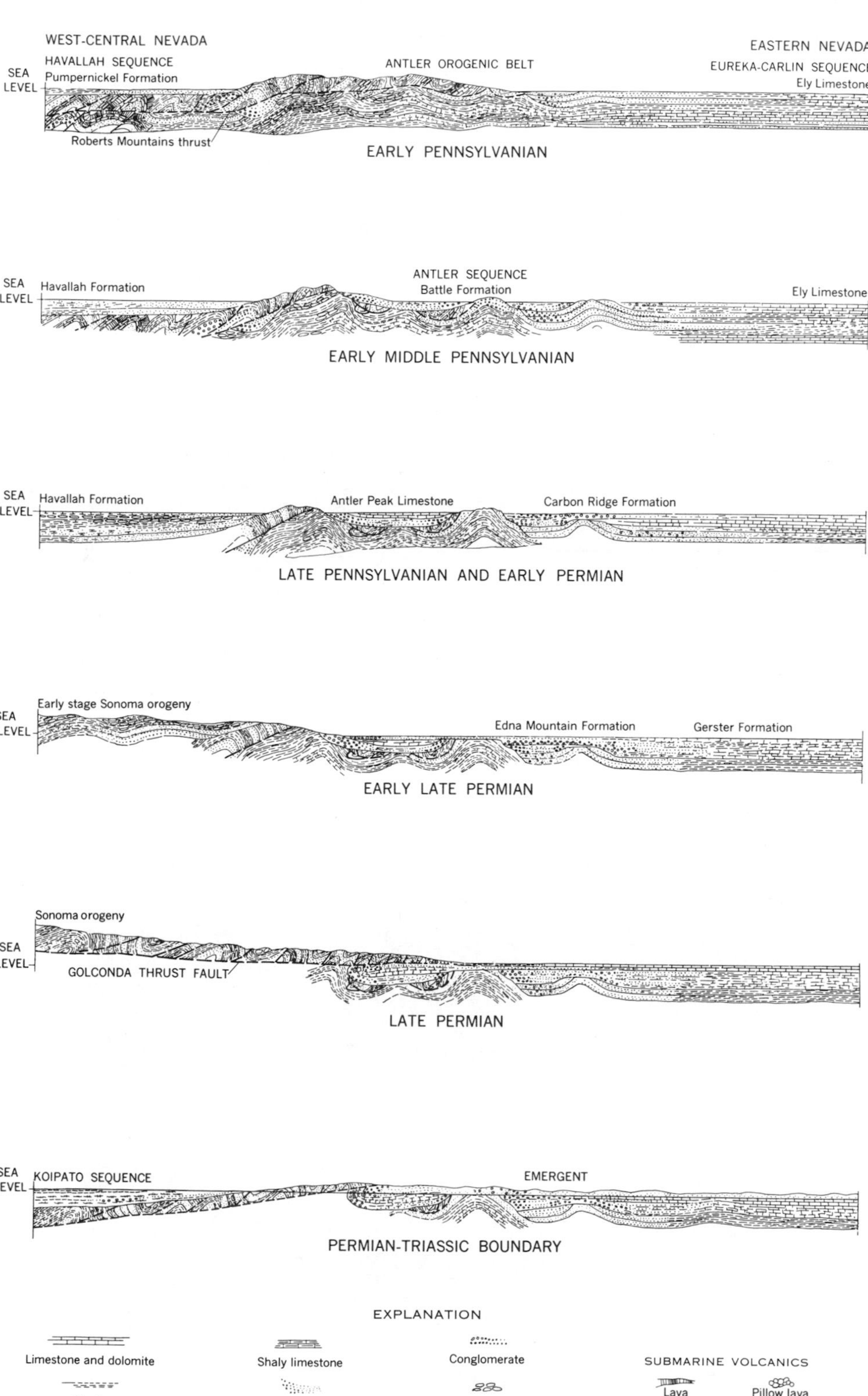
WEST-CENTRAL NEVADA
HAVALLAH SEQUENCE
Pumpernickel Formation
SEA LEVEL
ANTLER OROGENIC BELT
EASTERN NEVADA
EUREKA-CARLIN SEQUENCE
Ely Limestone
Roberts Mountains thrust
EARLY PENNSYLVANIAN
SEA LEVEL
Havallah Formation
ANTLER SEQUENCE
Battle Formation
Ely Limestone
EARLY MIDDLE PENNSYLVANIAN
SEA LEVEL
Havallah Formation
Antler Peak Limestone
Carbon Ridge Formation
LATE PENNSYLVANIAN AND EARLY PERMIAN
Early stage Sonoma orogeny
SEA LEVEL
Edna Mountain Formation
Gerster Formation
EARLY LATE PERMIAN
Sonoma orogeny
SEA LEVEL
GOLCONDA THRUST FAULT
LATE PERMIAN
SEA LEVEL
KOIPATO SEQUENCE
EMERGENT
PERMIAN-TRIASSIC BOUNDARY
EXPLANATION
Limestone and dolomite
Shaly limestone
Conglomerate
SUBMARINE VOLCANICS
Shale
Sandstone
Chert
Lava
Pillow lava
Intermediate and siliceous volcanic rocks

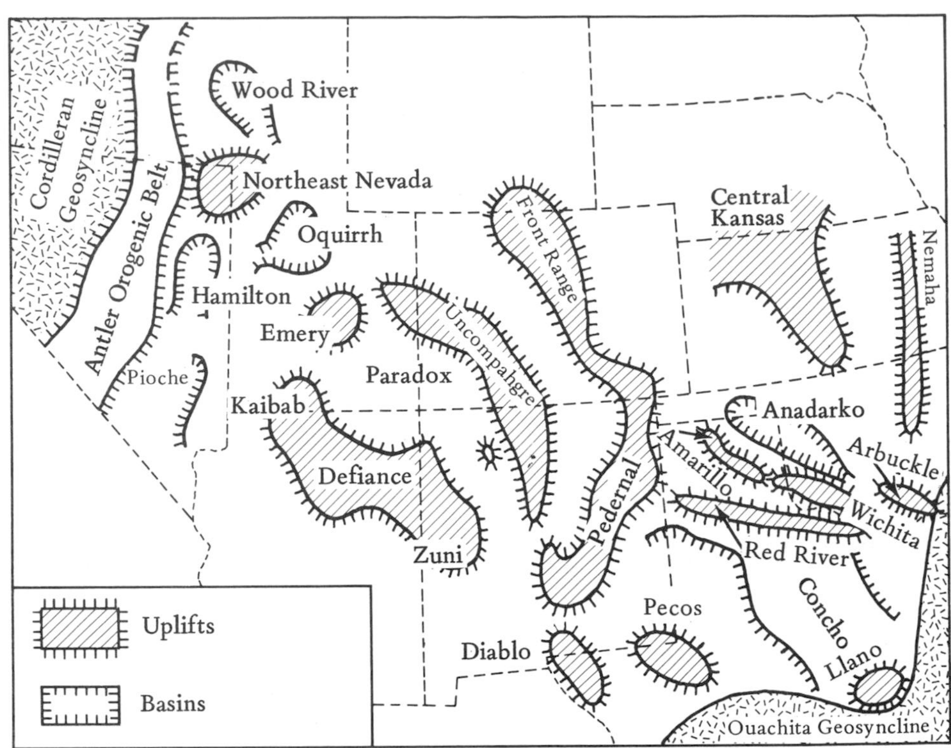

FIGURE 17.16A Present-Day Distribution of Colorado-Oklahoma Cratonal Mountains and Associated Basins.

FIGURE 17.16B Wichita Mountains near Lawton, Oklahoma. (Courtesy Oklahoma Industrial Development and Park Department.)

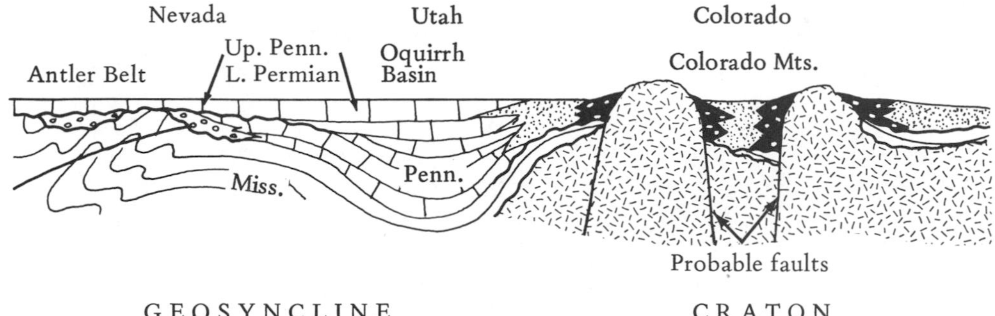

FIGURE 17.16C Cross-Section of the Colorado Mountains in the Pennsylvanian.

relatively flat-lying and exceedingly fossiliferous sequence of this system, particularly in western New York state (Figs. 17.28–29), is probably the finest Devonian section in the world.

The re-advance of the Kaskaskia Sea is represented by a basal black shale (Fig. 17.30), an atypical basal deposit, indicating that the sea progressively engulfed its swampy shoreline as it slowly covered the craton again. Geologists who have made the mistake of substituting rock-correlation for time-correlations have been puzzled by what conditions could cause a widespread black-mud-producing marine environment. The answer, of course, is that the environment was not widespread at any one time, but varied in age depending on the shifting position of the shore zone. The Early and Middle Mississippian deposits represent the culmination of the Kaskaskia Sea with widespread fossiliferous limestone (Figs. 17.31–32). The Kaskaskia Sea regression occurred in the Late Mississippian.

It was followed by the *Absaroka Sea* (Fig. 15.18), the last major epeiric sea invasion of North America in the Paleozoic. The Absaroka Sea reached its maximum extent near the close of the Pennsylvanian (Fig. 17.33), and then withdrew throughout the Permian and Triassic.

A peculiar characteristic of the transgressive phase of the Absaroka Sea is the presence of *cyclothems.* Cyclothems (Fig. 17.34) are cyclically repeated minor transgressive and regressive deposits that go upward from basal deltaic sandstones through nonmarine shale and freshwater limestone to underclay and coal to nearshore marine shale and offshore marine limestone, followed by nearshore limestone and shale, and finally nonmarine shale. To the east, as in Pennsylvania, the terrestrial parts of the sequence are more abundant; to the southwest, as in Kansas, the marine phases dominate; while in the mid-continent, such as Illinois, the complete sequence is best developed (Fig. 17.35). These sequences are repeated dozens of times indicating a constantly fluctuating sea level. Apparently cyclothems were caused by the coincidence of a broad, almost flat lowland over the North American interior with glacial-interglacial sequences of Gondwanaland which alternately withdrew and added large amounts of water from the oceans. A rise or fall of the sea of a few tens of meters could thus flood vast areas. Note that the coal formed just before the onset of marine conditions indicating that the coal swamp forests occurred in broad bands along or near the seashore. Thicker sections accumulated in the

FIGURE 17.17 The Garden of the Gods, Colorado. (Courtesy of William M. Mintz.)

A

B

FIGURE 17.18 Pennsylvanian Strata in the West. (A) Ouray, San Juan Mountains, Colorado. (Courtesy Colorado Dept. of Public Relations, (B) Goosenecks of the San Juan River, Utah. (Courtesy of William M. Mintz.)

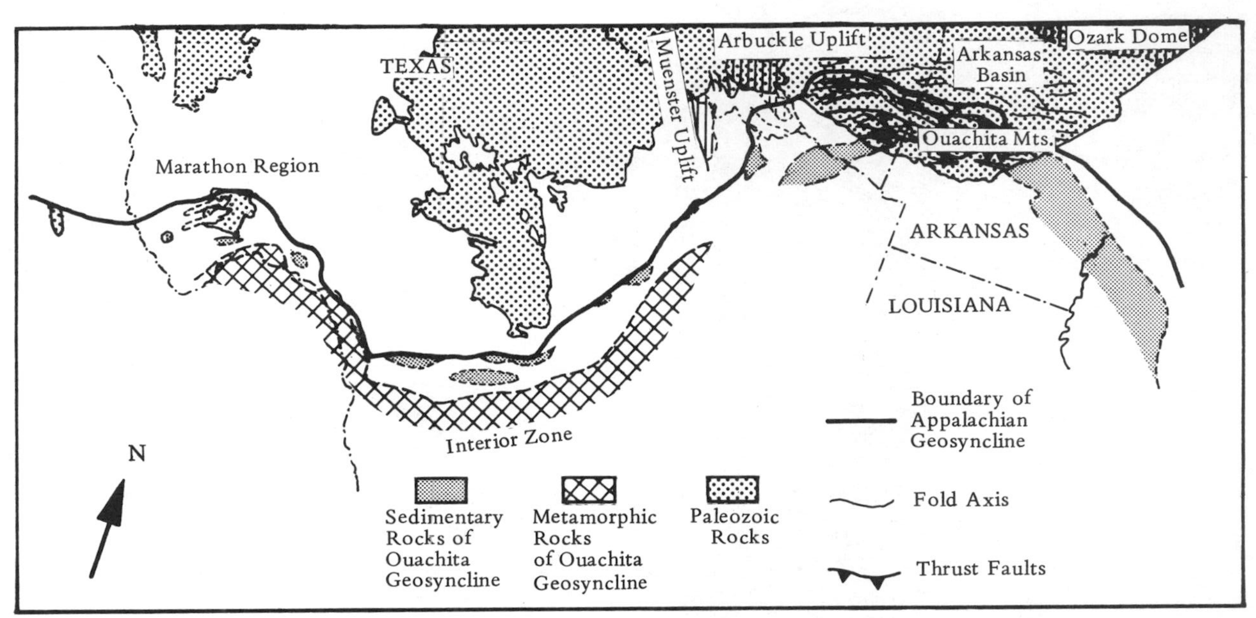

FIGURE 17.19 Present-Day Distribution of Ouachita-Marathon Orogenic Features.

FIGURE 17.20 The Ouachita Mountains, eastern Oklahoma. (Courtesy Mike Shelton, State of Oklahoma.)

more rapidly subsiding Illinois and Forest City Basins. The major coal beds of eastern North America today are of Pennsylvanian age (Fig. 17.36).

The regressive phase of the Absaroka Sea occupied the entire Permian. The sea gradually retreated to the southwest so that by Late Permian times it only remained in southwest Texas and adjacent New Mexico (Fig. 17.37). In the late Middle Permian great reefs grew in this region which have produced a wealth of fossils (Fig. 7.9) and one of the clearest demonstrations of intertonguing facies exposed anywhere (Fig. 17.38). At the same time in the present Rocky Mountains area of Idaho and adjacent states, another remnant of the regressing sea deposited the phosphate-rich Phosphoria Formation (Fig. 17.37). By the end of the Permian, virtually the entire North American craton was surrounded by mountains pouring clastics onto it. It was covered in many areas, particularly the present west, by great thicknesses of terrestrial redbeds that continue through the Triassic (Fig. 17.39).

In the Late Paleozoic, the equator ran roughly east-west through North America in the northern United States and southern Canada, thus making the coal forests tropical and the terrestrial redbed areas comparable to today's low latitude deserts (Fig. 17.40).

Europe

The European craton was flooded by terrestrial clastics from the Caledonian Mountains in the Devonian. Only by the Late Devonian did epeiric seas again return covering much of Russia (Fig. 17.41). In the Rhine-Meuse area of Ger-

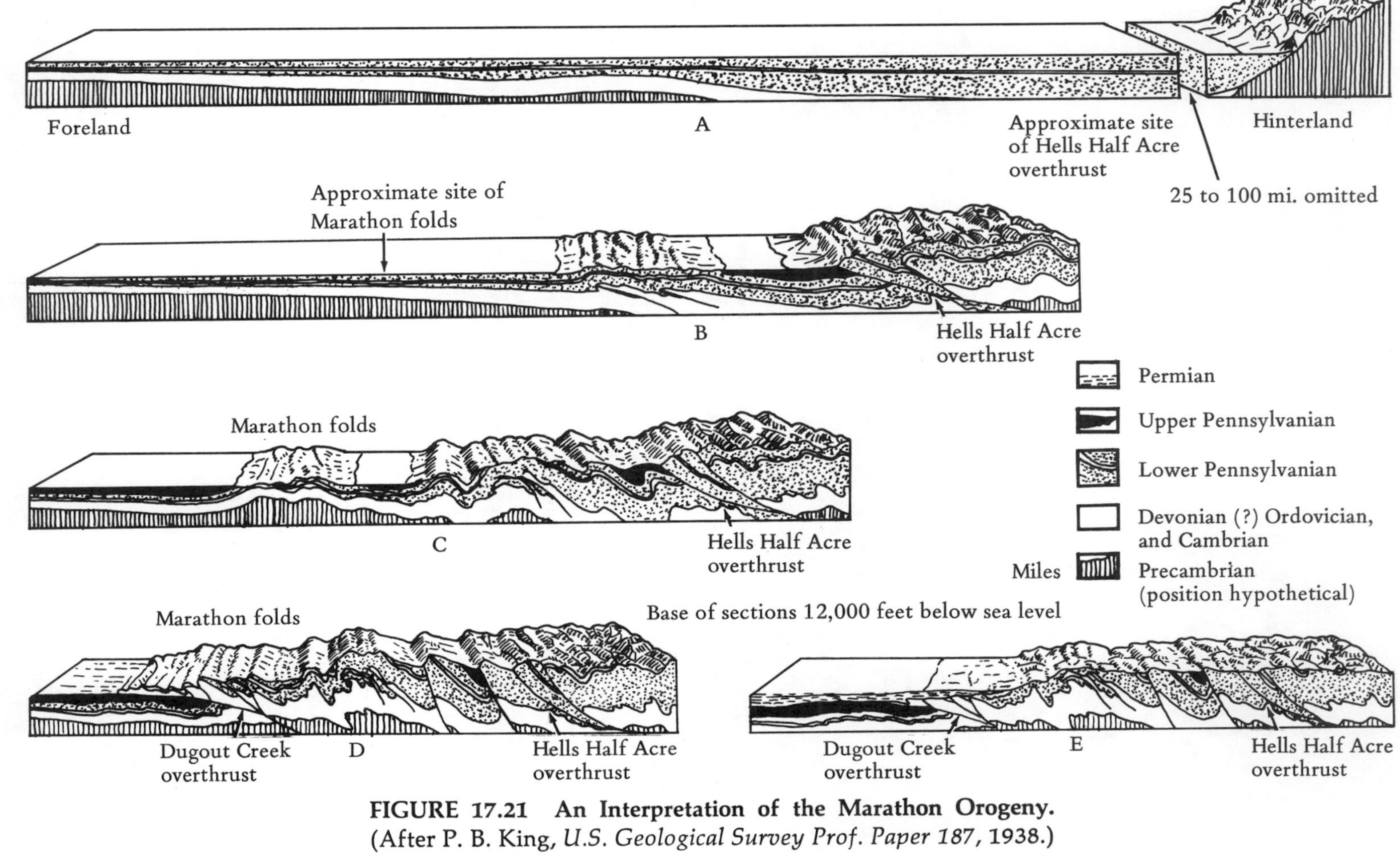

FIGURE 17.21 An Interpretation of the Marathon Orogeny. (After P. B. King, *U.S. Geological Survey Prof. Paper 187*, 1938.)

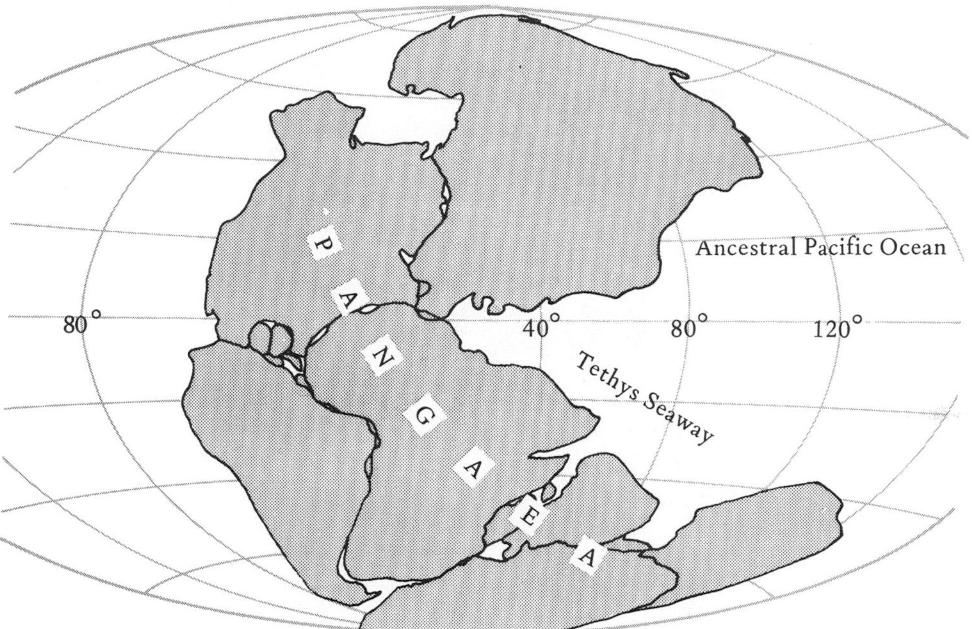

FIGURE 17.22 Permian Paleogeography. Epeiric seas not shown. (Data from works of Dietz and Holden, 1970.)

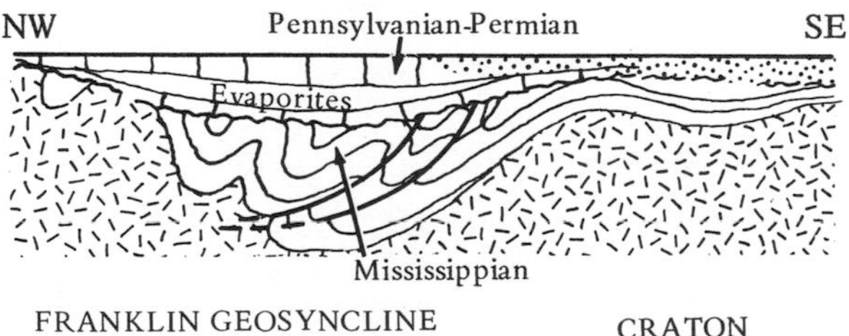

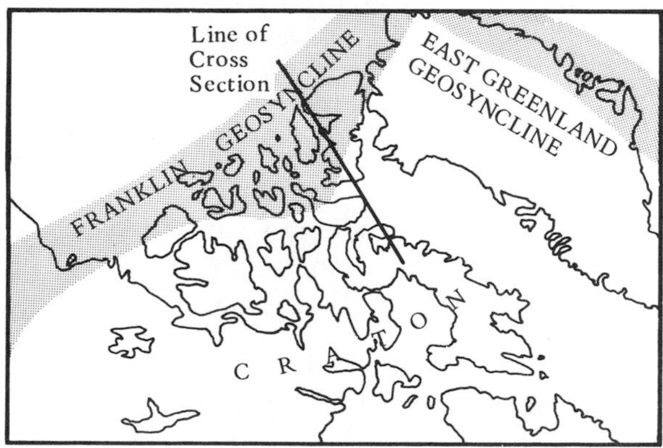

FIGURE 17.23 Restored Cross-Sections through Franklin Geosyncline Showing an Interpretation of the Innuitian Orogeny.

FIGURE 17.24 Innuitian Mountains. Castle Mountain at Yelverton Pass, Ellesmere Island, Canada. (Courtesy of R. L. Christie, Geological Survey of Canada, Ottawa.)

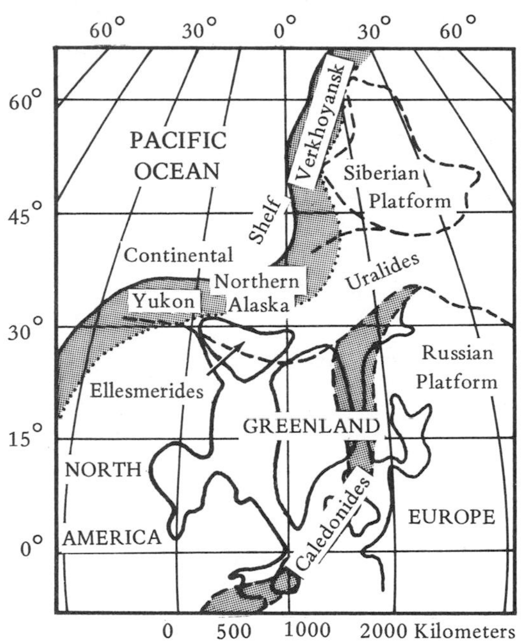

FIGURE 17.25 An Interpretation of Late Paleozoic Geography in the Present North Polar Region. (After Hamilton, *G.S.A. Bulletin*, vol. 81, September 1970. Used by permission of the author and G.S.A.)

many, in the miogeosyncline of the Hercynian Geosyncline, is an exceptionally complete and fossiliferous Devonian section that is better than the more highly deformed type area in England. The Late Devonian epeiric sea expanded in the Early Carboniferous even flooding the area of the former Caledonian Mountains (Fig. 17.42), which was then greatly reduced and, thereafter, was largely a stable region.

FIGURE 17.26 A Devonian Coral Reef in the Alpena Limestone, Thunder Bay Quarry, Alpena, Michigan. (Courtesy of Robert V. Kesling, University of Michigan.)

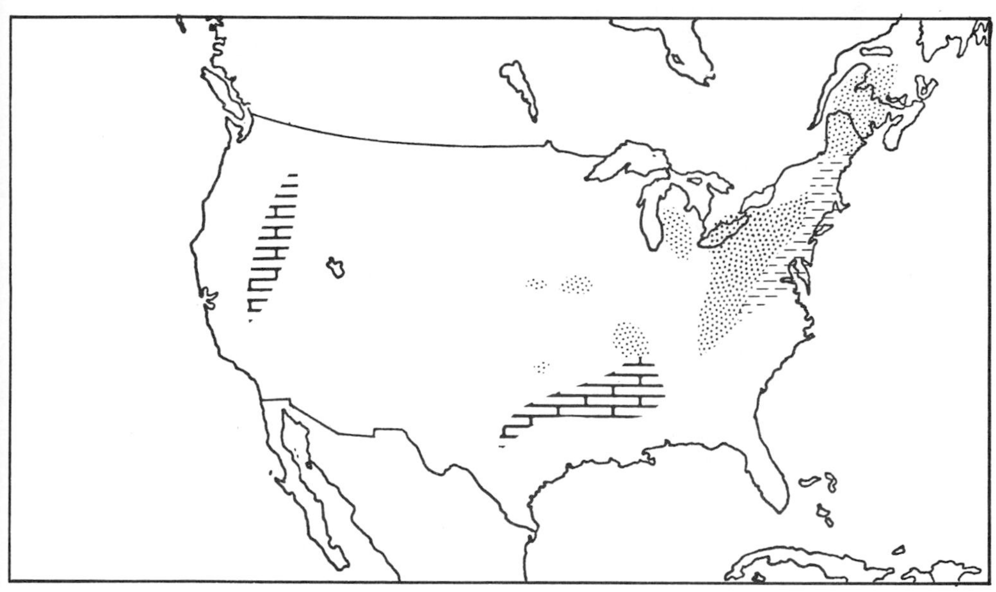

FIGURE 17.27 Present-Day Distribution of Basal Kaskaskia Sea Deposits.

In the Late Carboniferous, the culmination of the *Hercynian Orogeny* drove the epeiric sea from the European craton for the remainder of the Paleozoic. The terrestrial clastics (Fig. 17.43) that accumulated around these mountains and on the shelf to the north include the great coal beds of present-day Europe. Coal swamp forests flourished here as they did further to the present southwest in what is now North America. In the Permian, most of western Europe was the site of terrestrial redbed deposition broken only by the Late Permian hypersaline *Zechstein Sea* which flooded most of its northern margin. The Russian part of the craton remained flooded by an epeiric sea until the very close of

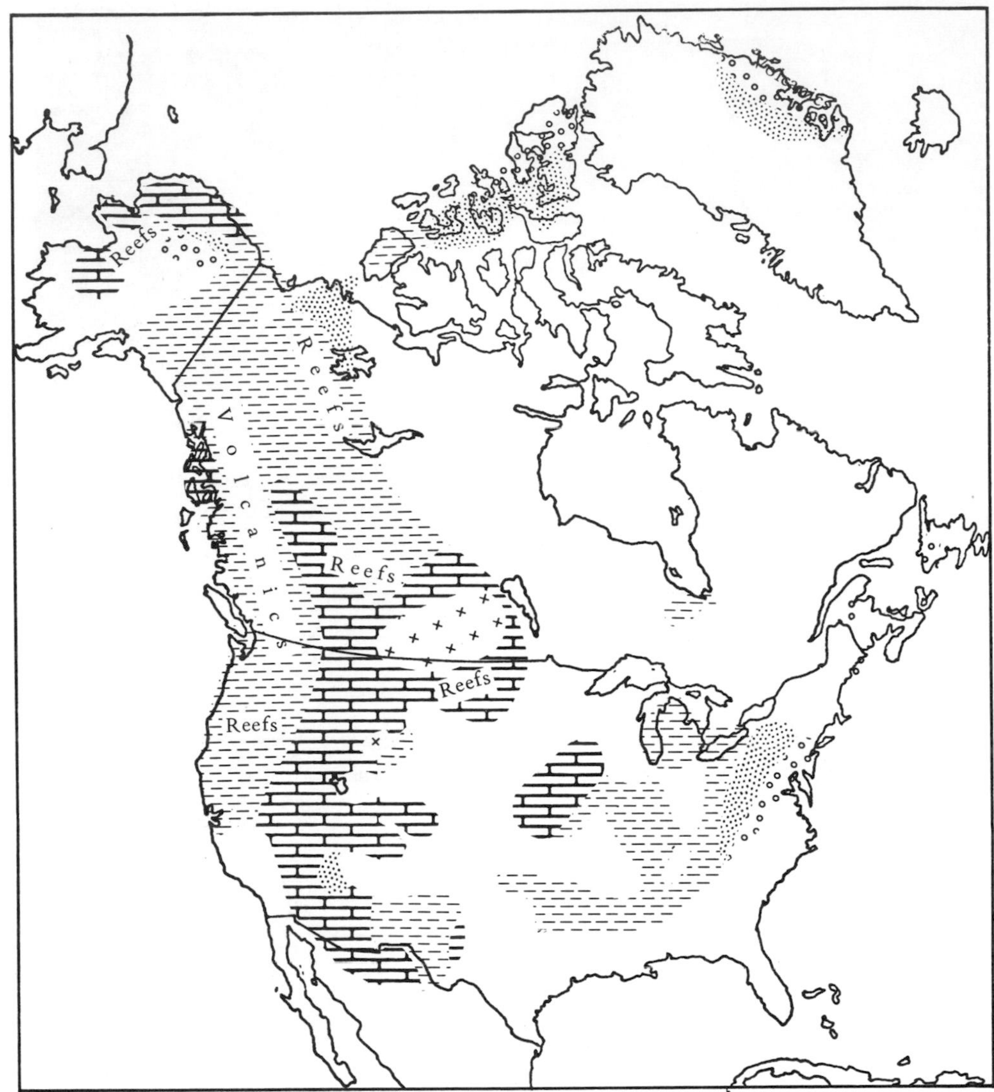

FIGURE 17.28 Upper Devonian Deposits of Present-Day North America.

the period when the influx of clastics from the rising Ural Mountains drove it out (Fig. 17.44).

Beginning in the Permian, the plate which contained Gondwanaland began to rotate counterclockwise toward present southern Eurasia. As it did so it began to close off the western end of the broad seaway which extended into western Pangaea (Figs. 17.22, 17.47). The trench-subduction zone systems of the former Hercynian and Mongolian Geosynclines became the site of continuing compression now called the *Tethyan Geosyncline*. This great unstable belt received sediments and underwent deformation throughout the Mesozoic, reaching a culmination of orogeny in the Cenozoic. In the Late Paleozoic, the equator ran diagonally from the present northwest to southeast across present-day extreme western Europe (Fig. 17.45) thus placing Tethys in the tropics.

FIGURE 17.29 Kaskaskia Sea (Devonian) Rocks of New York. (A-B) Watkins Glen State Park. Fine-grained sandstones and siltstones of Late Devonian age, (C-D) Letchworth State Park. Upper Devonian sandstones and shales in the Genesee River Gorge.

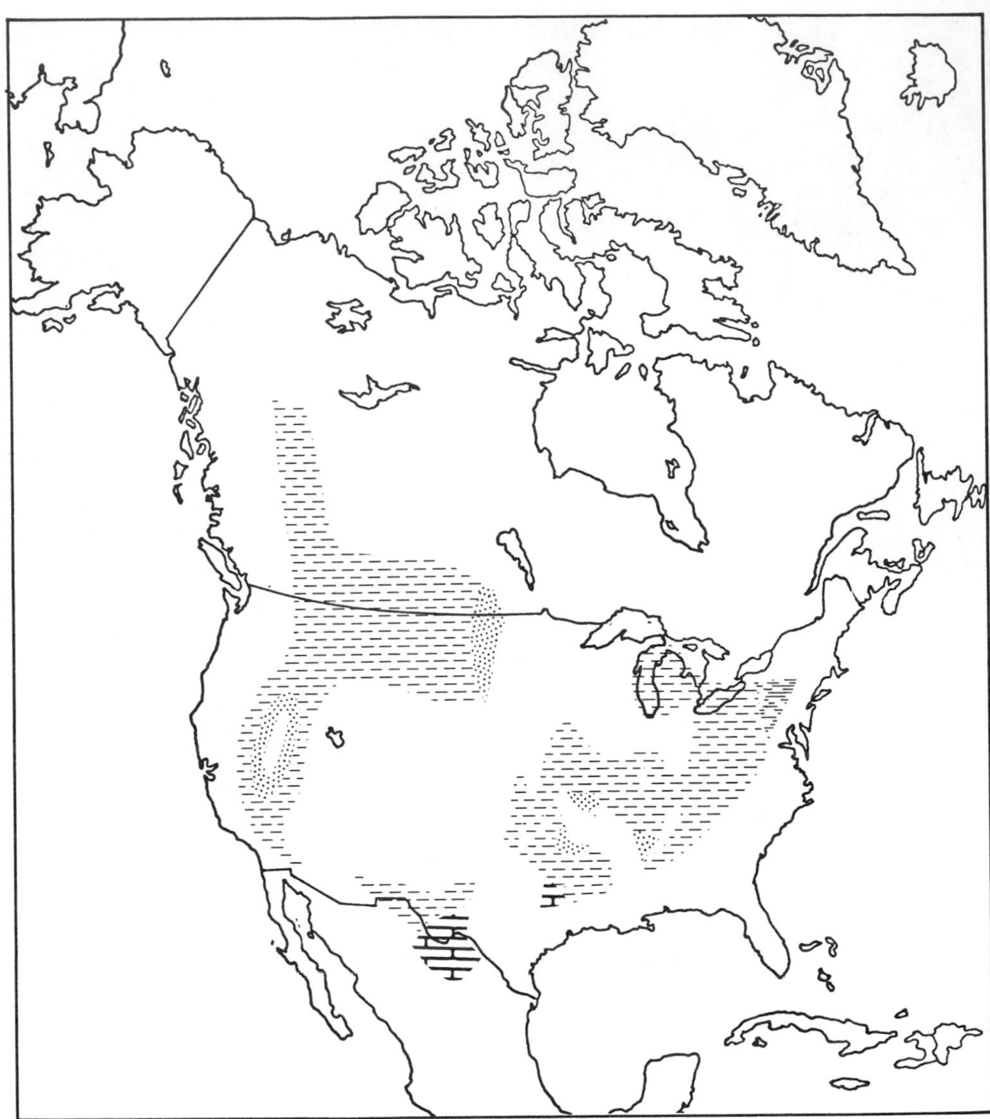

FIGURE 17.30 Present-Day Distribution of Basal Deposits of Second Advance of Kaskaskia Sea (Tamaroa Sea).

Asia

The Asian craton was largely emergent throughout the entire Late Paleozoic interval. Surrounded on all sides by mountain belts undergoing deformation, its deposits are of continental origin (Fig. 17.46). This part of Laurasia was in middle latitudes in the Late Paleozoic and at approximately right angles to its present position (Fig. 17.45). Eastern Siberia, which may have been a separate continental block, possibly contributed the deformation along the present eastern edge of Asia or that in present northwestern North America. The collision of the eastern edge of Laurasia with the Pacific plate was probably re-

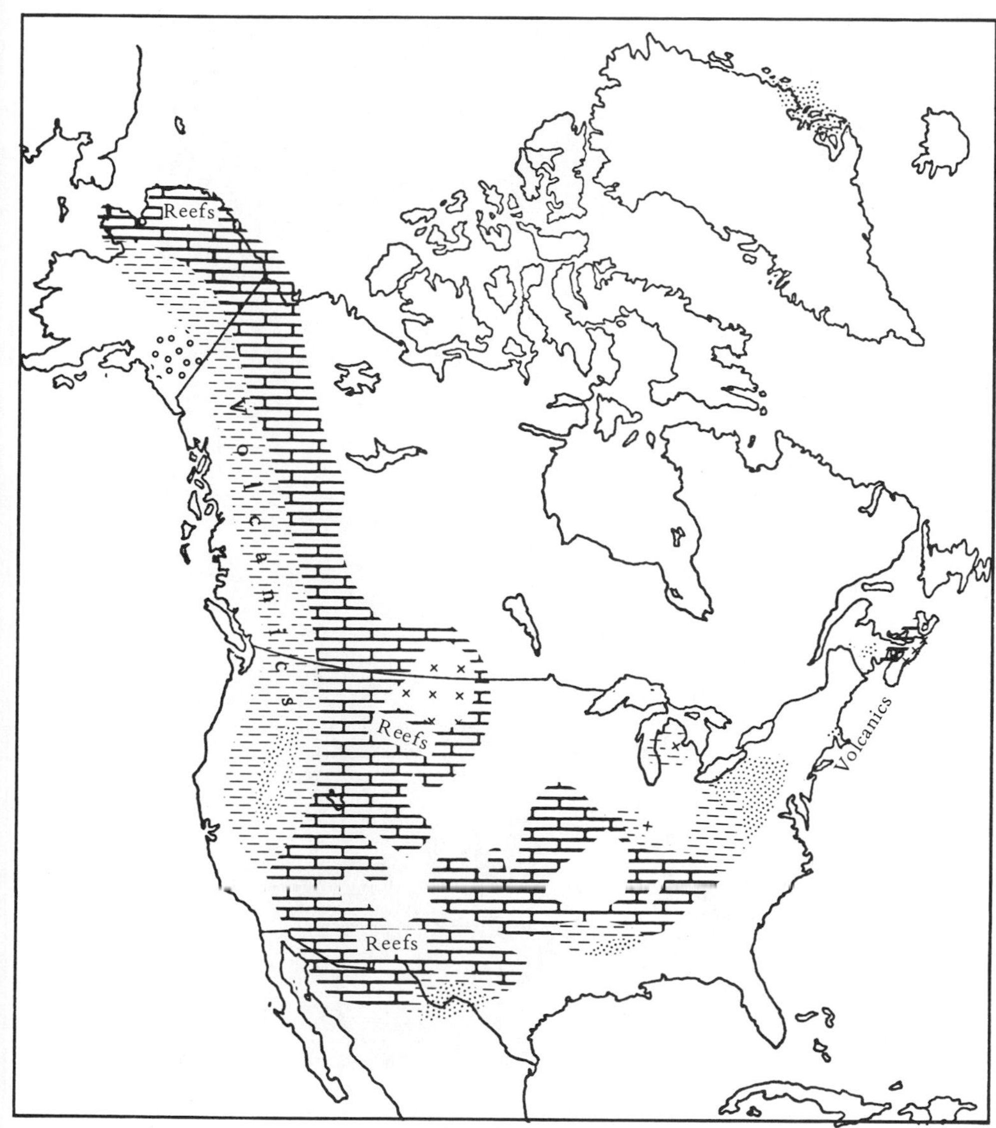

FIGURE 17.31 Present-Day Distribution of Middle Mississippian Sedimentary Rocks in North America.

sponsible for much of the Late Paleozoic instability in that zone. The Pacific or North Pole was situated in the area east of Japan in Late Paleozoic times (Fig. 17.45). The Chinese Craton apparently collided with the Asian one in the Late Paleozoic producing the culminating deformation of the eastern Mongolian Geosyncline. The craton itself was flooded by epeiric seas during much of this interval.

The Late Paleozoic was a time of extreme continental unification. The conditions produced by this unique geography had significant effects on the life of the Late Paleozoic as we shall see in the next chapter.

A

B

FIGURE 17.32 Mississippian Limestones in (A) Mississippi River Bluffs near Alton, Illinois, (B) Grand Canyon (Redwall Limestone forms prominent cliff in mid-canyon). (A) Courtesy of Illinois State Geological Survey, (B) Courtesy of William M. Mintz.

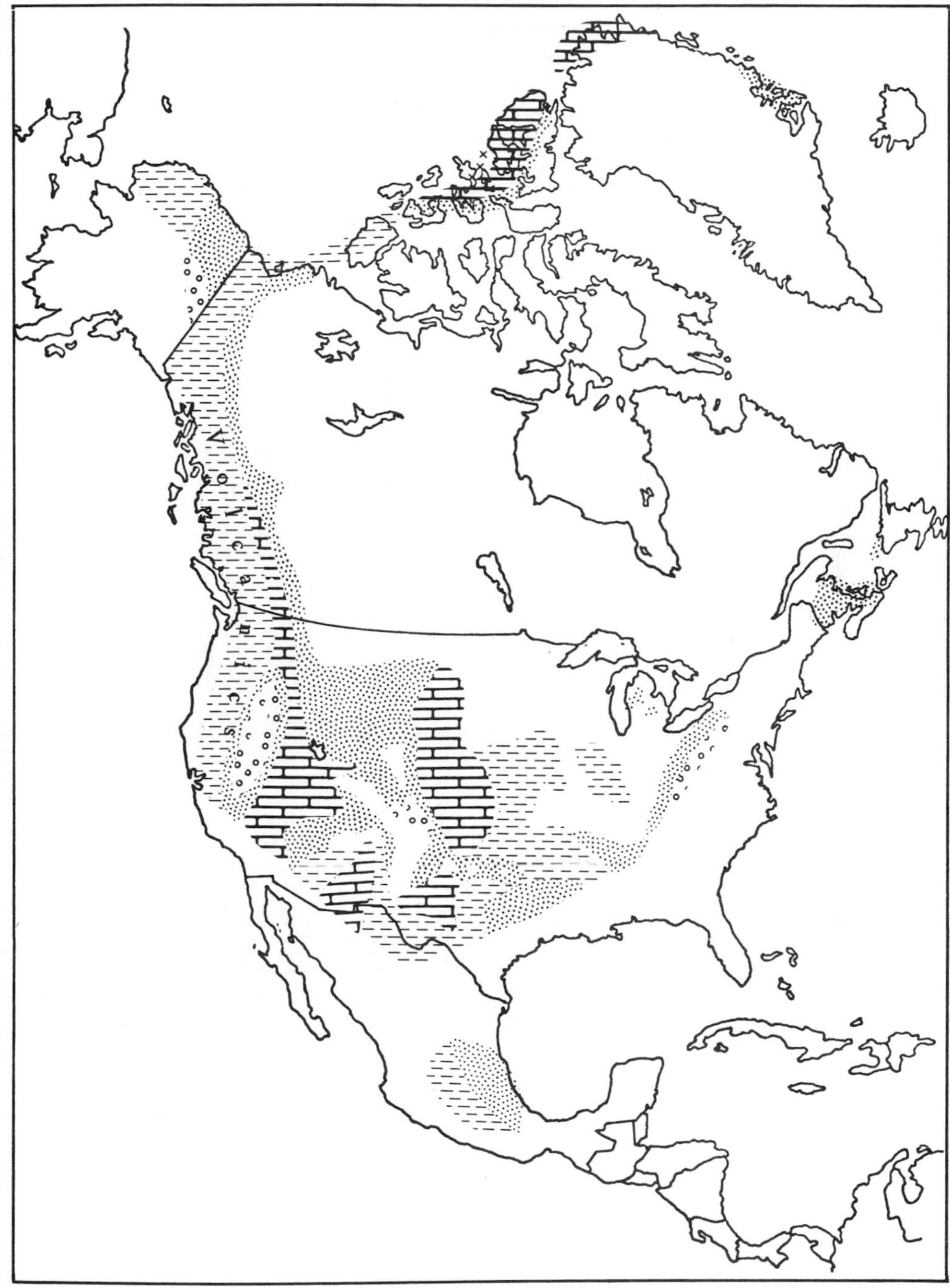

FIGURE 17.33 Present-Day Distribution of Middle Pennsylvanian Sedimentary Rocks in North America.

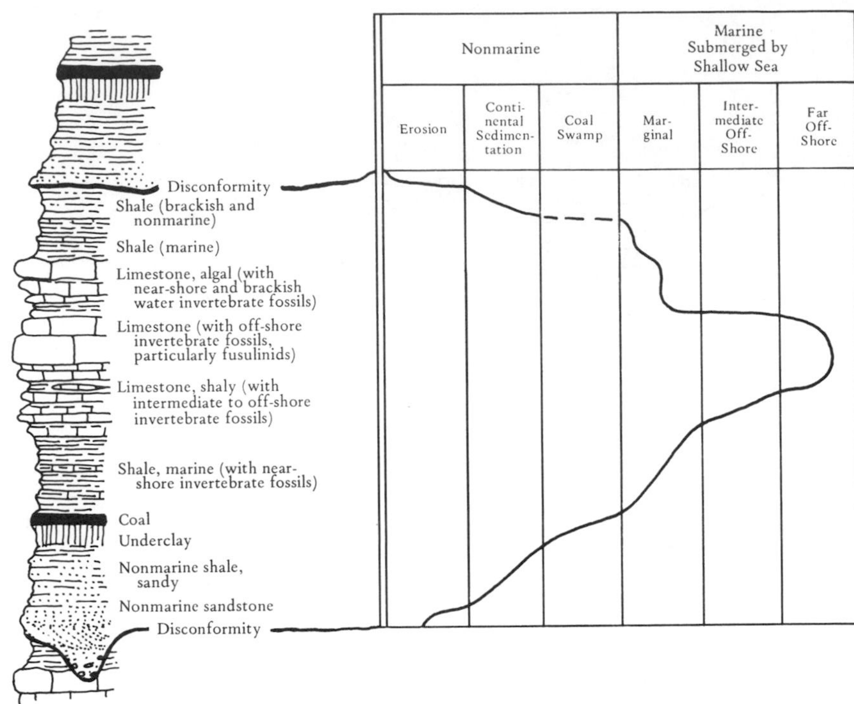

FIGURE 17.34 Cross-Section of a Complete Cyclothem.

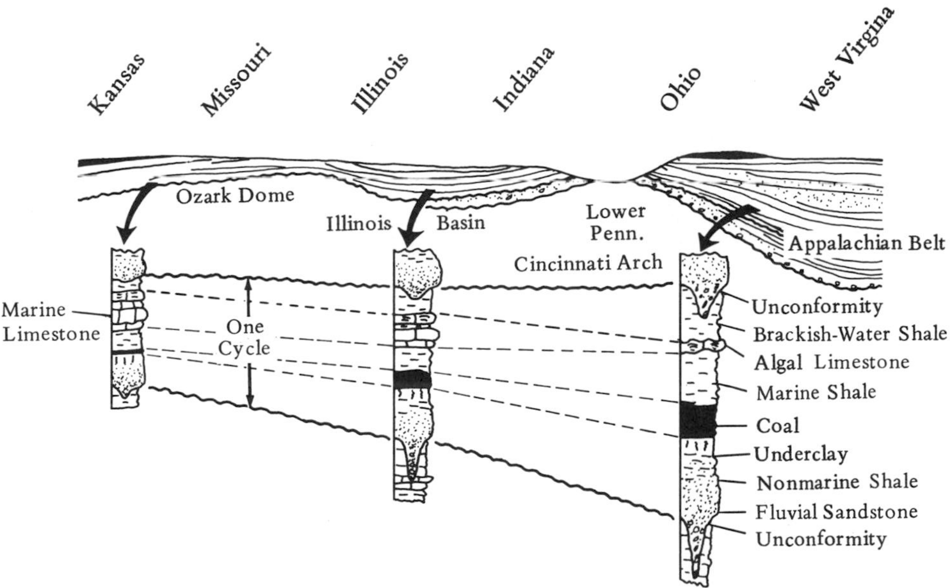

FIGURE 17.35 Cross-Section of Central North America Showing Lateral Variations in One Cyclothem.

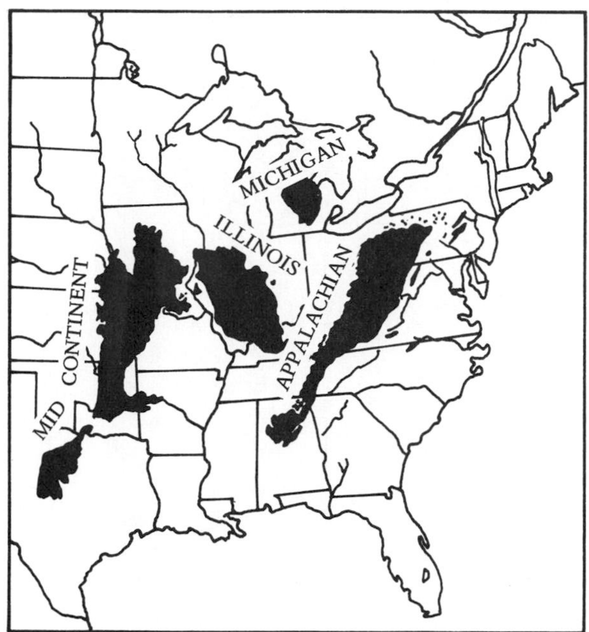

FIGURE 17.36 Pennsylvanian Coal Fields of the Eastern United States.

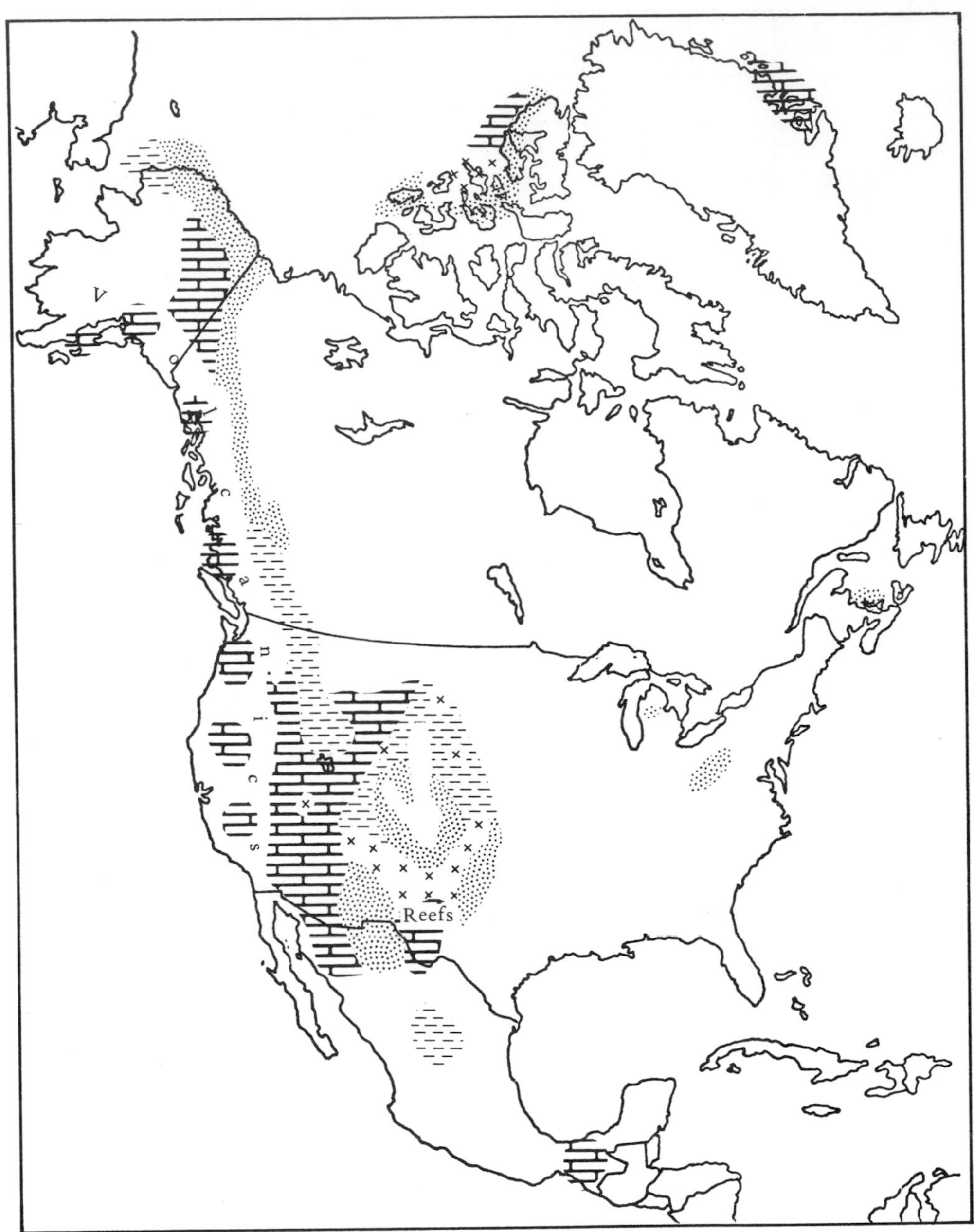

FIGURE 17.37 Present-Day Distribution of Middle Permian Rocks in North America.

FIGURE 17.38 Permian Reefs of West Texas. (A) Block diagram showing facies relationships. (B), (C) El Capitan, Guadalupe Mountains, National Park. The crest is composed of the Capitan reef limestone. (Courtesy Texas Highway Department.)

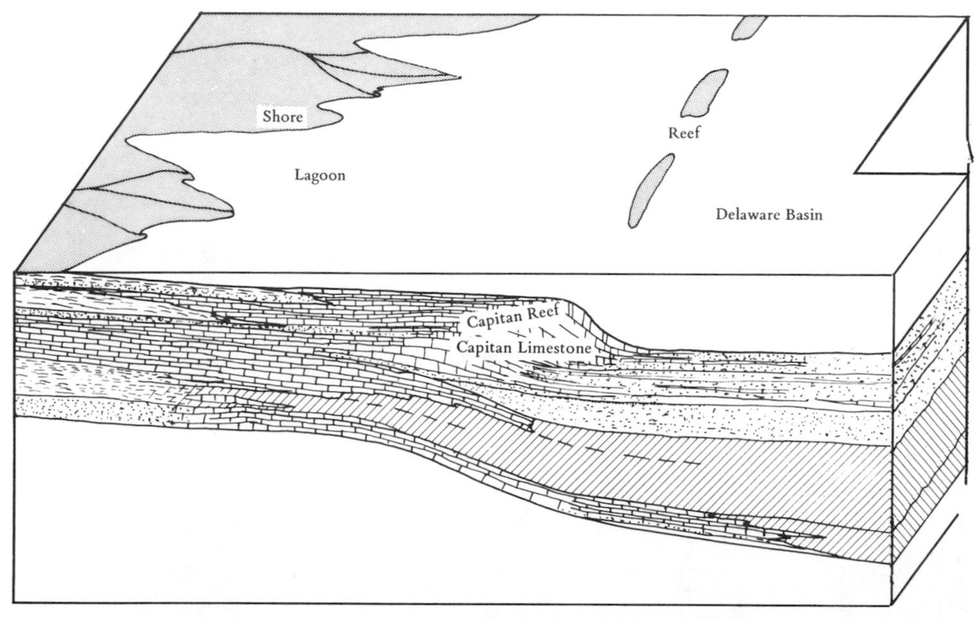
Shore
Reef
Lagoon
Delaware Basin
Capitan Reef
Capitan Limestone

A

B

C

FIGURE 17.39 Permian Redbeds in the Southwest. Monument Valley, Arizona-Utah. (Courtesy Utah Tourist and Publicity Council.)

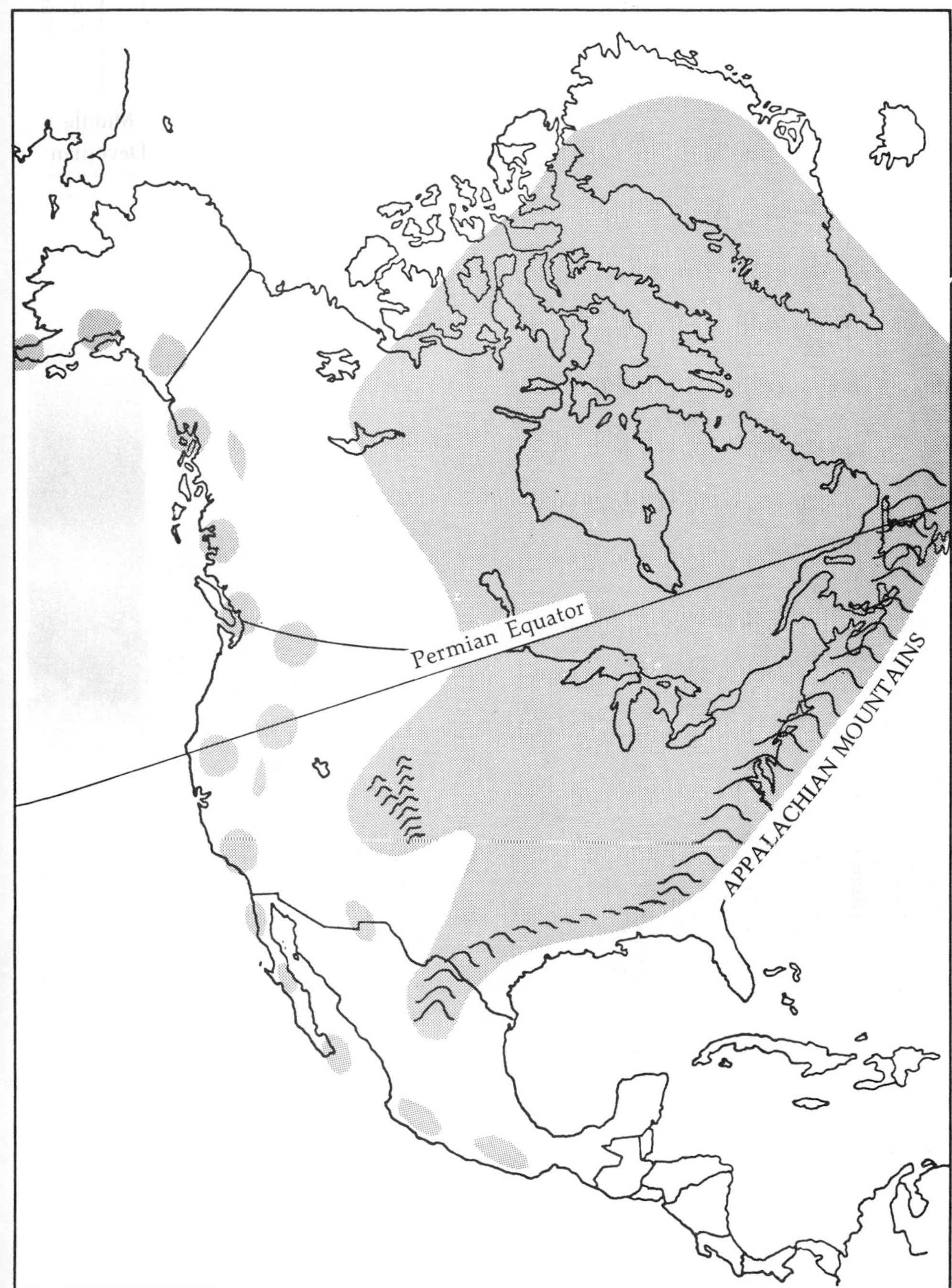

FIGURE 17.40 An Interpretation of Middle Permian Paleogeography on a Map of Present-Day North America.

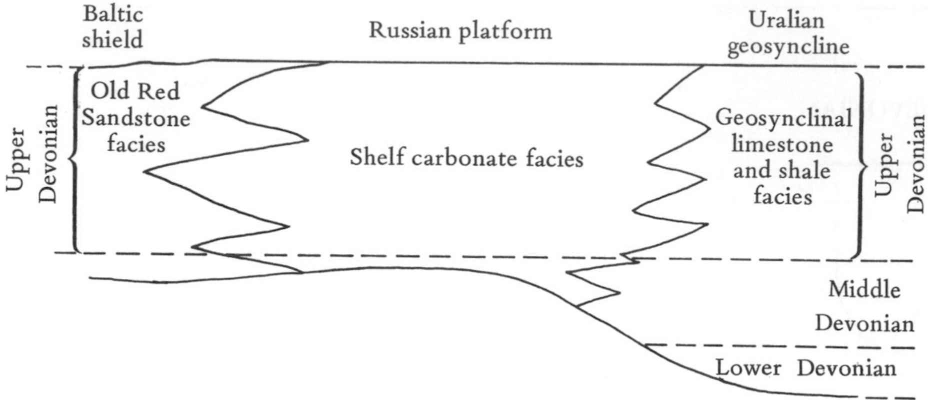

FIGURE 17.41 Cross-Section of European Devonian Rocks.

FIGURE 17.42 Lower Carboniferous Limestone in Great Britain near Llangollen, Wales. (Crown Copyright Geological Survey photograph. Reproduced by permission of the Controller of Her Britannic Majesty's Stationery Office.)

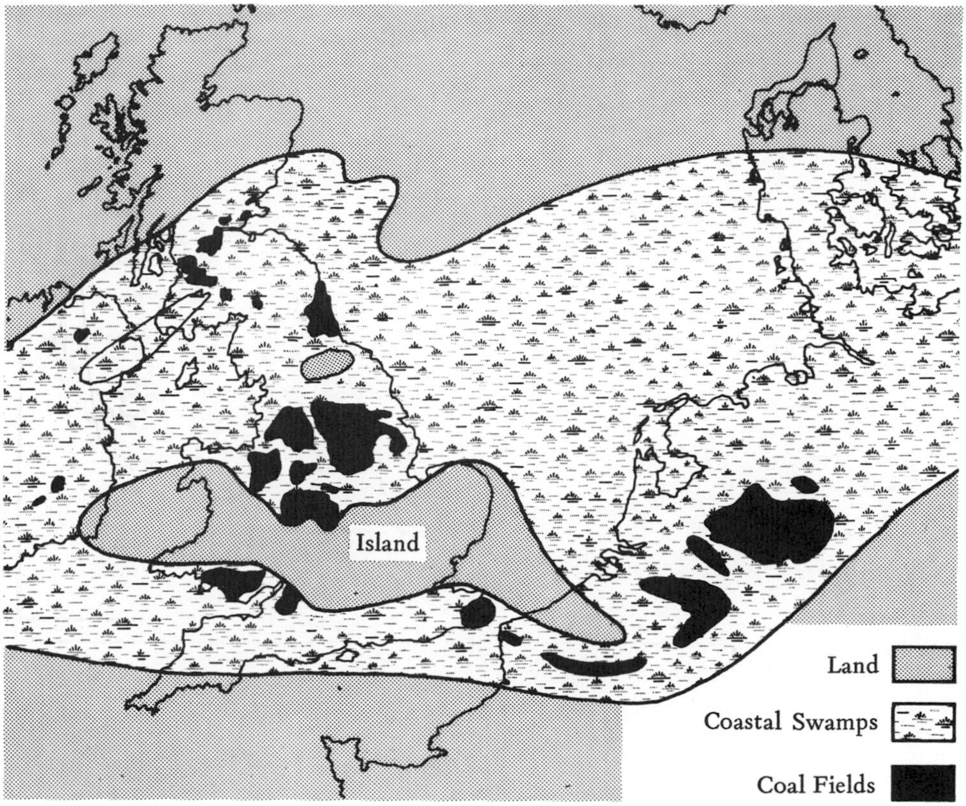

Island

Land

Coastal Swamps

Coal Fields

FIGURE 17.43 Paleogeography of Present-Day Northwest Europe in the Late Carboniferous.

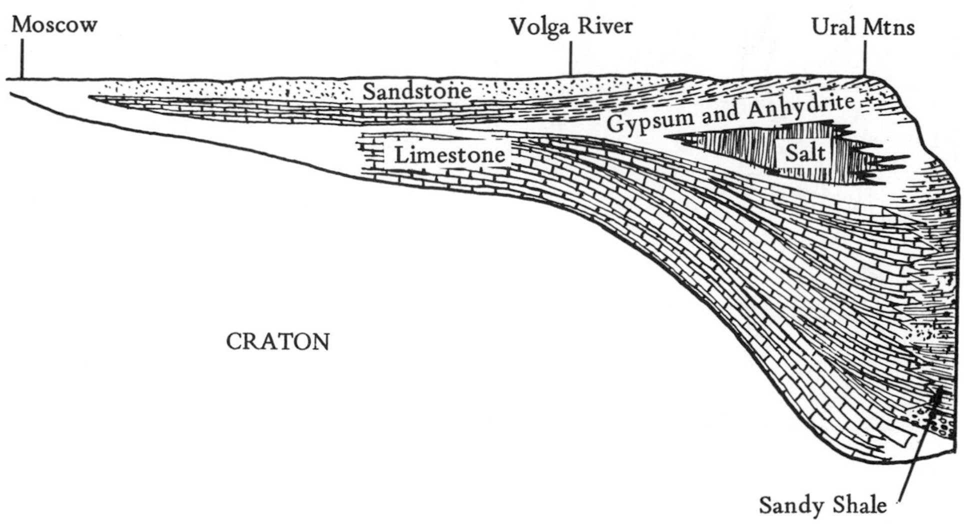

FIGURE 17.44 Cross-Section of Russian Permian Rocks. Vertical dimension highly exaggerated.

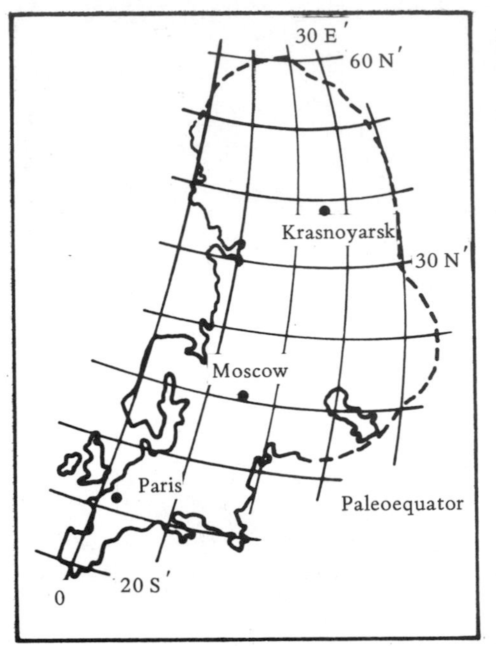

Devonian

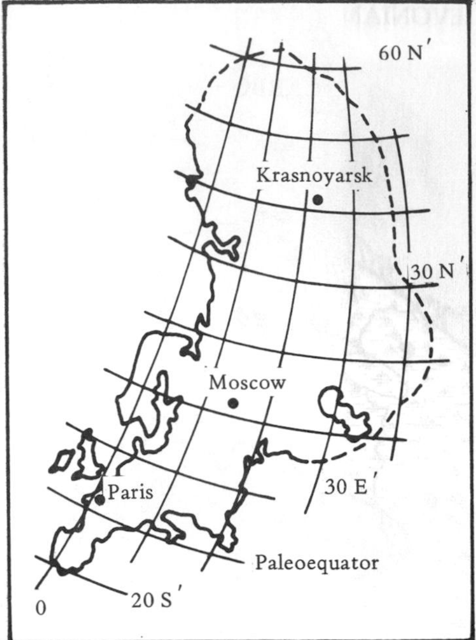

Carboniferous

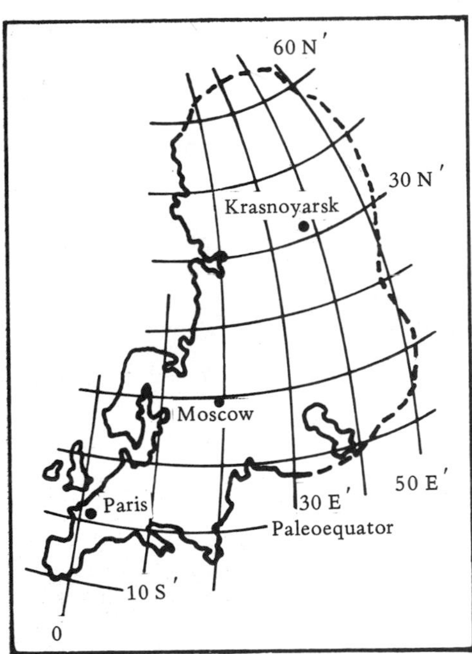

Permian

FIGURE 17.45 Paleolatitudes for Eurasia in the Late Paleozoic. (From E. Irving, *Paleomagnetism and Its Application to Geological and Geophysical Problems,* Copyright © 1964, by permission of John Wiley & Sons, Inc.)

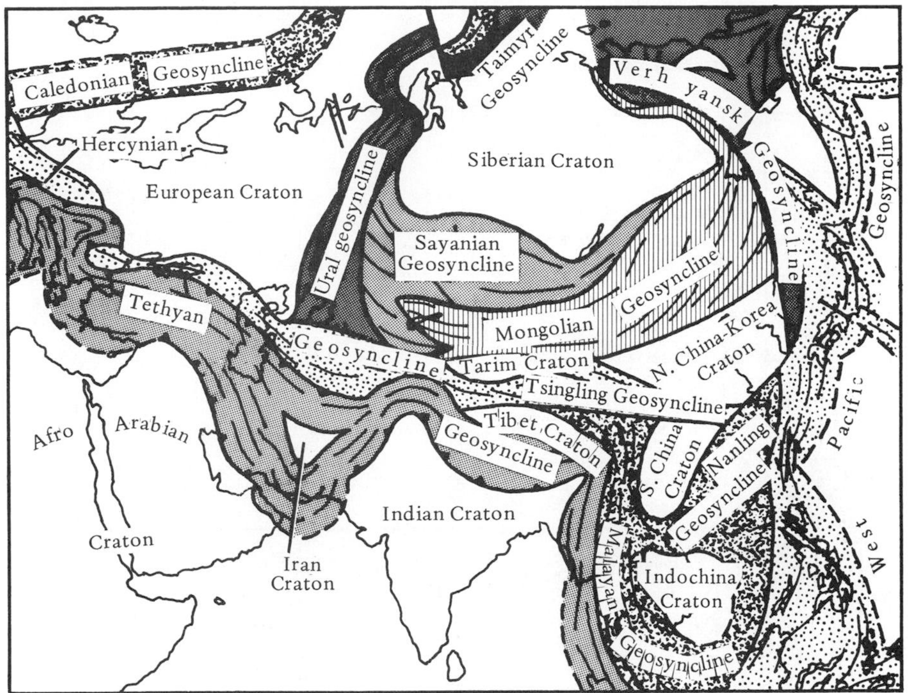

FIGURE 17.46 Map of Present-Day Asia Showing Ancient Geosynclines and Cratons.

N. American Pole-Path

Eurasian Pole-Path

Approx. -900 m. Contour

Mobile Belts

Trans-Atlantic Tertiary Volcanic Province

Mesozoic

Late Paleozoic

Early Paleozoic

Precambrian

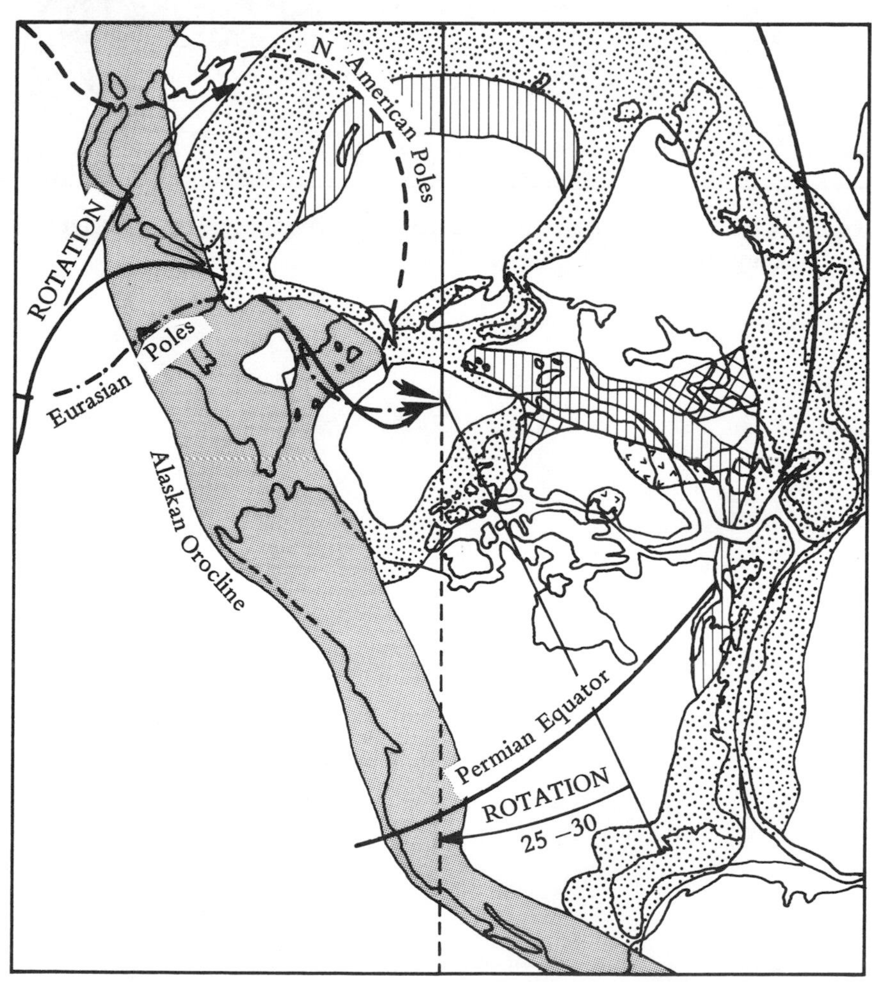

FIGURE 17.47 Reconstruction of Laurasia. Note position of Tethys in the Tropics. (From *Evolution of the Earth* by Dott and Batten. Copyright © 1971 by McGraw-Hill, Inc. Used by permission of the McGraw-Hill Book Company.)

Questions

1. Where was the Gondwana Pole situated in the Late Paleozoic according to paleomagnetic data and what geologic evidence supports these positions?
2. Where was the Samfrau Geosyncline and when did it undergo major deformation?
3. What is the significance of the Caledonian-Acadian Orogeny and when did it occur?
4. How are today's Catskill Mountains related to the Late Paleozoic Acadian Mountains?
5. What are two possible explanations for the Appalachian-Alleghenian Orogeny and when did it occur?
6. What is the significance of the Hercynian and Uralian Orogenies in the evolution of Eurasia?
7. What major orogenies affected the Cordilleran Geosyncline in the Late Paleozoic and what was their probable origin?
8. What is the significance of the Ancestral Rockies–Oklahoma Mountains uplifts and when did they occur?
9. What might the Ouachita-Marathon Orogenies indicate about plate positions in the Late Paleozoic?
10. When did the Kaskaskia and Absaroka epeiric seas advance, peak, and retreat?
11. What are cyclothems and what factors may have caused their formation?
12. Where and when did major redbeds form in the Late Paleozoic of Gondwanaland and Laurasia?
13. When was the European craton flooded by seas in the Late Paleozoic?

18

Invasion of the Land:

Life of the Late Paleozoic

Beginning with Devonian, there was widespread invasion and occupation of the earth's land masses by organisms. Among the invertebrates, mollusks and particularly arthropods successfully invaded the land, while the vertebrates succeeded with the amphibians, transitional forms between fish and typical tetrapods. Before the end of the Paleozoic, reptiles dominated the animal world on land. The plants covered the landscape in the Late Paleozoic and diversified into all the present phyla with the probable exception of the flowering plants.

Marine Life

This is not to say that life in the sea (Figs. 18.1–18.4) did not continue to flourish in the Late Paleozoic. Indeed it continued to diversify and multiply, but many of the orders and some of the classes are different from those of the Early Paleozoic. The major Late Paleozoic invertebrate groups were the *foraminifera, corals, bryozoans, brachiopods, ammonoids, ostracods, crinozoan echinoderms,* and *conodonts.* There was a high degree of cosmopolitanism in Late Paleozoic seas because of the close concentration of all the continental masses.

Foraminifera

The *foraminifera* became abundant for the first time in the Mississippian. The common types of that period were planispirally coiled, granular calcareous genera such as *Endothyra* (Fig. 18.5A). From these evolved the elongate, spindle-shaped *fusulines* (Fig. 18.5B) of the Pennsylvanian and Permian. These abundant and rapidly evolving foraminifera are major correlation fossils for rocks

FIGURE 18.1 Diorama of Devonian Marine Life. (Courtesy Field Museum of Natural History.)

FIGURE 18.2 Diorama of Mississippian Marine Life. (Courtesy Field Museum of Natural History.)

FIGURE 18.3 Diorama of Pennsylvanian Marine Life. (Courtesy U.S. National Museum.)

FIGURE 18.4 Diorama of Permian Marine Life. (Courtesy Field Museum of Natural History.)

of those systems. Common genera include *Fusulinella, Fusulina, Triticites,* and *Pseudoschwagerina* (Fig. 18.5B–C). No fusuline forams are known to survive the Permian.

Corals

The *coral* fauna of the Late Paleozoic was characterized by the abundance of *tetracorals,* an important group for regional correlation in the Devonian and Mississippian. The tabulate corals that dominated the Ordovician and Silurian were only abundant in the Early Devonian regressive places of the Tippecanoe Sea. Thereafter, the tetracorals underwent a great expansion reaching a Devonian peak and building reefs such as the one that today forms the Falls of the Ohio at Louisville, Kentucky. Common Devonian genera (Fig. 18.6, 18.7) include the solitary *Zaphrentis, Heliophyllum,* and *Hadrophyllum,* and the colonial *Hexagonaria* and *Pachyphyllum.* The tetracorals were somewhat diminished, but still common, in the Carboniferous (Fig. 18.8). Representative genera include the solitary *Caninia* and *Lophophyllidium,* and the colonial *Lithostrotionella.* The tabulate coral *Chaetetes* (Fig. 18.9), with its very slender corallites, is locally common in Carboniferous strata. Tetracorals diminished greatly in the Permian as widespread continental conditions caused the retraction of the epeiric seas in which they thrived. A few Permian tetracorals show the insertion of new septa into all six spaces between the original septa and may be transitional to hexacorals. Until recently, it was thought that the tetracorals failed to survive

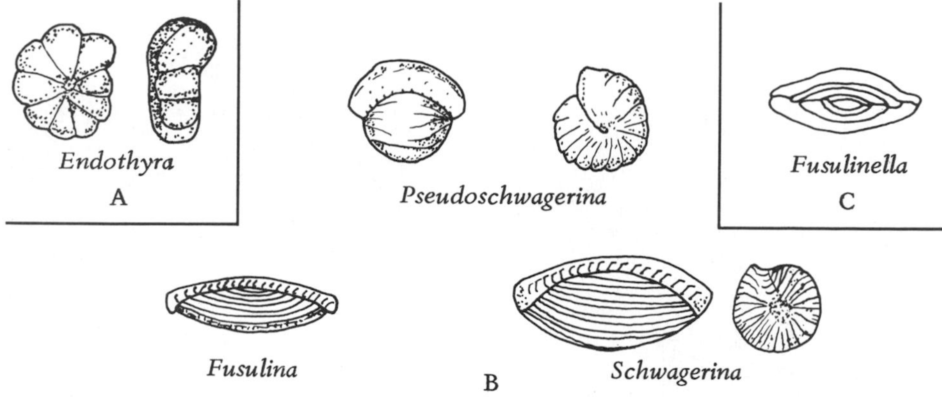

FIGURE 18.5 Late Paleozoic Foraminifera. (A) Mississippian Genus *Endothyra,* (B) Pennsylvanian-Permian Fusulinaceans, (C) Cross-section of Fusulinacean.

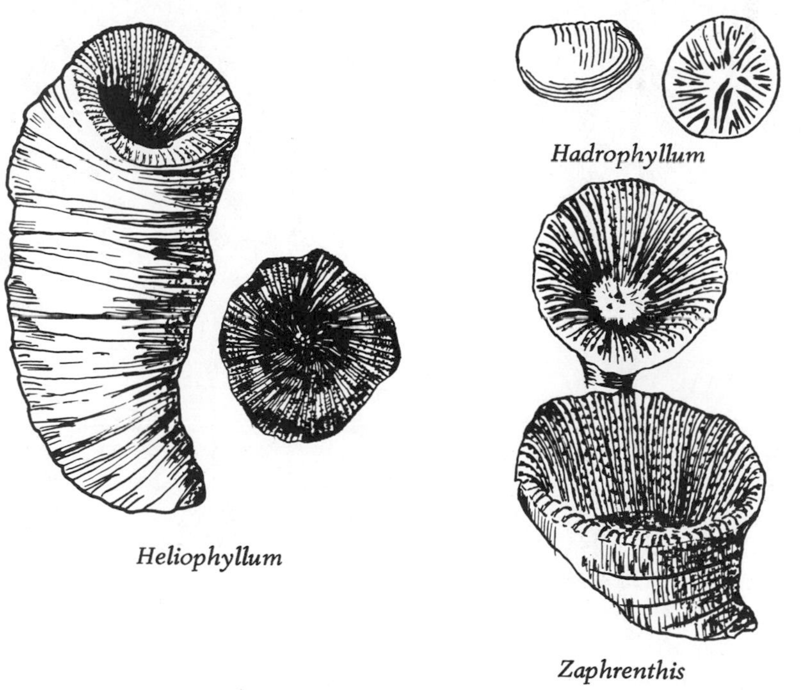

FIGURE 18.6 Devonian Solitary Tetracorals.

the Paleozoic, but recent finds in the Soviet Union and the Indian subcontinent indicate they survived into rocks containing a transitional Permian and Triassic fauna. The tabulate corals also apparently failed to survive the Paleozoic, although some questionable Mesozoic finds have been reported.

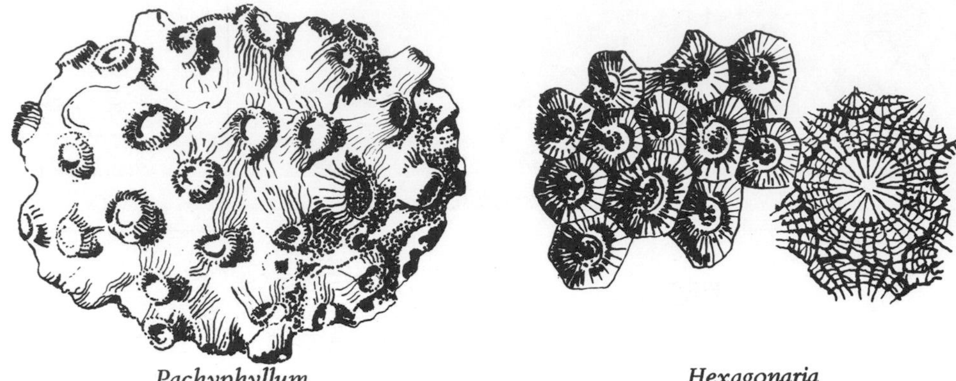

FIGURE 18.7 Devonian Colonial Tetracorals.

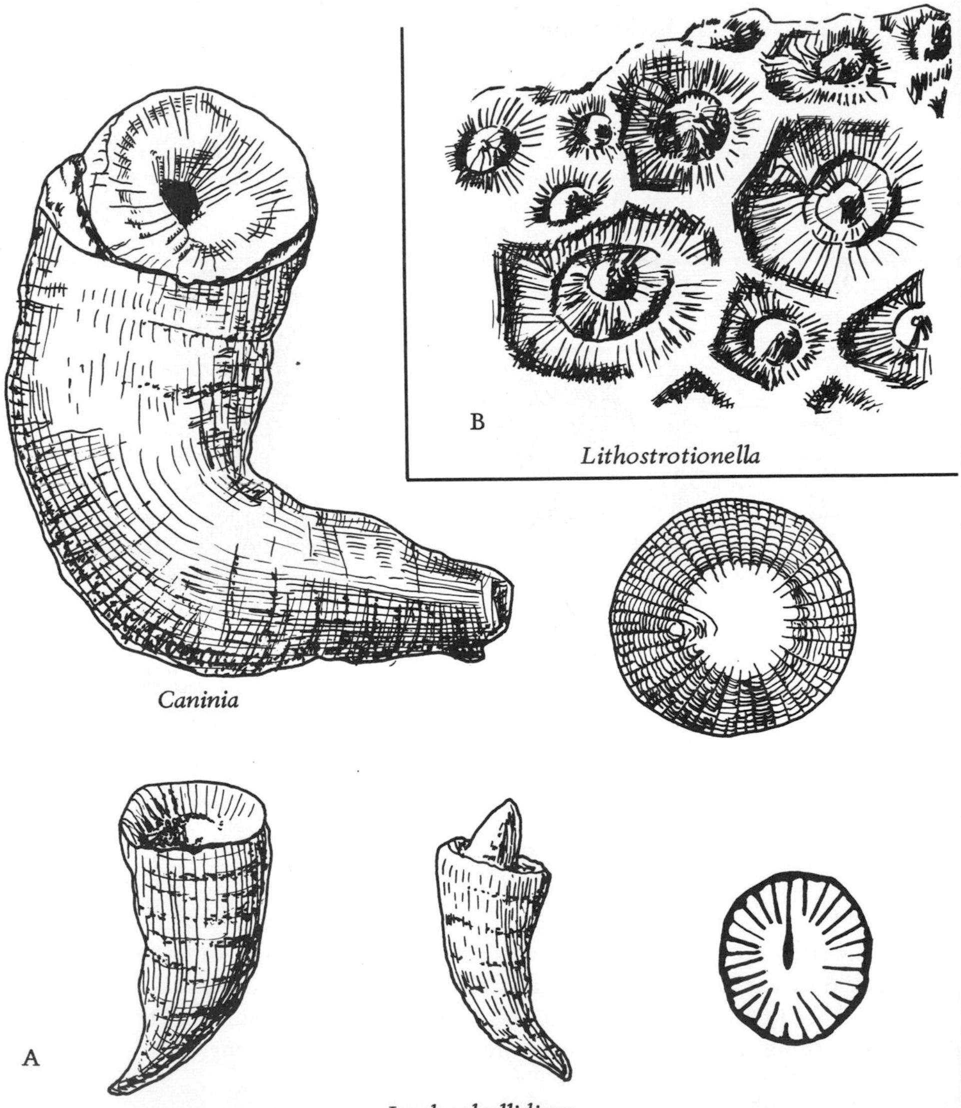

B

Lithostrotionella

Caninia

A

Lophophyllidium

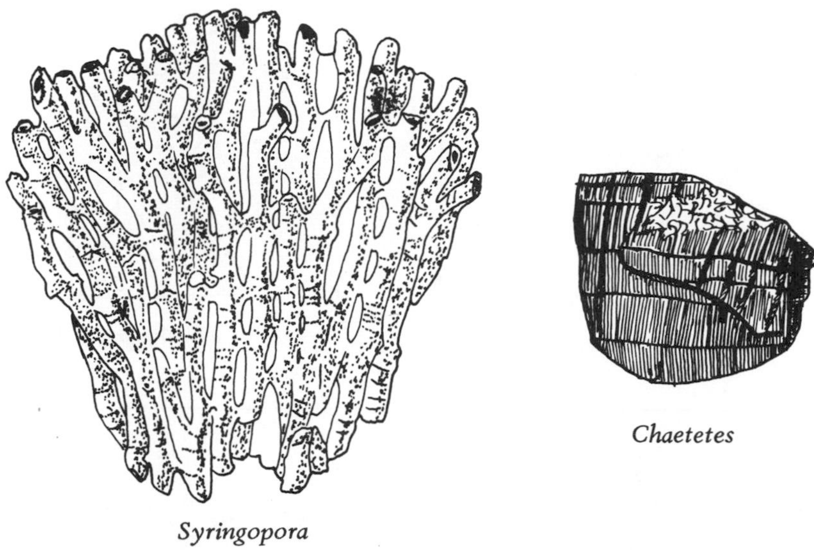

FIGURE 18.9 Late Paleozoic Tabulate Corals.

Bryozoans

The *bryozoans* are another significant faunal element that displays a change in the dominance of orders from Early to Late Paleozoic. The stony bryozoa or trepostomes, so abundant in the Ordovician and to some extent, the Silurian, decline to insignificance in Upper Paleozoic rocks. *Tabulipora* (Fig. 18.10A) was one of the few common Late Paleozoic genera. Cystoporoids also declined with only *Fistulipora* (Fig. 18.10B) being occasionally common. The *cryptostome* bryozoans, particularly the *lacy* fenestellids, diversified greatly and dominate Upper Paleozoic bryozoan assemblages. The lacy forms (Fig. 18.11) reached a Devonian-Carboniferous peak and include the abundant genera *Fenestrellina, Polypora,* and the chiefly Mississippian *Archimedes* which had a screw-shaped axis supporting its lacy fronds. Delicate twig-like cryptostomes (Fig. 18.10C) such as *Rhombopora* also were common. The cryptostomes declined greatly in the Permian and only a few survived into the transitional Permotriassic beds. The trepostomes and cystoporoids also appear not to have survived the Paleozoic, but some later bryozoans may be their descendants.

Brachiopods

Brachiopods flourished in the Late Paleozoic, but once again we note a shift in the abundance of orders. Of the Early Paleozoic brachiopod groups, the inarticulates declined; the orthoids (Fig. 18.12) diminished but still included

FIGURE 18.8 Carboniferous Tetracorals.
(A) Solitary, (B) Colonial.

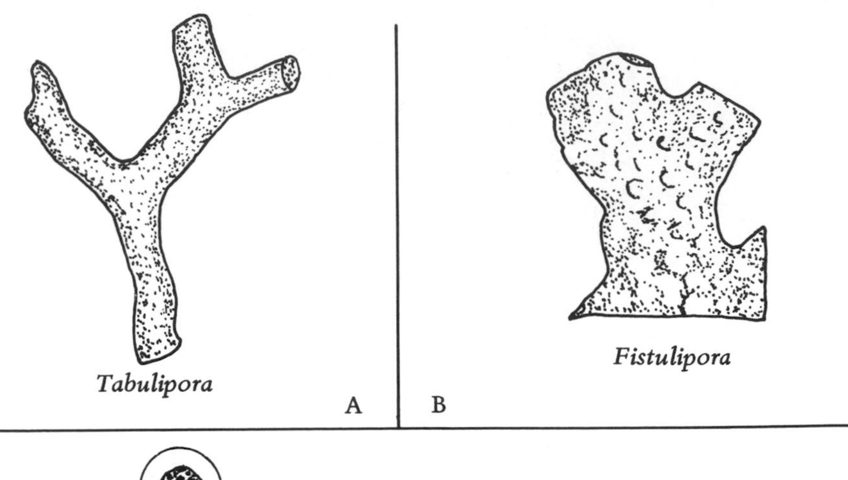

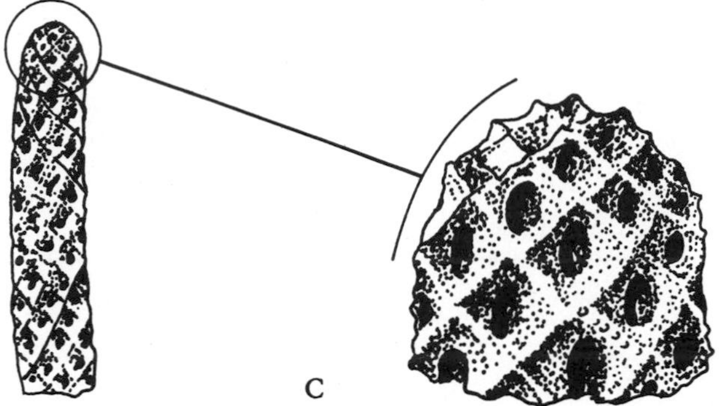

FIGURE 18.10 Late Paleozoic Twig-Like Bryozoans. (A) Trepostome, (B) Cystoporoid, (C) Cryptostome.

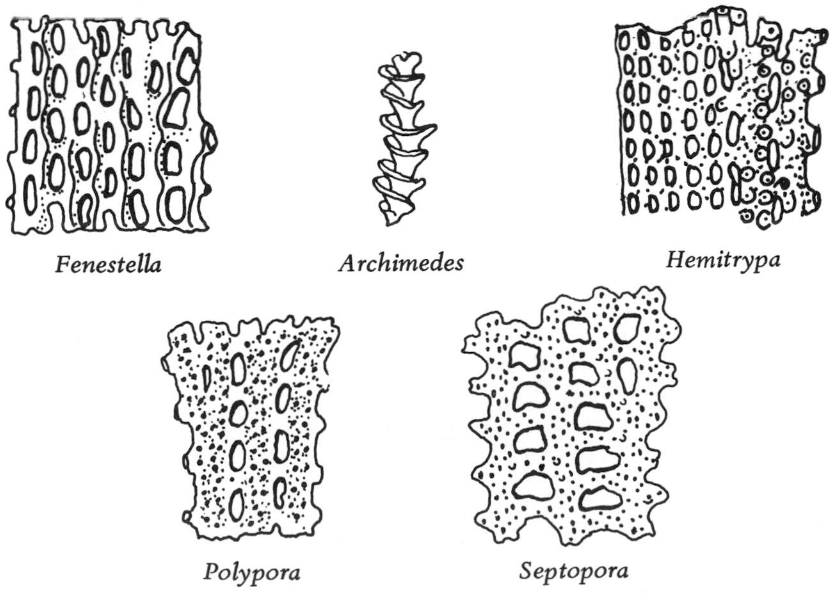

FIGURE 18.11 Late Paleozoic Lacy Cryptostome Bryozoans.

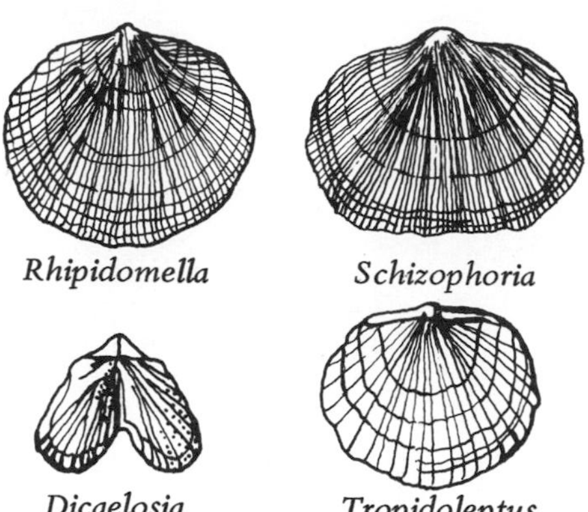

FIGURE 18.12 Late Paleozoic Orthoid Brachiopods.

the common genera *Tripidoleptus, Schizophoria,* and *Rhipidomella;* the strophomenoids (Fig. 18.13) did well only through the Devonian when they included the common genera *Douvillina, Leptaena,* and *Stropheodonta;* the pentameriods became extinct in the Devonian with only a few forms such as *Gypidula* (Fig. 18.14) common in that period; and the *rhynchonelloids* (Fig. 18.15) continued to flourish with such genera as *Eatonia, Camarotoechia,* and *Pugnoides.* The Late Paleozoic sees the dominance of the *chonetoids, productoids, davidsonioids, atrypoids, athyridoids,* and the *spiriferoids.* The chonetoids (Fig. 18.16), particularly abundant in the Devonian and Carboniferous, include genera such as *Devonochonetes, Mesolobus,* and *Neochonetes.* The spiny productoids appeared in the Devonian and became the dominant Carboniferous and Permian brachiopod group. Common genera of this very diverse order were *Dictyoclostus, Retaria, Juresania,* and *Linoproductus* (Fig. 18.17). In the Permian, genera such as *Prorichthofenia* converged on corals as they inhabited the same reef environment (Fig. 18.17). Another peculiar brachiopod order, the *oldhaminoids* (such as *Leptodus*), with their peculiar slotted shells (Fig. 18.18) also inhabited these reefs. The davidsonioids (Fig. 18.19) were common in the

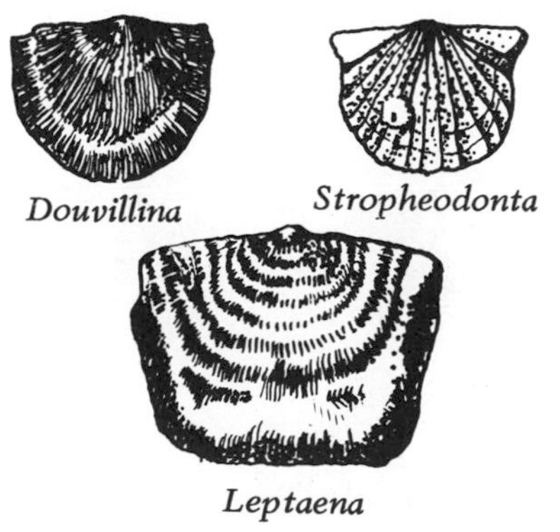

FIGURE 18.13 Late Paleozoic Strophomenoid Brachiopods.

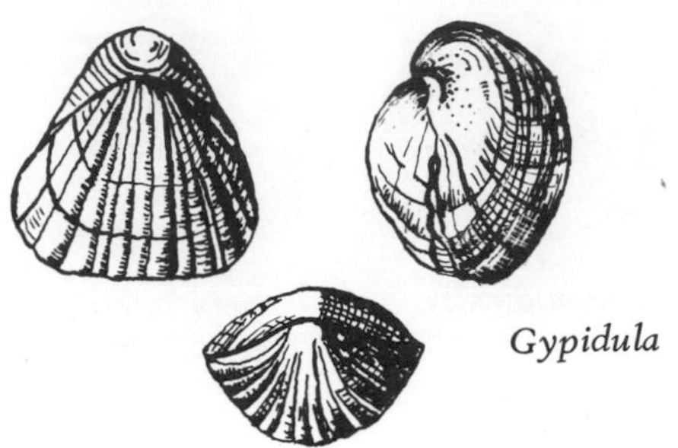

FIGURE 18.14 Devonian Pentameroid Brachiopods.

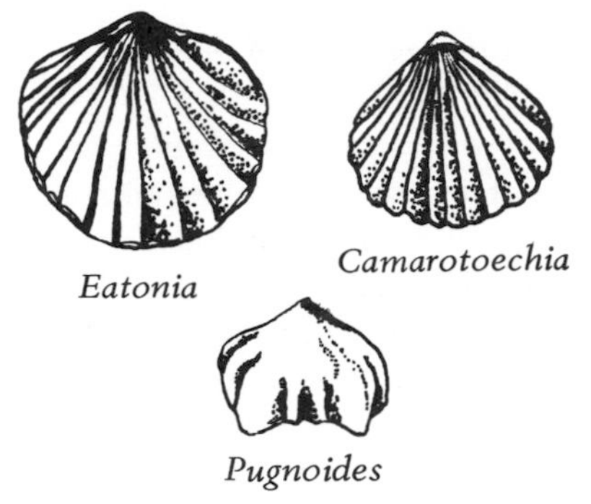

FIGURE 18.15 Late Paleozoic Rhynchonelloid Brachiopods.

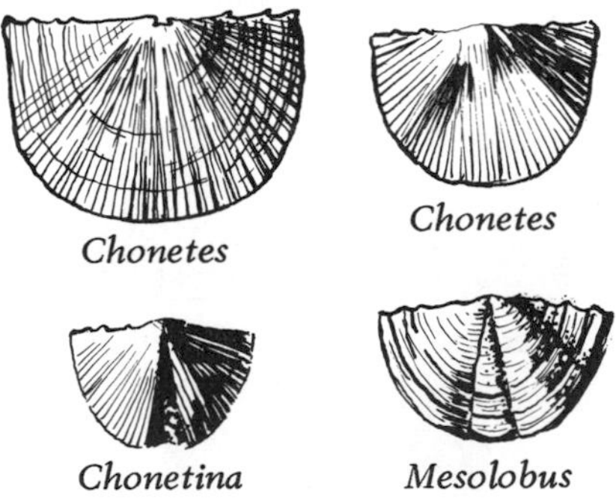

FIGURE 18.16 Late Paleozoic Chonetoid Brachiopods.

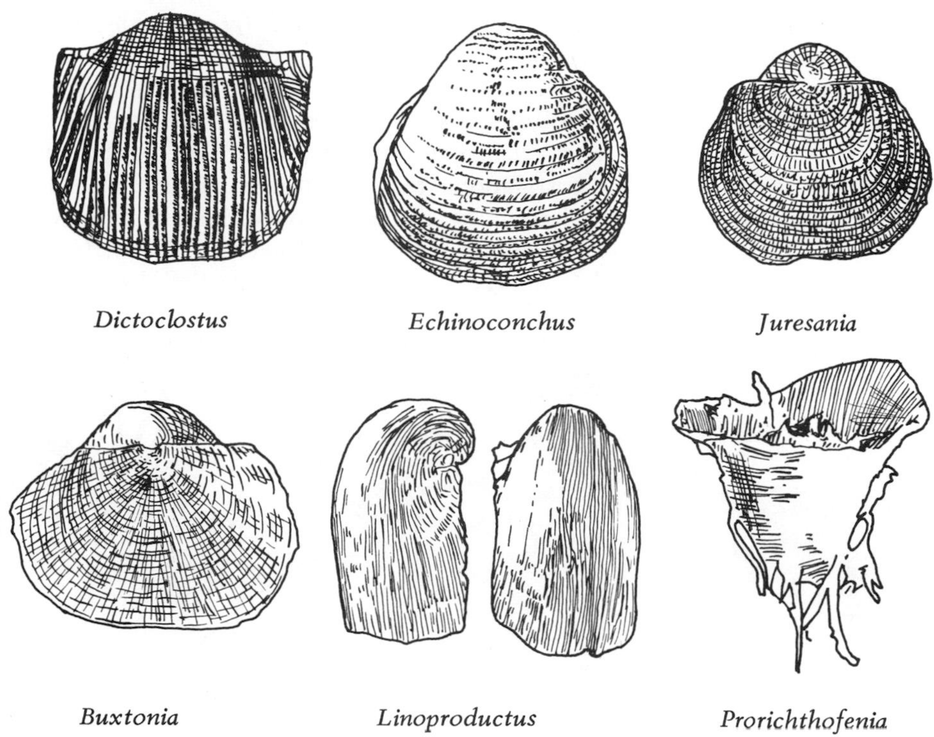

FIGURE 18.17 Late Paleozoic Productoid Brachiopods.

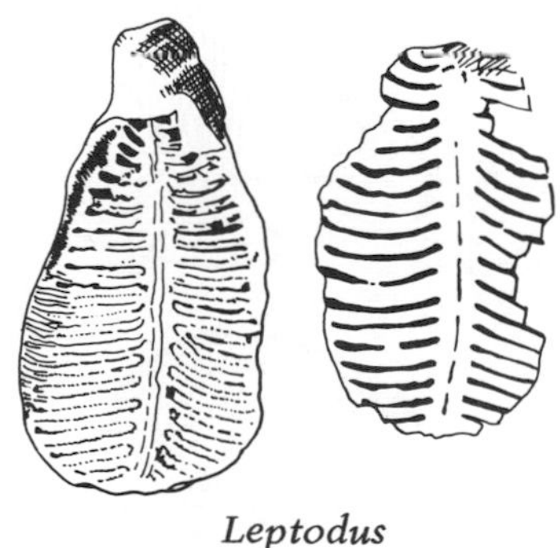

FIGURE 18.18 Peculiar Slotted Oldhaminoid Brachiopod.

Permocarboniferous, especially the genera *Derbyia* and *Meekella*. Atrypoids (Fig. 18.20) reached a Devonian climax and then became extinct. *Atrypa* and *Leptocoelia* were both very abundant in the period. Athyridoids (Fig. 18.21) were common throughout the Late Paleozoic and include such common genera as *Athyris, Composita,* and *Meristella*. The spiriferoids (Fig. 18.22) reached a Devonian peak, but continued to be important through the remainder of the era. The Devonian forms include *Mucrospirifer, Paraspirifer, Brachyspirifer,* and *Cyrtospirifer,* while *Spirifer* and *Neospirifer* were common in the Carboniferous-Permian interval. The dominant modern order of articulate brachiopods, the terebratuloids (Fig. 18.23) appeared in the Devonian, but only a few genera such as *Rensselaeria,* and *Dielasma* were common. Brachiopods are important for making regional correlations throughout the Upper Paleozoic.

Most of the abundant Late Paleozoic brachiopod groups did not fare well thereafter. As noted above, atrypoids did not survive the Devonian. The orthoids, declining for a long time, failed to survive the Permian; while the also declin-

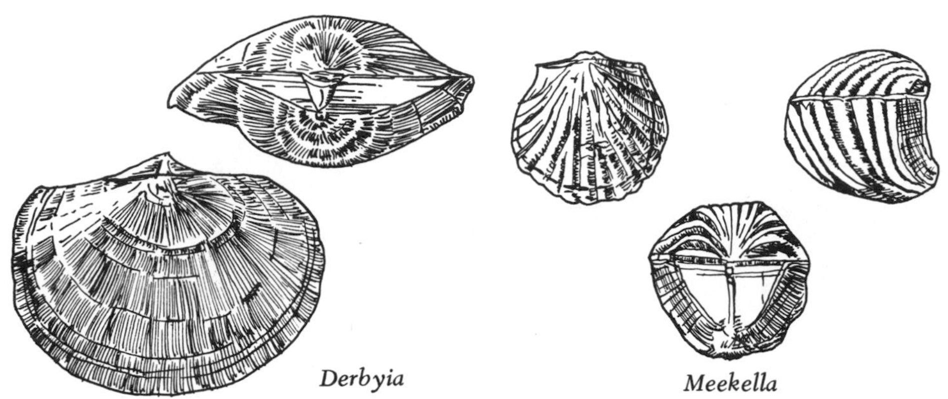

FIGURE 18.19 Late Paleozoic Davidsonioid Brachiopods.

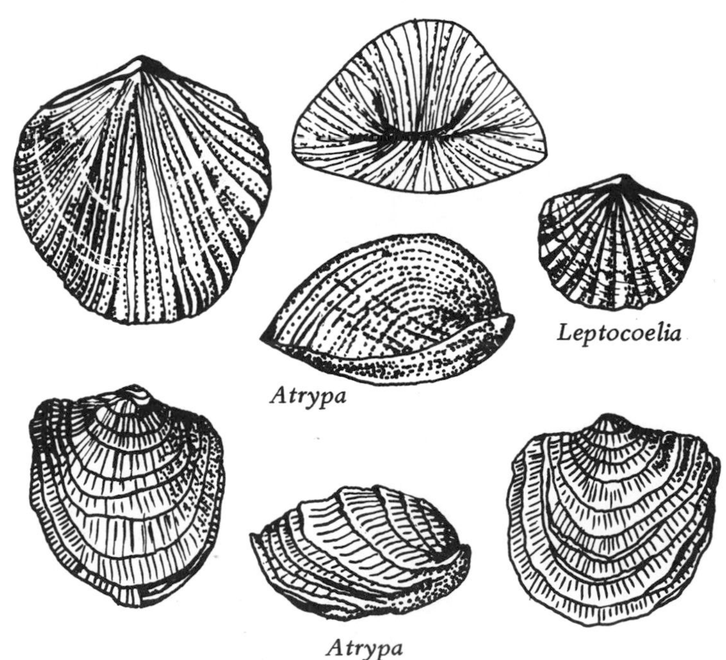

FIGURE 18.20 Devonian Atrypoid Brachiopods.

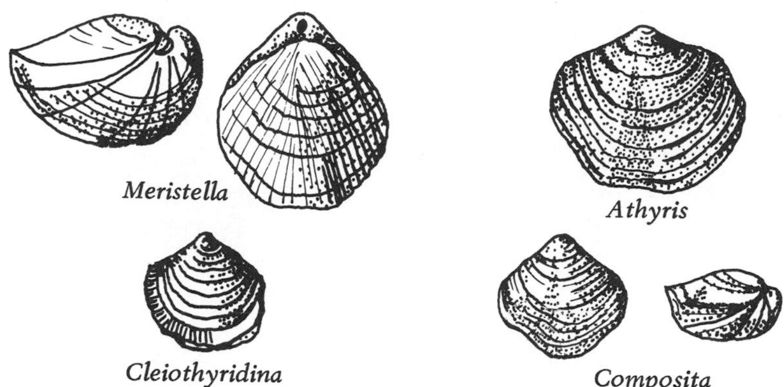

FIGURE 18.21 Late Paleozoic Athyridoid Brachiopods.

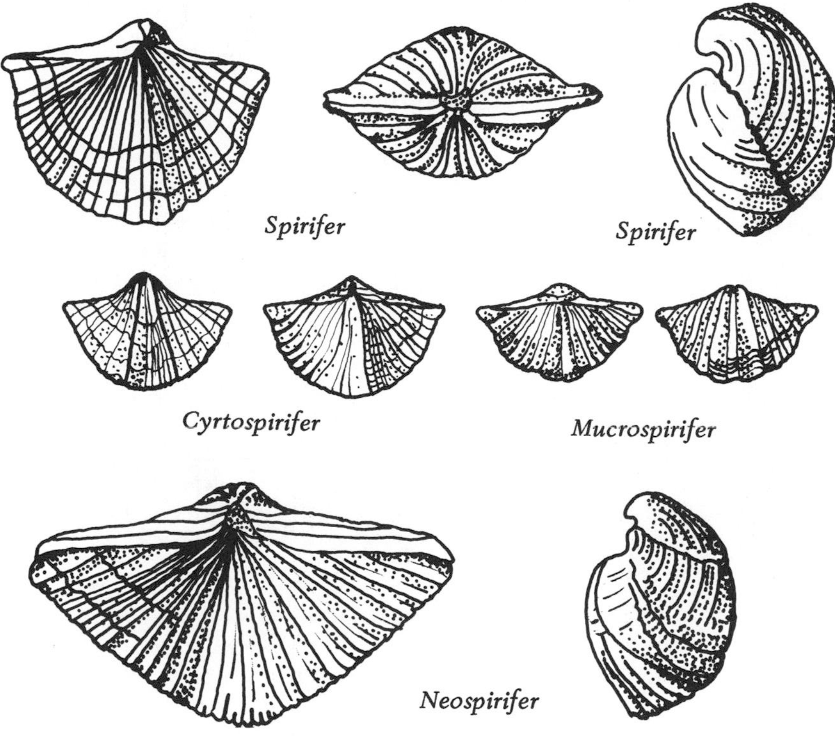

FIGURE 18.22 Late Paleozoic Spiriferoid Brachiopods.

ing strophomenoids had a few Triassic descendants. Except for some questionable Jurassic descendants, the chonetoids did not live beyond the Permian. Productoids survived into the Permo-Triassic transition beds and a few peculiar Lower Mesozoic brachiopods may have been their descendants. Davidsonioids and athyridoids had some Triassic survivors and the spiriferoids even lasted into the Jurassic, but all three groups had only rare, atypical Mesozoic descendants. Only the rhynchonelloids and terebratuloids, the two surviving articulate groups, continued to diversify in the Mesozoic.

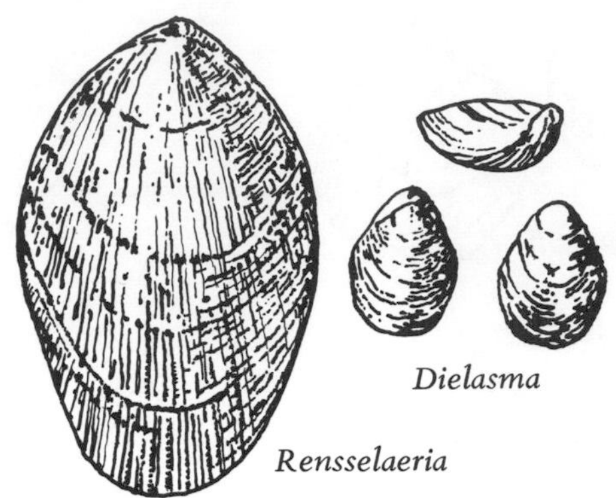

FIGURE 18.23 Late Paleozoic Terebratuloid Brachiopods.

Mollusks

The *mollusks* became more important in the Late Paleozoic, chiefly because of the appearance and radiation of the *ammonoids* (Figs. 18.24–25). Ammonoids are relatively common fossils from the Late Devonian onward, though a diminution appears to occur near the end of the era. Most Paleozoic ammonoids are *goniatites* (Fig. 18.24) such as *Tornoceras* (Devonian), *Muensteroceras* (Mis-

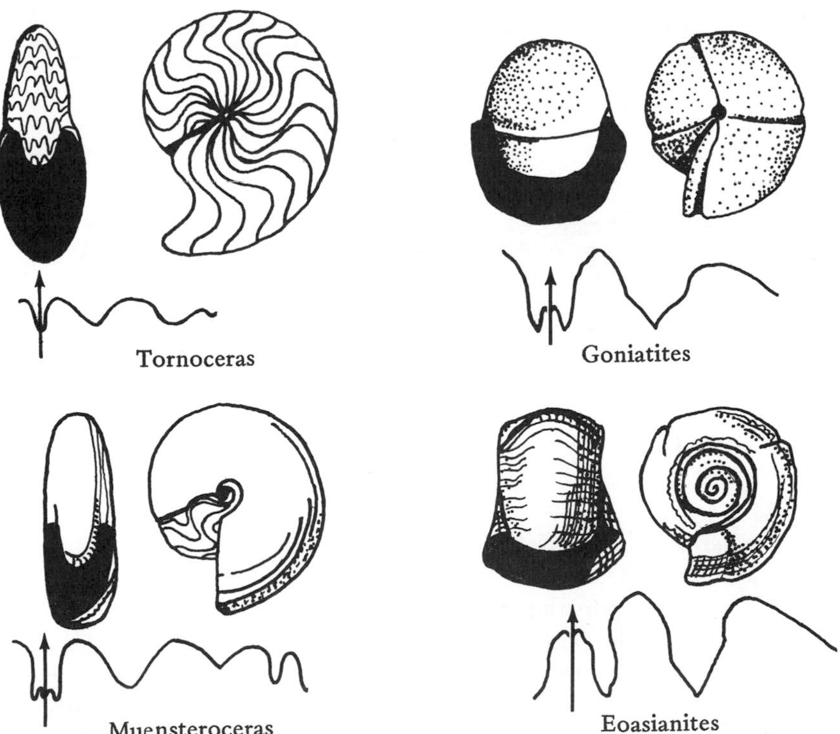

FIGURE 18.24 Late Paleozoic Goniatite Ammonoids. Sutures shown below specimens.

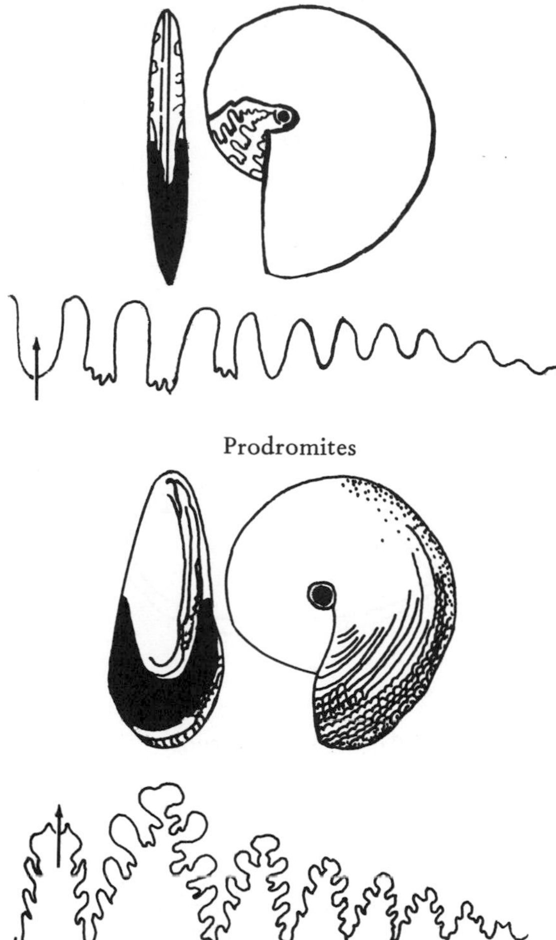

FIGURE 18.25 Late Paleozoic Ammonoids with Ceratite and Ammonite Sutures.

sissippian), and *Eoasianites* (Pennsylvanian-Permian). Ammonoids with ceratite sutures appear in the Mississippian and with ammonite sutures in the Permian (Fig. 18.25). Because their pelagic life habits enabled them to spread widely, goniatites are major fossils for making long-distance correlations of Upper Paleozoic rocks. Goniatites nearly all disappeared before the close of the Permian. Only a few survived into the Permotriassic transition rocks.

Ostracods

The tiny, bivalved *ostracod* crustaceans (Fig. 18.26) increase in abundance in the Upper Paleozoic strata. Such highly ornamented genera as *Dizygopleura, Hollina,* and *Kirkbyella* are common, along with the smoother forms such as *Thlipsura* and *Bairdia.* In common with most marine groups, ostracods declined in the later Permian when the seas were contracted by the extreme emergence of continents.

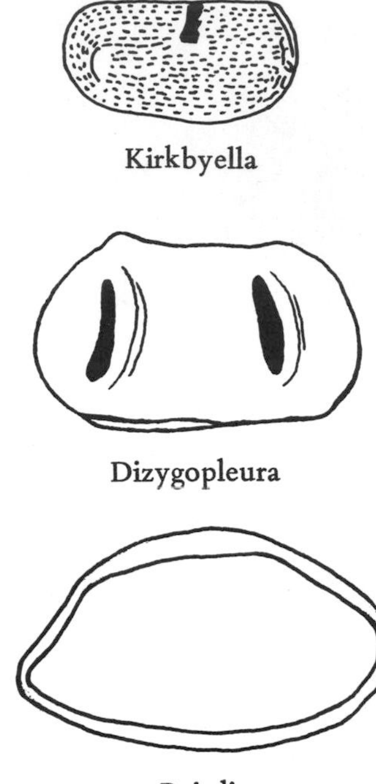

FIGURE 18.26 Late Paleozoic Ostracods.

Echinoderms

The stalked or *crinozoan echinoderms* reached a peak in the Late Paleozoic seas. Most abundant were the *crinoids*. Crinoids were common in the Devonian seas where genera such as *Arthroacantha, Dolatocrinus,* and *Gennaeocrinus* occurred (Fig. 18.27). But the great peak of the crinoids occurred in the widespread limestone-depositing epeiric seas of the Mississippian (Fig. 18.2). *Agassizocrinus, Taxocrinus, Dizygocrinus, Rhodocrinites,* and numerous others abound (Fig. 18.28). The Pennsylvanian saw a decline in crinoids and only a few members of one of the Paleozoic subclasses survived the Permian. The

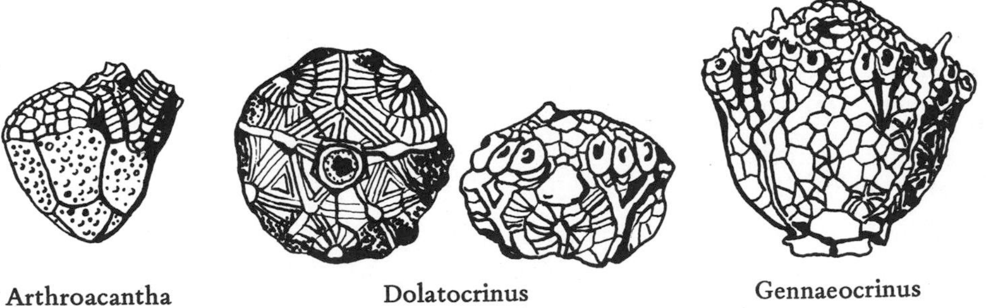

FIGURE 18.27 Devonian Crinoids.

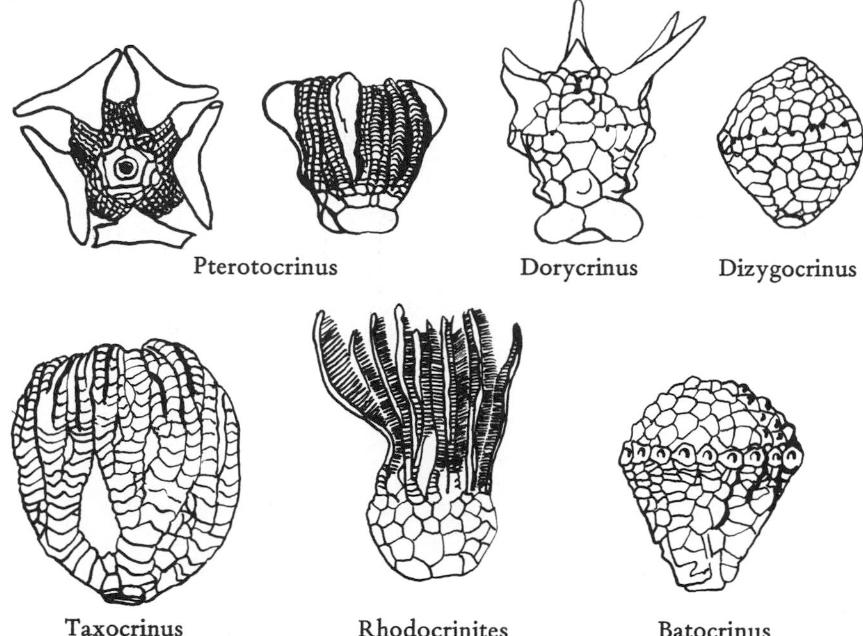

FIGURE 18.28 Mississippian Crinoids.

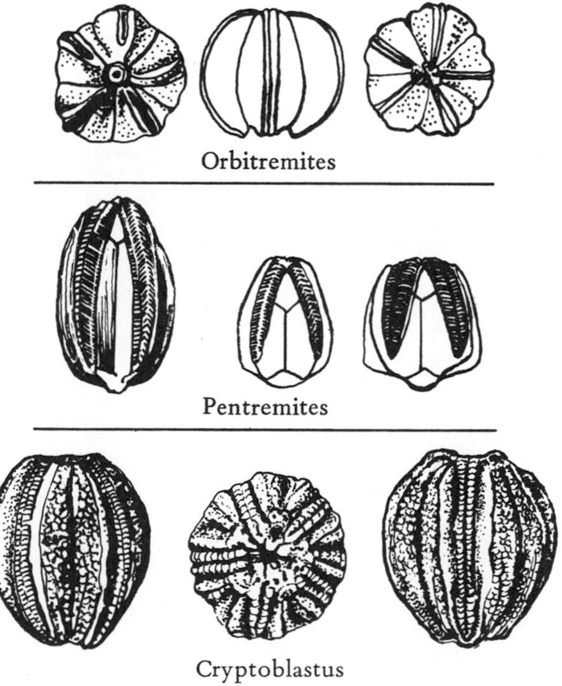

FIGURE 18.29 Late Paleozoic Blastoids.

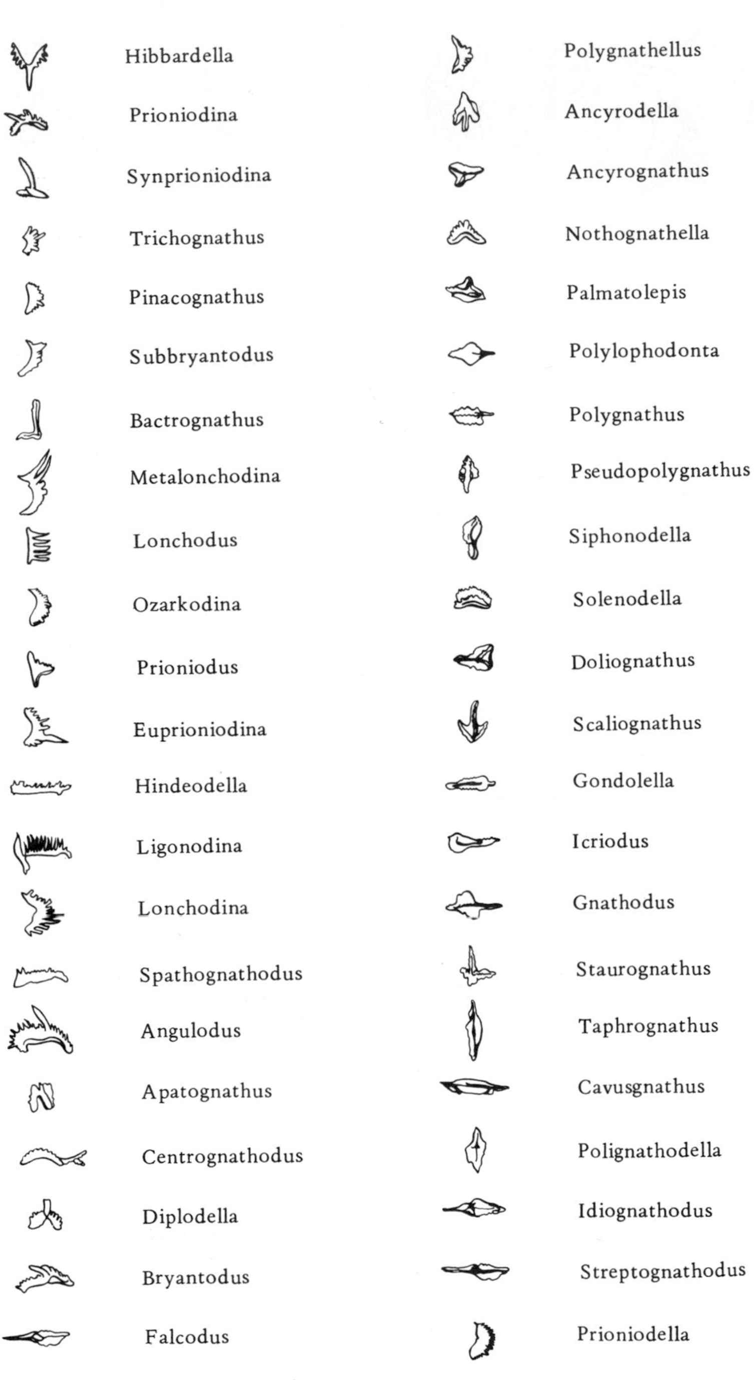

FIGURE 18.30 Late Paleozoic Conodonts.

blastoids (Fig. 18.29) also reached a Mississippian peak with genera such as *Pentremites,* and *Orbitremites* common. Blastoids declined greatly in the Pennsylvanian and Permian, and failed to survive the latter period.

Conodonts

Conodonts (Fig. 18.30), the tiny tooth or plate-like structures belonging to a pelagic worm-shaped animal, are among the most common and geologically useful fossils of the Upper Paleozoic. They are particularly abundant in Devonian and Carboniferous rocks where they have been used to zone the strata. Common genera are *Hindeodella, Ozarkodina, Spathognathodus,* and *Palmatolepis.* Conodonts were of lesser importance in Permian and Triassic times when the shallow seas were very restricted.

Other Groups

Other marine invertebrate groups were occasionally common in the Late Paleozoic. *Glass sponges* (Fig. 18.31), siliceous forms with an interlocking gridwork of spicules, were very common in shallow Devonian seas such as occurred in western New York. A peculiar group of *calcareous sponges* which resemble beads on a string (Fig. 18.32) were common in the Pennsylvanian of many areas. The *stromatoporoids* (Fig. 18.33) were equally as abundant in the

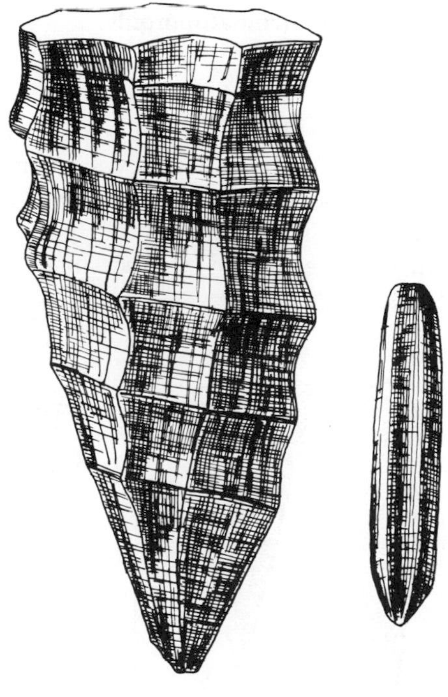

Hydnoceras *Prismodictya*

FIGURE 18.31 Devonian Glass Sponges.

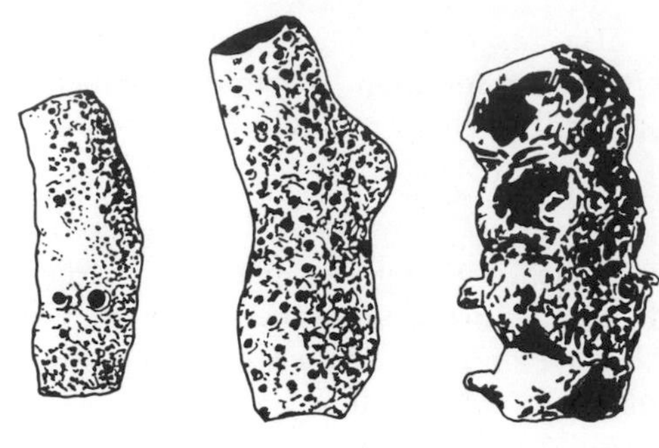

FIGURE 18.32 Carboniferous Calcareous Sponges.

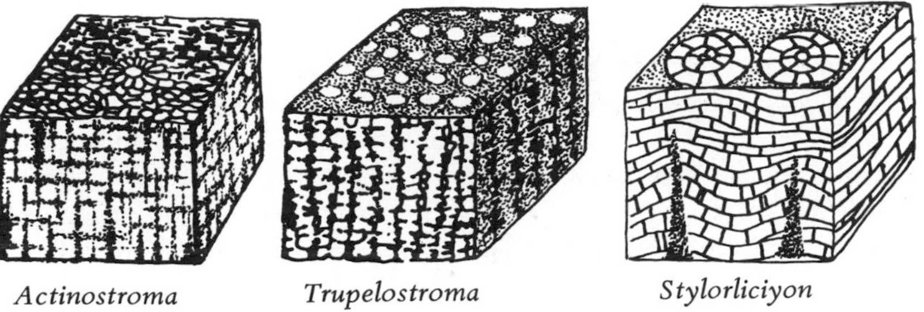

FIGURE 18.33 Devonian Stromatoporoids.

Devonian as in the Silurian, but thereafter are represented only by atypical forms. *Conularoids* (Fig. 18.34A), possible scyphozoan coelenterates, were at their peak in the Late Paleozoic, but declined to extinction shortly thereafter. *Tentuculites* (Fig. 18.34B) were still common in the Devonian. *Gastropods* and *pelecypods* continued to increase and produce many common genera. Im-

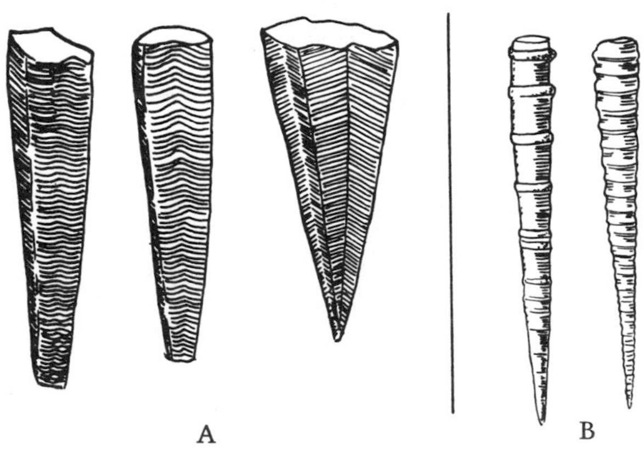

FIGURE 18.34 Enigmatic Conical Fossils of the Late Paleozoic. (A) Conularoids, (B) Tentaculites.

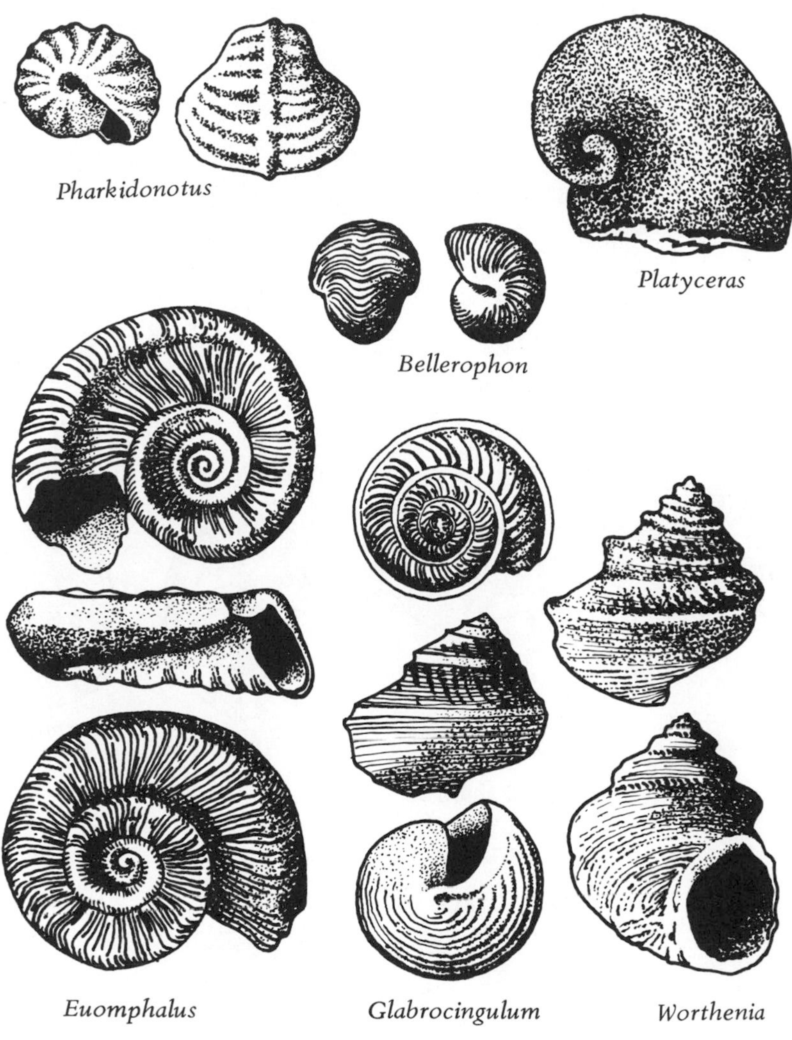

FIGURE 18.35 Late Paleozoic Gastropods.

portant Late Paleozoic gastropod genera include *Bellerophon, Euomphalus, Platyceras,* and *Worthenia* (Fig. 18.35). Significant pelecypods were *Nuculopsis, Myalina, Aviculopecten,* and *Wilkingia* (Fig. 18.36). The *trilobites* (Fig. 18.37) are common for the last time in the Devonian with the *phacopoids,* such as *Phacops* and *Greenops,* the major group. Local correlation is often done with phacopoid trilobites. Thereafter, trilobites were only represented by occasional illaenoids and these too finally passed from the scene in the Late Permian. Trilobite decline and extinction was a long process undoubtedly related to replacement by superior competitors such as mollusks, other arthropods, echinoderms, and fish. Finally, certain large globular sea urchins called *melonechinoids* (Fig. 18.38) are common in some Mississippian strata.

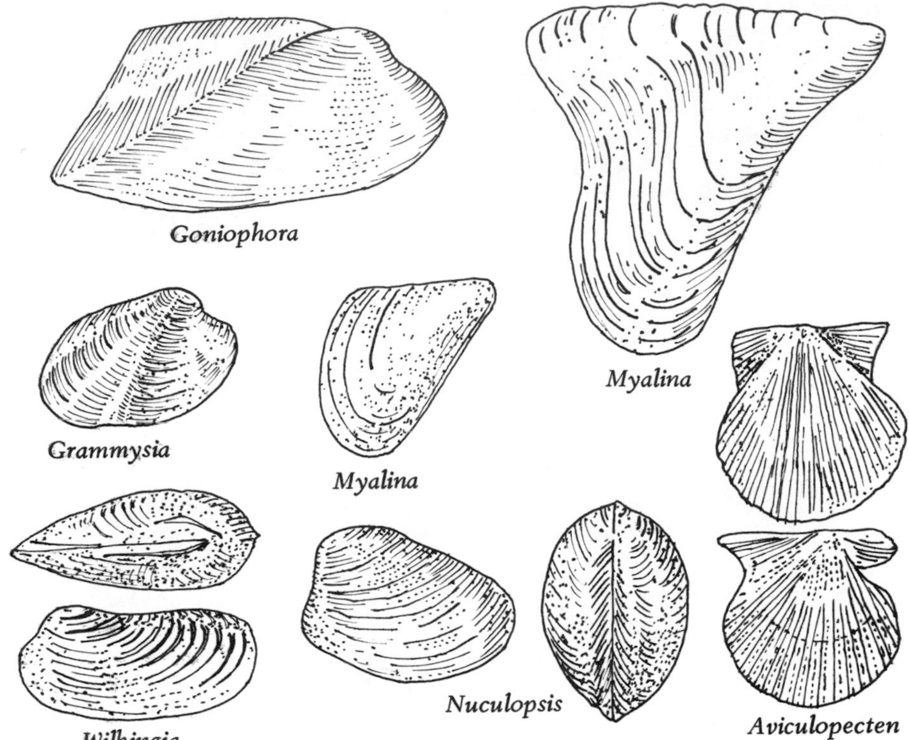

FIGURE 18.36 Late Paleozoic Bivalves.

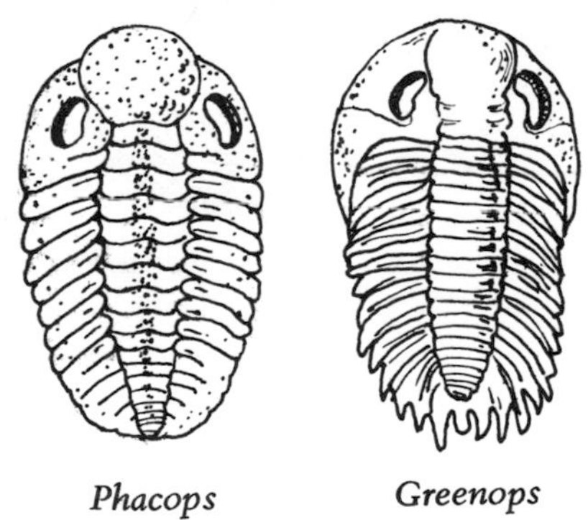

FIGURE 18.37 Devonian Phacopoid Trilobites.

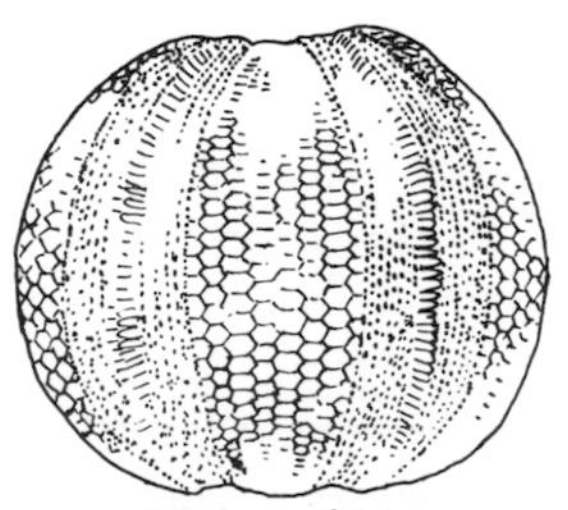

FIGURE 18.38 Mississippian Melonechinoid.

Biogeography

The marine invertebrate biogeography of the Late Paleozoic generally agrees with the inferences about continental positions deduced from physical evidence. The Gondwanaland faunas are poorly known, but largely distinctive from those of the present northern continents. The Laurasian continents show varying degrees of similarity with one another throughout the Late Paleozoic depending on the position of local mountain-building disturbances and other effects of plate movements such as latitudinal position. Particularly striking is the great faunal similarity between present western North America and Siberia during much of the Late Paleozoic. The nearness of these two areas and their similar latitudes were discussed in the last chapter (see Fig. 17.25).

Extinctions

From the preceding discussion of Late Paleozoic marine invertebrates, it becomes apparent why early geologists were struck by the number of disappearances of Paleozoic groups at or near the close of the Permian. Extinctions occurred in many important lineages of forams, corals, bryozoans, brachiopods, arthropods, and stalked echinoderms. Triassic faunas contained representatives of other lineages in these phyla and the first great abundance of the mollusks. It was these profound differences between typical Permian and Triassic faunas that originally led to the separation of the Paleozoic and Mesozoic Eras at this boundary. It is apparent that the major groups suffering a decrease in diversity were inhabitants of shallow epeiric and marginal seaways. As these became contracted, their environment shrank and they either retreated into the deeper waters of the ocean basins or became extinct. Many became extinct such as the fusulines, trilobites, and blastoids. Others must have survived because they have Triassic descendants, either as struggling survivors or new groups that evolved from them. We have no direct record of most of these events because of the destruction of ocean basin floors by plate movements. Another possible cause of the decrease in diversity of life in the Permian–Triassic boundary interval was the existence of Pangaea. Not only did the seas which formerly intervened between continents disappear, but the remaining shelves were in continuous contact around the supercontinent. This brought formerly separate faunas into competition with one another with a resultant loss of species.

In recent years, transitional rocks have been found, chiefly in geosynclinal areas where deposition was more continuous, between the Permian and Triassic. The record shows, as we might expect from our knowledge of modern life, that there was no mass extinction or catastrophe, just a gradual disappearance of Permian forms and replacement by Triassic forms. Groups formerly thought to have become extinct in the Permian such as tetracorals, cryptostome bryozoans, productoid brachiopods, goniatite ammonoids, and certain ostracods occur in the same strata with characteristic Triassic pelecypods. When we keep in mind that a large interval of 25–35 million years from Late Permian to Late Triassic, has a very poor record of shallow seas, it is not surprising that a great amount of evolution and extinction is largely unrecorded. Once again we have evidence that abrupt faunal replacement does not mean mass extinction, but instead indicates a shift in facies and/or a gap in the record. The fact that land life which has a good fossil record for this interval shows no effects of catastrophic changes further indicates that the great Paleozoic-Mesozoic differences in the sea are probably functions of the nature of our geologic record.

The Marine Ecosystem

The lack of the present phytoplankton producers, except for occasional dinoflagellates, in the Late Paleozoic seas poses the same problem about the ecosystem as it did in the Early Paleozoic. Dinoflagellates occur, but are uncommon fossils. Perhaps they were abundant, but most lacked preservable hard parts. The enigmatic acritarchs were abundant in the Devonian, but declined greatly thereafter (Fig. 16.40).

Fresh-Water Invertebrates

Fresh-water invertebrate life expanded in the Upper Paleozoic also as the *pelecypods* invaded nonmarine waters in the Devonian and the *gastropods* in the Carboniferous (Fig. 18.39). Some of the gastropods may well have been terrestrial. The *branchiopod* crustaceans (Fig. 18.40), a predominantly fresh-water group, appeared in the Devonian. *Ostracods* invaded fresh waters in the

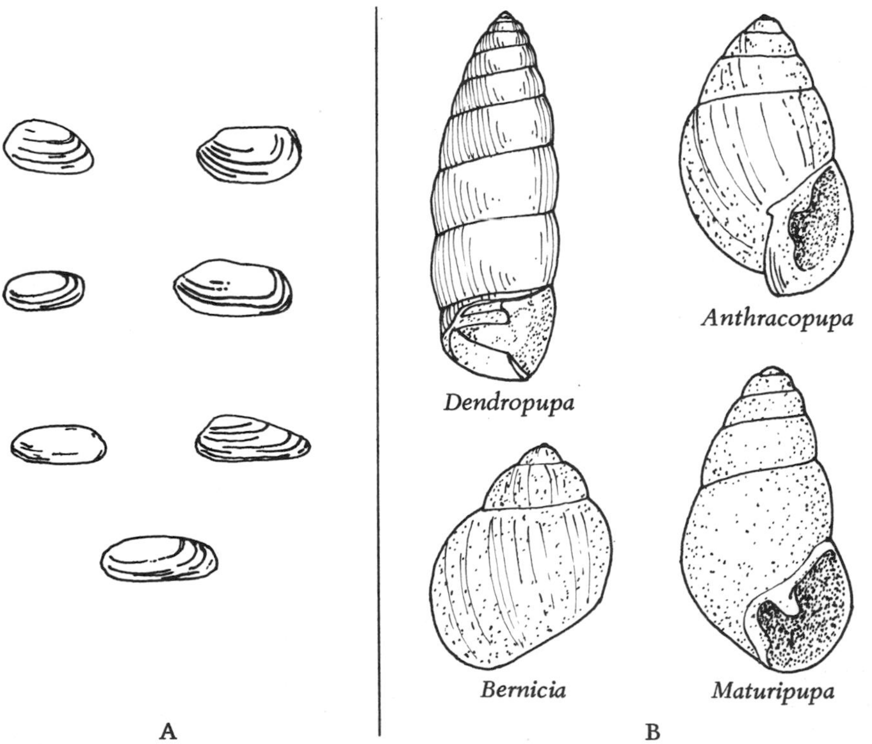

FIGURE 18.39 Late Paleozoic Non-Marine Mollusks. (A) Variation in a species of Carboniferous Bivalve, (B) Carboniferous Gastropods.

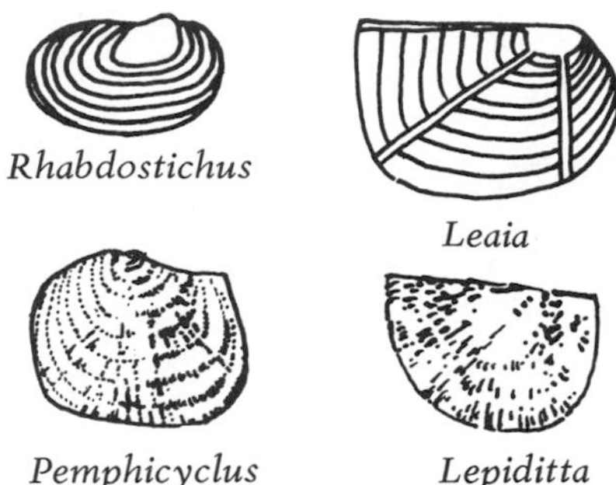

FIGURE 18.40 Late Paleozoic Branchiopod Crustaceans.

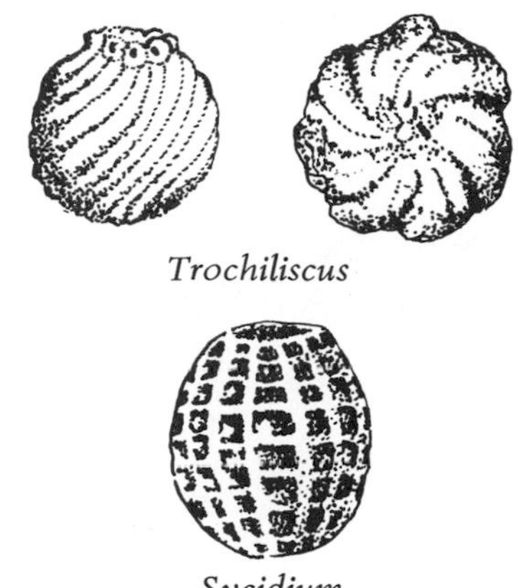

FIGURE 18.41 Late Paleozoic Charophyte Oogonia.

Pennsylvanian (some questionable Devonian forms have been reported). Some of the Paleozoic malacostracans may also have been fresh water. The protistan *charophytes* first appeared in the Devonian (Fig. 18.41).

Fish

The aquatic vertebrates, represented chiefly by the *fish,* are important denizens of Late Paleozoic waters. At the close of the Lower Paleozoic, fish became abundant for the first time in the fresh-water and nearshore marine areas of the Old Red Sandstone "Continent." This expansion continued through the Devonian, often called the "age of fishes" because every fish class was present and common during the period. The chiefly fresh-water *ostracoderms* (Fig. 18.42) occurred, but were unimportant after the Early Devonian. *Placoderms* (Fig. 18.43) reached their peak of abundance and diversity in the Devonian and left only a few last Carboniferous survivors. Placoderms include a host of types such as the largely marine *arthrodires,* shark-like *petalichthyoids,* skate or ray-like *rhenanoids,* and the fresh-water *antiarchs* which resembled ostracoderms. The *sharks* (Fig. 18.44), which arose in the Devonian and are very common in the Carboniferous, are known chiefly from teeth and spines because of their cartilage skeletons. Except for one group they were all marine. The largely

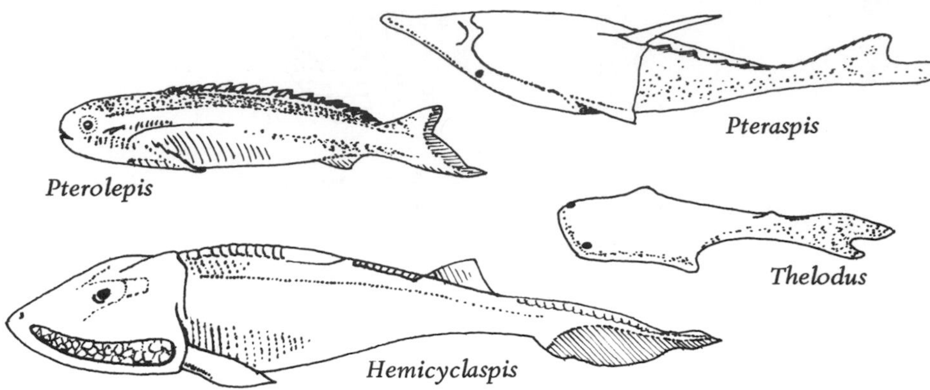

FIGURE 18.42 Devonian Ostracoderms.

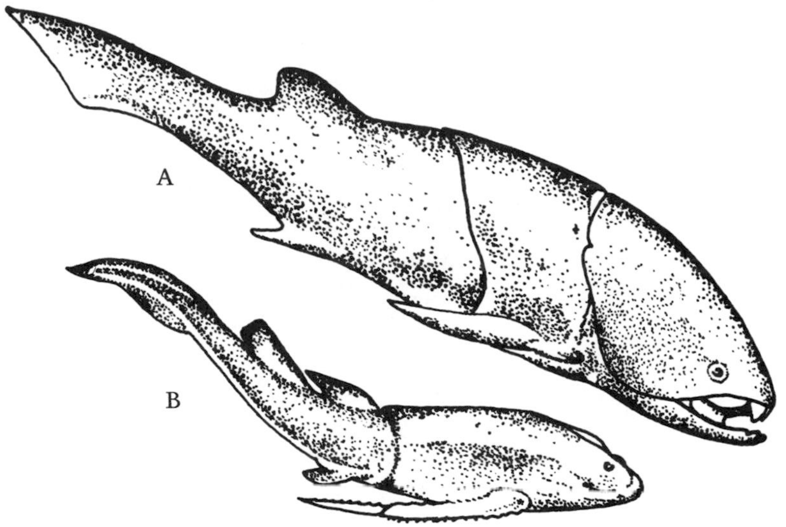

FIGURE 18.43 Devonian Placoderms. (A) *Coccosteus,* an Arthrodire, (B) *Bothriolepis,* an Antiarch.

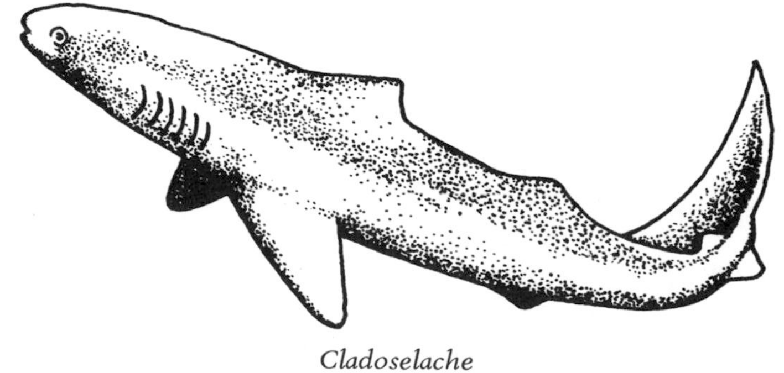

FIGURE 18.44 *Cladoselache,* **a Primitive Devonian Shark.**

fresh-water *acanthodeans* (Fig. 18.45) culminated in the Devonian and rapidly declined thereafter. They were the probable ancestors of the *bony fish* (Fig. 18.46), a group which first appeared in Devonian fresh waters. Both ray-fins (actinopterygians) and lobe-fins (sarcopterygians) occurred, but whereas the ray-fins diversified greatly thereafter, chiefly by invading the sea, the lobe-fins were unimportant after the Devonian and barely survived the Paleozoic (Fig. 7.109). *Cheirolepis* is a representative Devonian ray-finned bony fish and *Osteolepis*, a typical lobe-finned form.

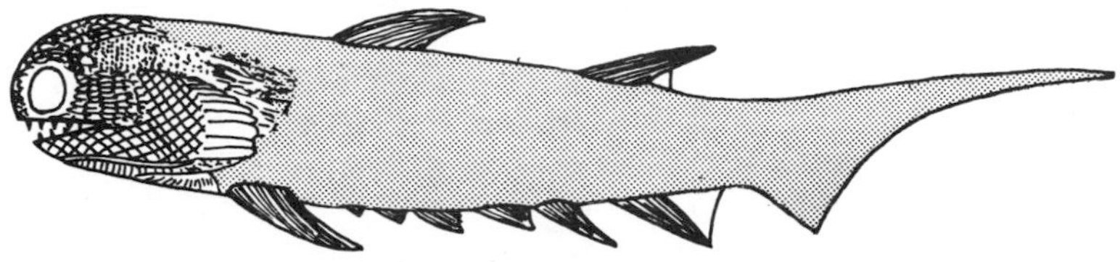

FIGURE 18.45 *Climatius*, **a Devonian Acanthodean.**

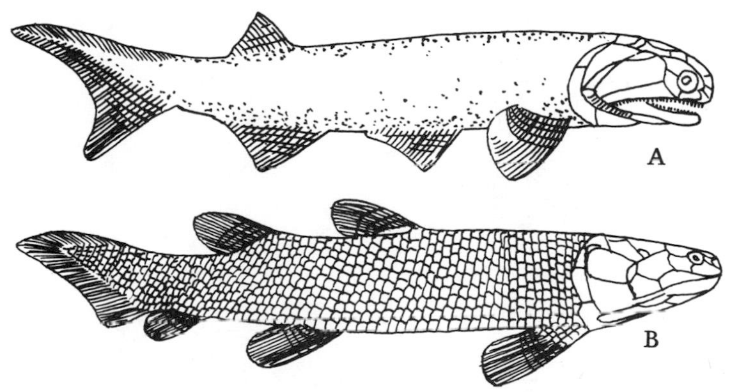

FIGURE 18.46 Devonian Bony Fish. (A) *Cheirolepis*, a ray-finned form, (B) *Osteolepis*, a lobe-finned form.

Land Life

Life on land is characterized by an explosive expansion in Late Paleozoic times. The *arthropods* are the chief invertebrate group to exploit this environment. The *arachnean* chelicerates evolved from the aquatic eurypterans via the scorpions. By the Carboniferous Period, unequivocal terrestrial scorpions occur in the coal swamp deposits and associated terrestrial or near-shore beds (Fig. 18.47). *Spiders* and similar forms (Fig. 18.48) are even more common in these rocks. The oldest known member of the tick and mite group is occasionally common in the Devonian. It is the mandibulate arthropods, chiefly the *insects*, that are the real success on land, however. Flightless insects evolved in the Devonian and by the time of the great Pennsylvanian coal swamps, flying insects of numer-

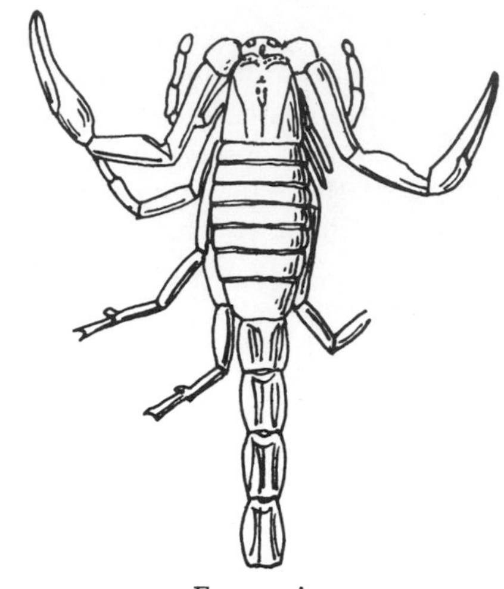

FIGURE 18.47 A Pennsylvanian Scorpion.

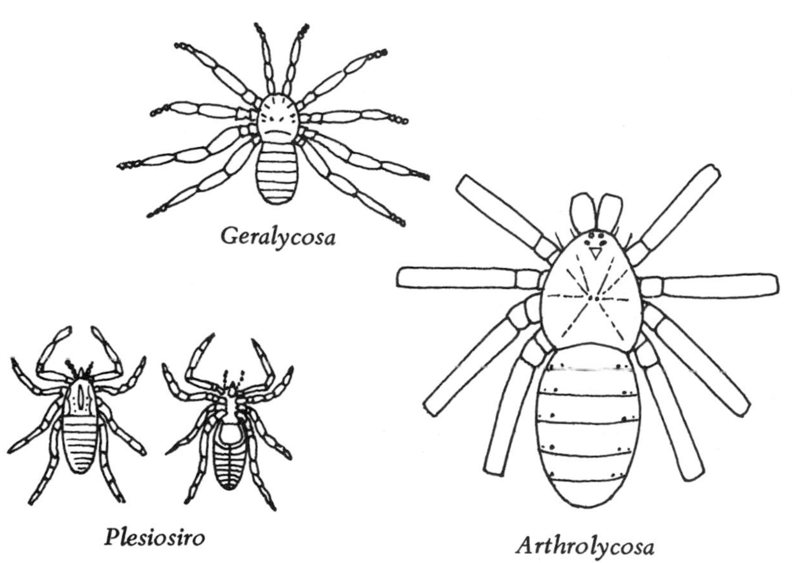

FIGURE 18.48 Pennsylvanian Spiders and Spider-Like Forms.

ous orders (Fig. 18.49) had evolved, many of large size. Dragon fly-like forms and cockroaches were among the most spectacular of these. Some of the former group had two-foot wing spans. Flying insects are also abundant in some of the Permian terrestrial and fresh-water beds such as those in central Kansas. The myriapods, another terrestrial mandibulate group, are represented by archipolypods and millipedes (Fig. 18.50) in the Pennsylvanian coal swamp deposits.

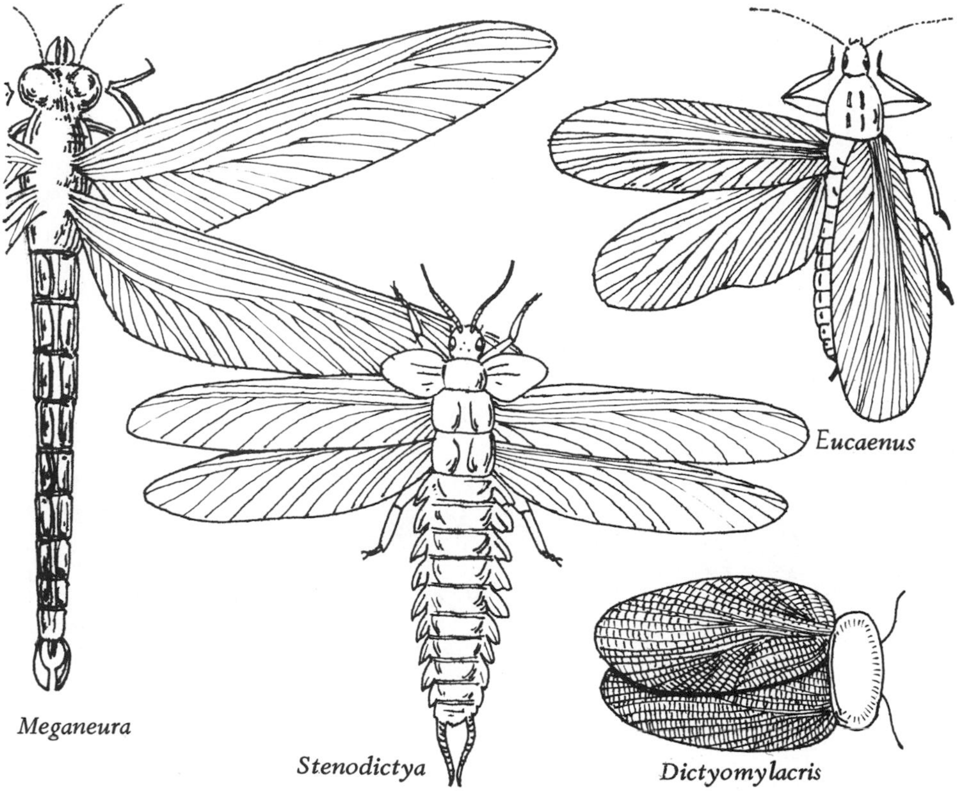

Eucaenus
Meganeura
Stenodictya
Dictyomylacris

FIGURE 18.49 Pennsylvanian and Permian Insects.

Pleurojulus
Glomeropsis
Acantherpestes

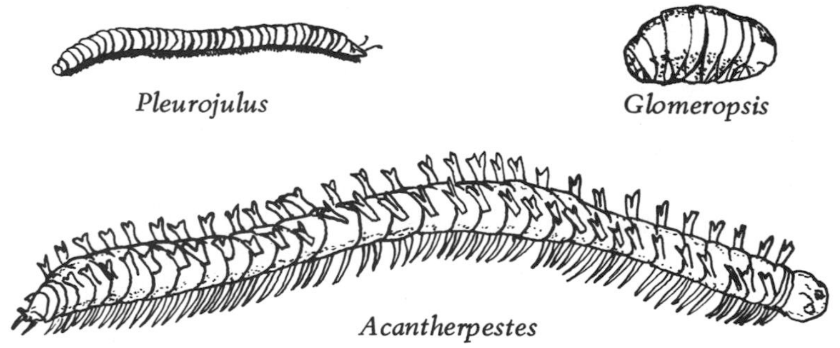

FIGURE 18.50 Late Carboniferous Myriapods.

Land Vertebrates

The *vertebrates* invaded the land for the first time in the Late Devonian. Many Devonian *fish* display adaptations fitting them for the land. Most fresh-water forms had lungs and several, particularly the *rhipidistian sarcopterygians* (Fig. 18.51), had bones which attached to supporting shoulder and hip girdles and

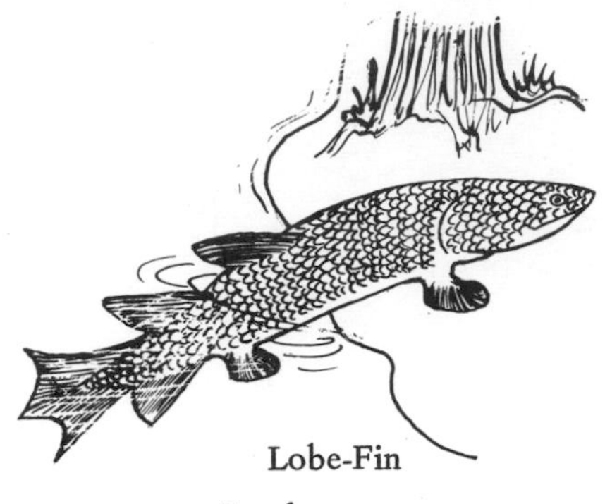

FIGURE 18.51 Rhipidistian Lobe-Finned Fish, Precursor of the Tetrapods.

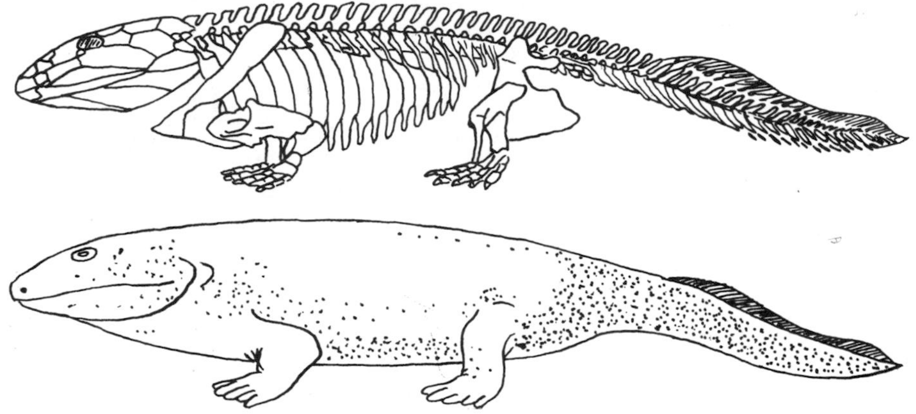

FIGURE 18.52 *Ichthyostega,* **the Oldest Known Amphibian.** Upper Devonian, Greenland.

supported their paired fins. The rhipidistians also had internal nostrils connecting their nasal passages with the passageway to the lungs. These features were probably very useful to the fish in crawling from pool to pool in the dry season, either in search of food or to escape a drying pond. These fish, such as *Eusthenopteron,* are best known from the Old Red Sandstone "Continent" which was probably a low latitude savanna-like area with a pronounced dry season as indicated by the paleomagnetic and lithologic data. Selection favored the gradual strengthening of the limbs and girdles and the improvement of the lungs until ultimately the fish-amphibian boundary was bridged.

The oldest known *amphibian, Ichthyostega* (Fig. 18.52), from the Late Devonian of the east Greenland part of the Old Red Sandstone region, still has a fish-like tail and weak limbs. The bones of the skull and the teeth of the earliest amphibians are very similar to those of the most amphibian-like fish. In the Pennsylvanian coal swamps of Laurasia, the amphibians (Fig. 18.53) underwent a major radiation including several types of large, bulky, flat-headed *labyrinthodonts* and small, salamander-like *lepospondyls.* Most of these forms still spent

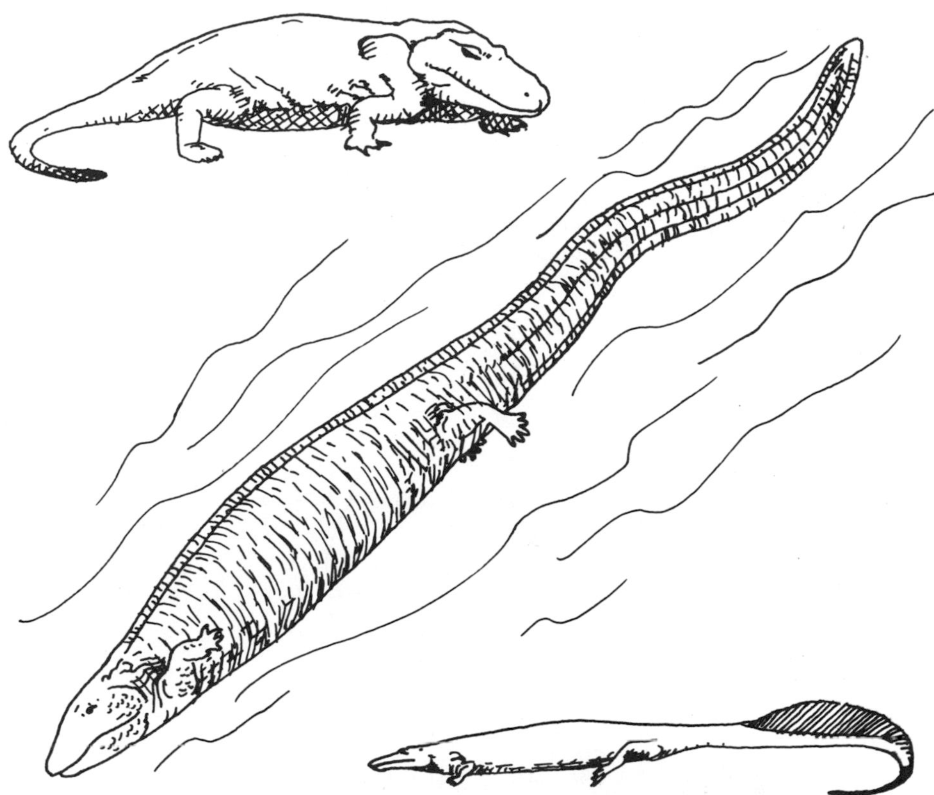

FIGURE 18.53 Late Paleozoic Labyrinthodont Amphibians.

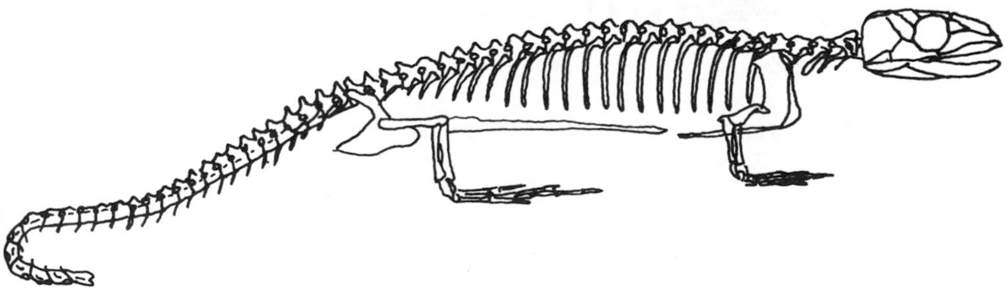

FIGURE 18.54 *Hylonomus*, **a Pennsylvanian Cotylosaur.** One of the oldest known reptiles.

most or all of their lives in the water. Some larval forms with fish-like gills are known. Both types of Paleozoic amphibians continued into the Permian with some reduction as the coal-producing swamps gradually disappeared. The labyrinthodonts succeeded in invading Gondwanaland in the Permian as Pangaea formed. The lepospondyls did not survive the Permian.

Some time in the Pennsylvanian, the *reptiles* arose from the labyrinthodont amphibians and began to diversify. Only a few imperfect specimens of Pennsylvanian reptiles are known, chiefly of the stem-reptile or *cotylosaur* group (Fig. 18.54), though a few *mammal-like* reptiles are also known. The latter group

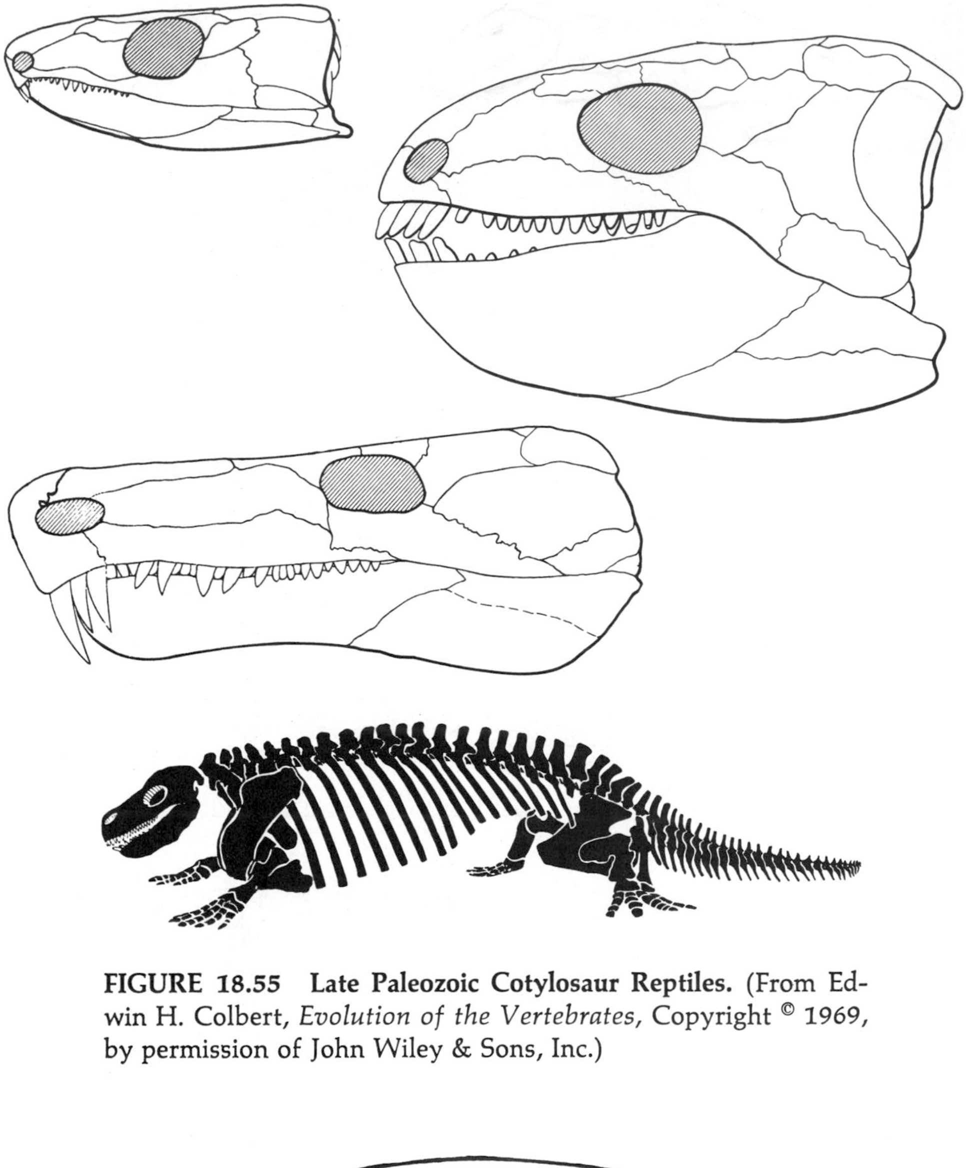

FIGURE 18.55 Late Paleozoic Cotylosaur Reptiles. (From Edwin H. Colbert, *Evolution of the Vertebrates,* Copyright © 1969, by permission of John Wiley & Sons, Inc.)

Seymouria

FIGURE 18.56 *Seymouria,* **a Permian Tetrapod Intermediate in Structure Between Amphibians and Reptiles.**

FIGURE 18.57 **Permian Life of West Texas.** Fin-backed pelycosaurs (mammal-like reptiles) include carnivores and one herbivorous form. Small pelycosaurs are seen on the left, while small aquatic amphibians appear on the right. Conifers (left) and calamites (right) dominate vegetation. (Mural by C. R. Knight, courtesy American Museum of Natural History.)

includes the first herbivorous reptiles. The cotylosaurs (Fig. 18.55) did not differ greatly from their labyrinthodont ancestors. In the Permian, the reptile record greatly improves with the widespread occurrence of terrestrial redbeds. In these rocks, survivors of the transitional amphibian-reptile forms occur (Fig. 18.56), forming such perfect connecting links that it is impossible to decide in which class to place them. Fossil reptilian amniotic eggs are first known in the Permian as well. Cotylosaurs (Fig. 18.55) were still common in the Permian and spread from Laurasia to Gondwanaland. It is the mammal-like reptiles, however, that dominate Permian land faunas. The Permian forms of *pelycosaurs* include the spectacular fin-backed genera *Dimetrodon* and *Edaphosaurus* (Fig. 18.57). Originating in the North American part of Laurasia, they gradually spread to Gondwanaland by the later Permian via present-day Europe. The later Permian and Triassic forms belong to the more advanced group, the *therapsids* (Fig. 18.58). Large faunas are known from Russia and particularly South Africa where a sequence of abundant forms occurs right across the Permo-triassic boundary. Thus, in contrast to oceanic life, which was greatly restricted at the close of the Paleozoic because of the large decrease in the area of shallow seas, terrestrial life radiated greatly at this time because of the large, interconnected terrestrial environment. The existence of the Gondwanaland-Laurasia

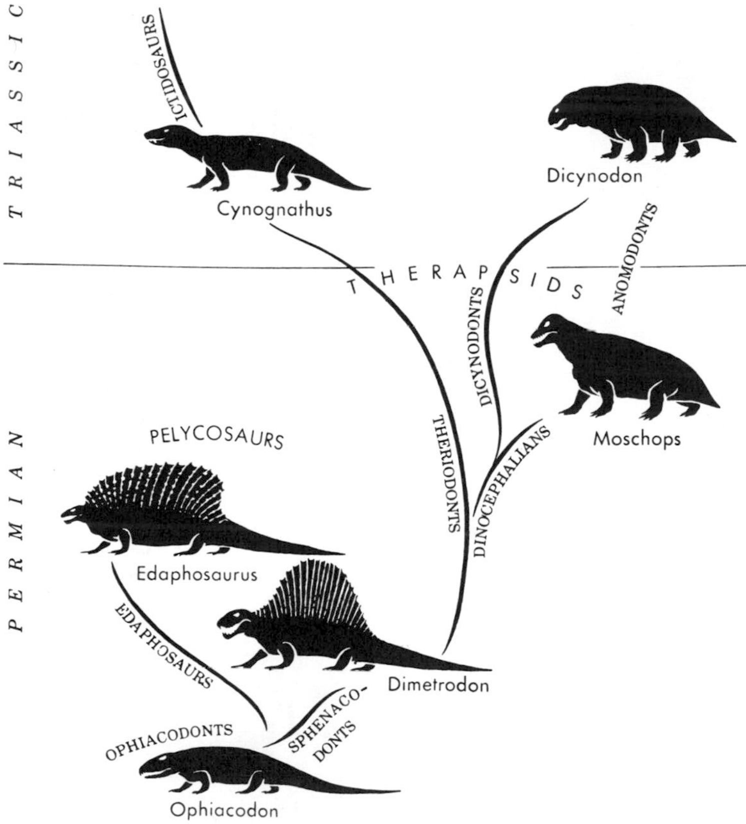

FIGURE 18.58 Mammal-like Reptiles. (From Edwin H. Colbert, *Evolution of the Vertebrates*, Copyright © 1969, by permission of John Wiley & Sons, Inc.)

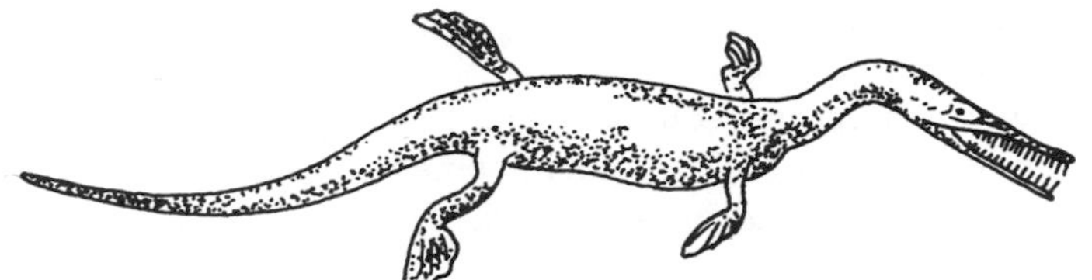

FIGURE 18.59 *Mesosaurus.* A small, fresh-water reptile of Late Paleozoic age.

connection in the Permian permitted a great deal of interchange of reptilian faunas between the two areas.

A few other Permian reptilian types (Fig. 7.116) are known. These are the protorosaurs, perhaps ancestral to the sauropterygians, and the eosuchians, perhaps ancestral to the lizards. Both had lizard-like proportions and were relatively small. The former arose in Laurasia and the latter probably did. Finally, no discussion of Late Paleozoic reptiles would be complete without mention of the enigmatic *mesosaurs* (Fig. 18.59) of South America and South Africa. The present disjunct distribution of this Late Carboniferous or Early Permian form was one of the lines of evidence that originally led to the formulation of ideas on continental drift.

Land Floras

The first great land floras developed in the Late Paleozoic. The simple *psilophytes* (Fig. 18.60), which first appeared near the close of the Silurian, diversified in the Devonian, their peak of abundance. At least three major lineages existed,

FIGURE 18.60 Early Devonian Flora. The low, dichotomous plants dominating the vegetation are chiefly psilophytes.

represented by *Rhynia, Zosterophyllum,* and *Trimerophyton.* Before the close of the Devonian, however, lycopods, articulates, and ferns each evolved separately from the psilophytes and became common (Fig. 18.61). Although most were low-growing herbaceous forms in the Devonian, some of them, such as the lycopod *Archaeosigillaria* and the fern *Eospermatopteris,* reached tree size.

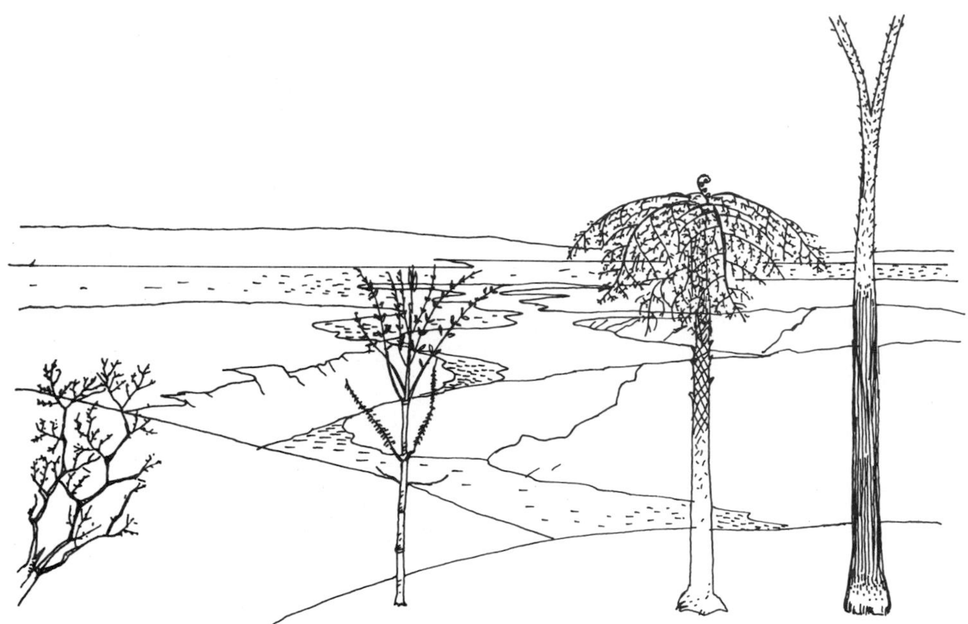

FIGURE 18.61 Late Devonian Flora. Lycopods, Articulates, and Ferns dominate.

In the great Pennsylvanian coal swamps of Laurasia (Fig. 18.62), the *lycopods* such as *Lepidodendron* and *Sigillaria, articulates* such as *Calamites* and *cordaitean* conifers such as *Cordaites,* all reached the size of large trees. The *seed ferns* such as *Medullosa,* which also reached a peak in the Pennsylvanian, were common low trees. The undergrowth consisted of herbaceous lycopods, articulates, and ferns. The lack of growth lines in the coal swamp plants and the paleomagnetically determined continental positions for the Pennsylvanian leave little doubt that this vegetation lived in tropical latitudes. Distinctive temperate floras flourished in present western North America–northern Asia and present southeast Asia (Fig. 17.4).

In the Permian, as the coal swamps gradually disappeared, the great tropical Laurasian plant groups of the Paleozoic declined. *Conifers* became the dominant form of plant life in the Permian (Fig. 18.63).

The Gondwanaland flora of the Late Paleozoic is dominated by the seed fern *Glossopteris* and its allies (Fig. 18.64). Apparently low shrubs, these plants thrived in the cool climates near the Late Paleozoic glaciers and accumulated to sufficient thicknesses in the interglacials to form the major coal beds of the present Southern Hemisphere continents. The barrier of the Ouachita-Appalachian-Hercynian Mountains and the Tethys seaway kept the Laurasian floras distinct from those of Gondwanaland.

FIGURE 18.62 Pennsylvanian Coal-Swamp Flora. Large-scale trees (lycopods), calamites (articulates), and cordaites (conifers) dominate, while seed ferns are the major shrubs and low trees.

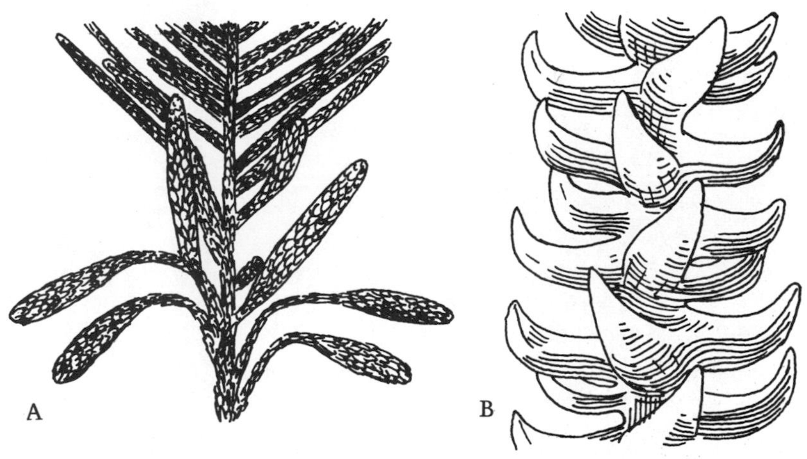

FIGURE 18.63 Late Paleozoic Conifer. (A) Leafy branch with cones. (B) Leaves.

As with land animals, there is no profound Permotriassic floral break, only the continuing rise of the conifers and also the cycadophytes and gingkos which had appeared in the Late Paleozoic but were uncommon. The large-scale trees, calamites, and cordaites, were essentially gone well before the Mesozoic began. In Gondwanaland, the *Glossopteris* flora continued without interruption into the Triassic.

FIGURE 18.64 Glossopteris Flora. *Glossopteris* and the larger *Gangamopteris* dominate.

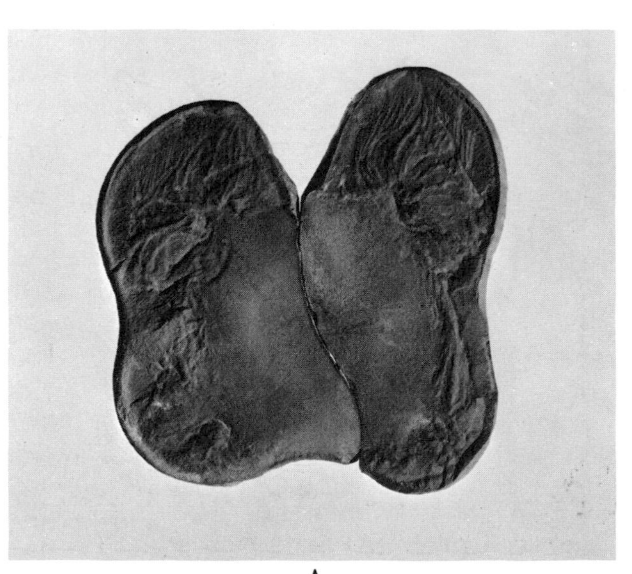

A

B

FIGURE 18.65 The Mazon Creek Fauna. (A) Medusoid coelenterate *Antracomedusa,* (B) Tully Monster, *Tullimonstrum,* a worm-like form of unknown affinities, (C) Xiphosuroid Arthropod *Palaeolimulus,* (D) Malacostracan (Eocarid) Crustacean *Belotelson,* (E) Malacostracan (Syncarid) Crustacean *Palaeocaris,* (F) Larval fish, (G) Paleoniscoid ray-finned bony fish. (Courtesy of Ralph G. Johnson, University of Chicago.)

Mazon Creek Fossils

A significant fossil locality of the Late Paleozoic is the Mazon Creek area of Illinois. In Pennsylvanian concretions (Fig. 18.65) are preserved impressions of a sample of nearly the entire biota of near-shore marine and terrestrial communities of the area. The marine fossils include numerous soft-bodied worm and arthropod groups, a tadpole larva of the tunicates (possible vertebrate relatives) and the oldest fossil lamprey (the soft-bodied parasitic survivor of the Agnathea). Also included are enigmatic creatures such as the Tully monster, *Tullimonstrum*. The terrestrial biota includes insects, arachneans, small vertebrates, and plant leaves. With discoveries such as these, the paleontologist can reconstruct evolutionary histories and ecosystems which could hardly be guessed at otherwise.

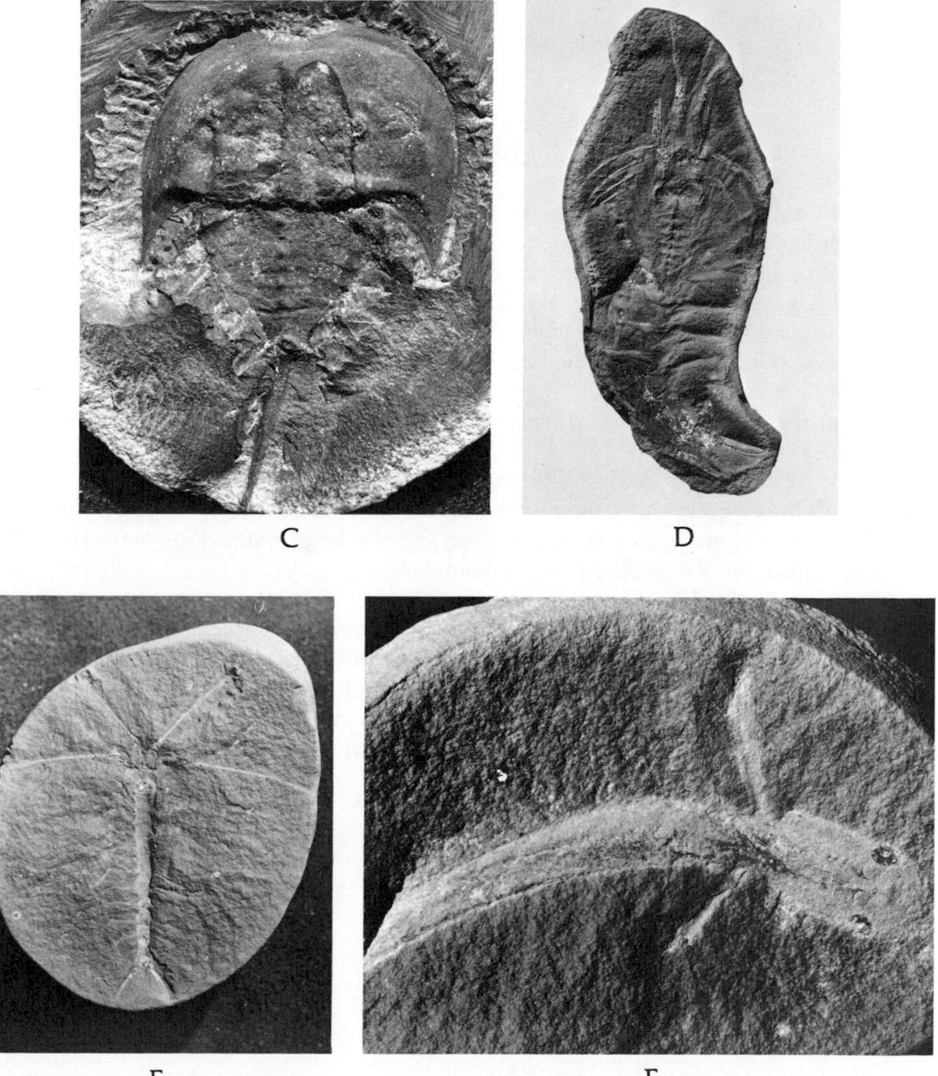

G

Questions

1. When did foraminifera become important in the geologic record, and what is the major Late Paleozoic group?
2. What coelenterates were the major reef builders of the Late Paleozoic?
3. Which group of bryozoans dominated Late Paleozoic faunas?
4. What were the six major brachiopod groups of the Late Paleozoic?
5. What significant group of mollusks arose and became common in the Late Paleozoic?
6. What two major types of crinozoan echinoderms reached a peak in the Mississippian?
7. Why are conodonts geologically important in the Late Paleozoic?
8. Which major groups of Paleozoic organisms died out at or near the close of the era?
9. Give at least one plausible explanation for the large number of extinctions that occur near the Paleozoic-Mesozoic boundary.
10. What major fresh-water groups appeared in the Late Paleozoic?
11. Why is the Devonian known as the Age of Fishes?
12. Which groups of organisms invaded the land in the Late Paleozoic?
13. Discuss the Late Paleozoic evolution of the amphibians.
14. What were the major Late Paleozoic reptile groups?
15. What were the dominant plant types of the Carboniferous coal swamps?

19

The Fountainhead of Stratigraphy:

The Mesozoic Erathem

THE THREE PERIODS OF THE MESOZOIC ERA, THE TRIASSIC, JURASSIC, AND CRETACEOUS, encompass approximately 160 million years of earth history from about 225 to 65 million years ago. The type area of the *Triassic* is in the post-orogenic terrestrial redbed sequence that occupies the site of the old Hercynian Geosyncline in central Germany; the type area of the *Jurassic* is in the Jura Mountains of northwest Switzerland, a shelf area along the northern margin of the Tethyean Geosyncline; and the type *Cretaceous* is on the shelf in the present Paris Basin of northern France (Fig. 2.19). At its beginning the Mesozoic is characterized by the extreme concentration of the continental masses into Laurasia and Gondwanaland, themselves interconnected forming Pangaea; at its close, it is characterized by extreme fragmentation as a new pattern of plates and their movements develop (Fig. 19.1–19.3).

Triassic Events

The pattern of magnetic bands in the present North Atlantic Ocean basin suggests that it may have begun to form off the present eastern United States as early as the Triassic, though no oceanic rocks of that age have been found. The oldest known sediments, from the basin margins as expected, date from the Jurassic, but there may be older material beneath them. Triassic basalts are common in present eastern North America indicating the rift was in existence in that period and the Late Paleozoic Appalachian disturbances may indicate an even earlier initiation of rifting. The paleomagnetic data suggest a Late Jurassic opening for South Atlantic and the oldest rocks thus far found are Late Cretaceous in age. (The extreme southern Atlantic may have opened even earlier.) Thus, before the end of the Mesozoic, North America and South America were separating from Gondwanaland (Fig. 19.3).

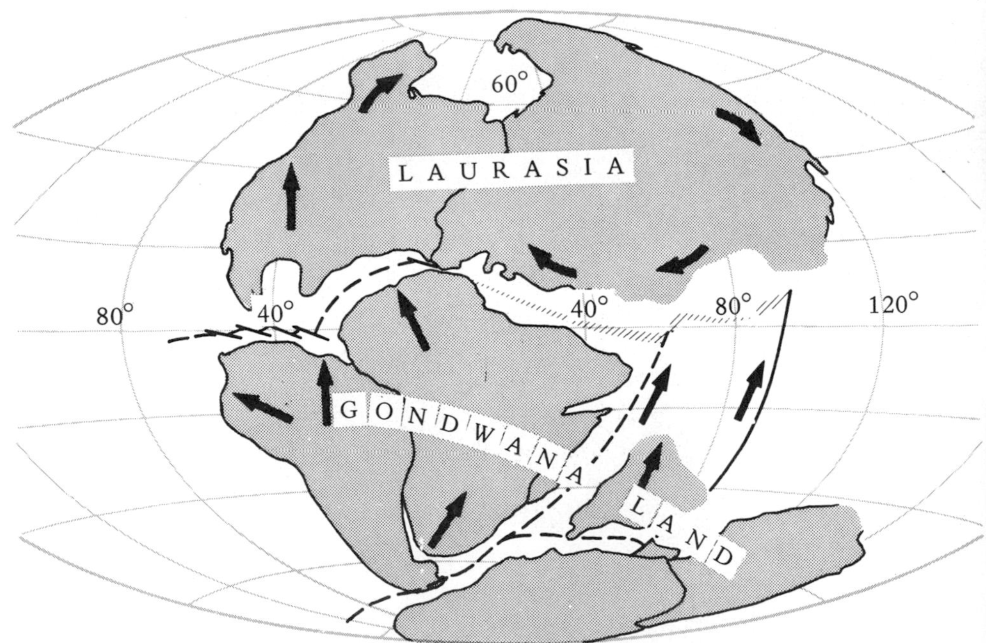

FIGURE 19.1 Late Triassic Paleogeography. Epeiric seas are not shown. (Data from works of Dietz and Holden, 1970.)

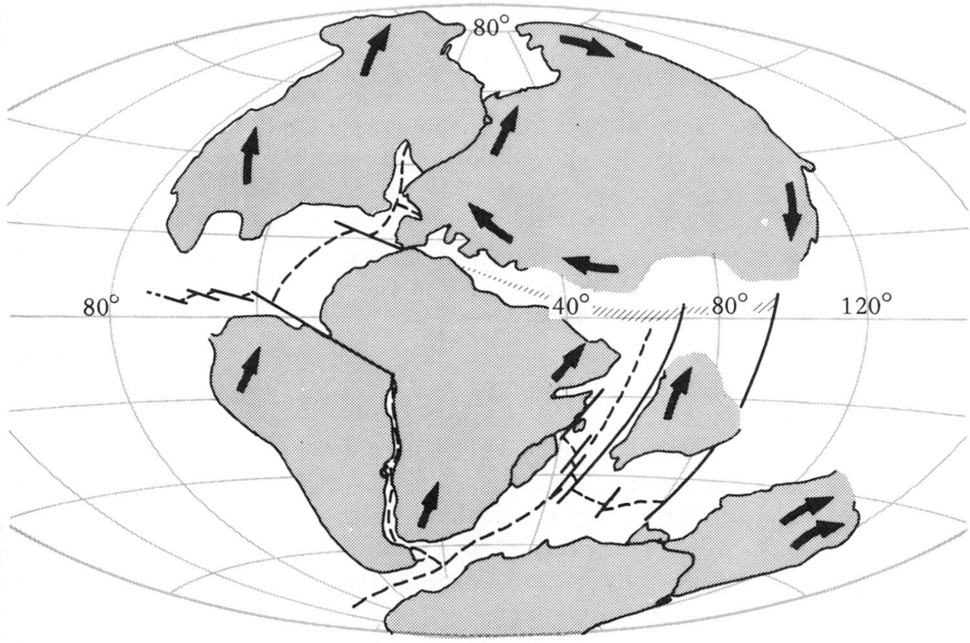

FIGURE 19.2 Late Jurassic Paleogeography. Epeiric seas are not shown. (Data from works of Dietz and Holden, 1970.)

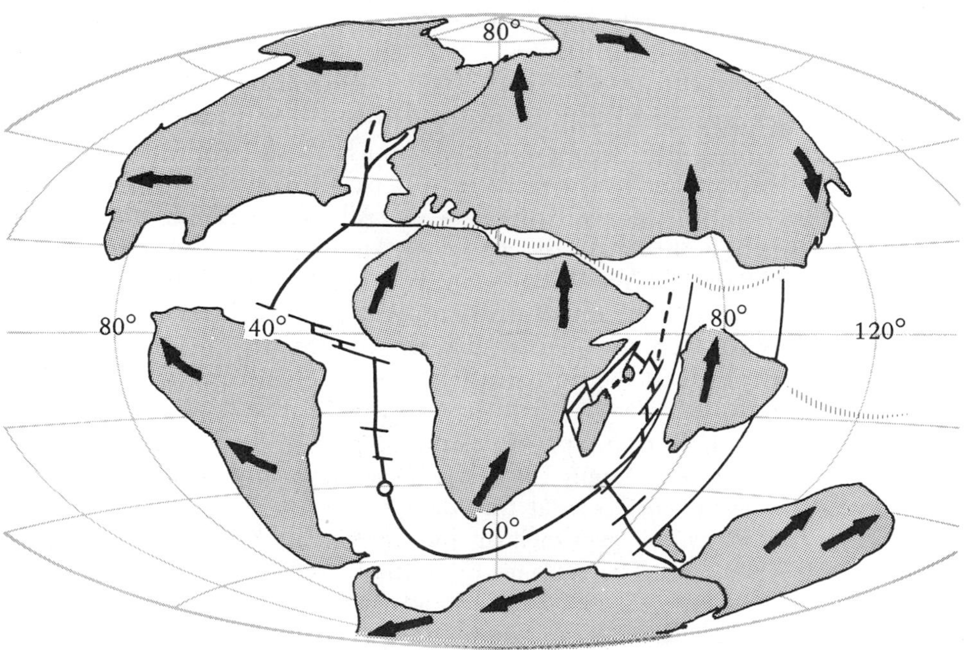

FIGURE 19.3 Late Cretaceous Paleogeography. Epeiric seas are not shown. (Data from works of Dietz and Holden, 1970.)

Gondwanaland suffered further fragmentation as new rifts developed and another new ocean basin, the Indian, began to form. Present-day Africa split off from the remainder of Gondwanaland, present-day India, Australia, New Zealand, and Antarctica, beginning in the Late Triassic (Fig. 19.1). An extension of this rift continued up the present east side of India separating it from the other former Gondwanaland continents. In the Mesozoic, the former western Gondwanaland continents, South America, Africa, and India, moved into equatorial regions, while the Antarctic-Australian piece remained in high latitudes. The more oceanic conditions in the South polar area apparently moderated conditions as no continental glaciations are known.

As the North Atlantic rift began to form, what appear to be large-scale Permotriassic transform faults developed in the continents. These are now preserved on the presently widely separated sides of the Atlantic as the *Brevard Fault* in the southern Appalachians and the *Cabot Fault* of the northern Appalachians, whose continuation in Europe is the *Great Glen Fault* of Scotland (Fig. 19.4). The area of the old Appalachian–West African Mountains became a zone of tension with its attendant block faulting, intrusion and terrestrial deposition characterizing the Triassic as the Atlantic rift opened. In North America, the surviving terrestrial redbeds and associated igneous rocks, the Newark Group (Figs. 19.5, 19.6), are preserved in downfaulted basins from the Canadian Maritime Provinces through the Carolinas. The Triassic equator passed through the present southern United States and Mexico (Fig. 19.7) and the Pacific Pole was situated in the area presently east of Siberia.

The extreme continental conditions that characterized the later Permian continued through much of the Triassic as well. The cratons were all emergent except for some marginal seas. There were no epeiric seas that covered North America in this period, only the final marginal seas in the Idaho-Wyoming-Utah areas of the regressing Absaroka Sea (Fig. 19.8). By the Late Triassic, the terrestrial redbeds (Fig. 19.9) covered all the Southern Rocky Mountains–Colorado Plateau area except for the remnant peaks of the Ancestral Rockies. This sequence, which has been carved into much of the present spectacular scenery of the Colorado Plateau (Fig. 19.10), passes without interruption or evidence of a boundary right into the Jurassic. All over northern and western Europe, a terrestrial redbed or Germanic facies (Figs. 19.11, 19.12) characterizes this system. The Permian-Triassic boundary is difficult to establish in the nearly continuous and unfossiliferous sequence of this New Red Sandstone "Continent." A minor marine invasion, the *Muschelkalk Sea,* occurred in the Middle Triassic (Fig. 19.11) and at the close of the period another incursion began. The Asian continental block was also emergent in the Triassic. In the Gondwanaland continents, the Karroo-type terrestrial deposition continued from the Permian without break. Marginal Triassic Seas in western Australia reflect the opening of the Indian Ocean at that date.

The three major Mesozoic geosynclines, the *Cordilleran, Andean,* and *Tethyan,* contain marine strata of Triassic age, in contrast to the almost total lack of such rocks on the cratons. The Cordilleran and Andean unstable belts developed from the collision of east Pacific plates with the continental margins of the new American plates (Fig. 19.1) along a trench. The *Palmer Peninsula Geosyncline* of present west Antarctica is a continuation of the Andean belt. Consumption of Pacific crust and flysch, volcanism, and orogeny continued

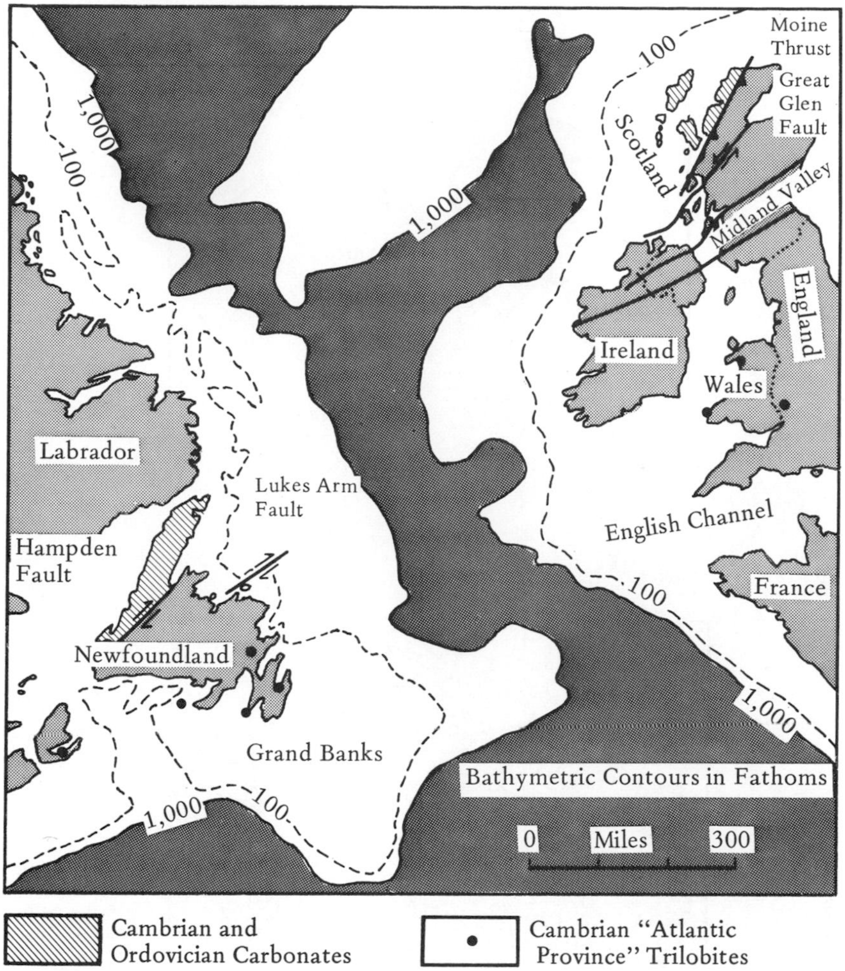

A

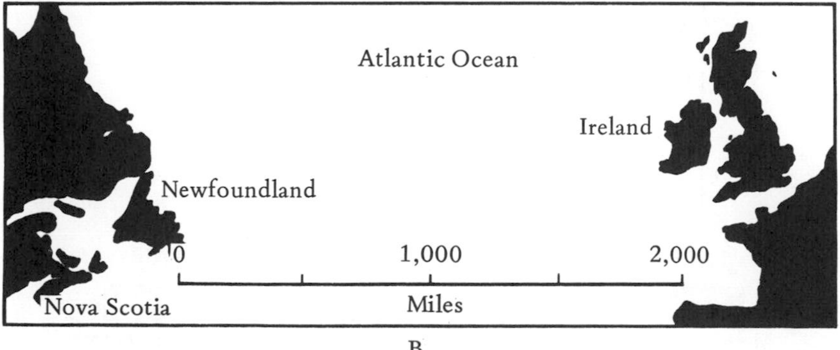

B

FIGURE 19.4 Match of Major Faults Across North Atlantic. (A) Late Paleozoic Reconstruction, (B) Present Relationships. (From Kay and Colbert, *Stratigraphy and Life History*, Copyright © 1965, by permission of John Wiley & Sons, Inc.)

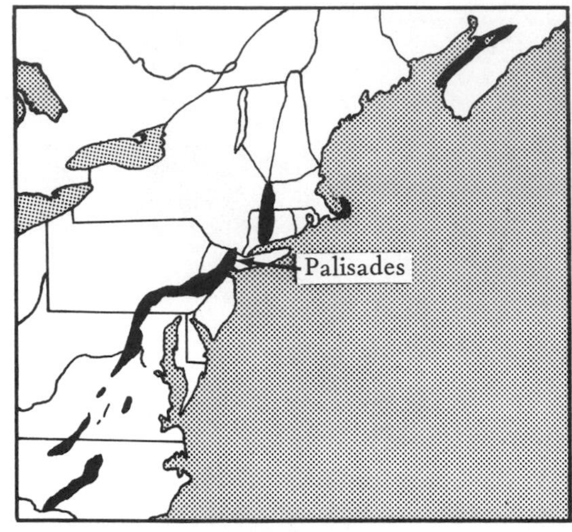

A

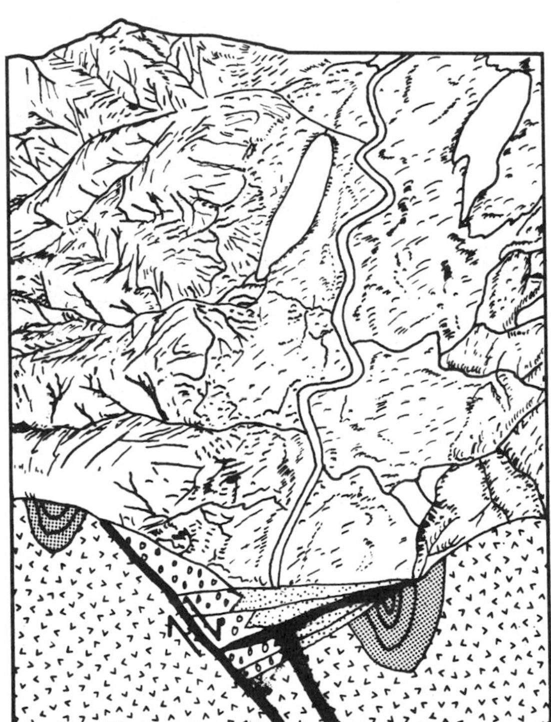

PALEOZOIC ROCKS

NEWARK GROUP

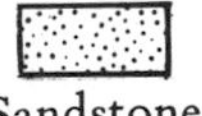

Metamorphic Rock

Granites and other Intrusives

Sandstone

Conglomerate

B

FIGURE 19.5 Triassic Fault-Block Basins of Eastern North America. (A) Map of Distribution, (B) Block Diagram of Typical Triassic fault-block basin.

FIGURE 19.6 Palisades of the Hudson River, Palisades Interstate Park, New Jersey. A sill intruded in Newark Group rocks. (Courtesy State of New Jersey.)

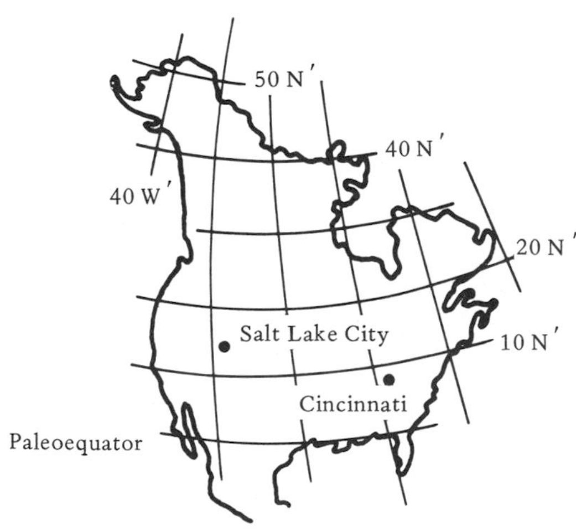

FIGURE 19.7 Triassic Paleolatitudes for North America. (From Irving, *Paleomagnetism and Its Application to Geological and Geophysical Problems.* Copyright © 1964, by permission of John Wiley & Sons, Inc.)

throughout the Mesozoic. In the present North Pacific area in the Late Triassic, the small east Siberian plate must have been moving toward Asia. The Tethyan Geosyncline (Fig. 19.1), which resulted from the impingement of plates containing Africa and India (and probably some microcontinents as well) in a trench along present southern Eurasia, developed excellent sequences for all

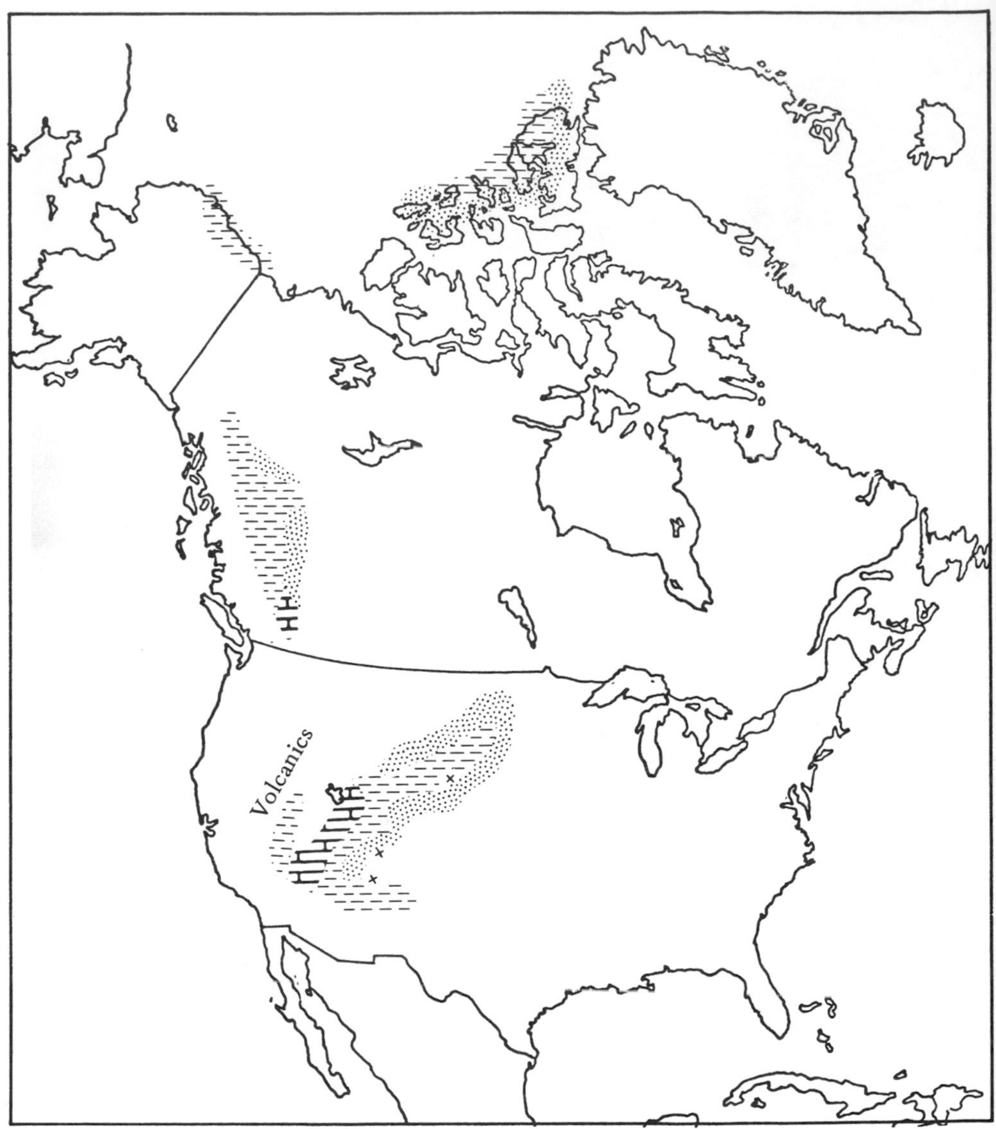

FIGURE 19.8 Present-Day Distribution of Lower Triassic Sedimentary Deposits of North America.

the Mesozoic systems. The abundantly fossiliferous Triassic section, best known in the Alps, is the world standard for the system which is chiefly represented by terrestrial rocks in its type area. Tethys was in the Triassic tropics and contained abundant reefs of hexacorals and dasycladaceous green algae which today form the great stone blocks of the Dolomite and other limestone Alps (Fig. 19.13).

Jurassic Events

In the Jurassic, as the Atlantic and Indian Ocean basins developed further (Fig. 19.2) and the Late Paleozoic mountains were largely reduced, there was a return

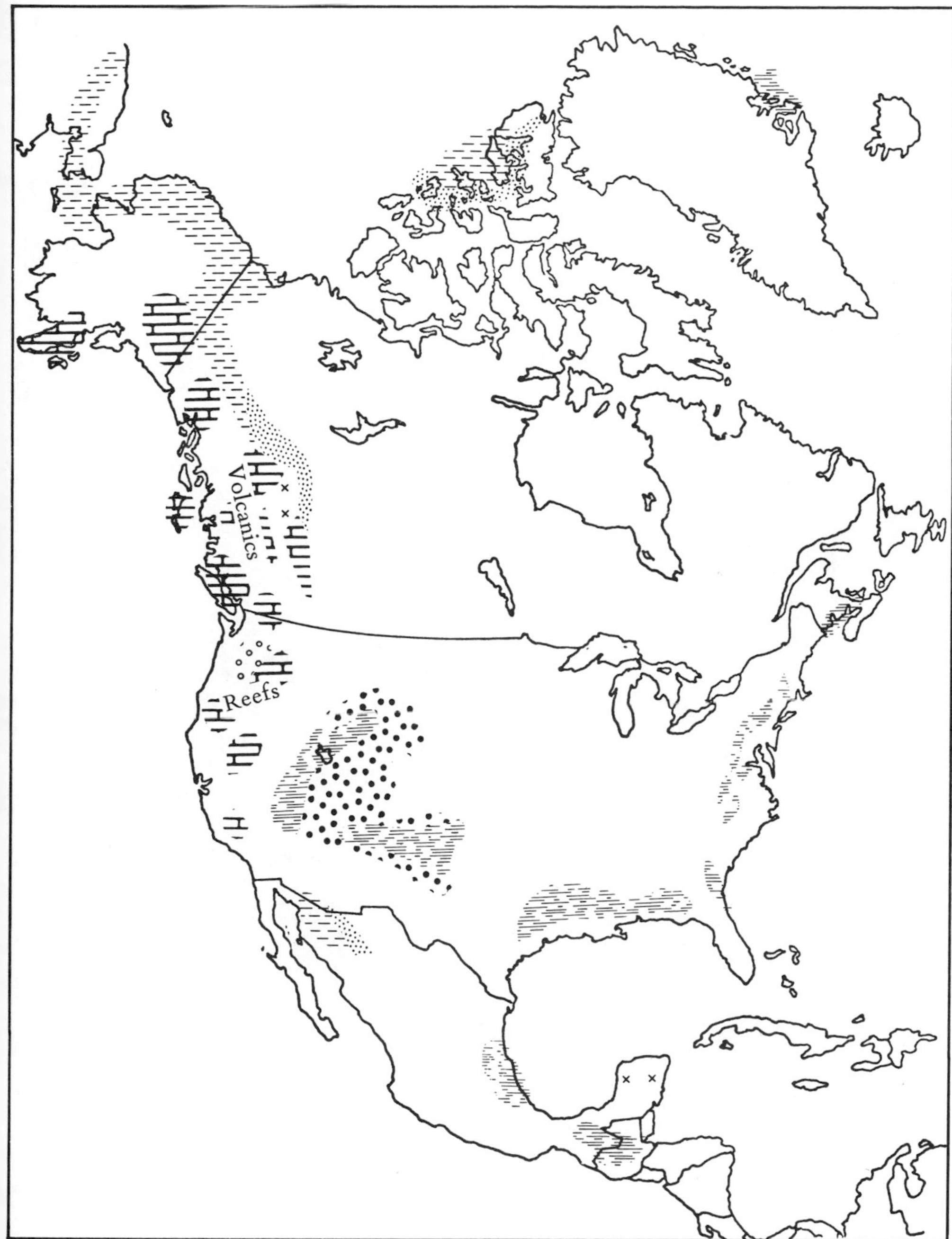

FIGURE 19.9 Present-Day Distribution of Upper Triassic Sedimentary Deposits of North America.

to more widespread epicontinental seas. In North America, the *Zuni Sea* (Figs. 15.18, 19.14) began transgressing in the Late Triassic-Early Jurassic interval, from both the Cordilleran Geosyncline on the west and the newly formed Gulf of Mexico part of the Atlantic on the southeast. As the Atlantic rift opened, the crust in the rift area thinned and the trailing edge of the continent foundered. Thus the site of the old Ouachita-Appalachian Geosyncline became flooded by

A

B

C

FIGURE 19.10 Triassic-Jurassic Redbeds and other Continental Strata in Colorado Plateau Country. (A) Echo Cliffs, northern Arizona, (B) Deadhorse Point, eastern Utah, (C) Checkerboard Mesa, Zion National Park, southwest Utah, (D) Colorado National Monument, western Colorado, (E) Arches National Monument, eastern Utah. (Courtesy of William M. Mintz.)

D

E

seas in the Mesozoic. In the Jurassic, eastern Mexico, southern Texas and Louisiana, and perhaps additional areas to the east were flooded, and the sea probably connected with the one coming from the Cordilleran region.

The northern incursion of the Zuni Sea flooded much of the present Great Plains and southern Rocky Mountain area of the craton in the Late Jurassic. The deposits of this sea (Fig. 19.15), often called the Sundance, are chiefly clastics reflecting the instability of the Cordilleran Geosyncline to the west. Near the close of the Jurassic, compression from an east Pacific plate (some authorities believe it contained a microcontinent) produced major deformation of this geosyncline, the *Nevadan Orogeny*. A great flood of clastics, the *Morrison* clastic wedge, temporarily drove the Zuni Sea off much of the craton.

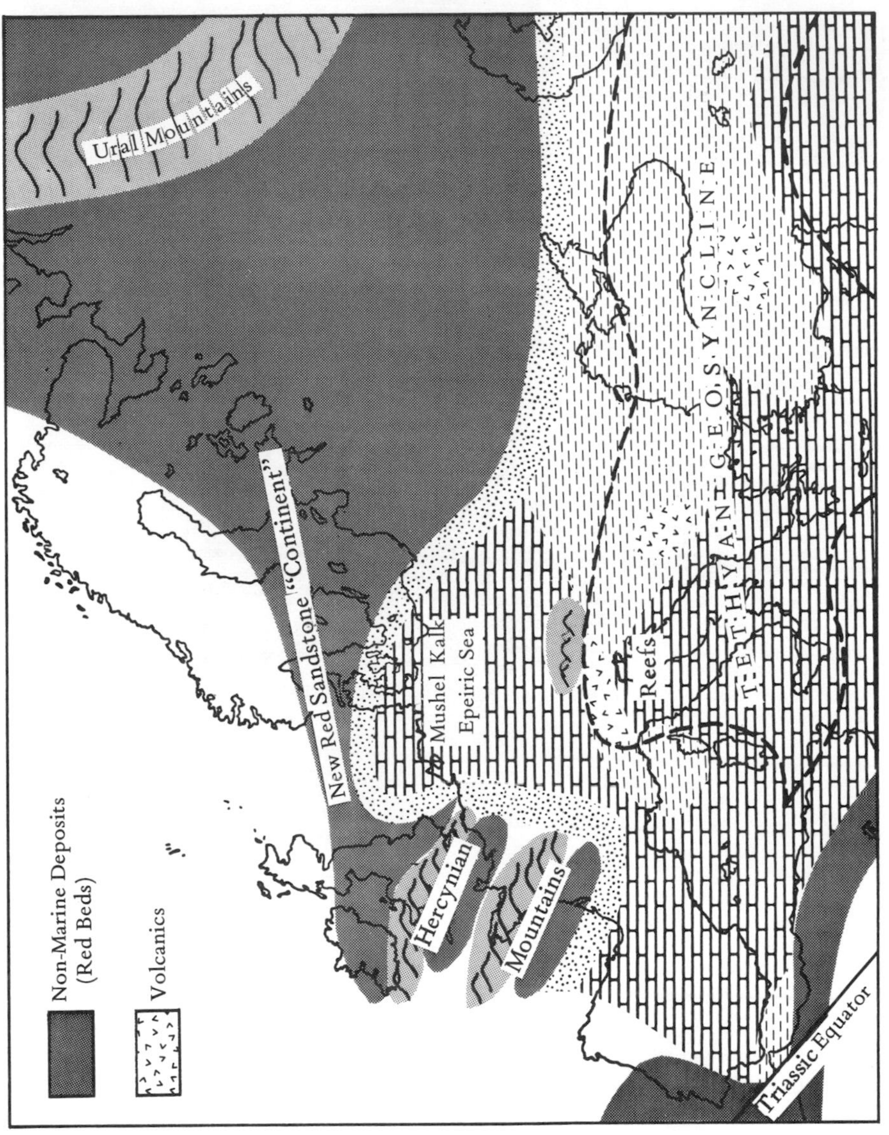

FIGURE 19.11 Middle Triassic Muschelkalk Sea Deposits of Present-Day Europe.

FIGURE 19.12 New Red Sandstone of Europe. Permotriassic redbeds in Scotland. (Crown Copyright Geological Survey photograph. Reproduced by Permission of the Controller of Her Britannic Majesty's Stationery Office.)

A

B

FIGURE 19.13 The Dolomite Alps, Italy. (A) The Tre Cime di Lavaredo, (B) The Sassolungo. (Courtesy Italian State Tourist Office.)

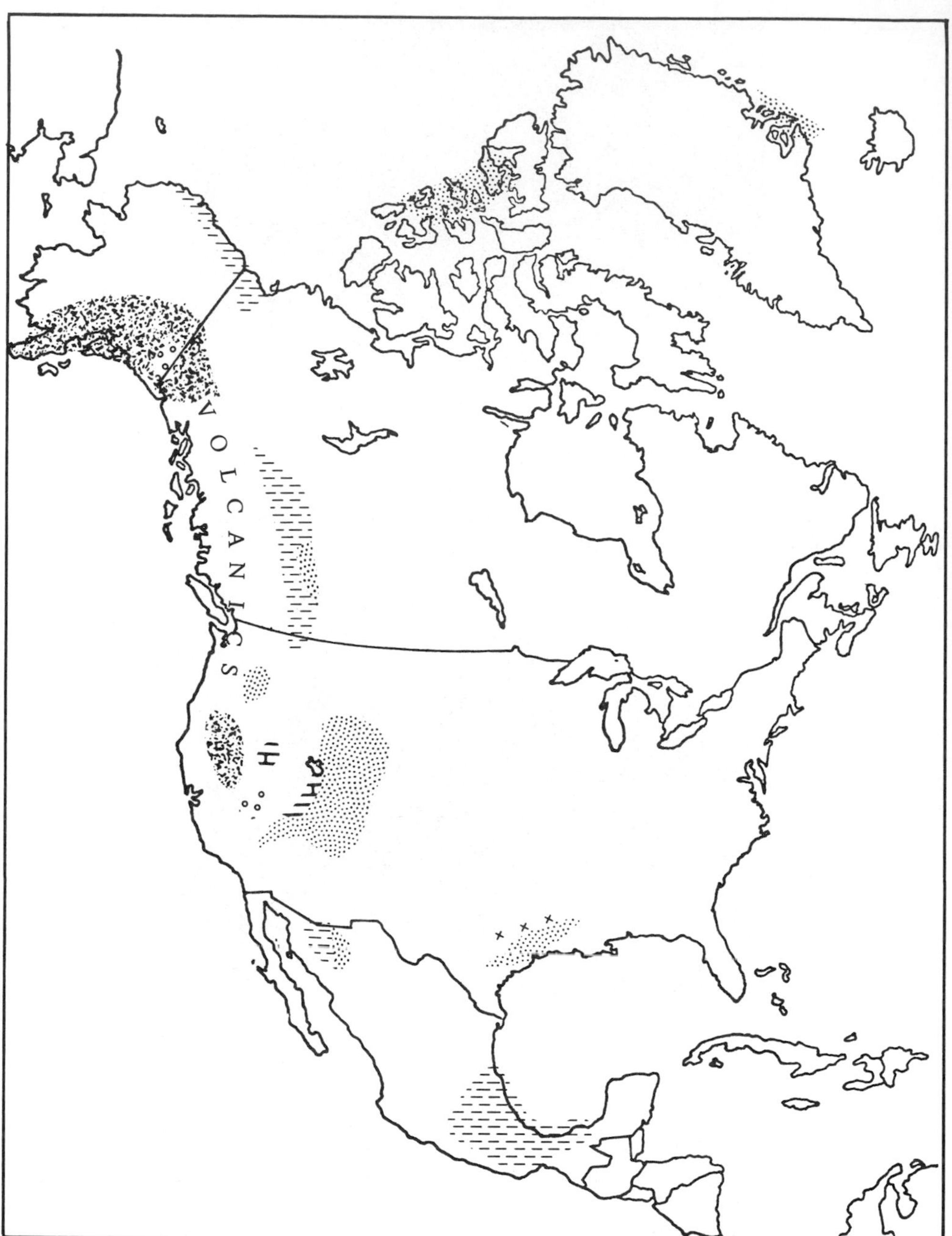

FIGURE 19.14 Present-Day Distribution of Lower Jurassic Sedimentary Deposits in North America.

Great granite intrusions of Jurassic-Cretaceous age occurred in the eugeosyncline from Alaska through Baja California (Fig. 19.16) including the huge Sierra Nevada batholith of California whose name is given to the orogeny. (In Nevada and Utah this deformation is also known as the *Sevier Orogeny.*) As the east Pacific plate continued to be thrust under the western continental edge of the North American plate, a great melange of oceanic and continental

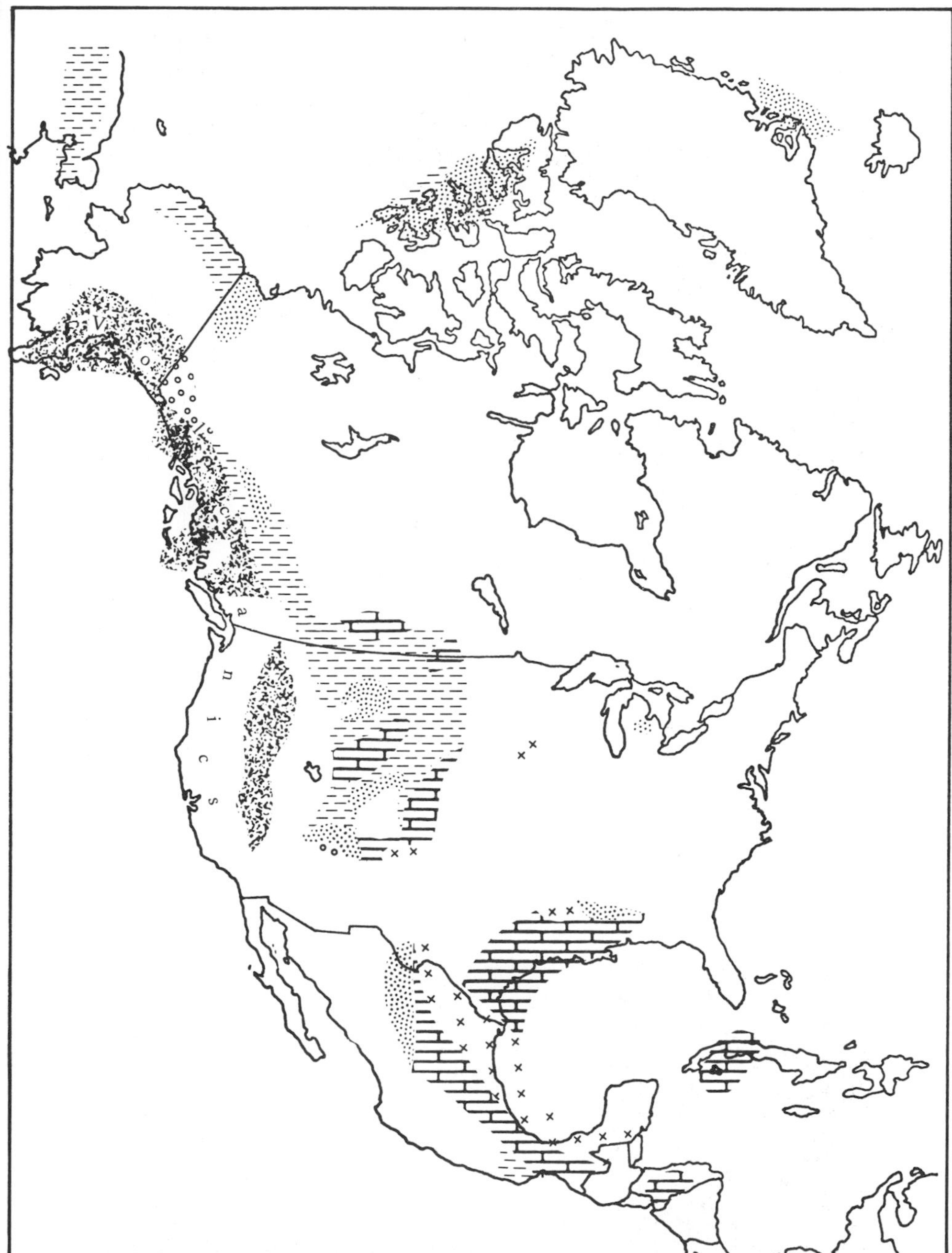

FIGURE 19.15 Present-Day Distribution of Upper Jurassic Sedimentary Deposits in North America.

rocks called the Franciscan Formation developed in California (Figs. 19.17, 19.18). Compare these events with those discussed in Chapter 9 and Figure 9.6.

In Europe, an epeiric sea (Fig. 19.19) that began at the end of the Triassic spread over the continent in the Jurassic from the new northerly extension of the Atlantic Ocean to the Tethyan Geosyncline except for Hercynian remnants

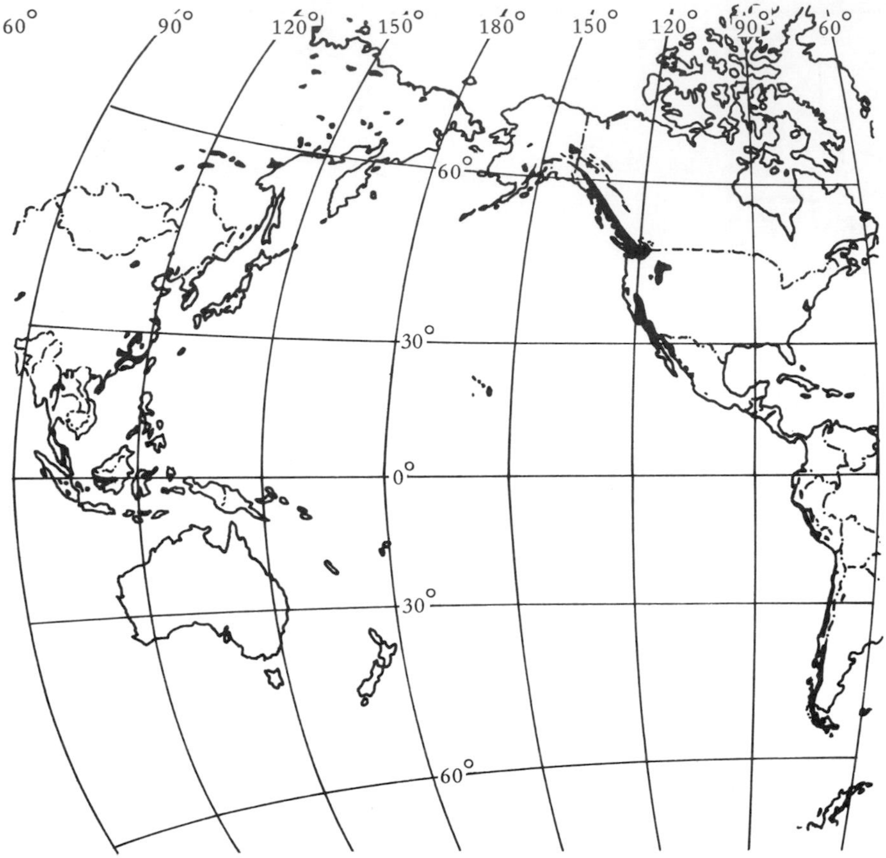

60° 90° 120° 150° 180° 150° 120° 90° 60°

60°

30°

0°

30°

60°

A

B

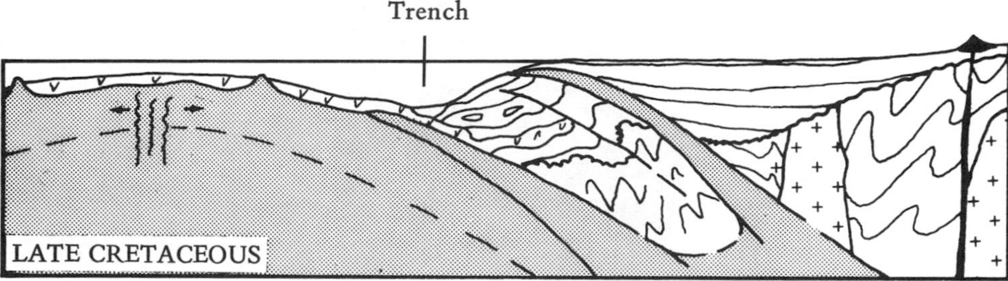

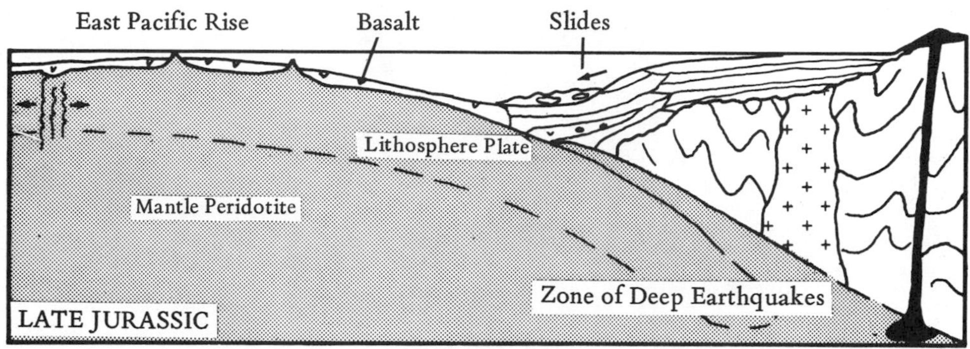

FIGURE 19.17 An Interpretation of the Formation of the Franciscan Formation. (From *Evolution of the Earth* by Dott and Batten. Copyright © 1971 by McGraw-Hill, Inc. Used by permission of the McGraw-Hill Book Company.)

in present Eastern Europe. This study of these widespread, abundantly fossiliferous Jurassic rocks (and the Cretaceous strata of the area to some extent also) by early geologists led to the formation of many fundamental stratigraphic principles such as faunal succession, facies, and stage and zone terminology and correlation. Thus, the Mesozoic Erathem and the Jurassic System in particular are really the wellsprings of much of modern geologic thinking. Toward the close of the Jurassic, partial regression occurred.

The African craton (Fig. 19.19), the interior of Gondwanaland for so much of earth history, received its first major marine flooding in the Jurassic as the Atlantic and Indian Oceans opened. The present north coast, particularly Algeria, and the east coast were flooded. Marine embayments developed in western Australia as the Indian Ocean developed further in the Jurassic. As the rifts opened in Gondwanaland, great amounts of basaltic lava poured from fissures.

FIGURE 19.16 Mesozoic Intrusions of the Circumpacific Area. (A) Map of distribution. The North American ones are chiefly Jurassic-Cretaceous age. (B) The Sierra Nevada Batholith, Yosemite Valley, Yosemite National Park, California. ((A) From Bateman and Eaton, *Science*, vol. 158, 15 Dec., 1967, Copyright © by AAAS, used by permission of the authors and AAAS. (B) Courtesy of William M. Mintz.)

A

B

C

FIGURE 19.18 Franciscan Rocks. (A) Weathered serpentine, Diablo Range (Courtesy of G. Oakeshott), (B) Pillow basalts, Point Bonita (Courtesy of C. W. Jennings), (C) Chert, Wolfback Ridge (Courtesy of Mary Hill), (D, E) Blueschists (Courtesy of Edgar H. Bailey). From *California's Changing Landscapes*, by Gordon Oakeshott. Copyright © 1971 by McGraw-Hill, Inc.

D

E

These Jurassic-Cretaceous basaltic rocks accumulated to great thicknesses on all the present southern continents. In some areas such as South Africa, these flat-lying rocks have subsequently been elevated to great heights and deeply eroded producing spectacular mountains such as the *Drakensberg* (Fig. 19.20). The formation of the Walvis and Chagos hot spots caused these flood basalts and their locations have left nemataths that allow accurate positioning of the continents from the Late Jurassic onward. The Gondwana pole was near the present eastern margin of the Antarctic-Australia continent in the Jurassic (Fig. 19.21).

The Jurassic also saw the collision of the small east Siberia plates with present east Asian Eurasia producing the culminating deformation of the *Verkhoyansk Geosyncline,* the *Kimmerian Orogeny.* Batholiths generated by the Kimmerian orogenic welt abound in eastern Siberia (Fig. 19.16). Batholiths also occur in eastern and southeastern Asia probably reflecting the impingement of the Indian and Pacific Plates, bearing microcontinents, against the cratons.

Although the Pacific Ocean clearly must have been in existence at least since the Precambrian to take up all the crustal space left by the bunching

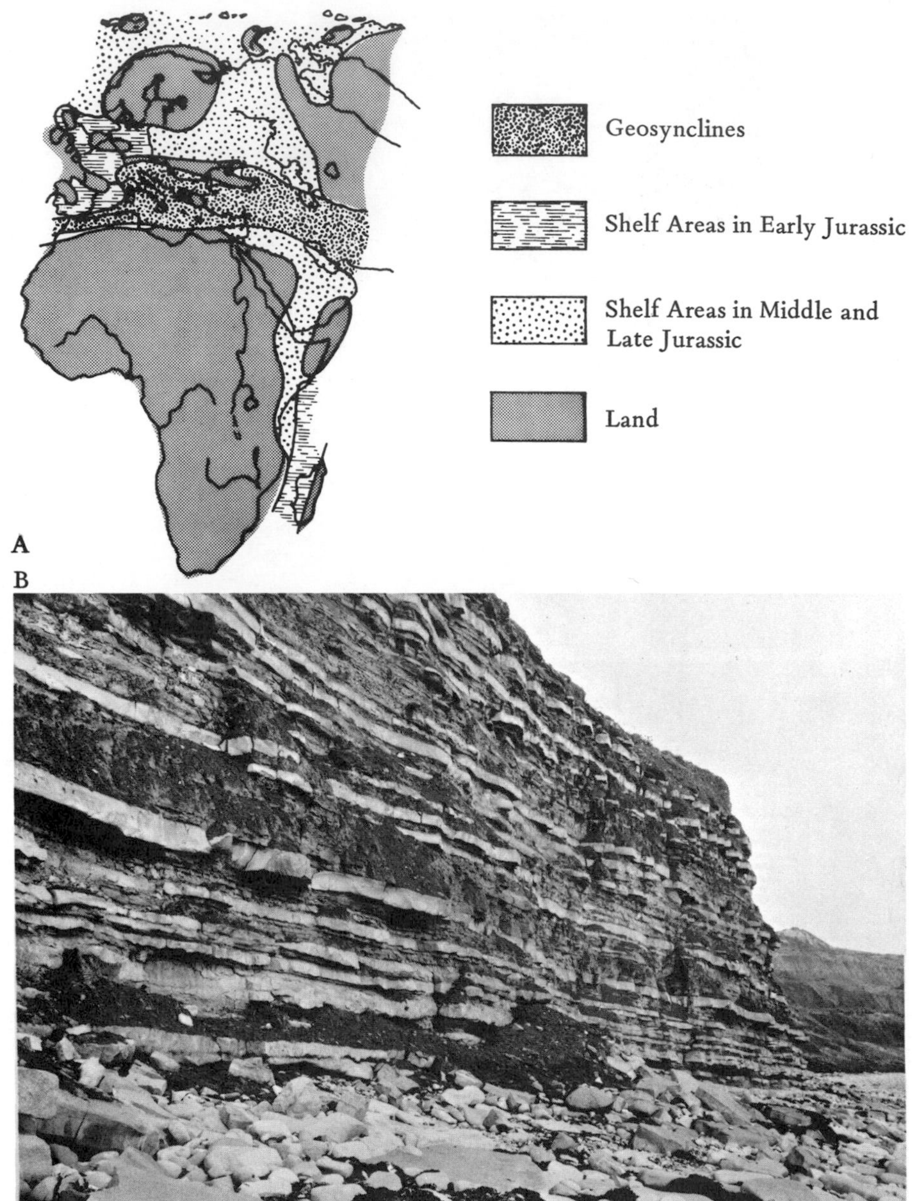

FIGURE 19.19 Jurassic Deposits of Present-Day Europe and Africa. (A) Map showing distribution, (B) Limestone and shale strata of Early Jurassic age near Lyme Regis, England. ((B) Crown Copyright Geological Survey photograph. Reproduced by Permission of the Controller of Her Britannic Majesty's Stationery Office.)

together of the continents in the Paleozoic, none of its crust of that age is currently known to have survived to the present in the ocean basin. It has been deformed and/or destroyed as its margins have descended into the various geosynclinal trenches along the present-day west coasts of the Americas and Antarctica and the present-day east coasts of Asia and Australia. The oldest rocks yet known from the Pacific Ocean basin are uppermost Jurassic carbonate

FIGURE 19.20 Drakensberg Escarpment, South Africa. (Courtesy South African Tourist Corporation.)

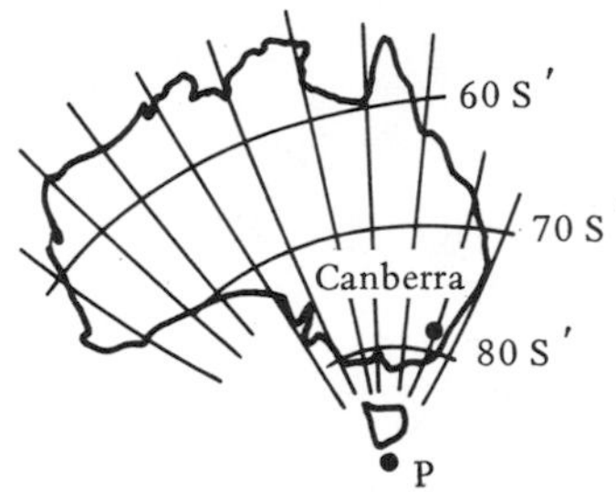

FIGURE 19.21 Paleolatitudes of Australia for the Jurassic. (From E. Irving, *Paleomagnetism and its Application to Geological and Geophysical Problems*, copyright © 1964, by permission of John Wiley & Sons, Inc.)

oozes from the western margin of the present Pacific plate along the Mariana and Japan Trenches (Fig. 11.16). Older rocks may occur beneath these sediments.

Cretaceous Events

The Cretaceous was a long (60–70 million years) and eventful interval in earth history. It began (Fig. 19.2) with only the incipient Atlantic and Indian Oceans and, hence, the continents still largely interconnected, and ended with a time of extreme fragmentation and isolation of land masses (Fig. 19.3).

North America

In North America, which was becoming more and more separated from the Eurasian part of old Laurasia, the Cretaceous was a time of great flooding by the Zuni epeiric sea (Fig. 2.19). After a slight withdrawal in the latest Jurassic and Early Cretaceous, the sea began advancing again from both the Cordilleran Geosyncline and the Gulf of Mexico. The two branches eventually joined in the Late Cretaceous, flooding the entire present Great Plains and Gulf Coastal Plain areas. Marginal flooding continued around the eastern edge of the continent to present Long Island. The part of the sea toward the Gulf of Mexico produced limy deposits including chalk (Fig. 19.22), but clastics became prevalent in the Great Plains area adjacent to the Cordilleran Geosyncline (Fig. 19.23).

A

B

FIGURE 19.22 Cretaceous Chalk Deposits of North America. (A) Castle Rock, Kansas, (B) The Sphinx, Kansas. (Courtesy Kansas Dept. of Economic Development.)

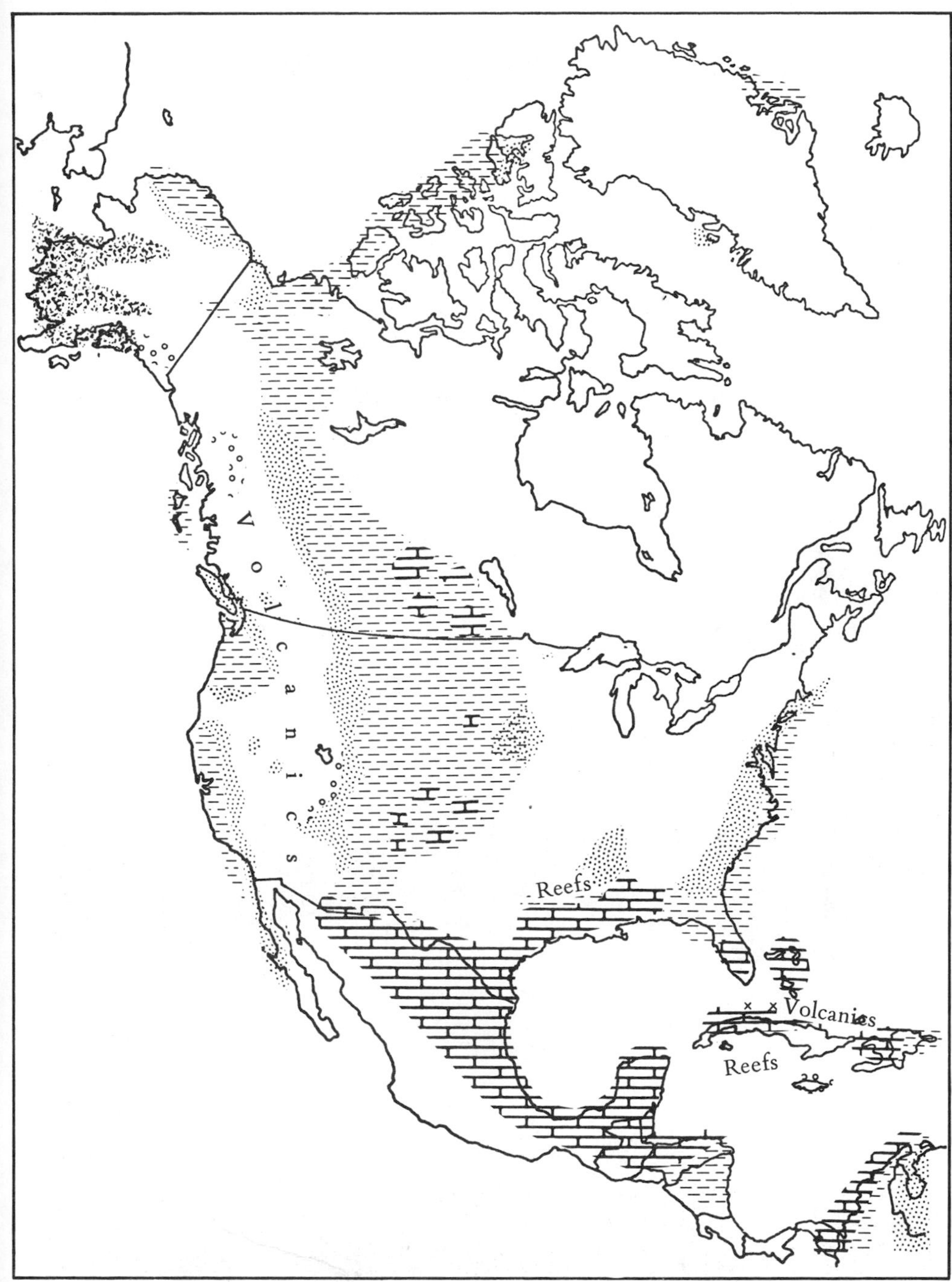

FIGURE 19.23 Present-Day Distribution of Lower-Upper Cretaceous Sedimentary Deposits in North America.

These reflect the beginning stages of the great *Laramide* or *Cordilleran Orogeny* that deformed the entire geosyncline in the Late Cretaceous and the Paleogene. Collision of an east Pacific plate continued to produce underthrusting of the western margin of North America (Fig. 19.17), mirror image thrust faults

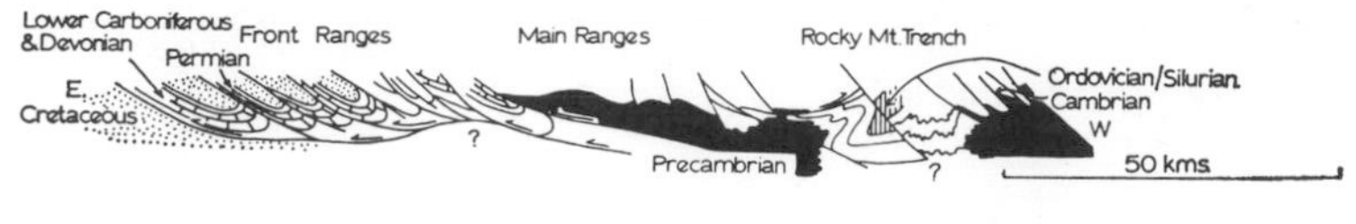

A

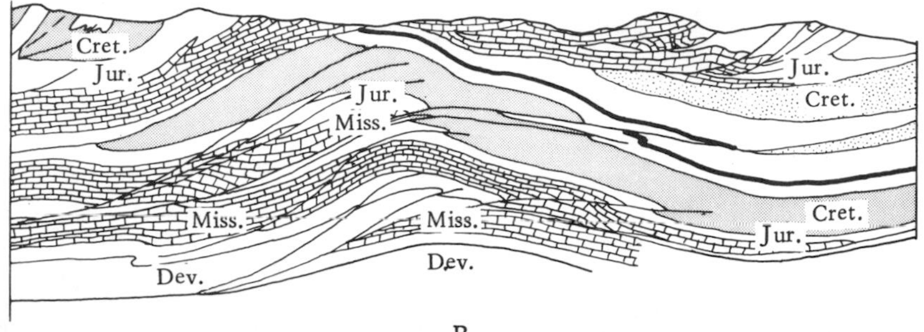

C B

D

FIGURE 19.24 The Cordilleran Orogeny. (A) Cross-section of deformed belt today, (B) Detailed cross-section of deformed miogeosynclinal deposits, (C-D) Thrust-Faulted ranges of Canadian Rockies, Banff National Park. ((A) from Dewey and Bird, *Journal of Geophysical Research*, vol. 75, no. 14, May 10, 1970. By permission of authors and AGU.) (C–D) Courtesy of Alberta Dept. of Industry & Tourism.)

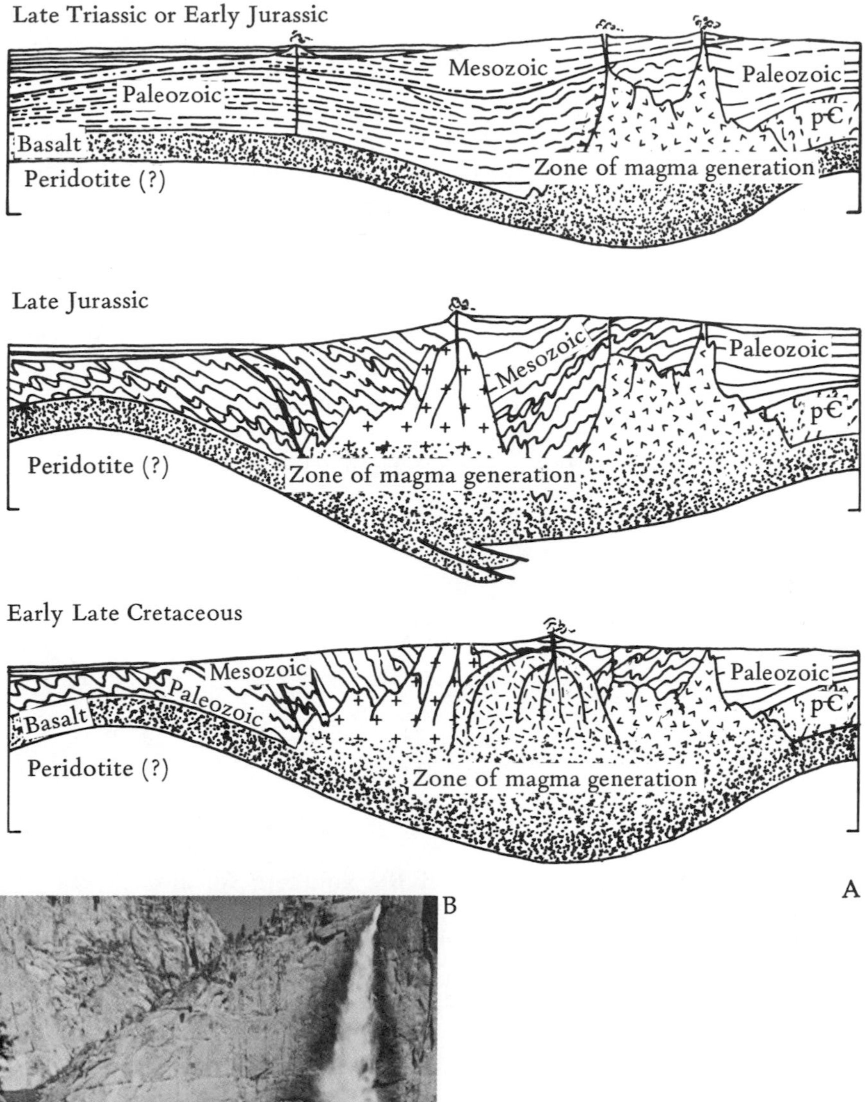

Late Triassic or Early Jurassic

Mesozoic
Paleozoic
Paleozoic
pЄ
Basalt
Zone of magma generation
Peridotite (?)

Late Jurassic

Mesozoic
Paleozoic
pЄ
Peridotite (?)
Zone of magma generation

Early Late Cretaceous

Mesozoic
Paleozoic
Paleozoic
pЄ
Basalt
Peridotite (?)
Zone of magma generation

A

B

FIGURE 19.25 The Nevadan-Cordilleran Orogeny. (A) Orogenic Welt of Cordilleran Eugeosyncline. The effects of plate tectonics are not shown on the diagram. (B) Yosemite Falls, Yosemite Valley, Yosemite National Park, California. Valley walls are composed of Late Mesozoic granite intrusives. ((A) from *California State Division of Mines Bulletin, 190,* (B) Courtesy of William M. Mintz.)

in the Cordilleran Geosyncline from the Canadian Rockies through western Utah and eastern Nevada (Fig. 19.24), extensive igneous intrusion forming a rising orogenic welt along the present west coast (Fig. 19.25), and buckling of the craton to produce the Southern Rocky Mountains stretching from southern Wyoming through New Mexico (Fig. 19.26). A great clastic wedge, often called the *Mesa Verde* (Figs. 19.27, 19.28), built eastward, driving out the Zuni Sea by the close of the Cretaceous. By the end of the Cretaceous, North America was still connected with present Europe via Greenland. North America was also still close to South America, probably with a narrow isthmus or island chain in between, but the connection was getting more tenuous as the two continents were rotating away from one another and the Central America–Gulf of Mexico–Caribbean Oceanic area expanded (Fig. 19.3). North America was in the middle latitudes in the Cretaceous, but tipped from its present position so that the pole was just to the present west of Alaska and the latitude lines ran diagonally across the continent from what is currently southwest to northeast.

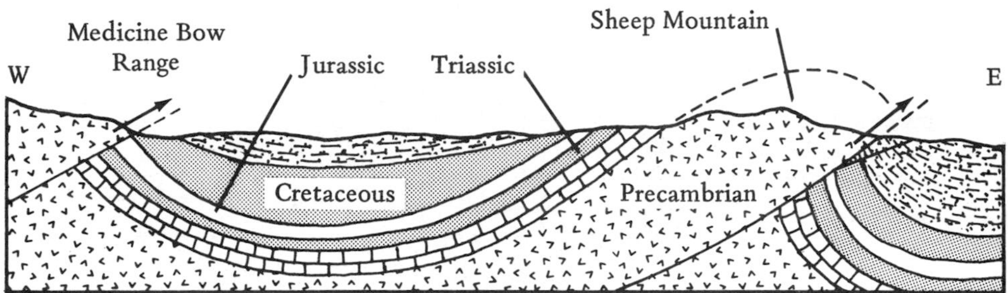

FIGURE 19.26 Cross-Section of the Southern Rockies. A deformed part of the Craton.

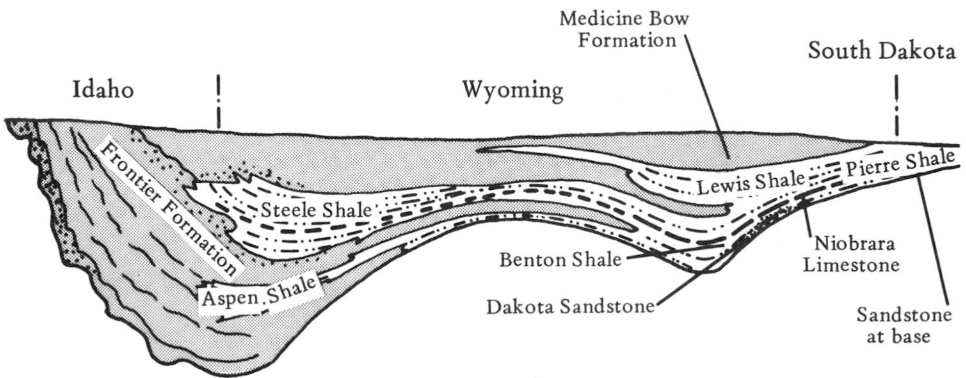

FIGURE 19.27 Cross-section of the Mesa Verde Clastic Wedge.

A

B

FIGURE 19.28 Mesa Verde Group Rocks. (A) Point Lookout, Mesa Verde National Park, Colorado, (B) Book Cliffs, eastern Utah. (Courtesy of William M. Mintz.)

Eurasia

In present Europe, the epeiric seas also returned after an Early Cretaceous regressive phase. Spreading northward from Tethys, they engulfed nearly all of Europe except the Baltic shield by the Late Cretaceous (Fig. 19.29). Limy deposits were widespread particularly in the area of northern France and southern England where the famous Cretaceous chalk formed (Fig. 19.30). The seas regressed at the close of the Cretaceous. In present-day Asia, the craton was flooded for the first time since the Early Paleozoic Seas spread south from the Arctic Ocean along the present Ob River Valley to join the Tethyan Geosyncline (Fig. 19.31). The southern margin of Eurasia was the site of the great Tethyan Geosyncline trench which continued to receive sediments and volcanics and undergo intermittent disturbances in the Cretaceous as Africa, India, and small microcontinents closed in from the south (Fig. 19.3). The seaway

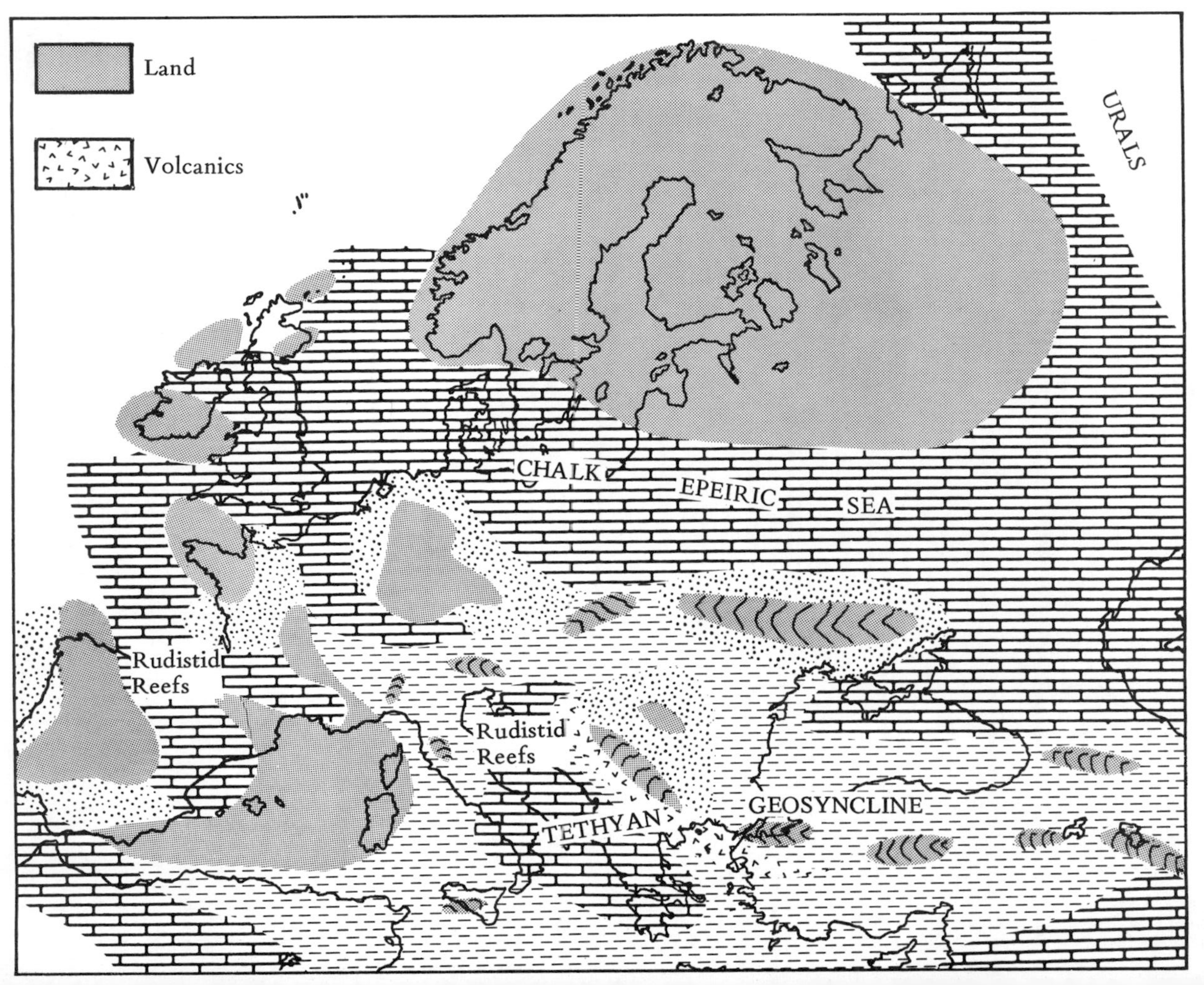

FIGURE 19.29 Cretaceous Chalk Sea Deposits of Present-Day Europe.

FIGURE 19.30 The White Cliffs of Dover. (Crown Copyright Geological Survey photograph. Reproduced by Permission of the Controller of Her Britannic Majesty's Stationery Office.)

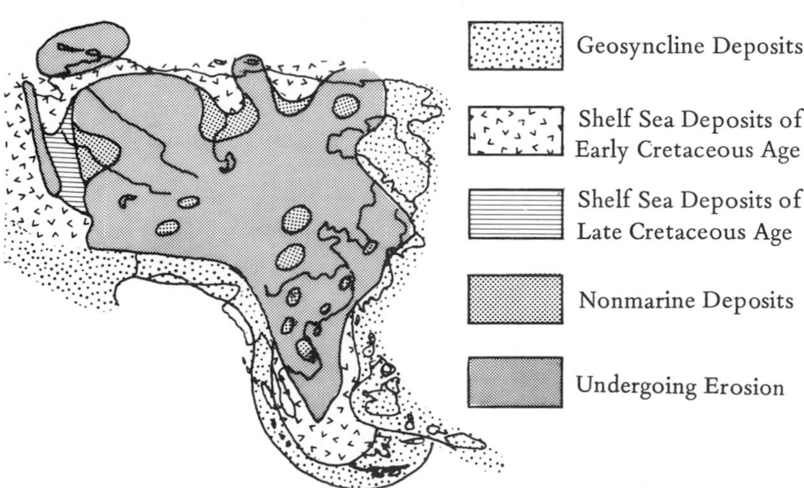

FIGURE 19.31 Cretaceous Deposits of Present-Day Asia.

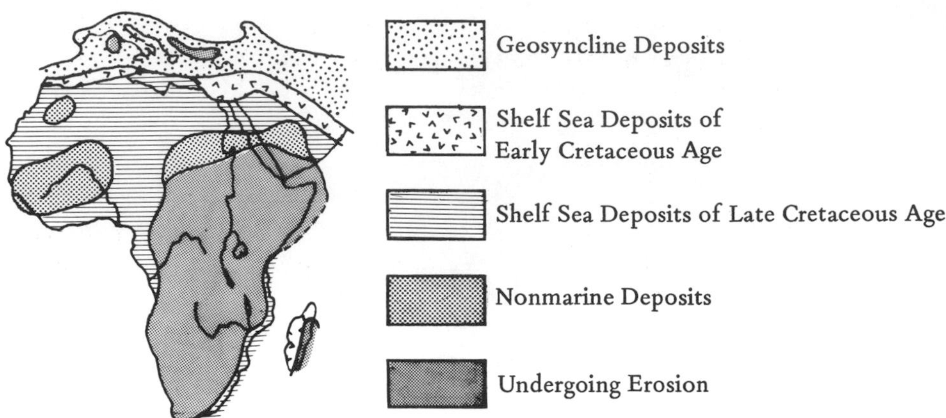

FIGURE 19.32 Cretaceous Deposits of Present-Day Africa.

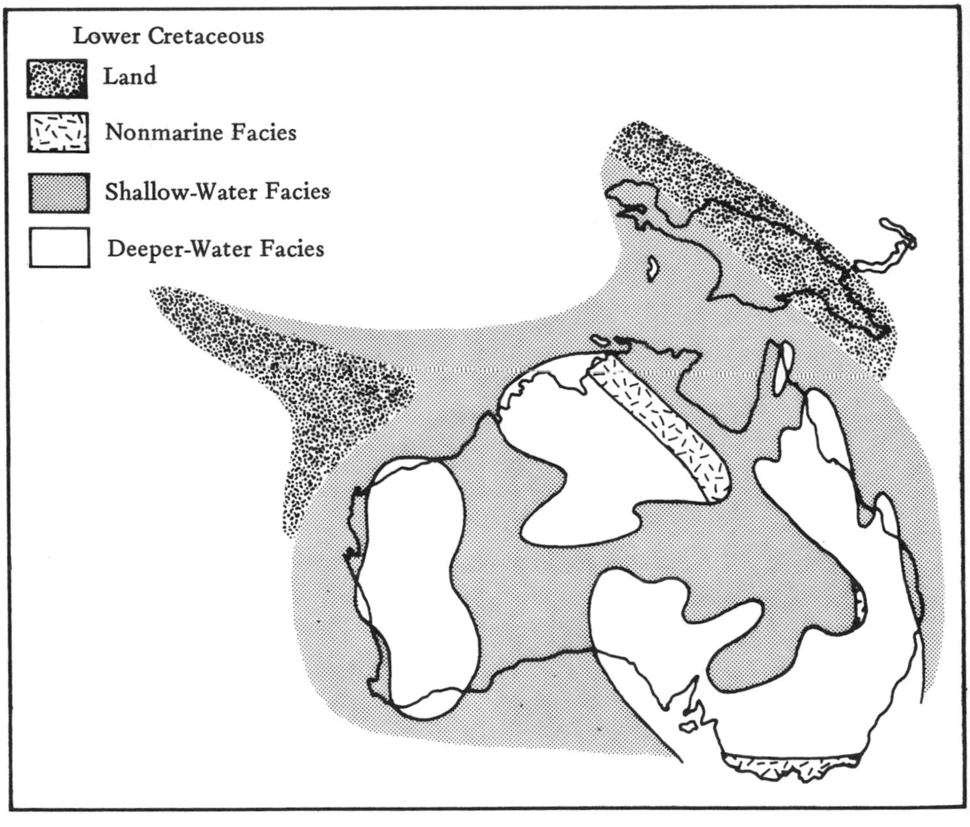

Lower Cretaceous
Land
Nonmarine Facies
Shallow-Water Facies
Deeper-Water Facies

FIGURE 19.33 Lower Cretaceous Deposits of Present-Day Australia.

was narrower in the present west where Europe and Africa had long been in proximity, but widened to the present east where the continental edges of what is now Arabia and India had not yet encountered Asia. Tethys was in tropical latitudes just north of the equator in the Cretaceous.

Gondwanaland

The fragmentation of Gondwanaland continued (Fig. 19.3). South America continued to move away from Africa toward the present southwest as the South Atlantic basin expanded. Cretaceous marginal seas flooded portions of the eastern edge of South America. In the Late Cretaceous, the impinging east Pacific plate strongly deformed the Andean Geosyncline just as it did the western edge of North America. This belt of deformation continued into the Palmer Peninsula Geosyncline of Antarctica which was still in close proximity.

Africa, a separate continent at this time also, was extensively submerged in the Cretaceous. A large epeiric sea covered most of the western bulge of Africa from the Tethyan Geosyncline southward to the Gulf of Guinea and also extended eastward to Arabia (Fig. 19.32). Marginal embayments from the Indian Ocean flooded east Africa. India was an isolated continent and marginally flooded. Western Australia was flooded by an epeiric sea from the Indian Ocean in the Middle Cretaceous (Fig. 19.33). The Pacific plate was colliding with the northern edge of the plate including Australia, for New Guinea was deformed in the Cretaceous. New Zealand separated from Gondwanaland at this time also and was deformed. Australia and Antarctica were still one continent in the Cretaceous and the Gondwana Pole was near the present eastern edge of Antarctica, just southeast of present Australia (Fig. 19.34).

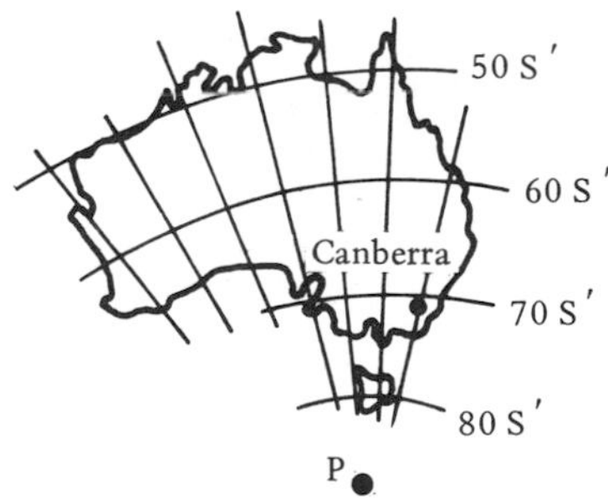

FIGURE 19.34 Paleolatitudes for Australia in the Cretaceous. (From E. Irving, *Paleomagnetism and its Application to Geological and Geophysical Problems*, copyright © 1964, by permission of John Wiley & Sons, Inc.)

The Cretaceous Earth

In the late Cretaceous, we see the reverse conditions of those that closed the Paleozoic Era. The Late Cretaceous was a time of widespread flooding of continental masses and of numerous ocean basins separating the continents and, in turn, separated by them. The major land areas of North America, South America, Europe, Asia, Africa, India, and Australia-Antarctica were all isolated at this time. The oceans were interconnected in the equatorial regions by Tethys and the Central American seaway, a condition which provided a unique circumglobal tropical seaway. The geography of the Mesozoic, with its fragmentation as opposed to the consolidation of the Paleozoic, certainly must have had significant effects on the life of the era. We shall now turn to the fossil record of the Mesozoic.

Questions

1. What is the evidence for the time of opening of the Atlantic Ocean?
2. Discuss the Mesozoic fragmentation of Gondwanaland.
3. What is the significance of the Newark Group fault basins in terms of plate tectonics?
4. What are the dominant types of Triassic deposits on the Laurasian cratons?
5. What was the origin of the Tethyean Geosyncline and what was its paleoclimatology in the Mesozoic?
6. Discuss the geologic history of the Zuni Sea.
7. What was the significance of the Morrison and Mesa Verde clastic wedges?
8. What is the geologic evidence for the Nevadan and Laramide Orogenies?
9. When was Europe flooded by epeiric seas in the Mesozoic?
10. What significant events affected the African craton in the Mesozoic?
11. What was the distribution of the continents at the close of the Mesozoic and how did it differ from that at the beginning of the era?

20

The Age of Reptiles and Ammonoids:

Mesozoic Life

THE DRAMA OF MESOZOIC LIFE IS CHARACTERIZED BY THE DOMINANCE OF REPTILES among land animals and mollusks, particularly the cephalopods, among the larger marine invertebrates. The fossil record of this era closes with another apparent "sudden disappearance" or "mass extinction" whose explanation is more difficult to ascertain than the Permian one.

Marine Life

Life in the sea flourished again in the Mesozoic Era (Figs. 20.1, 20.2) after the temporary contraction of epeiric seas induced by the extreme continental conditions of the Middle Permian through Middle Triassic. As the Atlantic and Indian Oceans formed and epeiric seas spread from them over the now-lowered continents, marine life returned with a flourish. Its appearance in most areas is quite dramatic, however, because it is so different from that of the preceding Paleozoic. This is not surprising in view of the long missing interval and the fact that a major life habitat, the epeiric sea, had been virtually eliminated in the Permotriassic boundary interval. Many lineages of organisms were eliminated by the loss of this habitat so that when the environment returned it was open to a new adaptive radiation and colonization. In the geosynclines, such as the Tethyan, where marine sediments continuous across the boundary zone are preserved, the faunal changeover is the normal gradual replacement. In the Jurassic we get our first good look at life in the ocean basins, for the oldest known oceanic rocks not destroyed or deformed by subduction date from that period.

FIGURE 20.1 Diorama of Jurassic Marine Life. (Courtesy, Field Museum of Natural History.)

FIGURE 20.2 Diorama of Cretaceous Marine Life. (Courtesy, Field Museum of Natural History.)

Protistans

The major marine protistan groups of the Mesozoic were the *dasycladaceous green algae, coccolithophores, diatoms, dinoflagellates, foraminifera, radiolaria,* and *tintinnoids.* Most coccoliths, diatoms, dinoflagellates, radiolaria, and the planktonic forams appear late in the era once the oceanic basin record is preserved.

The benthonic dasycladaceous *green algae* (Fig. 20.3) were abundant in the warm, tropical waters of Tethys. They were major contributors to the reefs which today stand as the isolated blocks of the Dolomite Alps after the less resistant intervening facies had been stripped away. The group reaches a peak in the Triassic-Jurassic interval. Common genera include *Triploporella, Goniolina,* and *Diplopora.*

Coccolithophores (Fig. 20.4) first appeared in the Jurassic and expanded greatly in the Cretaceous. The calcareous platelets of these microscopic pelagic protistans collected in tremendous numbers on the ocean floors giving rise to the great Cretaceous chalk beds. They are also used in correlating Mesozoic ocean basin sediments. Common genera include *Coccolithus, Braarudosphaera,* and *Discoaster.* Coccolithophores have continued to be abundant down to the present day and are major producers supporting all life in modern oceans. Their

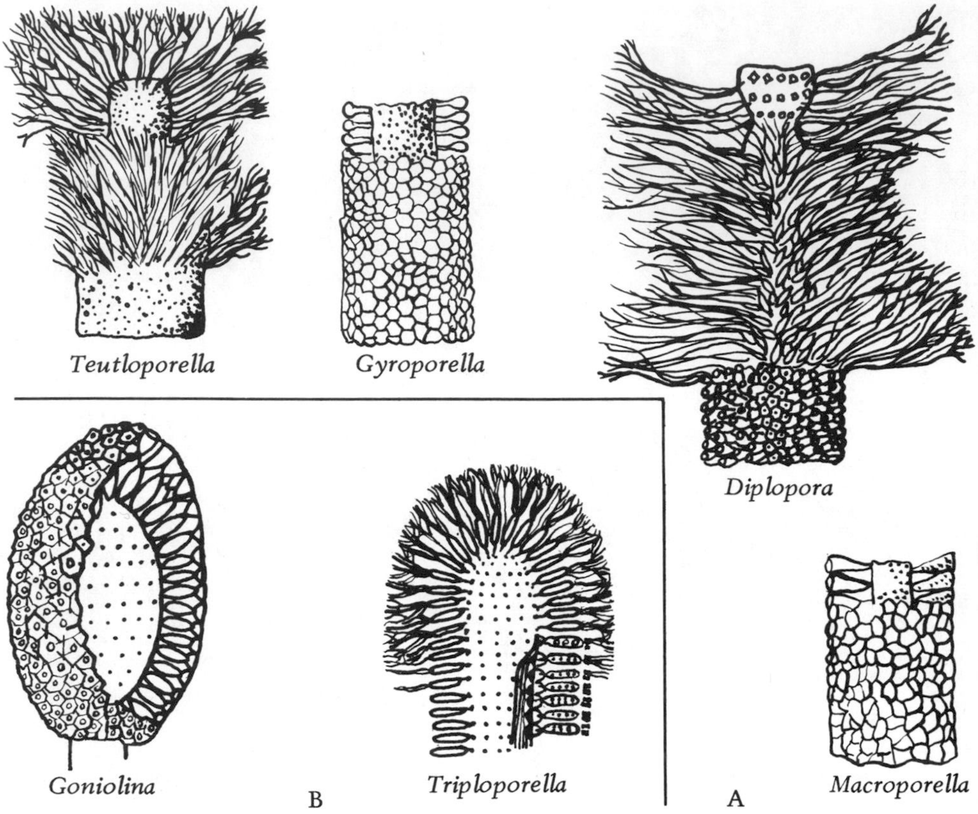

Teutloporella
Gyroporella
Diplopora
Goniolina
B
Triploporella
A
Macroporella

FIGURE 20.3 Mesozoic Dasycladaceous Green Algae. (A) Triassic, (B) Jurassic.

Coccolithus
Discoaster
Broarudosphaera
Discoaster

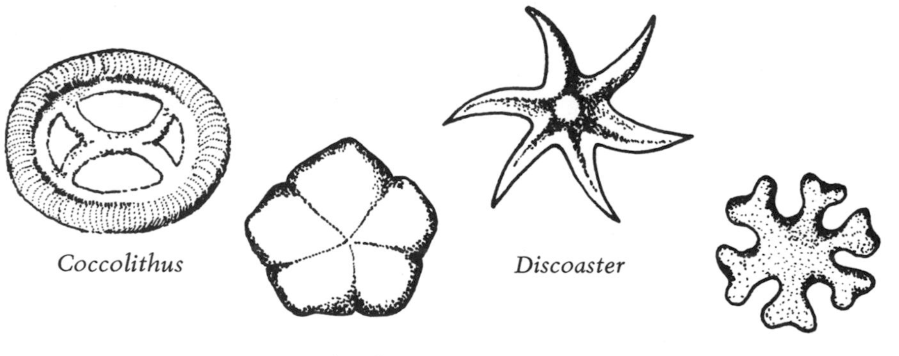

FIGURE 20.4 Late Mesozoic Coccoliths.

lack before the Jurassic is hard to explain unless they lacked the ability to secrete hard parts before this, or their niche was occupied by some currently unknown shell-less microorganism. If either of the two hypotheses is true, the prodigious amounts of CO_2 they extracted from the sea once they began to secrete coccoliths surely must have had repercussions in the earth's ecosystem.

If they really did not exist or had no ecologic predecessors before the Late Mesozoic, their appearance and spread, particularly with the widespread Cretaceous seas, may have triggered a quantum jump in the earth's oxygen supply at that time. Considering that diatoms, the second greatest group of modern photosynthesizers, did not appear in the fossil record until the Cretaceous, this increase may have been so significant that it somehow may have controlled the extinctions at the end of the era. The *silicoflagellates* (Fig. 20.5), also important producers today and near relatives of the coccolithophores, appeared in the Cretaceous.

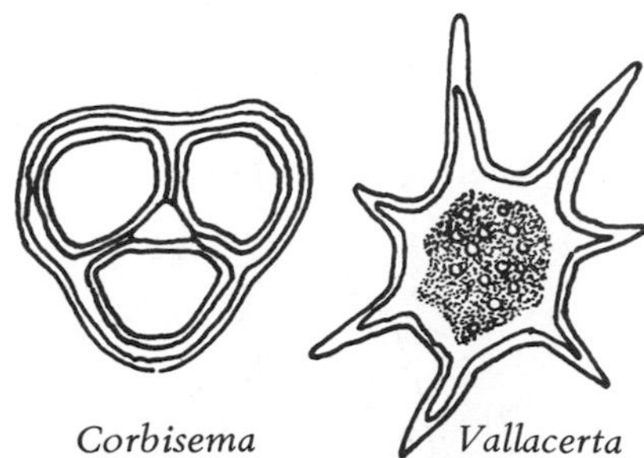

FIGURE 20.5 Cretaceous Silicoflagellates.

The siliceous *diatoms* (Fig. 20.6) are not known until the Cretaceous, but were abundant enough by the end of that period to form siliceous shales such as the Moreno of California. The group is chiefly planktonic and the Mesozoic forms were all marine. Common genera are *Actinoptychius and Biddulphia.*

Dinoflagellate cysts such as *Gymnodinium* (Fig. 20.7) become common for the first time in the Upper Mesozoic. This pelagic group is known from rare finds back into the Silurian and is the only known group of major producers in the oceans today that appears in the fossil record before the Jurassic.

Foraminifera (Figs. 20.8–20.10) undergo a great radiation in the Jurassic and Cretaceous that continues down to the present day. The small, benthonic types (Fig. 20.8), such as *Spiroplectammina, Frondicularia,* and *Marginulina,* diversified greatly and gave rise to the pelagic *Superfamily Globigerinacea* in the Jurassic. This group (Fig. 20.9), including *Heterohelix* and *Globotruncana,* multiplied so profusely in the Cretaceous that the tests of individuals were major contributors to marine sediments. These include not only the ocean basin

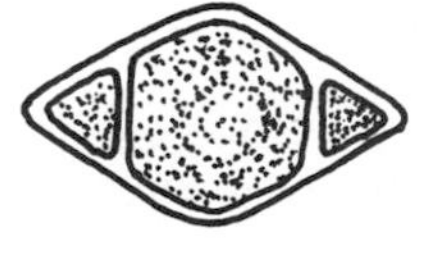

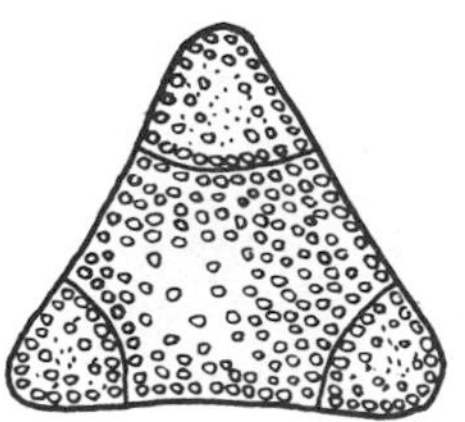

FIGURE 20.6 Cretaceous Diatoms.

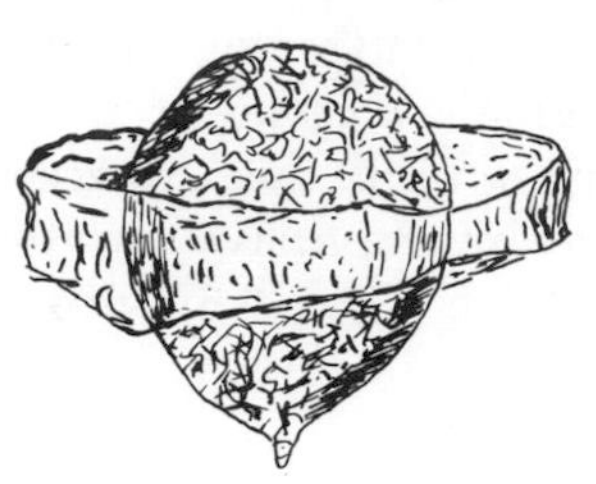

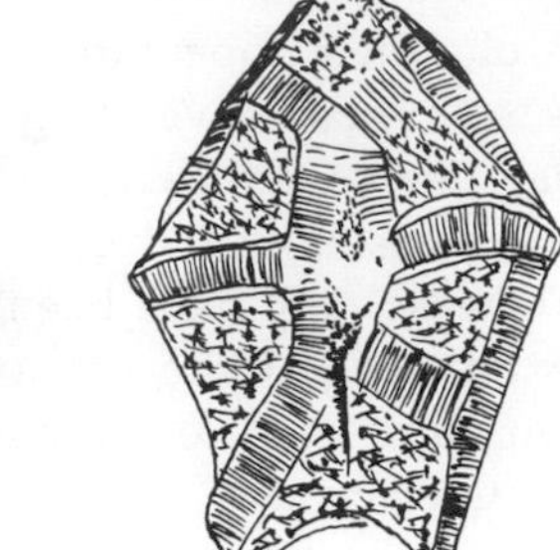

Stephodinium *Gymnodinium* *Peridinium*

FIGURE 20.7 Mesozoic Dinoflagellates.

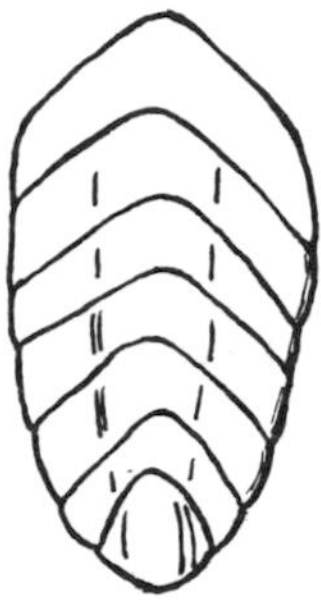

Sprioplectammina *Marginulina* *Frondicularia*

FIGURE 20.8 Mesozoic Small Benthonic Foraminifera.

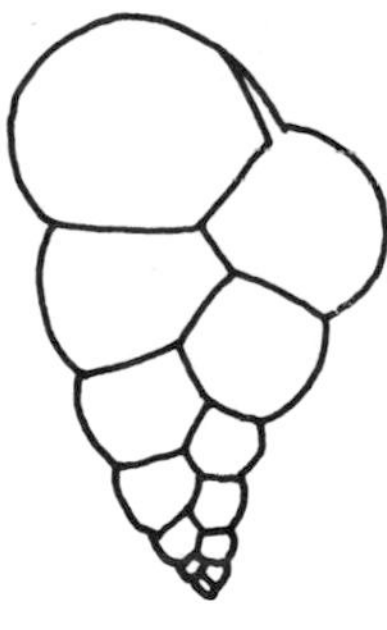

Heterohelix *Globotruncana*

FIGURE 20.9 Mesozoic Planktonic Foraminifera.

oozes, but also the great shelf chalks that accumulated in the widespread epeiric seas of the Cretaceous. From the Late Cretaceous onward, pelagic forams are important in long-range correlations. The large, disc-shaped orbitoids (Fig. 20.10A) such as *Cymbalopora* and the large spindle-shaped alveolinids (Fig. 20.10B) such as *Praealveolina* appeared and became common in the shelf seas of the Late Cretaceous. They are probably the ecologic successors of the fusulines which disappeared with the last Paleozoic epeiric sea.

The *radiolaria* (Fig. 20.11) were abundant for the first time in the Jurassic and Cretaceous. Chiefly open-ocean pelagic forms, they are not common before this except in areas where oceanic crust had been incorporated in geosynclines by underthrusting. In the Jurassic and Cretaceous, there are large-scale examples of this occurrence as well as an abundance of oceanic siliceous oozes where they are used for correlation. Common genera are *Hexadoridium* and *Lithomitra*.

The pelagic ciliate group of *tintinnoids* (Fig. 20.12) reaches a Jurassic-Cretaceous peak in the fossil record. *Tintinnopsis* is a typical genus.

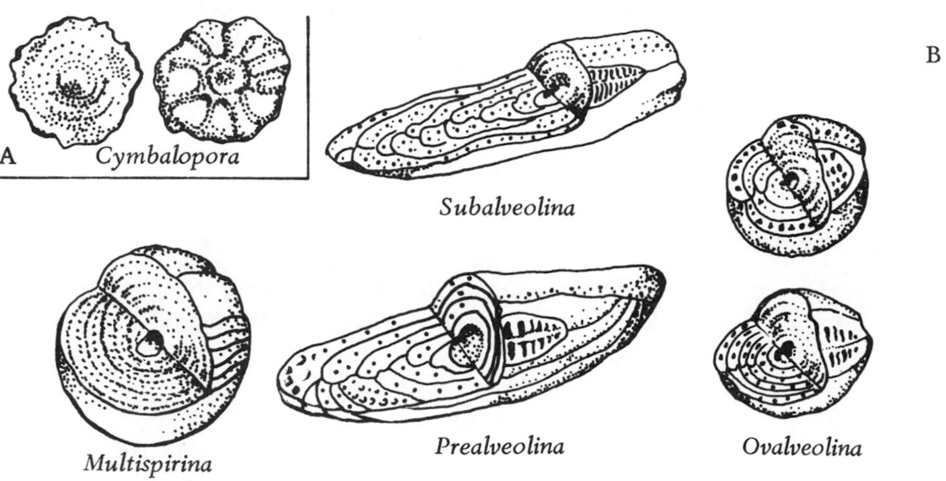

FIGURE 20.10 Cretaceous Large Foraminifera. (A) Orbitoid, (B) Alveolinids.

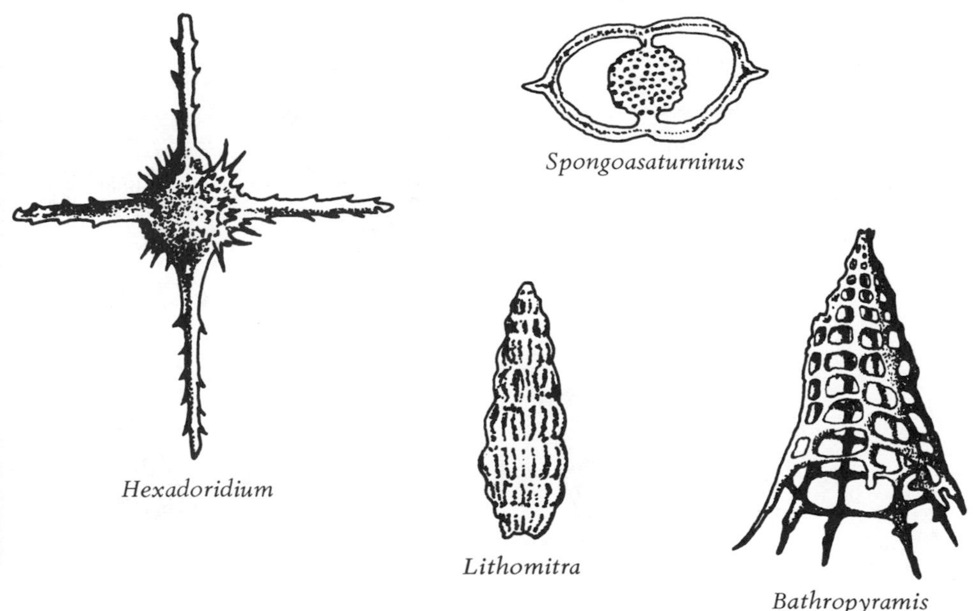

FIGURE 20.11 Mesozoic Radiolaria.

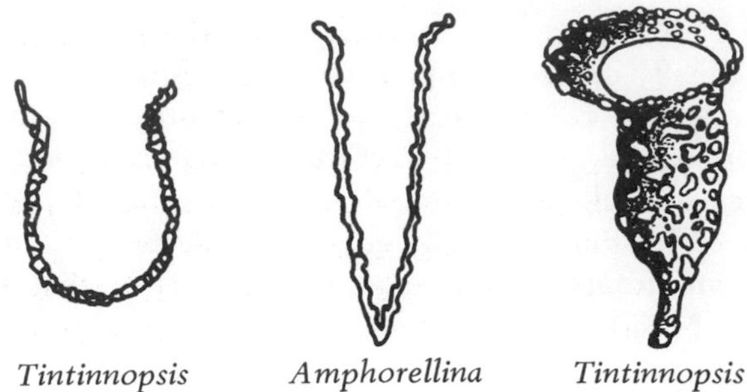

FIGURE 20.12 Late Mesozoic Tintinnoids.

Invertebrate Animals

The major marine invertebrate groups of the Mesozoic were the *sponges, hexacorals, bryozoans, brachiopods, gastropods, pelecypods, ammonoids, belemnoids, ostracods,* and *echinoids.* Almost all Mesozoic time-correlations are made with the abundant, nektonic, and rapidly evolving ammonoids, the single most important group of Mesozoic fossils to geologists.

The *sponges* (Fig. 20.13) were common in the tropical seas of Tethys in the Jurassic and particularly the Cretaceous. Both calcareous and siliceous types occurred with the latter predominating. Common genera include *Raphidonema* (calcareous), *Ventriculites* (siliceous), and *Siphonia* (siliceous).

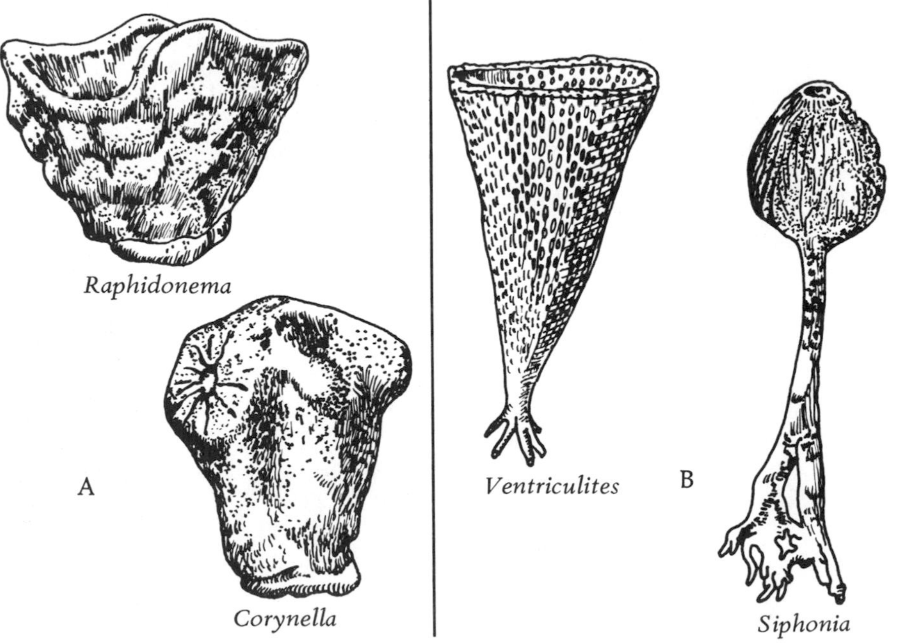

FIGURE 20.13 Late Mesozoic Sponges. (A) Calcareous forms, (B) Siliceous forms.

Hexacorals (Fig. 20.14) first appeared in the Middle Triassic from unknown ancestors in tropical Tethys. By the Late Triassic, they extended worldwide and were building reefs. From the Middle Jurassic onward hexacorals are important faunal elements, particularly in reefs, with their greatest abundance and diversity in Tethys, that great east-west seaway near the equator. The Atlantic was not very wide at that time and the fauna extended to southern North America. Common genera were *Montlivaltia, Thamnastria, Montastrea* and *Trochocyathus.*

Bryozoans, like most shallow-water marine invertebrates, were rare in the Triassic so we lack the necessary evolutionary links between the last Paleozoic types and the new Mesozoic lineages. The surviving *cyclostomes* (Fig. 20.15A) underwent a Jurassic and Cretaceous expansion and, until the Late Cretaceous, dominated bryozoan faunas. *Entalophora* was a typical form. The *cheilostomes* (Fig. 20.15B) appeared in the Cretaceous and by the latter part of the period had become the major group of bryozoans, a position they maintain today. *Membraniporidra* was a typical encrusting form.

Montlivaltia

Thamnasteria

Trochocyathus

Montastrea

FIGURE 20.14 Mesozoic Hexacorals.

Meliceritites

A Entalophora

Membraniporidra

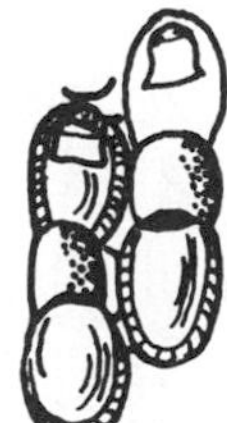
Mollia

B

FIGURE 20.15 Mesozoic Bryozoans. (A) Cyclostomes, (B) Cheilostomes.

Brachiopods had dwindled considerably since the Paleozoic and most of the dominant orders of that era are either gone or nearly so. The *terebratuloids* and *rhynchonelloids* (Fig. 20.16), the two living articulate orders, continue to diversify, however. Both groups reached a Jurassic peak and declined thereafter. *Goniorhynchia* is a common Jurassic rhynchonelloid and *Kingena*, an abundant Cretaceous terebratuloid.

The *mollusks* were the major phylum of large invertebrates in the Mesozoic. *Gastropods* (Fig. 20.17) increased steadily throughout the era, being most abundant in the Cretaceous, the period when the modern carnivorous types appeared. Especially noteworthy was a group of huge, high-spired forms with internal plications called *Nerineids* found only in Jurassic and Cretaceous strata, chiefly the latter. The *Turritella* group of high-spired snails that is abundant in the Paleogene becomes common in the Cretaceous along with many other snails. The *pelecypods* are even more abundant in the Mesozoic, though they too increase in importance as the era goes on. In the Triassic, a group of scallops or pectens including *Daonella, Halobia,* and *Monotis* (Fig. 20.18) are widespread and used for correlation within regions. They may have attached themselves to floating organisms such as seaweed or logs to achieve their wide distribution. In the Jurassic and Cretaceous, clams such as *Trigonia, Pholadomya,* the wing "oyster" *Inoceramus,* the scallops such as *Buchia* and *Camptonectes,* and the true oysters such as *Gryphaea* and *Exogyra* were extremely abundant (Fig. 20.19). Most of these genera were greatly reduced or disappeared by the end of the Cretaceous, but the clam group as a whole, particularly the more advanced burrowing clams, such as *Cardium,* continued to expand down to the present. A very significant group of reef-dwelling clams, the *rudists* or *hippuritoids* is restricted to Upper Jurassic and Cretaceous rocks. These large bivalves, such as *Durania* and *Monopleura,* converged on corals by developing a conical right or lower valve and a lid-like left or upper valve (Fig. 20.20). Gregarious tropical forms, they lived in the same environment as reef corals which accounts for the similar shape. None survived the Cretaceous and there is no apparent reason for their extinction.

The *cephalopods* are the most geologically important group of mollusks because their chiefly nektonic mode of life ensures the wide geographic distribution so vital for long-distance correlations. Far and away the most abundant

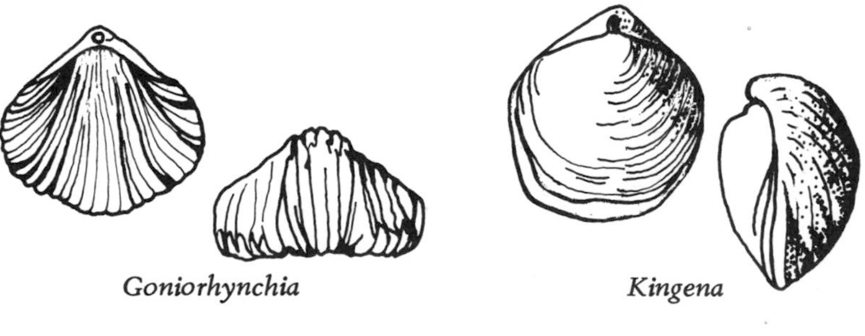

FIGURE 20.16 Mesozoic Articulate Brachiopods. (A) Rhynchonelloid, (B) Terebratuloid.

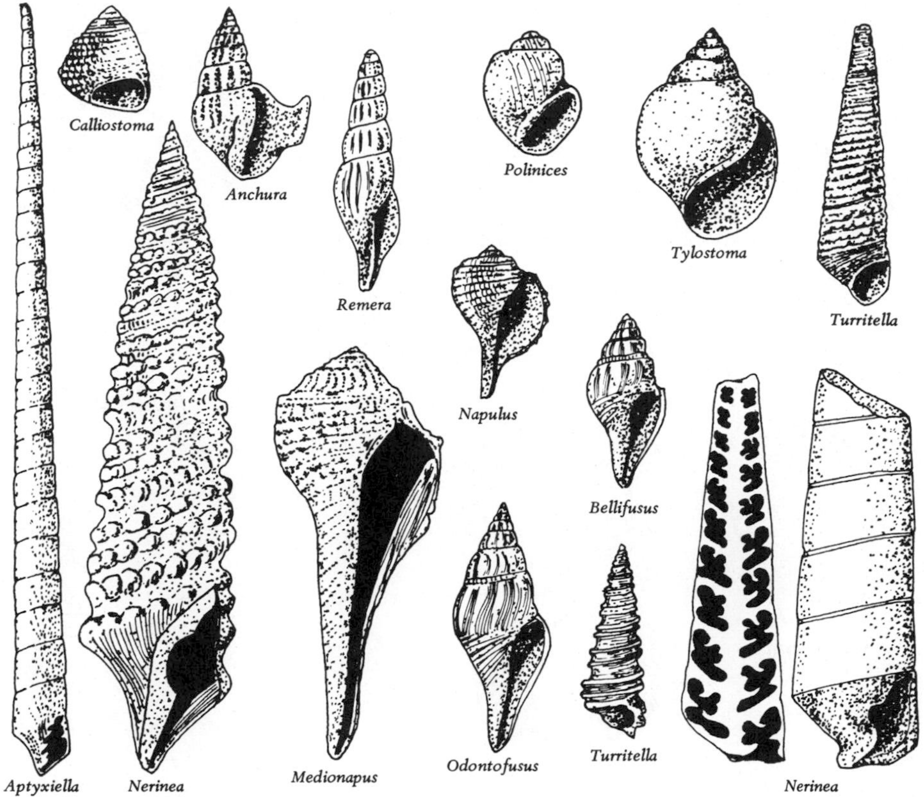

FIGURE 20.17 Mesozoic Gastropods.

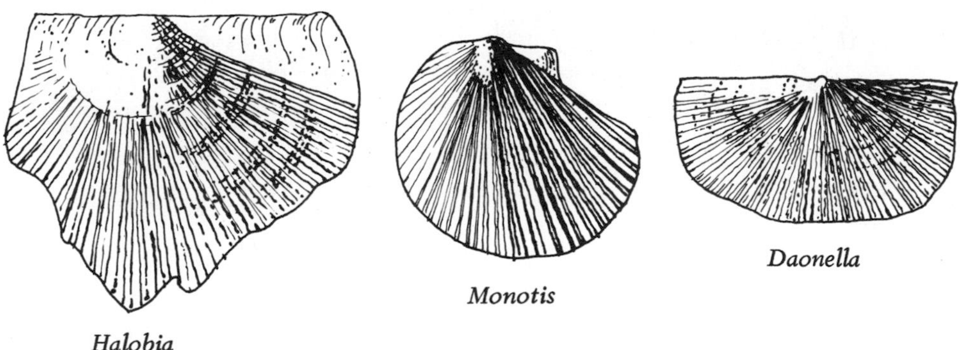

FIGURE 20.18 Triassic Bivalves.

Mesozoic cephalopods were the forms with complex sutures, the *ammonoids*. In the Triassic, *ceratites* dominated, but they did not survive the period. Common genera were *Meekoceras* and *Ceratites* (Fig. 20.21). In the Jurassic and Cretaceous, the *ammonites* (Fig. 20.22) underwent the greatest radiation in ammonoid history. In addition to the typical planispiral forms such as *Amaltheus* and *Perisphinctes* of the Jurassic and *Oxytropidoceras*, *Placenticeras*, and *Du-*

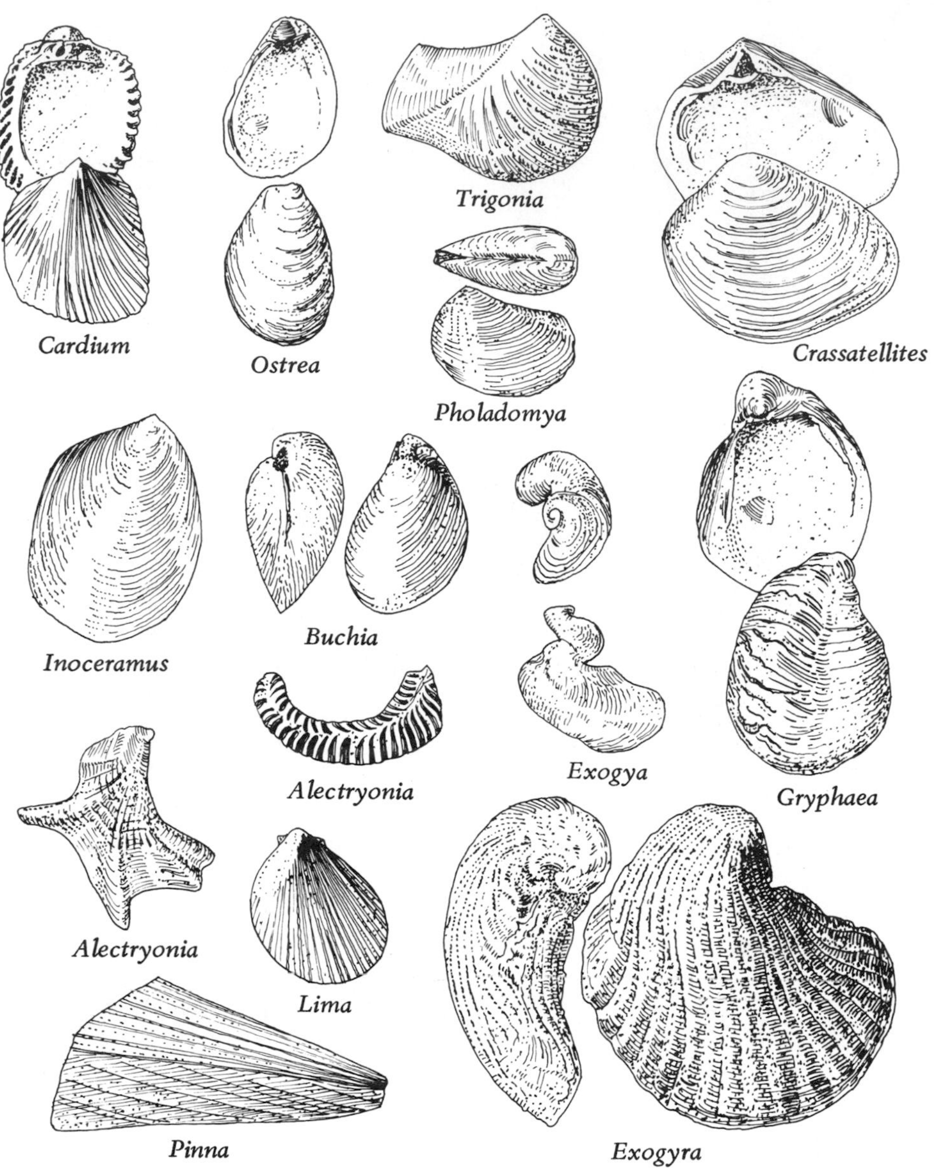

FIGURE 20.19 Jurassic and Cretaceous Bivalves.

frenoyia of the Cretaceous, there were numerous uncoiled or strangely coiled lytocerines (Fig. 20.23) such as *Hamites, Turrilites,* and *Baculites.* Ammonoids disappeared by the end of the Cretaceous.

It is their extinction and the approximately simultaneous demise of the dinosaur lineages on land that led to the original separation of the Mesozoic and Cenozoic Eras. The extinction of these two major fossil groups of the Mesozoic is certainly very real and puzzling. Although both groups declined toward the close of the Mesozoic, some members survived in fair abundance until the very end. Late Cretaceous seaways were widespread and ammonites were largely pelagic anyway so we cannot invoke restriction of the seas as a cause for their

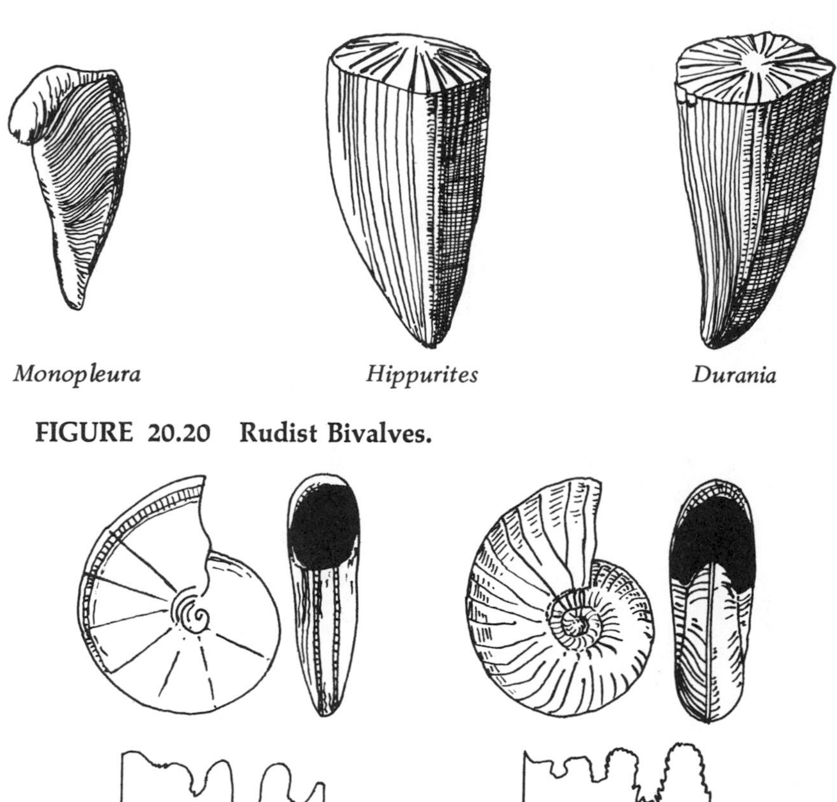

FIGURE 20.20 Rudist Bivalves.

Meekoceras *Ceratites*

FIGURE 20.21 Triassic Ceratite Ammonoids. Sutures shown below specimens.

extinction. Two hypotheses for their extinction appear reasonable. One is the possible change in the proportions of gaseous constituents of the earth's atmosphere and oceans alluded to earlier. The pelagic ammonoids, like the living nautilus, depended on a delicate adjustment of gases and fluids in their chambers for the proper bouyancy. Anything that could drastically alter the balance, such as a rapid increase or decrease in the amount of dissolved oxygen or carbon dioxide in sea water, could send them bobbing to the surface or sinking like a lead balloon. The other hypothesis points to the fact that modern types of ray-finned bony fish, the teleosts, underwent a great expansion at the same time the ammonites declined. Because both were nektonic carnivores, it is possible that the fish were superior competitors and thus, gradually eliminated the ammonoids. A few workers have suggested that predation by fish eliminated the ammonoids. It is true that certain fish ate ammonoids, but it is impossible for any group to literally consume another group to extinction as studies of the checks and balances of modern ecosystems show. Besides, the fish went right on expanding into the Cenozoic, a feat they hardly could have accomplished if they had eliminated their own food supply. Modern cephalopods, such as the squid, octopus, and cuttlefish, also arose in the Late Mesozoic and perhaps also were superior competitors that contributed to the fall of the am-

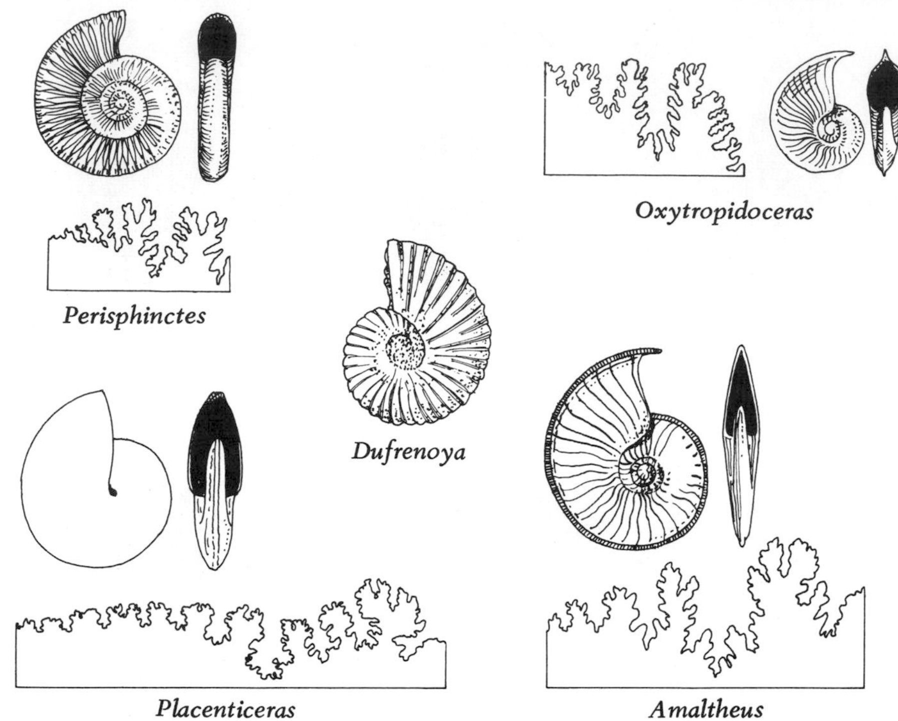

FIGURE 20.22 Jurassic-Cretaceous Ammonitic Ammonoids. Sutures shown below specimens.

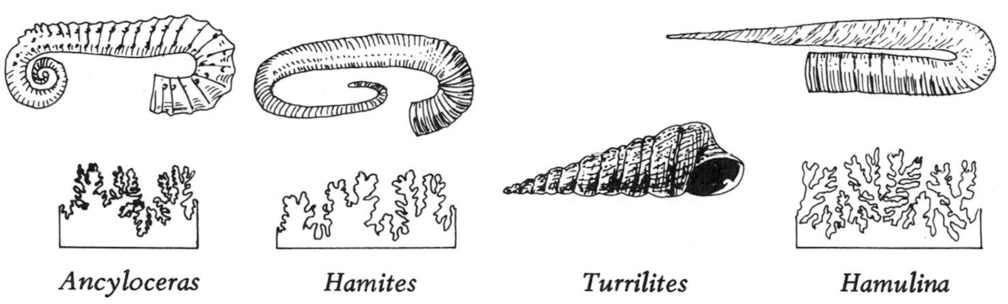

FIGURE 20.23 Lytocerine Ammonoids. Sutures shown below specimens.

monoids. These forms have delicate shells or lack shells and, hence, have left few fossils.

The cigar-shaped *belemnoids* (Fig. 20.24A) reached a peak in the Jurassic and Cretaceous. The abundant remains of these extinct nektonic squid-like forms are used for correlations within regions. *Pachyteuthis* was a common Jurassic genus; *Belemnitella,* a significant Cretaceous form. Only a very few belemnoids survived the Cretaceous. A few fossils of their chiefly soft-bodied relatives, the squids and cuttlefish, occur in the Upper Mesozoic indicating that these lighter, perhaps more agile, forms were superior competitors and eliminated the *belemnoids.* In the Triassic, slender relatives of the belemnoids known as *aulacoceroids* were common. *Atractites* (Fig. 20.24B) is a representative genus.

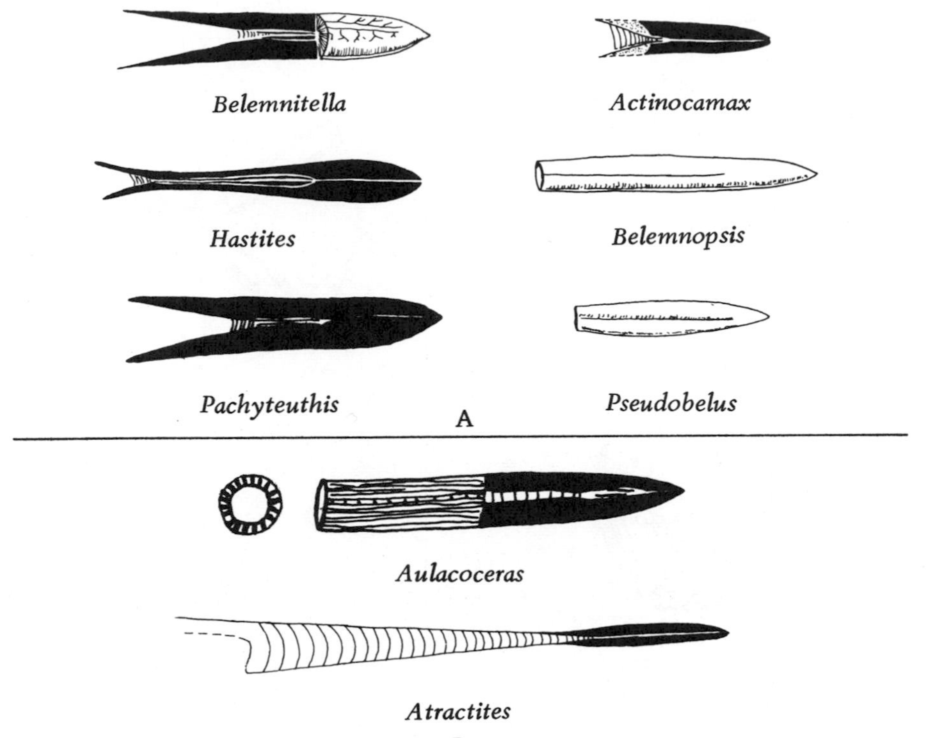

FIGURE 20.24 Mesozoic Coleoid Cephalopods. (A) Belemnoids, (B) Aulacoceroids.

The *ostracods* continued to increase in abundance in the Mesozoic after the Permotriassic decline in which the typical highly ornamented Paleozoic types largely disappeared. *Brachycythere, Bythocypris,* and *Cytherella* are common Mesozoic genera (Fig. 20.25).

Echinoids (Figs. 20.26, 20.27) became abundant for the first time in the Jurassic with the appearance of the burrowing irregulars. The sea-urchin-like, yet irregular, *holectypoids* such as *Holectypus* were common in the Jurassic and Cretaceous, but have declined to obscurity since. The elongate, dome-shaped *cassiduloids* with their star-shaped floscelle around the mouth are also common in rocks of these ages. *Pygurus* is a typical form. The most important irregulars, however, are the toothless, deposit-feeding *heart urchins.* Several

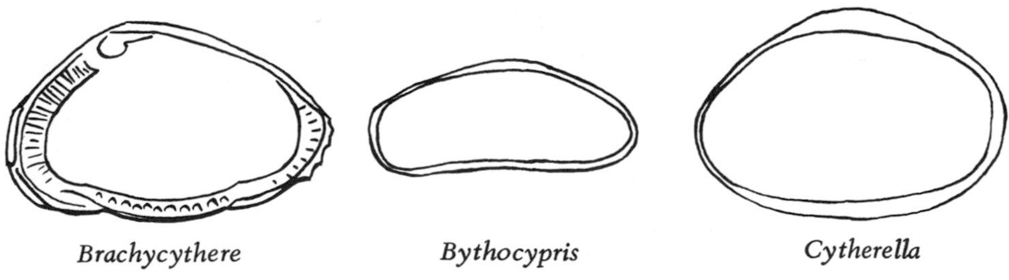

FIGURE 20.25 Mesozoic Ostracods.

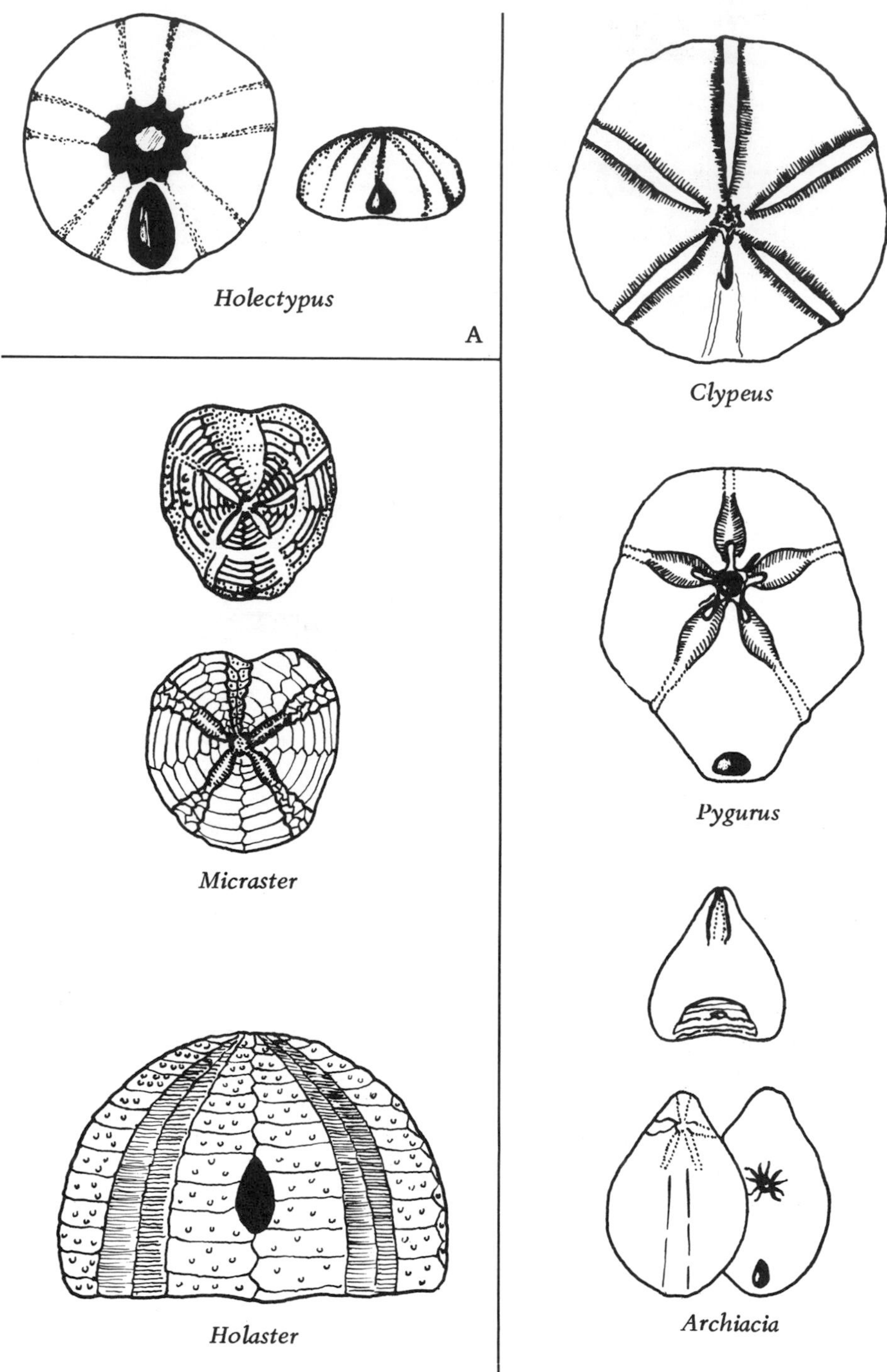

FIGURE 20.26 Mesozoic Irregular Echinoids. (A) Holectypoid, (B) Cassiduloids, (C) Heart Urchins.

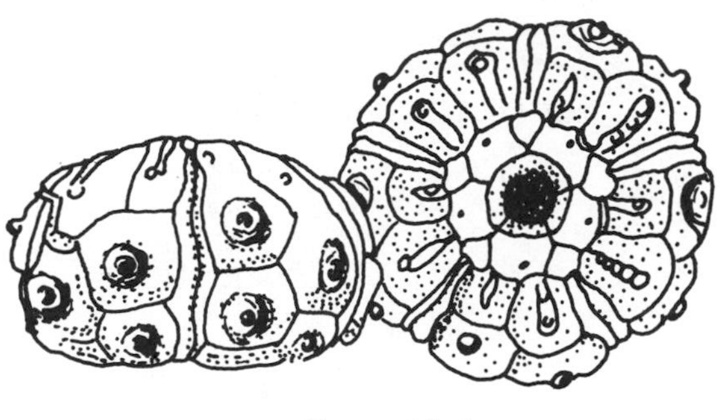

Stereocidaris Phymosoma

FIGURE 20.27 Mesozoic Regular Echinoids.

major types evolved in the Jurassic and the modern orders became dominant in the Cretaceous. *Holaster, Hemiaster,* and *Micraster* are abundant genera, the last being well-represented in chalk facies. The regular, sea-urchin types of echinoids were occasionally common in the Late Mesozoic. *Stereocidaris* and *Phymosoma* are two of the best known genera.

Other marine invertebrate groups are common for shorter intervals in the Mesozoic Era. The *spongiomorphoids* (Fig. 20.28), an extinct group related to the stromatoporoids and thus probably hydrozoan coelenterates or sclerosponges, were common reef builders in the Triassic of Tethys. *Malacostracan crustacean* fragments are abundant and complete specimens are occasionally encountered in Jurassic and Cretaceous rocks (Fig. 20.29). *Annelid* tubes (Fig. 20.30) are particularly common in the Cretaceous. The star-shaped columnals of some *crinoids* (Fig. 20.31) are occasionally common in Triassic and Jurassic strata, and the modern floating-swimming crinoids (Fig. 20.31) underwent a short, but widespread evolutionary burst in the Late Cretaceous seas that makes them useful for long-distance time-correlation of a thin stratigraphic interval.

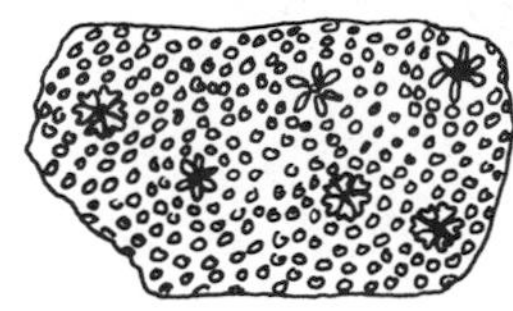

FIGURE 20.28 Spongiomorphoids.

Marine Biogeography

Marine biogeography of the Mesozoic is characterized by a broad tropical east-west Tethyean fauna between the Laurasian and Gondwanaland continents and a boreal fauna of the somewhat cooler epeiric seas of northern Laurasia. The Tethyean fauna showed occasional provincialism along its east-west length. In particular, the Pacific margins tended to develop differences (Fig. 20.32).

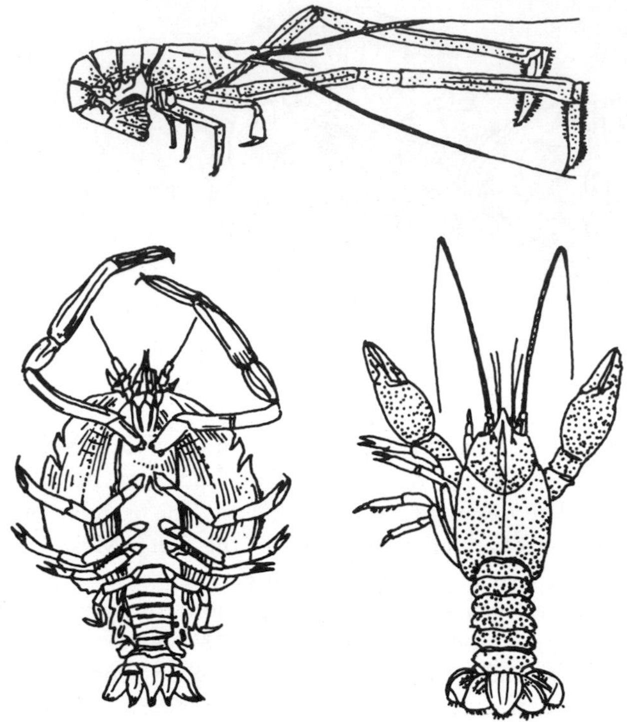

FIGURE 20.29 Mesozoic Malacostracans.

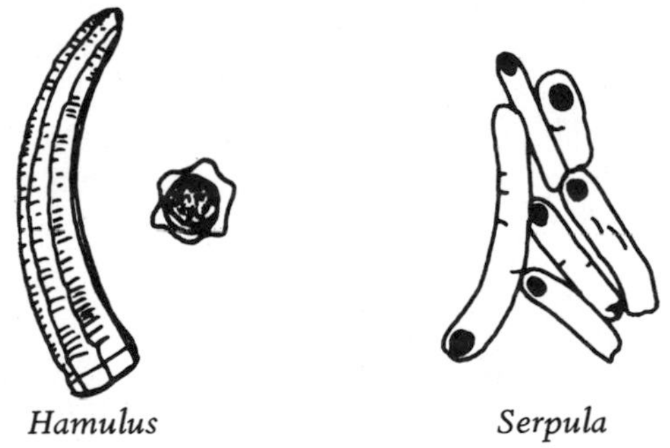

FIGURE 20.30 Mesozoic Annelid Tubes.

Marine Vertebrates

As noted earlier the ray-finned bony fish underwent a great radiation in Mesozoic seas, particularly in the Cretaceous. *Xiphactinus* is a well-known Cretaceous genus (Fig. 7.109).

The reptiles, so abundant on the land in the Mesozoic, returned to the seas and achieved great success there also. The fish or dolphin-like *ichthyosaurs* (Figs. 20.33, 20.34) were common in the Triassic and reached a Jurassic peak. These viviparous forms have an unknown ancestry and died out by the Middle Cretaceous. The *sauropterygian* transition to the sea is better documented. The terrestrial, lizard-like protorosaurs gave rise to the amphibious Triassic

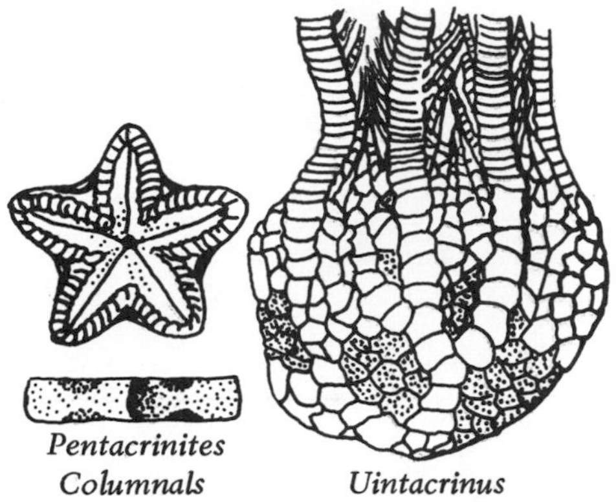

FIGURE 20.31 Mesozoic Crinoids.

nothosaurs (Fig. 20.34). From these evolved the Triassic seacow-like, but mollusk-crushing *placodonts* and the chiefly Jurassic-Cretaceous sea-serpent-like plesiosaurs (Figs. 20.33, 20.34). Both long- and short-necked varieties of these huge marine reptiles lived during their Late Cretaceous peak. All died out by the end of the Cretaceous. The *lizards* and *turtles* (Figs. 20.34, 20.35) went to sea in the Cretaceous. The *mosasaurs* or huge marine lizards did not survive the end of the Cretaceous, but marine turtles of smaller size than their Cretaceous relatives persist to the present.

Fresh-Water and Land Life

Protistans and invertebrates continued to diversify in fresh waters and on land. The *charophytes* became common in fresh-water continental deposits of the Jurassic and Cretaceous (Fig. 20.36). Their widespread occurrence helps to date the generally isolated dinosaur finds in rocks of these systems. *Atopochara, Clavator,* and *Perimneste* were representative genera. Fresh-water clams (Fig. 20.37) continue and the oldest unequivocal fresh-water and terrestrial snails (Fig. 20.38) are found in Late Mesozoic strata; aquatic forms, in the Jurassic; and terrestrial, in the Cretaceous. The *branchiopod crustaceans* (Fig. 20.39) such as *Cyzicus* continue to be common in Triassic fresh-water deposits just as they are in Carboniferous and Permian non-marine strata. Fresh-water *ostracods* (Fig. 20.40) also carried through from the Late Paleozoic and increased in numbers. The terrestrial arthropods continued their evolution, particularly the *insects* (Fig. 20.41). Most of the modern orders appeared in the Mesozoic. Among the lesser arthropod groups, the terrestrial centipedes appeared in the Cretaceous.

Among the vertebrates, ray-finned *fish* continued to be common in fresh waters. The large, flat-headed stegocephalian or labyrinthodont *amphibians* (Fig. 20.42) were still common in many areas in the Triassic as continental deposition remained widespread. Similar genera occurred in such distant areas as the Spitsbergen Island, at the present northern limits of old Laurasia, and South Africa and Australia, at the present southern limits of Gondwanaland. (Both were in temperate zones in the Mesozoic, however.) The recent find of a labyrinthodont

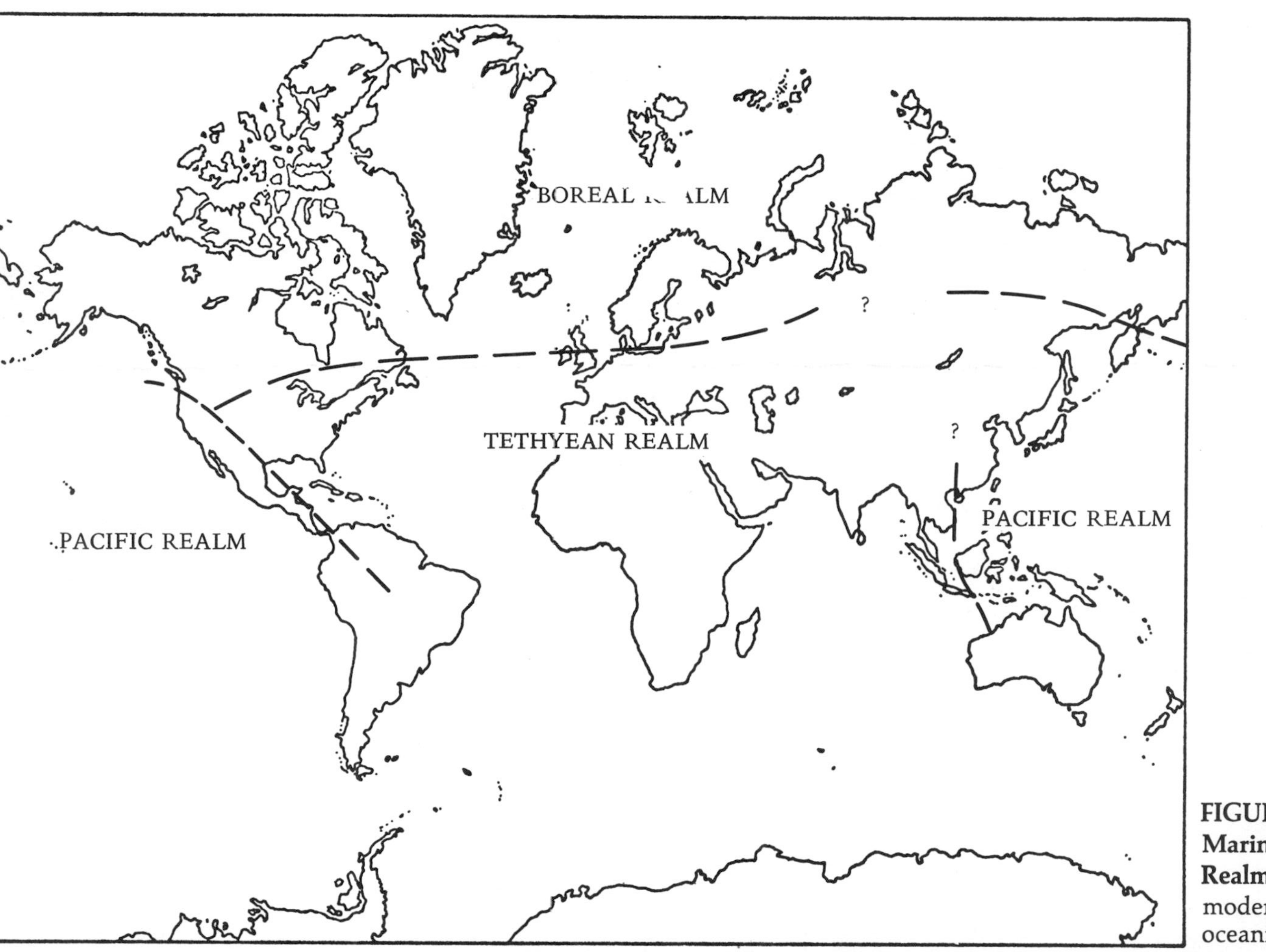

FIGURE 20.32 Jurassic Marine Faunal Realms plotted on modern continental-oceanic distribution.

FIGURE 20.33 Mesozoic Marine Reptiles. (A) Jurassic Plesiosaurs and Ichthyosaurs. (C. R. Knight mural, courtesy Field Museum of Natural History.)

FIGURE 20.33 Mesozoic Marine Reptiles. (B) Fossil Ichthyosaur with young. (Courtesy of the American Museum of Natural History.)

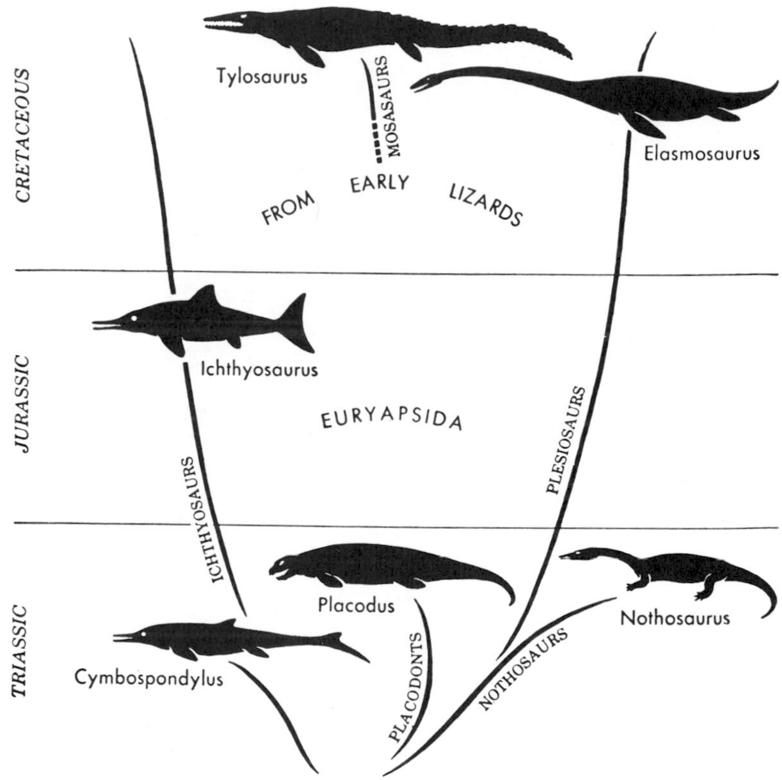

FIGURE 20.34 Evolution of Mesozoic Marine Reptiles. (From Edwin H. Colbert, *Evolution of the Vertebrates,* Copyright © 1969, by permission of John Wiley & Sons, Inc.)

(and a mammal-like reptile) in Antarctica was a factor that won over most vertebrate paleontologists to the idea of moving continents. The ancestor of the frogs has been found in the Triassic of Gondwanaland on the present island of Madagascar, and ancestral salamanders are known from the Jurassic and Cretaceous of North America.

FIGURE 20.35 Mesozoic Marine Reptiles and Flying Reptile. Marine turtle and lizard (mosasaur) and a flying reptile or pterosaur. Cretaceous forms from North America. (C. R. Knight mural, courtesy American Museum of Natural History.)

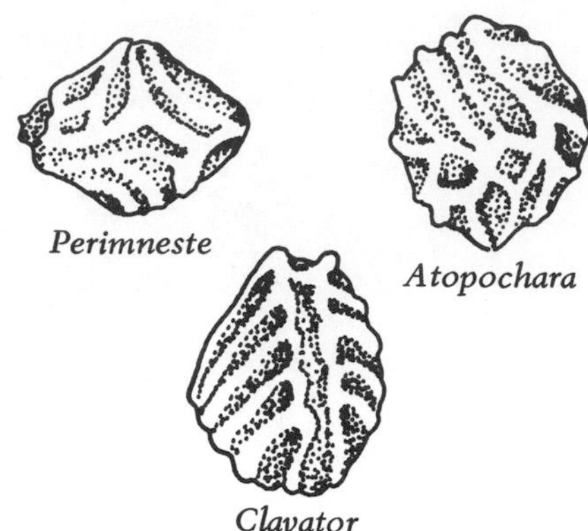

FIGURE 20.36 Mesozoic Charophytes.

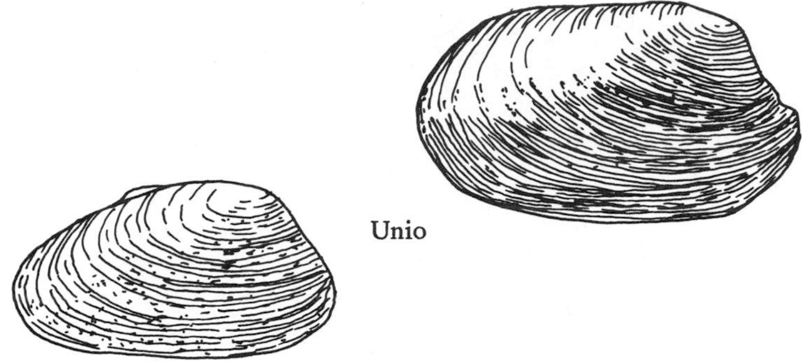

FIGURE 20.37 Mesozoic Fresh-Water Bivalves.

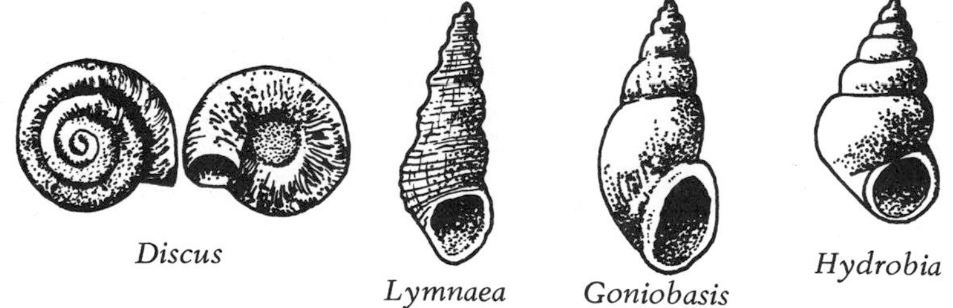

FIGURE 20.38 Mesozoic Fresh-Water Gastropods.

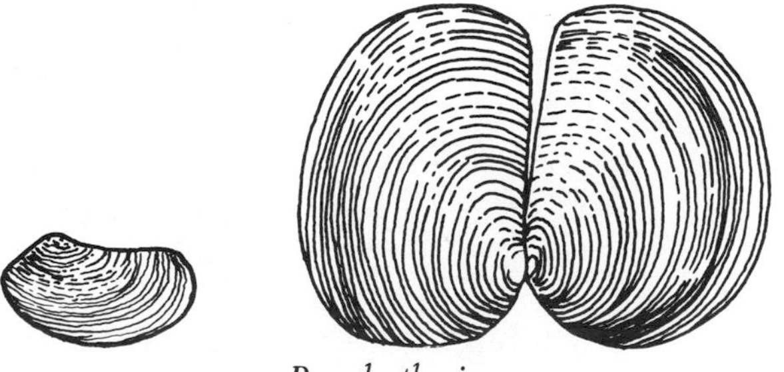

FIGURE 20.39 Mesozoic Branchiopods.

FIGURE 20.40 Mesozoic Fresh-Water Ostracod.

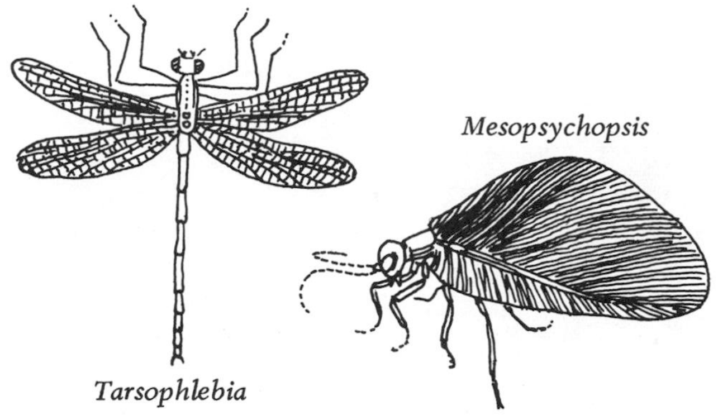

FIGURE 20.41 Mesozoic Insects. These forms are from the famous Solenhofen area.

FIGURE 20.42 A Triassic Amphibian from Europe.

FIGURE 20.43 Mammal-Like Reptiles from South Africa. (C. R. Knight mural, courtesy Field Museum of Natural History.)

Reptiles

In the terrestrial environment, the *reptiles* dominated the faunas. The early Mesozoic reptile assemblage consisted mainly of the *mammal-like* or synapsian group (Fig. 20.43) which continued without interruption from the Paleozoic. These reptiles, which were abundant until the Late Triassic, became more and more mammal-like as the period progressed and near its close a carnivorous lineage crossed the boundary and became the first small mammals. Many lineages so closely approximate the mammalian condition that not all the so-called primitive "mammals" may actually be mammals, but just independent offshoots of the synapsian reptiles. It is probable that some of the early Mesozoic mammal-like reptiles had a furry covering and were able to internally regulate their own temperature. In the Late Triassic, the mammal-like reptiles, which included both carnivores and herbivores and had successfully invaded all the present continental areas, were ecologically replaced by other orders of reptiles, chiefly the dinosaur groups (Fig. 20.44). The synapsian reptiles persisted into the Jurassic in declining numbers and then became extinct, unless some of the so-called primitive Mesozoic and Cenozoic mammal orders are not ancestral to other mammals. In that case, the group may actually survive down to the present day in the form of monotreme "mammals," the platypus and spiny "anteaters." The true mammals, which descended from the synapsian reptiles, also may have contributed to the demise of the reptile group by ecologically replacing the smaller carnivorous, insectivorous, and herbivorous types.

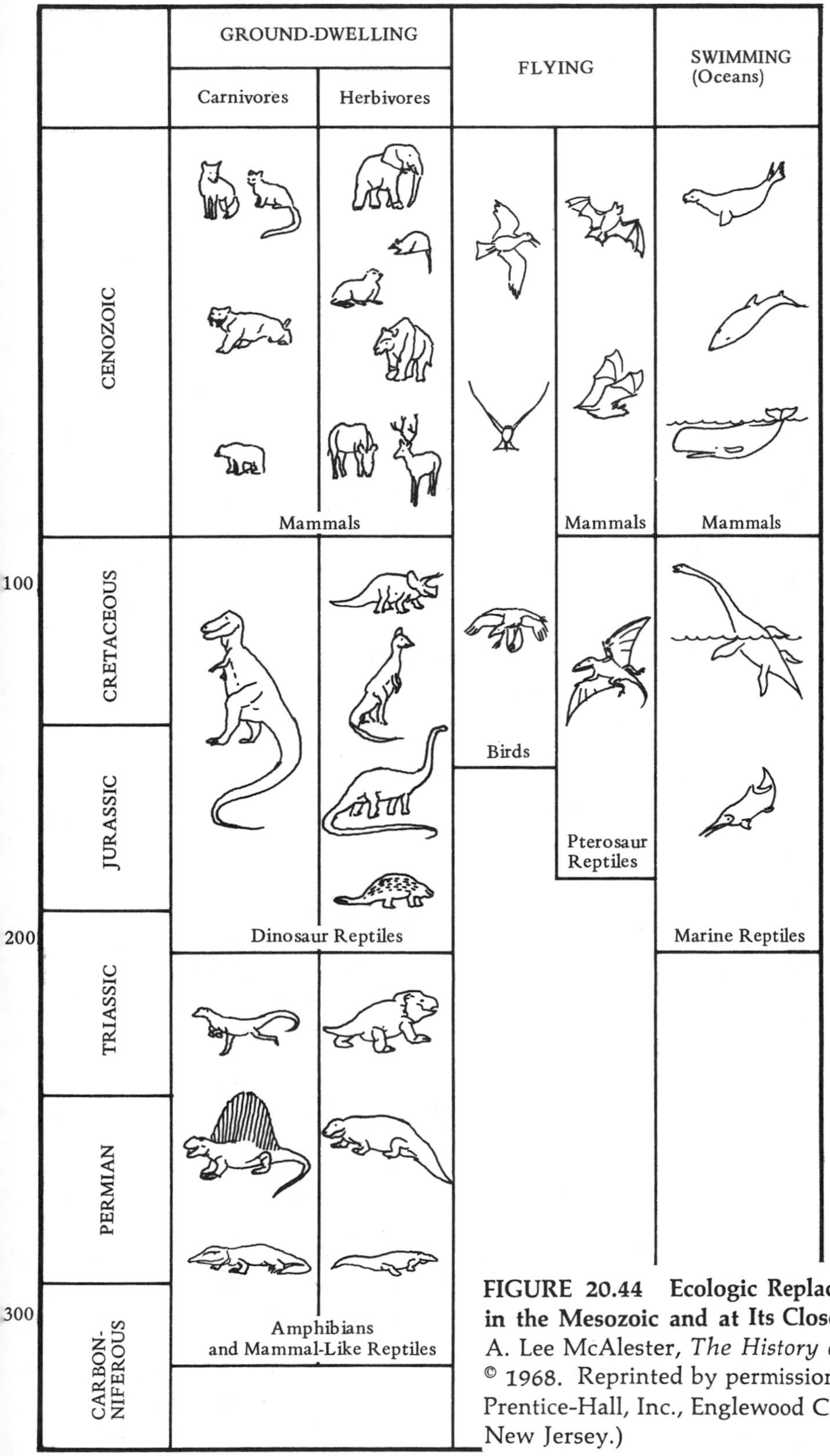

	GROUND-DWELLING		FLYING		SWIMMING (Oceans)
	Carnivores	Herbivores			
CENOZOIC	Mammals			Mammals	Mammals
100 CRETACEOUS	Dinosaur Reptiles		Birds	Pterosaur Reptiles	Marine Reptiles
JURASSIC					
200 TRIASSIC	Amphibians and Mammal-Like Reptiles				
PERMIAN					
300 CARBON-NIFEROUS					

FIGURE 20.44 Ecologic Replacements in the Mesozoic and at Its Close. (From A. Lee McAlester, *The History of Life*, © 1968. Reprinted by permission of Prentice-Hall, Inc., Englewood Cliffs, New Jersey.)

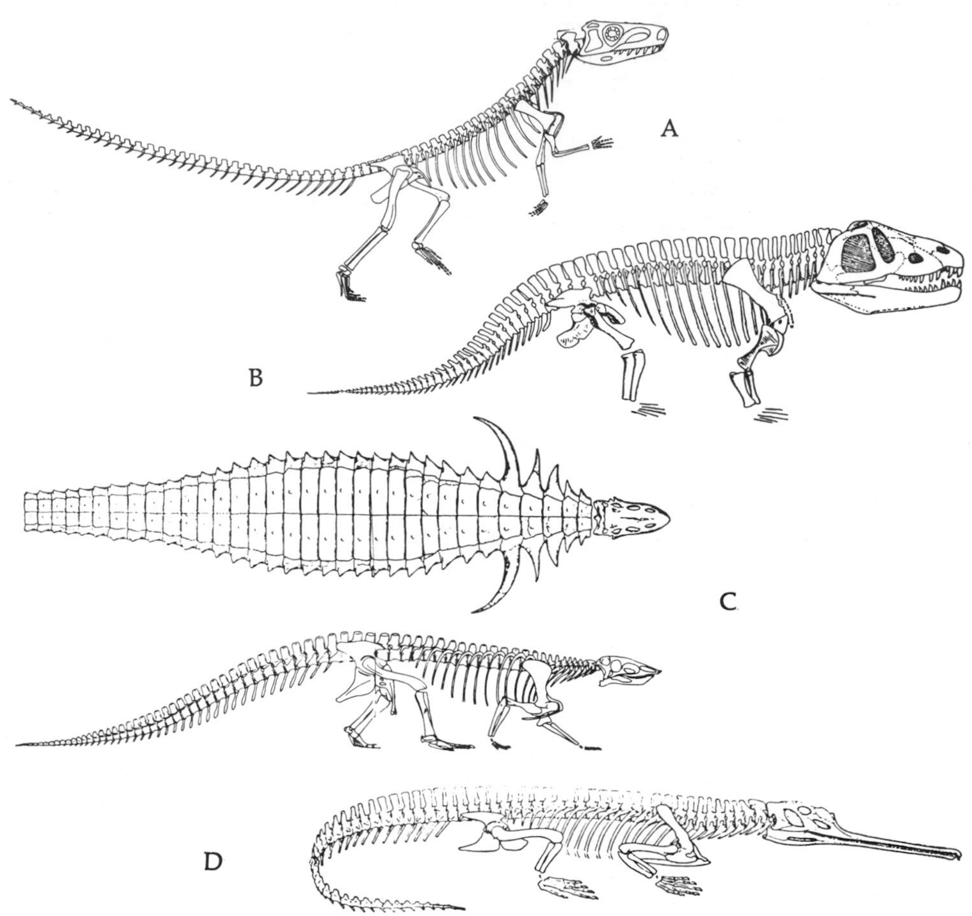

FIGURE 20.45 Thecodonts. (A) Pseudosuchian (Dinosaur ancestor), (B) Proterosuchian, (C) Aetosaurs, (D) Phytosaur. (From A. S. Romer, *Vertebrate Paleontology*, Copyright © 1966 by the University of Chicago Press. Used by permission of the publisher.)

The dinosaurs actually belong to two (or perhaps three) separate orders of reptiles. They arose independently from a group of reptiles called the *thecodonts* (Fig. 20.45). Thecodonts were common in the Triassic and included several adaptive types including the phytosaurs, the ecologic but not evolutionary predecessors of the crocodiles. Thecodonts, a diverse lot, gave rise to many reptilian groups in the Mesozoic including the two dinosaur orders and the crocodiles in the Triassic and the pterosaurs and the true birds in the Jurassic. The Triassic group of bipedal thecodonts called *pseudosuchians* (Fig. 20.45A) probably gave rise to the dinosaur orders as both types are primitively bipedal and the ancestral condition shows in the large hind limbs and pelvis of the later quadrupedal types (Fig. 20.46). The oldest members of both kinds of dinosaurs occur in Gondwanaland indicating they originated there, but dinosaurs spread to all continents before there was any appreciable fragmentation of the early Mesozoic continental unity.

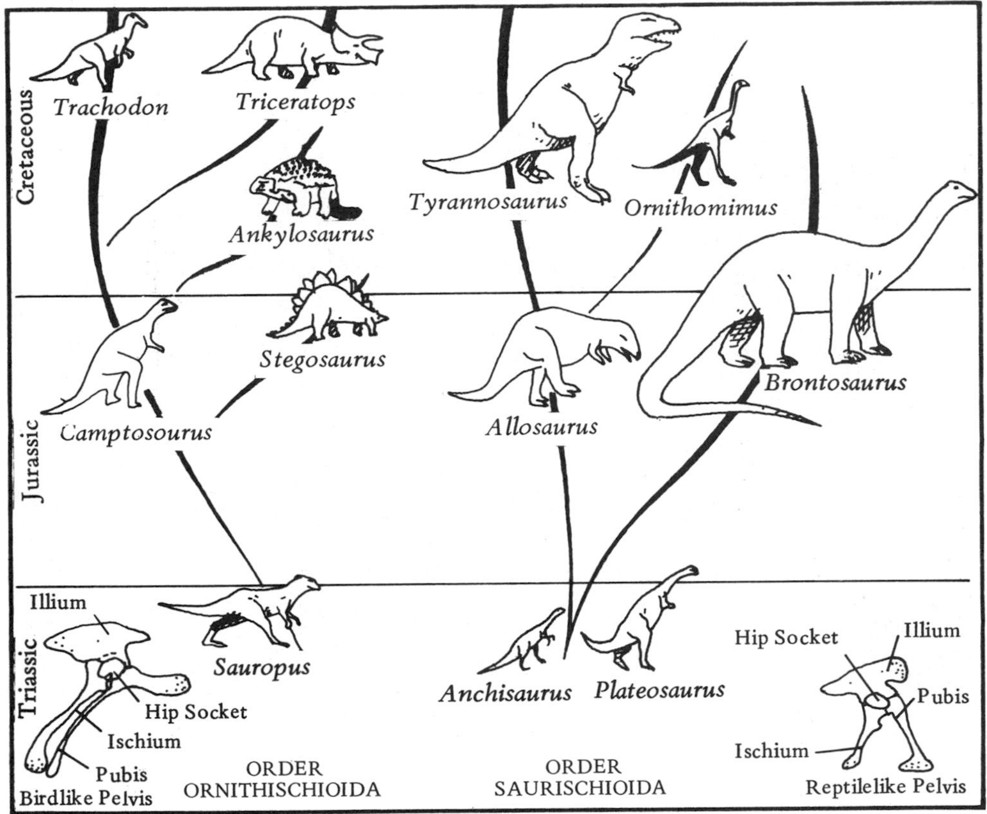

FIGURE 20.46 Dinosaur Evolution.

One group of dinosaurs which may not be a natural unit is called the *lizard* or *reptile-pelvis* type and is characterized by the typical triradiate reptilian pelvis (Fig. 20.46). Its early, Triassic members were the bipedal *prosauropods,* most of which were herbivorous, but some were carnivorous. From these arose in the Late Triassic–Early Jurassic interval the huge, quadrupedal, amphibious herbivores called *sauropods* and the large bipedal carnivores known as *theropods.* The sauropods were most abundant in the Jurassic and include the familiar genera *Apatosaurus, Brachiosaurus, Brontosaurus,* and *Diplodocus* (Fig. 20.47). The theropods were common in both the Jurassic and Cretaceous with *Allosaurus* a common Jurassic form and *Tyrannosaurus,* a significant Cretaceous genus (Fig. 20.48). There were also smaller and more slender theropods of ostrich-like proportions including *Coelurus* and the toothless, egg-eating *Ornithomimus* (Fig. 20.46).

The other dinosaur order is characterized by the possession of a *bird-like-pelvis* (Fig. 20.46) and probably arose from the same group of thecodonts that gave rise to the birds. There were four kinds of bird-pelvis dinosaurs, all apparently herbivorous. The bipedal *ornithopods* culminated in the amphibious Cretaceous hadrosaurs or duck-billed dinosaurs (Fig. 20.49). The armored *stegosaurs* (Fig. 20.50) were common in the Jurassic and the armored *ankylosaurs* (Fig. 20.49) in the Cretaceous. Both forms were quadrupeds. Also quad-

FIGURE 20.47 Jurassic Sauropods. (C. R. Knight mural, courtesy American Museum of Natural History.)

FIGURE 20.48 Cretaceous Ceratopsians and Theropods. (C. R. Knight mural, courtesy Field Museum of Natural History.)

FIGURE 20.49 Cretaceous Ornithopods, Ankylosaurs, Ostrich-Like Theropod Dinosaurs. (C. R. Knight mural, courtesy Field Museum of Natural History.)

rupedal were the great horned dinosaurs or *ceratopsians* (Fig. 20.48) of the Late Cretaceous. *Triceratops* is a common North American genus.

All the dinosaur groups became extinct by the end of the Cretaceous with theropods and ceratopsians persisting the longest. The Mesozoic-Cenozoic boundary is marked in terrestrial deposits by the remains of the highest dinosaurs. Their extinction still remains a mystery because there were no immediate, large mammalian successors, thus ruling out superior competitors as a possible cause of the extinction. In a few areas, such as the Bug Creek region of Montana, a transitional sequence across the boundary occurs. This sequence reveals a gradual reduction in dinosaur abundance and a concomitant increase in mammalian remains. There is also a progressive change in the plant communities in these strata. Such evidence suggests that the food chain that dinosaurs dominated may have been gradually replaced by a food chain based on different plants which interacted with the rapidly increasing modern types of insects and small seed- and insect-eating mammals. The diminution of the dinosaurs through the whole Late Cretaceous interval suggests that some sort of ecologic change such as this was occurring. On the other hand, it is hard to reconcile the complete disappearance of a diversified group such as the dinosaurs with the slow replacement of plant communities. It also fails to explain the simultaneous extinction of all the successful groups of large reptiles that invaded the seas in the Mesozoic. Only something that would cut across all environments and food chains could cause the Late Mesozoic extinction.

A change in the atmospheric O_2 and CO_2 ratio could have triggered such a reaction and is also an attractive hypothesis because it could explain the simultaneous extinction of the ammonoids. As stated earlier, the appearance and great increase in abundance of the large groups of modern photosynthesizers or their acquisition of the ability to secrete shells in the Late Mesozoic could have caused large-scale changes in the ratio of the gases. The exact effect on the reptiles is uncertain, but it could have affected them because their sluggish, cold-blooded metabolism was not suited to high oxygen levels, or it could have altered the earth's heat budget through changes in the greenhouse effect. A cooling trend at the end of the Mesozoic is indicated by several lines of geologic, paleontologic, and geochemical evidence and this could have been initiated by a large decrease in the CO_2 content of the atmosphere which controls the greenhouse effect.

Other, chiefly terrestrial, reptilian groups such as the turtles, lizards, snakes, rhynchosaurs, and crocodiles arose in the Mesozoic and were occasionally common but less spectacular than the preceding groups. It is notable, however, that these smaller reptiles are the ones that survive today.

Aerial Vertebrates

The thecodonts also gave rise to two groups of flying vertebrates. The flying reptiles or *pterosaurs* (Figs. 20.35, 20.51, 20.52) appeared in the Jurassic, their exact ancestry unknown, and culminated in the great seagoing toothless *Pteranodon* of the Cretaceous. Some individuals of this species had a wing span in excess of 25 feet. The pterosaurs died out near the close of the Cretaceous. In the Jurassic, another group of thecodonts gave rise to the *birds*. The birds, with their warm-blooded, feathered condition, underwent a great radiation that con-

FIGURE 20.50 Jurassic Stegosaurs. (C. R. Knight mural, courtesy American Museum of Natural History.)

FIGURE 20.51 Jurassic Life in Solenhofen Area of Germany. Pterosaurs or flying reptiles above, earliest birds in middle, small dinosaurs and arthropod below. Plants are chiefly cycadeoids. (C. R. Knight mural, courtesy Field Museum of Natural History.)

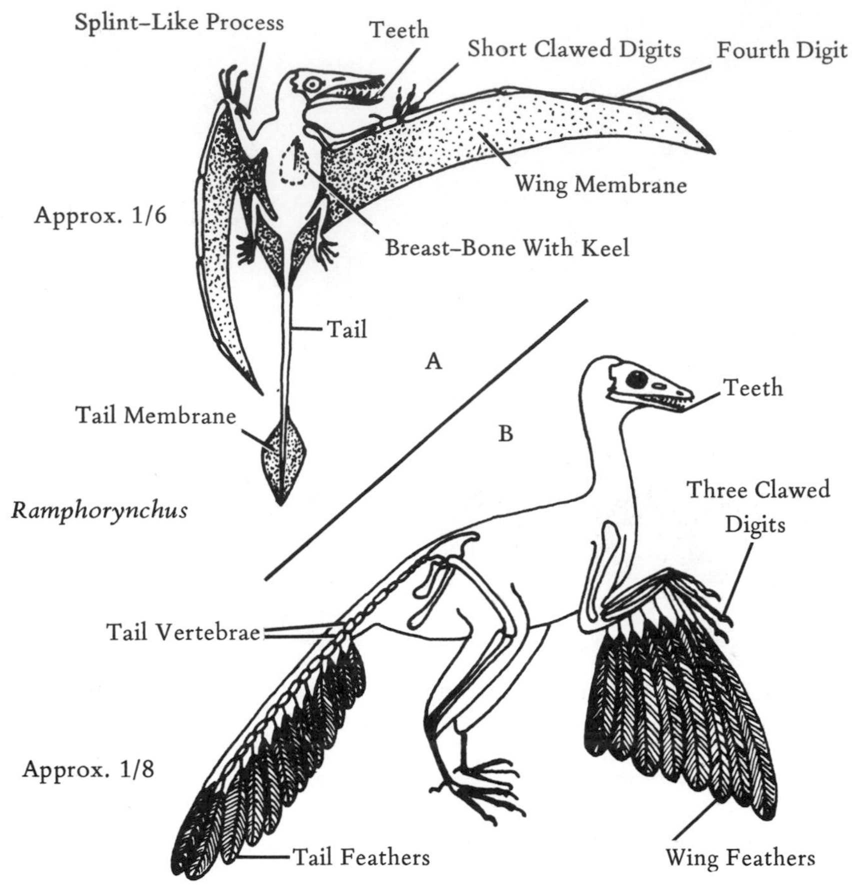

FIGURE 20.52 Mesozoic Flying Vertebrates. (A) Jurassic Pterosaur. (B) Jurassic Bird. (From R. M. Black, *The Elements of Paleontology*, Copyright © 1970, The Cambridge University Press.)

tinues to the present day, thus entitling them to rank as a separate vertebrate class. The Jurassic "bird," *Archaeopteryx* (Figs. 20.51, 20.52), is actually a connecting link between the two classes. With its teeth, claws, long bony tail, and lack of hollow bones and a beak, its skeleton is that of a reptile and not a bird. But the impressions of feathers covering the body and forming most of the wings indicate that this gliding creature was well on its way to becoming a typical bird. In the Cretaceous, a few toothed birds persisted (Fig. 20.53), but the radiation of modern orders had begun. This event probably caused the extinction of the pterosaurs which were unable to compete with the better-adapted birds.

Mesozoic Mammals

The *mammalian* class began to diversify after its Late Triassic origin from small, carnivorous, mammal-like reptiles. Carnivores, insectivores, and seed-eaters ap-

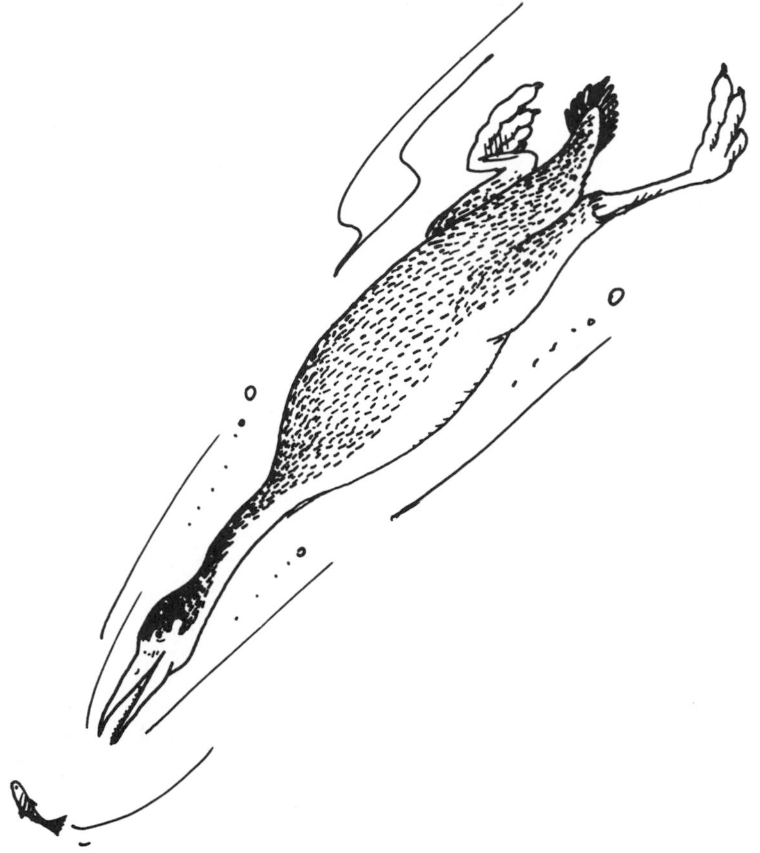

FIGURE 20.53 Cretaceous Toothed Bird *Hesperornis.*

peared, but most of Jurassic and Early Cretaceous forms were members of extinct orders representing early radiations (Figs. 20.54, 20.55). By the Late Cretaceous, however, both the pouched *marsupials* and the *placental mammals* were diversifying into their modern orders at least five of which appeared before the end of the period (Fig. 20.56). The condition of the earth at the time of the mammalian radiation was quite different from that of the Late Paleozoic and Early to Middle Mesozoic when the reptiles underwent their radiation. In that interval, the continents were all together or in close proximity allowing the rapid and easy intercommunication between distant areas that resulted in the great cosmopolitanism of Carboniferous through Early Cretaceous terrestrial life (Fig. 20.57). As we saw in the preceding chapter, however, the Late Cretaceous was a time of extreme fragmentation and isolation, a condition which persisted throughout the Paleogene and into the Neogene (Fig. 20.58). The Early to Middle Cretaceous mammals were able to reach all the continents, but thereafter their evolution was able to occur in isolation producing several distinct provincial faunas such as those of South America, Africa, Australia, and the three northern continents, North America, Europe, and Asia, which were occasionally in contact and, hence, had faunal interchange.

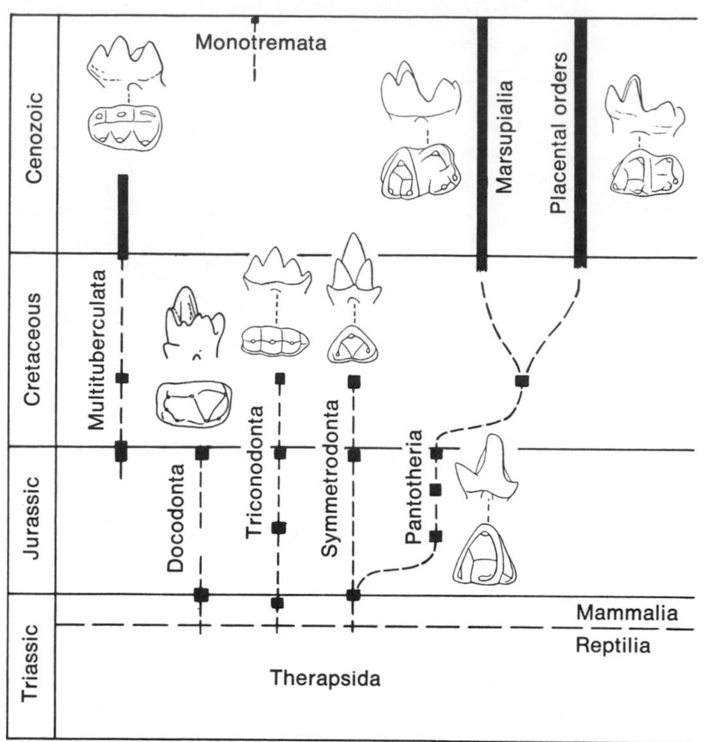

FIGURE 20.54 Teeth and Inferred Relationships of Mesozoic Mammals. (From B. Kummel, *History of the Earth*, W. H. Freeman and Co., Copyright © 1970, Second Edition.)

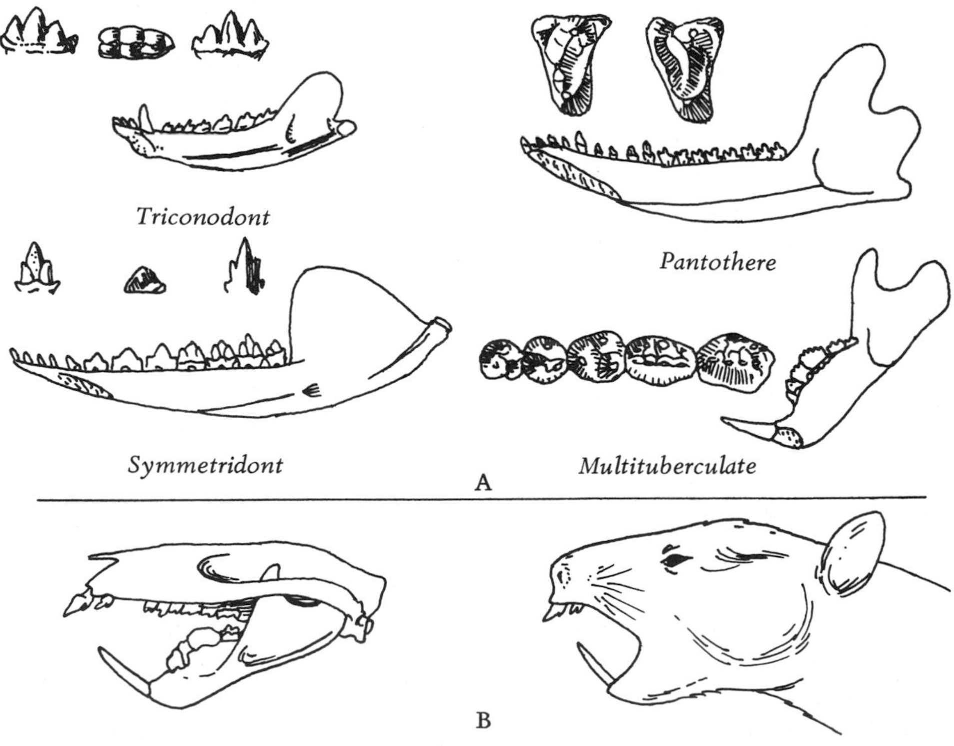

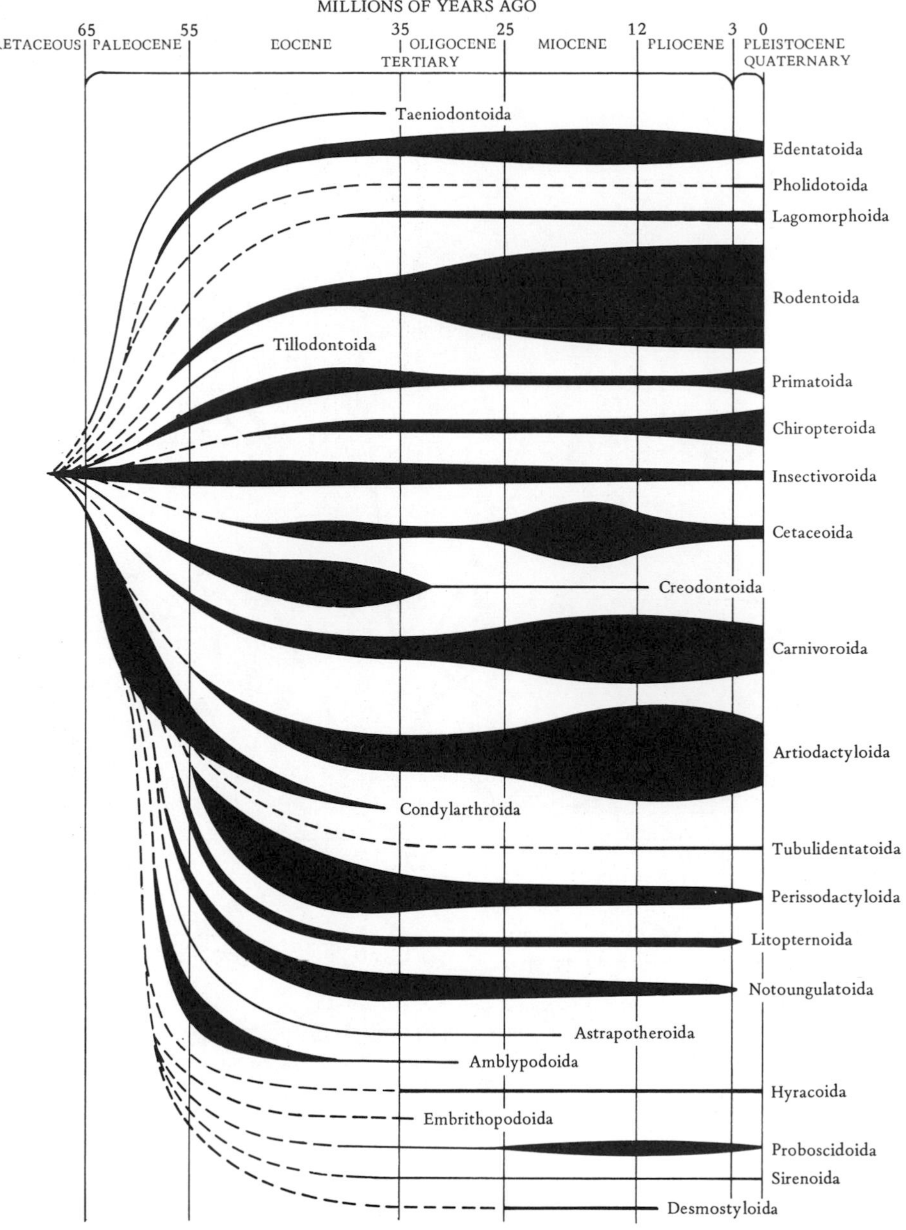

FIGURE 20.56 The Mammalian Radiation. Note the beginnings of most mammalian orders in the latest Mesozoic or earliest Cenozoic. (After Kurten.)

FIGURE 20.55 Mesozoic Mammals. (A) Teeth and jaws of major Mesozoic mammal groups, (B) Skull and restoration of a Mesozoic mammal.

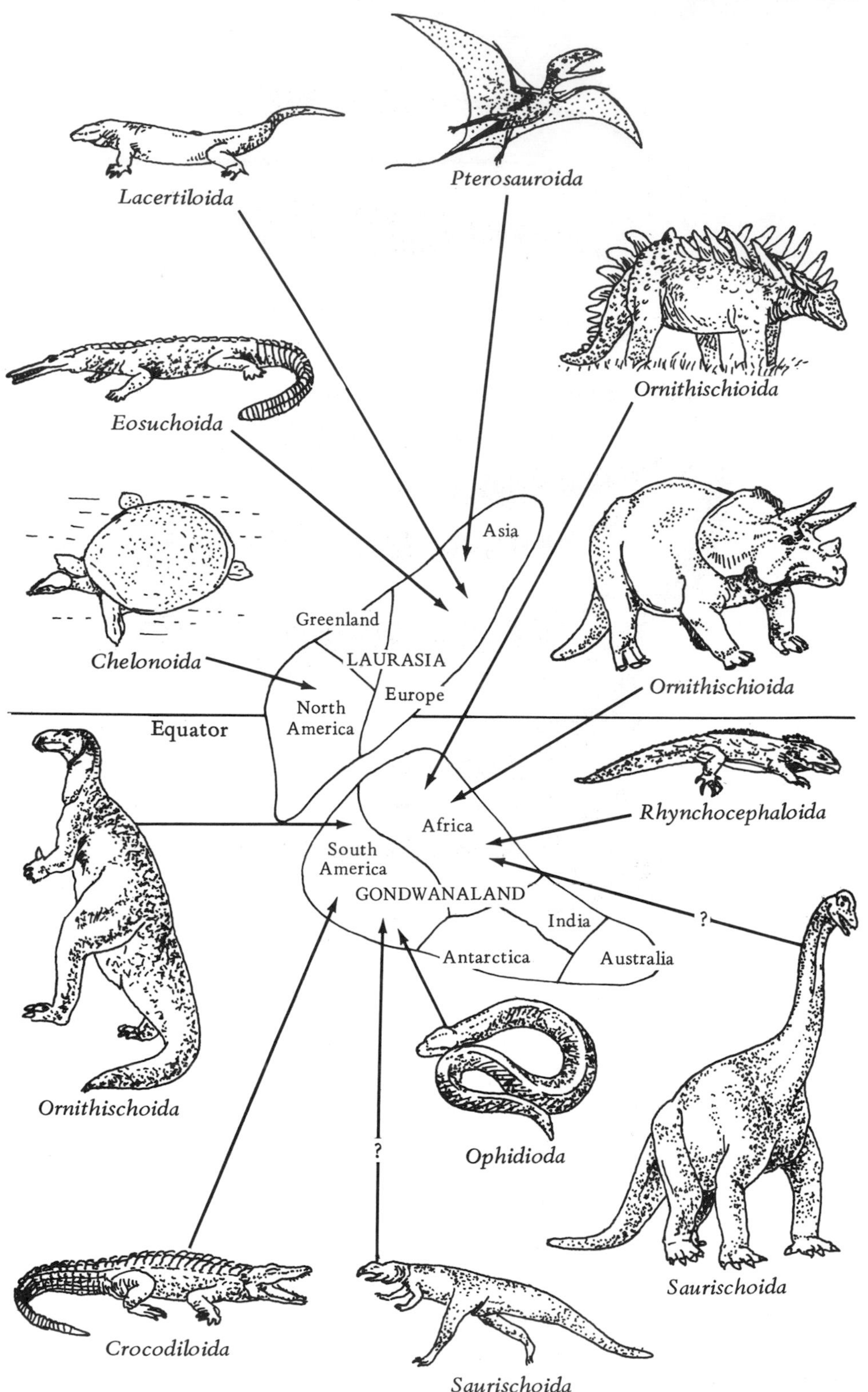
Lacertiloida
Pterosauroida
Ornithischioida
Eosuchoida
Asia
Greenland
LAURASIA
Europe
North America
Chelonoida
Ornithischioida
Equator
Africa
Rhynchocephaloida
South America
GONDWANALAND
India
Antarctica
Australia
?
Ornithischoida
Ophidioda
Saurischoida
Crocodiloida
Saurischoida

Plants

The plant life of the Mesozoic Era was characterized by two distinct floras, the gymnosperm flora which continued from the Permian and persisted into the Early Cretaceous, and the angiosperm flora which became dominant in the Middle Cretaceous and has persisted to the present. The gymnosperms or naked-seed plants have no fruit covering their seeds. They include the *conifers, cycadophytes,* and *gingkos,* the major elements of the plant world through much of the Mesozoic (Figs. 20.59, 20.60). The conifers had attained dominance of the plant world in the Permian and continued in this status through much of the Mesozoic. The large Triassic *Araucarioxylon* (Fig. 20.61), so abundant in the Petrified Forest National Park of Arizona, represents a group now restricted to the southern continents by continental drift. This freed it from competition with the more advanced conifers such as the pines and sequoias that evolved in the northern continents later in the Mesozoic. Cycadophytes, which arose in the Carboniferous, became very abundant in the Mesozoic. Both the surviving cycads and the totally Mesozoic cycadeoids were present. *Williamsonia* is a common Triassic-Jurassic genus of cycadeoid and *Cycadeoida* is a characteristic Jurassic-Cretaceous form. *Bjuvia* is a Triassic cycad. The gingkos also arose in the Late Paleozoic, but reached a Mesozoic peak. *Baiera* is a typical Mesozoic genus of these large trees. The undergrowth of this Mesozoic flora consisted chiefly of ferns and some peculiar seed ferns. The latter group died out in the Cretaceous.

The origin of *angiosperms* is uncertain. Some Carboniferous and Jurassic pollen resembles that of angiosperms, a Triassic leaf is apparently that of a palm (Fig. 20.62), and palm wood is known from rocks of possible Jurassic age in Utah. Because flowers and fruits, two of the most diagnostic parts of angiosperms, are rarely preserved as fossils, it is virtually impossible to identify the ancestors of the group. In addition, it appears that angiosperms may have arisen in upland areas where deposition is not the rule and, hence, could only be represented in the preserved lowland record by pollen. Unquestioned angiosperm leaves, apparently belonging to modern genera, appeared in the Early Cretaceous (Fig. 20.63). By the Middle and Late Cretaceous, angiosperms dominated the earth's floras (Fig. 20.64), either replacing or forcing into more vigorous climatic regions the other plant divisions. The rise of the angiosperms was undoubtedly closely coupled with that of the pollenizing insects. Indeed, the angiosperm-insect relationship is probably the most important ecologic interaction in terrestrial ecosystems today. The Late Mesozoic floral change may have contributed to the extinction of the dinosaurs, but it apparently can't explain the extinction of marine reptiles and ammonoids unless it affected atmospheric gas ratios. The floral changeover on land may have either triggered or been triggered by changes in the O_2–CO_2 ratio. It certainly appears to be more than coincidence that the major

FIGURE 20.57 Conditions at the Beginning of Reptilian Radiation. The continents were concentrated in two large supercontinents which in themselves were interconnected. The faunas were cosmopolitan, but the inferred continent of origin is indicated.

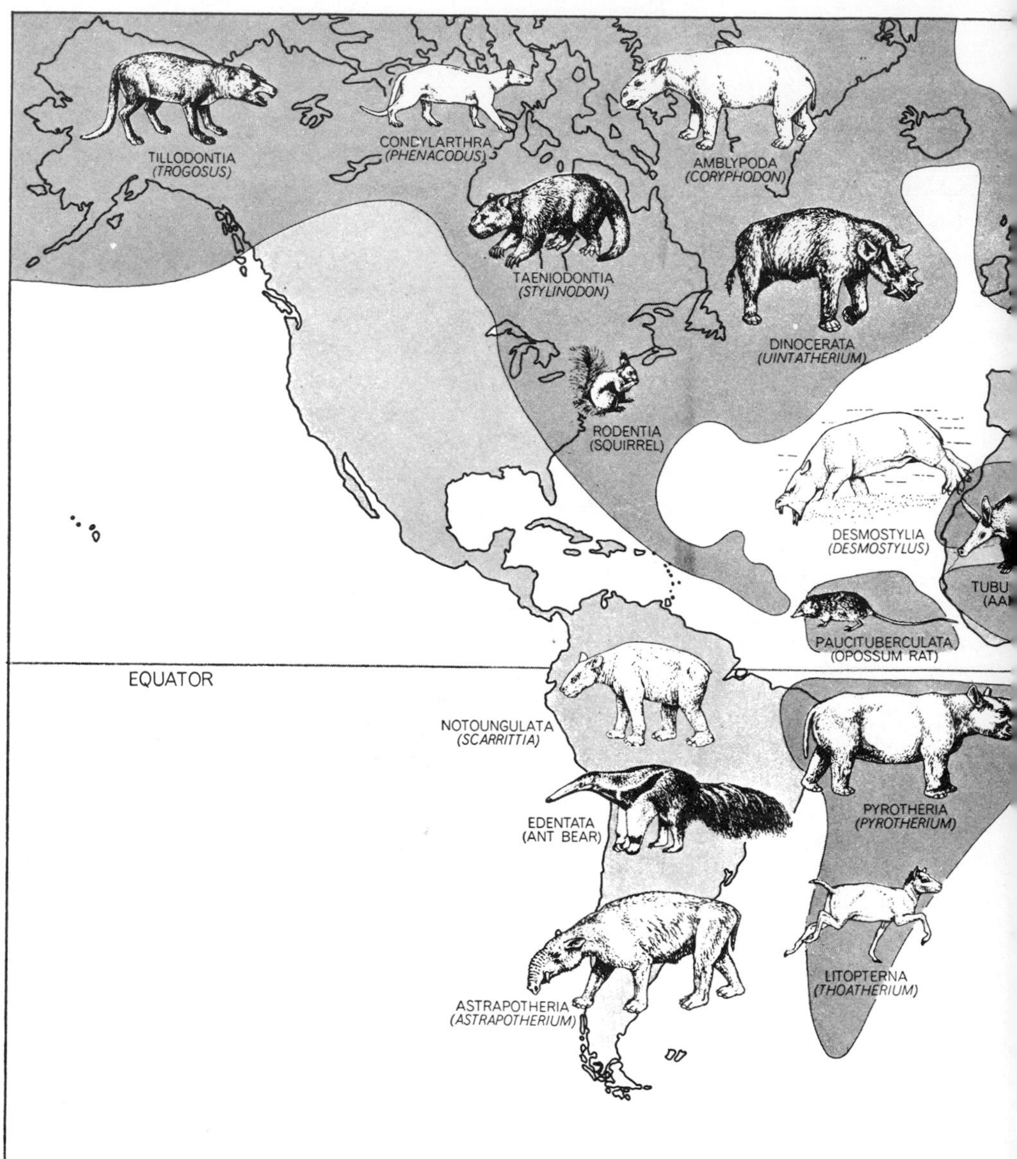

FIGURE 20.58 Continental Drift and Mammalian Evolution. (From *Continental Drift and Evolution* by Bjorn Kurten,)

producers of the earth today, both on land and in the sea, reached their present dominance in the Late Cretaceous.

Palynology is a subscience of paleontology that deals with fossil spores and pollen. Fossil pollen became common in the Late Jurassic and has continued to increase down to the present (Fig. 20.65) making palynology an important discipline.

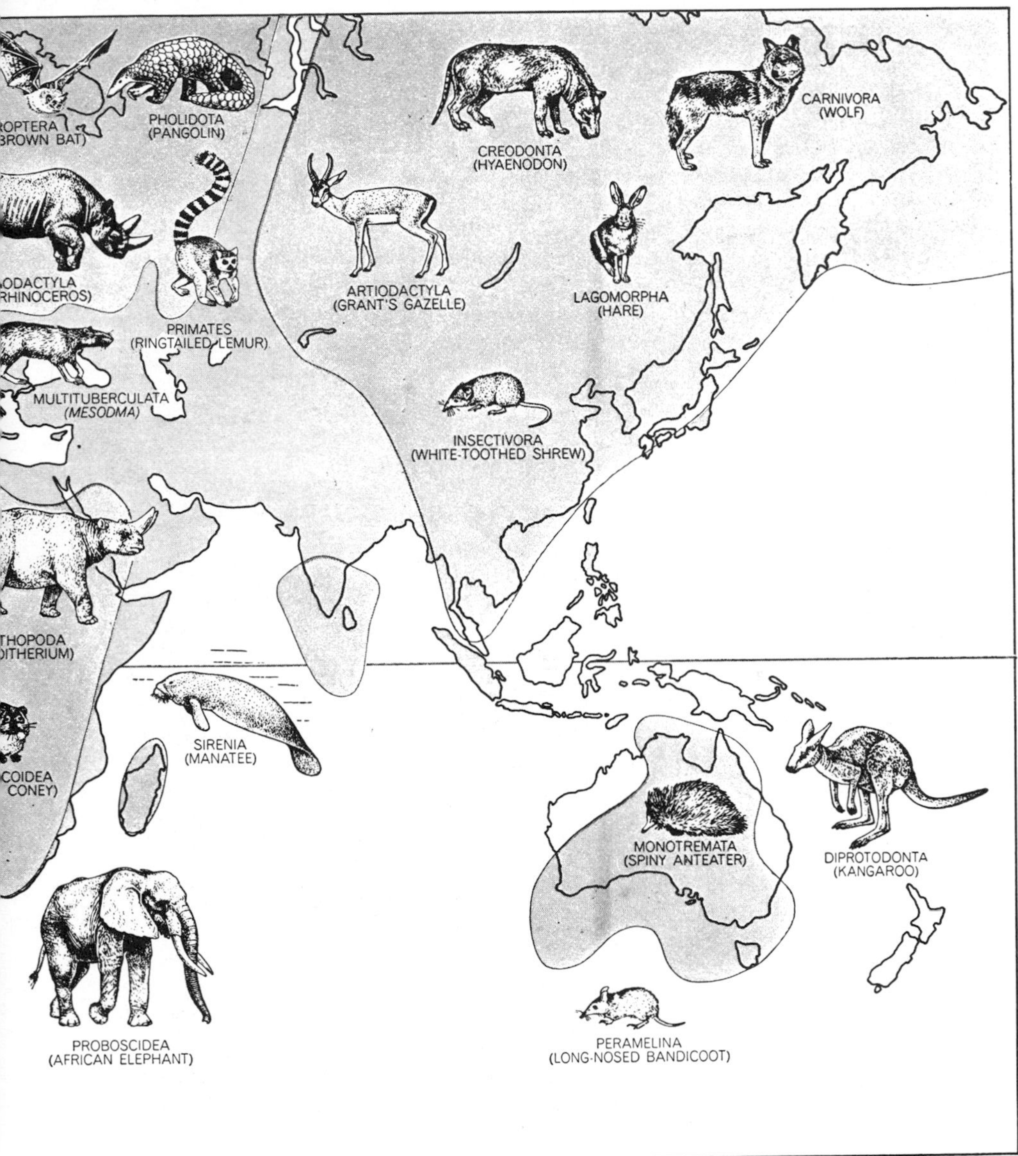

Solnhofen Fossils

Before closing a discussion of Mesozoic life, it is necessary to mention one of these rare fossil occurrences that gives us a glimpse of the soft-bodied life of the past so rarely preserved in the geologic record. Near Solnhofen, in Southern Germany, are deposits of a very fine-grained lithographic limestone that originally formed as a chemical precipitate in quiet lagoons along the shores of the regressing Late Jurassic epeiric sea (Fig. 20.51). Into this quiet lime-mud environment settled the remains of jellyfish, insects, pterosaurs, birds, and many other generally unpreserved forms. The body outlines showing jellyfish tentacles, insect wings, pterosaur wing membranes, and the feathers of the oldest birds are preserved as impressions (Fig. 20.66) in this remarkable deposit.

FIGURE 20.59 Mesozoic Flora. Note dominance of cycadeoids.

FIGURE 20.60 Mesozoic Flora of Late Triassic Age. Note gingkos, conifers, cycadeoids and ferns.

FIGURE 20.61 Petrified *Araucarioxylon* **logs.** Petrified Forest National Park, Arizona. (Courtesy of William M. Mintz.)

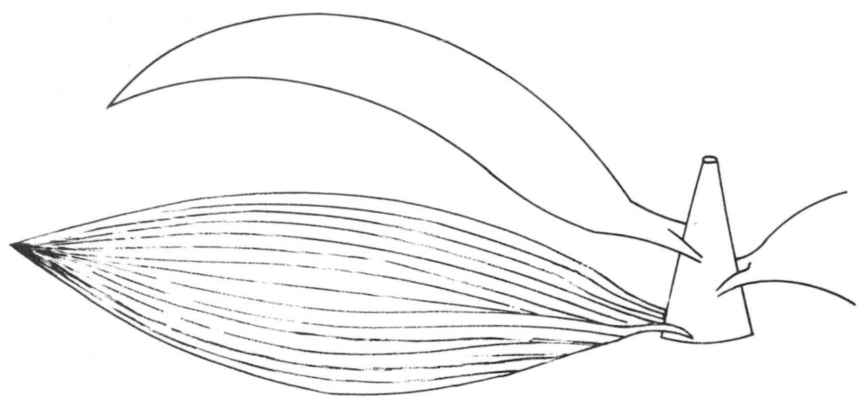

FIGURE 20.62 *Sanmiguelia,* **a possible Triassic palm.**

FIGURE 20.63 Cretaceous Angiosperm Leaves.

FIGURE 20.64 Late Cretaceous Angiosperm Flora. Based on Greenland fossils.

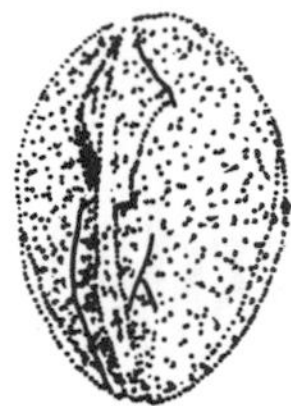

U. Juras.

U. Cretaceous

FIGURE 20.65 Mesozoic Pollen Grains.

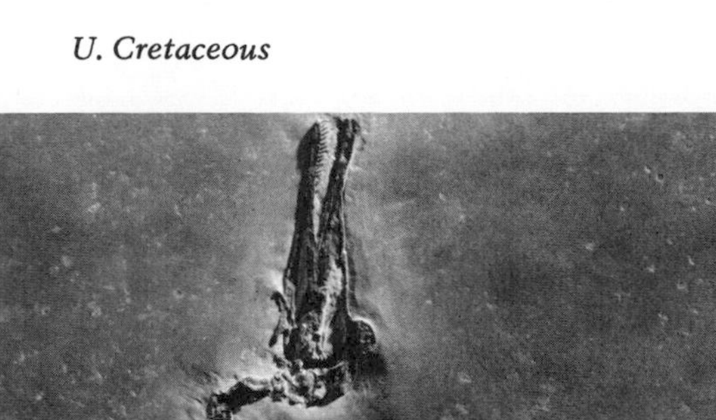

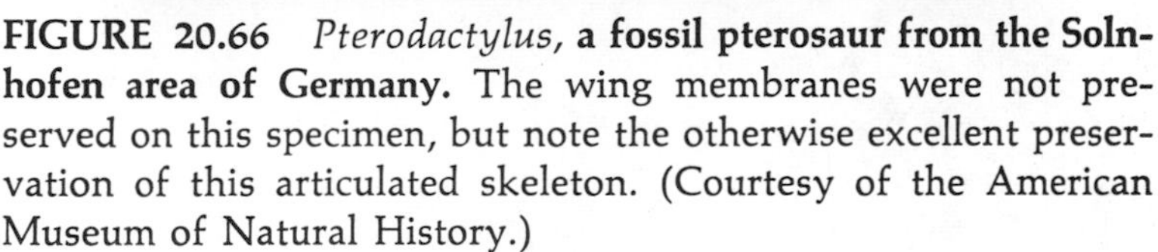

FIGURE 20.66 *Pterodactylus,* **a fossil pterosaur from the Solnhofen area of Germany.** The wing membranes were not preserved on this specimen, but note the otherwise excellent preservation of this articulated skeleton. (Courtesy of the American Museum of Natural History.)

Questions

1. What major modern groups of oceanic producers appeared in the Mesozoic Era?
2. Where and when were dasycladaceous green algae significant in the Mesozoic?
3. What significant ecologic breakthrough was made by the foraminifera in the Late Mesozoic?
4. What was the major group of reef-building corals in the Mesozoic?
5. Which three classes of mollusks dominate Mesozoic faunas?
6. Why are the ammonoids so biostratigraphically significant?
7. What types of sutures do most Mesozoic ammonoids possess?
8. What is the major Mesozoic group of echinoderms?
9. Why is the Mesozoic Era called the Age of Reptiles?
10. What significant extinctions occurred near the end of the Mesozoic and what are some possible causes?
11. Discuss the evolution of the major dinosaur groups.
12. What were the major plant groups of the Mesozoic and what major changeover occurred in the Cretaceous?
13. What two major vertebrate classes originated in the Mesozoic?

21

Just Yesterday:

The Cenozoic Era

The Cenozoic Era encompassed the last 65 million years of earth history, the time when the present physical features, plate distribution, and organisms of the planet developed. The Cenozoic Era is usually subdivided into seven epochs, from oldest to youngest, *Paleocene* (11 million years), *Eocene* (17 million years), *Oligocene* (11 million years), *Miocene* (16 million years), *Pliocene* (7 million years), *Pleistocene* (3 million years), and *Holocene* (10,000 years). The type areas of the first four are situated on the sites of marginal shelf seas in northern Europe. The Paleocene type area is in northern France; that of the Eocene, in the Paris and London Basins; the Oligocene, in the Hanover Basin of northern Germany; and the Miocene, in western France. The Pliocene and Pleistocene type areas are situated in the Tethyan Geosyncline of Italy, the former in the north and the latter in the south. The Holocene or Recent has no type area; it encompasses the whole present surface of the earth.

The distribution of these epochs into periods, however, is not universal (Fig. 21.1). Most American and many European geologists use the older terms *Tertiary* and *Quaternary* with the boundary located between the Pliocene and Pleistocene. The type area of the Tertiary is that of the Pliocene in northern Italy, while the type Quaternary is in northern France. This separation is very unequal, for the Quaternary only encompasses about 3 million years, but it does emphasize the striking worldwide Late Cenozoic glaciations. Geologists using these terms emphasize that the whole era is so short, including less time than the Cretaceous Period, that other period subdivisions are not needed. Most European and some American geologists on the other hand, prefer the *Paleogene* (or *Nummulitic*) and *Neogene* Periods as subdivisions of the Cenozoic. In this case, the boundary falls between the Oligocene and Miocene and is near the climax of the orogeny in the Tethyan Geosyncline. This subdivision gives periods of more equal length, 39 million and 26 million years respectively. There are no type areas of these two systems, only the type areas of the epochs assigned to them. Recently some geologists have developed a compromise that incorporates the best features of both. In this scheme, Paleogene, Neogene, and Quaternary Periods are used, the first including the Paleocene, Eocene, and Oligocene, the second, the Miocene and Pliocene, and the third, the Pleistocene and Holocene. Thus, the long Tertiary, which was almost synonymous with Cenozoic, is subdivided and yet the unique-

CENOZOIC TIME SCALES

<table>
<tr><th>Periods</th><th>Epochs</th><th>Periods</th></tr>
<tr><td rowspan="2">Quaternary
3 Million
Years Long</td><td>Holocene</td><td rowspan="4">Neogene
26 Million
Years Long</td></tr>
<tr><td>Pleistocene</td></tr>
<tr><td rowspan="5">Tertiary
62 Million
Years Long</td><td>Pliocene</td></tr>
<tr><td>Miocene</td></tr>
<tr><td>Oligocene</td><td rowspan="3">Paleogene
39 Million Years
Long</td></tr>
<tr><td>Eocene</td></tr>
<tr><td>Paleocene</td></tr>
</table>

FIGURE 21.1 Cenozoic Time Scales.

ness of the late Cenozoic ice age is recognized. In our discussion, we will discuss the history as it unfolds epoch by epoch and leave the debate over periods to the specialists.

Plate Positions and Movements

In the Cenozoic Era, the plates gradually assumed their present form and positions (Fig. 21.2). The northern continents of North America and Eurasia clustered

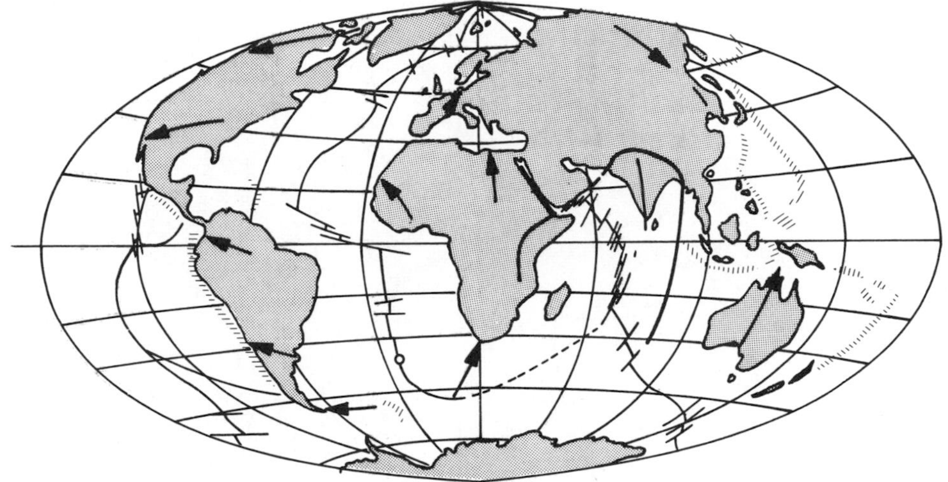

FIGURE 21.2 The Present Plate Configuration.

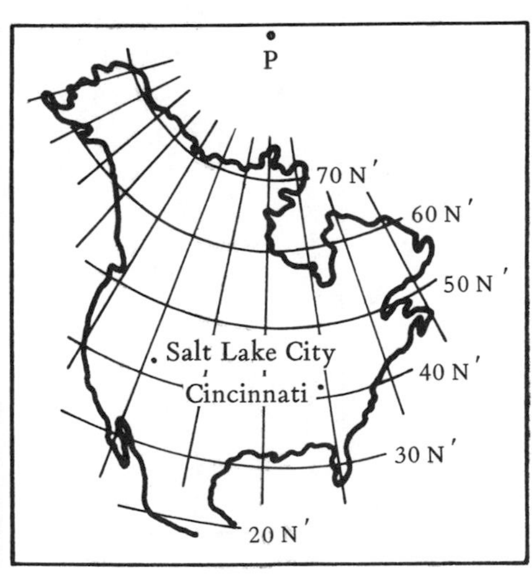

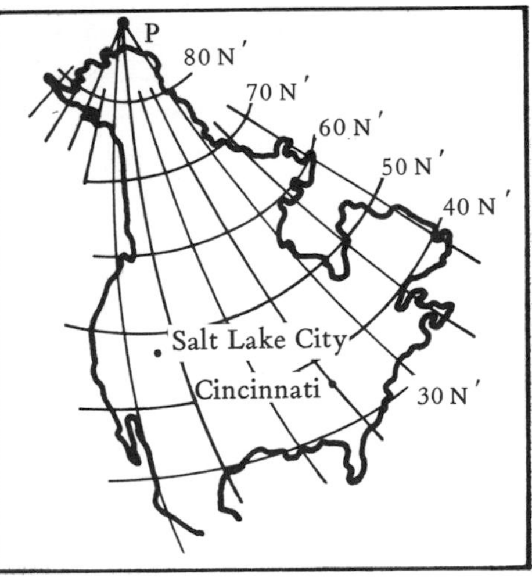

FIGURE 21.3 Paleolatitudes for the Cretaceous and Late Tertiary. (From E. Irving, *Paleomagnetism and its Application to Geological and Geophysical Problems,* Copyright © 1964, by permission of John Wiley & Sons, Inc.)

around the North or Pacific Pole which became situated in the Arctic Sea (or Ocean). More of western North America was rotated northward into cooler zones as the Cenozoic progressed (Fig. 21.3). Greenland, part of the American plate, began to separate from Norway in the Paleocene, thus opening an oceanic connection between the Arctic Sea and the Atlantic Ocean. Eventually, as this rift widened and the rotation of North America largely closed off the Arctic from the

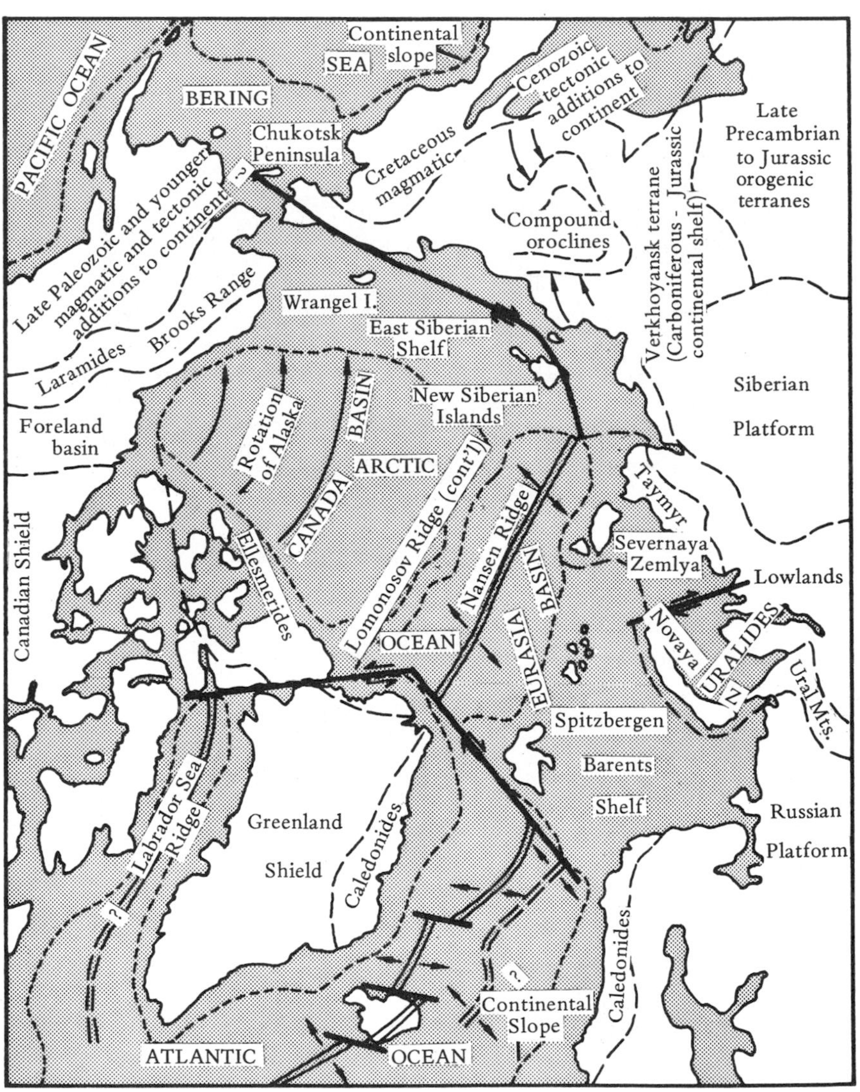

FIGURE 21.4 Tectonic Elements and Presumed Plate Motions in the Arctic Ocean Region. (From Hamilton, *G.S.A. Bulletin*, vol. 81, Sept. 1970. Used by permission of the author and G.S.A.)

Pacific, the Arctic Sea became a northward extension of the Atlantic (Fig. 21.4). Large-scale strike-slip faulting has also apparently affected the Arctic. North America probably collided with Asia somewhere in the Alaska–East Siberian region, but geologists still disagree as to the boundary between the North American and Eurasian Plates. The geology of this region is very difficult to decipher as it is the point of collision of at least three plates, the Eurasian, American, and Pacific, and possibly another, called the Kula, which has disappeared into the Aleutian Trench (Fig. 21.5). Toward the close of the Eocene, the present Bay of Biscay opened in western Europe and the Iberian Peninsula (present Spain and Portugal) rotated counterclockwise to the southeast. The collision of Iberia with Europe produced the present Pyrenees Mountains. This opening allowed the seas to spread into western France in the type area of the Miocene (Fig. 21.6). The island of Iceland (Fig. 21.2), between Greenland and Norway, was born from volcanic eruptions on the Mid-Atlantic Ridge in the Miocene. In the Early Cenozoic, the western boundary of North America was a trench in which the small Farallon Plate, spreading eastward from the East Pacific Rise, was being consumed (Fig. 21.7). In the Late Oligocene, western North America began encountering the spreading ridge (Figs. 21.7, 21.8) and the boundary between the Pacific and North American Plates became a transform fault, the present San Andreas. This fault expanded with time as the Farallon Plate disappeared. The formation of a transform fault boundary in western North America apparently converted the whole western Cordillera into a broad zone of transform faulting which triggered new unrest throughout the area. One of the most spectacular pieces of evidence of these latter features was the opening of the Gulf of California as Baja California split off North America.

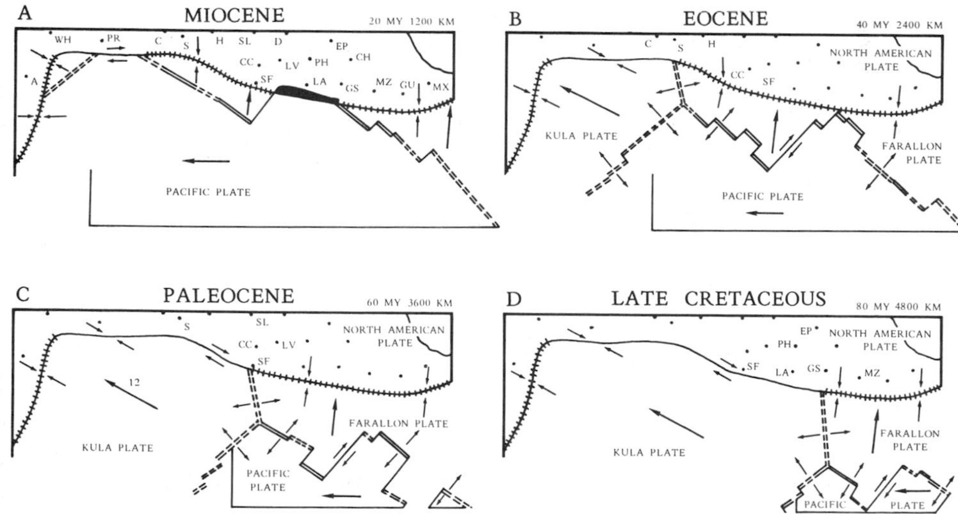

FIGURE 21.5 Plate Movements in the Northeast Pacific in the Cenozoic Assuming Constant Motions. (From Atwater, *G.S.A. Bulletin*, vol. 81, December, 1970. Used by permission of the author and G.S.A.)

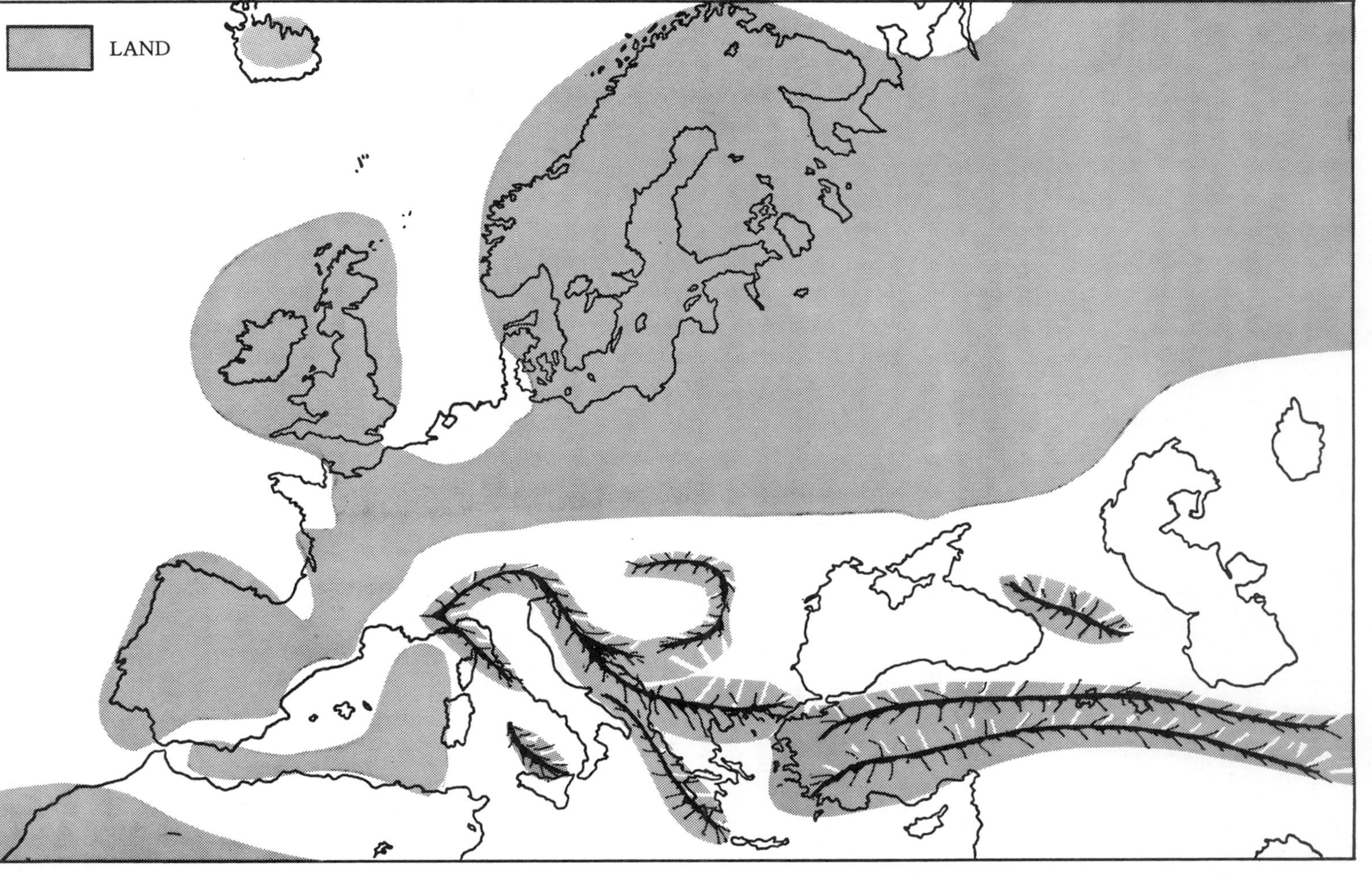

FIGURE 21.6 Paleogeographic Map of Europe in the Miocene. (Plate motions not considered in preparation of map.)

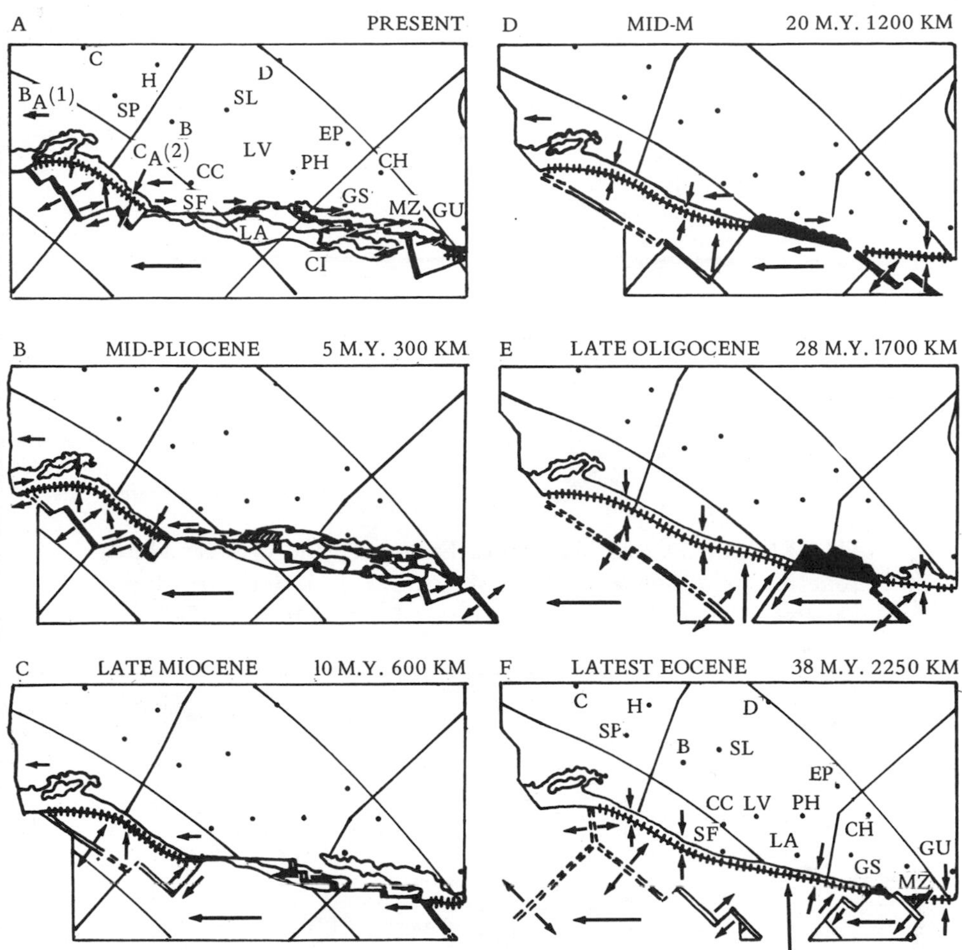

FIGURE 21.7 Plate Movements Off Western North America in the Cenozoic. Constant motion of 6 cm/year is assumed. (From Atwater, *G.S.A. Bulletin*, vol. 81, December, 1970. Used by permission of the author and G.S.A.)

Gondwanaland continued its fragmentation in the Cenozoic (Fig. 21.2). South America and Africa further separated as the Atlantic Ocean basin widened. Africa itself started to be sundered by the development of the great rift valleys, beginning in the Oligocene. Arabia has already split off and an incipient ocean basin, the Red Sea, has formed. Eastern Africa is now becoming separated from the remainder of the continent as the rift valleys, marking a new plate junction, develop there. Australia and Antarctica began their separation in the Eocene with Australia shifting northward into temperate and tropical climates, while Antarctica became centered on the Gondwana or South pole. The opening of the direct Indian and Pacific Ocean link in the Cenozoic completed the development of the present world-encircling ocean in high latitudes of the Southern Hemisphere. The collision of the African, the incipient Arabian, and the Indian plates with Eurasia in the Middle Cenozoic caused the great deformation of the Tethyan Geosyncline known as the *Alpine Orogeny.*

A high rate of ocean spreading in the southeastern Pacific not only continued the Andean deformation but caused the development of the Central American

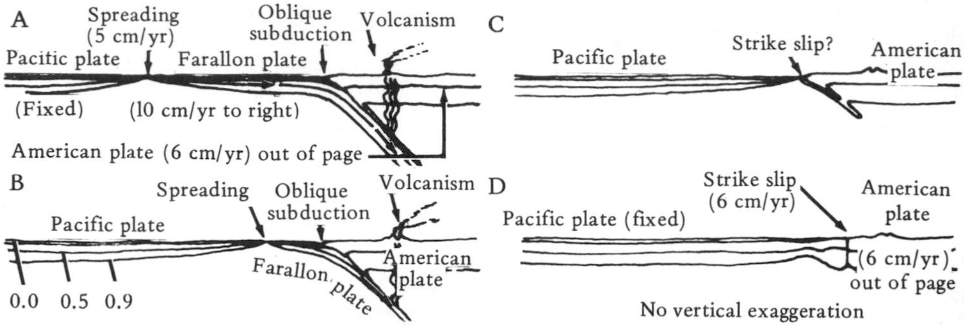

FIGURE 21.8 Cross-Sections of Eastern Pacific-Western North America in the Cenozoic. (From Atwater, *G.S.A. Bulletin,* vol. **81,** December, 1970. Used by permission of the author and G.S.A.)

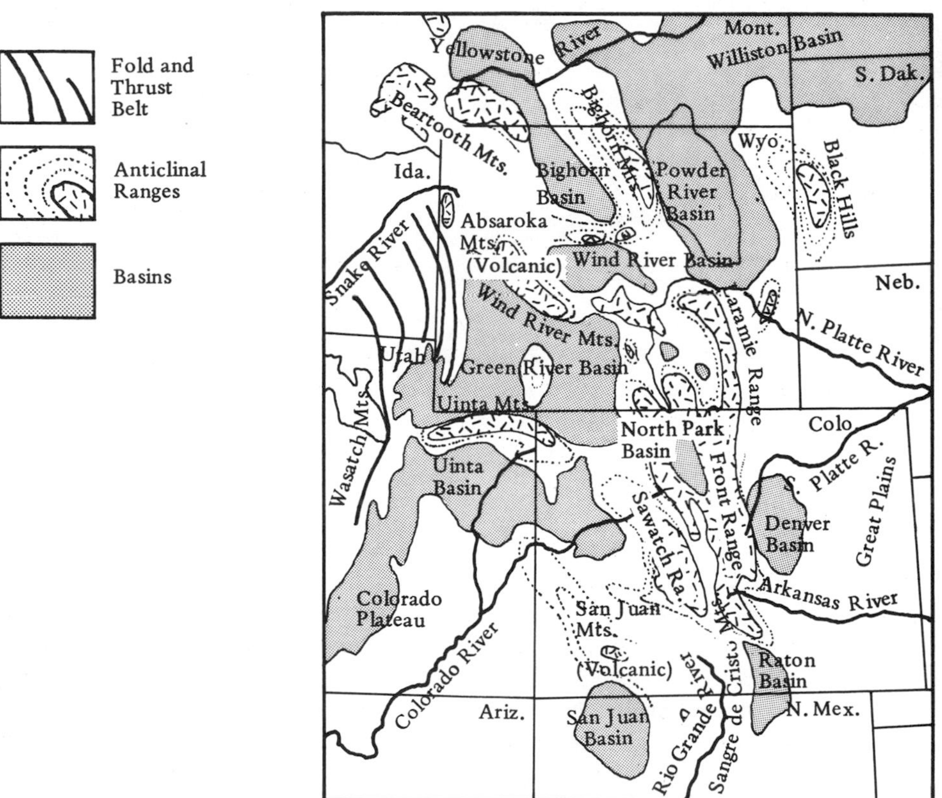

FIGURE 21.9 Early Cenozoic Mountains and Intermontane Basins of the Central and Southern Rocky Mountains Area.

and West Indian arcs connecting North and South America (Fig. 21.2). The Pacific Plate continued to impinge against eastern Asia producing continuous deformation in this area.

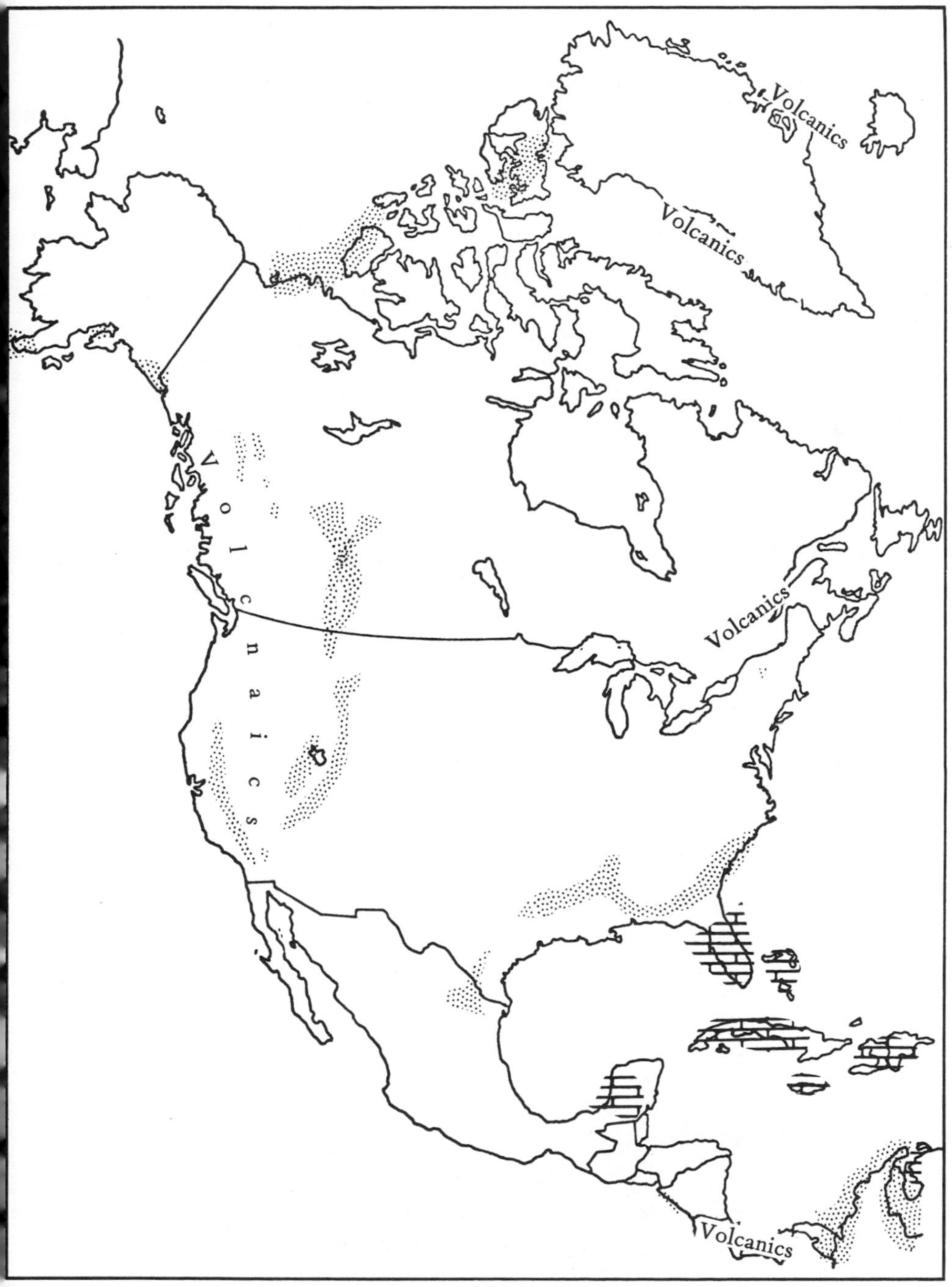

FIGURE 21.10 Present-Day Distribution of Paleogene Sedimentary Deposits in Western North America.

North America

Within this general framework, let us examine some of the details of Cenozoic physical history. In North America, the Laramide or Cordilleran Orogeny (Fig. 21.9) continued well into the Paleogene. The last epeiric sea to invade the craton, the *Tejas Sea* (Fig. 15.18), made only a brief Paleocene incursion into the Great Plains area, but was soon driven out by the Fort Union clastic wedge of terrestrial sediments coming from a strong pulse of the Cordilleran mountain-building (Fig. 21.10). The Tejas Sea gradually withdrew out through the Gulf Coastal region during the remainder of the Cenozoic.

Intermontaine basins within the rising Rocky Mountains (Fig. 21.9) contain numerous terrestrial deposits of Paleocene and Eocene age. One of the best known is the Green River Formation (Fig. 21.11) of southwestern Wyoming and adjacent states, an ancient lakebed shale famous for its fish fossils. Another of these terrestrial units, the colorful Wasatch Formation, has been subsequently uplifted and carved into the fantastic pillars of Bryce Canyon National Park (Fig. 21.12). In the Oligocene, another phase of deformation in the mountains sent a sheet of clastics, the White River clastic wedge, over the Great Plains. This deposit, well known for its mammalian fossils, has been carved by erosion into the noted Badlands of South Dakota (Fig. 21.13). Great outpourings of lava flows occurred in the Oligocene of some areas of the Rockies. These have since been uplifted and carved into the jagged San Juan Mountains of Colorado and the Absaroka Mountains of Wyoming and Montana (Figs. 21.14, 21.15). In the Miocene, the Rocky Mountains had been reduced to low relief by erosion and burial of the mountains in their own debris. Rivers ran across the countryside without regard for the underlying rocks. There was broad uplift and block faulting (Fig. 21.16) in the Late Pliocene and Pleistocene. During this rise of the present Rockies, the major rivers were able to maintain their courses across the exhumed mountain ranges forming numerous spectacular canyons (Figs. 21.17, 21.18).

FIGURE 21.11 Eocene Green River Strata, Wyoming.

FIGURE 21.12 Bryce Canyon, Bryce Canyon National Park, Utah. (Courtesy William M. Mintz.)

FIGURE 21.13 The White River Badlands of South Dakota, Badlands National Monument. (Courtesy of William M. Mintz.)

(A)

(B)

(C)

FIGURE 21.14 Yellowstone National Park in the Absaroka Mountain area of Wyoming. The dying effects of volcanism are shown by (A) The Norris Geyser Basin and (B) Old Faithful Geyser. Volcanic deposits are displayed in (C) the Grand Canyon of the Yellowstone. (A) Courtesy of William M. Mintz, (B–C) Courtesy of National Park Service.

FIGURE 21.15 The San Juan Mountains, Colorado. Two scenes in the Mt. Sneffels area near Ridgeway. (Courtesy Colorado Dept. of Public Relations.)

FIGURE 21.16 The Grand Tetons, Wyoming. (Courtesy of William M. Mintz.)

Vulcanism was widespread in the Pacific Northwest throughout the Cenozoic. Submarine lava flows abounded in the Paleogene. There were great outpourings of basalt forming the Columbia River Plateau (Fig. 21.19) chiefly in the Miocene. In the Late Neogene, the high volcanoes of the Cascade Mountains developed (Fig. 21.20). There was also extensive vulcanism in the Colorado Plateau and Sierra Madre regions in the Late Cenozoic (Fig. 21.21).

In California the geology is extremely difficult to decipher because of the great dislocations caused by the San Andreas fault system, a great transform fault or fracture (Fig. 21.22). This fault system connects the offset portions of the East Pacific Rise in the Gulf of California and off Northern California

FIGURE 21.17 Physiographic Diagram of a Part of the Western States. Note canyons cut through the ranges of the Rockies, block-fault mountains, and volcanic mountains.

(A) (B)

FIGURE 21.18 Canyons Cut Through the Rocky Mountains. (A) Black Canyon of the Gunnison, Colorado, (B) Royal Gorge, Colorado. (Courtesy of William M. Mintz.)

FIGURE 21.19 Columbia River Plateau Basalts at Dry Falls State Park, Washington. (Courtesy of Washington State Department of Commerce and Economic Development.)

FIGURE 21.20 High Cascade Volcanoes. (A) Mt. Shasta, California, (B) Mt. Mazama, Oregon, caldera now occupied by Crater Lake, (C) Mt. Hood, Oregon, (D) Mt. Rainier, Washington. ((A) Courtesy of William M. Mintz, (B–C) Courtesy Oregon State Highway Travel Division, (D) Courtesy Washington State Dept. of Commerce & Economic Development.)

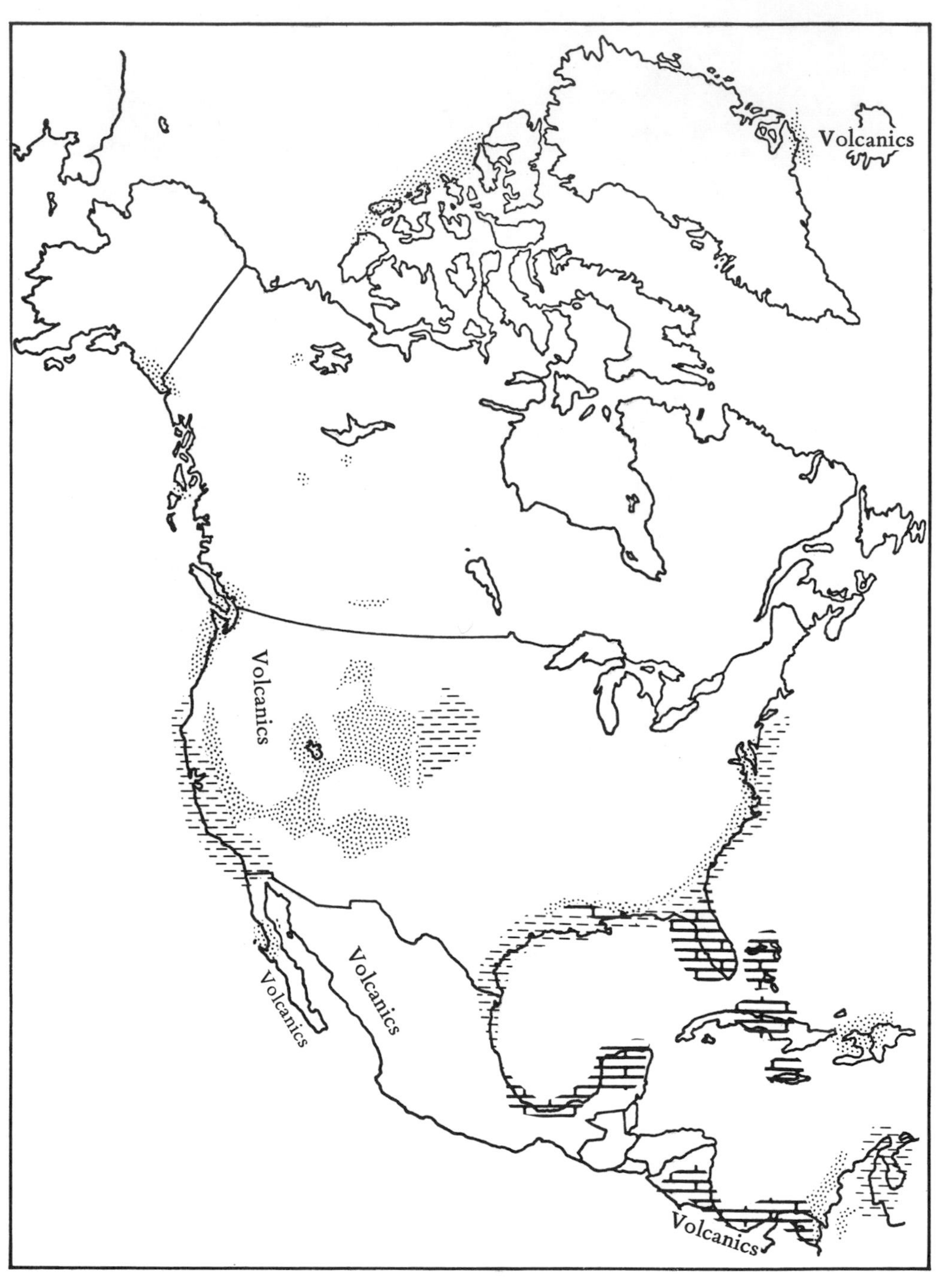

FIGURE 21.21 Present-Day Distribution of Early Neogene Sedimentary Deposits of North America.

A

B

FIGURE 21.22 The San Andreas Fault, California. (A) Northern California. Rift zone from Tomales Bay (foreground) to Bolinas Lagoon in the distance. (B) Southern California. Elkhorn Hills and Carrizo Plain area. (Courtesy of T. W. Dibblee, Jr., *California State Division of Mines Bulletin 190.*)

(Fig. 21.23). The missing pieces of granite (and associated metamorphics) connecting the southern Sierra Nevada and northern Baja California batholiths were moved northwestward so that they now lie in the Coast Ranges near and to the south of San Francisco (Fig. 21.21). The entire west coast was the scene of marginal embayments and seas through much of the Cenozoic with major incursions occurring in the Eocene and Miocene (Fig. 21.21). The shearing engendered by North America's colliding with the Pacific rift zone forming a transform fault zone, however, produced folding, faulting, and uplift along the west coast in the Neogene (Fig. 21.24). This deformation gradually drove the seas out of all but Southern California by the present. This disturbance is called the *Cascadian "Orogeny"* in the northwest and the *Pasadenan* or *Coast Range "Orogeny"* in California (Figs. 21.25, 21.26). Black faulting occurred on a grand scale from the Sierra Nevada to the Rockies in the Late Pliocene and Pleistocene. This elevated the great two-mile high east scarp of the Sierra (Fig. 21.27) and the numerous high ranges of the Great Basin (Fig. 21.28).

Eastern North America had been foundering since the Atlantic rift opened. The crust thins beneath a rift zone and the trailing edges of plates subside. The Atlantic and Gulf Coastal Plains (Figs. 21.29, 21.32) of North America subsided throughout the Late Mesozoic and Cenozoic. Most of the Mesozoic and many of the Cenozoic sediments were marine, but as the Rockies and even the Appalachians underwent Late Cenozoic rejuvenation, the flood of clastics was so great that they all eventually became terrestrial (Fig. 21.29). (Cenozoic rejuvenation of the Appalachians has probably been caused by isostatic adjustment.) Thus the broad uplift that produced today's topographic Cordilleran and Appalachian Mountains (Figs. 21.30, 21.31) is of the same age, although the earlier times of accumulation and deformation differed. The isostatic rise of salt domes from Mesozoic evaporate deposits has also characterized the Gulf Coast in Cenozoic times (Fig. 21.32).

North America and South America had been separated by the opening of the Atlantic rift in the Early Mesozoic. It not only opened the ocean basin, but rotated the two continents away from one another in the Late Mesozoic and the Paleogene (Figs. 19.2, 19.3). In the Neogene, however, the competing compression of the westward-moving western Atlantic and eastward-moving

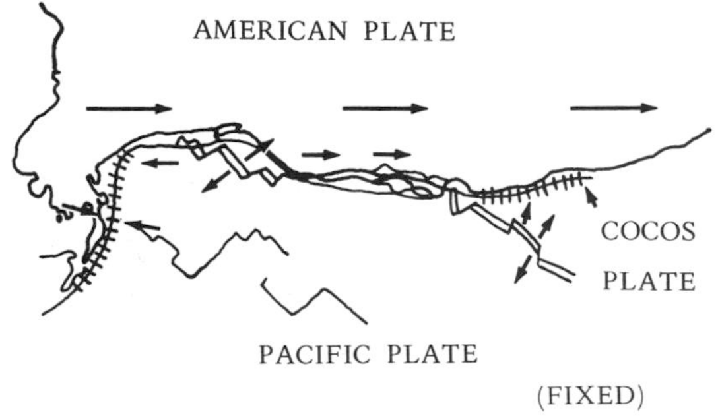

FIGURE 21.23 Relationship of the San Andreas Fault to Other Plate Boundaries in the Northeast Pacific Area. (From Atwater, *G.S.A. Bulletin*, vol. 81, December, 1970. Used by permission of the author and G.S.A.)

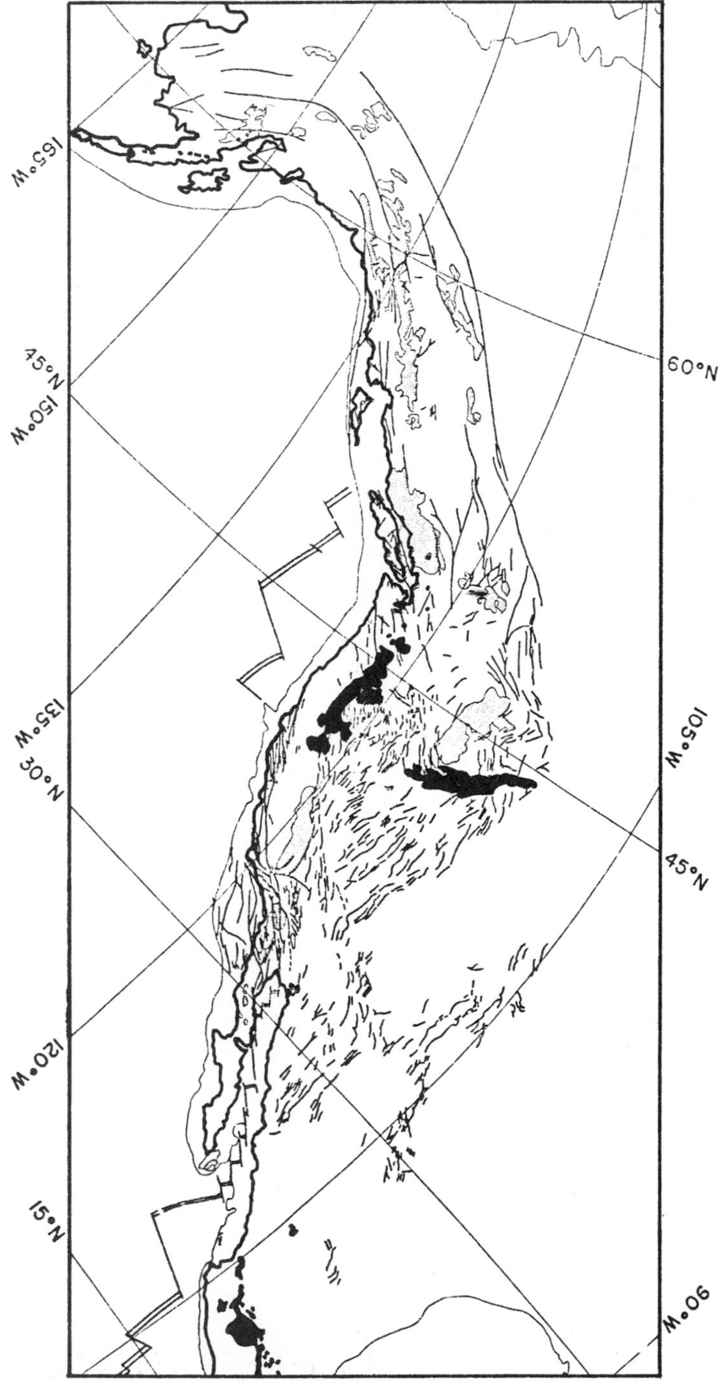

FIGURE 21.24 Major Tectonic Features of Western North America Related to the Motion Between the American and Pacific Plates. (From Atwater, *G.S.A. Bulletin*, vol. 81, December, 1970. Used by permission of the author and G.S.A.)

east Pacific produced a great deal of deformation in the region between the two continents. Previous to this time, an island arc trench system developed from Honduras through the West Indies into northern Venezuela (Fig. 21.33). The Atlantic part of the North American plate still dives beneath the trench on the eastern side of the arc today (Fig. 11.33). In the Cenozoic, the East Pacific Plate spreading from the East Pacific Rise impinged on a continental side of the South American Plate in South America and triggered renewed mountain-building in the Andes (Figs. 9.7, 21.34). In the narrow sea between

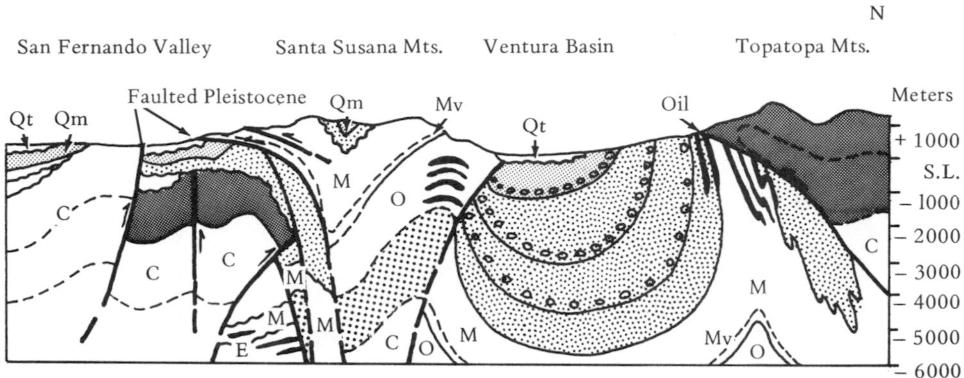

FIGURE 21.25 Cross-Section Through Part of California Coast Ranges Showing Late Cenozoic Deformation. (From *California State Division of Mines Bulletin 17*, 1954.)

FIGURE 21.26 California Coast Ranges. Peachtree Valley in southern Gabilan Range, Monterey County. (Courtesy of Mary Hill.

the Americas, however, two oceanic plate edges collided. In this compression, the Caribbean became an isolated plate with the East Pacific Plate diving beneath its western edge. A volcanic arc above this subduction zone formed the Panama land bridge linking the two continents for the first time since the Mesozoic. It also interrupted, along with the Tethyean mountain-building of the Old World, the great east-west flow of water in the equatorial regions which had occurred in the Late Mesozoic and Early Cenozoic.

FIGURE 21.27 Sierra Nevada Eastern Scarp. Mt. Williamson and Mt. Barnard. (Courtesy of William M. Mintz.)

FIGURE 21.28 The Basin-Ranges of Nevada. Wheeler Peak in the Snake Range. Upfaulted Lower Paleozoic Sedimentary Rocks. (Courtesy Nevada Dept. of Economic Development.)

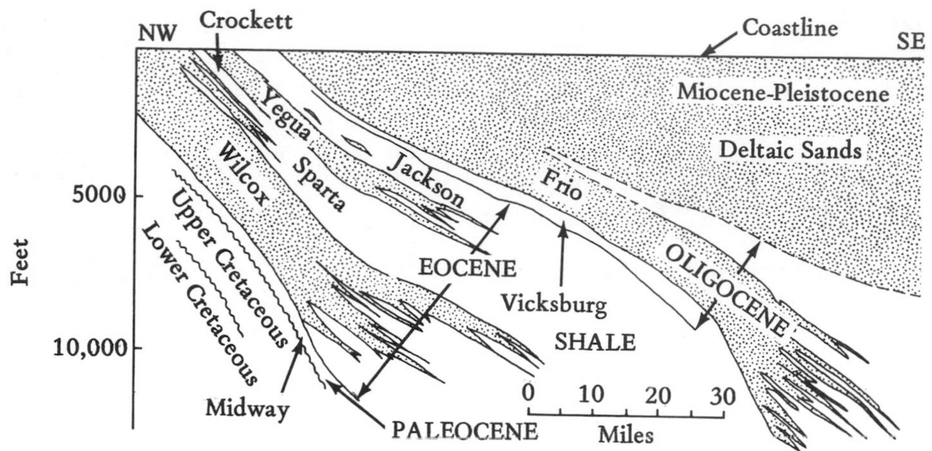

FIGURE 21.29 Cross-Section of Gulf Coastal Plain Showing Regressive Sequence.

(A)

(B)

FIGURE 21.30 The Rocky Mountains. (A) Hallett Peak from Bear Lake, Rocky Mountain National Park, Colorado. Rocks are Precambrian Granites, (B) Garden Wall, Glacier National Park, Montana. Precambrian Belt Supergroup Sedimentary Rocks. (Courtesy of William M. Mintz.)

FIGURE 21.31 The Appalachian Mountains. (A) The Adirondacks of New York, (B–D) The Great Smoky Mountains of North Carolina and Tennessee. ((A) Courtesy of William M. Mintz, (B–D) Courtesy of National Park Service.)

Europe

In Europe, the most important Cenozoic physical event was the *Alpine Orogeny*. After the Late Cretaceous-Paleocene regression, the craton was marginally flooded in the Eocene both in the northwest and southeast (Fig. 21.35). In the Oligocene, the major compression resulting from the close approach of Africa to Eurasia occurred. The great folding and thrust faulting of the Alps and other southern European ranges (Fig. 21.36) reached its peak at this time, though deformation had been continuous since the Mesozoic and continues today in many areas of the Mediterranean region (Fig. 21.37). There was also much flooding of the northern craton in the Oligocene (Fig. 21.38). In the Neogene, the flood of clastics from the Alps, Appennines, Dinaric Alps, Balkan Mountains, and Caucasus gradually drove the seas from the craton and eliminated them from much of the geosyncline (Figs. 21.39, 21.40). The Mediterranean is a shrinking ocean basin caught in a vise between two approaching continents. A long eastward extension of Tethys encompassing the present Black, Caspian, and Aral Seas (and several large Balkan lakes) became isolated from the remainder and broke up into separate basins. Only the Black Sea has regained a tenuous connection with the Mediterranean.

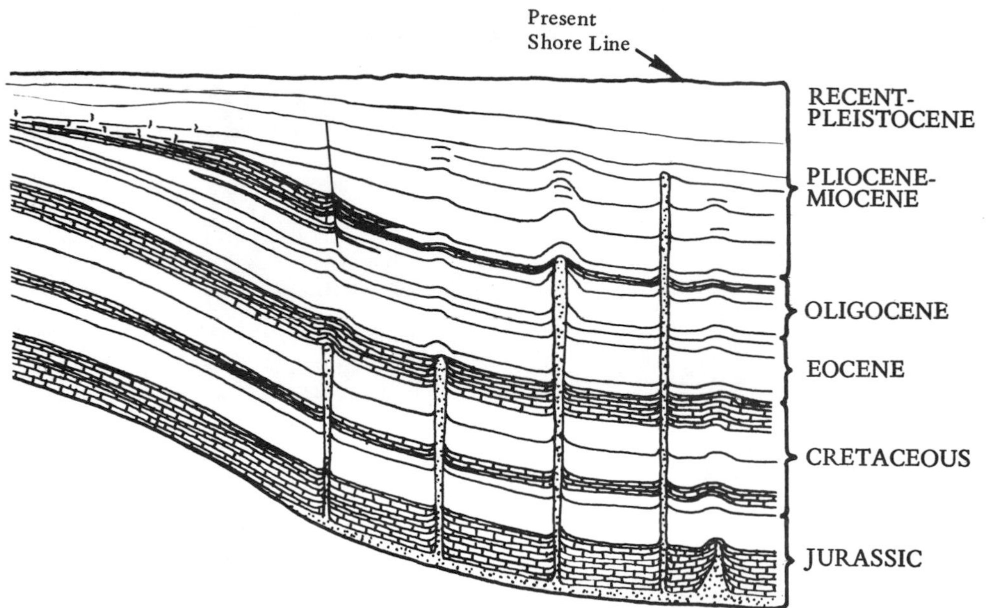

FIGURE 21.32 Cross-Section of the Gulf Coastal Plain Showing Salt Domes.

FIGURE 21.33 The Caribbean Region. (A) Present-Day Distribution of Middle Cenozoic Deposits, (B) Tectonic Setting of the Region Today.

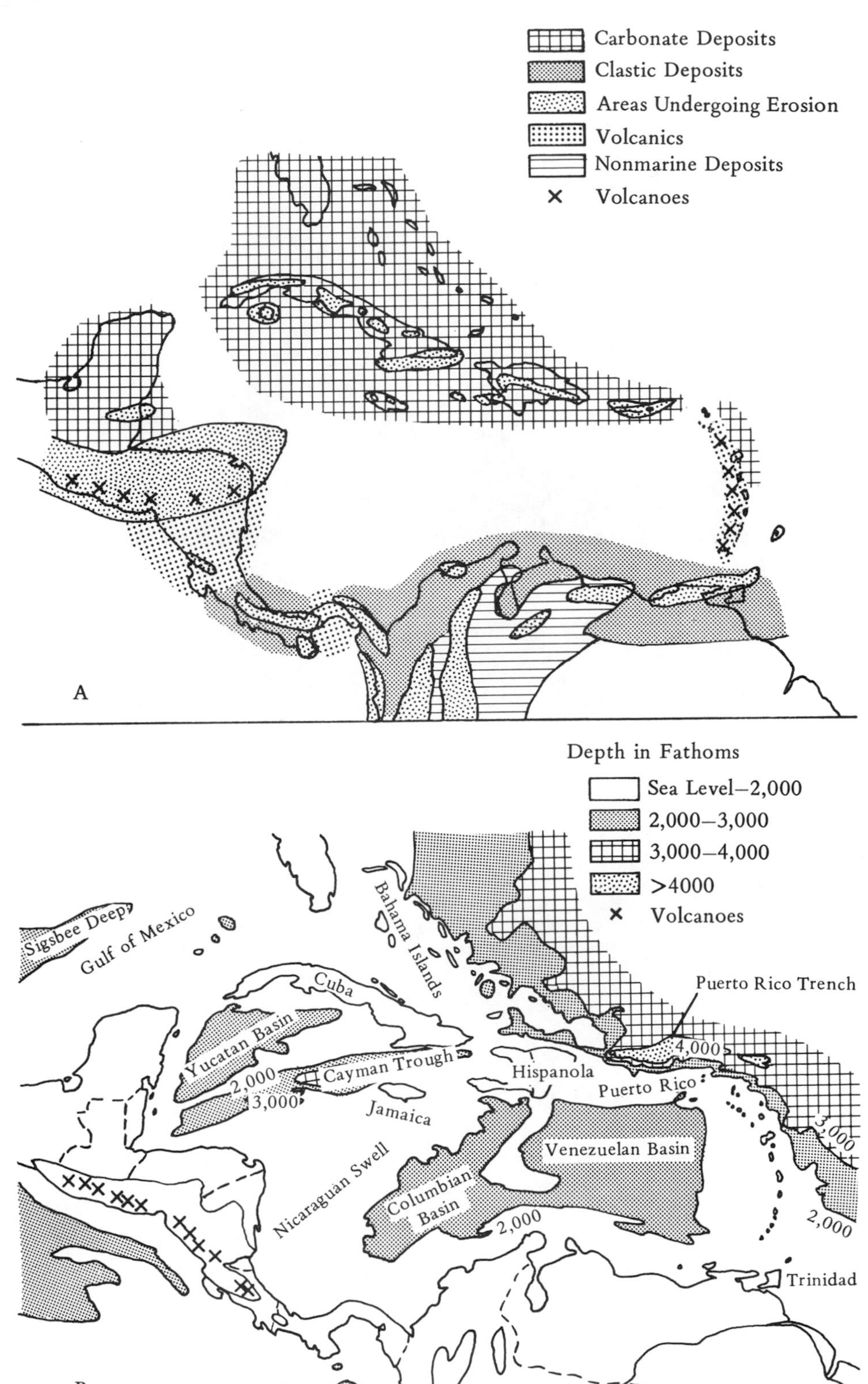
Carbonate Deposits
Clastic Deposits
Areas Undergoing Erosion
Volcanics
Nonmarine Deposits
Volcanoes
A
Depth in Fathoms
Sea Level–2,000
2,000–3,000
3,000–4,000
>4000
Volcanoes
Sigsbee Deep
Gulf of Mexico
Bahama Islands
Cuba
Yucatan Basin
2,000
3,000
Cayman Trough
Hispanola
Jamaica
Puerto Rico Trench
4,000
Puerto Rico
3,000
Venezuelan Basin
Nicaraguan Swell
Columbian Basin
2,000
2,000
Trinidad
B

FIGURE 21.34 The Andes Mountains of South America. The Cordillera de Huayhuash, Peru. Strata are of Mesozoic and Cenozoic age. (Courtesy of Peter J. Coney, from *G.S.A. Bulletin*, vol. 82, July 1971.)

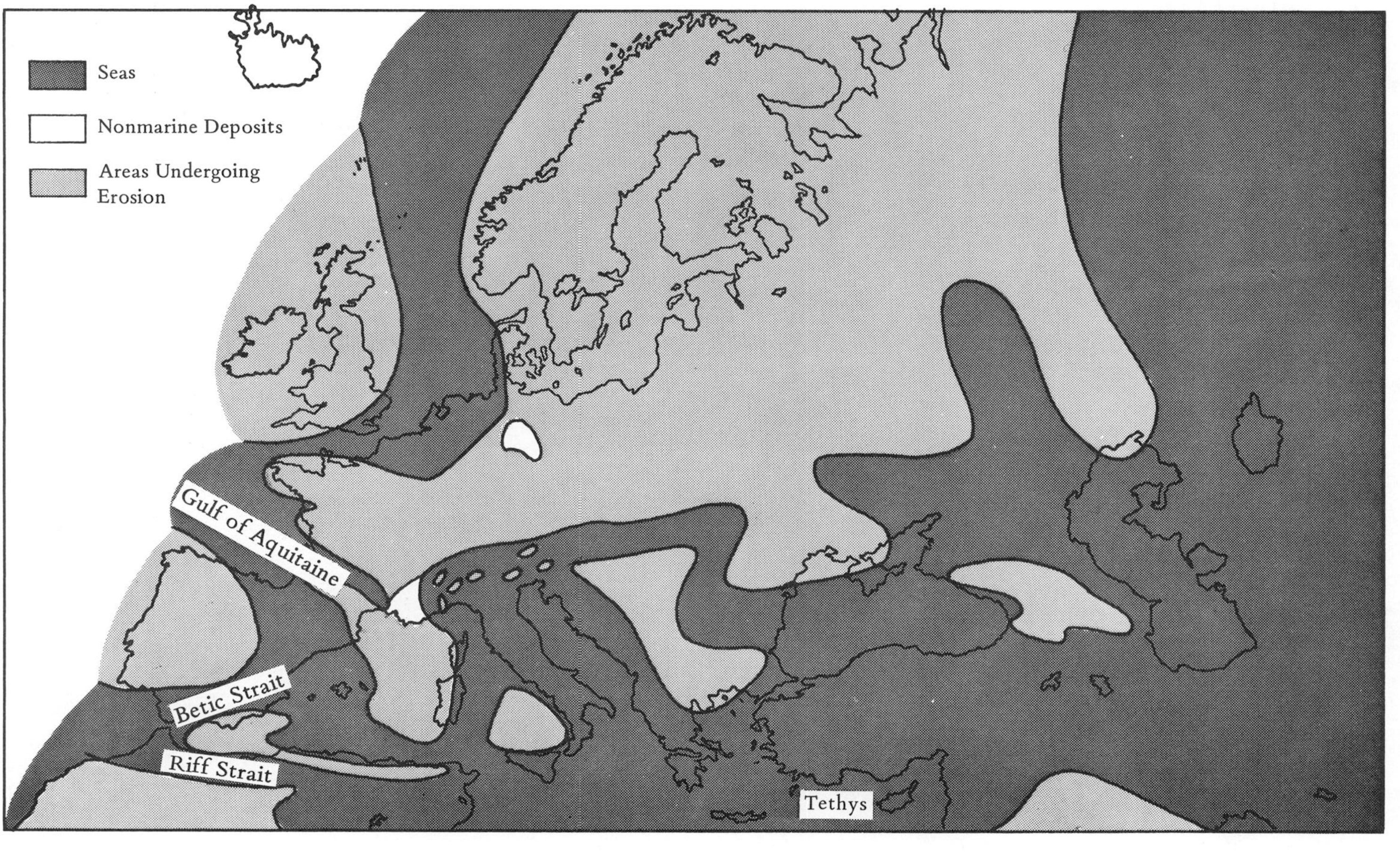

FIGURE 21.35—Paleogeographic Map of Europe in the Eocene. Effects of plate tectonics not shown.

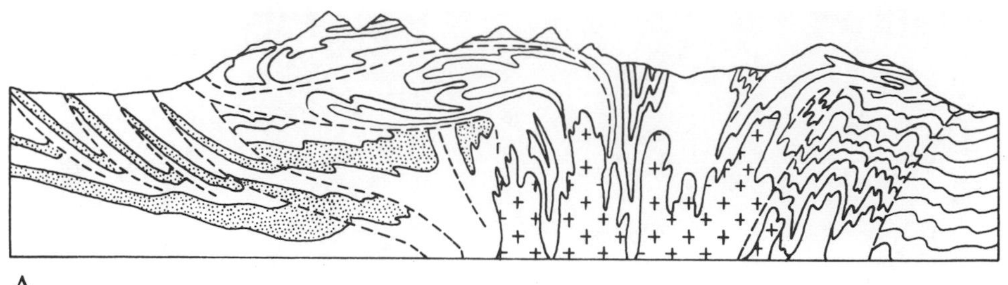

A

B

C

FIGURE 21.36 Cross-Sections of the Alps Showing Effects of Alpine Orogeny. (A) Details of folds and thrust faults, (B) Diagrammatic view of major tectonic elements particularly the nappes or large overthrust sheets. (From Dewey & Bird, *Journal of Geophysical Research*, vol. 75, no. 14, May 10, 1970. By permission of authors and AGU.) (C) Pennine Alps, Weisshorn in foreground, Matterhorn in distance, (D) The Matterhorn, (E) Pilatus from Rigi, Lake Lucerne, (F) Jungfraujoch. (Courtesy Swiss National Tourist Office.)

D

E

F

Asia and the Gondwana Continents

In the Mesozoic, India had separated from Africa when the Carlsberg Ridge developed forming the Andaman Sea of the Indian Ocean (Figs. 19.1, 19.3). In the Oligocene, Arabia was split off from Africa by the Red Sea rift (Fig. 21.41). Both the Indian and Arabian plates moved to the present northeast and collided with the Eurasian plate in the Middle to Late Cenozoic. Where the boundaries of the plates were oceanic in the present East Indies area (Fig. 11.33), an island arc and trench system developed with the Indian diving beneath the Eurasian. Along most of their lengths, however, the colliding plates were continental and this produced the great mountain ranges of the Near East and the Himalayan and associated ranges (Figs. 21.42–21.44). In the Himalayan region, the Indian plate has been overrun by the Eurasian one and a double thickness of continental crust has resulted (Fig. 21.45). This is why the Himalayas are so much higher than any other mountain range on earth today and why the vast Tibetan Plateau lies at such a great elevation. The great influx of clastics from these highlands has completely filled the sinking trough that developed between the crumpled edges of the two approaching continents. The interior of Asia

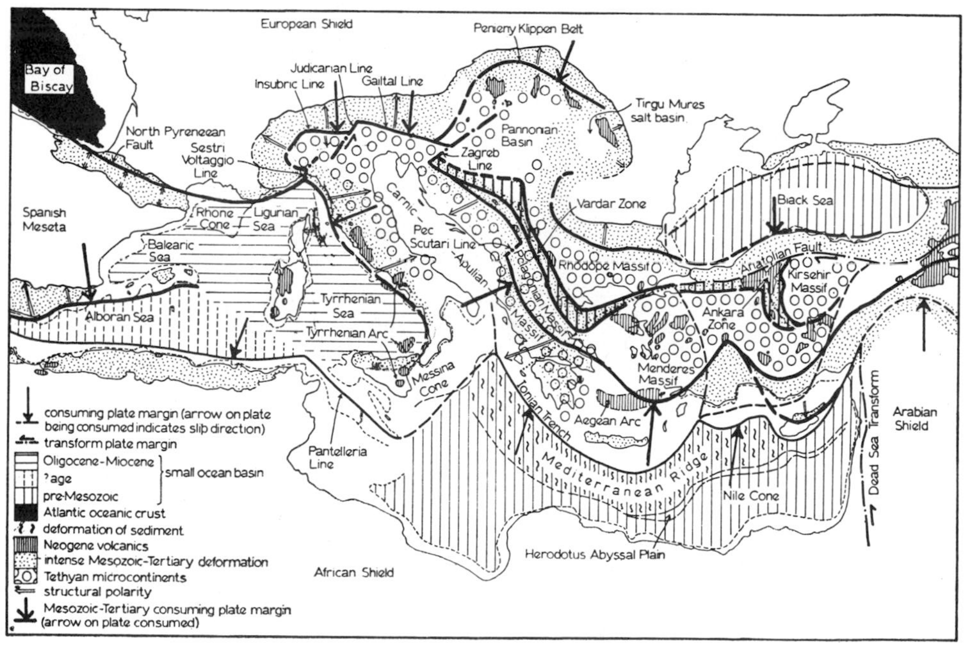

FIGURE 21.37 Major Tectonic Elements of Mediterranean Region Today. (From Dewey & Bird, *Journal of Geophysical Research*, vol. 75, no. 14, May 10, 1970. By permission of authors and AGU.)

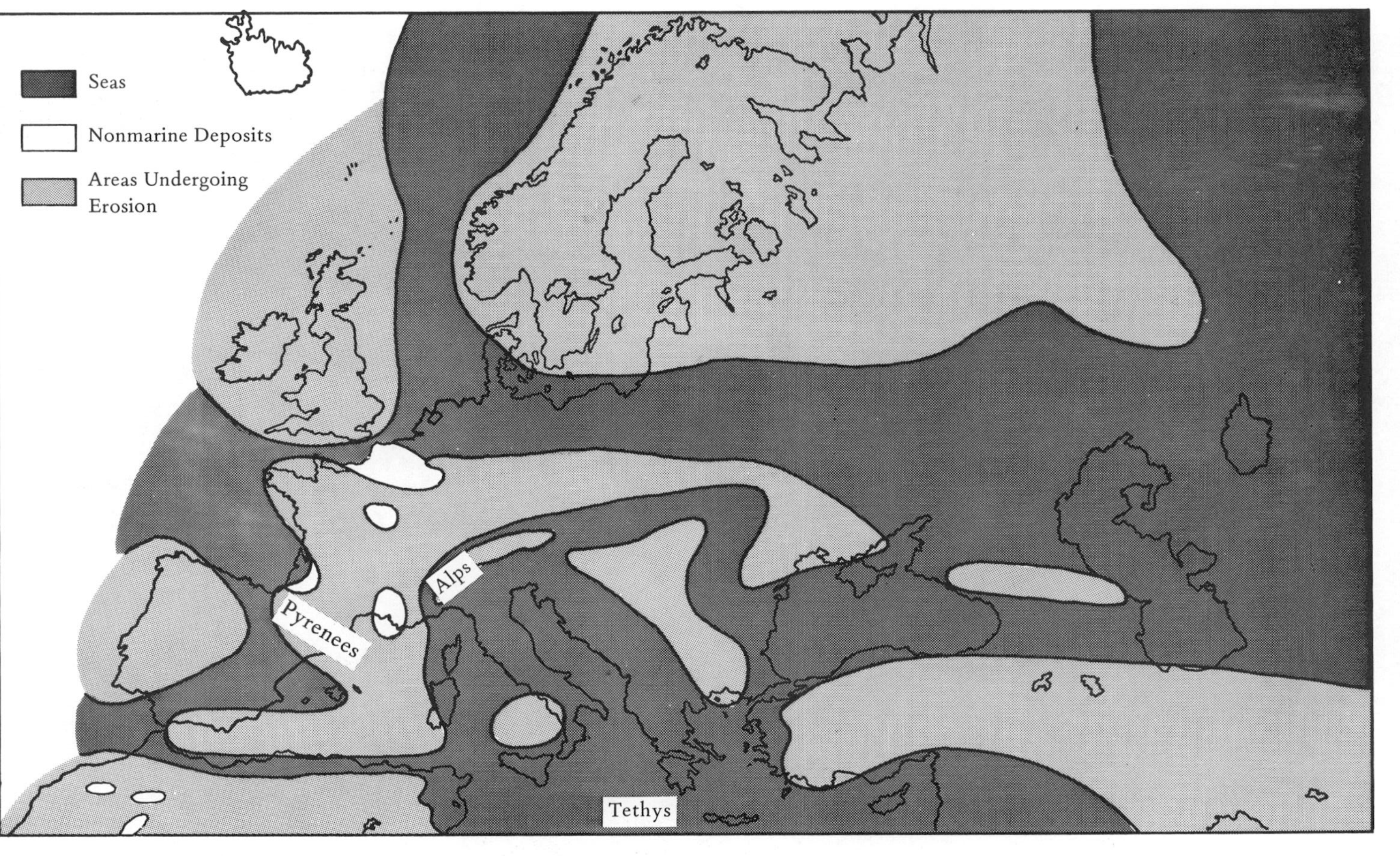

FIGURE 21.38 Paleogeographic Map of Europe in the Oligocene. Effects of plate tectonics not shown.

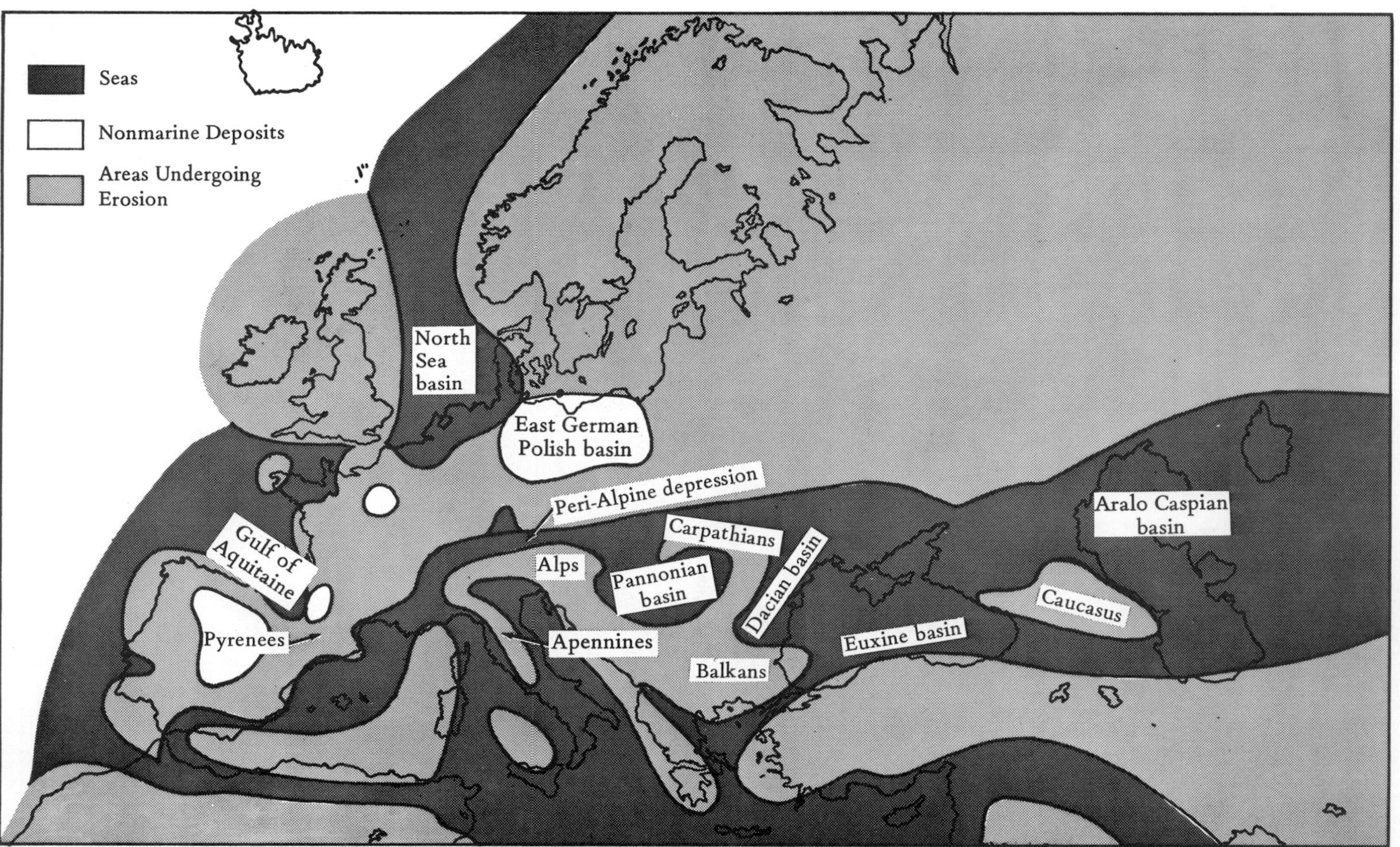

FIGURE 21.39 Paleogeographic Map of Europe in the Miocene. Effects of plate tectonics not shown.

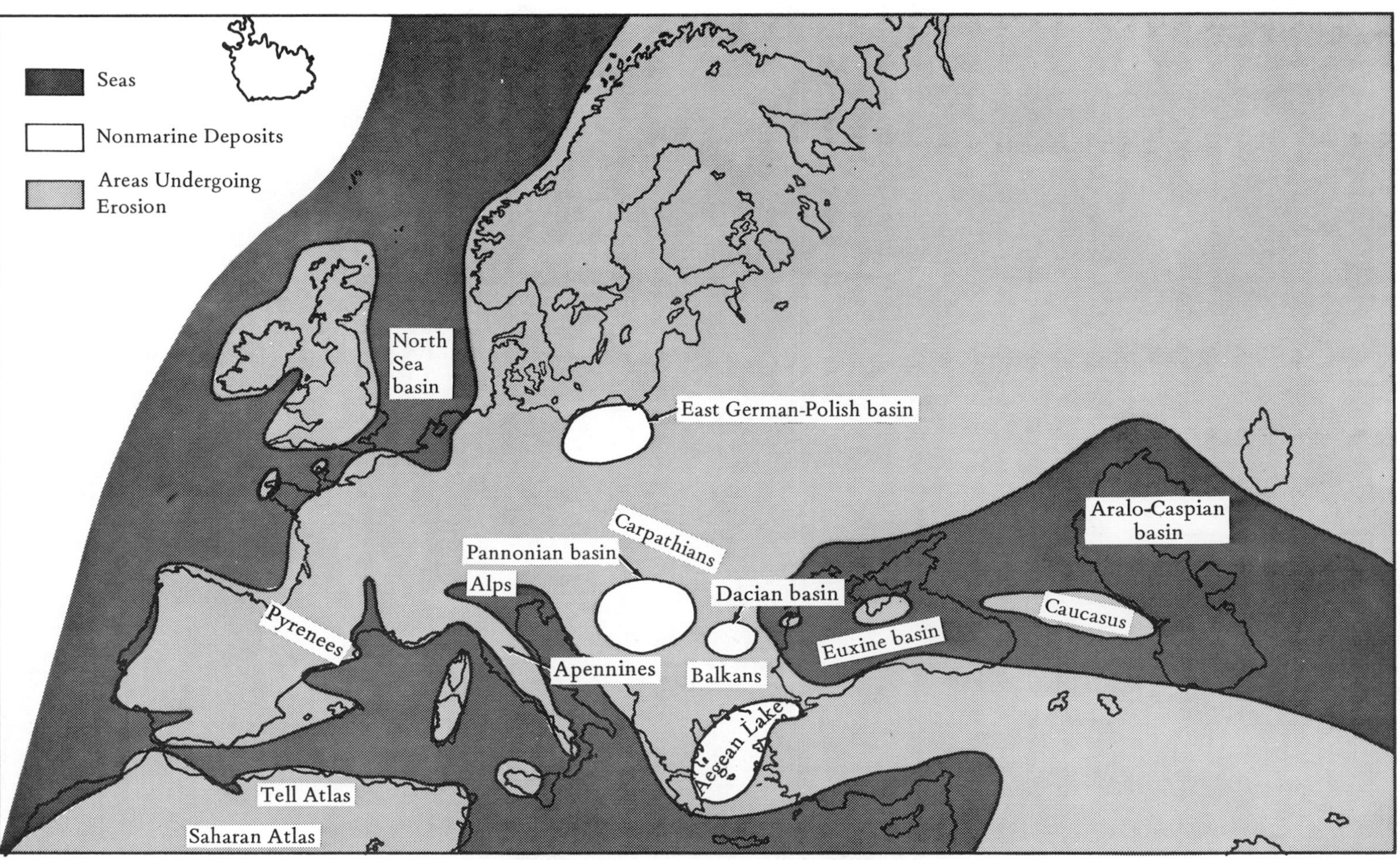

FIGURE 21.40 Paleogeographic Map of Europe in the Pliocene. Effects of plate tectonics not shown.

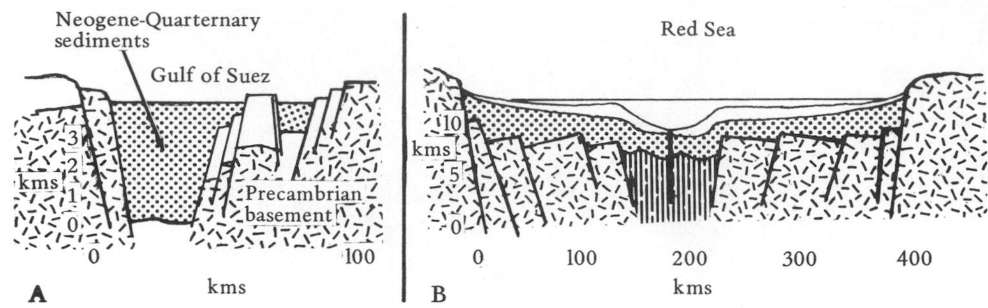

FIGURE 21.41 Cross-Section of the Red Sea Today Showing Its Rift Zone Character. (A) Gulf of Suez, an incipient rift, (B) Red Sea proper, a mini-ocean basin. (From Dewey & Bird, *Journal of Geophysical Research*, vol. 75, no. 14, May 10, 1970. By permission of authors and AGU.)

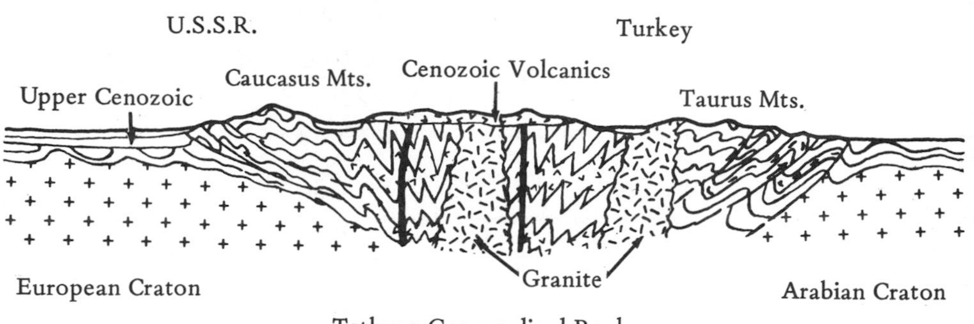

FIGURE 21.42 Cross-Section of Near Eastern Ranges produced by the collision of Arabia with Eurasia.

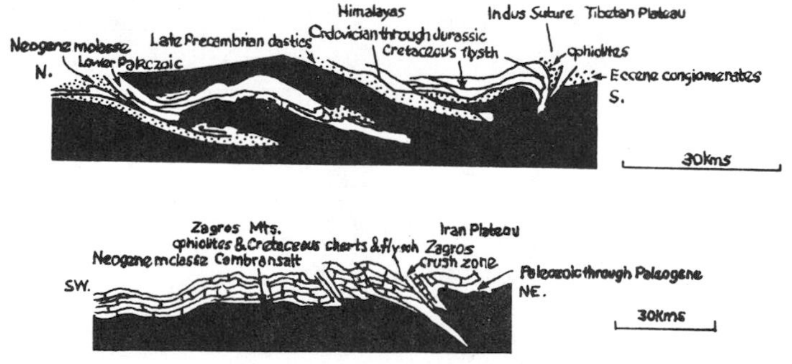

FIGURE 21.43 Cross-Sections of (A) Zagros and (B) Himalayan Mountain Ranges produced by the collision, respectively, of Arabia and India with Eurasia. (From Dewey & Bird, *Journal of Geophysical Research*, vol. 75, no. 14, May 10, 1970. By permission of authors and AGU.)

FIGURE 21.44 The Himalayas. (Courtesy Sierra Club)

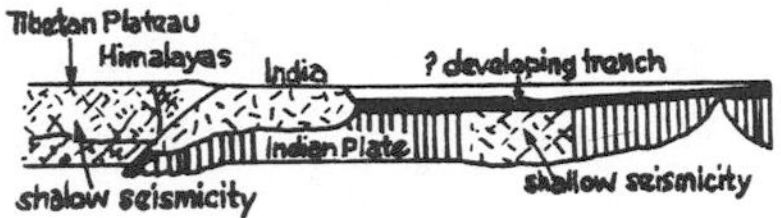

FIGURE 21.45 Cross-Section of Indian-Eurasian Collision Zone. (From Dewey & Bird, *Journal of Geophysical Research,* vol. 75, no. 14, May 10, 1970. By permission of authors and AGU.)

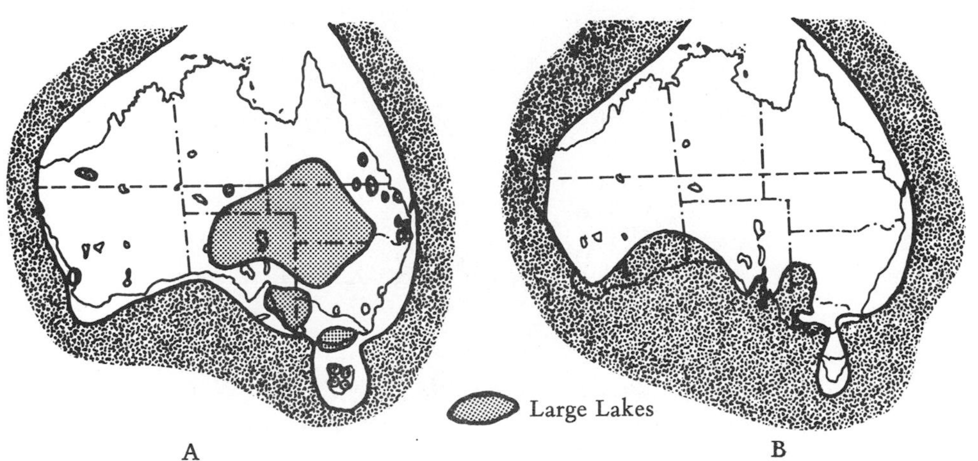

FIGURE 21.46 Paleogeographic Maps of Australia. (A) Oligocene, (B) Miocene.

is characterized by terrestrial Cenozoic deposits. Eurasia and southeast Asia seem to be coming apart in the Late Cenozoic along the Baikal rift zone (Fig. 11.33). Eastern Asia is bounded by trenches where the oceanic Pacific Plate and associated minor plates plunge beneath the continental Eurasian plate (Fig. 9.7). The compression in this contact zone of island arcs has produced the continual instability and deformation which characterizes the region even today (Fig. 11.33).

(A)

(B)

FIGURE 21.47 The Southern Alps of New Zealand. (A) Mt. Sefton, Mt. Cook National Park, (B) Milford Sound, Fiordlands National Park. (Courtesy of New Zealand Tourist Office.)

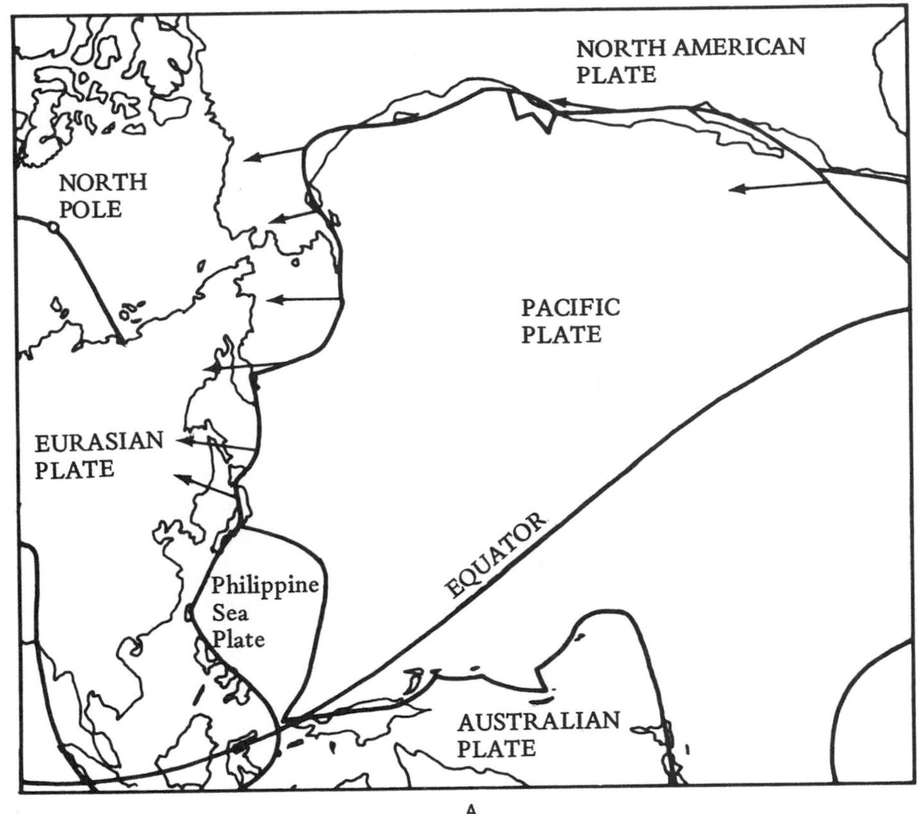

A

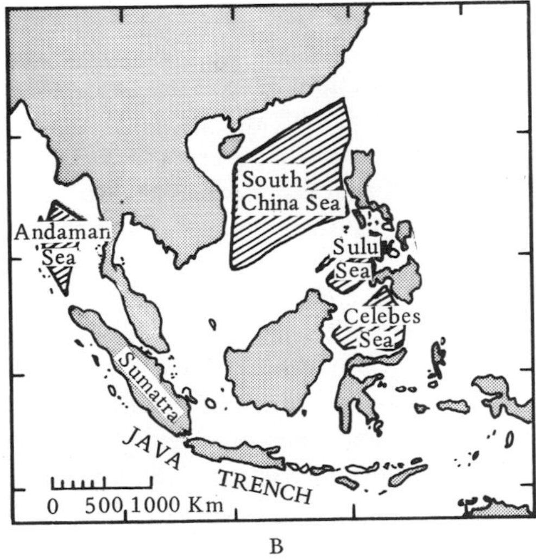

B

FIGURE 21.48 Development of Small Oceanic Plates near Southeast Asia. (A) Philippine Sea Plate, (B) South China, Sulu, Celebes, Andaman Sea Plates.

Australia and Antarctica became separate continents in the Cenozoic (Figs. 11.26, 11.38). Both were largely emergent and characterized by terrestrial and fresh-water deposits with volcanics common in some areas. There were marginal marine embayments in the south and west of Australia in the Early and Middle Cenozoic (Fig. 21.46). The small continental block of New Zealand became situated at the colliding edges of the Indian and Pacific plates in the Cenozoic and the ensuing deformation produced the Southern Alps of that country (Fig. 21.47).

The Ocean Basins

The ocean basins continued to receive a sedimentary record during the Cenozoic Era (Figs. 11.14–11.16). The calcareous deposits formed on the ridge flanks moved outward and downward gradually dissolving away as they went into the deep-water oceanic waters. The older Cenozoic strata, situated near the margins of the present basins, are less calcareous and contain more residual clays. The large Pacific Plate, entirely oceanic, is notable because its western edge is broken into a smaller plate occupied by the Philippine Sea (Fig. 21.48). Apparently pieces of plates can be broken off major plates in zones where several plates impinge such as southeast Asia (or the Caribbean area mentioned earlier) and undergo a separate development (Fig. 11.33).

The Pleistocene Ice Age

An event of profound importance for the physical and biological history of the entire earth is the glacial ice age that occurred in the most recent part of the Cenozoic. It was not until the 1820s and 1830s that geologists first began to perceive the connection between various geological features of areas distant from present glaciers and the action of glaciers. It was the Swiss, naturally enough, who noted that the erosional and depositional features produced by present glaciers extend far beyond the present extent of ice, even onto the plains. Previous to that time, European geologists attributed the poorly sorted clastic debris called "drift" found over much of the surface of northern Europe to Noah's Flood. A noted skeptic, the naturalist *Louis Agassiz,* went to the Alps to prove *Venetz-Sitten, de Charpentier,* and the other Swiss wrong, but came away convinced of the reality of an ice age in the not distant past. Agassiz discovered even more abundant evidence of glaciation in the British Isles, Scandinavia, and the north European Plain (Fig. 21.49). In 1846, he immigrated to America and found that an even greater area, virtually all of the continent north of the Ohio and Missouri Rivers, had been glaciated (Fig. 21.49).

Also in 1846, *Edward Forbes* suggested that Lyell's Pleistocene "Period," originally based on the percentage of species of living mollusks it contained, was synonymous with this recent glacial age, the Quaternary. Most geologists, unfortunately, ultimately accepted this argument, including Lyell himself in 1873. As a result, they have been perpetually plagued by the problem of

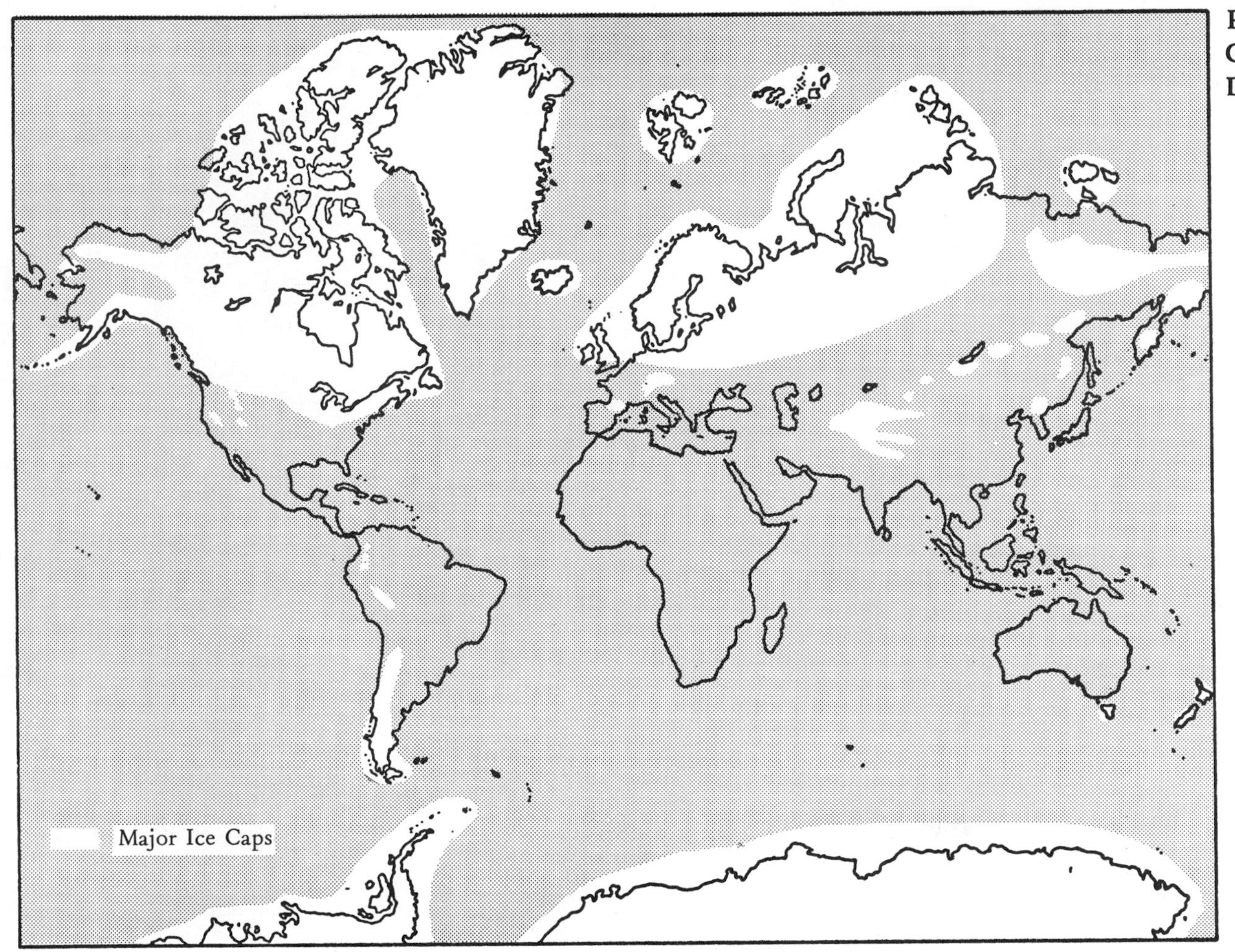

FIGURE 21.49
Glaciated Regions
During the Pleistocene.

assigning boundaries to an epoch or period originally defined in two entirely different ways. The type Pleistocene of Lyell is in the marine rocks of southern Italy, while the type Quaternary is the terrestrial glacial debris of northern France. Recently obtained radiometric dates indicate both the type Pleistocene and the onset of widespread continental glacial conditions began approximately 3 million years ago.

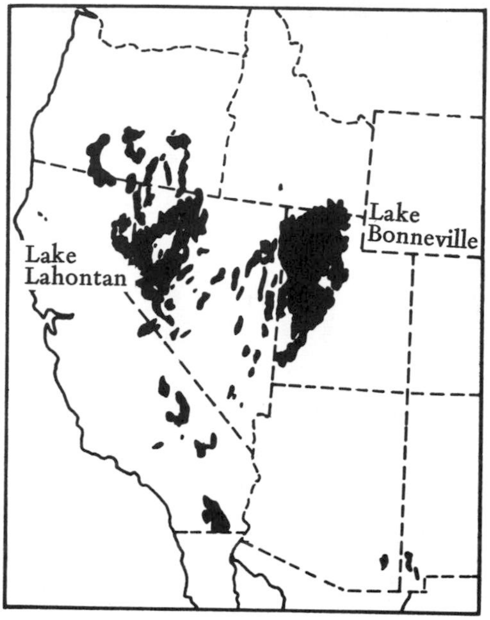

Lake Bonneville
Lake Lahontan

FIGURE 21.50 Pleistocene Pluvial Lakes of the Southwest.

FIGURE 21.51 Pluvial Lake Shoreline. Lake Bonneville terrace near Salt Lake City, Utah. (Courtesy Utah Travel Council and Hal Rumel.)

Geologists now know that Quaternary continental ice caps (Fig. 21.49) covered not only North America (including Greenland) north of the Ohio and Missouri Rivers, northern Europe, and the Alps, but also western and southern South America, all of Antarctica, northern Siberia, the central Asian mountains and plateaus, Tasmania, and New Zealand. In addition, high mountains all over the earth either had much larger glaciers than now or had glaciers where none exist today.

Other indirect evidences of glaciations and interglaciations were found. These include the elevated shorelines cut by seas when sea level was high because the icecaps had melted during an interglacial, the pluvial lakes that existed

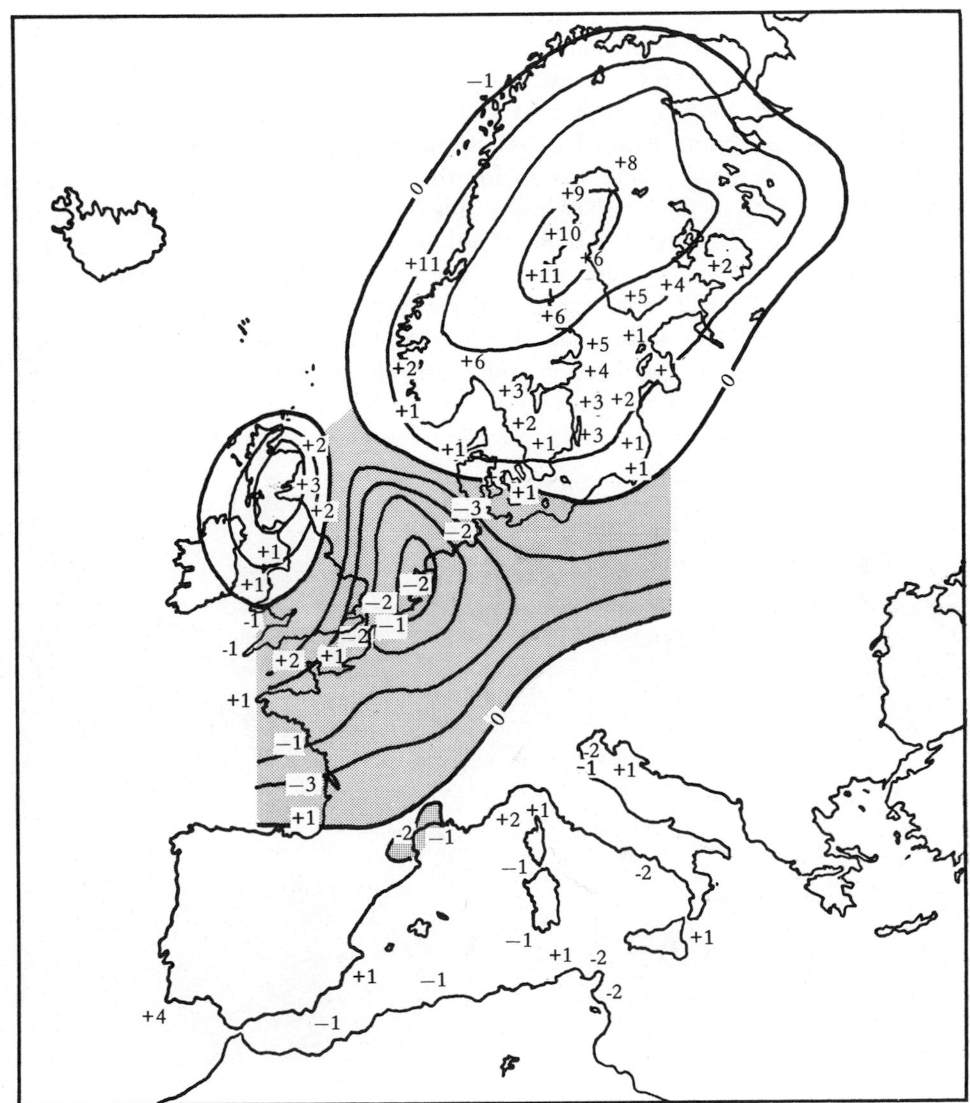

FIGURE 21.52 Crustal Rebound in Scandinavia Following Deglaciation.

during glaciations in present arid regions such as the American southwest and the African Sahara (Figs. 21.50, 21.51), large canyons cut by large amounts of glacial meltwater but now occupied by puny streams, and the rebounding of land areas after removal of the ice load (Figs. 21.52, 21.53).

Geologists also have discovered that there was not a single glaciation, but several, separated by intervals that were warmer than present and during which the ice largely disappeared. The evidence (Fig. 21.54) includes multiple layers of glacial deposits separated by zones of deep weathering, soils, loess (wind-blown dust), and fossiliferous sediments bearing the remains of warm-climate animals and plants. In the oceans, layers of alternately warm and cold water faunas succeed one another in the sedimentary record (Fig. 21.55) offering further proof. The traditional pattern (Fig. 21.56) was four glaciations called Nebraskan, Kansan, Illinoisan, and Wisconsinan from oldest to youngest in North America, and three interglacials called Aftonian, Yarmouthian, and Sangamonian from oldest to youngest in North America. Europeans called the glacial stages Gunzian, Mindelian, Rissian, and Wurmian from oldest to youngest and had no separate names for the interglacials. (More recently, an earlier fifth or Donau Glacial has been added.) The system soon broke down, however, and the Wisconsinan tills proved to be multiple and the stage was further subdivided into smaller glacial advances and retreats. Work in the ocean basins utilizing temperature sensitive foraminifera, geochemical techniques, and radiometric dating has shown that rather than four glacials and three interglacials, there has been a continuous succession of fluctuating glacial

FIGURE 21.53 Elevated Beaches South of Prince of Wales Island, Northwest Territories, Canada. (Courtesy Geological Survey of Canada.)

and interglacial intervals back into the Pleistocene (Fig. 21.57). Some of these appear to have been more or less glacial than others and perhaps correspond to the original seven subdivisions.

Debates rage about whether or not we are still in the Pleistocene and whether or not we are still in an interglacial epoch. The answer to the first question is no because Lyell originally established the Pleistocene on the basis of 90 percent living species. When we reach 100 percent living species, we are in his Recent (or Gervais' Holocene) Epoch. The different question about whether the present is or is not an interglacial is intimately involved in understanding the causes of glaciation.

Causes of Glaciation

Glaciers have probably always existed in higher mountains at least in high latitudes. But continental glaciation, in which large, thick masses of ice override vast lowland areas of continents, has not always existed. It occurred at

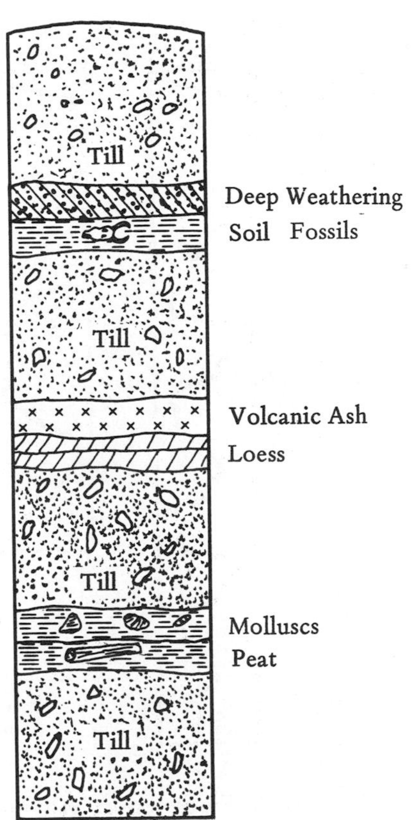

FIGURE 21.54 Evidences of Multiple Glaciations in Pleistocene.

least twice in the Precambrian, once in the Early Proterozoic (Aphebian) of Canada and again in the Late Precambrian-Eocambrian of nearly all the present continents. It was more or less continuous in Gondwanaland throughout the Paleozoic, but best known in the Ordovician of Africa and the Permocarboniferous of the whole supercontinent. And finally, the great Quaternary glaciation occurred in the Late Cenozoic. Various hypotheses have been advanced to explain the phenomenon of continental glaciation, but most have been either non-uniformitarian or lacking in supporting evidence. The lack of any periodicity in the occurrence of glaciations and the alternation of glacial and warm interglacial stages have always been the biggest stumbling blocks to a coherent theory, if indeed all glaciations had the same cause. Clearly, some cyclical phenomenon must be responsible for the alternations of glacial and interglacial stages. This cannot be simply the starving of glaciers for lack of precipitation because the interglacials were warm intervals, not just cold times without glaciers.

There are three astronomical phenomenon that do repeat themselves regularly and may influence temperatures. These are variations in the ellipticity of the earth's orbit, the obliquity of the earth's axis, and the wobble or precession

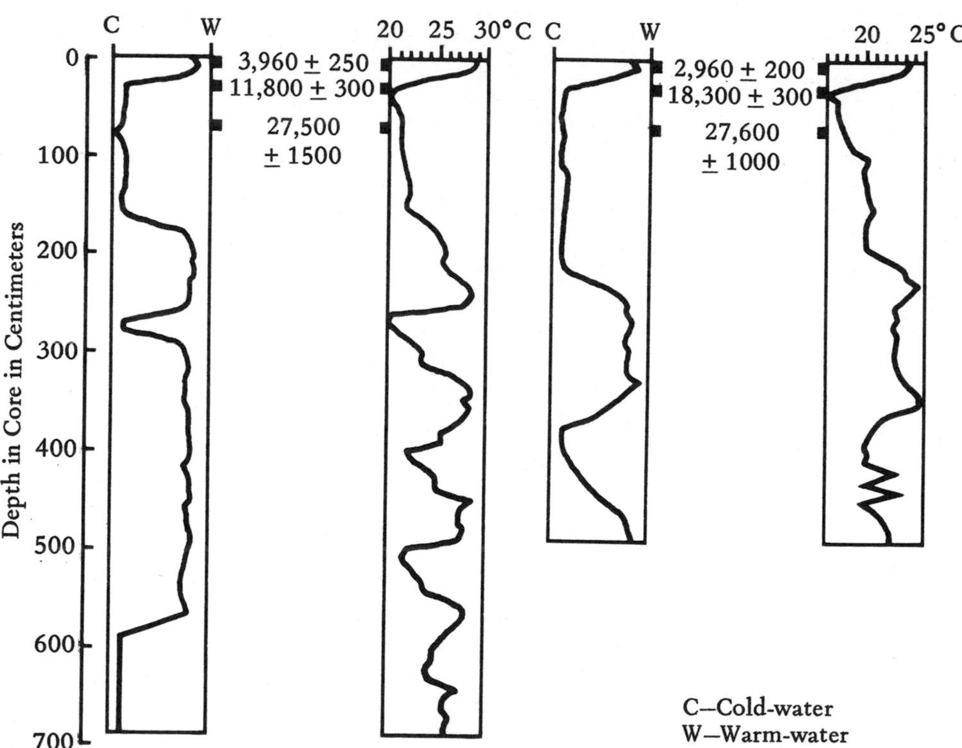

FIGURE 21.55 Oceanic Evidence of Alternating Glacials and Interglacials. Note alternating warm water and cold water foram assemblages. (From Leo F. Laporte, *Ancient Environments* © 1968, Prentice-Hall, Inc., Englewood Cliffs, New Jersey.)

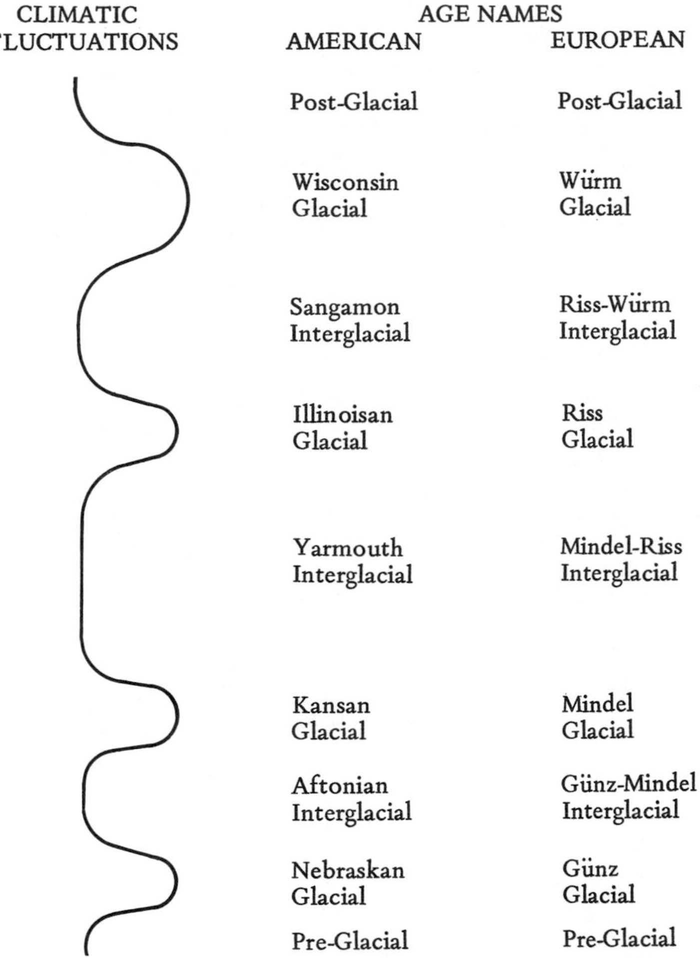

FIGURE 21.56 Traditional Pattern of Four Glacials and Interglacials.

of the earth's axis. The Yugoslavian meteorologist *Milankovitch* calculated the variations and their periods and the German climatologist *Koppen* suggested that they might correspond with the glacial-interglaciation sequence. The combined effect of the three factors produces a greater cooling every 40,000 years and corresponds well with the Late Pleistocene cold and warm spells (Fig. 21.58).

Kukla noted that the earth's albedo ratio, that is the ratio of the light reflected to the light received at its surface, varies in high latitudes according to the tilt of the earth's axis. Hence, as the earth wobbles, its albedo ratio in high altitudes changes and the amount of heat received at the earth's surface there varies. In his view, the precession of the axis provides the most plausible cause for the rhythmic alternation of glaciations and interglacials. Glaciers form when in that part of the wobble when the albedo ratio is high and they melt when it is low. A recent study of dated, sea-cut terraces in New Guinea

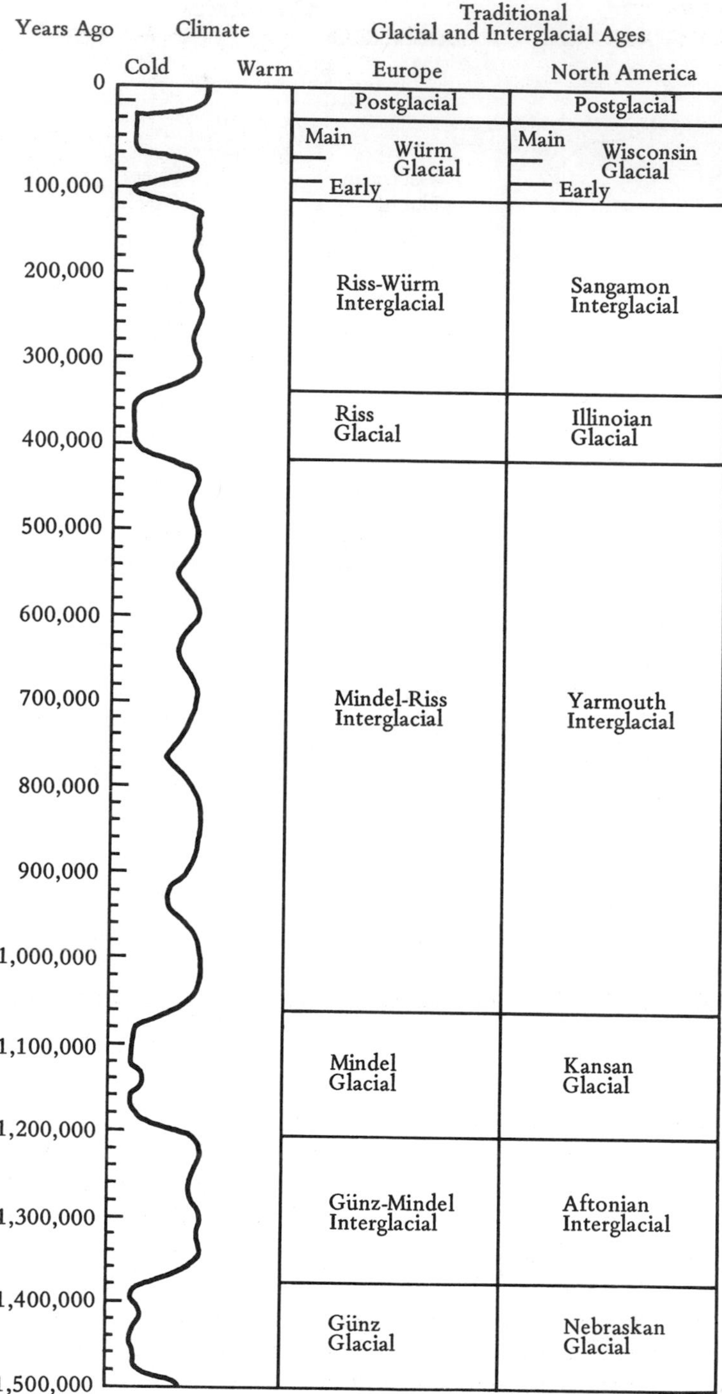

FIGURE 21.57 Recent Evidence of Numerous Glacials and Interglacials in Oceanic Cores. (From Leo F. Laporte, *Ancient Environments,* © 1968, Prentice-Hall, Inc., Englewood Cliffs, New Jersey.)

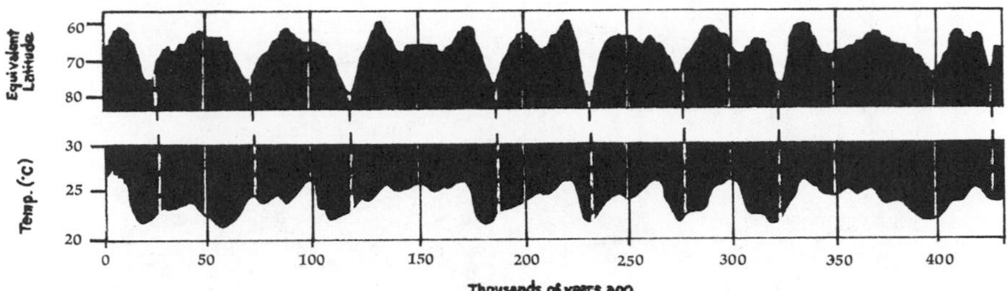

FIGURE 21.58 Correspondence of Milankovitch Curve (upper) with Glacial-Interglacial Sequence (lower). Determined from oceanic cores using foraminifera. (From C. Emiliani, *Science*, vol. 154, 18 Nov., 1966. Copyright 1966 by the American Association for the Advancement of Science. Used by permission of the author and AAAS.)

has indicated that the timing of the glacial-interglacial sequences coincides with albedo fluctuations predicted astronomically.

But the precession of the axis and the other astronomical phenomena have occurred continuously since the earth's formation; why haven't glaciations? Perhaps paleogeography offers the answer, a suggestion made in the 1800s by *Lyell* and *Andrew Ramsay*. When the circulation of the oceans is upset by the situation of the continents, the earth's entire heat balance can be disturbed. Furthermore, high continents increase the albedo ratio and extensive ones lower temperatures. Certainly during the Paleozoic Gondwana glaciations and the Late Cenozoic glaciations at both poles, the continents were high and very extensive because they were all very near. And, they were in the polar regions. In the Paleozoic and Mesozoic, the Pacific Pole was in the Pacific basin and there was no glaciation near it. In the Mesozoic, the Gondwana Pole was on the margin of Gondwanaland and the vast Pacific Ocean. Again, there was no glaciation. So we can say that perhaps glaciations occur when the three astronomical phenomenon noted above coincide to reduce the earth's heat budget or during the high albedo ratio parts of the wobble cycle when large, high continental masses cover or surround the polar regions. Much further testing and observation is necessary, however, before we can confidently state that we have found the explanation for continental glaciations. If we are right though, we can say that the present is only an interglacial, because all the conditions are right for glaciation except the current astronomical positions which will return to glacial levels in the near geologic future.

The Future?

It may be wise to close this final chapter on the earth's physical history with another prediction for the future. Based on extrapolations of present plate motions and amounts, the world 50 million years from now will probably look something like that portrayed in Figure 21.59.

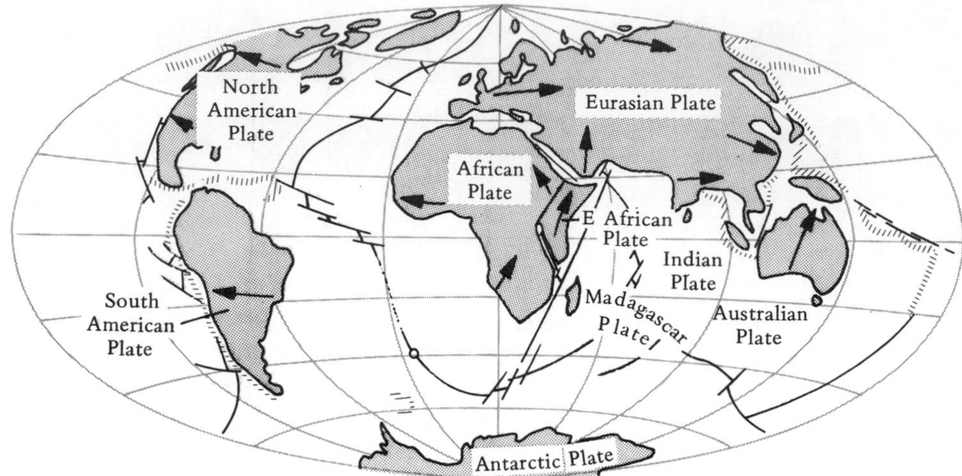

FIGURE 21.59 The Surface of the Earth as It May Appear in 50 Million Years. Epeiric seas not shown. The present direction and rate of plate movements are assumed to remain constant. (Data from Dietz and Holden.)

Questions

1. What are the two major ways in which the Cenozoic epochs are distributed into periods?
2. What has been the main motion of the North American Plate in the Cenozoic Era?
3. What events produced the major deformation of the Tethyean Geosyncline?
4. What is the relationship of the Fort Union clastic wedge to the Tejas Sea?
5. What is the origin of the spectacular canyons of the Rocky Mountains?
6. What is the historical significance of the San Andreas Fault of California?
7. How and when did the present Panamanian land bridge develop?
8. What major orogeny affected Europe in the Cenozoic?
9. How does the original definition of the Pleistocene differ from the concept of it as the great Ice Age?
10. What is the evidence for multiple Pleistocene glaciations?
11. Discuss at least one plausible theory for the origin of continental glaciations.

22

The Rise of the Modern Biota:

Cenozoic Life

In the Cenozoic Era, the story of present-day life unfolds. On land the mammals were the dominant large animal group, giving rise to the common name for the era, the age of mammals. Insects and angiosperms, so vital to the support of all terrestrial life, were just as significant, however. In the seas, the same groups that dominated the Mesozoic, the protistans and the mollusks, continued to expand and diversify with the significant exception of the extinct ammonoids.

Marine Life

The major marine fossil organisms of the Cenozoic are the *coccolithophores, diatoms, dinoflagellates, red algae, foraminifera, radiolaria, hexacorals, bryozoans, gastropods, pelecypods, annelids, crustaceans, echinoids,* and *fish.* The foraminifera, gastropods, and pelecypods are the major correlation fossils for shallower water environments, while the coccolithophores, planktonic forams, and radiolaria are important in dating ocean basin deposits. The increasing provincialism that occurs in the Cenozoic because of the fragmented oceans and continents and the strong latitudinal gradients of the final ice age, make correlation from region to region more difficult than at any other time in the geologic past since the Precambrian.

Protistans

The pelagic golden algae (Figs. 22.1, 22.2), particularly the *coccolithophores,* but also the *silicoflagellates,* were the major photosynthesizers and, hence producers, on earth during the Cenozoic. Their fossils are most abundant in the calcareous and siliceous oozes that blanket large areas of the ocean basins. One important variety of plate or coccolith is the star-shaped *discoaster* which is important for correlation and died out in the Pleistocene. *Corbisema* is a common silicoflagellate fossil.

The siliceous *diatoms* (Fig. 22.3) are second only to the golden algae as producers in the Cenozoic. Their remains are abundant in oceanic ooze, nearshore basin deposits, and they even invaded fresh water in the Neogene. Several formations are composed chiefly of diatom frustules forming the economically important rock diatomite which is used in filters, insulation, etc. Common diatom genera are *Coscinodiscus, Melosira, Navicula,* and *Nitzschia.*

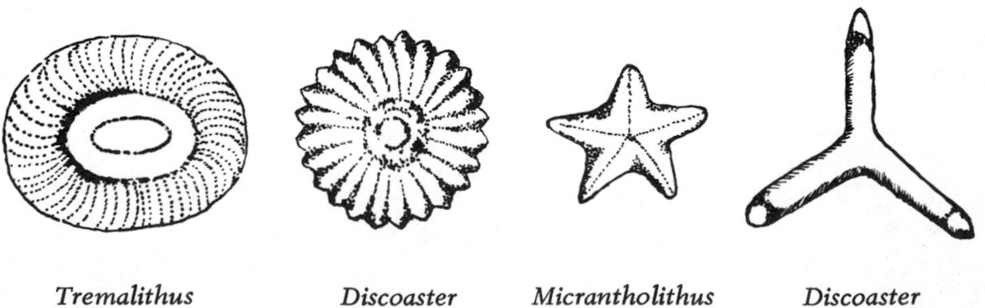

Tremalithus *Discoaster* *Micrantholithus* *Discoaster*

FIGURE 22.1 Cenozoic Coccolithophore Plates.

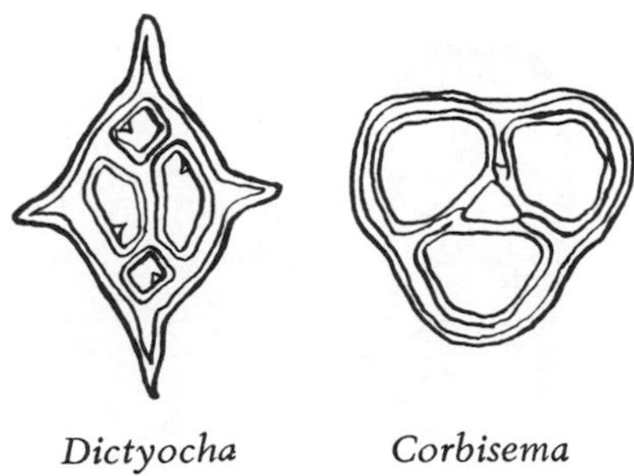

FIGURE 22.2 Cenozoic Silicoflagellates.

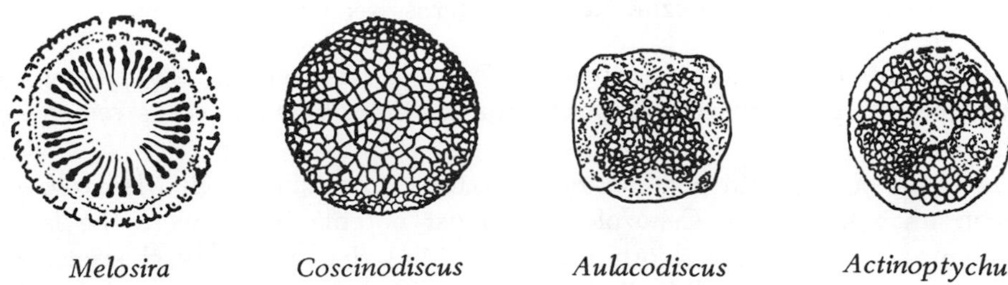

FIGURE 22.3 Cenozoic Marine Diatoms.

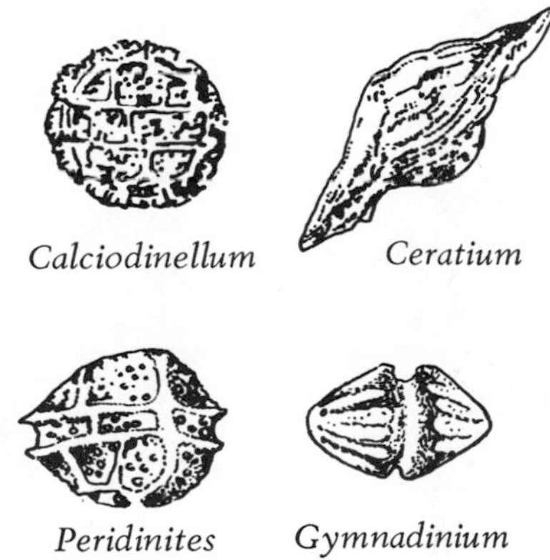

FIGURE 22.4 Cenozoic Marine Dinoflagellates.

The cellulose cysts of *dinoflagellates* (Fig. 22.4) including the *hystrichospheres* are another common Cenozoic fossil group. These pelagic producers are very important in the marine ecosystem also. Representative genera are *Gymnodinium, Hystrichosphaeridium,* and *Peridinium.*

The *red algae,* like the green and the brown, include numerous seaweed types ill-suited for preservation and therefore probably much more abundant in the past than our record shows. More so than the green or brown, however, the red algae have evolved lineages with calcareous remains. The encrusting sheets of *Lithothamnion* and the delicate, jointed branches of the coralline red algae

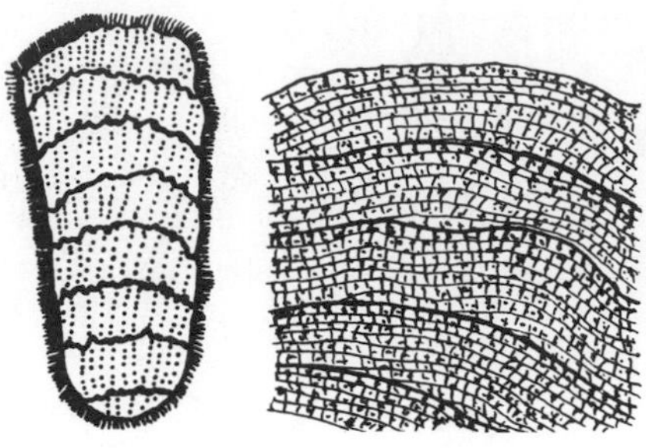

FIGURE 22.5 Cenozoic Red Algae (cross-sections).

are significant, but usually overlooked, Cenozoic fossils (Fig. 22.5). They are major elements in coral reefs of the Cenozoic where they served as cementing agents.

The *foraminifera*, both benthic and planktonic, continued their Mesozoic expansion unabated in the Cenozoic. The most notable forms were the large, disk-shaped nummulites and orbitoids (Fig. 22.6) that especially characterized the Paleogene or Nummulitic Period. *Nummulites, Discocyclina,* and *Lepidocyclina* were common genera. These forams were very numerous in the shallow,

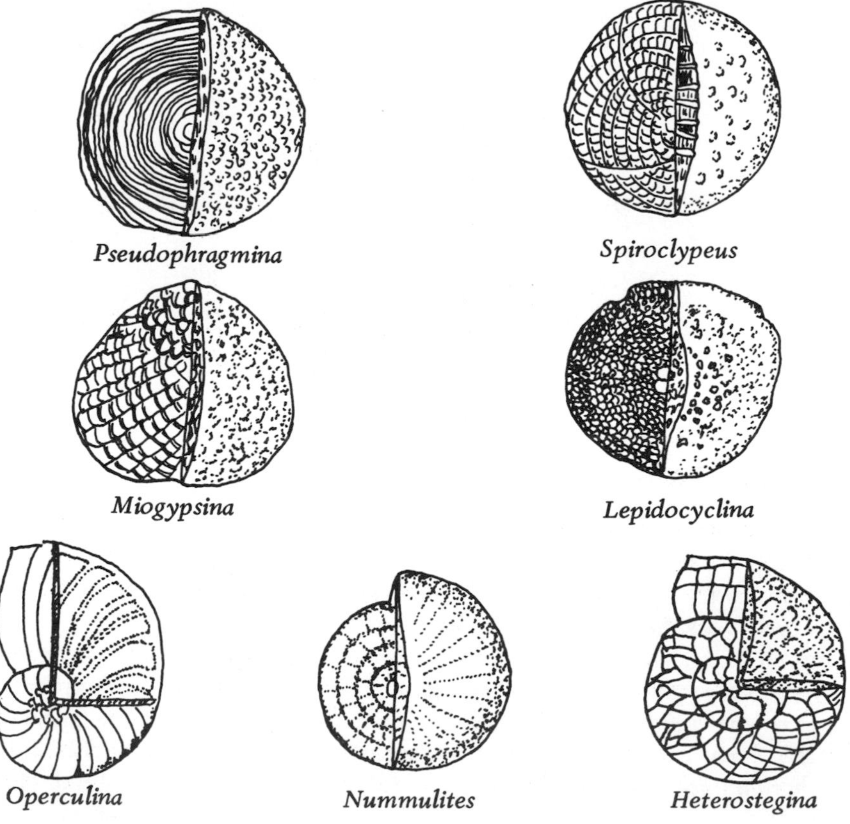

FIGURE 22.6 Cenozoic Large Disc-Shaped Forams.

tropical seas of that time, particularly Tethys. Their remains are very abundant in the limestone used to face the Egyptian pyramids and the Greek historian Heredotus mistook them for food lentils carried by the slaves who toiled on the tombs. These large foraminifera, and the similarly adapted but elongate wheat-grain shaped *alveolinids* (Fig. 22.7), were drastically reduced and geographically restricted by the cooling climates of the Late Cenozoic. The abundant small forams (Fig. 22.8) include genera such as the benthonic *Quinqueloculina, Bulimina,* and *Eponides* and the planktonic *Globigerina.*

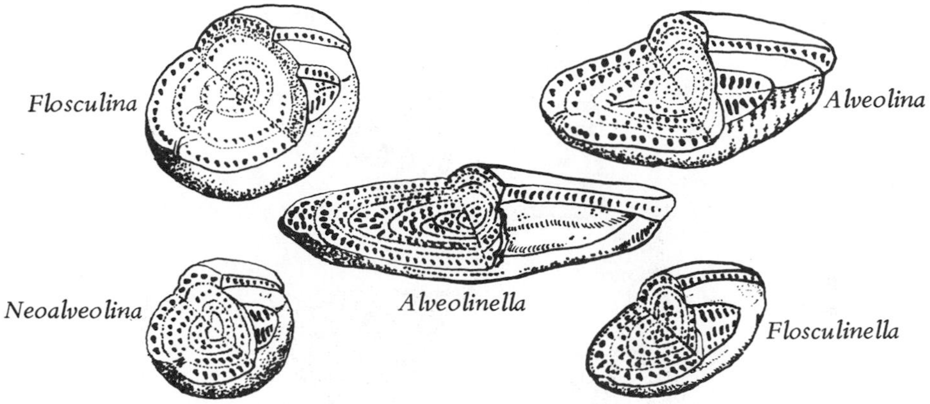

FIGURE 22.7 Cenozoic Alveolinid Forams.

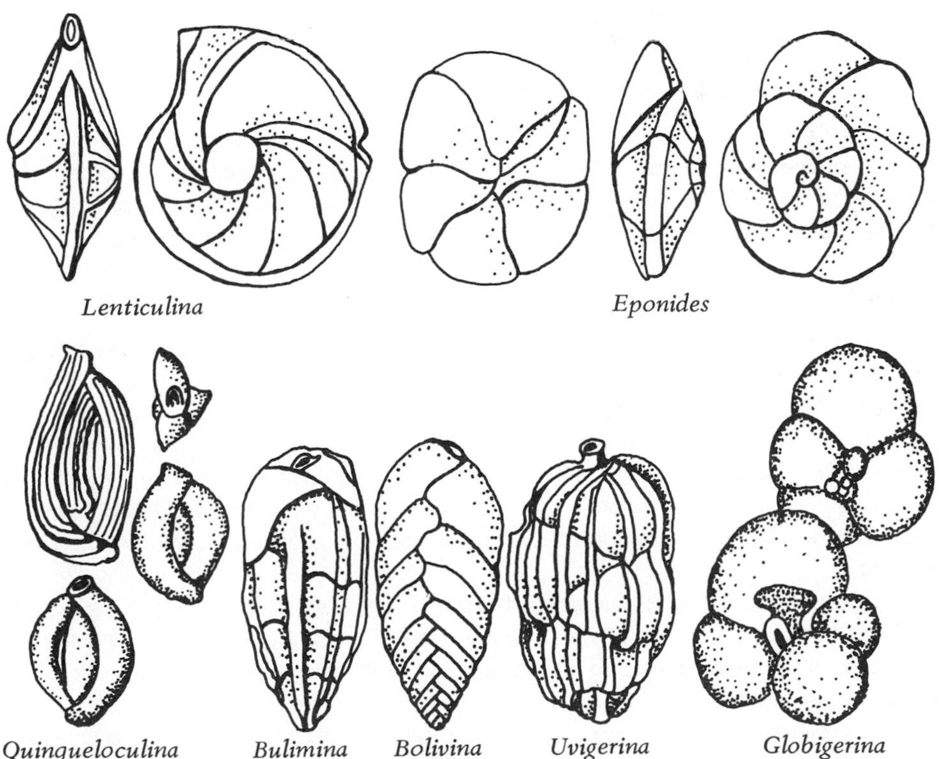

FIGURE 22.8 Cenozoic Foraminifera.

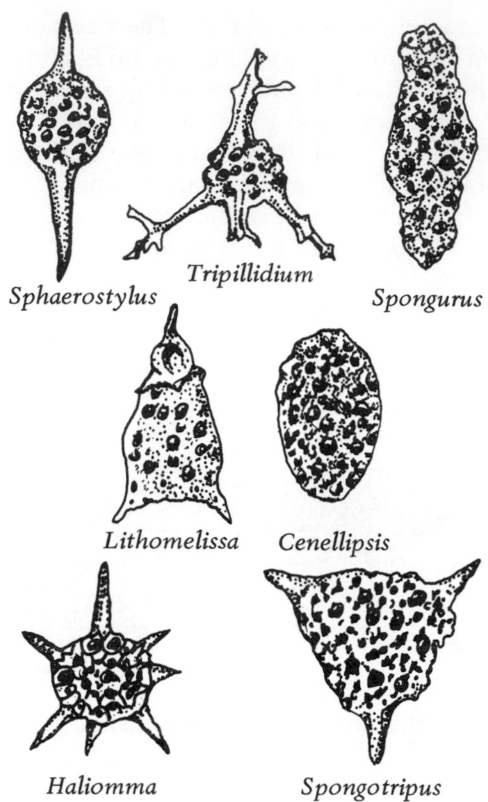

FIGURE 22.9 Cenozoic Radiolaria.

The siliceous *radiolaria* (Fig. 22.9) reach a peak in Cenozoic deposits. They are abundant in both the deposits of marine basins marginal to continents and the oceanic oozes. *Sphaerostylus* and *Tripilidium* are common genera.

Animals

Hexacorals, particularly the reef formers, were abundant and widespread in the Early Cenozoic epochs when seas were more widespread and climates milder. They became reduced and more restricted geographically with the cooling trend of the Late Cenozoic. Also, the mountain-building that severed the worldwide east-west oceanic connections in the Middle Cenozoic led to provincialism of the coral fauna. Common Cenozoic genera (Fig. 22.10) include the solitary *Flabellum* and *Balanophyllia* and the colonial *Astrhelia* and *Septastrea*.

Bryozoan faunas were dominated by the delicate *cheilostomes* (Fig. 22.11) such as *Flustra* and *Pliophloea*. Cyclostomes (Fig. 22.12), such as the twig-like *Idmonea*, remained second in importance. Cenozoic bryozoans are very abundant, but so small and delicate they are too often overlooked.

The mollusk faunas were dominated by the gastropods and pelecypods, though the conical *scaphopods* (Fig. 22.13) were always common. *Gastropods*

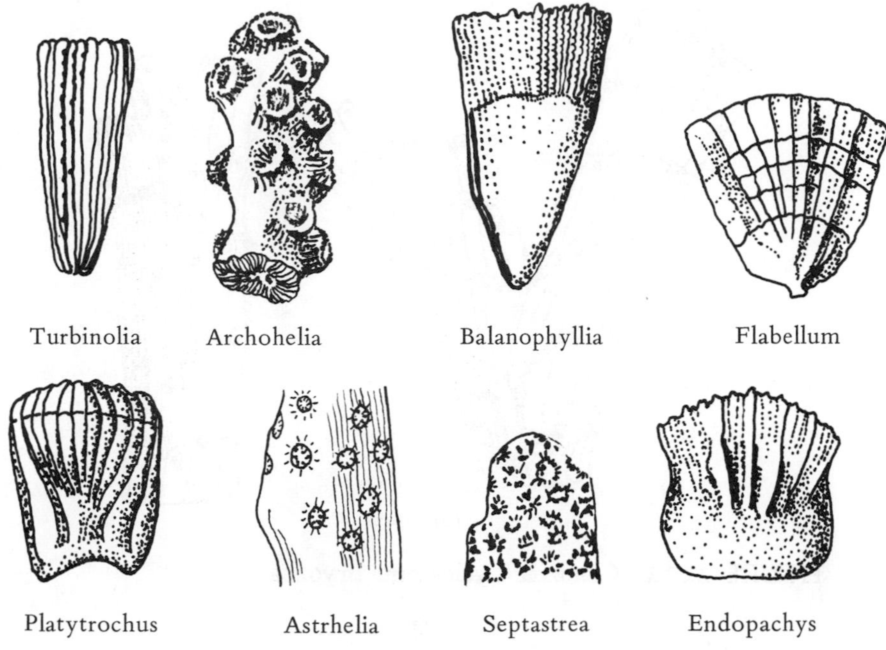

FIGURE 22.10 Cenozoic Hexacorals.

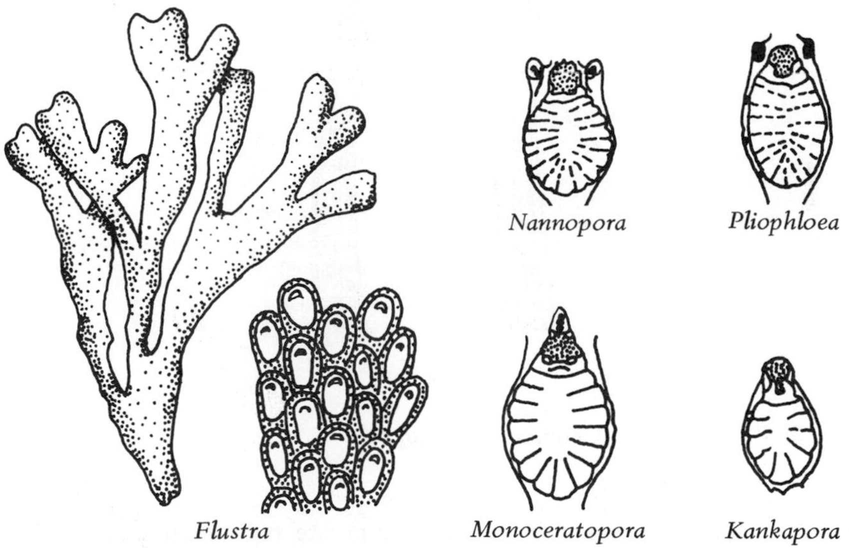

FIGURE 22.11 Cenozoic Cheilostome Bryozoa.

(Fig. 22.14) continued to increase in numbers and variety throughout the Cenozoic and are at their peak at the present when they are second only to insects in diversity. Numerous genera of gastropods such as *Calliostoma, Crepidula, Polinices, Siphonalia,* and *Conus* occur. Worthy of special mention, however, is the high spired genus *Turritella* and its relatives. This subtropical to tropical genus is very abundant and widespread through the Miocene when climates

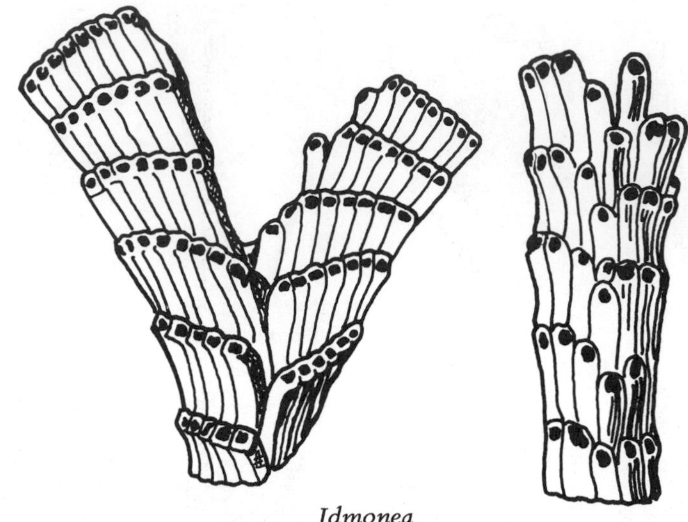

Idmonea

FIGURE 22.12 Cenozoic Cyclostome Bryozoa.

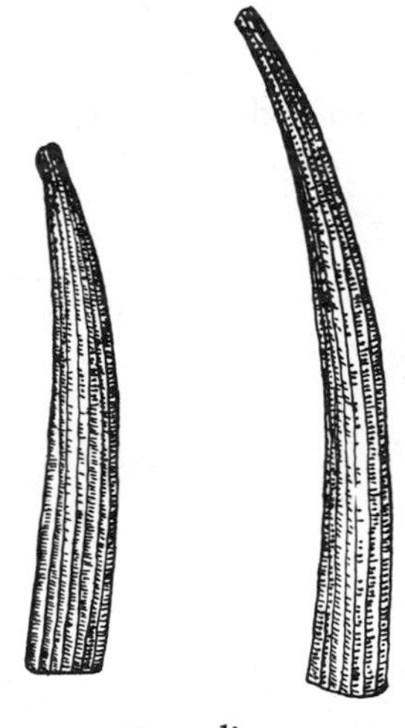

Dentalium

FIGURE 22.13 Cenozoic Scaphopods.

were milder than at present in current temperate regions and is widely used for correlation. Like most other warm-water marine forms, its range and numbers decreased in the cooling Late Cenozoic. *Pelecypods* (Fig. 22.15) also continue their great expansion in the Cenozoic. The modern burrowing types, such as *Chione, Mercenaria,* and *Tellina,* dominated, but the arks such as *Anadara,* the scallops such as *Lyropecten,* and the oysters such as *Ostrea* continued to be important. A significant Paleogene group of large bivalves is represented by the genus *Venericardia.* In the Neogene, many of the scallops and oysters reached very large size.

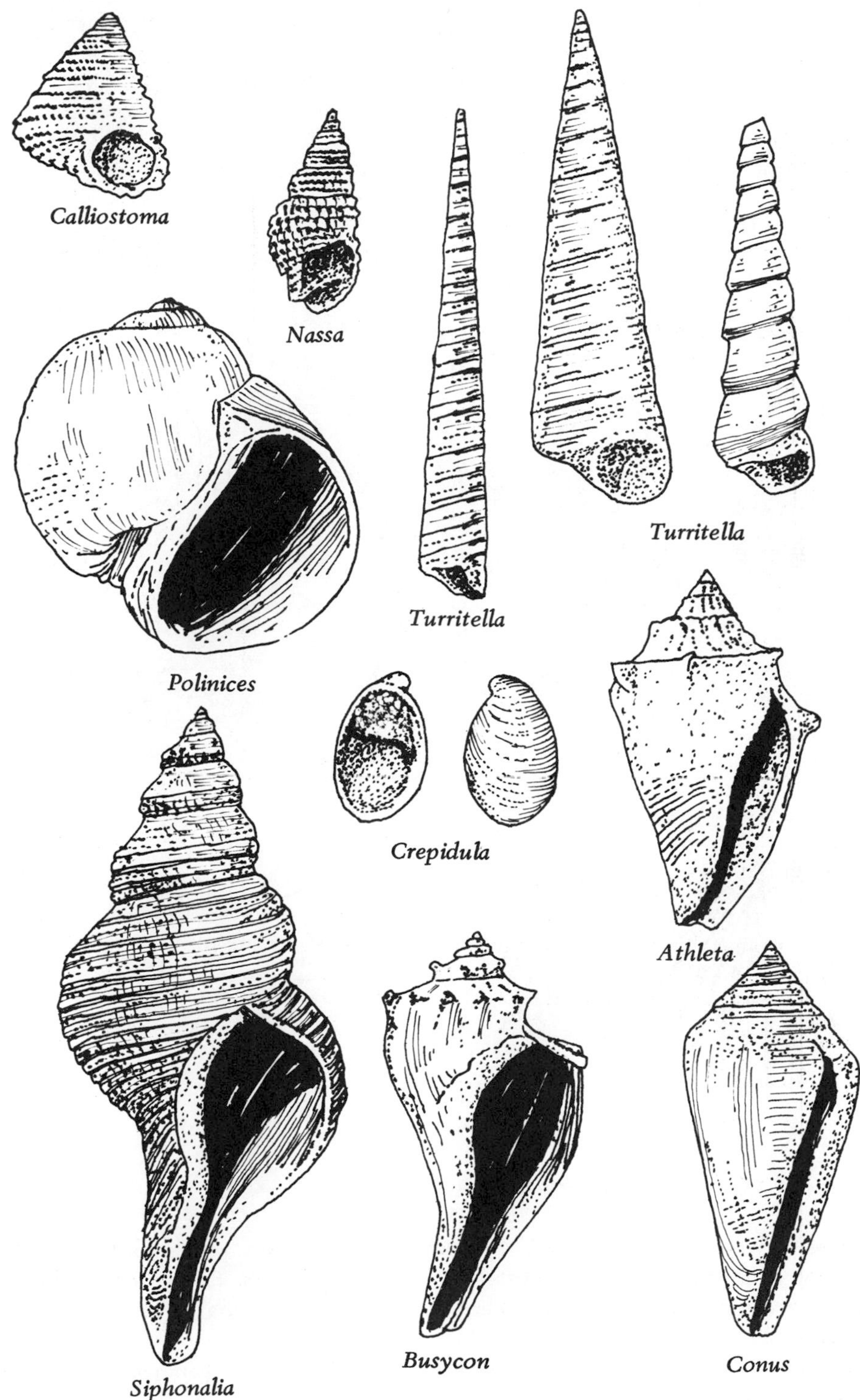

FIGURE 22.14 Cenozoic Gastropods.

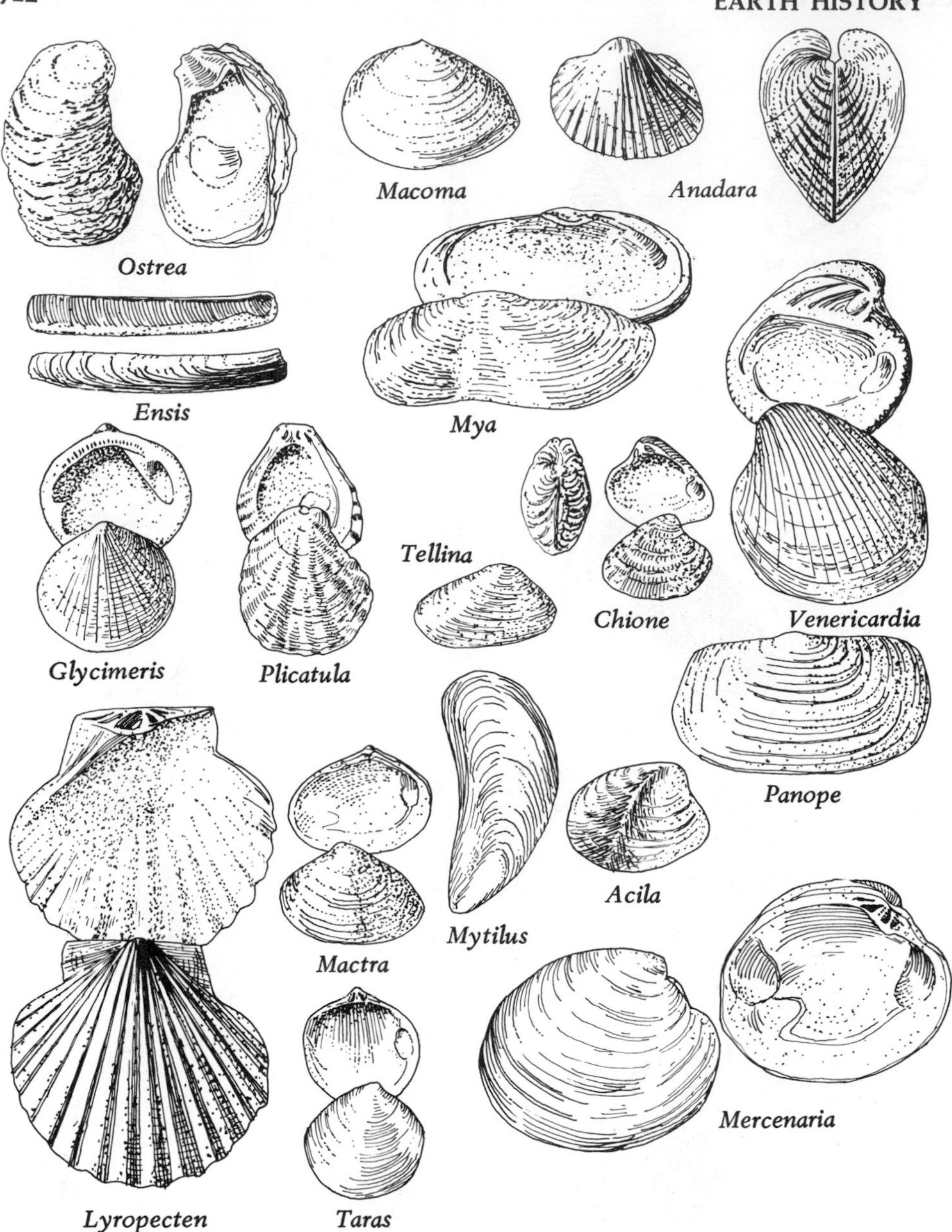

FIGURE 22.15 Cenozoic Bivalves.

Annelid worm tubes (Fig. 22.16) such as *Serpula* and the coiled *Spirorbis* are very common Cenozoic fossils, particularly as encrustations on mollusks. They are often overlooked, however, because of their small size and general simplicity.

The *crustacean* arthropods have several notable Cenozoic groups. The marine *ostracods* (Fig. 22.17) continue in abundance. *Hemicythere* and *Cytheridella* are typical forms. The *barnacles* (Fig. 22.18) become important in the Cenozoic. Some species of *Balanus* reach truly gigantic size in the Neogene. Finally, the

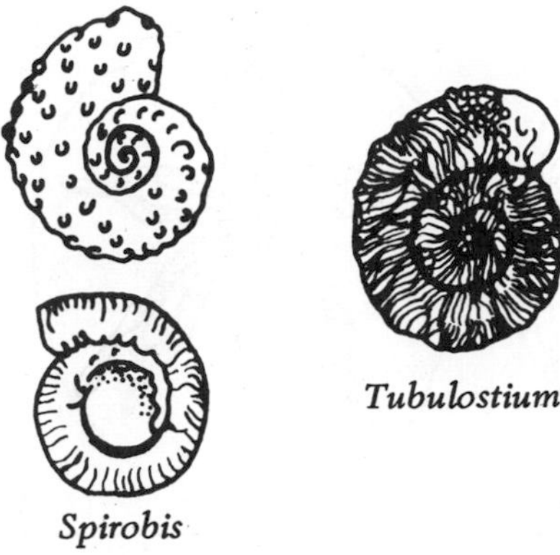

FIGURE 22.16 Cenozoic Annelid Worm Tubes.

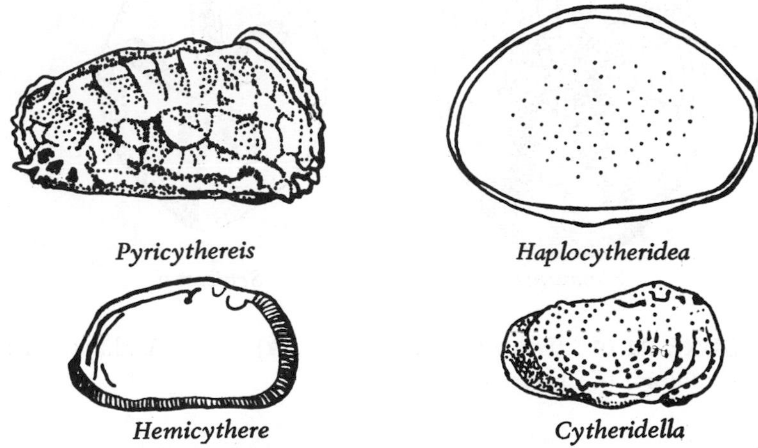

FIGURE 22.17 Cenozoic Ostracods.

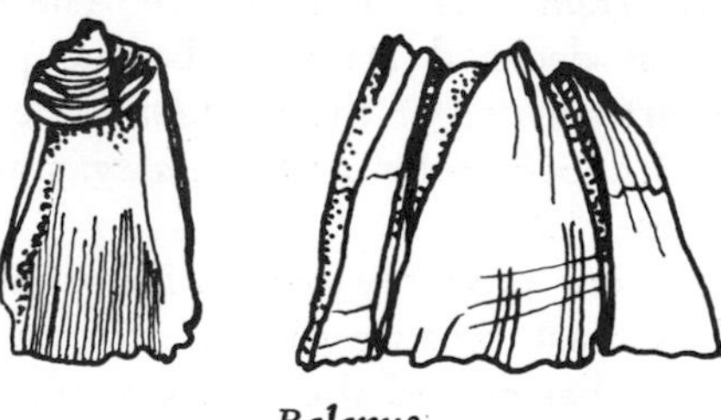

FIGURE 22.18 Cenozoic Barnacles.

malacostracan crustaceans are well represented by broken bits of appendages and carapaces and occasionally more complete fossils.

The *echinoids* (Fig. 22.19) are another group that reached their peak in the Cenozoic. The irregular burrowing forms dominated, but this is probably a function of their living in a favorable environment for fossilization. The *sand dollars* were the most successful Cenozoic echinoids and their internally sup-

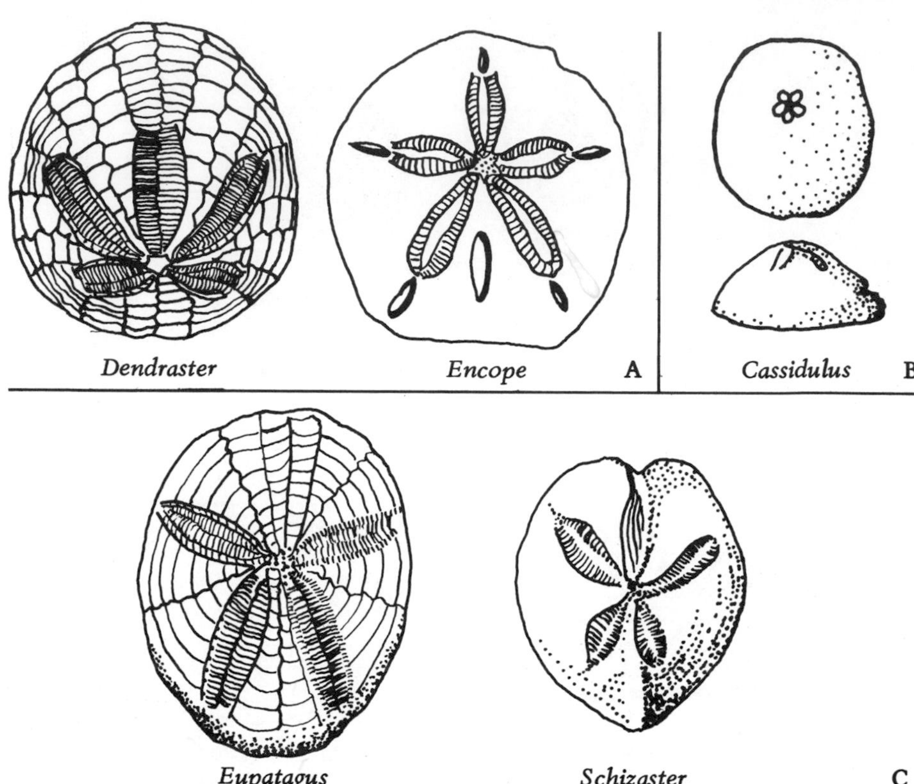

FIGURE 22.19 Cenozoic Echinoids. (A) Sand Dollars, (B) Cassiduloid, (C) Heart Urchins.

ported tests are well suited for preservation. They were common both in the tropical Early Cenozoic faunas and the cooler ones of the Late Cenozoic. Common genera include *Dendraster* and *Encope.* The cassiduloids are chiefly a warm-water group that declined after an Eocene peak. *Cassidulus* was a common genus. The *heart urchins* of the modern Order Spatangoida are common in Cenozoic deposits. *Schizaster* and *Eupatagus* were common members of this thin-shelled group.

Marine invertebrates were probably more diverse in the Cenozoic than at any time in the geologic past because of the extreme fragmentation of continental shelves and ocean basins and the strong north-south climatic gradients that developed in the Neogene (Figs. 3.11, 22.20).

The *fish* (Fig. 22.21), mostly modern ray-finned bony types, continued their success in the sea. Their scales, otoliths (ear bones), and disarticulated skeletons are abundant fossils (Fig. 22.22), but complete, articulated remains are only locally common. The cartilaginous fish are represented by numerous shark, skate, and ray teeth (Fig. 22.22C). In some areas, these teeth were heaped together in such amounts that they are major elements of the sediment.

Mammals succeeded in the sea with the evolution of the whales (Fig. 22.23) in Africa in the Eocene and in the pinniped carnivores (Fig. 22.24) in the North Pacific in the Miocene.

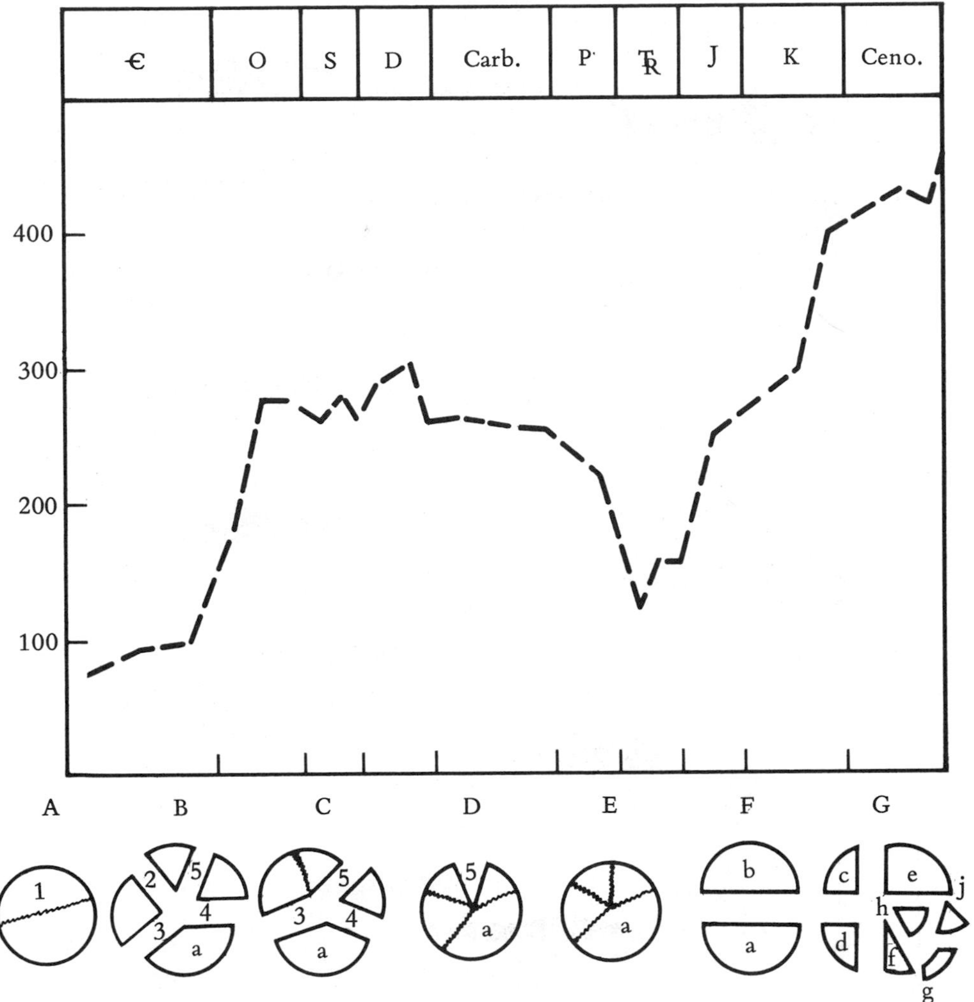

FIGURE 22.20 Correlation of Organic Diversity with Continental Assembly and Fragmentation. (A) Eocambrian suturing of Pan-African-Baikalian Orogeny and the Formation of Pangaea I; (B) Fragmentation of Pangaea in Cambrian-Ordovician and the development of (1) Caledonian-Acadian, (2) Appalachian, (3) Hercynian, and (4) Uralian Geosynclines; (C) Silurian-Devonian suturing of the Caledonian-Acadian Orogeny; (D) Permocarboniferous suturing of Appalachian-Hercynian Orogeny; (E) Permotriassic suturing of Uralian Orogeny and the formation of Pangaea II; (F) Triassic-Early Jurassic development of Tethyean Geosyncline; (G) Cretaceous-Recent closing of Tethys, opening of Atlantic, fragmentation of Gondwanaland. (a) Gondwanaland, (b) Laurasia, (c) North America, (d) South America, (e) Eurasia, (f) Africa, (g) Antarctica, (h) India, (j) Australia. (From Valentine & Moores, *Nature*, vol. 228, Nov. 14, 1970. Used by permission.)

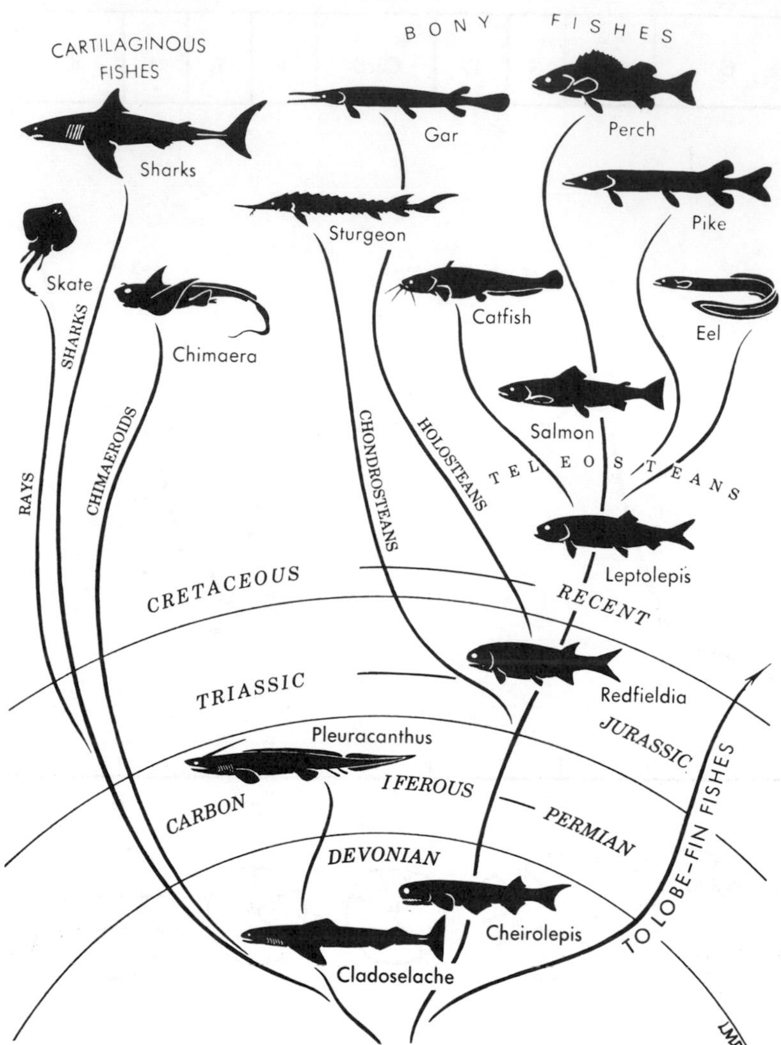

FIGURE 22.21 Evolution of the Cartilaginous and Bony Fishes. Note great expansion in the Cenozoic. (From Edwin H. Colbert, *Evolution of the Vertebrates*, Copyright © 1969, by permission of John Wiley & Sons, Inc.)

Fresh-Water Life

The fresh-water environment is abundantly represented by Cenozoic lake and stream deposits. The fossilizable denizens of this environment, *charaphytes, diatoms, gastropods, pelecypods, ostracods,* and *fish,* are preserved in great abundance. Representative genera are illustrated in Figure 22.25. The Eocene Green River lake beds are a particularly rich North American locality for fresh-water fossil organisms (Fig. 22.26).

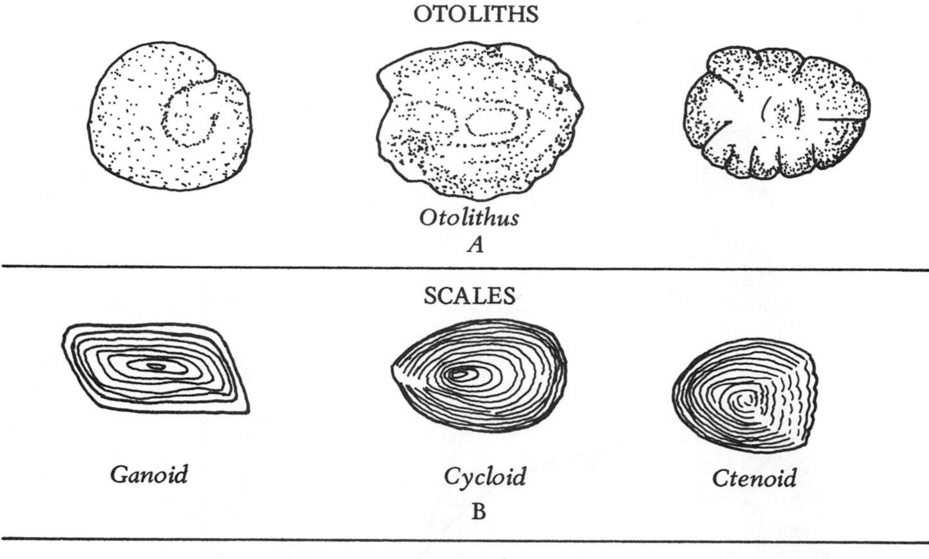

FIGURE 22.22 Cenozoic Fish Fossils. (A) Bony fish otoliths, (B) Bony fish scales, (C) Shark teeth.

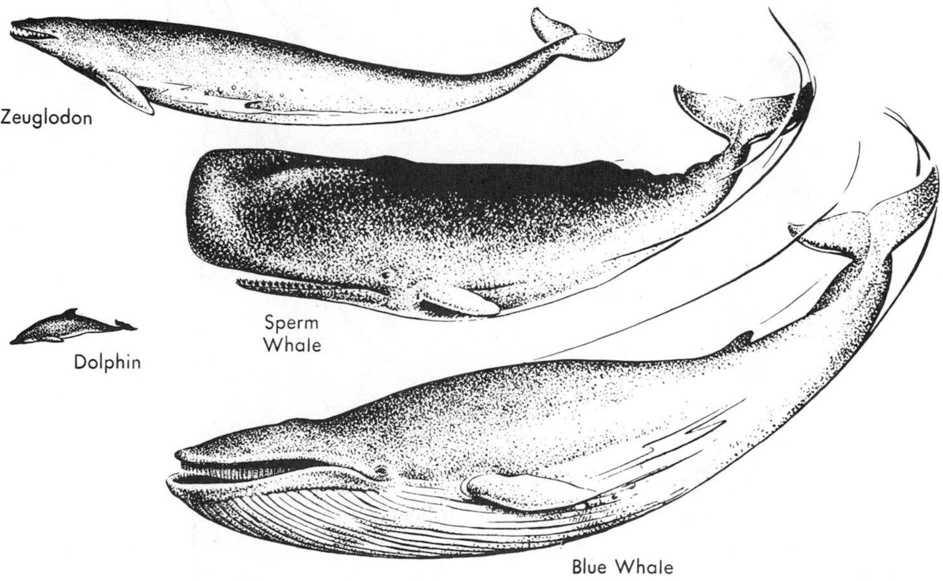

FIGURE 22.23 Cenozoic Whales. (From Edwin H. Colbert, *Evolution of the Vertebrates*, Copyright © 1969, by permission of John Wiley & Sons, Inc.)

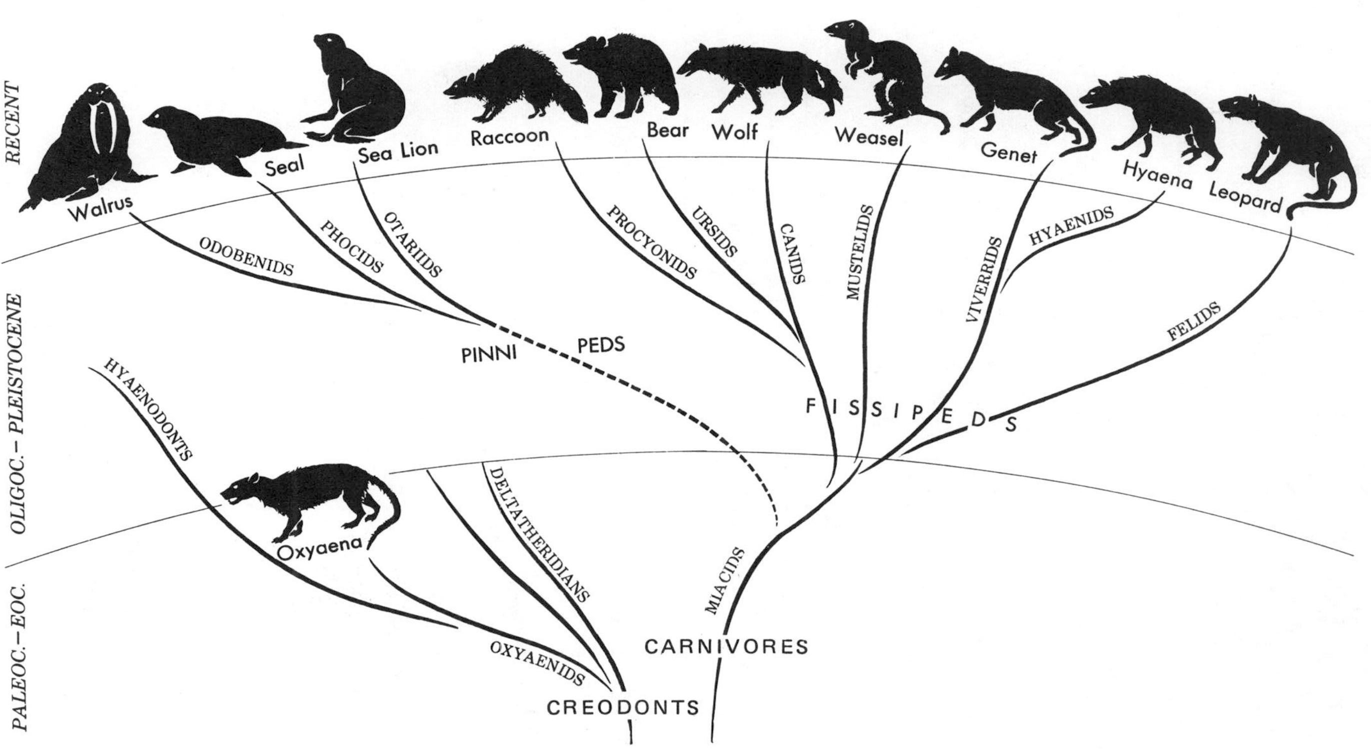

FIGURE 22.24 Evolution of the Carnivorous Mammals. (From Edwin H. Colbert, *Evolution of the Vertebrates*, Copyright © 1969, by permission of John Wiley & Sons, Inc.)

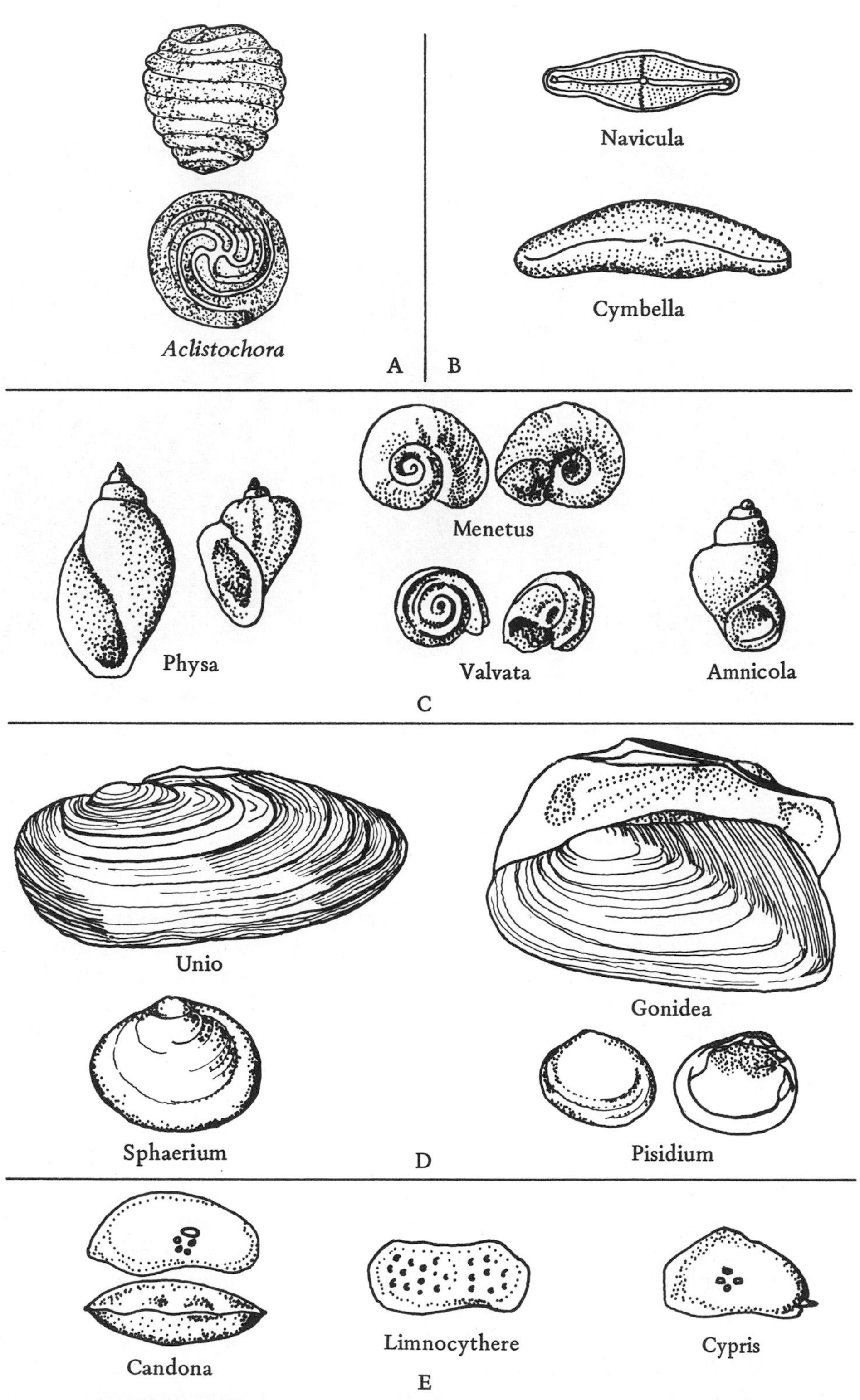

FIGURE 22.25 Cenozoic Fresh-Water Protistans and Invertebrates. (A) Charophyte, (B) Diatoms, (C) Gastropods, (D) Bivalves, (E) Ostracods.

FIGURE 22.26 Diorama of Life in the Eocene Green River Lake. (Courtesy Field Museum of Natural History.)

Land Life

On land, the major invertebrate group continued to be the ever-expanding arthropod class of *insects* (Fig. 22.27). Well-preserved remains occur in the Eocene Green River lake beds of Wyoming, the Oligocene Baltic amber beds, the Oligocene Florissant lake beds of Colorado, the Miocene lake beds of the California Mojave Desert, and several other localities. Arachnean and myriapod arthropods also occur in these deposits, but are less common. Gastropods (Fig. 22.28) were also successful on land in the Cenozoic and are occasionally common fossils.

Mammals

Vertebrate life on land was dominated by the *mammals*. Amphibians, reptiles, and birds occurred, but most forms were small and found chiefly as disarticulated and broken pieces. Mammalian evolution was strongly influenced by the shifting pattern of the continents in the Cenozoic (Fig. 20.58). In the Late Cretaceous, you may recall, the old Laurasia and Gondwanaland Supercontinents had fragmented into most of their present pieces, and even some of these were separated into smaller entities by epeiric seas. Early marsupial and/or placental mammals invaded all the major continental blocks (with the possible

FIGURE 22.27 Oligocene Fly in Amber. *Sphegina carpenteri* from the Baltic amber beds. (Courtesy of Frank M. Carpenter.)

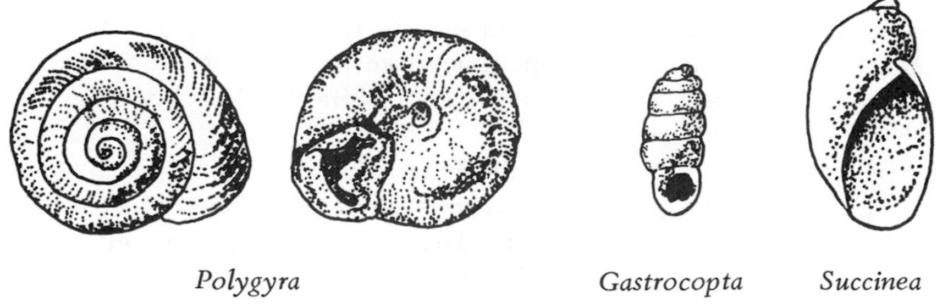

FIGURE 22.28 Cenozoic Terrestrial Gastropods.

exception of India) before this occurred, but then were free to evolve in isolation as the pieces moved apart.

North America and Eurasia remained in intermittent contact throughout the Cenozoic and there was much interchange between them. In the Early Cenozoic, the Greenland-Norway rift was still narrow and permitted communication. In the Late Cenozoic, North America and Eurasia came into contact in the Alaska-East Siberia area. The carnivorous marsupials, insectivores, bats, primates, edentates, rodents, rabbits, creodonts, carnivores, condylarths, amblypods, both odd- and even-toed hoofed animals or ungulates (perissodactyls and artiodactyls, respectively), and some other minor groups evolved on the northern continents in the Late Cretaceous or Early Paleogene (Fig. 20.58). Only the marsupials, edentates, condylarths and creodonts were able to invade the southern continents before they were too isolated for immigration to occur, and only the marsupials made it to Australia-Antarctica. Bats, because of their flying habits, were able to spread to all continents. Through the remainder of the Paleogene, South America, Africa, and Australia were each isolated and had their own mammalian evolution separate from that of the northern continents. (Nothing is known of Paleogene faunas in India or Antarctica.)

Australia was inhabited only by some surviving, egg-laying *monotremes*, which perhaps should be classified as surviving mammal-like reptiles, and the *marsupials* (Fig. 20.58). The marsupials diverged from their original carnivorous stock into herbivorous and omnivorous types. In so doing, all three varieties converged in many ways on the placental orders filling the same niches elsewhere in the world (Fig. 7.119A). For example, there are Australian marsupials that closely resemble moles, squirrels, groundhogs, mice, and wolves, and others which do not closely resemble any northern continent animal, but are the ecologic equivalents of mammals found there. For example, the kangaroo is the ecological equivalent of the large herbivores of the northern continents. Only as Australia approached southeast Asia in the Late Pleistocene were a few northern mammals such as the ubiquitous rats and the ingenious primate man able to invade the continent. The recent arrival of western man has, unfortunately, flooded the continent with northern mammals that are superior competitors to the native marsupials which they are rapidly exterminating.

South America has, as original mammalian inhabitants, the carnivorous marsupials, the edentates, and the herbivorous condylarths. The *marsupials* filled the carnivore role, as no creodont or true carnivore had appeared before its isolation, and also evolved into some rat-like forms (Fig. 20.58). The carnivorous forms were eliminated by competition from the northern continents' carnivores such as cats, dogs, weasels, and raccoons after the Mio-Pliocene reunion of North and South America.

The *edentates* (Fig. 22.29) had their center of evolution in South America for the few ancestral Early Cenozoic forms of North America soon became extinct. The two main lineages, the armadillos, and the sloths, underwent spectacular evolution, particularly in the Neogene. The giant armadillo-like glyptodonts of the Pleistocene reached 9-foot lengths and the great ground sloths such as *Megatherium* reached 20-foot lengths. After the Neogene rejoining of North and South America, both the armadillos and the ground sloths successfully invaded North America, the latter becoming extinct in the Early Holocene, perhaps by the hand of man. The South American *condylarths* diversified and gave rise to three very successful orders of South American ungulates (Fig. 22.30), the bulky rhino-like or hippo-like *toxodonts* (a few northern forms may belong here indicating the group arose before isolation, but did not survive in the north), the elephant-like *astrapotheres,* and the horse and camel-like *litopterns.* Once again, we see convergent evolution of isolated forms in the absence of the typical placental ungulate orders. These South American groups reached the Oligocene-Early Miocene peak, but when the Panama land bridge formed in the Late Miocene-Early Pliocene interval, the competition

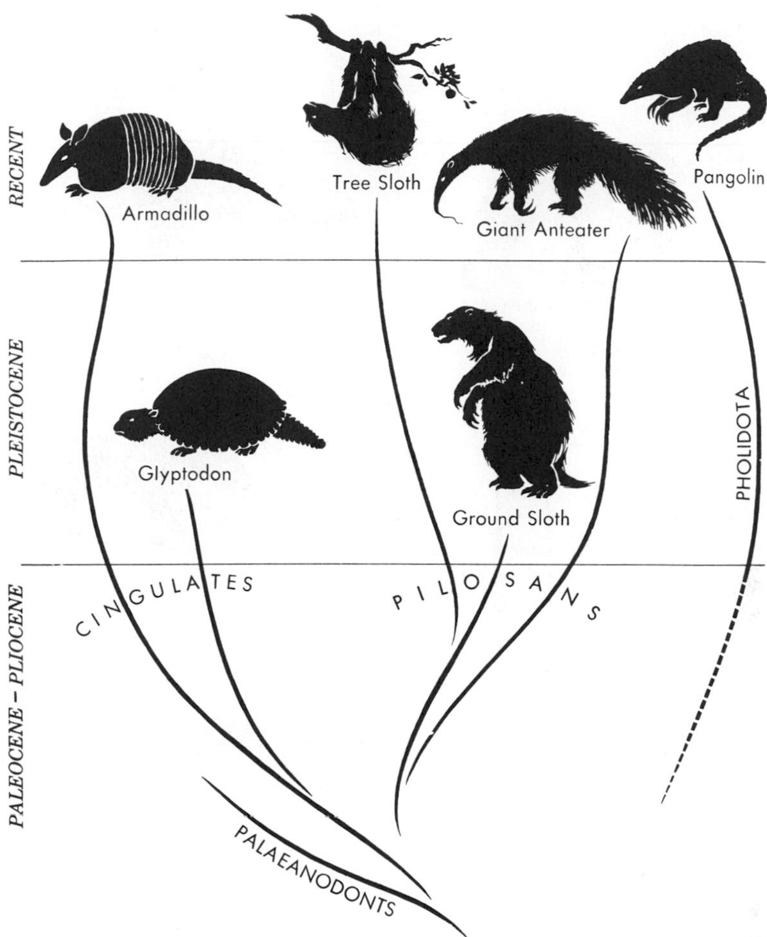

FIGURE 22.29 Evolution of the Edentate Mammals. (From Edwin H. Colbert, *Evolution of the Vertebrates,* Copyright © 1969, by permission of John Wiley & Sons, Inc.)

of the invading northern herbivores such as the perissodactyls, artiodactyls, and proboscideans drove them all to extinction before the close of the Pleistocene. In the Oligocene, there was apparently close enough proximity of North America and South America for a few small mammals to get into the southern continent. Some *rodents* entered South America and, in their isolation, evolved a whole host of unique forms such as guinea pigs, chinchillas, pacas, and the New World porcupine (Fig. 22.31). They did not suffer attrition after the two continents joined, and a few, such as the porcupine, successfully invaded the north. The other successful Oligocene invaders were the small *primates* that gave rise to the unique South American monkey assemblage (Fig. 22.32). There being no present-day equatorial jungles in North America, the group has never migrated out of South America or come in contact with Old World monkeys.

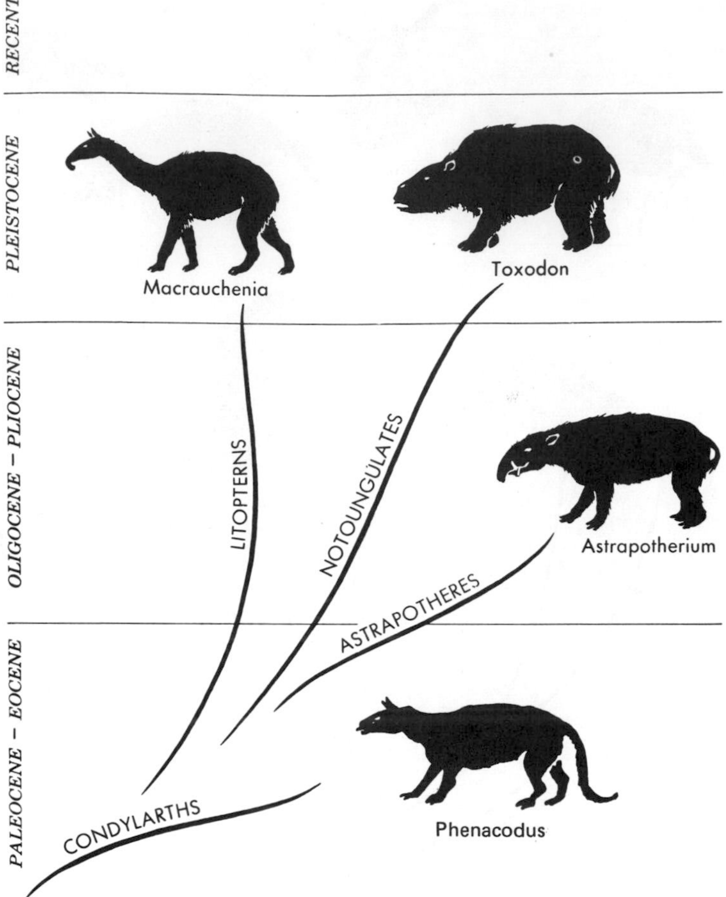

FIGURE 22.30 Evolution of the South American Ungulate Animals. (From Edwin H. Colbert, *Evolution of the Vertebrates,* Copyright © 1969, by permission of John Wiley & Sons, Inc.)

Africa was isolated for much less time than the other southern continents. Its original complement of pre-isolation mammals consisted of the *creodonts* and *condylarths.* The carnivorous creodonts were not abundant, but are the apparent ancestors of the *whales,* whose first carnivorous members occur in the Eocene of North Africa (Fig. 22.23). Likewise, the condylarths amounted to very little themselves in Africa, but gave rise to several orders of herbivore subungulates, some of which converged on the mammals of the other continents. These groups (Figs. 20.58, 22.33) were the large herbivorous *proboscideans* including the elephants and mastodons, the small rabbit-like *hyraxes* or true conies, the *embrithopods* which resembled huge rhinos, the *aardvarks* which resemble the South American edentate anteaters, and the *sirenians* or *sea cows* which took to the shallow seas. When Africa and Eurasia came into contact in the Oligocene and Miocene, the proboscideans not only survived, but eventually spread to all the continents except Australia; the hyraxes stayed

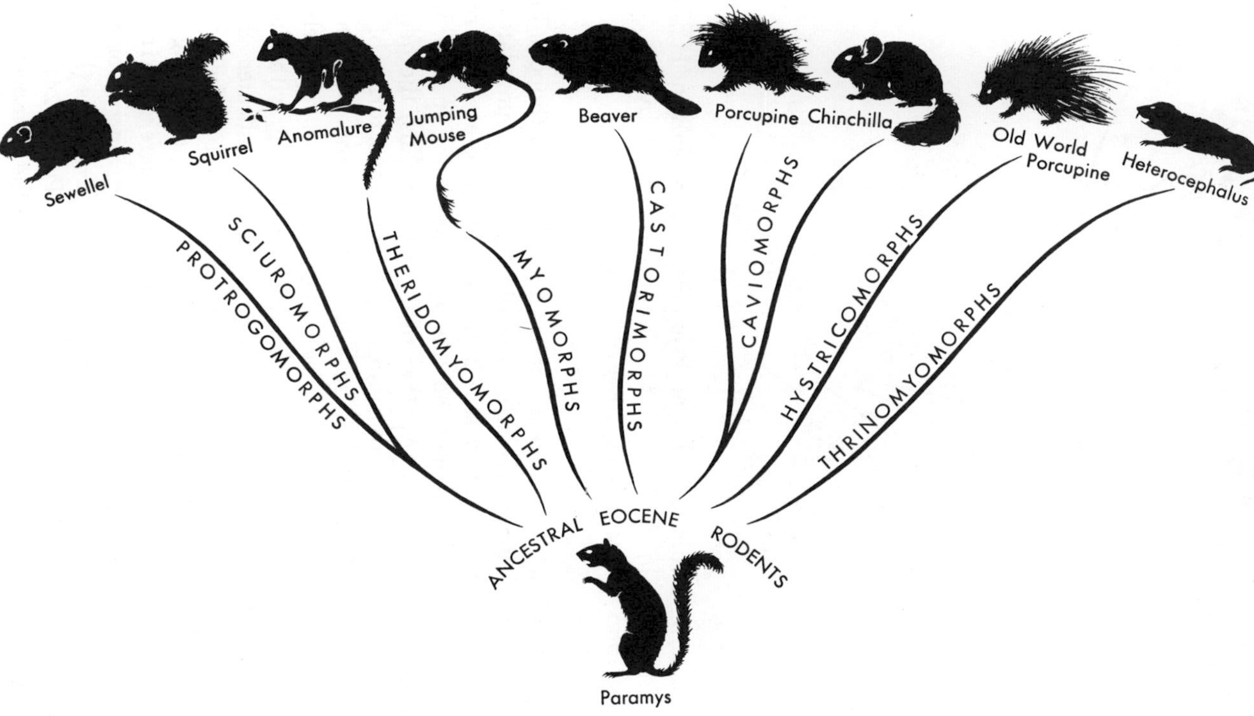

FIGURE 22.31 Evolution of the Rodents. (From Edwin H. Colbert, *Evolution of the Vertebrates*, Copyright © 1969, by permission of John Wiley & Sons, Inc.)

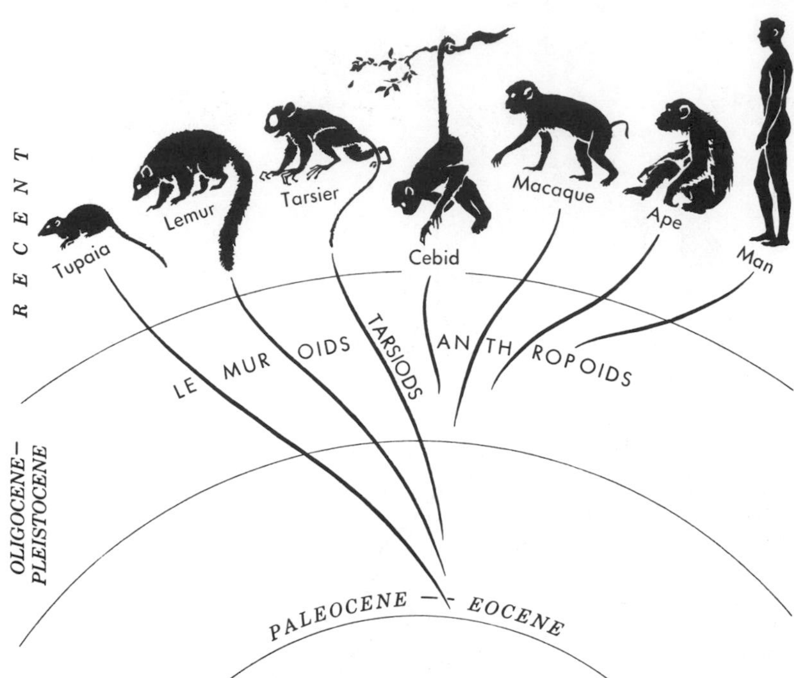

FIGURE 22.32 Evolution of the Primates. (From Edwin H. Colbert, *Evolution of the Vertebrates,* Copyright © 1969, by permission of John Wiley & Sons, Inc.)

at home save for one short-lived invader of Europe; the embrithopods became extinct, probably from competition with invading northern ungulates such as rhinos; the aardvarks remained in Africa and no edentate ever has reached that continent to compete with them; and the sea cows, being marine, rapidly spread through the world's oceans. Another order of mammals, the *desmostyloids,* were Miocene and Pliocene mollusk-eating, amphibious forms of hippo-like proportions inhabiting the shallow seas of east Asia and western North America. Their closely packed cylindrical teeth (Fig. 22.34) are relatively common fossils in some California nearshore marine beds along with the large oysters and pectens on which they fed. Their structure indicates they are related to the other subungulates but their African origin and their route to the North Pacific are unknown, though the migration is not surprising because the creatures were amphibious. The major northern invaders of Africa after the junction with Eurasia were the *primates* which underwent a considerable African radiation discussed in the next chapter (Fig. 22.32); *rodents* (Fig. 22.31); the true *carnivores* including the hyaenas, dogs and cats (Fig. 22.24); the *perissodactyls* including the claw-bearing chalicotheres and the rhinoceroses (Fig. 22.35); and the *artiodactyls* including pigs, hippos, camels, giraffes, and antelopes (Fig. 22.36). Africa has served as a Neogene refuge for many of the large members of these groups that did not survive the cooling climates of the northern continents.

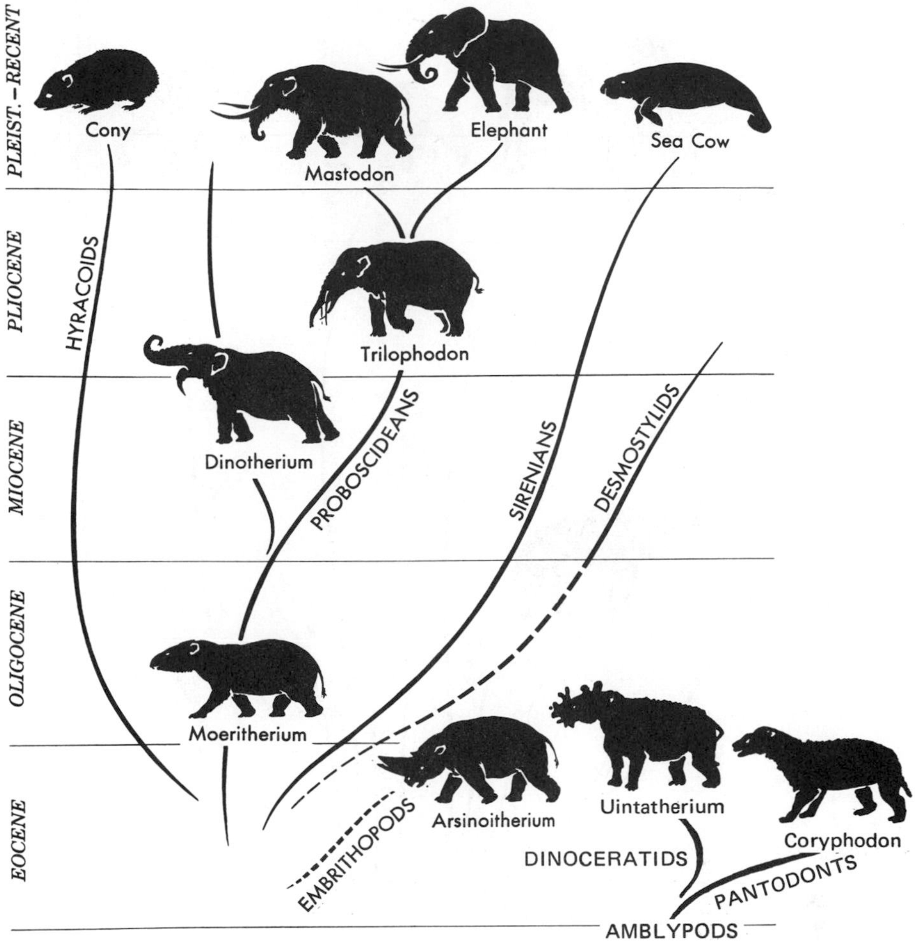

FIGURE 22.33 The Sub-Ungulate Mammals. (From Edwin H. Colbert, *Evolution of the Vertebrates*, Copyright © 1969, by permission of John Wiley & Sons, Inc.)

FIGURE 22.34 Restoration and Teeth of a Desmostylid. (After Stirton)

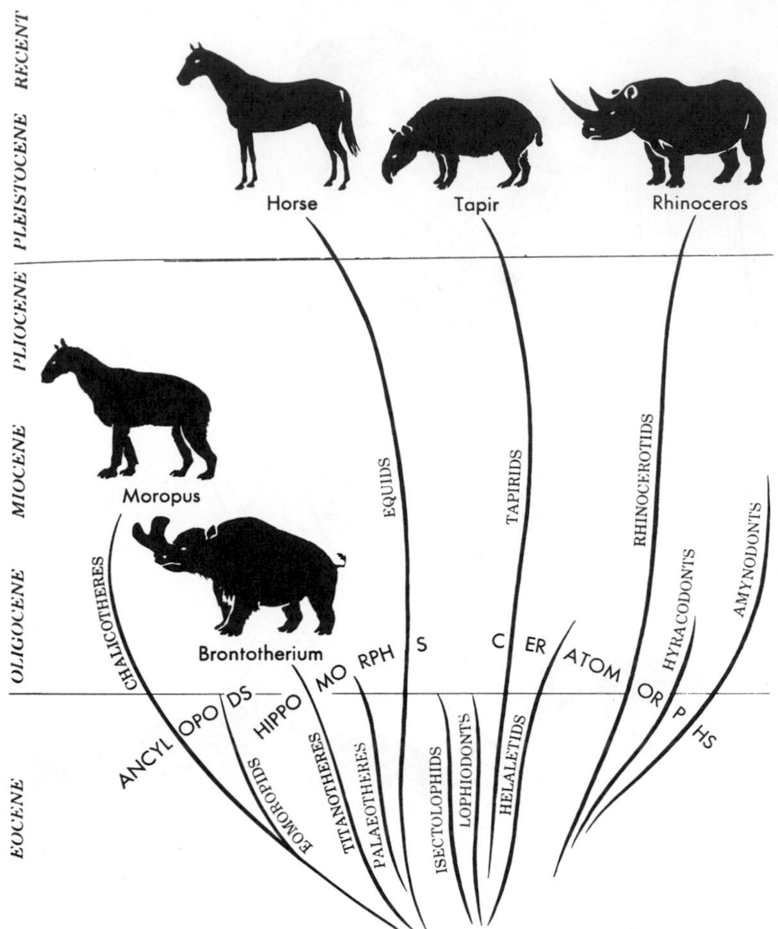

FIGURE 22.35 Evolution of the Perissodactyls or Odd-Toed Ungulates. (From Edwin H. Colbert, *Evolution of the Vertebrates*, Copyright © 1969 by permission of John Wiley & Sons, Inc.)

The interconnected northern land masses of *North America* and *Eurasia* were the main center of evolution of the insectivore, primitive primate, rodent, rabbit, creodont, carnivore, condylarth, amblypod, perissodactyl, and artiodactyl mammalian groups (Figs. 22.37–22.42). In addition, the flying bats originated there from insectivores, but have since spread to all continents. *Insectivores* include the Cenozoic shrews, moles, and hedgehogs, all rather small forms, but the nearest living relatives of the ancestral placental mammals. The primitive *primates* (Figs. 23.4, 23.5), such as the lemurs and tarsiers, were worldwide in the northern continents in the widespread moist forests of the Paleogene. They retreated to isolated jungle refuges in Africa, Madagascar, and southeast Asia in the Neogene. *Rodents* (Fig. 22.31) have become the most diverse of all mammalian orders and had spread to all continents by Late Neogene. The squirrel, rat-mouse, and beaver groups have centered their evolution in the northern continents, however. The *rabbits* (Fig. 22.43) were only slightly less

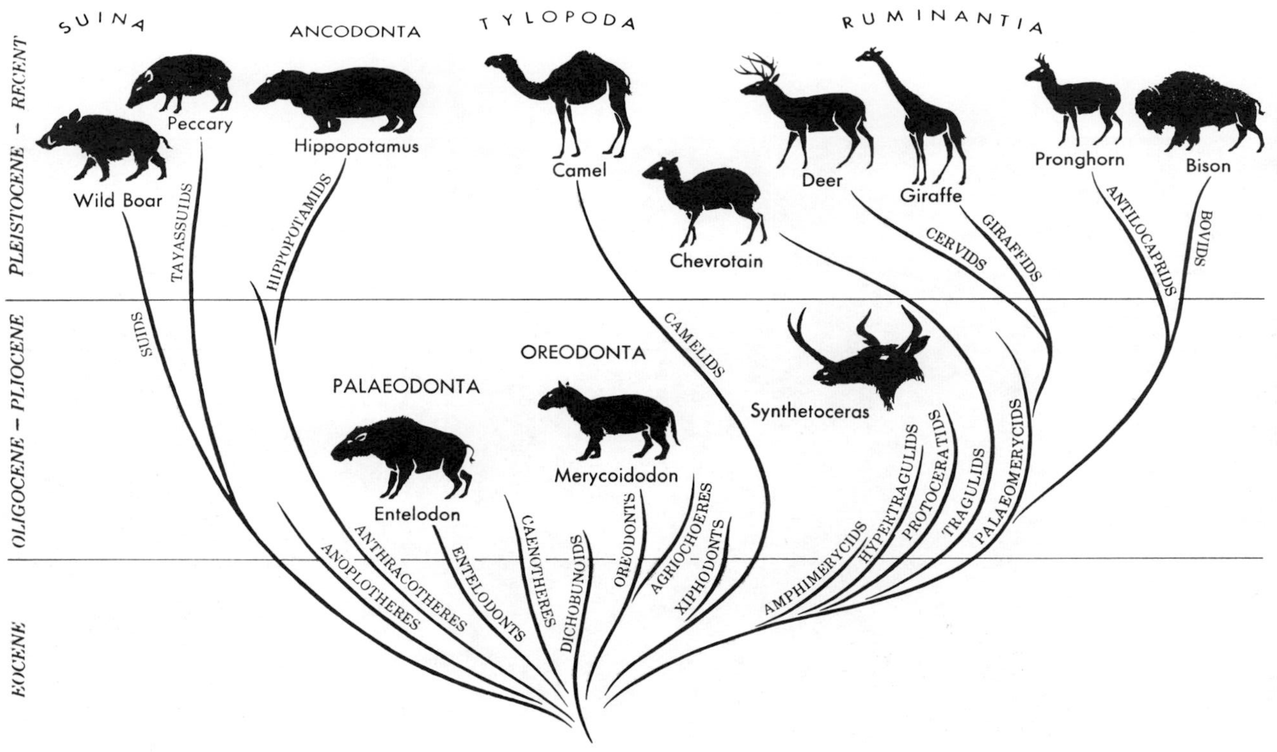

FIGURE 22.36 Evolution of the Artiodactyl or Even-Toed Ungulates. From Edwin H. Colbert, *Evolution of the Vertebrates,* Copyright © 1969 by permission of John Wiley & Sons, Inc.)

FIGURE 22.37 Paleocene Mammals. (Courtesy of Yale Peabody Museum.)

successful, not reaching Australia until their disastrous recent introduction by man. The small bones and teeth of both rodents and rabbits are among the commonest of mammalian fossils. The carnivorous *creodonts* and *carnivores* (Fig. 22.24) were almost exclusively northern continent forms until the Neogene when the true carnivores spread widely in Africa, India, and South America. The creodonts of the Paleogene were largely eliminated by competition with the superior true carnivores in the Neogene. The most spectacular true carnivores were the great stabbing cats such as *Smilodon* and the cave bears (*Ursus*) of the Pleistocene (Fig. 22.44). The fossil record is so complete for some lineages that we can trace the common ancestry of such groups as bears and dogs from bear-dogs. The *condylarths* and *amblypods* (including *uintatheres* and *pantodonts*), like the creodonts, radiated in the Paleogene, but were then gradually eliminated by competition from the more advanced herbivorous ungulates, the perissodactyls and artiodactyls. The odd-toed ungulates or perissodactyls reached an Early Neogene peak and have since declined nearly to extinction, while the even-toed artiodactyls have continued to expand indicating they may be superior competitors in the same environments.

Among the *perissodactyls* (Fig. 22.35) are five main lineages, the horses, brontotheres, chalicotheres (or ancylopods), tapirs, and rhinos. All were common in North America and indeed North America was the center for much of their evolution. Yet none have survived here until the recent and only the tapirs still live in the New World (excluding horses which were reintroduced by man). The horses form the best documented evolutionary lineages known in the fossil record (Fig. 4.18). The early small, many-toed, browsing forms lived in the sub-tropical forests of the Paleogene, but in the Neogene, with the expansion of grasslands, one lineage invaded that environment and eventually evolved into the large, one-toed grazing horses of today. The extinct chalicotheres were of horse-like build, but had large claws on their feet for digging plant roots and bulbs. The extinct, bulky brontotheres probably occupied the niche filled in the Late Cenozoic by the bison, an artiodactyl which evolved in Eurasia and did not reach America until the Pleistocene. (The grazing horses may have replaced the grazing brontotheres and may in turn have themselves been replaced by superior competitors, the bison.) The rhinoceroses underwent a significant radiation into many more adaptive types than their few Old World survivors. These included the agile running rhinos and the huge Asiatic Oligo-Miocene forms *Baluchitherium* (Fig. 22.45) and *Paraceratherium*, the largest land mammals known.

The *artiodactyls* (Fig. 22.36) include the swine, sheep, goat, cattle, and true antelope groups which underwent most of their evolution in the Old World and only sent a few, but very successful, invaders such as the peccaries, bighorn sheep, mountain goats, and bison into the New World in the Pleistocene. Other Old World artiodactyl groups such as the hippos, chevrotains, and giraffes never invaded the Americas. On the other hand, the camel, deer, and pronghorns underwent virtually all of their evolution in North America. The camels have since died out in North America, but survive in the Old World and South America which they invaded in the Late Cenozoic. The deer have spread to all continents except Australia, but the pronghorns have remained exclusively North American. The pig-like entelodonts, the sheep-like oreodonts, and the

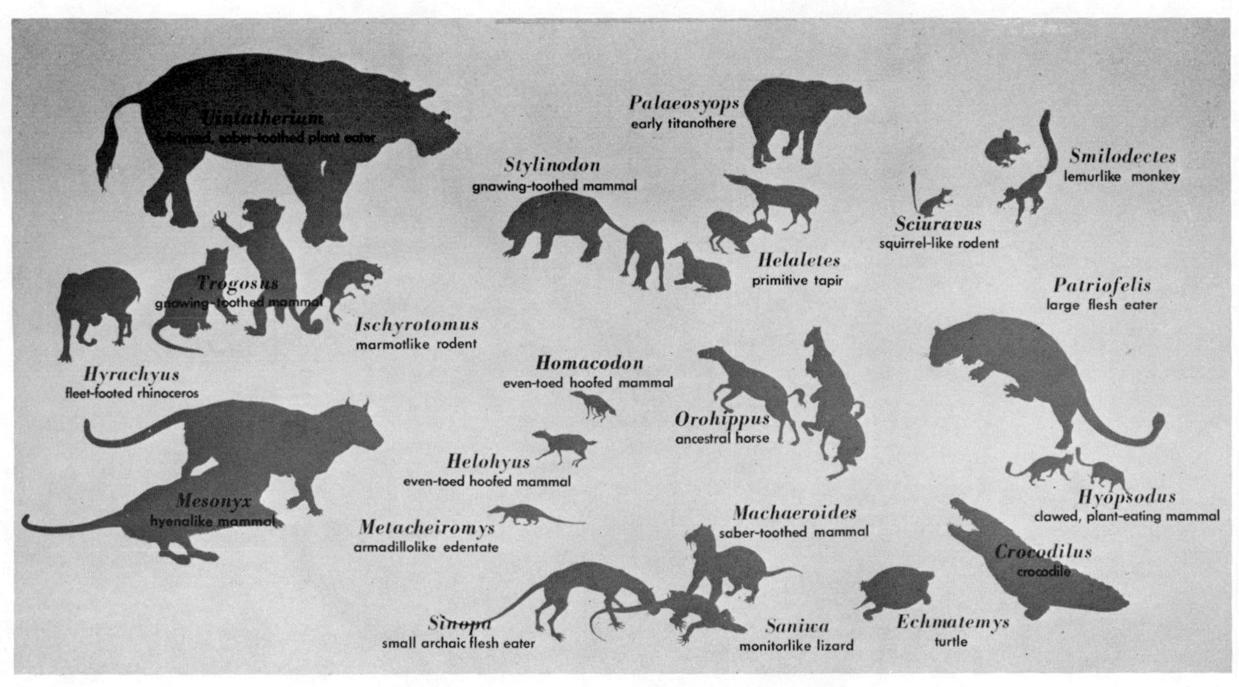

FIGURE 22.38 Eocene Mammals. Middle Eocene life of Wyoming. (Mural by J. H. Matternes, courtesy U.S. National Museum.)

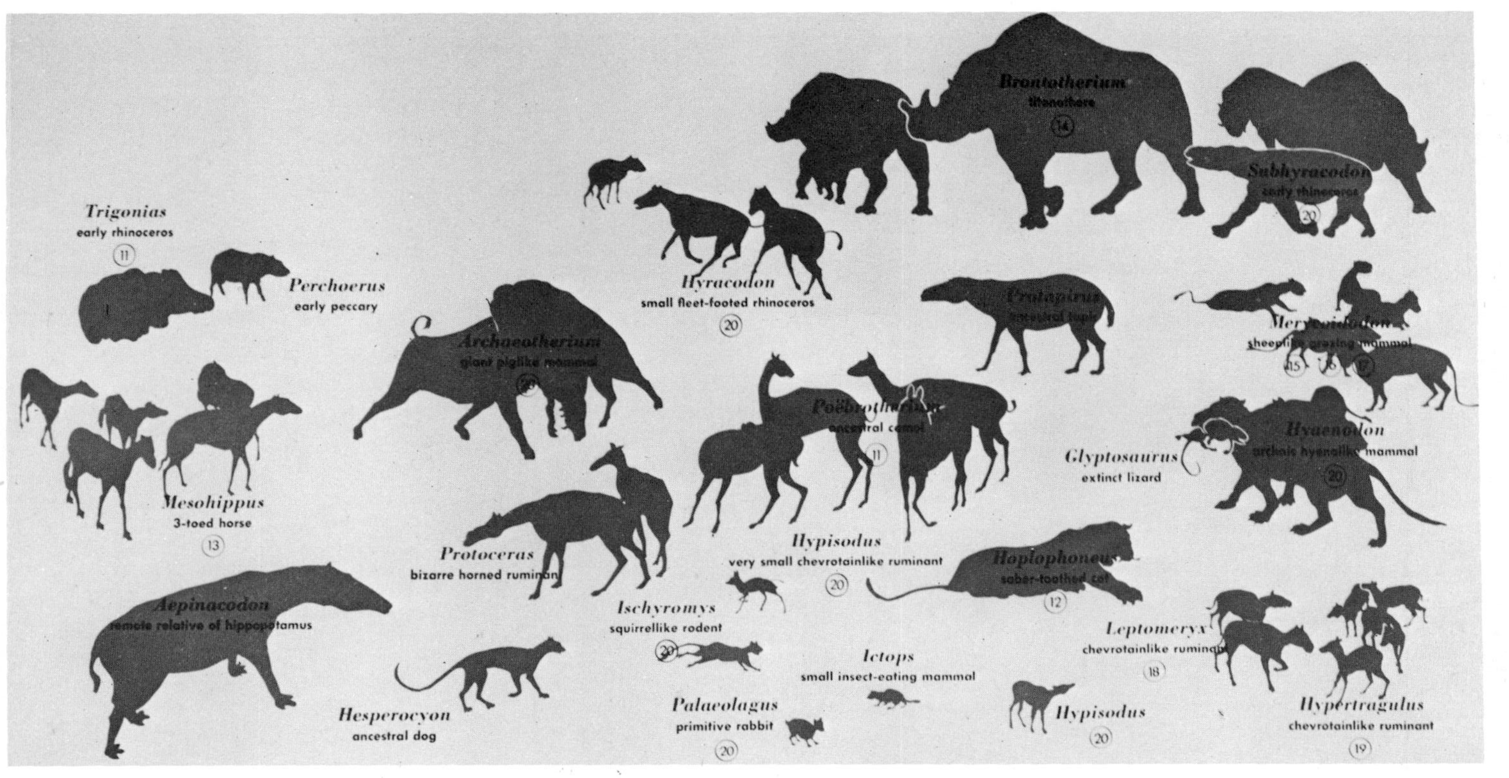

FIGURE 22.39 Oligocene Mammals. Early Oligocene life of South Dakota-Nebraska area. (Mural by J. H. Matternes, courtesy U.S. National Museum.)

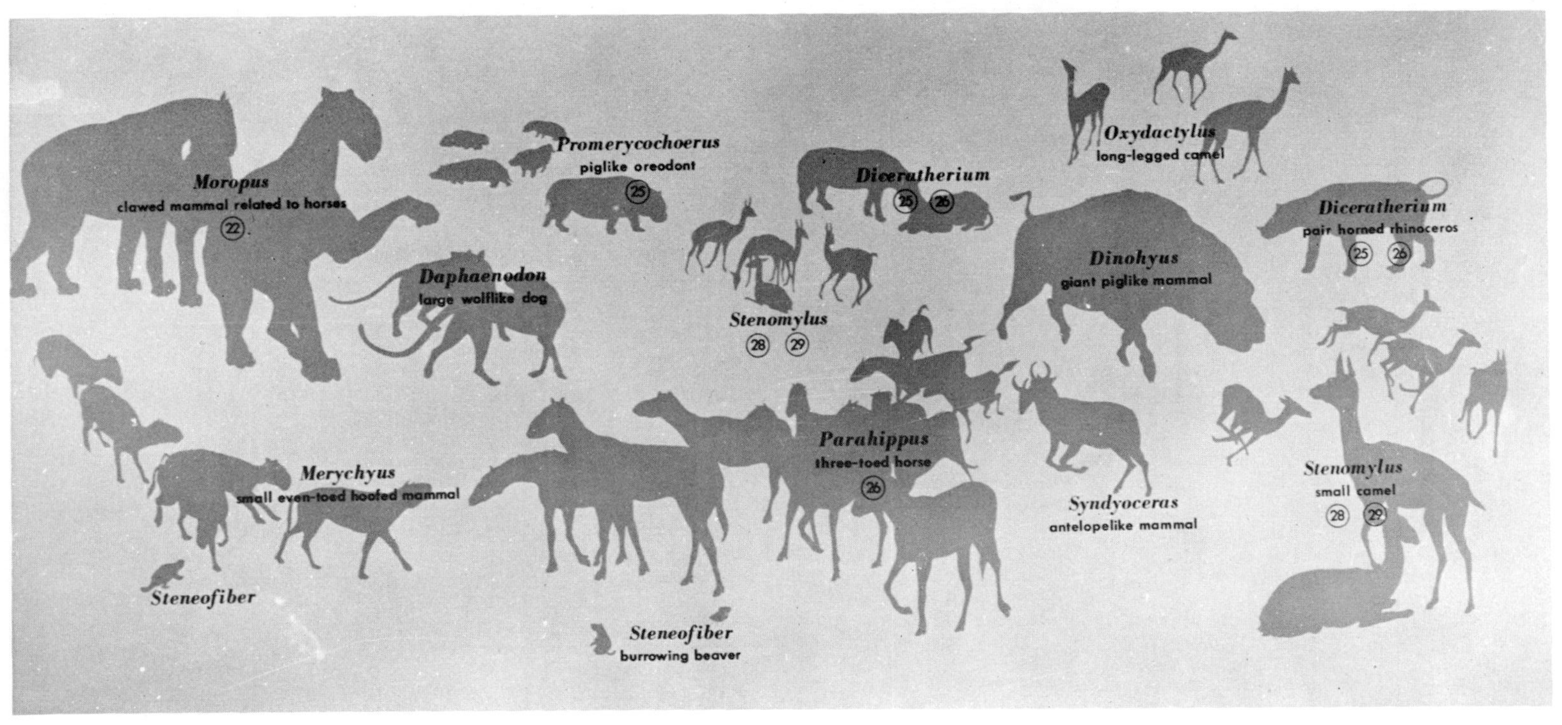

FIGURE 22.40 **Miocene Mammals.** Early Miocene life of Nebraska. (Mural by J. H. Matternes, courtesy U.S. National Museum.)

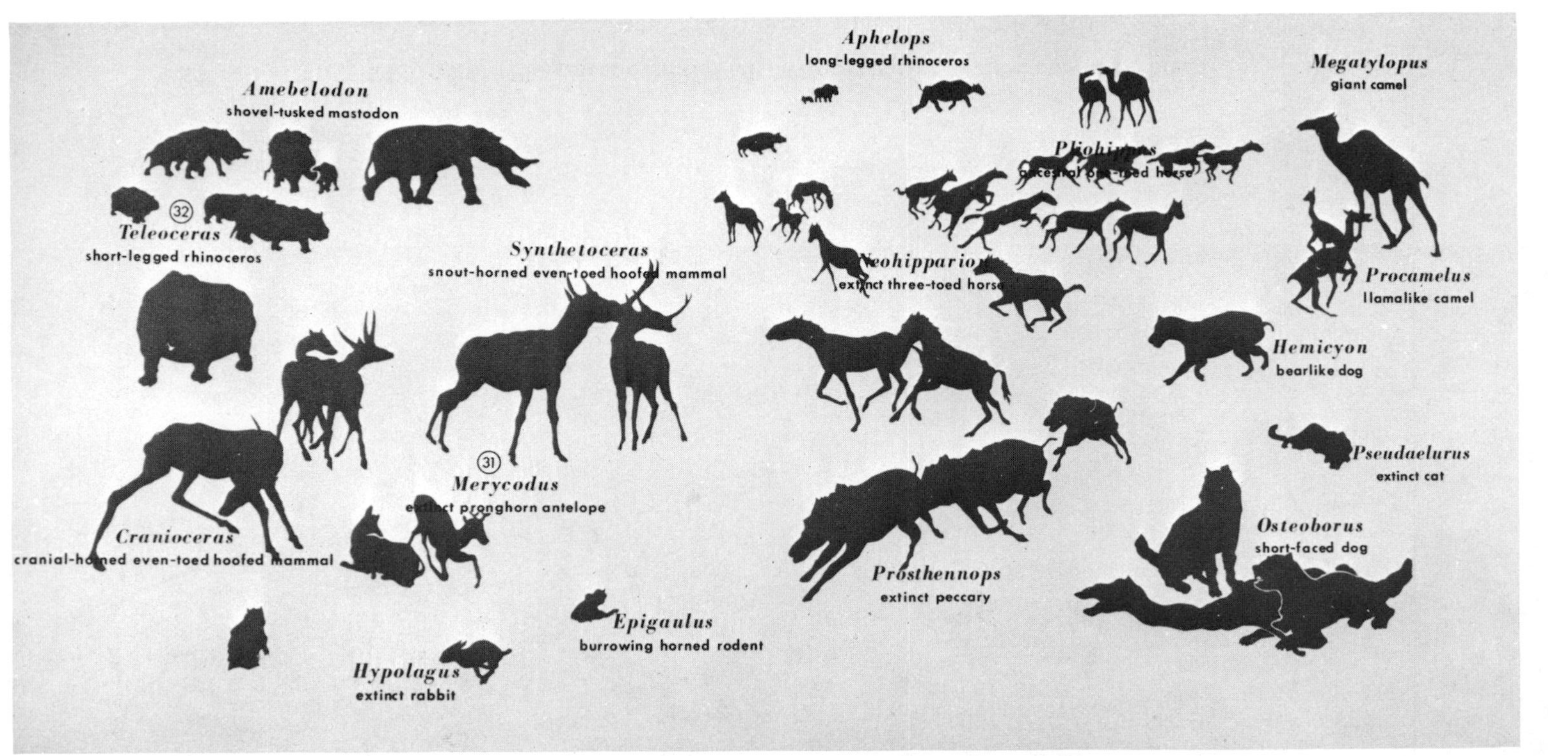

FIGURE 22.41 Pliocene Mammals. Early Pliocene life of the southern High Plains. (Mural by J. H. Matternes, courtesy U.S. National Museum.)

FIGURE 22.42 Pleistocene Mammals. (Courtesy of Yale Peabody Museum.)

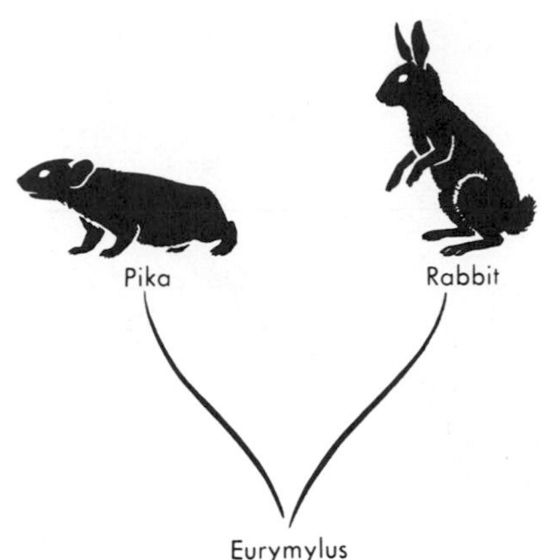

FIGURE 22.43 Lagomorph Mammals. (From Edwin H. Colbert, *Evolution of the Vertebrates,* Copyright © 1969, by permission of John Wiley & Sons, Inc.)

deer-like protoceratids, with their numerous cephalic horns, are extinct artiodactyl groups very abundant in North America in the Middle Cenozoic. The Oligo-Miocene rejoining of Africa with the northern continents permitted the successful invasion of the north by the proboscideans (Fig. 22.33), higher primates, and possibly the desmostylids.

Other notable features of mammalian evolution are the large members produced by almost every order living in the Pleistocene. These include giant marsupials, men, ground sloths, armadillos, beavers, elephants, and mastadons, rhinos, deer, and bison. Their size increase can probably be attributed to the cool climate of the Pleistocene for the larger an animal is, the lower the ratio of its surface area to volume, and, hence, the less heat loss. Almost all of these large forms are now extinct despite surviving numerous glacials and interglacials. Their extinction, along with that of many typical North American mammals around the Pleistocene-Holocene boundary (10,000 years ago), has been attributed to the rise of man, a Late Pleistocene invader, in the New World (Fig. 22.46). Men were known to have hunted these large forms, but whether they were actually able to completely eliminate whole groups of well-adapted mammals, and not others, remains a big question mark. Regardless of man's influence, there has certainly been an over-all reduction in mammalian diversity since the Miocene, a condition probably caused by the less favorable Late Neogene climate.

Birds

Before leaving the vertebrates, a few remarks must be made about the *birds.* This group is very significant in modern terrestrial faunas and certainly must have been so throughout the Cenozoic. The fragile skeleton of birds coupled with their aerial habitat makes them poor candidates for fossilization. Only scattered finds such as the Pleistocene La Brea Tar Pit of California (Fig. 22.44) give us a glimpse of the magnitude of bird evolution. One significant development in bird history was the evolution in the Early Cenozoic of large, flightless carnivores such as *Diatryma* and *Phororhacos* (Fig. 22.47). The former genus

FIGURE 22.44 La Brea Tar Pit Fauna. (Mural by C. R. Knight, courtesy American Museum of National History.)

FIGURE 22.45 Restoration of Baluchitherium. (C. R. Knight painting, courtesy of the American Museum of Natural History.)

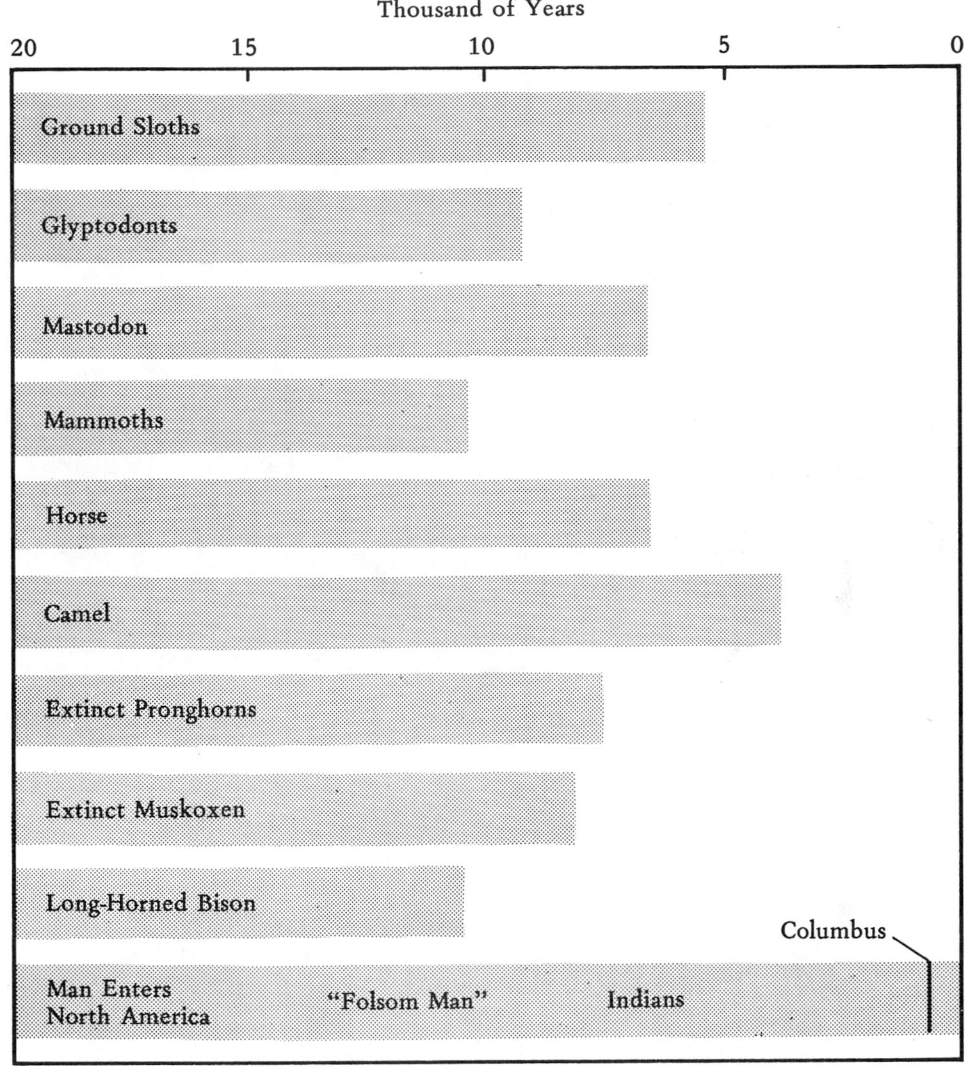

FIGURE 22.46 The Extinction of Pleistocene Mammals in North America and the Rise of Man.

evolved in North America in the Eocene, but did not last long, probably because of the presence of carnivorous mammals. *Phororhacos* and a host of relatives survived much longer in South America, probably because placental carnivores were lacking until the Mid-Neogene. Other, less ferocious, large flightless birds have survived on isolated islands and in some areas of continents down to the present (Fig. 22.47C).

Plants

The plant life of the Cenozoic Era was dominated by the *angiosperms* or flowering plants (Fig. 22.48). Their leaves, stems, roots, and particularly their

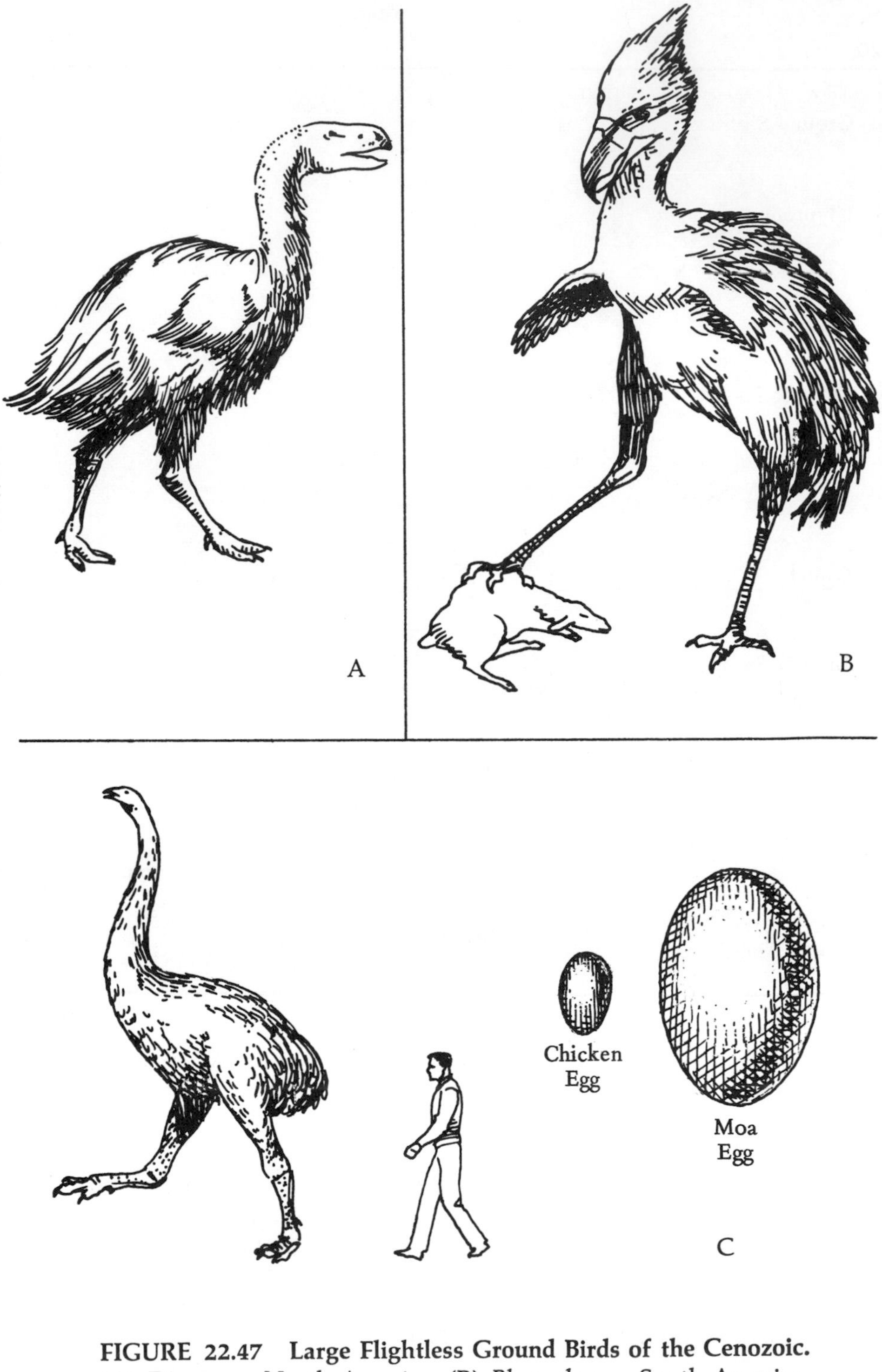

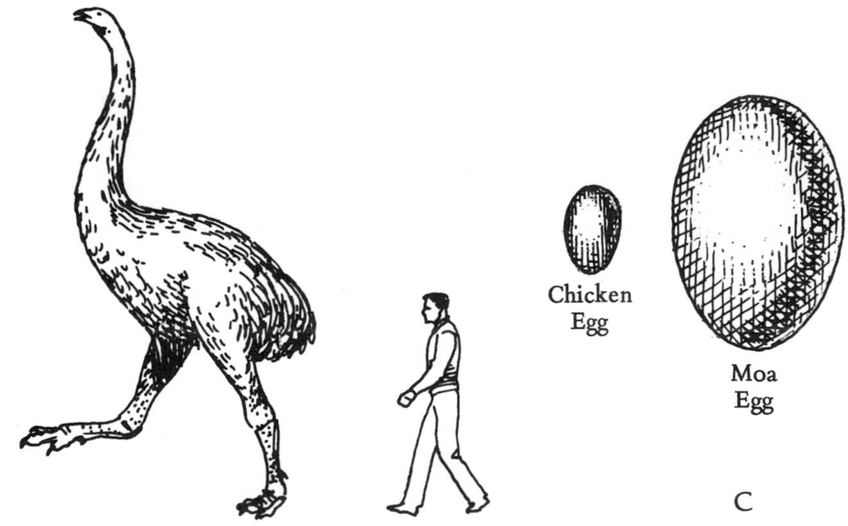

FIGURE 22.47 Large Flightless Ground Birds of the Cenozoic. (A) *Diatryma*, North America, (B) *Phororhacos*, South America, (C) *Moa*, New Zealand.

FIGURE 22.48 Angiosperm Flora of England in the Eocene.

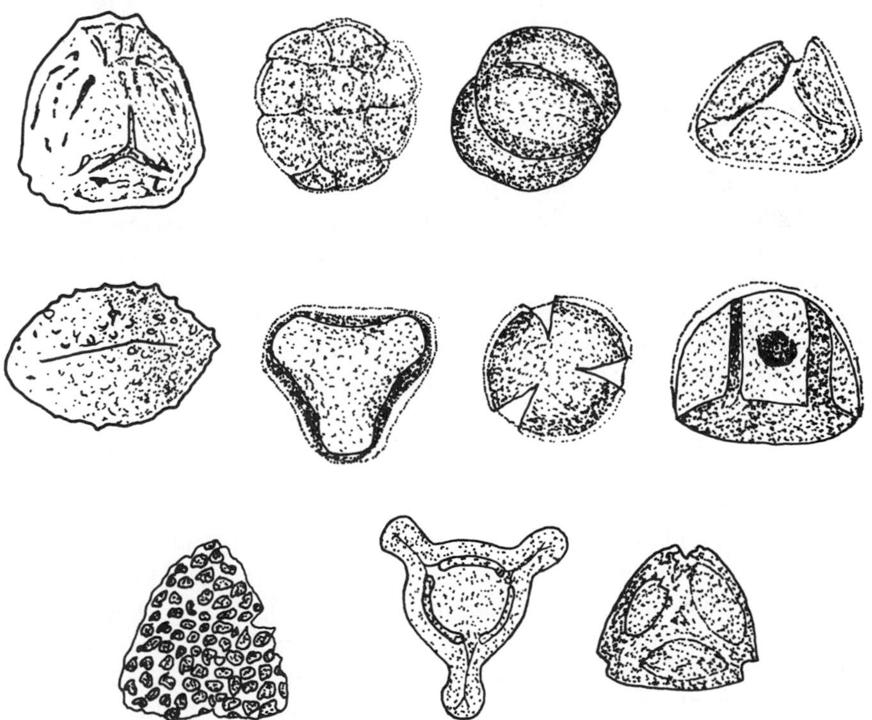

FIGURE 22.49 Cenozoic Pollen Grains.

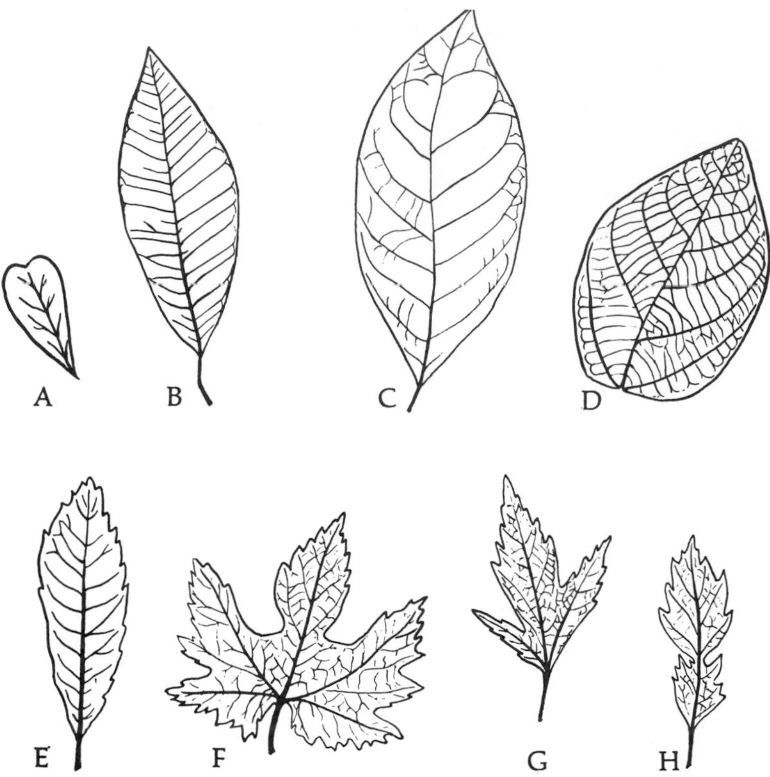

FIGURE 22.50 Cenozoic Angiosperm Leaves. (A–D) Entire margined leaves of early Cenozoic. (E–H) Dentate margined leaves of late Cenozoic. (From *Principles of Paleoecology* by D. Ager. Copyright © 1963. Used by permission of McGraw-Hill Book Co.)

pollen are abundantly preserved in continental strata (Fig. 22.49). The gymnosperms such as the conifers and gingkos declined to a few species common only in cooler and drier areas. Only the low-growing ferns remained common among the spore-bearing plants.

In the warmer and wetter conditions of the Early Cenozoic, most angiosperms had large, entire-margined leaves (Fig. 22.50) such as those of modern subtropical to tropical forms. As the climate became cooler because of the onset of glacial conditions and drier because of the rain shadows caused by the widespread rising mountains, these plants contracted their ranges to low latitudes. Furthermore, smaller leaves with serrate margins (Fig. 22.50), such as those characterizing present temperate to boreal angiosperms, became prevalent. In the dry climates of the Middle and Late Cenozoic, the *grasses* (Figs. 22.40, 22.41) became the dominant angiosperm group and carpeted vast plains areas on all the continents. Although their soft remains are uncommon fossils, their resistant seeds have left an abundant fossil record.

We now stand on the threshold of our modern world with protistans and mollusks abundant in the seas, and mammals, insects, and flowering plants dominating the land. Only one important element is necessary to complete the present day scene, the mammal called man.

Questions

1. What are the major Cenozoic groups of algae?
2. What significant group of large foraminifera reached a Paleogene peak?
3. What groups of mollusks dominate Cenozoic marine faunas?
4. What are the major echinoid groups of the Cenozoic?
5. Why is the Cenozoic Era known as the Age of Mammals?
6. Discuss the evolution of the fauna of South America in terms of its plate tectonic history.
7. What groups of mammals have been dominantly Laurasian in origin and development?
8. What affect has plate tectonics had on the evolution of African mammals?
9. What plant group is dominant in the Cenozoic?
10. What changes occurred in the floras of North America as the Cenozoic progressed?

23

The Psychozoic "Era":

The Rise of Man

THE RISE OF MAN, A VERY RECENT EVENT IN THE LONG HISTORY OF THE EARTH, IS deserving of special attention not only because the author and readers of this text belong to the lineage, but because it has had such a dramatic effect on the earth's entire ecosystem. Whether or not the human species survives, its span of existence on this planet will leave significant geologic reminders. Because of this, some scientists have suggested setting aside the last 5,000 years of earth history as a significant new chapter called the *Psychozoic Era*, characterized by man and his works. In this chapter we will use the term in an informal, expanded sense to discuss the entire rise of man.

Primate Characteristics

Man is a member of the mammal order *Primatoida* (Fig. 22.32) meaning the prime animals, a typical example of man's opinion of himself. The primate order arose from the insectivores in the Late Cretaceous. The probable ancestral group is that of tropical arboreal *tree shrews* (Fig. 23.1), but fossils of this family are not known before the Paleocene. Tree shrews have typical insectivore teeth and claws, but they have large brains with a diminished olfactory (sense of smell) region, the thumb is partially opposable, and the skull is lemur-like, all features of the primates (Fig. 23.2). Some workers classify the tree shrews as insectivores, some, as primates, and still others make them a separate order.

FIGURE 23.1 Representatives of the Major Living Groups of Non-Human Primates. (A) Tree Shrew, (B) Lemur, (C) Tarsier, (D) Old World Monkey, (E) Apes.

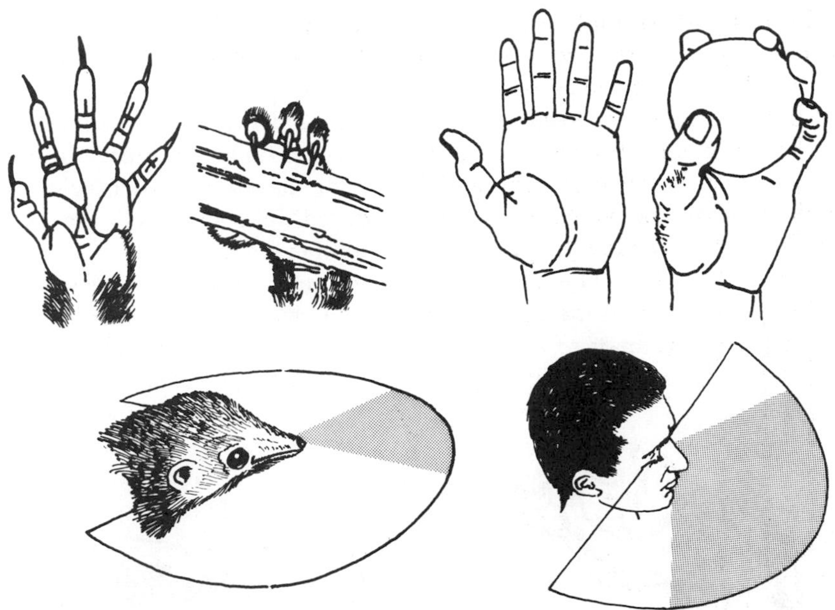

FIGURE 23.2 Contrasts Between the Grasping Ability and Vision of a Tree Shrew and Man. (From A. Lee McAlester, *The History of Life,* © 1968. Reprinted by permission of Prentice-Hall, Inc., Englewood Cliffs, N.J.)

However one classifies them, they appear to be related to the lineage from which the typical primates arose.

Plesiadapidines

A primitive, extinct group of primates with chisel-like front teeth, the *plesiadapidines* (Fig. 23.3) are known from Upper Cretaceous, Paleocene, and Lower Eocene rocks of North America and Europe indicating the primates arose on the northern continents. The oldest known tree shrew is also known from Europe, supporting this idea. The plesiadapidines already displayed the characteristic primate features of the grasping hand made possible by an opposable thumb, the large brain, and the shift of the eyes to the front of the face to permit overlapping stereoscopic vision. These features are all correlated with adaptation to a tree-living or arboreal existence, the mode of life of most primates. The teeth of most of the known plesiadapidines are different from those found in the main line of primates and indicate that they were not the immediate ancestors of the more advanced primates.

Lemurs

In the Late Paleocene and Eocene, the next step in primate evolution occurred with the appearance of the *lemurs* (Fig. 23.4). Lemurs are still rather small quadrupedal forms and resemble squirrels. They retain a muzzle and the eyes are not completely rotated to the front of the head. Most still have claw-like nails on at least some digits. The tail is usually long but not prehensile for grasping. The early lemurs were widespread in the Eocene, occurring in the subtropical to tropical forests of North America, Europe, and Asia. As these forests disappeared in the Oligocene because of continental movements and the widespread mountain-building that cooled climates and produced rain shadows, the range of lemurs contracted. Today they survive in southern Asia, eastern Africa, and Madagascar (Fig. 23.1). The island of Madagascar is their chief refuge because it split off from Africa in the Early Cenozoic and lacks carnivores and competing arboreal advanced primates.

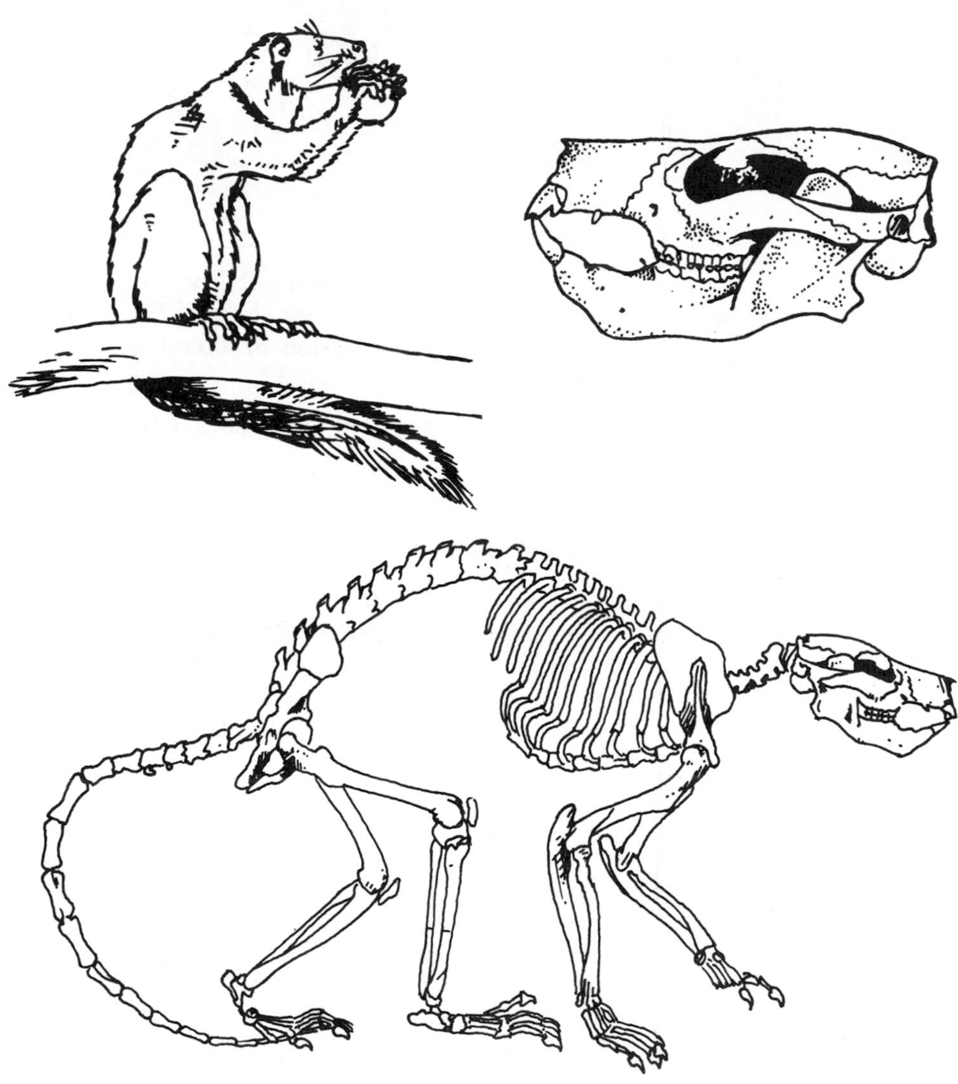

FIGURE 23.3 *Plesiadapis,* **A Paleocene Primate.** Restoration and skeleton. (After Simons.)

FIGURE 23.4 Restoration of an Eocene Lemur. *Notharctus osborni* of Middle Eocene age. (Painting by Ferguson, courtesy of the American Museum of Natural History.)

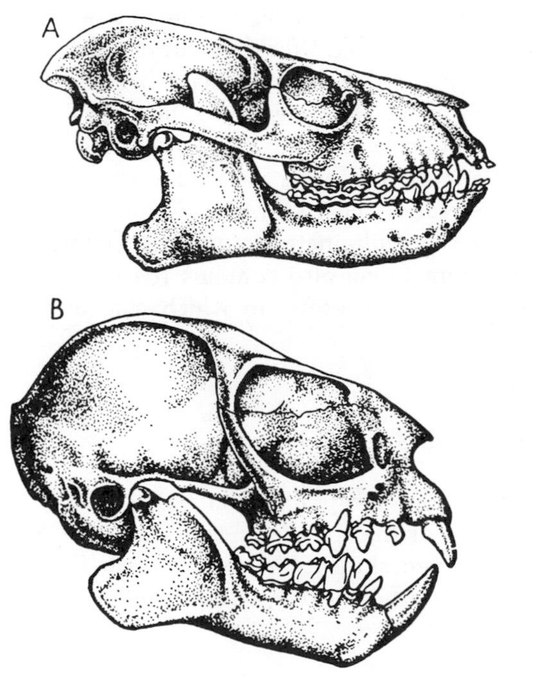

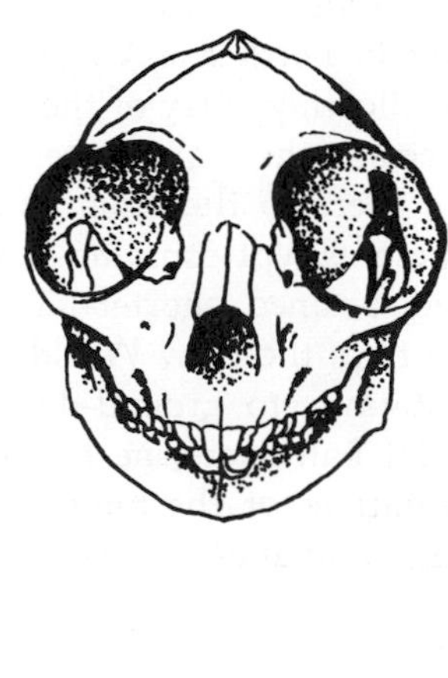

FIGURE 23.5 Eocene Primate Skulls. (A) Lemur, (B) Tarsiers. (From A. S. Romer, *Vertebrate Paleontology*, Copyright © 1966, by the University of Chicago Press. Used by permission of publisher.

Tarsiers

Also in the Late Paleocene and Eocene, the ancestors of the living hopping, rat-like, but arboreal *tarsiers* (Fig. 23.1) evolved. These ancestral forms had a transitional structure between the lemurs and the monkeys (Fig. 23.5). In particular, they, and the living tarsiers, have large, forward-directed eyes and have lost the muzzle. The brain, especially in the cerebral hemisphere area, is larger than that of lemurs. The tarsiers, like the lemurs, were widespread on the northern continents in the Eocene, but have since become restricted, probably for the same reasons, to isolated regions, in this case, the East Indies.

Monkeys

The *South American* or *platyrrhine* monkeys (Fig. 22.32) probably evolved from a Paleogene lemur or tarsier that found its way down the volcanic island arcs between North and South America. They are primitive in being small and possessing claw-like nails, an only slightly opposable thumb, a small brain, and a long tail which is prehensile in most. They are related to the Old World monkeys only through their common lemur or tarsier ancestor and the present similarities of the two groups are solely caused by convergent evolution in similar environments.

The *Old World* or *catarrhine monkeys* (Fig. 23.1) appear in the Oligocene of Egypt in northeast Africa. Africa was united with Eurasia in that region at the time allowing entry of the tarsier ancestor of the Old World monkeys. Africa was to become the main center of evolution for the remaining primates which spread from there to the other continents in the Late Neogene. The Oligocene Fayum fauna of Egypt contains fossils of several Old World monkeys. These are larger, bigger-brained, shorter-tailed and with a more opposable thumb and more typical nails than the New World monkeys. Several of these Old World monkeys show adaptations to ground-dwelling, a trait that culminates in the completely terrestrial Late Neogene baboons. The Fayum fauna also contains the oldest representatives of the Superfamily Hominoidea, the group to which man and the living great apes are assigned.

Hominoids

The Oligocene plexus of hominoids includes *Aegyptopithecus* (Fig. 23.6), the oldest known member, at 28 million years, of the specific lineage leading to man and the great apes. Because a well-preserved skull is the only known fossil of *Aegyptopithecus*, its mode of life, whether arboreal or ground-dwelling, cannot be determined.

Dryopithecines

Following this Late Oligocene genus, from the Middle Miocene through the Lower Pliocene (20 to 8 million years ago), is a group of fossils called the *dryopithecines* (Fig. 23.7). The dryopithecines were near the point of divergence of men and the surviving ground-dwelling apes, the gorilla and chimpanzee. At

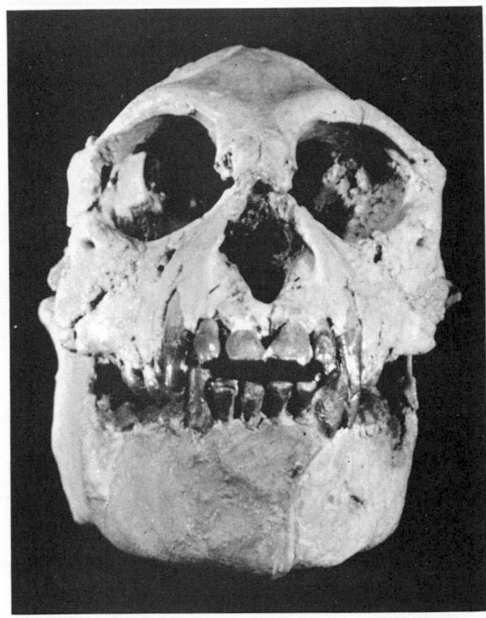

FIGURE 23.6 *Aegyptopithecus zeuxis* **Skull.** Oligocene, Fayum, Egypt. (Courtesy of Elwyn L. Simons, Yale University and *American Scientist.*)

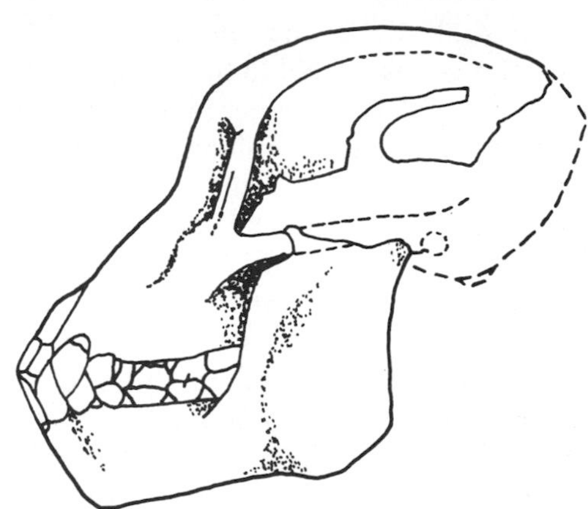

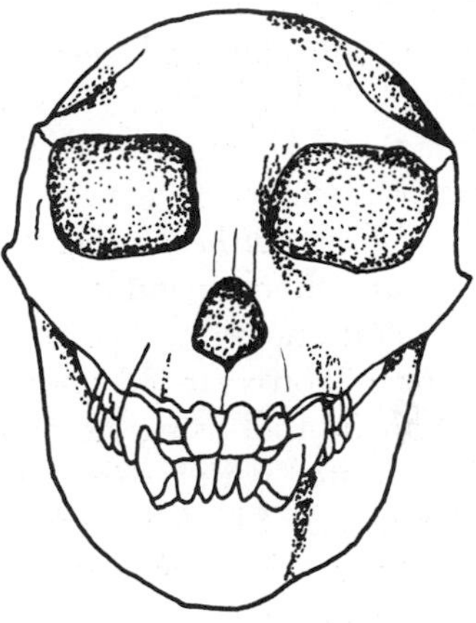

FIGURE 23.7 Dryopithecine Skull.

least the more advanced types, however, are more ape-like because the two sides of the lower jaw are nearly parallel as in that group, rather than slightly divergent as in man (Fig. 23.8). Dryopithecines are known from essentially complete skulls which display a low, rounded vault, moderate brow ridges, and a moderate forward protrusion of the jaws and face. Unfortunately the limbs are poorly known, but they do suggest that the creature was not highly adapted to brachiation and the foot is somewhat indicative of a bipedal posture. Perhaps *Dryopithecus* lived on the ground like the living baboon monkeys and occasionally reared up on its hind legs. The drying trend of the Neogene was restricting African forests at this time and the ape-man lineage was evolving in response to the spreading open grassland country. Ground-dwelling attributes were being selected for and one of these was surely the ability to rear up on the hind legs in

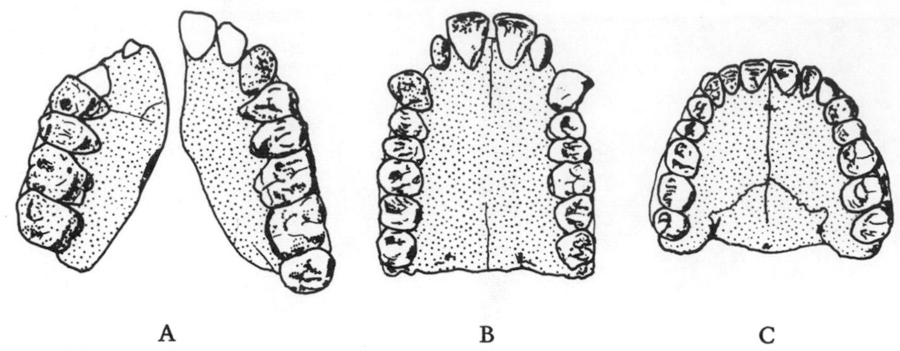

FIGURE 23.8 Comparisons of (A) *Ramapithecus* Jaw with jaws of (B) Ape (Orangutan) and (C) Man.

order to view more of the surrounding countryside. The dryopithecines were most common in east Africa, but spread to the north and through the Near East region to southern Asia including India, now joined to Eurasia and Europe (Fig. 23.9).

Ramapithecus and Kenyapithecus

In the Late Miocene or Early Pliocene of east Africa and southern Asia have been found some jaw and teeth fragments of a genus related to the dryopithecines. This genus, *Ramapithecus* (Figs. 23.8–11), has slightly divergent lower jaw bones as in man and is thus the oldest known fossil, about 10 million years old, unequivocally assigned to the *Family Hominidae,* the family of man. In east Africa, the genus *Kenyapithecus* (Fig. 23.9) has also been found. It ranges in age from Early to Late Miocene and may be synonymous with *Ramapithecus* or closer to its dryopithecine ancestors. The 20 million-year-old species *K. africanus* evolved into the 12–14 million-year-old species *K. wickeri.* This latter species is apparently the oldest known tool-using hominid. Blunt stones, not native to the site, were found associated with broken bones and crushed skulls of animals at Fort Ternan, Kenya. The use of tools indicates that the arms were in the process of being freed from walking and were being used for other purposes, perhaps indicating the onset of bipedalism.

Australopithecines

After the *Ramapithecus-Kenyapithecus* level of human evolution is that of the *australopithecines* (Fig. 23.11–15), chiefly found in Pleistocene rocks of 2 to .5 million years in age, but recently unearthed in Pliocene strata. *Australopithecus* was a long-lived genus consisting of several species, some of which tended to a more carnivorous diet and others of which were more vegetarian. *Australopithecus* is best known from South Africa where skulls, jaws, teeth, and limbs have been found (Fig. 23.12). More recently, good east African material and fragments from southern and eastern Asia have been unearthed (Fig. 23.13). The various species varied from chimpanzee to gorilla-sized, but all walked erect with limb bones essentially like those of modern men. The skull differed from modern man by its smaller brain case (half of that of modern man), more prominent brow ridges,

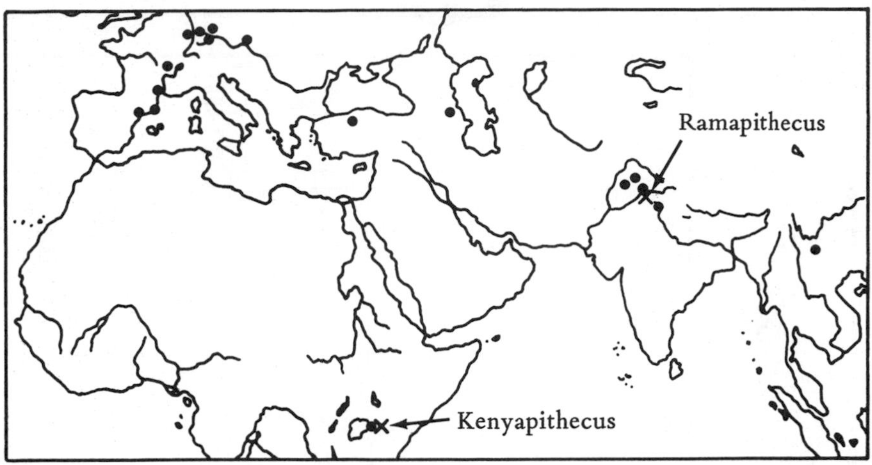

FIGURE 23.9 Distribution of *Dryopithecines* **and** *Ramapithecus-Kenyapithecus* **Complex.**

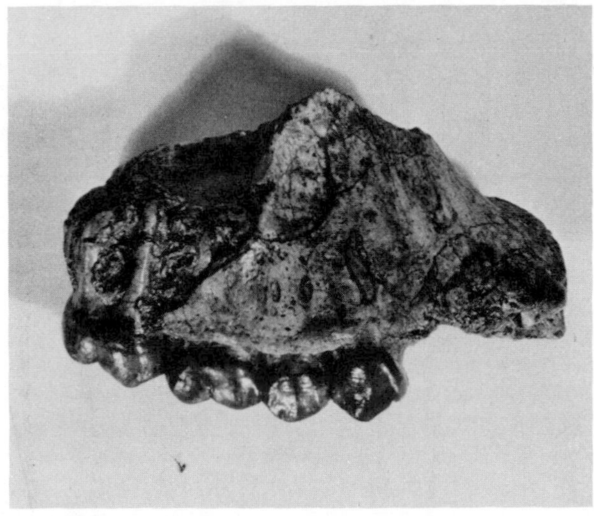

FIGURE 23.10 *Ramapithecus brevirostris.* (Courtesy of Elwyn L. Simons, Yale University.)

and slightly more protruding face. The bipedal posture freed the front limbs for other functions such as tool-making and using. Simple pebble tools called hand axes and probably used for cutting, scraping, and chopping have been found with the australopithecine fossils (Fig. 23.14, 23.25). These are assigned to the Oldowayan cultural level. Broken brain cases indicate that some australopithecines were cannibals.

About 1.75 million years ago, in east Africa, a form of man appeared that was transitional, in some respects, between the advanced australopithecines and the

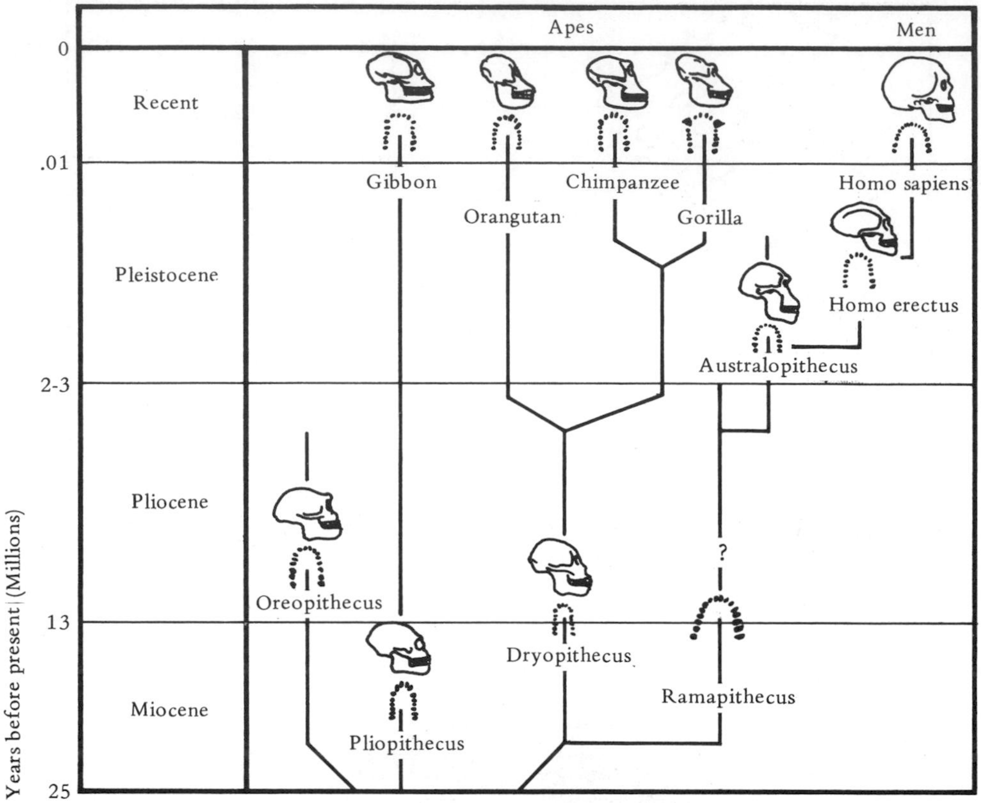

Apes
Men
Years before present (Millions)
0
Recent
.01
Pleistocene
2-3
Pliocene
13
Miocene
25
Gibbon
Orangutan
Chimpanzee
Gorilla
Homo sapiens
Homo erectus
Australopithecus
?
Oreopithecus
Dryopithecus
Ramapithecus
Pliopithecus

FIGURE 23.11 Evolution of the Apes and Men. (From A. Lee McAlester, *The History of Life*, © 1968. Reprinted by permission of Prentice-Hall, Inc., Englewood Cliffs, N.J.)

FIGURE 23.12 *Australopithecus* **Restoration.** (Augusta-Burian mural, courtesy of American Museum of Natural History.)

next step in human evolution. This creature, called *Homo habilis* (Fig. 23.15), is considered by some to be nothing more than an advanced australopithecine and not the direct ancestor of modern man.

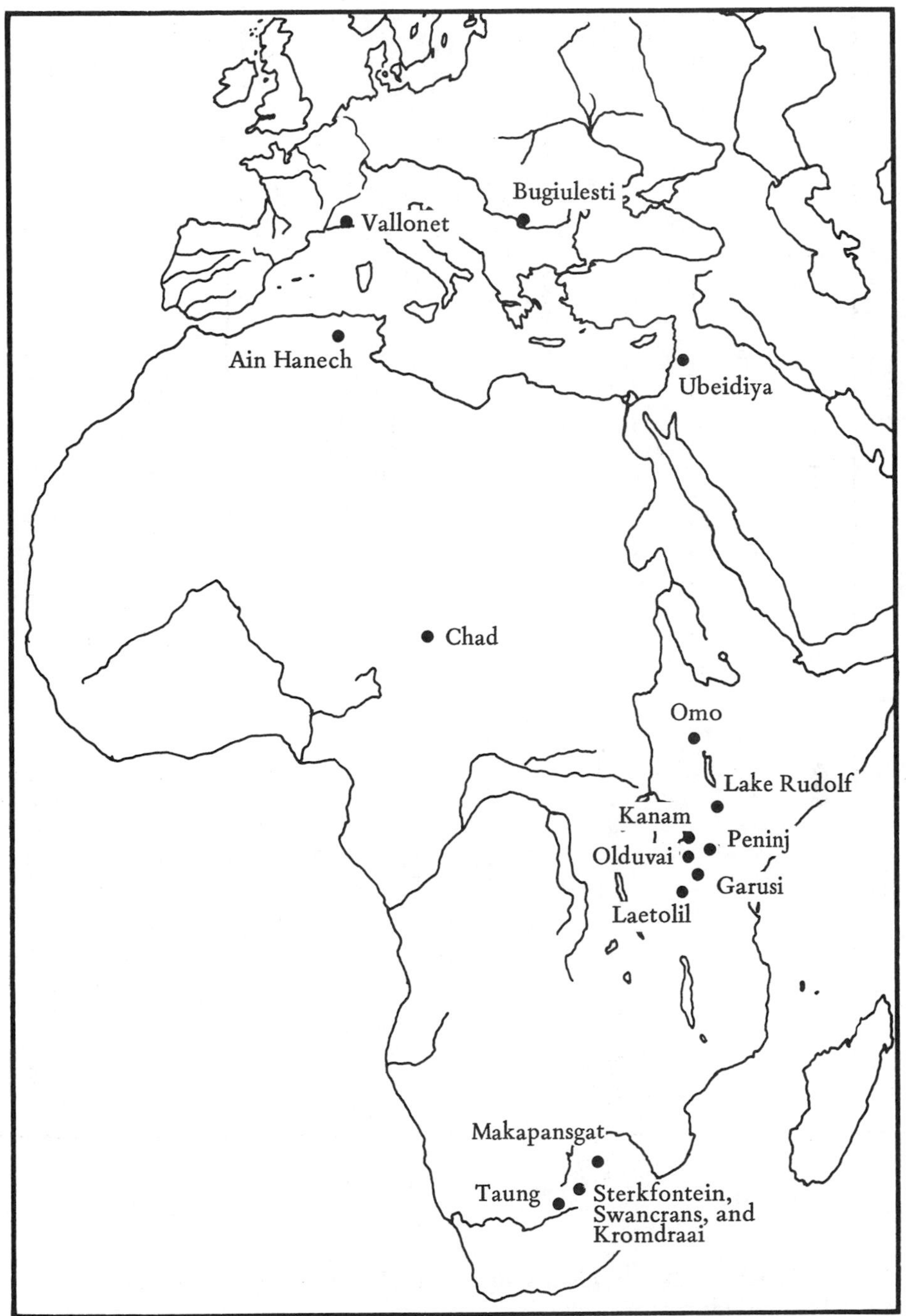

FIGURE 23.13 Localities Where *Australopithecus* Remains or Artifacts Have Been Found.

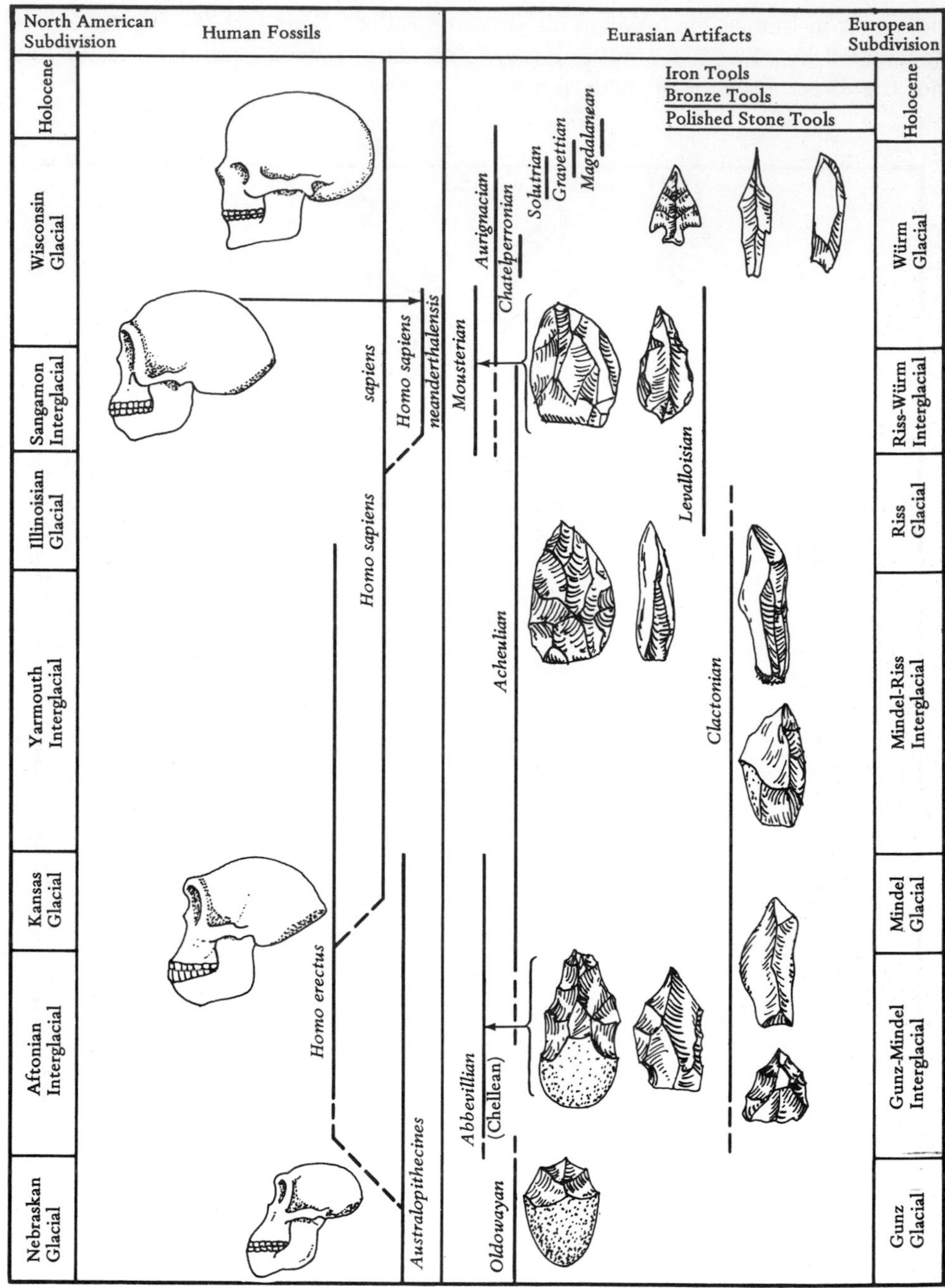

FIGURE 23.14 Skulls, Ranges, and Tools of Major Pleistocene Homonoids.

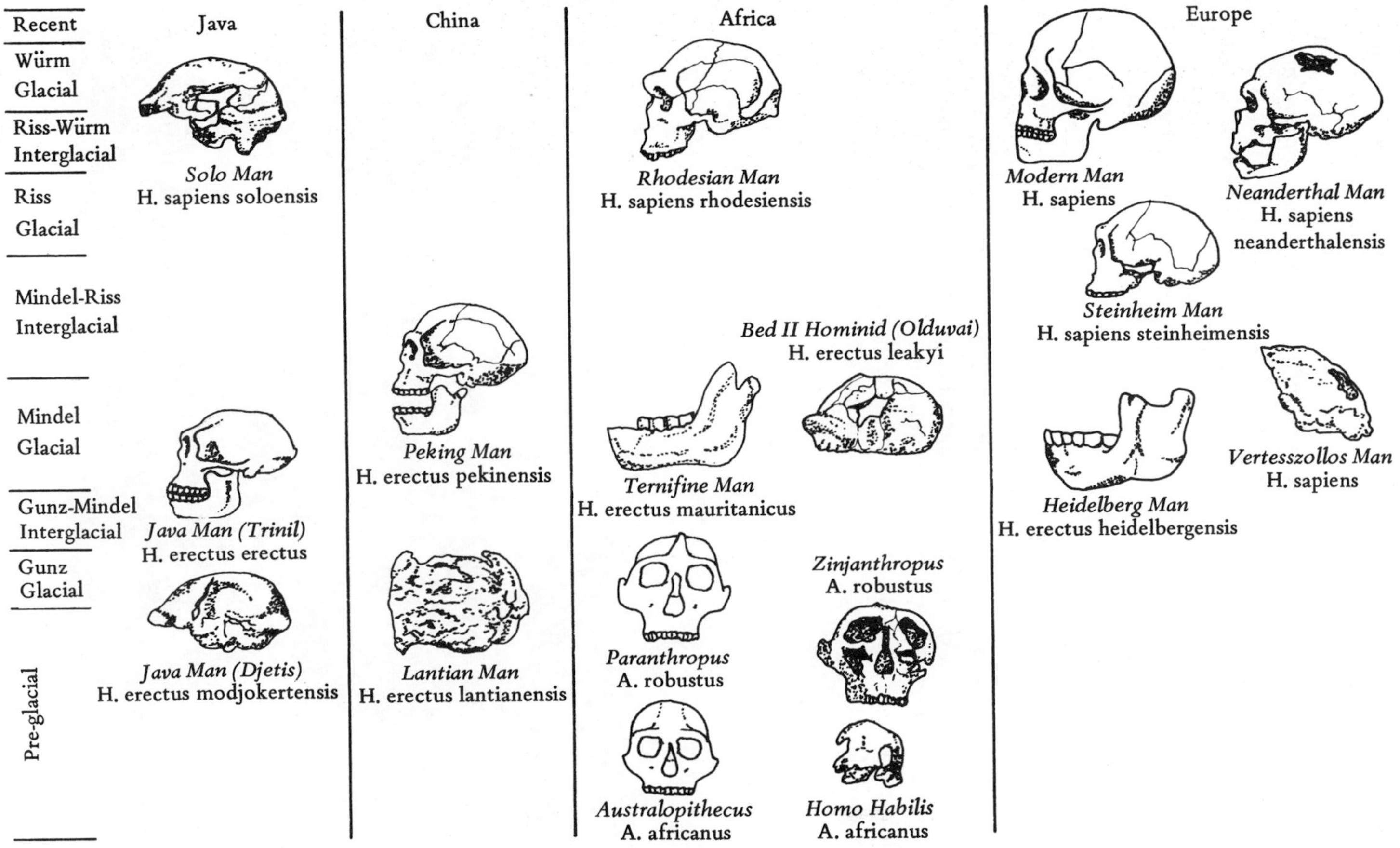

Recent
Würm Glacial
Riss-Würm Interglacial
Riss Glacial
Mindel-Riss Interglacial
Mindel Glacial
Gunz-Mindel Interglacial
Gunz Glacial
Pre-glacial

Java

Solo Man
H. sapiens soloensis

Java Man (Trinil)
H. erectus erectus

Java Man (Djetis)
H. erectus modjokertensis

China

Peking Man
H. erectus pekinensis

Lantian Man
H. erectus lantianensis

Africa

Rhodesian Man
H. sapiens rhodesiensis

Bed II Hominid (Olduvai)
H. erectus leakyi

Ternifine Man
H. erectus mauritanicus

Paranthropus
A. robustus

Zinjanthropus
A. robustus

Australopithecus
A. africanus

Homo Habilis
A. africanus

Europe

Modern Man
H. sapiens

Neanderthal Man
H. sapiens
neanderthalensis

Steinheim Man
H. sapiens steinheimensis

Heidelberg Man
H. erectus heidelbergensis

Vertesszollos Man
H. sapiens

FIGURE 23.15 Major Fossil Evidence of Human Evolution Known in 1966. (After Kummel.)

Homo erectus

From about 800,000 to 200,000 years ago, the first member of our genus, *Homo erectus* (Fig. 23.11, 23.14–16), lived. *Homo erectus* evolved from *Australopithecus,* was contemporaneous with it for 300,000 years, and eventually supplanted it. *Homo erectus* was first known from Java where it was called *Pithecanthropus* and China where it was known as *Sinanthropus.* Experts now agree that it is sufficiently like modern man to be assigned to the same genus. The skull of *Homo erectus* has a brain capacity intermediate between that of the australopithecines and modern man, but has a lower dome, more prominent brow ridges, and a more protruding face than that of modern man. It also lacks a chin which suggests to some that language had not yet evolved. Remains of *Homo erectus* are now known in eastern and northern Africa, Asia, and Europe. The tools of *Homo erectus* consisted of large, shaped hand axes of flint. Their quality of workmanship improved as the Middle Pleistocene progressed. The older hand axes, chipped only on one side, are assigned to the Abbevillian or Chellean cultural level (Figs. 23.17 A), while the younger, completely worked, hand axes are called Acheulian (Figs. 23.17 B). Scrapers and knives made from flakes of pebbles are called Clactonian (Figs. 23.17 C). The tools of these cultures are widely found in Europe, Asia, and Africa (Fig. 23.18). *Homo erectus* had also learned the use of fire.

FIGURE 23.16 *Homo erectus* **Restoration.** (Augusta-Burian mural, courtesy of American Museum of National History.)

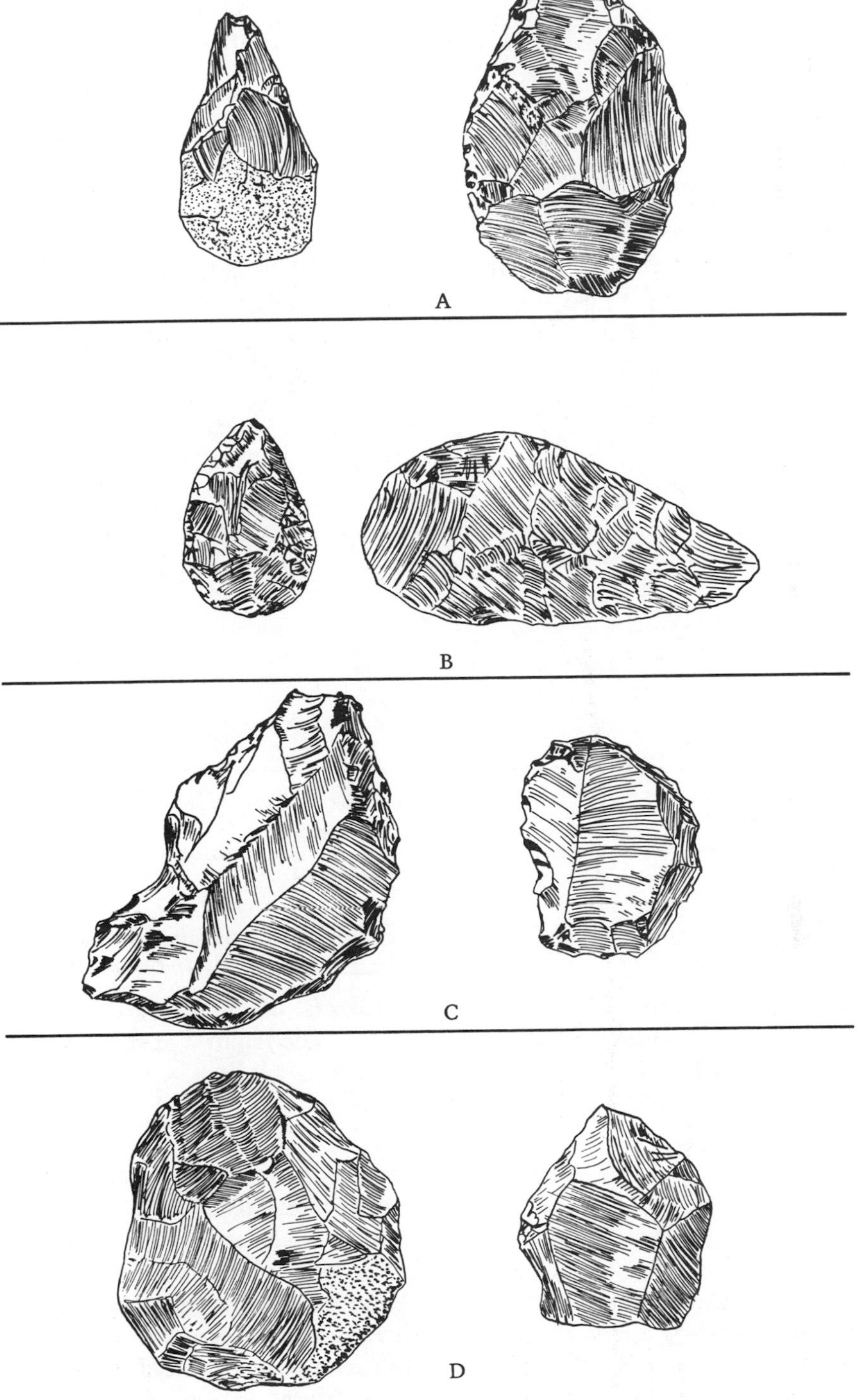

FIGURE 23.17 Artifacts of *Homo erectus*. (A) Early hand-axe or Abbevillian, (B) Late hand-axe or Acheulian, (C) European flake-tool or Clactonian, (D) Asian chopper-tool or Soan.

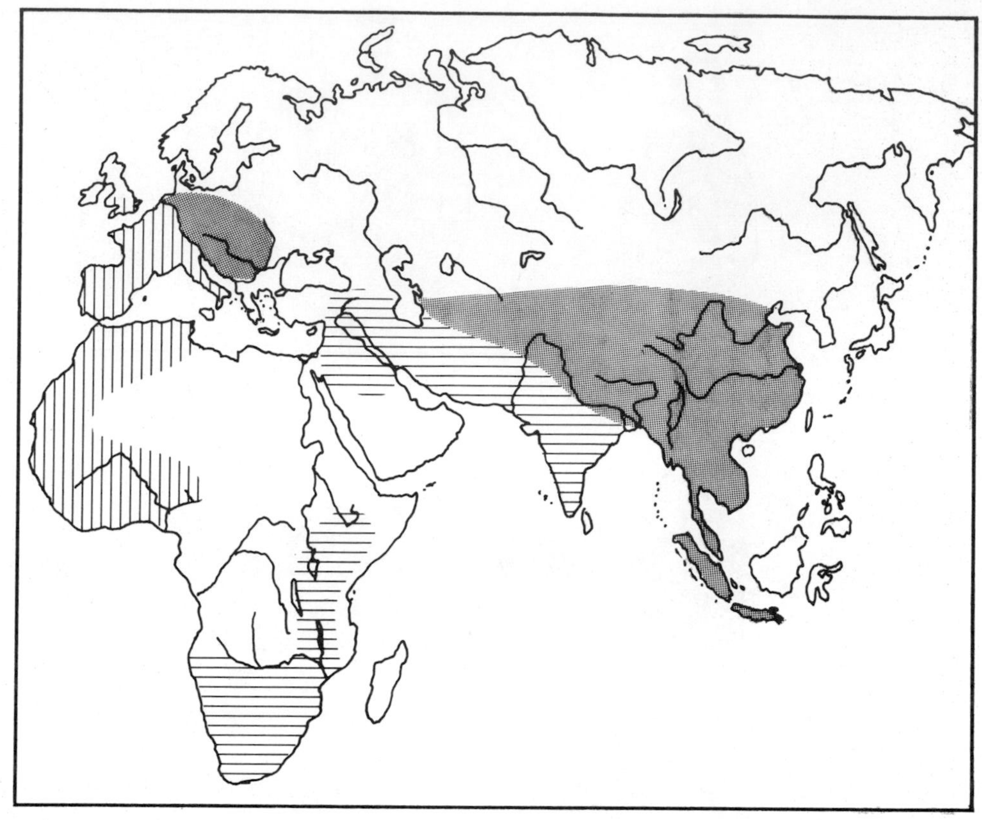

Culture Areas of Acheulean:

Distribution of Non-Acheulean

SW Europe—NW Africa

Distribution of Acheulean

Sub-Saharan Africa—Near East—India

FIGURE 23.18 Distribution of Artifacts of *Homo erectus.* (From C. S. Chard, *Man in Prehistory,* Copyright © 1969, by permission of McGraw-Hill Book Co.)

Neanderthal Man

The earliest records of our species, *Homo sapiens,* are scattered fragments from Morocco, France, Germany, and England of approximately 500,000 years in age. The earliest complete skeletons of members of our species are those of the so-called *Neanderthal Man* (Figs. 23.14, 23.19) who lived from about 110,000 to 35,000 years ago in Europe, Africa, and Asia. He differed from typical *Homo sapiens* by possessing a lower forehead, wider skull, more prominent brow ridges and by lacking a chin (Fig. 23.14). His body was modern in build, though many restorations picture him as a squat, slouching brute, an erroneous impression based on the skeleton of a diseased individual. He developed the cultures known as Levalloisian and Mousterian (Figs. 23.14, 23.20) with their well worked stone axes, scrapers, and points. The last were attached to shafts which then became

FIGURE 23.19 Neanderthal Man restoration. (Augusta-Burian mural, courtesy of American Museum of Natural History.)

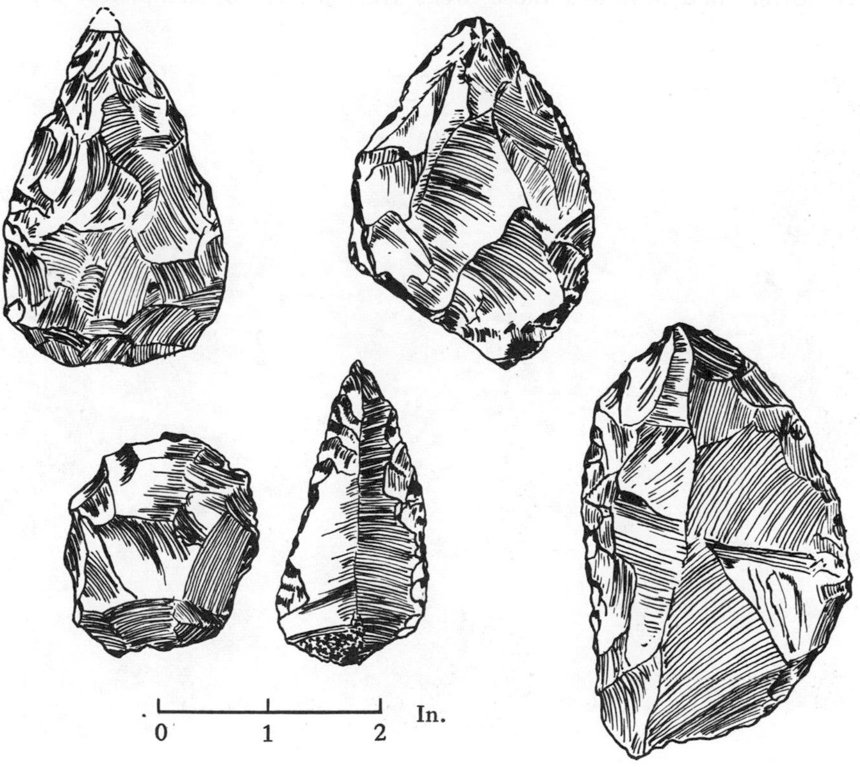

FIGURE 23.20 Artifacts of Neanderthal Man. Mousterian culture.

spears. There are gradations between Neanderthal Man and more typical *Homo sapiens*, even at the same site, such as Carmel, Israel, indicating that he was no more than a subspecies or race of our own species. The exact cause of his somewhat different skull features is uncertain, but their occurrence is not surprising in view of the great diversity of local populations and races that exists on earth today.

Homo sapiens

Completely modern skulls, belonging to what is called *Cro-Magnon Man* (Fig. 23.21), appear about 35,000 years ago. The complete skeletons of Cro-Magnon Man, his artwork (Fig. 23.22), and his stone tools (Figs. 23.23) are common throughout Europe, Africa, and Asia. He passed through a diversity of cultural levels characterized by the progressive refinement of projectile points, knives, chisels, and drills. These so-called blade traditions are the highest cultural levels of the Paleolithic or Old Stone Age and include the Perigordian, Chatelperronian, Aurignacian, Gravettian, Solutrean, and Magdalenian cultures (Fig. 23.23). The men of these last two cultures produced probably the finest stone tools and are also known for their striking cave paintings. *Homo sapiens* began to spread to other continents from his original African-Eurasian home. He emigrated to Australia via the East Indies and to the Americas via the Siberia-Alaska land bridge. This was particularly easy during the glacial stages when sea level was much lower than now and those areas were largely or completely land. The

FIGURE 23.21 Cro-Magnon Man restoration. (Augusta-Burian mural, courtesy of American Museum of Natural History.)

FIGURE 23.22 Artwork of Cro-Magnon Man. Reproductions of late Paleolithic cave art.

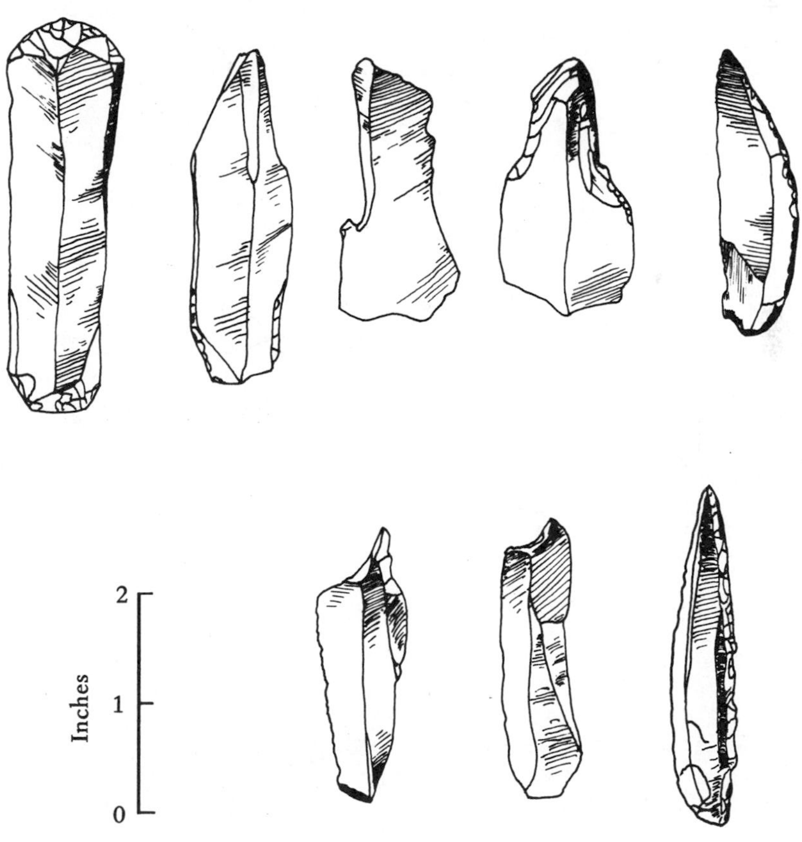

FIGURE 23.23 Artifacts of Cro-Magnon Man.

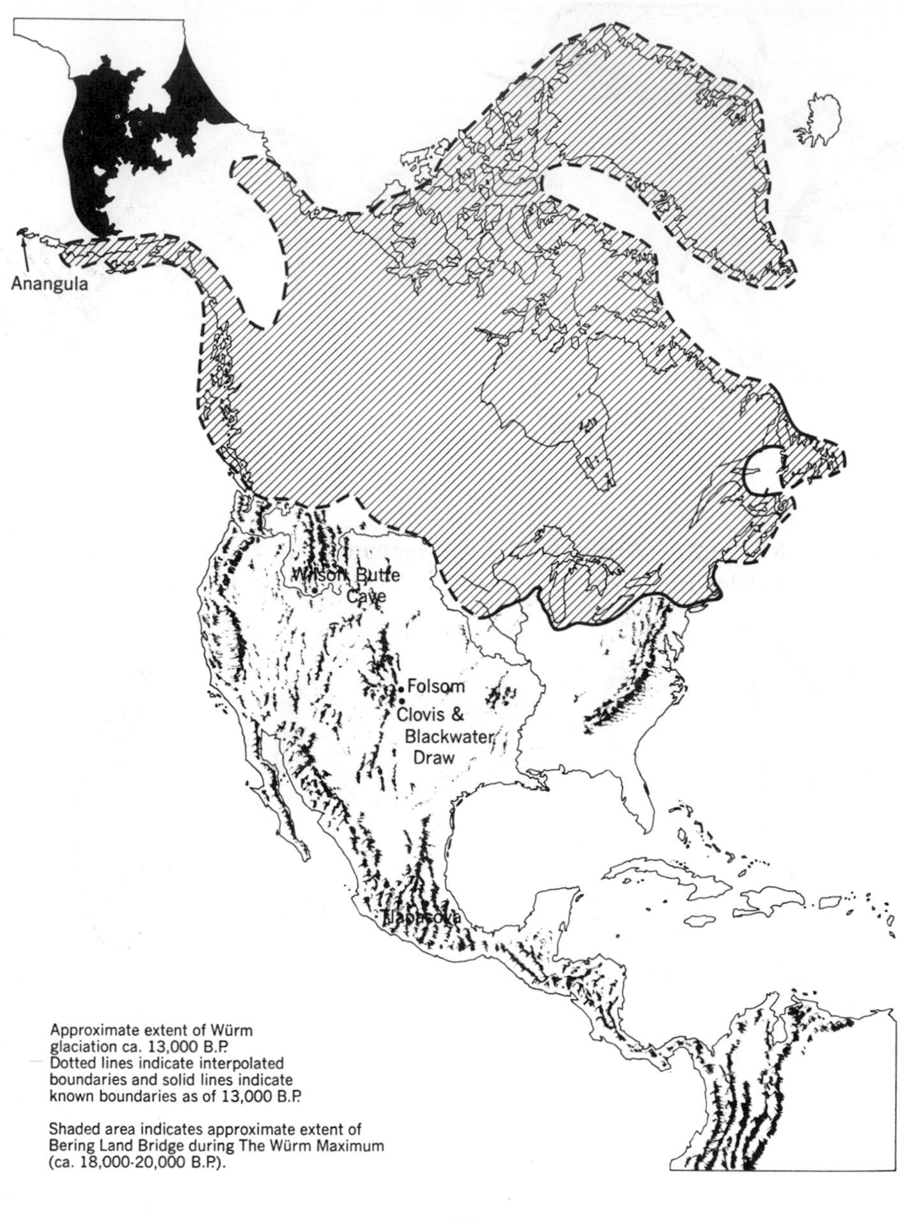

(A)

FIGURE 23.24 Early Man in North America. (A) Maximum extent of Würm (Wisconsin) Glaciation and Localities of Early Artifacts, (B) Postulated invasions of North America from Eurasia. (A) From C. S. Chard, *Man in Prehistory*, Copyright © 1969, by permission of McGraw-Hill Book Co, (B) From Muller-Beck, *Science*, vol 152, 27 May 1966, copyright by AAAS, used by permission of authors and AAAS.)

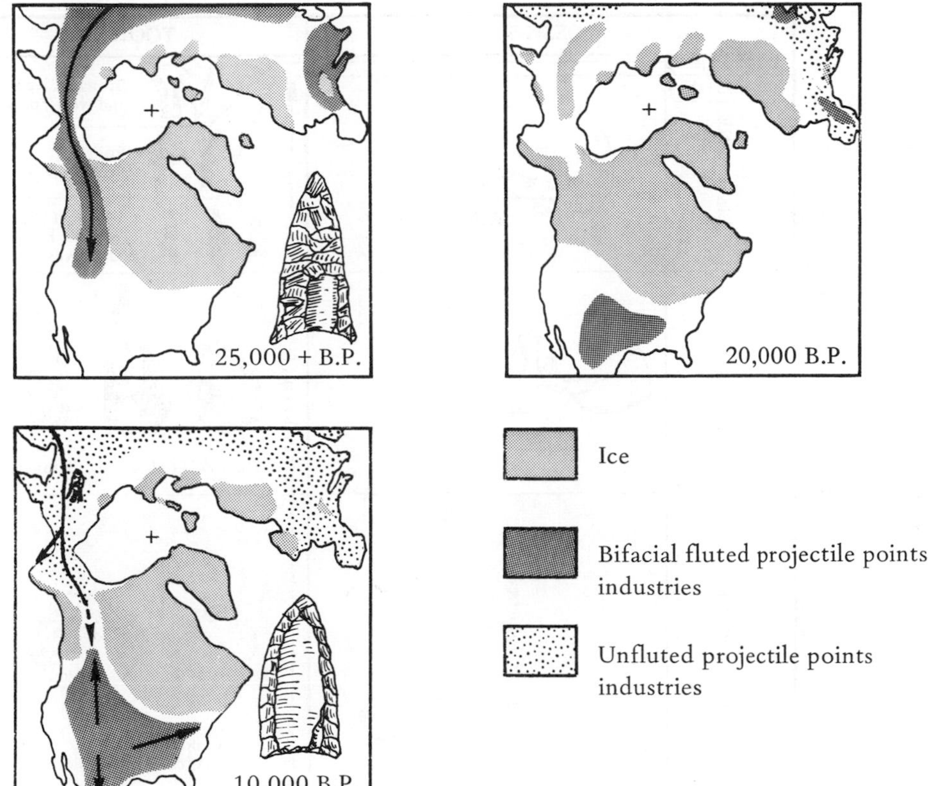

exact date of man's entry into North America is still uncertain. The possibility exists back as early as 35,000 years ago when typical *Homo sapiens* had evolved and the shelves were exposed, but other than some possible charred wood, bones, and ground, no evidence exists until the end of the Pleistocene around 13,000 years ago when fluted spear points appeared (Fig. 23.24). The extent of the glaciers at that time would have prevented man from entering North America so it is probable he arrived during the preceding interglacial.

About 15,000 to 10,000 years ago, near the Pleistocene-Holocene boundary, *Homo sapiens* man began to make polished tools by grinding and he moved into the Neolithic, or New Stone Age cultures (Fig. 23.25). The neolithic men also began to make pottery, domesticate animals, and cultivate plants. Man could then settle down in permanent sites which became the first towns. In these towns, some men could perform other functions than simply raising food. About 5,000 years ago the metal cultures replaced the stone age cultures, and man began to write his history. With this event, we pass from prehistory to history and thus from Historical Geology to History.

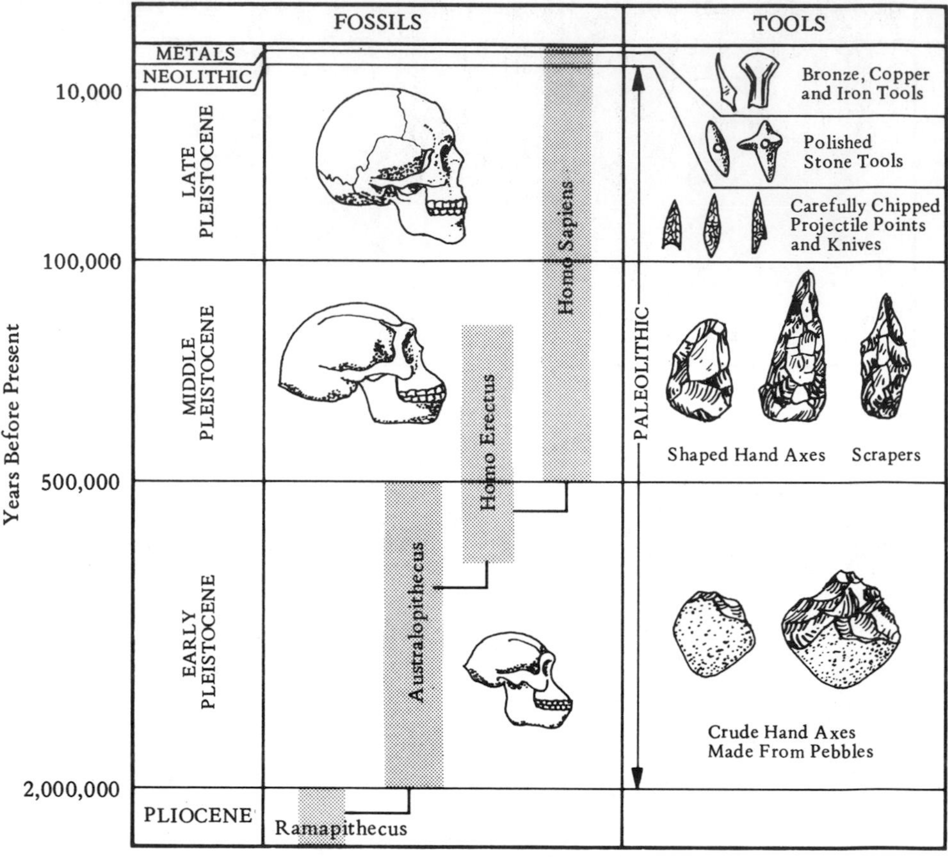

FIGURE 23.25 Summary of Man's Evolutionary and Cultural History showing the development of polished stone and metal tools at the beginning of recorded history. (From A. Lee McAlester, *History of Life*, © 1968, by permission of Prentice-Hall, Inc., Englewood Cliffs, New Jersey.)

Questions

1. To what order of mammals does man belong and what other groups are also members of this order?
2. What is the oldest known group of primates?
3. How do lemurs differ from more specialized primates?
4. Discuss the changes throughout the Cenozoic in the distribution of lemurs and tarsiers.
5. What is the oldest known member of the ape-man group of primates and when did it live?
6. What is the significance of *Dryopithecus* and *Ramapithecus*?
7. When did the Australopithecines live and what was their general physical appearance?
8. How did *Homo erectus* differ from the Australopithecines and from modern man?

9. How did Neanderthal man differ from Cro-Magnon man?
10. Which type of man was responsible for the following types of implements: Mousterian, Chellean, Aurignacian, Oldowayan?

Bibliography

General Geology Texts

Dott, R. H. and Batten, R. L., *Evolution of the Earth.* New York: McGraw-Hill, 1971.
Eardley, A. J., *General College Geology.* New York: Harper and Row, 1965.
Earth Science Curriculum Project, *Investigating the Earth.* New York: Houghton Mifflin, 1967.
Fagan, J. J., *View of the Earth.* New York: Holt, Rinehart, and Winston, 1965.
Foster, R. J., *General Geology.* Columbus, Ohio: Charles E. Merrill, 1969.
Mears, B., *The Changing Earth.* New York: Van Nostrand Reinhold, 1970.
Spencer, E. W., *Geology.* New York: Crowell, 1965.
Stokes, W. L., and Judson, S., *Introduction to Geology.* Englewood Cliffs, N.J.: Prentice-Hall, 1968.
Strahler, A. N., *The Earth Sciences.* New York: Harper and Row, 1971.

Historical Geology Texts

Clark, T. H., and Stearn, C. W., *The Geological Evolution of North America.* New York: Ronald, 1968.
Cloud, P. (editor), *Adventures in Earth History.* San Francisco: Freeman, 1970.
Dunbar, C. O., and Waage, K. M., *Historical Geology.* New York: Wiley, 1969.
Kay, M., and Colbert, E. H., *Stratigraphy and Life History.* New York: Wiley, 1965.
Kummel, B., *History of the Earth.* San Francisco: Freeman, 1970.
Moore, R. C., *Introduction to Historical Geology.* New York: McGraw-Hill, 1958.
Spencer, E. W., *Basic Concepts of Historical Geology.* New York: Crowell, 1962.
Stirton, R. A., *Time, Life, and Man.* New York: Wiley, 1959.
Stokes, W. L., *Essentials of Earth History.* Englewood Cliffs, N.J.: Prentice-Hall, 1966.
Wells, A. K., and Kirkaldy, J. F., *Outline of Historical Geology.* London: Murby, 1966.
Woodford, A. O., *Historical Geology.* San Francisco: Freeman, 1965.

Paleontology Texts

Ager, D. V., *Principles of Paleoecology.* New York: McGraw-Hill, 1963.
Andrews, H. N., *Studies in Paleobotany.* New York: Wiley, 1961.
Arnold, C. A., *An Introduction to Paleobotany.* New York: McGraw-Hill, 1947.
Banks, H. P., *Evolution and Plants of the Past.* Belmont, Calif.: Wadsworth, 1970.
Beerbower, J. R., *Search for the Past.* Englewood Cliffs, N.J.: Prentice-Hall, 1968.
Black, R. M., *The Elements of Paleontology.* Norwich, England: Cambridge University, 1970.
Clark, D. L., *Fossils, Paleontology and Evolution.* Dubuque, Iowa: Brown, 1968.
Colbert, E. H., *Evolution of the Vertebrates.* New York: Wiley, 1969.
Darrah, W. C., *Paleobotany.* New York: Ronald, 1960.
Delevoryas, T., *Morphology and Evolution of Fossil Plants.* New York: Holt, Rinehart, and Winston, 1962.
Easton, W. H., *Invertebrate Paleontology.* New York: Harper, 1960.
Fenton, C. L., and Fenton, M. A., *The Fossil Book.* Garden City, N.Y.: Doubleday, 1958.

Halstead, L. B., *The Pattern of Vertebrate Evolution.* San Francisco: Freeman, 1968.
Imbrie, J., and Newell, N. (editors), *Approaches to Paleoecology.* New York: Wiley, 1964.
Jones, D. L., *Introduction to Microfossils.* New York: Harper, 1956.
Laporte, L., *Ancient Environments.* Englewood Cliffs, N.J.: Prentice-Hall, 1968.
Mathews, W. H., *Fossils.* New York: Barnes and Noble, 1962.
McAlester, A. L., *The History of Life.* Englewood Cliffs, N.J.: Prentice-Hall, 1968.
Moore, R. C. (editor), *Treatise on Invertebrate Paleontology.* Lawrence, Kansas: University of Kansas, 1953–1972.
Moore, R. C., Lalicker, C. G., and Fischer, A. G., *Invertebrate Fossils.* New York: McGraw-Hill, 1952.
Raup, D. M., and Stanley, S. M., *Principles of Paleontology.* San Francisco: Freeman, 1971.
Rhodes, F. H. T., Zim, H. S., and Shaffer, P. R., *Fossils.* New York: Golden, 1962.
Romer, A. S., *Vertebrate Paleontology.* Chicago: University of Chicago, 1966.
Romer, A. S., *The Procession of Life.* Cleveland, Ohio: World, 1968.
Seward, A. C., *Plant Life Through the Ages.* New York: Hafner, 1959.
Shrock, R. R., and Twenhofel, W. H., *Principles of Invertebrate Paleontology.* New York: McGraw-Hill, 1953.
Weller, J. M., *The Course of Evolution.* New York: McGraw-Hill, 1969.

Stratigraphy Texts

Berry, W. B. N., *Growth of a Prehistoric Time Scale.* San Francisco: Freeman, 1968.
Dunbar, C. O., and Rodgers, J., *Principles of Stratigraphy.* New York: Wiley, 1957.
Eicher, D. L., *Geologic Time.* Englewood Cliffs, N.J.: Prentice-Hall, 1968.
Garrels, R. M., and MacKenzie, F. T., *Evolution of Sedimentary Rocks.* New York: Norton, 1971.
Gignoux, M., *Stratigraphic Geology.* San Francisco: Freeman, 1955.
Harbaugh, J. W., *Stratigraphy and Geologic Time.* Dubuque, Iowa: Brown, 1968.
Krumbein, W. C., and Sloss, L. L., *Stratigraphy and Sedimentation.* San Francisco: Freeman, 1964.
Shaw, A., *Time in Stratigraphy.* New York: McGraw-Hill, 1964.
Weller, J. M., *Stratigraphic Principles and Practice.* New York: Harper, 1960.

Index

Index